Handbook of Engineering Hydrology

Fundamentals and Applications

Handbook of Engineering Hydrology

Handbook of Engineering Hydrology: Fundamentals and Applications, Book I

Handbook of Engineering Hydrology: Modeling, Climate Change, and Variability, Book II

Handbook of Engineering Hydrology: Environmental Hydrology and Water Management, Book III

Handbook of Engineering Hydrology

Fundamentals and Applications

Edited by
Saeid Eslamian

CRC Press
Taylor & Francis Group
Boca Raton London New York

CRC Press is an imprint of the
Taylor & Francis Group, an **informa** business

CRC Press
Taylor & Francis Group
6000 Broken Sound Parkway NW, Suite 300
Boca Raton, FL 33487-2742

Printed on acid-free paper
Version Date: 20140114

International Standard Book Number-13: 978-1-4665-5241-8 (Hardback)

Visit the Taylor & Francis Web site at
http://www.taylorandfrancis.com

and the CRC Press Web site at
http://www.crcpress.com

Contents

Preface

Hydrological and ecological connectivity is a matter of high concern. All terrestrial and coastal ecosystems are connected with water, which includes groundwater, and there is a growing understanding that "single ecosystems" (mountain forest, hill forest, mangrove forest, freshwater swamp, peat swamp, tidal mudflat, and coral reef) that are actually the result of an artificial perception and classification can, in the long term, only be managed by a holistic vision at the watershed level. It is essential to investigate ecosystem management at the watershed level, particularly in a changing climate.

In general, there are two important approaches:

1. Adaption to hydrological events such as climate change, drought, and flood
2. Qualitative and quantitative conservation of water, thereby optimizing water consumption

The *Handbook of Engineering Hydrology* aims to fill the two-decade gap since the publication of David Maidment's *Handbook of Hydrology* in 1993 by including updated material on hydrology science and engineering. It provides an extensive coverage of hydrological engineering, science, and technology and includes novel topics that were developed in the last two decades. This handbook is not a replacement for Maidment's work, but as mentioned, it focuses on innovation and provides updated information in the field of hydrology. Therefore, it could be considered as a complementary text to Maidment's work, providing practical guidelines to the reader. Further, this book covers different aspects of hydrology using a new approach, whereas Maidment's work dealt principally with classical components of hydrologic cycle, particularly surface and groundwater and physical and chemical pollution.

The key benefits of the book are as follows: (a) it introduces various aspects of hydrological engineering, science, and technology for students pursuing different levels of studies; (b) it is an efficient tool helping practitioners to design water projects optimally; (c) it serves as a guide for policy makers to make appropriate decisions on the subject; (d) it is a robust reference book for researchers, both in universities and in research institutes; and (e) it provides up-to-date information in the field.

Engineers from disciplines such as civil engineering, environmental engineering, geological engineering, agricultural engineering, water resources engineering, natural resources, applied geography, environmental health and sanitation, etc., will find this handbook useful.

Further, courses such as engineering hydrology, groundwater hydrology, rangeland hydrology, arid zone hydrology, surface water hydrology, applied hydrology, general hydrology, water resources engineering, water resources management, water resources development, water resources systems and planning, multipurpose uses of water resources, environmental engineering, flood design, hydrometeorology, evapotranspiration, water quality, etc., can also use this handbook as part of their curriculum.

This set consists of 87 chapters divided into three books, with each book comprising 29 chapters. This handbook consists of three books as follows:

1. Book I: Fundamentals and Applications
2. Book II: Modeling, Climate Change, and Variability
3. Book III: Environmental Hydrology and Water Management

This book focuses mainly on the basic concepts of surface and groundwater hydrology and hydrometeorology, water resources, ecohydrology, and hydroecology in addition to hydrological data processing, flood monitoring, warning, and prediction in urban systems. The second book covers climate and hydrologic changes and estimation, mathematical modeling, risk and uncertainty, spatial and regional analysis, statistical analysis. The third book includes groundwater management, purification, sanitation and quality modeling, surface water management, wastewater and sediment management, water law and water resources management. The chapters in this book can be classified as follows:

- *Dam, reservoir, and hydroelectric*: Long-term generation of scheduling of hydro plants, check dam selection procedures in rainwater harvesting, and stochastic reservoir analysis
- *Ecohydrology*: Ecohydrology for engineering harmony in the changing world, concepts, and plant water use
- *Groundwater hydrology*: Conjunctive use of groundwater and surface water in a semiarid, hard-rock terrain; fundamentals of hydrodynamic modeling in porous media; groundwater exploration: geophysical, remote sensing, and Geographic Information Systems (GIS) techniques; and groundwater hydrology: saturation zone, groundwater–surface water interactions, hydrogeology of hard-rock aquifers, isotope hydrogeology, and karst hydrogeology
- *Hydroecology*: Hydrologic and hydraulic design of green infrastructure, hydrology–ecology interactions, and wetland hydrology
- *Hydrological data*: Data processing in hydrology, optimum hydrometric site selection and quality control, and homogenization of climatological series
- *Hydrometeorology*: Cold region hydrology and evapotranspiration and water consumption
- *Monitoring, warning, and prediction*: Modern flood prediction and warning systems and satellite-based systems for flood monitoring and warning
- *Surface hydrology*: Catchment water yield estimation, hydrograph analysis and baseflow separation, and low flow hydrology
- *Urban systems*: Sustainability in urban water systems and urban hydrology

About 200 authors from various departments and across more than 30 countries worldwide have contributed to this book, which includes authors from the United States comprising about one-third of the total number. The countries that the authors belong to have diverse climate and have encountered issues related to climate change and water deficit. The authors themselves cover a wide age group and are experts in their fields. This book could only be realized due to the participation of universities, institutions, industries, private companies, research centers, governmental commissions, and academies.

I thank several scientists for their encouragement in compiling this book: Prof. Richard McCuen from the University of Maryland, Prof. Majid Hassanizadeh from Utrecht University, Prof. Soroush Sorooshian from the University of California at Irvine, Profs. Jose Salas and Pierre Julien from Colorado State University, Prof. Colin Green from Middlesex University, Prof. Larry W. Mays from Arizona State University, Prof. Reza Khanbilvardi from the City College of New York, Prof. Maciej Zalewski from the University of Łódź, Poland, and Prof. Philip B. Bedient from Rice University.

In addition, Research Professor Emeritus Richard H. French from Las Vegas Desert Research Institute, who has authored the book *Open Channel Hydraulics* (McGraw-Hill, 1985), has encouraged me a lot. I quote his kind words to end this preface:

> My initial reaction to your book is simply WOW!
>
> Your authors are all well known and respected and the list of subjects very comprehensive. It will be a wonderful book. Congratulations on this achievement.

Saeid Eslamian
Isfahan University of Technology
Isfahan, Iran

Editor

Saeid Eslamian is an associate professor of hydrology at Isfahan University of Technology, Iran, where he heads the Hydrology Research Group in the Department of Water Engineering. His research focuses mainly on statistical and environmental hydrology and climate change. In particular, he specializes in modeling and prediction of natural hazards, including floods, droughts, storms, winds, and groundwater drawdowns, as well as pollution in arid and semiarid zones, particularly in urban areas.

Prof. Eslamian received his BS in water engineering from Isfahan University of Technology in 1986. Later, he was offered a scholarship for a master's degree at Tarbiat Modares University, Tehran. He completed his studies in hydrology and water resources in 1989. In 1991, he was awarded a grant for pursuing his PhD in civil engineering at the University of New South Wales, Sydney, Australia. His supervisor was Professor David H. Pilgrim, who encouraged him to conduct research on regional flood frequency analysis using a new region of influence approach. Soon after his graduation in 1995, Eslamian returned to Iran and worked as an assistant professor at Isfahan University of Technology (IUT). In 2001, he was promoted to associate professor.

Eslamian was a visiting professor at Princeton University, Princeton, New Jersey, in 2006 and at the University of ETH Zurich, Switzerland, in 2008. During this period, he developed multivariate L-moments for low flow and soil–moisture interaction.

Eslamian has contributed to more than 300 publications in books, research journals, and technical reports or papers in conferences. He is the founder and chief editor of the *International Journal of Hydrology Science and Technology* and the *Journal of Flood Engineering*. He also serves as an editorial board member and reviewer of about 30 Web of Science (ISI) journals. Recently, he has been appointed as the chief editor for a three-set book series Handbook of Engineering Hydrology by Taylor & Francis Group (CRC Press).

Prof. Eslamian has prepared course material on fluid mechanics, hydraulics, small dams, hydraulic structures, surface runoff hydrology, engineering hydrology, groundwater hydrology, water resource management, water resource planning and economics, meteorology, and climatology at the undergraduate level and material on evapotranspiration and water consumption, open channel hydraulics, water resources engineering, multipurpose operation of water resources, urban hydrology, advanced hydrology, arid zones hydrology, rangeland hydrology, groundwater management, water resources development, and hydrometeorology at the graduate level.

He has presented courses on transportation, Energy and Agriculture Ministry; and different university departments in governmental and private sectors: civil engineering, irrigation engineering, water engineering, soil sciences, natural resources, applied geography, and environmental health and sanitation.

Eslamian has undertaken national and international grants on "Studying the impact of global warming on the Kingdom of Jordan using GIS," "Study of the impact of different risk levels of climate change on Zayandehroud River Basin's climatic variables," "Feasibility of reclaimed water reuse for industrial uses in Isfahan Oil Refining Company," "Microclimate zoning of Isfahan city and investigation of microclimate effect on air temperature, relative humidity and reference crop evapotranspiration," "Feasibility of using constructed wetland for urban wastewater," "Multivariate linear moments for low flow analysis of the rivers in the north-eastern USA," and "Assessment of potential contaminant of landfill on Isfahan water resources." He has received two ASCE and EWRI awards from the United States in 2009 and 2010, respectively, as well as an outstanding researcher award from Iran in 2013.

Contributors

Hafzullah Aksoy
Department of Civil Engineering
Istanbul Technical University
Istanbul, Turkey

Anibal Tavares de Azevedo
Applied Science Faculty
State University of Campinas
Campinas, Brazil

Giovanni Barrocu
Department of Civil Engineering, Environmental Engineering and Architecture
University of Cagliari
Cagliari, Italy

Philip B. Bedient
Department of Civil and Environmental Engineering
Rice University
Houston, Texas

Saeid Khosrobeigi Bozchaloei
Department of Watershed Management,
Faculty of Natural Resources,
Tarbiat Modares University,
Tehran, Iran

Neil A. Coles
Centre of Excellence for Ecohydrology
Faculty of Engineering, Computing and Mathematics
University of Western Australia
Crawley, Western Australia, Australia

Jonathan Peter Cox
OTT Medioambiente Iberia S.L.
San Sebastian de los Reyes
Madrid, Spain

and

Department of Civil Engineering
Catholic University of Murcia
Murcia, Spain

Benoît Dewandel
Water Division
New Water Resources Unit
BRGN (French Geological Survey)
Montpellier, France

Ebru Eris
Department of Civil Engineering
Ege University
Izmir, Turkey

Faezeh Eslamian
Department of Civil Engineering
Isfahan University of Technology
Isfahan, Iran

Saeid Eslamian
Department of Water Engineering
Isfahan University of Technology
Isfahan, Iran

Zheng Fang
Department of Civil and Environmental Engineering
Rice University
Houston, Texas

Secundino Soares Filho
Department of Systems Engineering
State University of Campinas
Campinas, Brazil

Jonathan Gourley
National Severe Storms Laboratory
National Weather Center
Norman, Oklahoma

Jim Griffiths
Department of Geographical Sciences
University of Nottingham Ningbo China (UNNC)
Ningbo, People's Republic of China

Ove Tobias Gudmestad
Department of Mechanical and Structural Engineering
and Materials Science
University of Stavanger
Stavanger, Norway

José A. Guijarro
Department of Meteorological Studies of the Mediterranean,
State Meteorological Agency (AEMET),
Palma de Mallorca, Spain

Khaled H. Hamed
Faculty of Engineering
Department of Irrigation and Hydraulics
Cairo University
Giza, Egypt

Yang Hong
School of Civil Engineering and Environmental Sciences
The University of Oklahoma
Norman, Oklahoma

Ali Noor Islam
Parts Aviations Ltd
Dhaka, Bangladesh

Shafi Noor Islam
Department of Ecosystems and Environmental Informatic,
Brandenburg University of Technology
Cottbus, Germany

Pierre-Yves Jeannin
Swiss Institute for Speleology and Karst-Studies,
La Chaux-de-Fonds, Switzerland

Sandeep Joshi
Shrishti Eco-Research Institute
Pune, India

Rezaul Karim
Ministry of Local Government
Engineering Department
Agargaon, Bangladesh

Sara Shaeri Karimi
Dezab Consulting Engineers Company
Ahwaz, Iran

Sadiq Ibrahim Khan
School of Civil Engineering and Environmental Sciences
The University of Oklahoma
Norman, Oklahoma

Reza Khanbilvardi
NOAA-CREST Institute, City College
City University of New York
New York, New York

Patrick Lachassagne
Danone Waters
Evian Volvic World Sources
Evian-les-Bains, France

Shrikant Daji Limaye
Ground Water Institute (NGO)
Pune, India

Wenjuan Liu
School of Agriculture
University of Ningxia
Ningxia, Yinchuan, People's Republic of China

Leonardo Silveira de Alburqueque Martins
Applied Science Faculty
State University of Campinas
Campinas, Brazil

Matthew F. McCabe
School of Civil and Environmental Engineering
Water Research Centre
University of New South Wales
Kensington, Victoria, Australia

and

Water Desalination and Reuse Center
King Abdullah University of Science and Technology
Thuwal, Saudi Arabia

Miguel A. Medina, Jr.
Department of Civil and Environmental Engineering
Duke University
Durham, North Carolina

Seyed Sajed Motevallian
School of Civil Engineering
College of Engineering
University of Tehran
Tehran, Iran

Saumitra Mukherjee
Remote Sensing and Geology Laboratory
School of Environmental Sciences
Jawaharlal Nehru University
New Delhi, India

Tadanobu Nakayama
Center for Global Environmental Research
National Institute for Environmental Studies
Tsukuba, Japan

Saeid Okhravi
Department of Water Engineering
Isfahan University of Technology
Isfahan, Iran

Adam Porowski
Institute of Geological Sciences
Warsaw Research Centre
Polish Academy of Sciences
Warsaw, Poland

Antonia Sebastian
Department of Civil and Environmental Engineering
Rice University
Houston, Texas

Amit Singh
School of Environmental Sciences
Jawaharlal Nehru University
New Delhi, India

Chander Kumar Singh
School of Environmental Sciences
Jawaharlal Nehru University
and
Department of Natural Resources
TERI University
New Delhi, India

Satyanarayan Shashtri
School of Environmental Sciences
Jawaharlal Nehru University
New Delhi, India

Shaul Sorek
Department of Environmental Hydrology and Microbiology
Zuckerberg Institute for Water Research
and
Blaustein Institutes for Desert Research
and
Ben-Gurion University of the Negev
Beer Sheba, Israel

David Stephenson
Department of Civil Engineering
University of Botswana
Francistown, Botswana

Marouane Temimi
NOAA-CREST Institute City College
The City University of New York
New York, New York

Mehdi Vafakhah
Department of Watershed Management,
Faculty of Natural Resources,
Tarbiat Modares University,
Tehran, Iran

Lixin Wang
Department of Earth Sciences
Indiana University-Purdue University
Indianapolis, Indiana

and

School of Civil and Environmental Engineering
Water Research Centre
University of New South Wales
Kensington, Victoria, Australia

Hartmut Wittenberg
Faculty III, Environment and Technology
Leuphana University at Lüneburg
Lüneburg, Germany

Robert Wyns
Geology Division
BRGN (French Geological Survey)
Orléans, France

Ali Zahraee
NOAA-CREST Institute, City College,
City University of New York
New York, New York

Maciej Zalewski
European Regional Center for Ecohydrology
UNESCO

and

Department of Applied Ecology
University of Lodz
Lodz, Poland

Mônica de Souza Zambelli
Department of Systems Engineering
State University of Campinas
Campinas, Brazil

1

Catchment Water Yield

Jim Griffiths
University of Nottingham Ningbo China

AUTHOR

Jim Griffiths was born in South Wales (United Kingdom) and studied at both undergraduate and postgraduate level in the School of Geography at King's College London. His doctoral research involved spatial modeling of pore-water pressures in shallow translational landslides (in SE England and SE Spain), with respect to climate and land-use change. He spent 5 years as a hydrologist at the Centre for Ecology and Hydrology in Wallingford (formerly the Institute of Hydrology), where his research included development of continuous simulation rainfall–runoff models and investigation of surface water–groundwater interaction in permeable lowland catchments in the United Kingdom. From 2008 to 2011, he worked as senior hydrologist for a UK-based mining consultancy for which he conducted hydrological site investigation work in Sierra Leone, Congo Republic, Burkina Faso, Tanzania, Turkey, and Sweden and contributed to feasibility and prefeasibility level studies for a variety of mine developments in Northern Europe, Central Asia, Africa, South America, and Russia. He is a fellow of the Chartered Institute of Water and Environmental Management (CEng, CEnv, CSci) and a lecturer in environmental sciences at the University of Nottingham Ningbo China.

PREFACE

This chapter describes a number of different approaches for the estimation of catchment yield including water balance, reservoir, and tank models. Emphasis is given to the balance between precipitation and evapotranspiration (ET) by describing the role of different land-use and soil properties in determining the overall water budget. Attention is also paid to the importance of the spatial and temporal scale at which estimates are made and how this might affect their accuracy.

1.1 Introduction

In order to make an assessment of available water resources for a proposed or existing development, one of the first things an engineering hydrologist must do is to estimate the average water yield from surface water catchments. In addition to the estimate of the quantity of available water, the seasonality and interannual variability of catchment yield must be assessed in order to predict the probability of water deficits in drought years and the potential for surplus during wet years. This can be achieved using a range of hydrological models that exhibit varying degrees of complexity. This chapter reviews a number of procedures that can be used to predict catchment water yield, with particular emphasis on consideration of the role of soil and land-use type.

1.2 Definition of the Catchment

The catchment is the principle hydrological unit considered within the field of hydrology and fluvial geomorphology. Catchments can be represented or differentiated by a range of interrelated hydrological parameters including average climate characteristics (precipitation, temperature, and insolation), landform and drainage characteristics (topography, drainage density, channel length, and shape), and soil and land-use characteristics (soil structure and permeability and percentage of canopy cover).

Catchment area is sometimes referred to as drainage area or river-basin area. Catchment yield is the amount of water that will be transported to the catchment outlet from an area of land that lies up-gradient as defined by the surrounding topography. Each catchment is separated from neighboring catchments by a topographically defined drainage divide. Output from smaller catchments will drain into the larger catchments in a hierarchical pattern. A great number of smaller sub-catchments can be defined within any catchment and may be referred to as nested catchments.

In order to do derive catchment water yield, the catchment boundaries, or watershed, should first be identified from topographic maps. Although this can be done manually from contour maps (5–10 m), this is more easily achieved using a digital terrain model (DTM), which can be acquired from photogrammetry, land survey, or remote sensing. Commonly used sources of remotely sensed data include the Shuttle Radar Topography Mission (SRTM), Advanced Spaceborne Thermal Emission and Reflection Radiometer (ASTER), and Light Detection and Ranging (LiDAR) (see Nikolakopoulos et al. (2006) for a comparison of SRTM and ASTER elevation data and Harris et al. (2012) for description of the use of LiDAR data).

Figure 1.1 illustrates a DTM produced from manually surveyed data points. To achieve greater model accuracy in areas of increased topographic variability, the data are digitized using the triangular irregular network (TIN) method and then are converted into a rectangular grid (5 × 5 m). At this resolution,

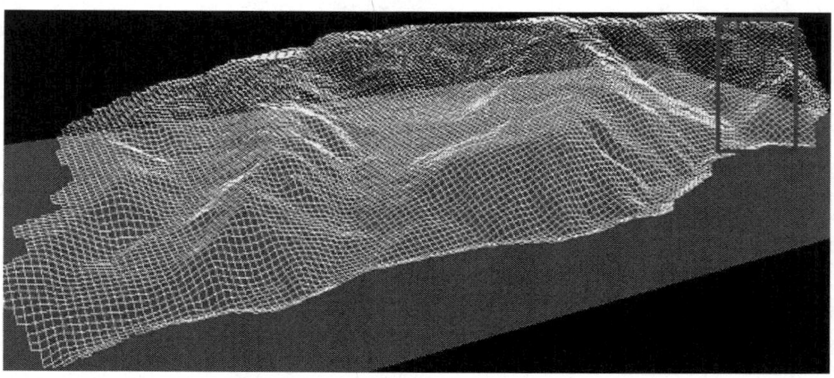

FIGURE 1.1 A 5 × 5 m resolution grid DTM derived from manually surveyed topographic data.

catchments as small as 50 m^2 are identifiable. For larger catchment areas, the use of remotely sensed data and a larger grid resolution may be necessary to reduce computational time. The use of digital topographic data for catchment delineation also allows calculation of fractional areas of different land-cover or soil type if suitable maps are available or if they can be derived from remote sensing imagery data (Rogan and Chen, 2004).

Many water resource models assume surface and groundwater catchment boundaries to be identical. While this is rarely the case (due to soil and geological heterogeneity), it is sometimes a useful assumption to make as it allows hydrological catchment delineation from surface topographic data alone. While both surface and groundwater catchment boundaries associated with a predefined catchment outlet point can be derived manually from paper contour maps or geological maps, this process is more frequently performed using a geographic information system (GIS) or computer-aided design (CAD) software.

1.3 Modeling Catchment Yield

In its most simple form, catchment yield can be defined as the precipitation (P) that leaves the catchment as surface water flow (Q) after Evapotranspiration (ET) losses and losses to the soil or groundwater. If the assumption of a closed groundwater system is made (i.e., there are negligible groundwater losses from or additions to the catchment), catchment yield may be described by a ratio of the difference between mean annual catchment precipitation and ET and catchment outflow (as represented in Equation 1.1):

$$\text{Catchment yield} = \frac{\text{P} - \text{ET}}{\text{Q}} \tag{1.1}$$

Estimation of catchment water yield can be more complex than that expressed in Equation 1.1, especially if a high level of precision and a relatively short time increment is required. Although there are a large range of different models and methods that can be employed by water resource planners and engineers, after definition of the catchment watershed, estimation of catchment water yield can be summarized by four generic steps that should be followed:

- Estimation of catchment inputs
- Estimation of catchment outputs
- Representation of transport processes
- Calculation of catchment yield at the catchment outlet

Assuming there are no upstream inputs, the largest hydrological input into a catchment is precipitation. However, there may be some difficulty in making reliable and representative estimates of catchment precipitation due to both the size of the catchment and the availability of historic data. Firstly, it is more difficult to estimate of precipitation in larger catchments as rainfall is less likely to be homogeneously distributed across the whole catchment or to occur at all locations within the catchment at the same intensity and at the same time. Secondly, the remoteness and extent of human development within a catchment can mean that there is little or no recorded rainfall at any location within the catchment. Both large and remote catchments therefore present a problem that must be solved through the use of the best available data and a number of statistical assumptions about the distribution of precipitation in the area.

It is acknowledged that some catchments will exhibit transboundary groundwater movements, but this is dealt with in more detail elsewhere within this volume (Chapter 22 of *Handbook of Engineering Hydrology: Fundamentals and Applications*). With the assumption of a closed groundwater catchment then, catchment output will consist exclusively of evaporation and transpiration. There are a range of methods to calculate evaporation, the choice of which will depend on available data and the temporal resolution of the required estimate. The maximum rate of water evaporation within a catchment is dependent on water availability. The potential evaporation rate of evaporation is therefore attenuated by the soil moisture deficit (SMD). Transpiration, the loss of water from plant stomata, can be calculated from species-specific physical properties.

Water transport processes that should be considered when making estimates of catchment yield include infiltration, throughflow, percolation, and runoff. The spatial and temporal resolution at which each process needs to be represented will depend on the resolution of required catchment yield estimates. For example, if monthly yield estimates are required, there is no need to calculate hourly infiltration rates. Conversely, if daily variation in catchment yield is required, some consideration in daily variation of SMDs, and thus infiltration, must be made. The dominant water transport process will be different in every catchment and will largely depend on variation in land cover, geology, and topography. Overlandflow (OF), for example, may be twice the magnitude of baseflow in permeable catchments, but less than half the magnitude of baseflow in permeable catchments.

If water transport processes are to be modeled at daily timesteps, then it is likely that quickflow and slowflow processes will be represented differently. The arrival of water from different parts of the catchment at the catchment outlet by different routes (overlandflow, throughflow, or baseflow) therefore needs to be summed in order to provide the end estimate of flow from the catchment per unit timestep. The rate at which water is transported to the catchment outlet will depend on both the nature of the process and catchment physiography, geology, and soil type.

1.3.1 Water Balance Models

The concept of the water balance was coined by C. Warren Thornthwaite in 1944 to represent the balance between hydrological inputs (precipitation and inter-catchment transfers of surface or groundwater) and outputs (evaporation, groundwater seepage, and streamflow from the catchment). The water budget can be calculated both at the soil profile or catchment scale and at any temporal interval (though most are calculated as daily or annual). The water budget of a small catchment with an impermeable bedrock is given by Equation 1.2. Catchment yield is represented by the term OF. The water balance approach is useful because it can represent a range of different catchment types and hydrological regimes:

$$P = I + ET_A + OF + \Delta SMD + \Delta GWS + \Delta GWR \tag{1.2}$$

where

- I is the canopy interception
- ET_A is the actual evapotranspiration
- ΔSMD is the change in SMD
- ΔGWS is the change in groundwater storage
- ΔGWR is the change in groundwater recharge

Effective precipitation (EP) can be defined as water from precipitation after losses to canopy interception, ET, and soil moisture storage (Equation 1.3). As such, it is closely related to catchment yield:

$$EP = P - I - ET_A - \Delta SMD \tag{1.3}$$

Figure 1.2 illustrates the difference between measured precipitation and calculated EP for a small catchment in SE England. It can be seen that EP is zero during the summer and autumn months. This was due to high canopy interception, ET, and SMD. As a result, surface water runoff (which consists predominantly of EP) was significantly reduced in summer months, and catchment yield was composed predominantly of baseflow (throughflow and groundwater seepage).

Thornthwaite and Mather (1957) suggested that in larger catchments, monthly EP might be reduced by up to 50% before it reaches the catchment outlet by natural storages such as lakes, drainage channels, and groundwater. The proportion of EP delayed in this way decreases with catchment size and when the water balance is calculated for periods smaller than a month.

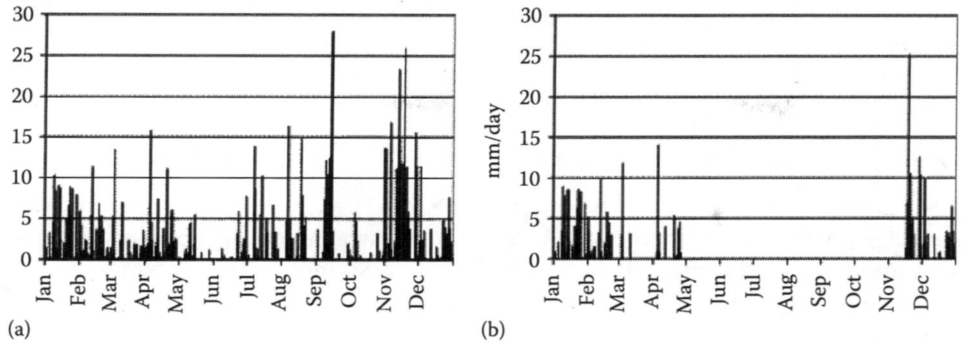

FIGURE 1.2 (a) Measured precipitation (P) and (b) calculated effective percipitation (EP) for a small catchment in SE England.

1.3.2 Reservoir Models

Horton (1938) suggested that catchment outflow can be represented as a simple linear or nonlinear storage reservoir. As the volume within the reservoir (or catchment) rises the overflow (or catchment yield) increases. The basic form of the linear or nonlinear function used to define that overflow at a point in time, Q(t), is given by Equation 1.4:

$$Q(t) = \frac{1}{k} s(t)^n \tag{1.4}$$

where

- s(t) is the volume of water in storage at time, t
- k is a constant (with units of time)
- n is the order of the reservoir

The response of the reservoir is determined by the exponent "n," such that a value of 1 produces a linear response and the values of 2 and 3 produce quadratic and cubic nonlinear responses, respectively. To increase the complexity of the catchment system response, multiple reservoirs may be used in series or in parallel cascades (see Sugawara et al., 1983). However, this also increases the task of parameter calibration and thus the requirement for observed data.

1.3.3 Tank Models

In order to simulate the hydraulic response of the catchment over long timescales (c. 30 years), it is necessary to simplify hydrological processes in order to reduce computing time. A simplified daily 1D water balance model for example, may be used to predict the movement of water within the soil profile.

The process of infiltration of water into the soil column is inherently complex due to the various stages of wetting or drying that might occur. As the water content of the soil changes, its structure may also change, as particles swell or simply settle against each other, thus altering pore space and reducing cracks and voids. Additionally, the hydraulic gradient within the soil will change, especially around the wetting front. Both of the previously mentioned phenomena are hysteretic in nature so that the magnitude of change will depend on whether the soil is drying out or becoming more wet.

Ideally, a description of the infiltration response of a soil to various rainfall patterns should be gained from site investigation. However, as this is not always possible, an approximation of the mechanism of moisture flow into the soil can be made. Rubin (1966) stated that the infiltration of water into the soil would occur in one of three possible circumstances:

- Nonponding infiltration: rainfall rate < infiltration rate
- Preponding infiltration: rainfall rate ≤ infiltration rate
- Rain ponding infiltration: rainfall rate > infiltration rate

For ponding and OF to occur, precipitation must be greater than the hydraulic conductivity of the soil. Once ponding occurs, whether immediately or after a period of pre-ponding, a positive pressure head acts upon the surface until rainfall rate declines. If precipitation is such that no surface ponding occurs, actual infiltration will tend toward the rate of precipitation with time. Precipitation rates will rarely be constant, and resulting infiltration may move intermittently between the previously mentioned modes in any one event. Antecedent moisture conditions will also affect a soil's ability to transport moisture, as will previous wetting and drying cycles that the soil has undergone, though the hysteretic nature of these is assumed to be negligible. Because of such intrinsic variability, any attempt to model infiltration must invariably simplify the process.

Figure 1.3 illustrates a schematic of how processes can be represented within a 1D tank model that was initially used to predict pore-water pressures within a landslide prone catchment in SE England (Collison et al., 2000). In this example, the vertical soil profile was represented by only three layers (though this can be more for soils of greater heterogeneity): a root zone, a colluvium layer, and an underlying impermeable layer. The rate of transfer of moisture from one soil layer to another was regulated by soil conductivity and its relationship to antecedent soil moisture using the van Genuchten (1980) method. The Green and Ampt method for moisture transfer (Green and Corey, 1971) is also regularly used for this purpose.

The actual potential difference between soil layers was represented using the ratio of the amount of water held in adjacent layers and their capacity to receive drainage from above. Total water content was then determined for each layer on a daily basis, accounting for rainfall, canopy interception, ET, bypass flow, and drainage. Total amount of water contained within each layer was calculated using the following mass-balance equations:

$$\text{Layer 1}: W_{t+1} = W_t + \left[P(1 - BP)\right] - ET_A - D_1 - OF \tag{1.5}$$

$$\text{Layer 2}: W_{t+1} = W_t + \left[P(BP)\right] + D_1 - D_2 \tag{1.6}$$

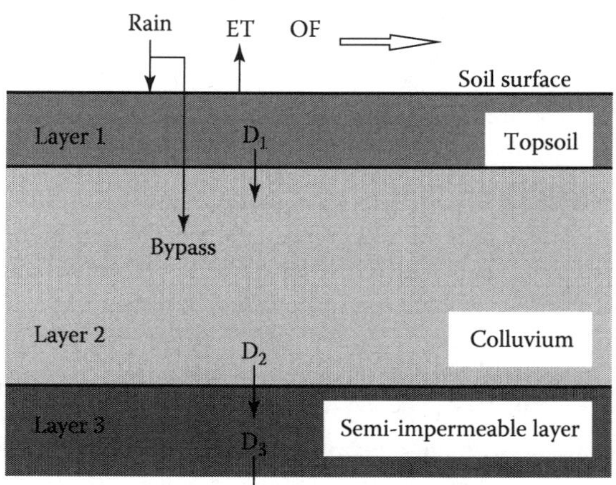

FIGURE 1.3 Schematic of *tank model* approach to representation of precipitation, evaporation, and infiltration.

$$\text{Layer 3}: W_{t+1} = W_t + D_2 - D_3 \tag{1.7}$$

where

W_t is the initial water content (mm)
W_{t+1} is the resulting water content (mm)
P is the net rainfall (mm)
BP is the bypass coefficient (mm)
OF is the overland flow (mm)
D_1 is the drainage from layerl (mm)
D_2 is the drainage from layer 2 (mm)
D_3 is the deep percolation (mm)
ET_A is the actual evapotranspiration (mm)

In this case, the model assumes that soil saturated hydraulic conductivity is always greater than rainfall intensity due to the presence of biopores in the topsoil (this means that runoff only occurred via saturation from a rising water table, rather than limiting infiltration). Each layer is assumed to drain to its field capacity at a rate dependent on its hydraulic conductivity or the capacity for drainage in the underlying layer. Total vertical drainage (D), saturation excess (Z), and water content of each layer at the end of the timestep (W_{t+1}) therefore depend on one of four possible conditions:

1. Amount of incoming water is insufficient reach fill micropores, resulting in no drainage from the layer.
2. Incoming water fills soil micropores, but water content remains less than field capacity, so limited drainage occurs.
3. Incoming water is sufficient to maintain field capacity and allow maximum drainage, but layer is not saturated.
4. Layer is saturated resulting in unconditional drainage, but saturation limits further inflow to the layer.

Table 1.1 illustrates how W_{t+1}, D and Z can be calculated with respect to layer properties, where

X is the available water in layer = initial water content + input (mm)
W_{fc} is the water content when soil is at field capacity (mm)
W_{sat} is the water content when soil is saturated (mm)
W_{t+1} is the water content at time t+1 (mm)
D_c is the drainage capacity: limited by available storage in layer below (mm/d)
D is the drainage (actual) (mm)
Z is the saturation excess (mm)

1.3.4 Land-Cover and Soil Properties

To predict daily catchment yield, it is necessary to represent processes that effect the movement and quantity of water within the catchment, such as evaporation and soil infiltration. Figure 1.4

TABLE 1.1 Possible Hydrological Conditions of Tank Model Soil Layers Defined by Water Content and Drainage Status

	1	2	3	4
Condition	$X \le W_{fc}$	$W_{fc} < X \le W_{fc} + D_c$	$W_{fc} + D_c < X \le W_{sat} + D_c$	$X > W_{sat} + D_c$
$W_{t+1} =$	X	W_{fc}	$X - D_c$	W_{sat}
D =	0	$X - W_{fc}$	D_c	D_c
Z =	0	0	0	$X - D_c - W_0$

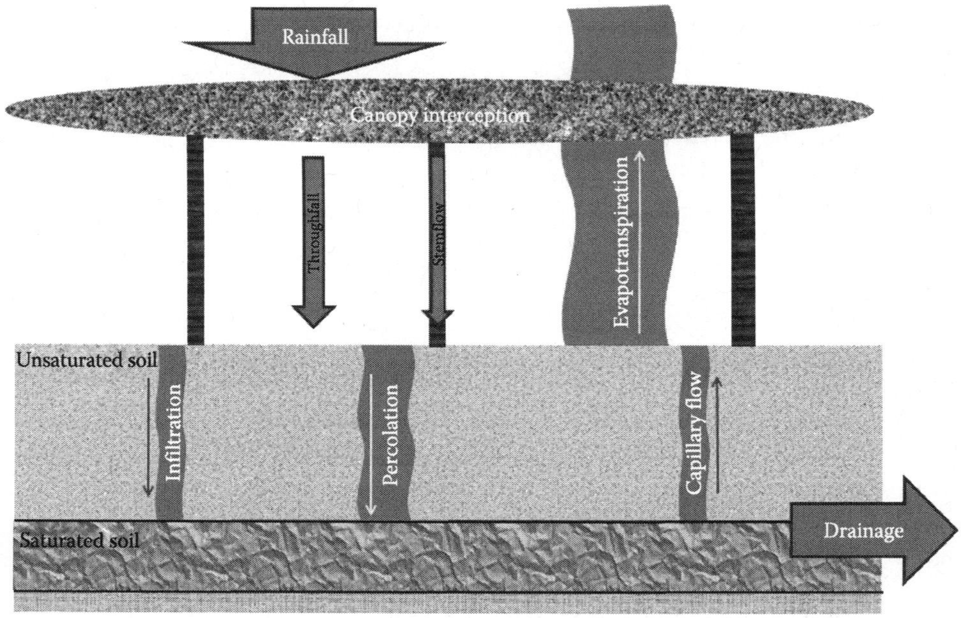

FIGURE 1.4 Schematic representation of catchment processes. Area of arrow indicates relative differences in flow, while width indicates relative differences rate of flow (thin = fast rate).

illustrates some of the processes that can be considered in order to achieve daily or sub-daily estimates of catchment yield, including canopy interception and ET, and infiltration and drainage. As the figure suggests, processes that control the movement of water within the catchment are dynamic and interrelated. Drainage outflow from the catchment will occur at a rate that is dependent not only on soil conductivity and the antecedent water content of the soil but also water availability, which will depend on both climate and land management practices.

1.4 Precipitation

Catchment precipitation is usually calculated from available historic data. This can be difficult where data are discontinuous or missing. Similarly, rainfall may not fall homogenously throughout a catchment, particularly if it is a large catchment, and number of techniques can be utilized to best represent the spatial and temporal distribution of catchment rainfall, a number of which are briefly discussed here and more thoroughly elsewhere within this volume.

1.4.1 Spatial Distribution of Precipitation

Patterns of rainfall are most often expressed in terms of return period, total magnitude, and maximum or mean intensity. For most European environments convective storms produced in anticyclonic conditions are generally of short duration but high intensity, whereas frontal storms associated with depressions are of lower intensity but longer duration (Brooks and Richards, 1994). Catchment outflow generally responds more quickly to convectional as opposed to frontal storms due to higher associated rainfall intensities and the positively skewed rainfall distribution within such events.

If catchment precipitation is to be derived from a network of different sources or gauges, it will be necessary to derive a representative mean of precipitation to use in calculation of yield estimates. To use a simple arithmetic mean of nearest gauges to the area of interest risks introducing bias into the estimate as some gauges will be closer and more representative than others. The use of Thiessen polygons to derive a weighted average, based on difference in distance or altitude between the target area and each gauge used, would produce a more realistic estimate. Alternatively, isohyets (contours of equal rainfall depth) may be drawn or interpolated if gauge density is sufficient.

1.4.2 Temporal Distribution of Precipitation

Long-term records of mean monthly and annual rainfall (and associated catchment yield) can be used to identify the likelihood of either prolonged wet or dry periods within a catchment. Such periods are often characterized by their frequency, magnitude, and duration of either above or below average rainfall. The beginning and end of an exceptionally dry or wet period can also be defined by a specific threshold limit. A drought period, for example, will begin when the cumulative precipitation deficit exceeds a specified threshold limit (as defined from historical records). Likewise, the drought ends when the cumulative precipitation deficit falls below the specified threshold. The intensity of a drought is generally described by the cumulative deficit during the drought divided by its duration.

Prolonged periods of low precipitation can be characterized by a number of different drought indices. The U.S. Department of Agriculture, for example, uses the Palmer Drought Severity Index that is based on monthly hydrological accounting of precipitation, ET, and changes in soil moisture storage, whereby a moisture anomaly index is first defined for each month based on long-term average monthly precipitation and ET (see Palmer, 1965). A number of other regularly utilized drought indices include the Surface Water Supply Index (Shafer and Dezman, 1982) and the Standard Precipitation Index (McKee et al., 1993).

1.4.3 Representative Measurement of Catchment Precipitation

In assessing how many rain gauges are required to realistically capture a reliable record of catchment precipitation, Bleasdale (1965) quoted that densities that ranged from a single gauge every 13 km^2 in smaller catchments (c. 25 km^2) to a single gauge every 325 km^2 in larger catchments (8000 km^2) for gathering representative monthly precipitation data. The Joint Committee of the Meteorological Office, Royal Meteorological Society, and the Institution of Water Engineers (1937) also indicated that density should decrease with catchment size but that the density should be greater in well-utilized areas. Similarly, a greater number of gauges may be needed in mountainous areas or where topography is known to increase the heterogeneity of regional rainfall patterns.

With the increased use of radar-derived precipitation data for hydrological modeling (see Krajewski and Smith, (2002) for an excellent review of related methodology), an increasing number of radar-derived precipitation data are becoming a realistic option for use in long-term catchment yield estimation.

1.5 Evapotranspiration

Potential evapotranspiration (ET_P) is the maximum rate of ET that can occur for an unlimited supply of water in the root zone. Actual evapotranspiration (ET_A) is a function of SMD and available water content (AWC). Hillel (1977) describes the relationship between ET_A and SMD in the following way:

$$ET_A = ET_P\left(1 - SMD/AWC\right) \tag{1.8}$$

when $\theta_{rz} > \theta_{pwp}$

$$ET_A = 0 \tag{1.9}$$

when $\theta_{rz} < \theta_{pwp}$

where

$$\text{SMD} = \text{Field capacity}(\theta_{fc}) - \text{Root zone soil moisture } (\theta_{rz}) \tag{1.10}$$

$$\text{AWC} = \text{Field capacity}(\theta_{fc}) - \text{Wilting point } (\theta_{pwp}) \tag{1.11}$$

The total amount of ET that will take place from the soil surface will depend on the effective surface area from which moisture may be extracted. Rates of ET_A within the vegetation canopy will vary relative to canopy density, leaf area index (LAI), and position within the canopy (the lower layer receiving lower airflow and incident shortwave radiation [see Jones, 1983]). Evaporation from forest floors is usually assumed to be minimal for the sake of simplicity (for more information see Roberts et al., 1980; Shuttleworth, 1979).

Evaporative losses from the catchment are derived from both evaporation from the ground surface and evaporation from the surface of vegetated surfaces. Unless calculating catchment yield from areas that have significant non-vegetated cover, for example, heavily urbanized or industrialized catchment, evaporation from non-vegetated surfaces within a catchment can be assumed to be negligible. Where tree cover exists, precipitation may first be intercepted by the canopy, thus preventing up to 50% of precipitation from reaching the ground surface; this may be further reduced by the presence of a deep litter cover (Helvey and Patric, 1965a).

1.5.1 Canopy Interception

The degree of canopy interception loss is dependent on two groups of variables:

1. Vegetation characteristics including canopy shape, size, distribution, stage of development, leaf size and shape, and canopy moisture storage capacity
2. Rainfall characteristics including intensity, duration, and frequency

In order to know how much rainfall reaches the soil surface, it is necessary to know the extent and capacity of the overlying vegetation canopy and the rate and magnitude of incident rainfall. As can be seen from Figure 1.5, net rainfall is composed of canopy throughfall and stemflow, that is, water that reaches the ground through the canopy and via the canopy structure, respectively, such that

$$\text{Net precipitation } (P_n) = \text{Rainfall } (P) - \text{Interception loss } (I_t) \tag{1.12}$$

$$\text{Interception loss } (I_t) = \text{Canopy interception loss } (I_c) + \text{Litter interception loss } (I_l) \tag{1.13}$$

and

$$\text{Canopy interception loss } (I_c) = \text{Rainfall } (P) - \text{Stemflow } (P_s) - \text{Throughfall } (P_t) \tag{1.14}$$

Thus

$$\text{Net rainfall } (P_n) = \text{Throughfall } (P_t) + \text{Stemflow } (P_s) - \text{Litter interception loss } (I_l) \tag{1.15}$$

Most of the precipitation that is incident on vegetation canopy will be stored until a maximum storage capacity is reached, at which point it "overflows" and becomes throughfall or stemflow. This means the relationship between precipitation and canopy interception will vary throughout a single storm, depending on the extent to which the capacity of the canopy has been filled. Accordingly, the interception resulting from two storms of similar water volume but of differing intensity and duration may vary considerably.

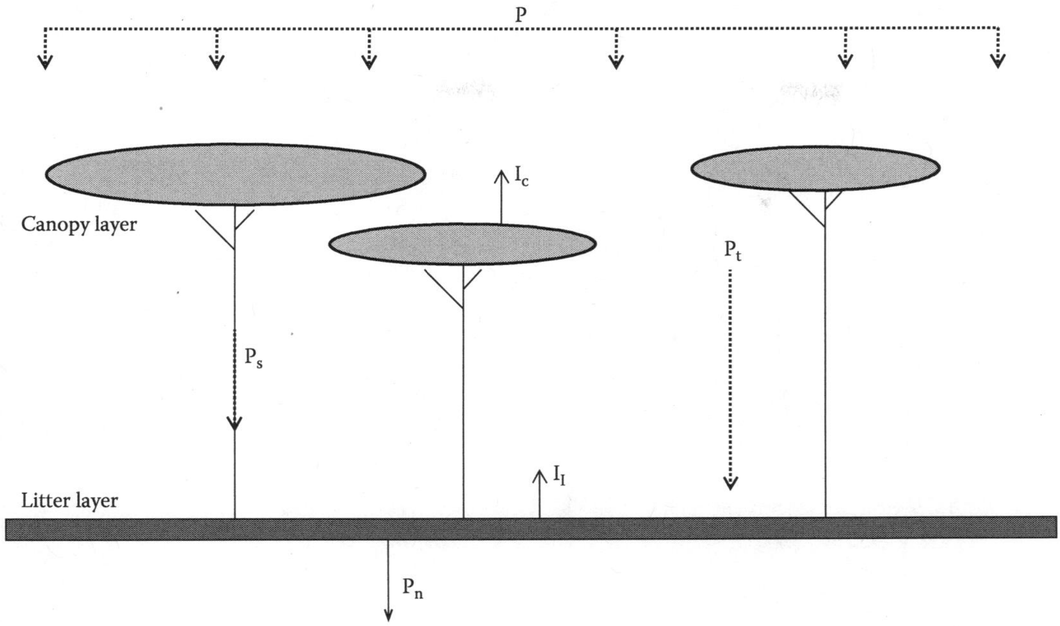

FIGURE 1.5 Schematic illustration of rainfall (P), canopy interception loss (I_c), throughfall (P_t), stemflow (P_s), litter interception loss (I_l), and net rainfall (P_n).

Once water has come to rest on the canopy, its rate of evaporation will depend on temperature, solar radiation, relative humidity, wind speed, and wind turbulence (which assist in removing saturated air from the canopy area). During intermittent rainfall, however, evaporation from the canopy will more become irregular to an extent that will depend on both canopy structure and LAI. For prolonged rainfall events, evaporation will be significantly reduced, as will the amount of water that would normally be lost from the plant through transpiration. This is especially true during rainfall that occurs within summer months (high temperature) and for plants of high LAI.

Robert Horton's 1919 method of estimating interception on a storm by storm basis determined both the water holding capacity of the canopy and the total evaporation occurring within a storm. Merriam subsequently developed the model by increasing canopy capacity as a function of total rainfall in order to better represent the downward movement of moisture through the canopy (Merriam, 1960). This approach is similar to the now frequently used "Rutter" interception model (Rutter et al., 1971a and b), which requires estimation of maximum canopy storage and trunk storage, in order to calculate total evaporation from the canopy. Empirical knowledge of throughfall and stemflow fractions is required to parameterize the Rutter model however. Rates of interception can then be calculated as functions of precipitation and canopy storage per timestep (see Figure 1.6). Net precipitation beneath the canopy is a function of incident precipitation and the amount of water stored within the canopy.

The Rutter model relies on the fact that actual evaporation from the canopy can be calculated as a function of wetness such that

$$E_A = \frac{C(E_P)}{S} \tag{1.16}$$

where

E_A is the actual evaporation
E_P is the potential evaporation
C is the actual canopy storage
S is the canopy storage capacity

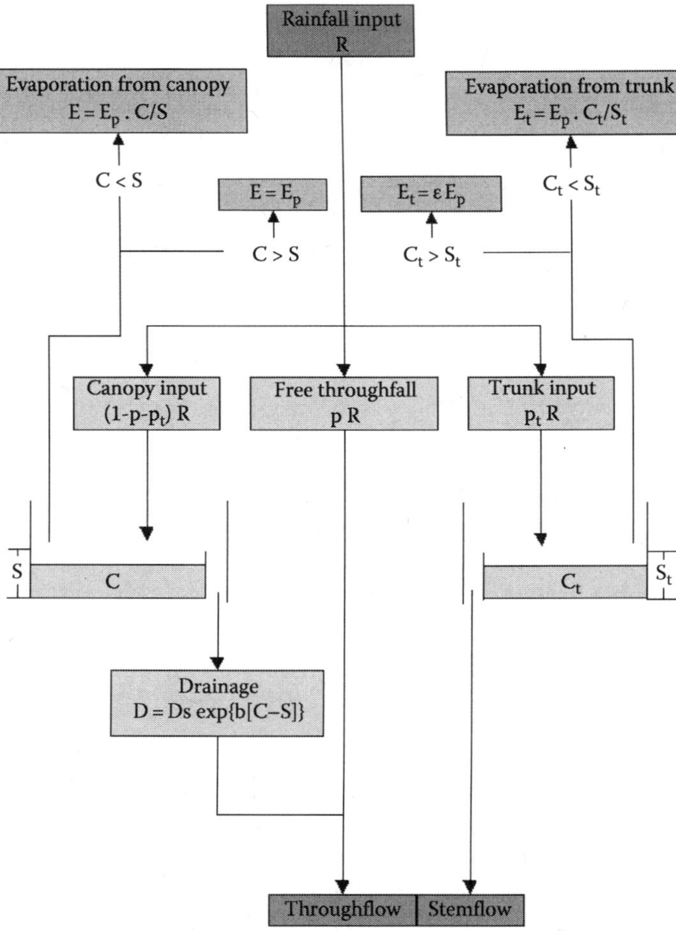

FIGURE 1.6 Schematic diagram of the Rutter dynamic interception model. (After Gash, J.H.C., *Quart. J. Roy. Meteor.*, 105, 43, 1979; Gash, J.H.C. and Morton, A.J., *J. Hydrol.*, 38, 49, 1978.) (R, rainfall; C, actual canopy storage; S, canopy storage capacity; Ct, actual trunk storage; St, trunk storage capacity; Ep, potential evaporation rate; E, actual evaporation from canopy; Et, actual evaporation from trunk; D, drainage from canopy; Ds, drainage from canopy when C = S; ε, empirical ratio of potential trunk evaporation to potential canopy evaporation; p, free throughfall coef.; pt, stemflow partitioning coef.; b, empirical const.)

While the Rutter model works well in situations where vegetation canopy is relatively continuous, Valente et al. (1997) have described modifications that are needed to account for discontinuous canopy, such as those that occur regularly in the more sparse forests of southern Europe. Calder (1985) also points out that problems may arise from using the Rutter model with vegetation species that exhibit larger canopy capacities, but not correspondingly large evaporation rates. Gash (1979) also suggested that by simplifying processes related to leaf drainage rates, more consistent results across species can be achieved.

As an alternative to the data-intensive method of using the Rutter model, percentage canopy interception can be approximated from measured throughfall and stemflow rates. Attempts were made by the Institute of Hydrology (now CEH, Wallingford, United Kingdom) to simplify calculation of annual interception losses at larger scales by considering only the percentage land cover of each vegetation type, the fraction of the year that the canopy is wet, and the average rate of potential evaporation (Calder, 1990). Though still essentially dynamic (albeit over longer timescales), such an approach is closer to empirically based interception models.

Characteristic regression equations of incident precipitation, to throughfall (Equation 1.17) and stemflow (Equation 1.18), may also be used to calculate net and intercepted rainfall:

$$P_t = m_t P + c_t \quad (1.17)$$

$$P_s = m_s P + c_s \quad (1.18)$$

where
P_s is the stemflow
P_t is the throughfall
m and c are empirically defined

This approach has the advantage of being simple and operable at any scale. Helvey and Patric (1965a and b) determined that for hardwoods, $m_t \cong 0.92$ and $m_s \cong 0.05$, while litter interception was assumed to be between 0.02 and 0.05. Numerous studies of interception characteristics have been made for a wide range of species, but as can be seen from Table 1.2, very few have included measurement of both net rainfall data and LAI, meaning that comparison with, or use in, subsequent study can be misleading. In addition, estimated values of throughfall and stemflow from regression equations for specific species tend to decrease as the sampling period increases. Likewise, if data from a single storm event are assessed, percentage interception predicted for storms of greater magnitude will tend to be higher than that predicted for smaller storms.

1.5.2 Evaporation and Evapotranspiration

The Thornthwaite original 1939 evaporation model (Thornthwaite and Holzman, 1939) relies on an empirically defined ratio between temperature and evaporation and is based on mean annual temperature. His 1944 model of monthly evaporation also considers number of daylight hours (Equation 1.19). Subsequent versions have been developed to determine both potential and ET_A, but remain empirical in nature:

$$Et_k = 1.6\left(\frac{10.ti}{TE_j}\right)^a \quad (1.19)$$

where
Et_k is the monthly ET_P (cm)
t is the mean monthly temperature in °C
TE_j is the monthly Thornthwaite temperature efficiency index

In a similar way, the Blaney and Criddle 1950 model can be employed using mean monthly temperature, fraction of daylight hours, and an empirical "crop factor." Although popular because of their simplicity and low data requirements (see Equation 1.20), both the Thornthwaite and Blaney–Criddle methods offer representation of a number of meteorological variables. The Thornthwaite equation, for example, would be of less use in conditions where relative humidity is a limiting factor on evaporative losses and similarly might overestimate losses if used in conditions where cloud cover frequently reduces incoming radiation:

$$ET_P = p\left(aT_{mean} + b\right) \quad (1.20)$$

where
p is the percentage fraction of daylight hours
T_{mean} is the mean daily temperature in °C $\left(\left(T_{max} - T_{min}\right)/2\right)$
a and b are empirically derived constants

TABLE 1.2 Canopy Parameters from Previous Studies (Seasonal Range Indicated where Known)

Source	Species/ Community	Net Rain	Interception	Throughfall	Stemflow	Storage Cap. (mm)	C*Leaf (cm/s)	LAI
		(as % of Total Rainfall)						
Dingman (1994)	Corsican pine	65	35				0.32	
Lee (1980)	Hornbeam	64	36					
	Douglas fir	61	39				0.8	
	Oak forest	78	22				0.3	
	Norway spruce	52	48					
	Sitka spruce	71	29				0.7	
	Scots pine	58	42					
Dolman (1987)	Oak forest			30–80		0.3–0.8		
López-Bermúdez (1996)	*Pinus halepensis*	76	24	74	2	5		
	Shrub cover	79	21	65	14	1.5		
	Shrub + pine	77	23	72	5	2		
Morgan and Rickson (1995)	Heather					1.5		
	Bracken					1.3		
	Meadow grass/ clover					2		
	Cereal					1.2–3		
	Temp deciduous forest					0.5–1		
	Pine/spruce forest					1.5		
Neal et al. (1991)	Beech plantation (hardwood)	89	11	84.6	4.4			
Parker (1995)	Deciduous broadleaf							4–6
	Evergreen broadleaf							7–12
	Deciduous conifer							5–7
	Evergreen conifer							15–20
	Pinus							7–12
Woodward (1987)	Deciduous hard forest							3–7
	Temp coniferous forest							10–47
Valente et al. (1997)	*Eucalyptus globulus*		10.8	87.5	1.7			3.2
	Pinus pinaster		17.1	82.6	0.3			2.7

Figure 1.7 illustrates calculated ET_P for a catchment in northern England using the Blaney–Criddle method. ET_A, estimated using the Food and Agriculture Organization (FAO) methodology for calculating crop water requirements (Allen et al., 1998), is also shown. It can be seen that ET_A falls below the potential rate from June to September due to increased SMD, shown in Figure 1.8 that has resulted from lower precipitation and higher ET_P.

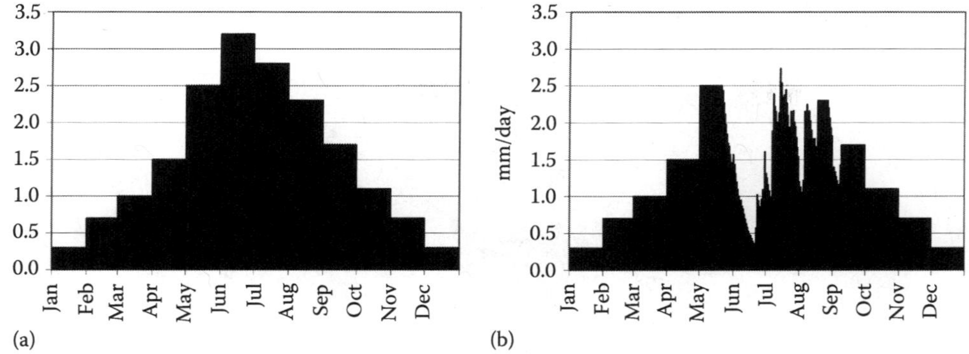

FIGURE 1.7 Estimated (a) ET_P and (b) ET_A for a 250 km² mixed vegetation, rural catchment in the United Kingdom.

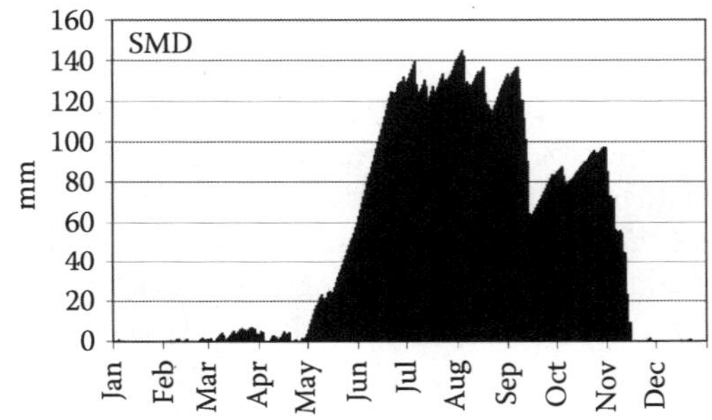

FIGURE 1.8 Estimated variation in SMD calculated using the FAO methodology for calculating crop water requirements. (From Allen, R.G. et al., *Crop Evapotranspiration: Guidelines for Computing Crop Water Requirements.* FAO Irrigation and Drainage Paper 56, Rome, Italy, 1998.)

While net radiation is the dominant controlling variable on ET, temperature is a more convenient surrogate measure of ET_P (Ayoade, 1971). This is because while variation in net solar radiation in response to global warming can be more accurately described than temperature (as it is largely a function of the level of CO_2 in the atmosphere), the potential for change in cloud dynamics makes its prediction much more difficult. Though direct estimates of expected changes to rates of ET in response to global warming scenarios are available (HadCM3), these relate only to standard vegetation types.

The accuracy of the prediction of the response of ET to temperature change will largely depend on the form of the empirical equation used, the season within which the empirical model is based, and the accuracy of parameters that represent ground conditions. As opposed to using the Thornthwaite or Blaney–Criddle methods, site- and vegetation-specific empirically based relationships between temperature and ET_P can be. Figure 1.9 illustrates the relationship between observed temperature and ET calculated using the Penman–Monteith model (parameterized to represent different types of vegetation cover).

The relationship internalizes independent climatic variables, which control the dependent variable (ET), and as such may be inaccurate under conditions of climate change. If such an approach is to be used within a climate impact assessment therefore, the assumption would have to be made that changes in radiation, humidity, and wind speed would not significantly affect the relationship between temperature and ET. Other studies that have produced similarly strong relationships include by McKenny and Rosenburg (1993) and Thornthwaite and Mather (1957) who identified a relationship between ET_A and SMD, which is dependent on available water capacity (AWC), which can also be represented by regression (see Dunne and Leopold, 1978).

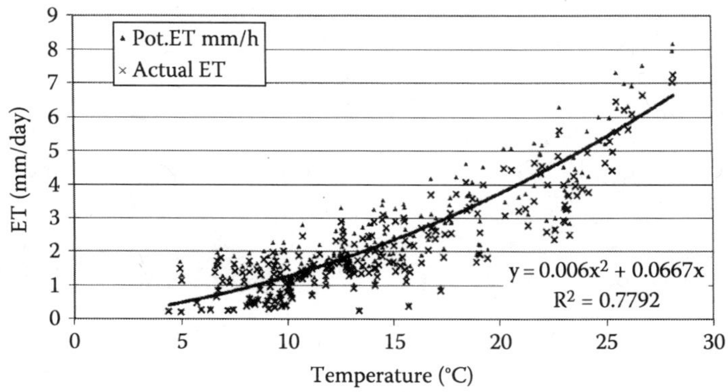

FIGURE 1.9 Relationship of temperature and ET calculated using the Penman–Monteith method.

A physically based representation of evaporation and transpiration processes is offered by the Penman–Monteith model that incorporates Monteith's (1965) estimation of canopy conductance into the Penman model for evaporation from a free surface (Penman, 1948). The main advantage of the Penman–Monteith model is that it allows representation of ET_P from various land-cover types. The structure of the model as described by Dingman (1994) is illustrated in Figure 1.10. This version of the model allows good representation of vegetation characteristics of height, roughness, albedo, as well as leaf and canopy conductivity. The model is derived from the mass and energy balance methods for calculation of ET. The combined equation (Equation 1.21) assumes that no change occurs in vegetation heat storage and no energy is lost through water advection or ground conduction:

$$ET_P = \frac{S(T_a)(K+L) + \gamma\rho_a c_a C_{at}\left[e_{sat}(T_a)\right](1-W_a)}{\rho_w \lambda_v \left\{s(T_a) + \gamma(1 + C_{at}/C_{can})\right\}} \tag{1.21}$$

where

ET_P is the potential evapotranspiration (cm/s)
$(K+L)$ is the net radiation (cal/cm²/s)
$S(T_a)$ is the slope of $e_{sat}(T_a)$ with temperature
$e_{sat}(T_a)$ is the saturation vapor pressure (mb)
W_a is the relative humidity (%)
C_{at} is the atmospheric conductance (cm/s)
C_{can} is the canopy conductance (cm/s)
and the following constants:
λ_v is the latent heat of vaporization (assuming full canopy) = 590 cal/g
ρ_w is the mass density of water = 1 g/cm³
ρ_a is the density of air = 0.00122 g/cm³
c_a is the heat capacity of air = 0.24 cal/g
γ is the psychrometric constant = 0.66 mb/°C

The rate of moisture release from the canopy to the air is represented in terms of canopy (C_{can}) and atmospheric (C_{at}) conductance, respectively, such that

$$C_{can} = L_t \cdot C_{leaf} \tag{1.22}$$

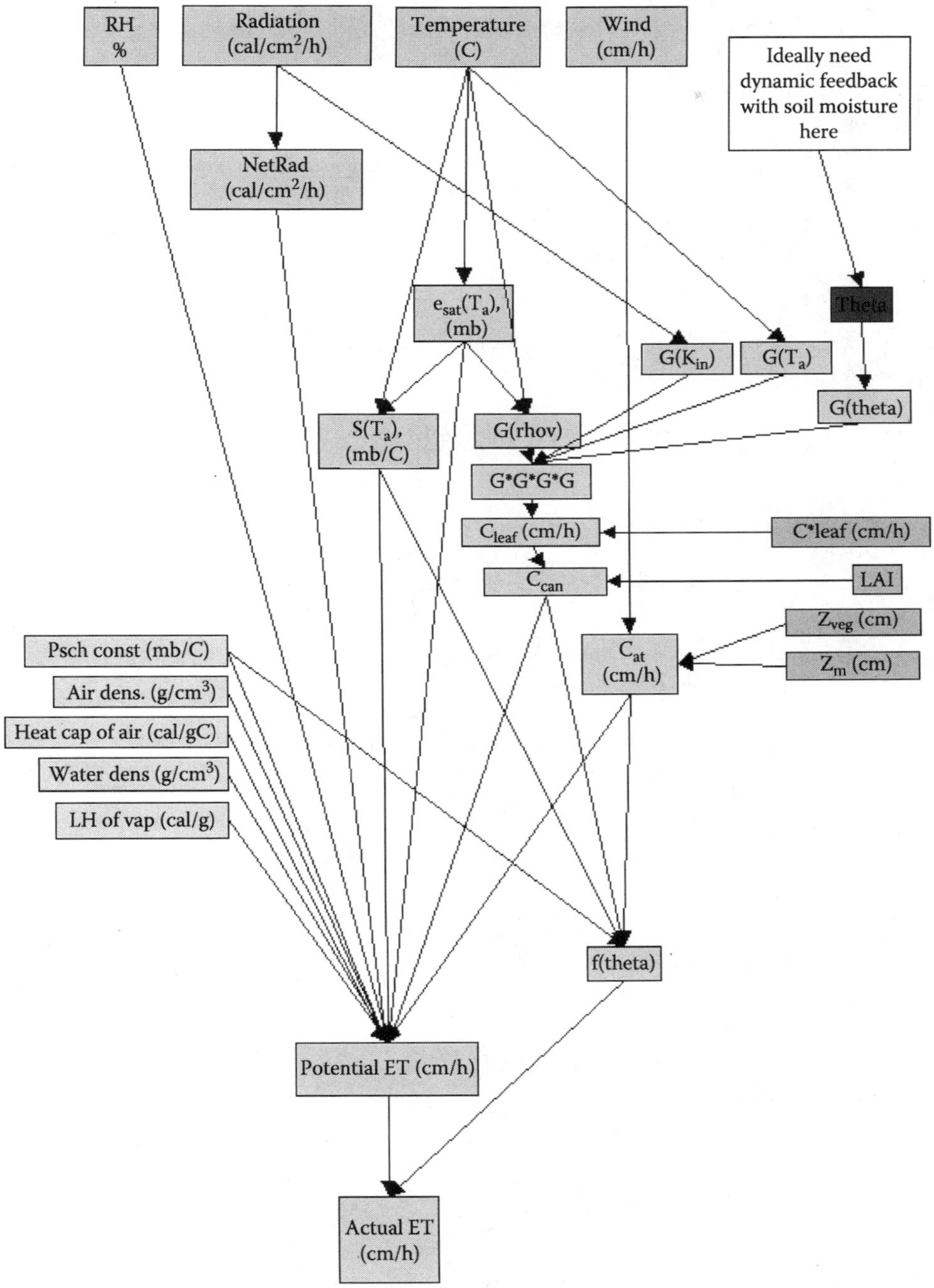

FIGURE 1.10 Schematic diagram of the "Penman–Monteith" ET model. Meteorological variables and soil moisture form the dynamic inputs to the model.

and

$$C_{at} = \frac{K^2U}{\left(\ln[z_m - z_d]/z_o\right)^2} \quad (1.23)$$

where

- U is the wind speed (cm/s)
- z_m is the height above ground at which U is measured (cm)
- z_d is the zero plane displacement = 0.7 z_{veg} (cm)
- z_o is the roughness height = 0.1 z_{veg} (cm)
- L_t = LAI (i.e., area of leaves per unit area of ground)
- C_{leaf} is the leaf conductance (cm/s)
- K is the von Kármán's constant = 0.41

Equation 1.24 describes the FAO of the United Nations approach to solving the Penman–Monteith model (see Hess and Lovelace, 1991; Smith, 1991):

$$ET_o = ET_{rad} + ET_{aero} \quad (1.24)$$

where

$$ET_{rad} = \frac{\delta(R_n - G)}{\lambda(\delta + \gamma^*)} \quad (1.25)$$

and

$$ET_{aero} = \frac{900\gamma U_2(e_a - e_d)}{(\delta + \gamma^*)(T + 275)} \quad (1.26)$$

where

- ET_o is the reference crop ET (mm/day)
- ET_{rad} is the radiation term (mm/day)
- ET_{aero} is the aerodynamic term (mm/day)
- δ is the slope of vapor pressure temperature curve (kPa/°C)
- R_n is the net radiation (MJ/m²/day)
- G is the soil heat flux (MJ/m²/day) = 0.01 Rn
- λ is the latent heat of evaporation (MJ/kg)
- γ is the psychometric constant = 0.066 (kPa/°C)
- γ^* is the modified psychometric constant (kPa/°C)
- T is the air temperature (°C)
- U_2 = wind speed at 2 m (ms^{-1})
- e_a = saturated vapor pressure at T (kPa)
- e_d = saturated vapor pressure at dew point (kPa)

The FAO method is less explicit about canopy structure and conductivity. For example, the modified psychrometric constant described in Equation 1.26 can be defined as a function of canopy and aerodynamic conductivity such that

$$\gamma^* = \gamma\left(1 + \frac{C_{at}}{C_{can}}\right) \quad (1.27)$$

This ratio between atmospheric (C_{at}) and canopy (C_{can}) conductivity has a strong influence on predicted values of ET (Bevan, 1979). In circumstances where one or neither of the variables is known, working estimates can be obtained through empirical study. For example, $C_{at}/C_{can} = 0.33.U_2$ (m/s) has been used to give acceptable results for many cereal-type crops (Hess and Lovelace, 1991).

1.6 Summary and Conclusions

Canopy interception and ET are complex processes, and their calculation requires some simplification for use in estimation of catchment yield. This chapter has taken a brief look at different approaches to estimate catchment yield (water balance, reservoir models, tank models, and process models), but has addressed canopy interception and ET in more detail as these estimates are required by all of the approaches. There are many other approaches for estimation of catchment ET and thus yield that have not been discussed. In particular, the use of crop coefficients in the FAO method of ET calculation is worth further investigation for catchments with a high percentage of agricultural land and thus markedly different growing seasons (see Allen et al., 1998).

Water resource engineering requires practical and robust modeling approaches as data availability and time constraints mean the use of a fully deterministic modeling approach is not possible. The focus on more accurate estimation of rainfall and ET as a way of improving and simplifying the catchment water balance has been noted by Zhang et al. (1999), who also see potential in the development of a parsimonious model for annual ET based only on rainfall and canopy cover fraction. In a similar way, the use of empirical regression models between ET and temperature has been highlighted in this chapter.

The use of 1D tank or reservoir models to allow a dynamic feedback between ET and soil moisture, also has the potential to provide data for models of water transport mechanisms within the catchment. However, such routing models are often considered outside the remit of catchment yield estimation unless daily resolution estimates are required: for which purpose the reader is referred to studies by Flügel (1995), Bevan and Kirkby (1979), and Young et al. (2008). Similarly, while catchment-specific unit hydrographs and flow duration curves are also measures of catchment yield, a more comprehensive description of their use and derivation is given elsewhere within this volume and can also be found in Boorman and Reed (1981) and Fennessey and Vogel (1990), respectively.

References

Allen, R.G., Pereira, L.S., Raes, D., and Smith, M. 1998. *Crop Evapotranspiration: Guidelines for Computing Crop Water Requirements*. FAO Irrigation and Drainage Paper 56, Rome, Italy.

Ayoade, J. 1971. Rainfall, evapotranspiration and water balance in Nigeria, Unpublished PhD thesis, University of London, Houston, TX.

Bevan, K. 1979. A sensitivity analysis of the Penman-Monteith actual evapotranspiration estimates, *Journal of Hydrology*, 44: 169–190.

Bevan, K. and Kirkby, M. 1979. A physically based, variable contributing area model of basin hydrology, *Hydrological Sciences Bulletin*, 24: 43–69.

Bleasdale, A. 1965. Rain gauge networks development and design with specific reference to the United Kingdom, *International Association of Scientific Hydrology Symposium on Design of Hydrological Networks*, Montreal, Quebec, Canada.

Boorman, D.B. and Reed, R.W. 1981. *Derivation of a Catchment Average Unit Hydrograph*. Wallingford, U.K., Institute of Hydrology, 50pp IH Report No.71.

Brooks, S.M. and Richards, K.S. 1994. The significance of rainstorm variations to shallow translational hillslope failure, *Earth Surface Processes and Landforms*, 19: 85–94.

Calder, I.R. 1985. What are the limits on forest evapotranspiration?, *Journal of Hydrology*, 82: 179–184.

Calder, I.R. 1990. *Evaporation in the Uplands*, Wiley, Chichester, U.K.

Collison, A.J.C., Wade, S., Griffiths, J., and Dehn, M. 2000. Modelling the impact of predicted climate change on landslide frequency and magnitude in SE England, *Engineering Geology*, 55: 205–218.

Dingman, S.L. 1994. *Physical Hydrology*, Prentice Hall, Englewood Cliffs, NJ.

Dolman, A.J., 1987. Summer and winter rainfall interception in an oak forest. Predictions with an analytical and a numerical simulation model, *Journal of Hydrology*, 90: 1–9.

Dunne, T. and Leopold, L.B. 1978. *Water in Environmental Planning*, WH Freeman and Co., San Francisco, CA.

Fennessey, N. and Vogel, R.M. 1990. Regional flow duration curves for ungauged sites in Massachusetts, *Journal of Water Resources Planning and Management*, 116: 530–549

Flügel, W.-A. 1995. Delineating hydrological response units by geographical information system analyses for regional hydrological modelling using PRMS/MMS in the drainage basin of the River Bröl, Germany, *Hydrological Processes*, 9(4): 423–436.

Gash, J.H.C. 1979. An analytical model of rainfall interception by forests, *Quarterly Journal of Royal Meteorological Society*, 105: 43–55.

Gash, J.H.C. and Morton, A.J. 1978. An application of the Rutter model to the estimation of the interception loss from Thetford Forest, *Journal of Hydrology*, 38: 49–58.

van Genuchten, M.Th. 1980. A closed-form equation for predicting the hydraulic conductivity of unsaturated soils, *Soil Science Society of America*, 48: 892–898.

Green, R.D. and Corey, J.C. 1971. Calculation of hydraulic conductivity: A further evaluation of some Predictive Methods, *Soil Science Society of America Proceedings*, 35: 3–8.

Harris, B., McDougall, K., and Barry, M. 2012. Comparison of multi-scale digital elevation models for defining waterways and catchments over large areas, ISPRS Annals of the Photogrammetry, *Remote Sensing and Spatial Information Sciences*, I-2: 75–80.

Helvey, J.D. and Patric, J.H. 1965a. Canopy and litter interception of rainfall by hardwoods of eastern United States, *Water Resources Research*, 1: 193–206.

Helvey, J.D. and Patric, J.H. 1965b. Design criteria for interception studies, *IAHS Publ. No. 67,* pp. 131–137, Wallingford, WA.

Hess, T.M. and Lovelace, G. (eds.) 1991. *Soil-Plant-Water Relationships: Practical Work Handbook* (7th edn.), Department of Agriculture and Water Management (Silsoe College), Cranfield, U.K.

Hillel, D. 1977. *Computer Simulation of Soil Water Dynamics*, International Development Research Centre, Ottawa, Southern Ontario, Canada.

Horton, R.E. 1938. The interpretation and application of runoff plot experiments, *Proceedings of Soil Science Society America,* 3: 340–349.

Joint Committee of the Meteorological Office, Royal Meteorological Society and the Institution of Water Engineers, 1937. Report on the determination of the general rainfall over any area, *Trans Institution of Water Engineers*, 42: 231.

Jones, H.G. 1983. *Plants and Microclimate*, Cambridge University Press, Cambridge, New York.

Krajewski, W.F. and Smith, J.A. 2002. Radar hydrology: Rainfall estimation, *Advances in Water Resources* 25: 1387–1394.

López-Bermúdez, F. 1996. *Field Site: Murcia, Spain, in Basic Field Programme Final Report (1991–1995)*, Medulus II: Project 1, EVSV–CT92–0128, pp. 38–60.

McKee, T.B, Doesken, N.J., and Kleist, J. 1993. The relationship of drought frequency and duration to timescales, *Proceedings of Eighth Conference on Applied Climatology*, Anaheim, CA.

McKenny, M.S. and Rosenburg, N.J. 1993. Sensitivity of some potential evapotranspiration methods to climate change, *Agricultural and Forest Meteorology*, 64: 81–110.

Merriam, R.A. 1960. A Note on the interception loss equation, *Journal of Geophysical Research*, 65(11): 3850–3851

Monteith, J.L. 1965. Evaporation and environment, in *Proceedings of the 19th Symposium of the Society for Experimental Biology,* Cambridge University Press, Cambridge, New York.

Morgan, R.P.C. and Rickson, R.J. (eds.) 1995. *Slope Stabilization and Erosion Control: A Bioengineering Approach*, E & FN Spon, London, U.K.

Neal, C., Robson, A.J., Hall, R.L., Ryland, T., Conway, T., and Neal, M. 1991. Hydrological impacts of hardwood plantation in lowland Britain: Preliminary findings on interception at a forest edge, Black Wood, Hampshire, southern England, *Journal of Hydrology*, 127: 349–365.

Nikolakopoulos, K.G., Kamaratakis, E.K., and Chrysoulakis, N. 2006. SRTM vs ASTER elevation products. Comparison for two regions in Crete, Greece, *International Journal of Remote Sensing*, 27: 21.

Palmer, W.C. 1965. Meteorological drought, *Technical Research Report Paper No. 45*, US Department of Commerce, Weather Bureau, Washington, DC.

Parker, G.G. 1995. Structure and microclimate of forest canopies, in Lowman, M.D. and Nadkarni, N.M. (eds.), *Forest Canopies*, Academic Press, CA.

Penman, H.L. 1948. Natural evaporation from open water, bare soil and grass, *Royal Society of London Proceedings, Series A*, 193: 120–145.

Roberts, J., Pymar, C.F., Wallace, J.S., and Pitman, R.M. 1980. Seasonal changes in leaf area, stomatal and canopy conductance and transpiration from bracken below a forest canopy, *Journal of Applied Ecology*, 17: 409–422.

Rogan, J. and Chen, D.-M., 2004. Remote sensing technology for mapping and monitoring land-cover and land-use change, *Progress in Planning*, 61: 301–325.

Rubin, J. 1966. Theory of rainfall uptake by soils initially drier than their field capacity, and its applications, *Water Resources Research*, 2: 739–749.

Rutter, A.J., Kershaw, K.A., Robins, P.C., and Morton, A.J., 1971b. A predictive model of rainfall interception in forests; 1. Derivation of the model from observations in a plantation of Corsican pine, *Agricultural Meteorology*, 9: 367–384.

Rutter, A.J., Morton, A.J., and Robins, P.C., 1971a. A predictive model of rainfall interception in forests; 2. Generalization of the model and comparison with observations in some coniferous and hardwood stands, *Journal of Applied Ecology*, 12: 367–380.

Shafer, B.A. and Dezman, L.E. 1982. Development of a Surface Water Index (SWSI) to assess the severity of drought conditions in snowpack runoff areas, *Proceedings of the Western Snow Conference*, Colorado State University, Fort Collins, CO.

Shuttleworth, W.J., 1979. Below canopy fluxes in a simplified one dimensional theoretical description of the vegetation-atmosphere interaction, *Boundary Layer Meteorology*, 17: 315–331.

Smith, M. 1991. *Expert Consultation on Revision of FAO Methodologies for Crop Water Requirements*, FAO-Land and Water Development Division, Rome, Italy.

Sugawara, M., Watanabe, I., Ozaki, E., and Katsuyame, Y. 1983. *Reference Manual for the Tank Model*, National Research Centre for Disaster Prevention, Tokyo, Japan.

Thornthwaite, C.W. and Holzman, B. 1939. The determination of evaporation from land and water surfaces, *Monthly Weather Review*, 67: 4–11.

Thornthwaite, C.W. and Mather, J.R. 1957. *Instructions and Tables for Computing Evapotranspiration and the Water Balance*, Laboratory of Climatology, Publication No. 10, Centerton, NJ.

Valente, F., David, J.S., and Gash, J.H.C. 1997. Modelling interception loss for two sparse eucalypt and pine forests in central Portugal using reformulated Rutter and Gash analytical models, *Journal of Hydrology*, 190: 141–162.

Woodward, F.I. 1987. *Climate and Plant Distribution*, Cambridge University Press, Cambridge, New York.

Young, A.R., Grew, R., Keller, V., Stannett, J., and Allen, S. 2008. Estimation of river flow time-series to support water resources management: The CERF model, *BHS 10th National Hydrology Symposium*, Exeter, England.

Zhang, L., Dawes, W.R., and Walker, G.R. 1999. Predicting the effect of vegetation changes on catchment average water balance, *Technical Report 99/12, Cooperative Research Centre for Catchment Hydrology*, CSIRO Land and Water.

2

Cold Region Hydrology

Ove Tobias Gudmestad

University of Stavanger

AUTHOR

Ove Tobias Gudmestad has since September 2008 been a full-time professor of marine technology at the University of Stavanger, Norway. From 1994 to 2008, he was an adjunct professor at the university, teaching courses on marine technology and offshore field development. He has a PhD in wave force analysis and experience from engineering, field development studies, oil and gas development projects, and research in Statoil from 1975 to 2008. When he left Statoil, he was the company's advisor for Marine and Arctic technology.

Gudmestad has published papers on the actions from waves and earthquakes, on the risk involved in marine operations, and on Arctic field development challenges. He has filed several patent applications related to offshore technology. From 2005 to 2013, he has also been working with the Norwegian University of Technology and Science in Trondheim, Norway, as adjunct professor of Arctic offshore civil engineering and from 2013 as adjunct professer of Cold Climate Technology at University of Tromsø, Norway. He has been awarded honorary doctoral degrees from the Gubkin State University of Oil and Gas in Moscow in 2002 and from Murmansk State Technical University in 2008.

PREFACE

With the needs for clean water for all biological processes and for most industrial processes, engineering hydrology is a most important subject that is of concern for all groups in the society, be it the public, the nongovernmental organizations, and the politicians.

The aspect of engineering hydrology in cold regions is of special character as the water takes two phases: the flowing phase and the solid phase. The solid ice phase is of particular concern as this phase lasts for more than half the year in certain areas, and this situation provides special challenges for the engineers. Furthermore, the frozen water encapsulates any pollution that will be released in the thawing season. The special challenge related to northward flowing rivers on the northern hemisphere also represents a specific challenge as the thawing starts upstream, which creates a large force on the downstream part of the ice. Huge flooding does result with potential for widespread distribution of pollution.

It is the aim of this chapter to give an introduction to engineering hydrology challenges, which are met in cold regions, and it is the intention to cover all the most relevant aspects of these challenges.

2.1 Introduction

This chapter on "Cold Region Hydrology" puts emphasis on the main theme of the book: engineering hydrology. The chapter is intended to serve as a checklist for engineers and scientists. The objective is to ensure that important aspects of hydrology be covered properly by those working with engineering and scientific aspects related to the cold regions of the world. The chapter is not aiming to reproduce important aspects on hydrology presented elsewhere in the book; however, the cold climate poses additional challenges that have to be solved by the humans working and living in this region. A systematic description of engineering challenges related to hydrology in cold climate is, as far as the author is aware of, not produced elsewhere. It is, furthermore, believed that the book will serve as inspiration for scientists studying specific aspects of cold region engineering hydrology.

It is, however, to be emphasized that wildlife and vegetation serve as "thermometers" on the soundness of the environment. The necessity to ensure that the land is clean and habitable is a key driver for the author's efforts to prepare the chapter. In this respect, traditional knowledge should be consulted and preserved. No one but those living in the cold climate really knows the promises, the challenges, and the limitations of living there.

The hydrology in cold regions is governed by the cold climate and the presence of permafrost that may be continuous to large depths. During wintertime, the ground is continually frozen from the surface down to the permafrost, and rivers and freshwater lakes are covered by ice, which at some locations can be several meters thick. Shallow freshwater ponds are frozen to the bottom, whereby no water is accessible from these ponds for the communities. The frozen rivers and lakes are used to transport goods to cold climate areas (on ice roads), to transport raw materials, and for the preparation of construction works. For calculation of the bearing capacity of the ice, see [11].

During the summer season, which may start from mid- to late June and continue to mid- to late August, the upper layer of the permafrost melts, resulting in the ground becoming nonaccessible by vehicles, and pollution might easily be carried with the water flow. Shallow tundra ponds become accessible during this season. The melting of glaciers is dependent upon the temperature, and the sum of the product of the average hourly temperature multiplied by the number of hours with this temperature will determine the melting rate. A tendency of increased flooding will be seen with an increase in the "melting degrees hours." On the other hand, whether the glaciers will be growing or shrinking will depend upon the net mass flux: snowfall minus melting. Where the glaciers are shrinking, there is a danger of imbalance in the future hydrology budget.

The upper soil under all roads, permanent access areas, and other civil engineering works/buildings must be cleared, and all construction activities must build on hard ground, that is, must be anchored down to the permafrost. In doing so, it is very important that heat transfer bridges are not created that will cause local melting of the permafrost. Civil engineering works must, therefore, be insulated from the permafrost. Typically, buildings are built above ground, and recently new materials, such as fiber clothing, are used for roads. Piles are being used to anchor civil engineering works, and where the annual precipitation is low, wooden piles having low heat transfer capacity can be used. It is also very important to map water flow paths to avoid buildings, roadways, etc., being flooded during the snowmelt and thaw period.

In between the summer and winter, there is the freeze-up period and the snowmelt period. In the freeze-up period, the ground freezes gradually and all ponds freeze. At some locations (such as in Barrow, Alaska [11]), precipitation is considerable during this period of the year, with a soaked surface layer as the freezing continues into deeper layers until the upper layers above the permafrost are entirely frozen. During rainfall, the ground becomes almost nonnavigable, and there is a considerable surface water flow.

During the snowmelt period, the runoff is considerable at the surface, but the flow does not penetrate deeply, as the lower layers are still frozen. Considerable flow can occur during this season, depending upon the temperature and the speed of the melting. Water can find new access ways and can cause considerable local damage. Vegetation will insulate the ground layers and be a binding material so the flow will find a route outside of vegetation. Any damage due to the use of heavy equipment can lead to increased surface damage during this period.

The rivers in cold climate areas (and there are huge rivers running north to the Arctic seas [1]) are exposed to enormous changes in the water level from the dry winter, where the surface freezes, and there is limited water transport for the huge amount of water that will be transported in the snowmelt season. In this period, the ice breakup also represents challenges, as the ice may pile up and jam the flow. Huge areas can be flooded, and the broken ice can substantially damage landscape, vegetation, and engineered structures. Furthermore, the rapid flow of water in this period also causes riverbank erosion, extensive sediment transport, and scouring around bridge supports, which can cause bridge collapse.

Last, but not least, the new challenges caused by less sea ice than in the past (an effect of the potential global warming) and, therefore, stronger wave actions cause more erosion of the shoreline than in the past. This influences the hydraulic regime in a way that traditional ponds used for drinking water might be eroded and filled with saltwater, the engineered structures might be undermined by the shrinking shoreline, or saltwater might penetrate deeper into estuaries, causing damage to wildlife and vegetation.

In the following, the specific challenges to the public caused by the hydraulic regime will be reviewed. The challenges and present conceptual solutions will be described, and in some cases, the solutions implemented by communities located in cold climate regions will be referred. The challenges discussed relate to:

- River flow during the snowmelting season
- The specifics during the river ice breakup
- The effect of the shrinking shoreline
- Spreading of pollution
- Drinking water availability
- Sewage transport
- The effect of hydrocarbon production activities

The specifics of cold climate hydrology will be emphasized and will refer to the subjects being discussed in the other chapters. The discussion will be qualitative as the variations are substantial. The chapter will, therefore, serve more as a reminder for persons involved in cold region hydrology challenges. The known can be solved. Those who know they have little knowledge know they need to know more. The dangerous situation is represented by those who do not know that the knowledge is limited and are ignorant about the potential unknown.

Another aspect that will finally be discussed in this introductory section is traditional knowledge. Predictions for the future are based upon sophisticated models, including models where global warming is foreseen. Often, however, information about the past is lacking. Many communities have knowledge about past's extreme events, and their elders know the legends. It is suggested that the first matter a community developer should consider is the relevance of the traditional knowledge in order to provide a design basis for any engineering project. This will in particular involve information about the hydrology and conditions of the ground and erosion due to actions of water. These data will also involve the behavior of animals, migration of mammals and birds as well as whales, and a consideration related to types of and changes in vegetation.

2.2 Public Challenge: River Flow during the Snowmelting Season

2.2.1 Description

The flow of water in Arctic rivers has a seasonal character (see Figure 2.1) with a low flow during the dry and cold winter where the river freezes, a very high flow during the snowmelt season, and a gradual decrease in the flow during the summer, in particular in areas with little summer precipitation, while the flow increases again before the freeze-up period when some rainfall is normal. Figure 2.1 shows the variation in the water flow in the Mackenzie River, one of the large Arctic rivers.

Some of the largest rivers in the world discharge into the Arctic seas. The major Arctic rivers are as listed in Table 2.1.

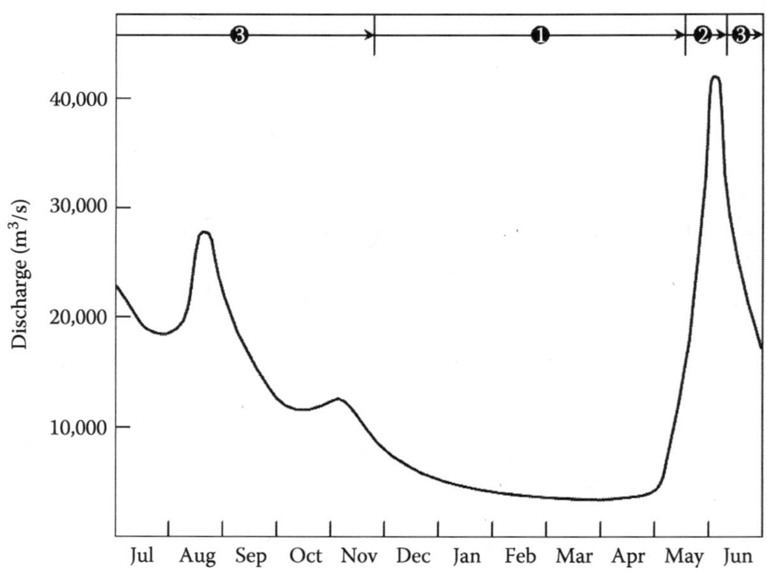

FIGURE 2.1 Typical seasonal variation of the flow of water in the Mackenzie River. (From Owens, E. et al., Field guide for oil spill response in Arctic waters, prepared for *Arctic Council Emergency, Prevention, Preparedness and Response (EPPR) Working Group, Environment Canada*, Yellowknife, Northwest Territories, Canada, p. 348, Figure 6.5, 1988. With permission from Arctic Council.)

TABLE 2.1 Rivers Discharging into Arctic Seas

River	Country	Discharge in Cubic Kilometers per Year
Dvina	Russia	105
Pechora	Russia	108
Ob	Russia	402
Yenisey	Russia	580
Lena	Russia	525
Kolyma	Russia	103
Yukon	United States (Alaska)	203
Mackenzie	Canada	281

Source: Bachman, J., Special report: Oil and ice: Worse than the gulf spill? 2010, http://www.reuters.com/article/2010/11/08/us-russia-oil-idUSTRE6A71IL20101108

2.2.2 Engineering Challenges

The large variation in the flow provides a number of challenges. The primary challenge is the erosion of the riverbank, the damage caused by the flowing ice and flooding (see also Section 2.3). The erosion causes the riverbanks to become unstable, and engineered structures can be undermined. This applies in particular to bridges spanning the river and to port facilities. Flooding causes damage to all engineered structures in large areas along the river.

2.2.3 Mitigation Measures to Avoid Damage in Strong River Flow

In order to mitigate the effects of the flooding due to snowmelt, several measures must be taken to protect engineered structures.

Like in any harbor, the banks can be protected by breakwaters and rock or concrete block plastering. It is important to calculate the dimensions of the plastering to ensure that it can stand up to the current flow [24]. In addition, the plastering or breakwaters must withstand the actions from the drifting ice during the ice breakup [25]. Recently, the use of textiles has been suggested (Figure 2.2) for plastering [5]; however, the textile must be sufficiently robust to withstand the actions to which they are exposed. Annual replacement of lost bags is required. It is expected that more research work will be conducted in the future to develop suitable materials:

- With a trend of Arctic warming, there will be increased permafrost melting, and the speed of erosion could accelerate. In addition, erosion destroys the vegetation, so erosion is self-reinforcing as vegetation holds the soil in place. It should also be noted that extensive erosion will cause large sediment transport with accumulation of sand or mud banks in the rivers, potentially in the river estuaries, hindering access by vessels. Also, dredging might be necessary to avoid flooding upstream.
- Bridges must be supported in a manner that the riverbank and mid-river supports are not exposed to erosion. This might be achieved by a layer of scour protection around the supports. The scour protection would often consist of rock of sufficient size to avoid being transported with the river current [10].
- Furthermore, the bridge supports must be designed to withstand the actions from ice. Of special interest are the bridge piers designed for the St. Lawrence River crossing to Prince Edward Island, the Confederation Bridge (see Figure 2.3). The piers have coned geometry in the waterline so ice can break up when forced against the bridge pier [7,16].

FIGURE 2.2 Testing the use of geobags to reduce erosion at Svea Mine, Spitsbergen. (Reproduced from Caline, F., Coastal sea ice action on a breakwater in a microtidal inlet in Svalbard, PhD thesis, NTNU, Trondheim, No. 226, October 2010.)

FIGURE 2.3 Confederation bridge (Canada) still under construction. (From Bercha Group, Homepage of Bercha Group, Consulting Services Since 1975, http://www.berchagroup.com/, 2012. With permission from Frank Bercha, Bercha Group, Calgary, Canada.)

- Of particular concern is the height of the bridge deck. In cases where the water flows against the bridge deck, there is a large possibility for loss of the bridge. This calls for data collection of the current speed, maximum water level, and ice conditions. Traditional knowledge regarding past extremes should be consulted to help define the extreme design situation. Like offshore structures for oil and gas, a safety level should be selected to ensure that the probabilities of bridge damage and bridge collapse are within acceptable levels agreed by the society. For offshore structures for the oil and gas industry, these levels are set as annual exceedances of 10^{-2} and 10^{-4}, respectively [11].
- Pipelines for oil and gas produced from onshore fields will often have to cross rivers. They might cross rivers on bridges that should be designed according to the principles defined above, or they might cross rivers on the river bottom. In the last case, the pipelines will be exposed to strong transverse currents and must be designed to be stable on the bottom. Trenching might be required, although

scouring might undermine the pipeline and cause free span. The erosion of the riverbank might be even more dangerous and cause free spanning pipelines exposed to considerable currents, possibly causing pipeline breakage. For a discussion of the potential pollution effects of damaged and leaking pipelines in the Arctic, see Section 2.8. Utmost care must, therefore, be taken when designing pipeline river crossings to avoid large pollution effects.

2.3 Public Challenge: The Specific Conditions during the Ice Breakup

During the ice breakup in the spring, there are challenges associated with large loading on riverbanks, structures placed in the rivers, and there is a potential for flooding of large areas.

2.3.1 Conditions Caused by the Ice Breakup

Many of the rivers in the Arctic region flow toward the north, and ice begins to melt in the south, while the river is still frozen in the north. This situation creates flooding that may be excessive in the south and also creates huge water and ice pressure on the ice in the north, potentially causing the ice to pile up and jam the river flow, causing even larger flooding.

A series of images demonstrating this phenomenon is shown in Figure 2.4. The images are taken by the Moderate Resolution Imaging Spectroradiometer (MODIS) on NASA's Terra satellite, showing the spring flooding in May 2007.

The images were made with both infrared and visible light. Land, burned by wildfire in the past year or two, is marked burn scar. The first image was taken on May 14, 2007, while the river was still frozen [18].

The second image was taken on May 23. The southern extents of the Lena and its surrounding areas are dramatically flooded.

The last image was taken a week later on May 28, and the floods had moved north. Water had, however, spread far beyond the river's banks. The flood put 12 towns (1,000 houses) under water, damaged or destroyed 41 bridges, and affected more than 14,000 people [18].

Figure 2.5 shows the flooding situation in Yakutsk on May 20, 2010.

2.3.2 Mitigating Measures to Avoid Large Damages

Even if it may not be possible to avoid flooding caused by snowmelt and ice breakup, it is of importance to provide forecasts such that people living in the area where flooding might be expected can take necessary precautions. Evacuation should be properly planned (each household should have a boat available for evacuation), and the most important buildings should be located on higher grounds. In order to plan properly, historical knowledge of the extent of flooding must be consulted in order to provide a suitable design basis for all planning works.

Ice jamming downstream might worsen the flooding situation considerably. Potentially, the use of explosives will help to open the jammed section. Ice jam at a river crossing might damage the bridge and cause large losses, both in terms of money and the time it will take to rebuild the bridge. Figure 2.6 shows how the Kuskokwim River threatens to flood the Western Alaska village of lower Kalskag on May 7, 2012.

2.4 Public Challenge: The Effect on Hydrology of the Shrinking Shoreline

The shrinking shoreline damages the built structures placed on or near the shoreline and causes large problems for water supply organized in the vicinity of shorelines.

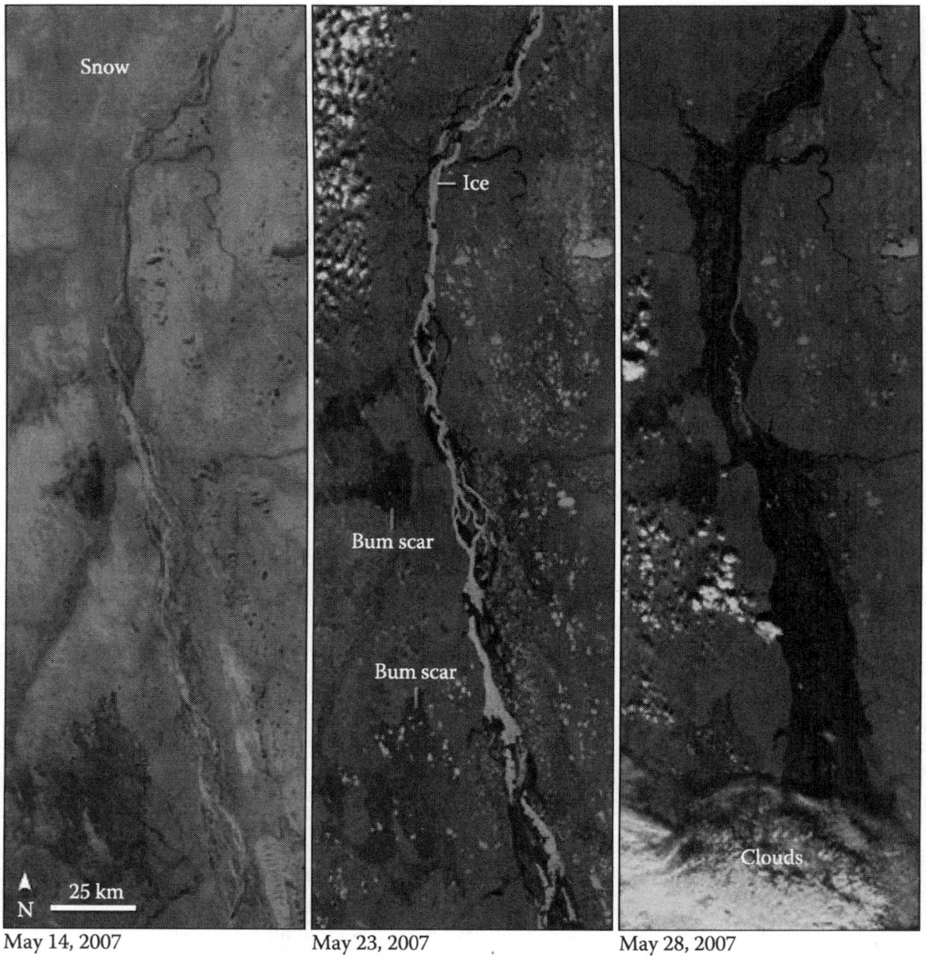

FIGURE 2.4 The ice breakup on River Lena, Siberia, 2007. (From NASA Earth Observatory, Flooding on the Lena River, http://earthobservatory.nasa.gov/NaturalHazards/view.php?id=18428. With permission from NASA's Earth Observatory according to their Image Use Policy.)

2.4.1 Shrinking Shoreline

The shrinking shoreline is an effect of the all-existing erosion of the Arctic shore. With increasing temperatures, the permafrost melts, resulting in increased shoreline erosion. This causes considerable problems for landfall of oil and gas pipelines. The erosion effect is reinforced by less ice offshore; the ice is arriving later and breaks up earlier. The open water area becomes larger as the ice border moves away from shore, providing larger fetch lengths where winds will increase the storm waves, leading to an increased erosion effect. Figure 2.7 shows the projected erosion of the Newtok village shoreline, Alaska [26]. The community of the village of Newtok has voted to relocate its 340 residents to new homes 9 miles away, up the Ninglick River.

In addition to erosion effects, the storm surge can create substantial flooding. Extreme storms combined with high tides can cause extensive flooding of low lands and damage engineering structures and freshwater reservoirs. On July 25, 2010, a large storm surge flooded the Varandey oil treatment terminal in Russia several hundred meters inland [8], causing massive damage to the terminal as the storm surge was combined with large waves. An airport built parallel to the shoreline was virtually destroyed during

FIGURE 2.5 Spring flood in Yakutsk. This picture was taken on May 20, 2010. (From Yakutsk, Spring Flood, The Lena River, Yakutsk, Yakutia/Siberia, http://www.flickr.com/photos/bolotbootur/http://nca2009.globalchange.gov/alaskasets/72157624105010964/, May 20, 2010. Reprinted with permission from photographer Bolot Bochkarev.)

FIGURE 2.6 Kuskokwim River threatening to flood an Alaskan village. Picture from the Alaska Division of Homeland Security's River Watch. (From DeMarban, A., Spring river flooding threat wanes across Alaska, save a few iffy villages, Alaska Dispatch, http://www.alaskadispatch.com/article/spring-river-flooding-threat-wanes-across-alaska-save-few-iffy-villages, 2012. Reprinted with permission from DeMarban.)

the flooding. With less Arctic ice, one can expect that such damaging storms to be more frequent and severe as the longer fetch length can cause the waves to be stronger than were in the past.

2.4.2 Effects on Hydrology

The shrinking shoreline could destroy settlements, and reports from the cold climate areas indicate extensive damage to built structures. Freshwater reservoirs can be destroyed, or flooding can cause

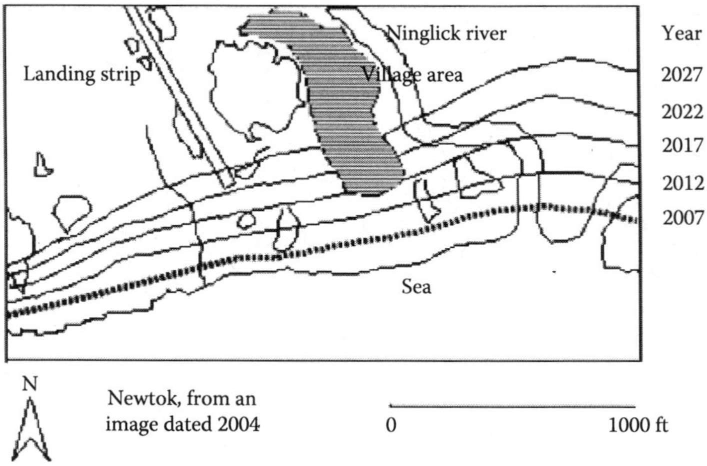

FIGURE 2.7 Newtok village; actual and projected shoreline erosion. (Redrafted from an image; see also US Global Change Research Program, Global climate change impacts in the United States, 2009, http://nca2009.globalchange.gov/alaska)

saltwater intrusion in the reservoirs. The hydrological effects can thus be large, and measures should be taken to prepare contingency plans in the event of flooding, and damage to freshwater reservoirs becomes a realistic situation. The damage to the shoreline will also cause changes to wildlife habitats, in particular, to bird nesting locations.

2.5 Public Challenge: Hydrology and Spreading of Pollution

The flow of water can cause severe spreading of pollution, and it is necessary to plan and implement mitigating measures to avoid such spreading.

2.5.1 Spreading of Pollution with the Water Flow

Noncontrolled flow of water can spread pollution from open dumps and dams holding waste, for example, waste from the mining or oil industries. The flow during the snowmelt season could, in particular, be noncontrollable, and poisoned water can find its way to water reservoirs or to settlement areas.

2.5.2 Mitigating Measures to Limit Spreading of Pollution

To avoid spreading of pollution with flowing water, in particular in cold climate areas where the snowmelting and river ice breakup could cause large flooding, the communities must map the older dumps and make every effort to cover the dumps properly. New dumps and waste areas must be located on higher grounds and properly be secured to avoid spreading of polluted water. However, seepage could still occur, in particular, if the permafrost melts and poisoned water finds its way from covered dumps to the water reservoirs. In the event that such situations are threatening, the dumps will have to be dug up, and the potentially polluting soil must be securely placed. In a worst case, massive pollution could make a settlement inhabitable. For reference, see the discussion of the effects of the large oil pollution in Komi Republic, Russia, in 1994, Section 2.8.

For the extraction of oil from tar sand, substantial amounts of water are being used. This water becomes contaminated and must be stored in large tailing ponds. This both depletes local water reserves and threatens to pollute the environment and nearby communities through leakages and during flooding [9]. For reference, see Figure 2.8.

FIGURE 2.8 View of tailing ponds looking south over the Syncrude processing facilities and upgrader north of Fort McMurray, Alberta. (From Greenpeace: Water pollution, http://www.greenpeace.org/canada/en/campaigns/Energy/tarsands/archive/threats/water-pollution. Reprinted with permission from Greenpeace.)

2.6 Public Challenge: Drinking Water Availability

In the Arctic region, the all-year availability of drinking water represents a large challenge, and there is a need to carefully plan for the provision of clean and nonpolluted drinking water to the communities.

2.6.1 Challenge to Provide Drinking Water

In the Arctic climate, permafrost is hindering natural groundwater from providing drinking water to the public. One is, therefore, during the winter dependent on water from frozen rivers or lakes. Shallow reservoirs, however, freeze, and much of the water is contained by the frozen ice. A primary concern is possible pollution, either from the ground (e.g., from dissolved metals) or from the public, in particular from older dumps that were common and that are leaking. Water treatment is, therefore, necessary to ensure acceptable water.

2.6.2 Example 1 Solution: Drinking Water to Longyearbyen, Svalbard

In winter, drinking water in Longyearbyen, 78°13′ N, 15°38′ E, comes from Isdammen located east of the settlement (Figures 2.9 and 2.10). Isdammen originally consisted of three small lakes and was established during the construction of the road to Mine 5. Today, Isdammen comprises a volume of 2.7 million m^3. The depth is on average 1.7 m, with a maximum of 5 m and water intake at 4 m. In winter, the ice thickness is typically less than 1.5 m. A large volume of the water reservoir is frozen to ice. The water source is basically meltwater from glaciers through the Endalen River [14] and possibly from groundwater [13] through an opening, "a talik" in the permafrost. The natural catchment area is 34.4 km^2, of which 15% is covered by glaciers [23] (Figure 2.11) [4].

The image shown in Figure 2.10 was acquired on July 12, 2003, and the area is located at 78.2° N latitude, 15.6° E longitude.

FIGURE 2.9 Isdammen, water drinking resource for Longyearbyen, Svalbard, during the winter period. (Printed with permission from Ivar Undheim, Leader Longyearbyen Lokalstyre, http://www.lokalstyre.no/)

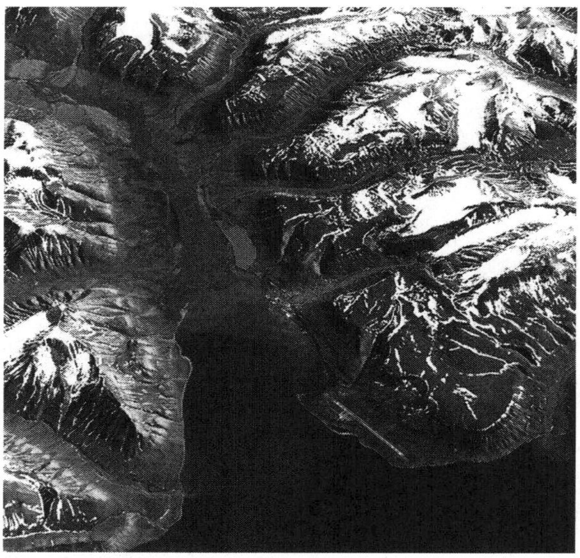

FIGURE 2.10 Longyearbyen, Svalbard, with airport and Isdammen. (From NASA: http://photojournal.jpl.nasa.gov/catalog/PIA10626. Published with permission from NASA's Earth Observatory according to their Image Use Policy.)

Before the water reaches the tap in Longyearbyen, it is filtered in three large sand filters and passes through UV radiation. The water treatment plant ensures that the water in Longyearbyen meets public drinking water requirements. The water supply net is 30 km long and covers the Longyearbyen settlement to the airport. The waterline is above ground and is insulated and equipped with heating cables to ensure the water does not freeze [15].

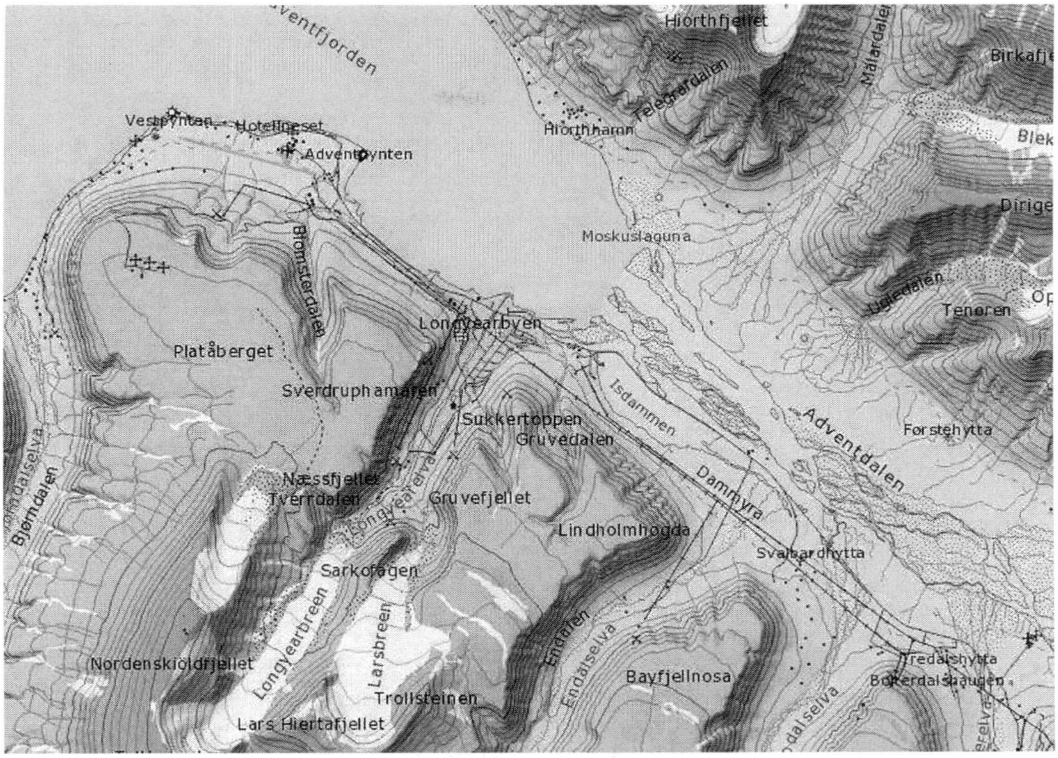

FIGURE 2.11 Isdammen and Endalen River, Longyearbyen, Svalbard. (From http://toposvalbard.npolar.no/. Published with permission from Norwegian Polar Institute, Tromsø.)

2.6.3 Example 2 Solution: Drinking Water to Barrow, Alaska

Isatkoak (Esatkuat) Lagoon is located centrally in the Barrow settlement (Figure 2.12). The lagoon remains partially unfrozen year round. The lagoon is separated into sections by artificial berms, and the upper lagoon serves as the primary source of drinking water for the city of Barrow, 71°23′ N and 156°28′ W. Water drawn from the lagoon is treated by filtration and distributed to the settlement.

2.7 Public Challenge: Sewer System and the Influence on the Hydrology

The design of a proper sewer system in an Arctic community represents a major challenge that needs to be given high attention. Of particular importance is the challenge to avoid pollution to the environment and to drinking water basins.

2.7.1 Challenge to Hydrology from the Sewer System

All modern societies must have a sewer system. The sewer will be mixed with water and could represent a source of pollution, in particular if the sewer comes in contact with the free flowing water. In areas where flooding occurs, the release of sewer could cause contamination of the freshwater system and a considerably unpleasant smell to the community.

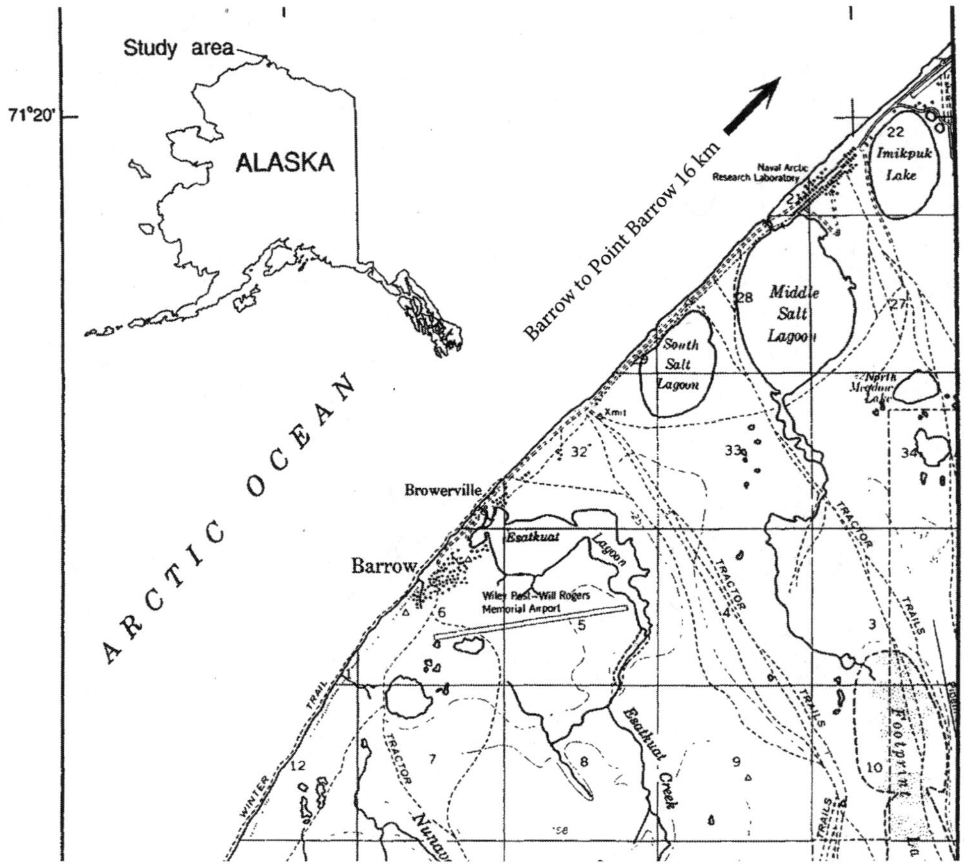

FIGURE 2.12 Location of Barrow, Alaska [17]; dashed lines represent 25-foot contours. (Courtesy of U.S. Geological Survey, Department of the Interior/USGS.)

2.7.2 Example Solution: Sewage System for Longyearbyen, Svalbard

Contained in Longyearbyen, there are exposed sewer pipes with swan-necks. These are connected to drain collection basins at each 50 m (see Figure 2.13). At the locations where the drain is being collected, there might be a little smell because the flow is dependent upon gravity and requires ventilation. These basins act as overflow basins if the drain becomes clogged or frozen. The drainage network is 23 km long. The drain is discharged at 50 m depth, 3 km into the Advent Fjord [14].

2.8 Effect from Hydrocarbon Production Activities

During the 1970s, oil production was increased in Russia due to an oil shortage in the world market, and the lack of proper technology (in particular pipeline technology and lack of leak detection technology) caused large pollution levels in some Russian rivers and lakes. In 1975, for example, several large West Siberian rivers that run north through Russia's biggest oil production region and empty into the Arctic Ocean had oil concentrations 21 times the maximum permissible level, according to a government report, "Status of Environmental Pollution in the USSR 1975–1976" [20].

A very large oil spill occurred in August 1994, when old pipelines in the northern Komi Republic began to leak. The oil spill was officially measured at 79,000 ton or 585,000 barrels; however,

FIGURE 2.13 Sewer collection system; isolated pipes with heating cables. (From Longyearbyen Lokalstyre/Svalbard Samfunnsdrift, www.norvar.no/content/download/9434/104875. Reprinted with permission from Ivar Undheim, Leader Longyearbyen Lokalstyre, http://www.lokalstyre.no/)

independent estimates put it at up to 2 million barrels [24]. The oil was stored in a large dam, as was the state of the art for containing pollution in the former Soviet Union. Two months after the spill began, heavy rains broke the dam, and a large amount of oil was released into rivers and across the tundra near the city of Usinsk [20].

The oil that did not immediately spill into the Arctic Ocean-bound Kolva, Usa, and Pechora rivers spread over 186 km^2 of marshland and tundra [22]. There it froze during winter months. The following spring, the oil from the frozen tundra washed back into the streams, seeped into the surrounding vegetation or traveled further down the Pechora to the Barents Sea.

2.9 Traditional Knowledge

The value of traditional knowledge should not be underestimated. This knowledge is critical when providing a design basis for any activity. In particular, in cold climates where written knowledge may be scarce, traditional knowledge going back hundreds of years will be extremely important [12]. Regarding hydrological aspects, large situations of flooding, large river ice situations with severe ice breakup, and large storm surge effects are some of the situations where traditional knowledge, even legends, might enlighten and constitute a design basis for engineering projects.

2.10 Summary and Conclusions

With reference to [12], the schematic of the Arctic hydrological cycle in a region of thick ice-rich permafrost can be presented as in Figure 2.14. Snowmelt is the most important event leading to large water flow and considerable challenges to engineered structures. In the event we do not respond properly to these challenges, large damage may occur. In this chapter, some descriptions of the challenges have been provided and pointed to solutions that may be implemented to minimize the damage.

Of particular recommendations in this chapter are measures to avoid environmental damages and consultation with the local society to identify past events that can be seen as dimensioning for future events. This will not, however, lead us to disregard models taking new trends into account, such as planning for a potentially future warmer climate. In a chapter on the hydrology of the cold climate regions, it must also be reminded that all life, vegetation, birds, animals, and humans are dependent upon clean, nonpolluted, and abundant volumes of water resources.

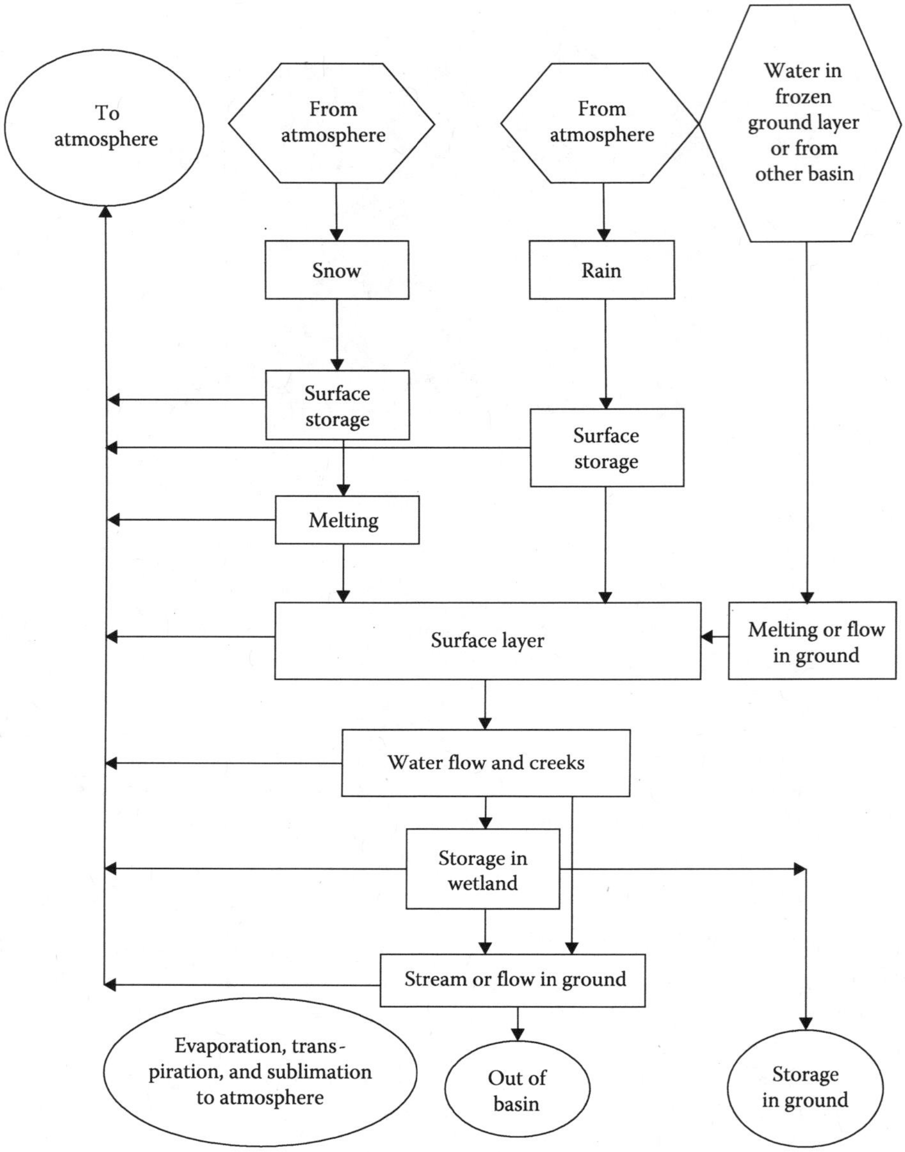

FIGURE 2.14 Schematic of the Arctic hydrological cycle in a region of thick ice-rich permafrost where we have added the interaction with frozen groundwater. We refer also to [28], where groundwater interaction is not considered.

References

1. Study of Environmental Artic Change (SEARCH) project, ACIA Map of Arctic River Basins, http://www.arctic.noaa.gov/detect/land-river-acia-image.shtml. Accessed 12 July 2013.
2. Bachman, J., 2010, Special report: Oil and ice: Worse than the gulf spill? http://www.reuters.com/article/2010/11/08/us-russia-oil-idUSTRE6A71IL20101108. Accessed 12 July 2013.
3. Bercha Group, 2012, Homepage of Bercha Group, Consulting Services Since 1975. http://www.berchagroup.com/. Accessed 12 July 2013.

4. Bogen, J. and Bønsnes, T., 2003, Erosion and sediment transport in High Arctic Rivers, Svalbard, *Polar Research,* 22(2), 175–189.
5. Caline, F., October 2010, Coastal sea ice action on a breakwater in a microtidal inlet in Svalbard, PhD thesis, NTNU, Trondheim, No. 226.
6. DeMarban, A., 2012, Spring river flooding threat wanes across Alaska, save a few iffy villages. Alaska Dispatch. http://www.alaskadispatch.com/article/spring-river-flooding-threat-wanes-across-alaska-save-few-iffy-villages. Accessed 12 July 2013.
7. Frederking, R., Kubat, I., and S. Prinsenberg, S., 2006, Response of two piers on confederation bridge to ice loading event of 4th April, 2003, *Proceedings of 18th International Symposium on Ice, IAHR'06,* Vol. 1, pp. 231–238, Sapporo, Japan.
8. Gazprom, May 2012, Private information provided by personnel from Gazprom during an Arctic Engineering Seminar arranged at the Gubkin University of Oil and Gas, Moscow, Russia.
9. Greenpeace, Water pollution. http://www.greenpeace.org/canada/en/campaigns/Energy/tarsands/. Accessed 12 July 2013.
10. Hughs, S.A., 2001, *Scour and Scour Protection,* U.S. Army Corps of Engineers, Trinidad, CO, September 24–28. http://www.oas.org/cdcm_train/courses/course4/chap_8.pdf
11. International Standardization Organization, ISO 19906, December 2010, *Arctic Offshore Structures,* ISO, Geneva, Switzerland.
12. Kane, D.L., September 2000, Arctic hydrological processes, In: Hinzman, L. and Vörösmarty, C., NSF-ARCSS Workshop on Arctic System Hydrology, Santa Barbara, CA. http://arcticchamp.sr.unh.edu/pdffiles/hinzmanvorosmarty2001.pdf
13. Killingtveit, Å., 2001, Er det grunnvannsinnstrømming til Isdammen, Svalbard i vinterhalvåret ?. Abstract in: Aune, T. and Cramer, J.: Program for Det 10. Seminar om hydrologi og miljøkjemi, NGU February 8–9, 2001, NGU Report 2001.010, (In Norwegian). See also: http://www.ngu.no/upload/Publikasjoner/Rapporter/2001/2001_010.pdf
14. Longyearbyen Lokalstyre/Svalbard Samfunnsdrift. http://norskvann.no/kompetanse/norsk-vann-bulletin. Bulletin nummer 3, oktober 2011. Accessed 12 July 2013.
15. Local Administration of Longyearbyen, Svalbard, Norway. http://www.lokalstyre.no/vann-og-avloep.5045471-247429.html. Accessed 12 July 2013.
16. Løset, S., Shkhinek, K., Gudmestad, O., and Høyland, K., 2010, *Impact of Ice on Offshore and Onshore Facilities,* LAN Publishing House, St. Petersburg, FL.
17. McCarthy, K.A., 1994, *Overview of Environmental and Hydrogeologic Conditions,* U.S. Geological Survey, Barrow, AK, Open-File Report 94–322.
18. NASA Earth Observatory. Flooding on the Lena River. http://earthobservatory.nasa.gov/NaturalHazards/view.php?id=18428. Accessed 12 July 2013.
19. NASA Jet Propulsion Laboratory, Califormia Institute of Technology. http://photojournal.jpl.nasa.gov/catalog/PIA10626. Accessed 12 July 2013.
20. OSS - Oil Spill Solutions Ltd. http://www.oilspillsolutions.org/contraversialspills.htm. Accessed 12 July 2013.
21. Owens, E., Solsberg, L.B., West, M.R., and McGrath, M., 1988, Field guide for oil spill response in Arctic waters, prepared for Arctic Council Emergency, Prevention, Preparedness and Response (EPPR) Working Group, Environment Canada, Yellowknife, Northwest Territories, Canada, p. 348. See also: http://www.fargisinfo.com/Referanser/LinkedDocuments/fldguide.pdf. Accessed 12 July 2013.
22. Russian Nature, 2012, Environmental problems of Northern Eurasia: Environmental Impact of Oil and Gas Development. http://rusnature.info/env/20.htm
23. Sund, M., 2008, Polar Hydrology, Report No 2, 2008, Norwegian Water Resources and Energy Directorate's Work in Svalbard. http://www.nve.no/global/publikasjoner/publikasjoner%202008/report%202008/report2-08.pdf

24. Tørum, A., Burcharth, H.F., Goda, Y., Kortenhaus, A., and Kriebel, D., 2007, A new international standard for actions from waves and currents on coastal structures. In: J.M. Smith (ed.), *Coastal Engineering Proceedings of the 30th International Conference*, World Scientific, San Diego, CA, September 3–8, 2006. Vol. 5, pp. 4290–4302.
25. Tørum, A., 2012, Coastal structures: Action from waves and ice, In: Y.C. Kim (ed.), *Series on Coastal and Ocean Engineering Practice*, World Scientific, Singapore, Vol. 1, pp. 95–252.
26. US Global Change Research Program, 2009, Global climate change impacts in the United States. http://nca2009.globalchange.gov/alaska. Accessed 12 July 2013.
27. Vilcheck, G.E. and Tishkov, A.A., 1997, Usinsk oil spill. In: R.M.M. Crawford (ed.), *Disturbance and Recovery in Arctic Lands*. Kluwer Academic, Dordrecht, the Netherlands, pp. 411–420.
28. Woo Ming-Kong et al., 2007, *Science Meets Traditional Knowledge: Water and Climate* in the Sahtu (Great Bear Lake) Region, Northwest Territories, Canada, Arctic, Vol. 60(1), pp. 37–46. http://pubs.aina.ucalgary.ca/arctic/Arctic60-1-37.pdf. Accessed 12 July 2013.
29. Yakutsk, May 20, 2010, Spring Flood, The Lena River,Yakutsk,Yakutia/Siberia. http://www.flickr.com/photos/bolotbootur/http://nca2009.globalchange.gov/alaskasets/72157624105010964/

3

Conjunctive Use of Groundwater and Surface Water in a Semiarid Hard-Rock Terrain

Shrikant Daji Limaye
Ground Water Institute (NGO)

AUTHOR

Dr. Shrikant Daji Limaye graduated in engineering from Poona University, India, and completed his master's in geophysics at the Indian Institute of Technology (Kharagpur) followed by his PhD in geology. For the past five decades, he has been working on national and international assignments in sustainable groundwater development for ensuring good quality and adequate quantity of groundwater. He has been a consultant to farmers for developing groundwater source in their farms for small-scale irrigation. He has also worked extensively on soil and water conservation, watershed development, forestation, and recharge augmentation. Dr. Limaye has widely traveled in over 35 countries and studied groundwater development. He has been associated with several organizations and has presented technical papers in international conferences. The website www.igcp-grownet.org of his UNESCO–IUGS–IGCP project GROWNET has attracted around 27,000 visits so far and is considered to be a unique source of information on groundwater development in low-income countries.

PREFACE

In semiarid, hard-rock regions in India, the surface water resources are scanty and available only for a few weeks after the monsoon rains. But if surface water is made to percolate into the shallow phreatic aquifer, then its velocity is reduced and it stays in the watershed for a few years, because of the low permeability of hard-rock aquifers. Monsoon rainfall takes place in about 3–4 months of the year (June–September) during which the fractured rock aquifer gets fully recharged. This storage gets depleted in the following prolonged dry period of winter and summer, due to pumpage from dug wells/bore wells and due to dry season flow and underflow in the beds of effluent streams/rivers. This depletion is a major portion of the total storage in the aquifer. The drawdown during the 8 months of dry season thus becomes almost equal to the total saturated thickness of phreatic aquifer. Groundwater flow does not take place across the surface water divides so that a subbasin or a mini-basin could be treated as a unit for water resources development and management. The groundwater body in a mini-basin or mini-watershed of 10–50 km^2 area is thus very small in size and is highly sensitive to the recharge during monsoon season. Activities related to water conservation and artificial recharge, during and after the rainy season, have therefore an important role to play. However, after the aquifer gets fully saturated, additional infiltration during the monsoons is rejected. In order to obtain sustainable supply in summer months, any increase in the pumpage for irrigation due to additional new wells in a mini-basin must be balanced by reduction in dry season streamflow and underflow and also by increasing the dry season recharge through artificial means. This chapter discusses the importance of "dry season artificial recharge" through percolation tanks (PTs). Side-by-side, it also emphasizes the necessity to adopt soil, water, and biomass conservation techniques in order to achieve sustainable supplies from the ever-increasing number of new wells and bores, due to population pressure in hard-rock areas. In future, the climatic pattern is going to be harsher and erratic. Monsoon rains may become more irregular. A watershed is the meeting point of climatology and hydrology. It is therefore necessary to manage our watersheds to absorb these climatic shocks. This can only be done through practicing soil and water conservation techniques for promoting recharge during rainy season and through construction of small PTs for artificial recharge during the dry season. These activities provide the foundation for conjunctive use of surface water and groundwater, in the rural scenario. This chapter has not considered the urban sector where conjunctive use is achieved by collecting runoff water from the streets or from the terraces of buildings and putting it into dug wells or bore wells after filtration. Although this is based on the author's experience in India, the concept of conjunctive use through PTs would be useful also in other countries having semiarid, hard-rock terrain.

3.1 Introduction

Groundwater is a natural resource, and its development could be defined as a concerted activity toward its sustainable use for human benefit. The concept of sustainable use is related to various factors like the volume of water storage in the aquifer, annual recharge or replenishment, volume of annual pumpage for the proposed use, benefit/cost ratio of the proposed use, and environmental impacts of the proposed use.

Hard-rock aquifers in this chapter include the noncarbonate, fractured rocks like the crystalline basement complex and metamorphic rocks, which cover an area of about 800,000 km^2 in central and southern India. Basalts of western India also known as the Deccan traps of late Cretaceous to early Eocene period are also included as a special case (Figure 3.1a). Deccan traps comprise hundreds of

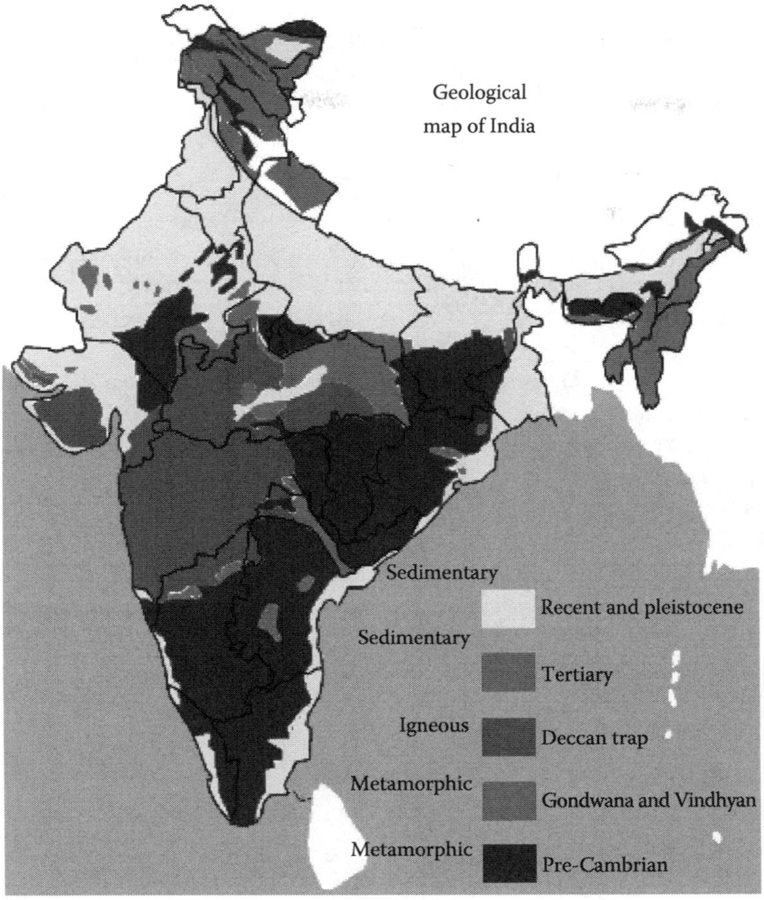

(a)

(b)

FIGURE 3.1 (a) Geological map of India showing basement complex and basalts of peninsular region (scale: 1 cm = 200 km). Basement complex (granite, gneiss, and other metamorphic rocks) is shown in dark gray. Basaltic area is shown in meduim gray. Black lines show state boundaries. The state of Maharashtra on the western coast is almost covered by basalt or Deccan trap (medium gray). Peninsular India is mostly covered by basalt and basement complex. (b) Nearly horizontal lava flows comprising the Deccan traps or basalts of western India. Location: Western Ghats hills between Pune and Mumbai, Maharashtra State.

nearly horizontal, basaltic lava flows in a thick pile and cover around 500,000 km^2 of western India (Figure 3.1b). The area covered by Deccan traps is the largest lava flow terrain in the world.

This pile was not tectonically disturbed after consolidation, and a hand specimen does not show any primary porosity due to the nonfrothy nature of the lava [1]. Hydrogeologically, the Deccan traps have low porosity and are therefore akin to fractured hard-rock aquifers.

The most salient features of the hard-rock aquifers are as follows:

1. The depth of groundwater occurrence, in useful quantities, is usually limited to 100 m or so.
2. A topographical basin or a subbasin generally coincides with groundwater basin. Thus, the flow of groundwater across a prominent surface water divide is very rarely observed. In a basin, the groundwater resources tend to concentrate toward the central portion, closer to the main stream and its tributaries.
3. The aquifer parameters like storativity (S) and transmissivity (T) often show erratic variations within small distances. The annual fluctuation in the value of T is considerable due to the change in saturated thickness of the aquifer from wet season to dry season. When different formulae are applied to pump-test data from one well, a wide range of values of S and T is obtained. The applicability of mathematical modeling is limited to only a few simpler cases within a watershed. But such cases do not represent conditions over the whole watershed.
4. The phreatic aquifer comprising the saturated portion of the mantle of weathered rock or alluvium or laterite, overlying the hard fractured rock, often makes a significant contribution to the yield obtained from a dug well or bore well.
5. Only a modest quantity of groundwater, in the range of 1–100 m^3 or so per day, is available at one spot. Drawdown in a pumping dug well or bore well is often almost equal to the total saturated thickness of the aquifer.

Groundwater development in hard-rock aquifer areas in India and many other countries has traditionally played a secondary role compared to that in the areas having high-yielding unconsolidated or semiconsolidated sediments and carbonate rocks. This has been due to the relatively poor groundwater resources in hard rocks, low specific capacity of wells, erratic variations and discontinuities in the aquifer properties, and the difficulties in exploration and quantitative assessment of the resource.

It should, however, be realized that millions of farmers in developing countries have their small farms in fractured basement or basaltic terrain. Whatever small supply available from these poor aquifers is the only hope for these farmers for upgrading their standard of living by growing irrigated crops or by protecting their rainfed crops from the vagaries of monsoon rainfall. It is also their only source for drinking water for the family and cattle. In many developing countries, like in India, hard-rock hydrogeologists have, therefore, an important role to play in promoting conjunctive use of surface water and groundwater.

3.2 Occurrence of Groundwater

Groundwater under phreatic condition occurs in the soft mantle of weathered rock, alluvium, and laterite overlying the hard rock. Under this soft mantle, groundwater is mostly in semi-confined state in the fissures, fractures, cracks, and joints [2]. In basaltic terrain, the lava flow junctions and red boles sandwiched between two layers of lava flows also provide additional porosity (Figure 3.2).

The ratio of the volume of water stored under semiconfined condition within the body of the hard rock to the volume of water in the overlying phreatic aquifer depends on local conditions in the miniwatershed. Dug-cum-bored wells tap water from the phreatic aquifer and also from the network of fissures, joints, and fractures in the underlying hard rock (Figure 3.3a and b).

The recharge to groundwater takes place during the rainy season through direct infiltration into the soft mantle overlying the hard rock and also into the exposed portions of the network of fissures and

FIGURE 3.2 Red bole (intertrappean bed) sandwiched between hard, fractured basalt flows.

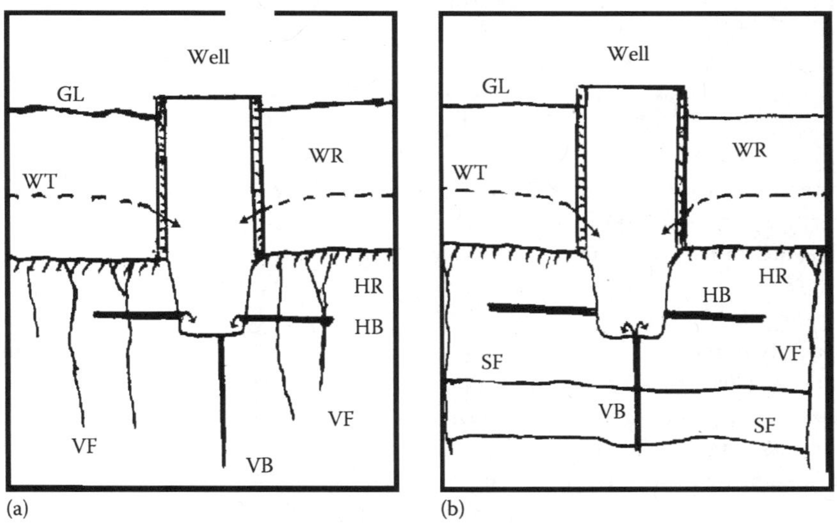

GL, Ground level; HB, horizontal bore; HR, hard rock;
SF, sheet fracture or joint; VB, vertical bore;
VF, vertical fracture; WR, weathered rock; WT, water table.

FIGURE 3.3 (a) and (b) Dug-cum-bored wells.

fractures. In India and other Asian countries in monsoon climate, the ratio of recharge to rainfall in hard-rock terrain is assumed between 3% and 15% [6]. This ratio depends upon the amount and nature of precipitation, the nature and thickness of topsoil and weathered zone, type of vegetation, evaporation from the surface of wet soil, profile of underlying hard rock, the topographical features of the subbasin, and the status of soil and water conservation activities adopted by villagers. Groundwater flow rarely occurs across the topographical water divides, and each basin or subbasin can be treated as a separate hydrogeological unit for planning the development of groundwater resources. After the rainy season, the fully recharged hard-rock aquifer gradually loses its storage mainly due to pumpage and effluent drainage by streams and rivers. The dry season flow of the streams is thus supported by groundwater outflow. The flow of groundwater is from the peripheral portions of a subbasin to the central-valley portion, thereby causing dewatering of the portions closer to topographical water divides. In many cases, the dug wells and bore wells yielding perennial supply of groundwater can only be located in the central-valley portion.

The annual recharge during monsoons being a sizable part of the total storage of the aquifer, the whole system in a subbasin or mini-basin, is very sensitive to the availability of this recharge. A couple of drought years in succession could pose a serious problem. The low permeability of hard-rock aquifer is a redeeming feature under such conditions because it makes small quantities of water available, at least for drinking purpose, in the dug wells or bore wells in the central portion of a subbasin. If the hard rocks had very high permeability, the groundwater body would have quickly moved toward the main river basin, thereby leaving the tributary subbasins high and dry. The low permeability in the range of 0.05–1.0 m/day thus helps in retarding the outflow and regulating the availability of water in individual farm wells. More farmers are thus able to dig or drill their wells and irrigate small plots of land without causing harmful mutual interference.

3.3 Groundwater Development

In the highly populated but economically backward areas in hard-rock terrain, governments in many developing countries have taken up schemes to encourage small farmers to dig or drill wells for small-scale irrigation. This is especially true for the semiarid regions where surface water resources are meager and seasonal. For example, in peninsular India, hard rocks such as granite, gneiss, schist, quartzite (800,000 km^2), and basalts (Deccan traps—500,000 km^2) occupy about 1.30 million square kilometers area out of which about 40% is in semiarid zone, receiving less than 750 mm rainfall per year. Over 4.50 million dug wells and bore wells are being used in the semiarid region for irrigating small farm plots and for providing domestic water supply.

The development of groundwater resources for irrigational and domestic use is thus a key factor in the economic thrift of vast stretches of semiarid, hard-rock areas. The basic need of millions of farmers in such areas is to obtain an assured supply for protective irrigation of at least one rainfed crop per year and to have a protected, perennial drinking water supply within a reasonable walking distance. The hard-rock hydrogeologists in many developing countries have to meet this challenge to impart social and economic stability to the rural population, which otherwise would migrate to the neighboring cities. Rapid and uncontrolled urbanization caused by exodus of rural population toward the cities is creating excessive stress and degradation in urban infrastructure. This is a common problem for many developing countries and could only be solved by providing assurance of at least one crop and increasing the opportunity for employment on farms for landless people.

Groundwater development in a subbasin results in increased pumpage and lowering of the water table due to the new wells, resulting in the reduction of the effluent drainage from the subbasin. Such development in several subbasins draining into the main river of the region reduces the surface flow and the underflow of the river, thereby affecting the function of the surface water schemes depending on the river flow. In order to minimize such interference, it is advisable to augment groundwater recharge by adopting artificial recharge techniques during rainy season and also during dry season. The measures for artificial recharge during monsoon rains include contour trenching on hillslopes, contour bunding of farms, gully plugging, farm ponds, underground stream bunds, and forestation of barren lands with suitable varieties of grass, bushes, and trees. Artificial recharge in dry season is achieved through the construction of PTs.

However, the increase in pumpage takes place through the initiative of individual farmers to improve their living standard through the irrigation of high-value crops, while recharge augmentation is traditionally considered as government's responsibility and always lags far behind the increase in pumpage. In many parts of the world, particularly in developing countries, groundwater is thus being massively over-abstracted. This is resulting in falling water levels and declining well yields, land subsidence, intrusion of saltwater into freshwater supplies, and ecological damages, such as drying out wetlands.

Groundwater governance through regulations has been attempted without much success, because the farmers have a strong sense of ownership of groundwater occurring in their farms. Integrated Water Resources Management (IWRM) is being promoted as a policy or a principle at national and international levels, but in practice at field level, it cannot be attained without the cooperation of rural community. NGOs sometimes play an important role in educating the villagers and ensure their cooperation.

3.4 Conjunctive Use through Dry Season Recharge from Percolation Tank

During the rainy season from June to September, the recharge from rainfall causes the recuperation of water table in a subbasin from its minimum level in early June to its maximum level in late September. This is represented by the equation

$$P = R + ET + r \tag{3.1}$$

where

P is the precipitation

R is surface runoff

ET is evapotranspiration during the rainy season

r is the net recharge, represented by the difference between the minimum storage and maximum storage in the aquifer

However, after the aquifer gets fully saturated, the additional infiltration during the monsoons is rejected and appears as delayed runoff.

During the dry season, the depletion of the aquifer storage in a subbasin, from its maximum value to minimum value, is represented by the following equation:

$$S = s + P + F + R \tag{3.2}$$

where

S is the aquifer storage at the end of rainy season, that is, maximum storage

s is the aquifer storage at the end of summer season, that is, minimum storage

P is the pumpage, mainly for irrigation, during the dry season from dug wells and bore wells

F is the dry season effluent streamflow and underflow supported by groundwater

R is the recharge, if any, available during the dry season, including the return flow from irrigated crops

The left-hand side of the aforementioned equation has an upper limit, as mentioned earlier. On the right-hand side, the minimum storage cannot be depleted beyond a certain limit, due to the requirement for drinking water for people and cattle. Dry season streamflow and underflow supported by groundwater have to be protected, as explained earlier, so that the projects depending upon the surface flow of the main river are not adversely affected. Any increase in the pumpage for irrigation during dry season due to new wells must therefore be balanced by increasing the dry season recharge.

The best way to provide dry season recharge is to create small storages at various places in the basin by bunding gullies and streams for storing runoff during the rainy season (Figure 3.4) and allowing it to percolate gradually during the first few months of the dry season. Such storages created behind earthen bunds put across small streams are popularly known as PTs.

In semiarid regions, an ideal PT with a catchment area of 10–50 km^2 or so holds maximum quantity by the end of September and allows it to percolate for the next 4–5 months of winter season. The excess of runoff water received in monsoon flows over the masonry waste weir constructed at one end of the earthen bund. By February or March, the tank is dry, so that the shallow water body is not exposed to

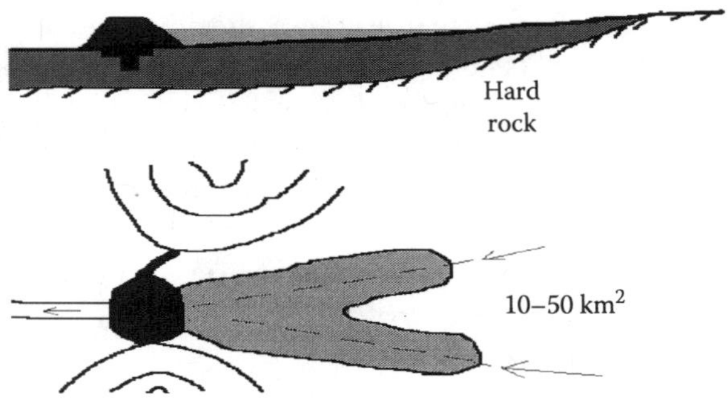

FIGURE 3.4 Cross section and plan of a typical PT on a stream between two hillocks. Hillocks are represented by black contours. Earthen bund of the PT is shown in dark gray. Cutoff trench below the bund is in black and accumulated rainwater is in light gray.

high rates of evaporation in summer months (Figure 3.6). Groundwater movement being very slow, whatever quantity percolates between October and March, is available in the wells on the downstream side of the tank even in summer months till June or the beginning of next monsoon season. The irrigation of small plots by farmers creates greenery in otherwise barren landscape of the watershed. Studies carried out in granite–gneiss terrain have indicated that about 30% of the stored water in the tank percolates as recharge to groundwater in the dry season. The efficiency is thus 30%. In basaltic terrain, if the tank is located at suitable site and the cutoff trench in the foundation of tank bund does not reach up to the hard rock, higher efficiencies up to 70% could be obtained [5]. However, more research is required for the estimation of the impact of PTs in recharge augmentation. In the state of Maharashtra in western India, over 10,000 PTs have been constructed so far [3] (Figure 3.5). They are beneficial to the farmers and are very popular with them.

The initial efficiency of a PT reduces due to silting of its bottom by receiving muddy runoff from the watershed. If the watershed is well forested and has a cover of grass, bushes, and crops, the silting is

FIGURE 3.5 Stone pitching on the face of the earthen bund of a PT under construction. Photo from village Hiware Bazar, District: Nagar, Maharashtra State.

FIGURE 3.6 A PT about to get dry toward the beginning of summer. Location: Village: Ralegan Siddhi, District: Nagar, Maharashtra State.

minimal. But in an average of 5–6 monsoon seasons, the tank bed accumulates about 0.20–1.00 m of silt. Silt reduces the storage capacity of the tank and also impedes the rate of vertical flow of recharge because of its low permeability. The efficiency gets reduced due to silting and desilting of tank bed when it dries in summer becomes necessary [4] (Figure 3.6).

Another type of recharge available during the dry season is the return flow or the percolation below the root zone of crops from irrigated farms. This return flow to groundwater is usually estimated at about 25%–30% of the volume of groundwater pumped in dry season and applied for irrigation. However, due to increasing popularity of more efficient irrigation methods like sprinkler or drip systems, this type of recharge has a declining trend.

3.5 Problems at Field Level

Although the construction of PTs and of soil and water conservation structures in a watershed promotes conjunctive use of surface water and groundwater, such recharge augmentation in dry season at several PTs in a river basin could create problems at field level. Suppose a major surface water supply project has been competed by building a big dam on a river and the large reservoir created behind the dam is supplying water to major cities and industrial areas. While designing such a major dam, certain calculations are made as to the "water crop" or "water harvest" coming into the reservoir for a given amount of rainfall in the large catchment area. If several PTs are constructed on small streams within this large catchment area of the river and water in these tanks makes a significant contribution to groundwater recharge in dry season, the "water harvest" at the big dam would get reduced, thereby causing water scarcity in cities and industrial areas. In semiarid regions, significant percentage of water accumulated at tiny field bunds constructed for soil conservation may be lost to evaporation and may not contribute to surface runoff or to groundwater recharge. The water supply department looking after the water supply to cities and industries from the reservoir at the dam is interested in getting more "water harvest" and would therefore object to the construction of too many PTs and field bunds. On the other hand, the farmers in the catchment area at higher level than the reservoir like to assert their right on the runoff that flows from their farms to the reservoir, by impounding the runoff at tiny farm bunds and PTs.

In such a conflict of interests, the NGOs usually take the side of dryland farmers in the catchment area and support their right to harvest the rainfall in farm ponds, field bunds, and PTs, so as to promote recharge to groundwater. Promoting such recharge is the only way for these poor farmers to get some assured crop by using groundwater for small-scale irrigation and to obtain safe quality drinking water. Moreover, the NGOs argue that the surface water collected in the large reservoir at the dam is never used efficiently. The farmers in the lower valley getting water for irrigation from canals coming from the large reservoir always over-irrigate their farms causing waterlogging in the command area of the canals. Such a gross wastage of water should be avoided first before complaining about the reduction in "water harvest" due to field bunds and PTs in the catchment area.

One more point in favor of the construction of field bunds and PTs to promote groundwater recharge in catchment area is that they reduce the silt load coming from the catchment into the reservoir of the big dam, thereby prolonging the useful life of the dam project.

It may, however, be noted that there is also a change of domain here. Surface water management is mostly in government sector, that is, public domain. Groundwater is traditionally owned by farmers and is in the private domain. Questions could be asked as to the productivity, that is, the contribution to the GDP from 1 m^3 of groundwater recharged from surface water for agricultural use by farmers, vis-a-vis its productivity if that cubic meter of water was used in surface water supply to city or industry. Although the contribution to GDP per cubic meter of water from urban sector is much more than that from the agricultural sector, the farmers' population has to survive and produce food. So, contribution to GDP cannot be the sole criterion. In other words, a balanced approach is needed at field level in order to maintain amicable relations between the competing stakeholders. Industrial and urban wastewater has to be recycled and reused, while irrigation has to be made more efficient so as to generate more output per cubic meter of water. IWRM through the conjunctive use of surface water and groundwater is thus a complicated matter at field level.

Equally complicated is the management or pumpage control of groundwater even in overexploited watersheds, because groundwater is traditionally the landowner's property. A farmer has therefore a right to dig or drill anywhere in his land and to pump as much water as possible, even if such pumping results in diminishing the yields of surrounding wells owned by others. There are no legal battles over this. Suppose a farmer has 1 ha of farm in a watershed of 500 ha and luckily has a very good well in his farm. Then, he pumps water for day and night and irrigates the whole farm with water-intensive crops like banana or sugarcane. Here, his equitable share in the groundwater of the watershed is only 0.2%. Because he is lucky to have a very good well in the farm, how much should he increase his share in groundwater resources of the watershed? Ten to fifty times? Such questions are difficult to answer. Moreover, through the intervention and social pressure from the village council, if the farmer reduces his pumping say by 50%, in hard-rock aquifers, it is difficult to assess where the unpumped 50% would flow. So a better alternative is to pump the high-yielding wells to full extent and give 50% to the well owner and 50% to neighboring farmers. The same high-yielding wells should be used for recharging during the rainy season thereby promoting the conjunctive use. Such activities would not be possible through any legislation. Only the village councils and active NGOs could follow this difficult path, leading to a kind of social revolution in which community's right on private groundwater body would be established and the recharge activity would be partly brought into the private domain.

3.6 Summary and Conclusions

A watershed is the meeting point of climatology and hydrology. It is therefore necessary to manage our watersheds so as to absorb the climatic shocks likely to come from the erratic climatic patterns expected in the near future. This can be done only through practicing soil and water conservation techniques for artificial recharge during rainy season and through construction of small PTs for artificial recharge during the dry season.

Basin or subbasin management begins with soil and water conservation activities taken up with people's active participation in several subbasins within a large basin. This improves the shape of hydrograph of the stream or river in the basin, from a "small time-based and sharp-peaked hydrograph" to a "broad time-based and low-peak hydrograph." Such a change also increases groundwater recharge.

Small water storages or tanks created in the subbasins by bunding streams and gullies store runoff water during the monsoon season and cause recharge to groundwater during the next few months of dry season. The residence time of water in the basins is thus increased from a few months to a few years, and the percolated water is available in the wells even during the summer season of a drought year.

After a few years of operation, silting of the tank bed reduces the volume of water stored and also the rate of vertical infiltration. Regular desilting of tanks by local people is, therefore, advisable.

A national policy for afforestation of degraded basins with proper species of grass, bushes, and trees should be formulated. Afforestation with eucalyptus trees should not be encouraged in low-rainfall areas as this effectively reduces groundwater recharge. The main aim of forestation of a degraded watershed with local spices of hardy trees, grasses, etc., should be to conserve soil, reduce velocity of runoff water, promote recharge to groundwater, and increase the biomass output of the watershed.

The involvement of NGOs should be encouraged in forestation schemes and soil and water conservation programs so as to ensure active participation of rural community in recharge augmentation. NGOs also motivate the farmers to maintain the soil and water conservation structures put in by government departments so as to ensure long-term augmentation of recharge to groundwater. Along with such management on supply side, demand management is also equally important. NGOs play a significant role in promoting the use of efficient irrigation methods and selection of crops with low water requirement.

The conjunctive use of surface water and groundwater is ideal in theory, but in practice, several problems arise at field level during implementation. NGOs play an important role in bringing competing stakeholder on one platform and reach an amicable solution. The problem in many developing countries is that although the Governments own most of the major sources of surface water they are not able to manage surface water provided for irrigation through canals. Several thousands of hectares get water logged every year. Governments also cannot effectively mange ground water because the it is owned by the farmers. NGOs therefore play a role in educating the farmers.

Although the discussion in the chapter refers to hard-rock terrain in India, it would be equally applicable to many other developing countries, having a similar hydrogeological and climatic setup.

References

1. Adyalkar, P.G. and Mani, V.V.S., 1971. Paleogeography, geomorphological setting and groundwater possibilities in Deccan Traps of western Maharashtra. *Bulletin of Volcanology*. 35: 696–708.
2. Deolankar, S.B., 1980. The Deccan Basalts of Maharashtra State, India: Their potential as aquifers. *Ground Water*. 18 (5): 434–437.
3. DIRD website www.dird-pune.gov.in/rp_PercolationTank.htm Efficiency of percolation tanks in Jeur sub Basin of Ahmednagar District, Maharashtra state, India. (Website Visited June 2012.)
4. Limaye, S.D., 2010. Review: Groundwater development and management in the Deccan Traps (Basalts) of western India. *Hydrogeology Journal*. 18: 543–558.
5. Limaye, D.G. and Limaye, S.D., 1986. Utility of percolation tanks in hard rock areas. *Lakes and Reservoirs: Research and Management*. 6: 269–271.
6. Limaye, S.D. and Limaye, D.G., 1986. Ground water management under stress conditions in small sub-basins in Basaltic terrain of western India. *Proceedings of International Conference on Groundwater Systems under Stress*. Australian Water Resources Conference Series, 13. pp. 277–282.

4

Data Processing in Hydrology

David Stephenson
University of Botswana

AUTHOR

David Stephenson is a professor emeritus at the University of the Witwatersrand and adjunct professor at the University of Botswana. He was a visiting professor at the University of Stuttgart, Tokyo Institute of Technology, and McMaster. He is the author of 11 books in water resources and over 200 papers. He is advisor to governments in Africa on water projects and consults internationally in water development and hydraulic structures.

PREFACE

The basis of all hydrological calculations is available data. Although time series may be extended theoretically, it requires some real data to base new model data on. The accuracy and extent of data are therefore of paramount importance. And location of data collecting stations influences answers. Here, the basis for planning data collection networks and methods of collecting data and processing it are described. The various parameters affecting hydrology and water quality are discussed, and the methodology for collecting the relevant data is explained.

4.1 Introduction

Hydrological data cover a wide spectrum. It can range from precipitation through surface runoff, streamflow, hydraulics, and groundwater, covering volume, rate, and quality. Data collection in some instances is very difficult and in others a very simple process. Data are required for planning the use of the resource, controlling it, and designing structures or systems to manage it. It could be the water depth; the water velocity or flow rate, which is required; or it could be water quality or even sustainability.

The format of the data and the accuracy and period of data collection depend on requirements, as well as the means of collecting it. In many situations, there is insufficient data for accurate management of the resource, and data series may have to be synthesized. Separate chapters cover mathematical generation of hydrological data while this chapter is confined to the collection of real records, mainly in the form of flow rates.

The spatial and temporal variation of the data is wide. Frequently, one will be required to focus on a specific point, which may or may not have a history of data collection. More often, and in fact in the majority of situations, any gauging that has taken place is not exactly what the designer or planner has in mind. Therefore, transposition of data in time or space or synthesis of data is required.

An understanding of the hydrological cycle is required to identify key data in the system. Often a mass balance is possible if the full hydrological cycle is understood, and missing data can be obtained by subtraction of one flow rate from another.

Apart from spatial variation, different flows are obtained by different models, longitudinally for 1D flow, laterally for 2D flow, and vertically for 3D flow. The data required for most hydrological systems are usually 1D. In other words, flow rate in a specific direction or most probably water depth from which flow rate and a variety of other parameters can be established.

Flow gauges have usually been established at specific points, which are amenable to measurement and not necessarily at the location at which a structure or project is required. Locations of bridges, culverts, drains, and dams will be sited from the point of view of topography, aesthetics, environment, and economics. These sites rarely coincide with the position of a flow gauge, if indeed ever. All too frequently there is no flow gauge on the river system being studied and data have to be synthesized. Again, extrapolation from neighboring catchments or even from related parameters, such as rainfall, is necessary. The more remote the available parameter is from the actual parameter to be estimated, the more uncertainty there is in transforming the data.

The time-span over which data is required is also highly variable depending on the type of data required. Whereas bridges and conduits may require peak flow rates, other structures such as dams and flood plain models may require flows over a period of time, and indeed continuous records enable the system to be modeled in time.

The extent in time over which data have to be collected is also often a limitation. Data measurements are usually confined to a specific time: either now, that is, real-time measurements, or over a period of time from when the gauge was installed. This is usually insufficient to give a complete picture. If a spectrum of floods is required for estimation of extreme events, then the period of recording should be over many years. Here again, there may be limitations on the amount of data available and extrapolation is often required. This could be by stochastic methods based on the statistics of the available data, or it could be synthesized by analogous models using other data such as rainfall data. Rainfall records in different forms are often available over a longer period of time than streamflow gauges. Rainfall data are generally more transposable than streamflow data because the rainfall pattern does not vary much from one point to another over a catchment or region. Streamflows obviously increase as the flow proceeds downstream, and therefore no point can be used in place of another point where specific data are required.

The length of record is also important from the point of view of risk or reservoir sizing. In other words, duration and intensity of droughts vary with a somewhat statistical basis, and the longer the record, the more likely it is that an extreme event has been picked up. However, planners prefer to work with extreme events, that is, with recurrence intervals covering 100 years, but data over that length of time are not often available. So the records have to be extended by stochastic processes or analogies with other data, such as rainfall or neighboring streamflows.

Another parameter in the selection of data is the confidence of data accuracy or risk of inaccuracy of data. Errors in data can be just as severe in giving incorrect results as lack of data or the error in processing data from one parameter to another.

4.2 Hydrometeorological Network Design

Hydrometeorological data are often easier to access and more transposable than streamflow data. Hydrological data are a particular problem as fairly sophisticated technology is needed, not just instrumentation. That is, weirs need to be built or river sections rated. Stage–discharge tables are needed, and these need periodic review as siltation or erosion can affect them. Channel roughness also changes with stage so that inaccurate flood estimates may result even though low flows are gauged accurately.

Often rainfall records are more comprehensive and reliable than streamflow record. Frequently, computer modeling may have to be used to obtain streamflow records. The accuracy of modeling is, however, limited so that a large inherent error in water resources planning may occur. The cost of overdesign or under-design can be severe and can produce high project costs. Another problem is the impatience of designers. At least 10, but preferably more than 20, years of reliable records are needed for major projects if risks are to be minimized, so planning of such projects should occur well ahead of construction.

Rainfall data are generally more readily available than streamflow data. The data can be transposed easier than stream data and generally can be patched if there are missing records, from adjacent gauges, with minimal extrapolation. Rain gauges are cheaper and less complicated to install than stream gauges. Rain gauges are used for many purposes, that is, agriculture, meteorology, tourism, and hydrology. It is therefore understandable that the chronology of readings may not always be what the hydrologist wants. But the depth of rain is always there. The most common type of rain gauge is the cone, which is used by householders and has accuracy consistent with the volume of precipitation. They are generally read once a day, but monthly totals are often all that is available. Tipping buckets or siphon tubes are more accurate and can be automated. The data can be logged on an EPROM (erasable programmable read only memory) or transmitted to a computer for direct processing.

The optimum spacing of rain gauges depends on the study area (see Figures 4.1 and 4.2). The larger the catchment, the longer the time interval to model the runoff and the more an average rain is adequate. Generally, about 5 or 6 gauges will ensure an accuracy of 5% [13] (see also Table 4.1). The precipitation type, that is, thunderstorm versus air transported, will affect the accuracy too, but unless a subdivision of the catchment is required, 10 rain gauges is an upper limit required.

Weather stations can be classified into groups:

- Weather stations. These will be for accurate meteorological gauging, and high standards of installation are adhered to. Masts for wind speed measurement will be 10 m high. Topography should

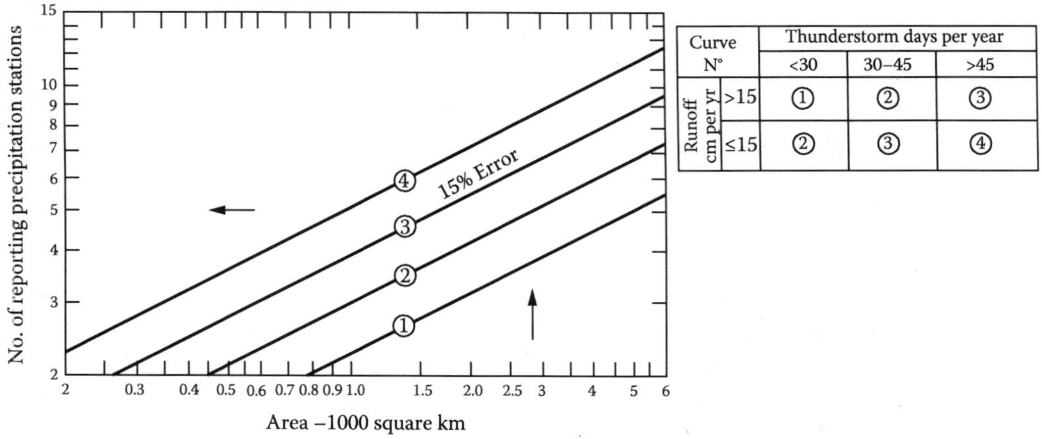

FIGURE 4.1 Minimum network requirement for continuous reporting stations. (From World Meteorological Organisation (WMO), *Casebook on Hydrological Networks*, WMO Publication, 324, Geneva, Switzerland, 1972.)

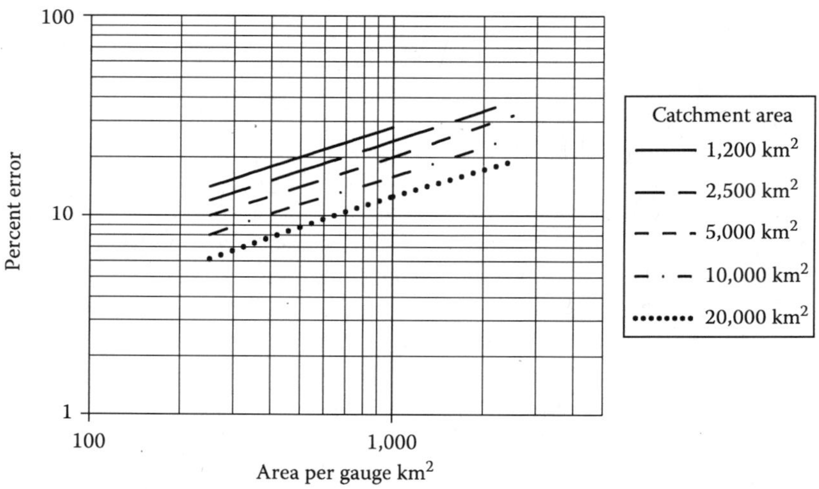

FIGURE 4.2 Gauge density–area–error graph. (Courtesy of U.S. Weather Bureau, 1947.)

be such that there is no effect on wind and radiation. Radiation, temperature, humidity, and atmospheric pressure are recorded. Snow measurement is measured separately from rain.

- Hydrometeorology stations. Installed for hydrological purposes and may include autographic precipitation gauge, evaporation pan, temperature, and humidity.
- Rain gauges. Ranging from cones to autographic tipping buckets.
- Atmospheric. From balloons for climate studies.
- Remote sensing. For example, radar and satellite imagery. These are subject to research as raindrop size can influence radar reflections. Thermal emissions and vegetation covers can be measured; thus, both of which contribute to evaporation [20].

Evaporation is an important component in the water balance of any catchment. It is generally the second biggest input/output after precipitation. Hydrologists tend to use evaporation pans, whereas climatologists calculate evaporation from radiation, humidity, and wind speed (e.g., [17]). Evaporation pan readings should be corrected by a factor of approximately 0.7 in transposing readings to reservoir or lake, to account for different radiation level. A correction is also made for precipitation.

TABLE 4.1 Minimum Numbers of Rain Gauges Required in Moorland Areas

Area		Rain Gauges		
Miles^2	km^2	Daily	Monthly	Total
0.8	2	1	2	3
1.6	4	2	4	6
7.8	20	3	7	10
15.6	41	4	11	15
31.3	81	5	15	20
46.9	122	6	19	25
62.5	162	8	22	30

Source: Joint Committee of the Meteorological Office, Royal Meteorological Society and the Institution of Water Engineers, *Trans. Inst. Water Eng. Lond.*, 42, 231, 1937.

TABLE 4.2 Minimum Numbers of Rain Gauges for Monthly Percentage of Average Rainfall Estimates

Area		Number of Rain Gauges
Miles^2	km^2	
10	26	2
100	260	6
500	1300	12
1000	2600	15
2000	5200	20
3000	7800	24

The method used to collect data can be manual in the case of remote poor regions or on clockwork driven charts that require subsequent processing or on battery-driven charts or EPROMS or transmitted to central computers by satellite link or radio transmitter. Obviously, if the data are required for real-time control, immediate direct transmission is necessary.

Table 4.2 indicates densities that are more appropriate for countrywide networks. In the application of the general guidance embodied in this table, it must be appreciated that any large river basin will almost invariably have within it a number of sub-basins for which the relatively dense networks would be recommended. Moreover, the minimum densities suggested would often be substantially increased in mountainous areas and would be closely followed only in areas of low or moderate elevation without complex topography.

Literature related to design of hydrometeorological networks is very extensive [7,8]. Several symposia have been convened concerning networks (e.g., [2,11,12]). In 1972, the Casebook on Hydrological Network Design Practice [21] was published, and supplements added in 1978 and 1981. Rodda [18] pointed out that the basic questions of station network design are simple:

1. How many stations?
2. When?
3. For how long?

The network density is determined primarily by the uses for which the data are intended. An early study of several storms in the Muskingum Basin, Ohio, shows that the standard error of rainfall averages for individual storm events varies with basin size and network density (see Figure 4.2).

There is a fine balance between expenditure on data, which may be of use, and over-expenditure on a gauging network. However, when one weighs up the cost of incorrect data, gauging networks generally under-expend. That is, the cost of lack of data or inaccurate data is far greater than the cost of the investment in getting better data to reduce incorrect expenditure on water resources development works.

Not only data accuracy is required but a lengthy record. Impatient politicians or hydrological engineers may plan dams based on scanty data and be unaware of extreme droughts or floods that occur, which were excluded from the limited records available. The incorrect location of gauging stations or the scarcity of gauging stations also means that the data have to be extrapolated to dam sites or river crossings and this also increases the uncertainty.

Many projects are designed from extrapolated time series. This is because the records may be short and statistical parameters obtained from the brief record. The mean and deviations are used to build a mathematical model of the flows. Figure 4.3 illustrates the effect of time step. The parameters used in stochastic models range from simply the mean and standard deviation such as in Markov chain models [5] to more sophisticated models including serial correlation, skewness, and trend. The unknown random component is generated with random numbers, but as we learn more about nature, we may be able to predict rainfall model river flow more accurately. That is indeed occurring with global climate modeling.

It should be noted from Figure 4.4 that a basin area of 20,000 km^2 will produce an error of 10% with a gauge density of 700 km^2 per gauge (28 gauges), while a 5,000 km^2 basin must have a gauge density of 250 km^2 per gauge (20 gauges) to yield the same error. Thus, it is apparent that the number of gauges required to estimate total or mean basin precipitation does not change significantly from small to large basins. These results are borne out by Johansen [12]. He explored the errors involved in simulating streamflow from a dense rain gauge network in central Illinois.

The calibration error is the ratio (percent) of the standard error of estimate to the standard deviation of the observed flow. The dispersion error is expressed as a percentage of the coefficient of variation of the historic record. It should be noticed that neither error is closely related to network density. It is interesting to note that if an error of 10% can be tolerated, 4–6 gauges appear adequate regardless of basin size.

Kohler [20] developed a relationship between thunderstorm days and runoff per year, with catchment area and the number of reporting precipitation stations necessary for flood forecasting. The standard

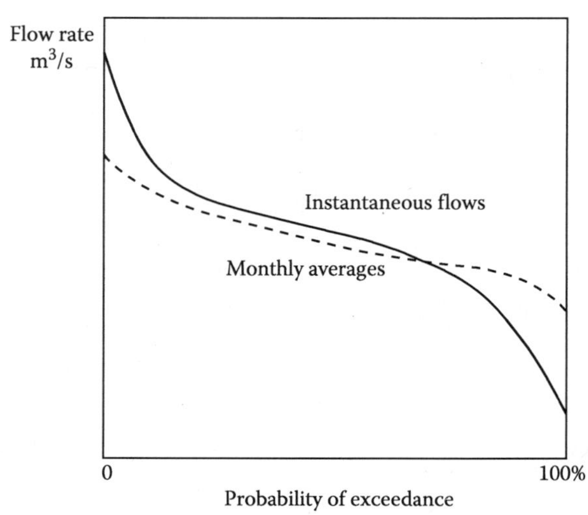

FIGURE 4.3 Comparison of instantaneous and monthly flows.

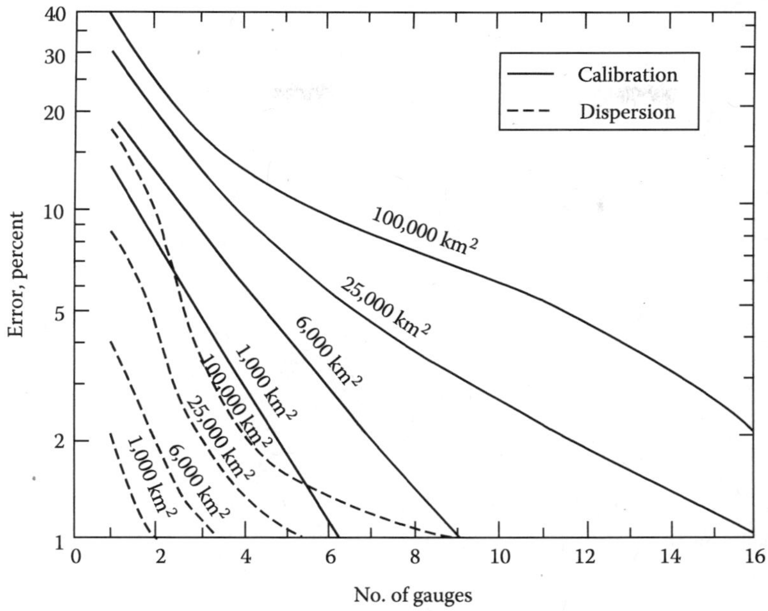

FIGURE 4.4 Error in streamflow for various gauges. (From Johansen, R.C., *Precipitation Network Requirements for Streamflow Estimation*, Department of Civil Engineering, Stanford University, Stanford, CA, Tech. Report 147, 1971.)

error *E* (percent) from his study of average storm rainfall for a catchment area *A* (in square kilometers) and *N* stations is given by

$$E = \frac{7.7(A)^{0.2}}{(N)^{0.48}} \tag{4.1}$$

4.3 Hydrological Networks

Hydrological data are a particular problem as fairly sophisticated technology is needed, not just instrumentation. That is, weirs need to be built or river sections rated. Stage–discharge tables are needed. And roughness also changes with stage so that inaccurate flow estimates may result even though low flows are gauged accurately. More important is the processing of the records and data collection, and conversion to flow records requires dedicated trained personnel.

Most streamflow networks tend to be developed for specific purposes and, frequently, appear to have little relevance to national concerns with water. For example, the networks for flood forecasting normally include gauges primarily at forecasting points (usually locations with high potential flood losses) and may not be appropriated for irrigation, municipal water supply, industrial uses, or navigation. Kovacs [15] demonstrated that runoff variability tends to increase with decreasing basin size. Also, the sampling interval is critical in determining data accuracy. Figure 4.5 relates drainage basin area to sampling interval. Nevertheless, for areas larger than 2000 km^2, sampling should probably be at least twice each day since the error can become fairly large in estimating daily runoff.

Moss [16] discussed a useful categorization in network design, which employs a use-orientated taxonomy as follows:

1. Project operations
2. Project design

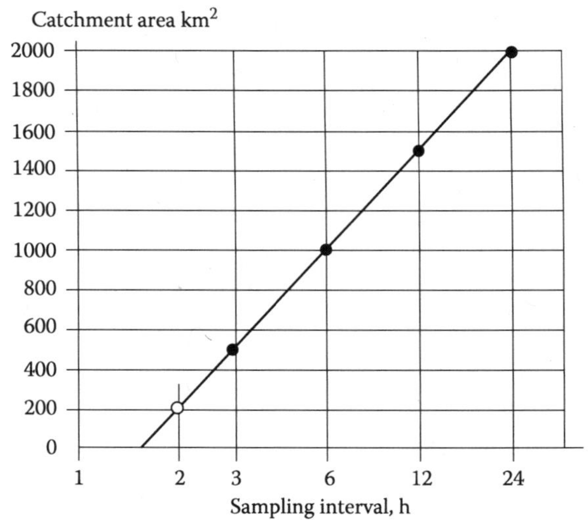

FIGURE 4.5 Relation between sampling interval and catchment area. (From Kovacs, G., *Proc. Budapest Symp. IAHS*, 158, 283, 1986.)

3. Resource planning
4. Water policy development and evaluation
5. Research

4.3.1 Density of Gauges

The World Meteorological Organisation (WMO) [20] has recommended that minimum gauging densities include

1. For temperate and tropical climates, plains regions: 1 gauge per 3000–5000 km^2
2. For mountainous basins in temperate and tropical zones: 1 gauge per 1000 km^2
3. For desert regions and polar zones: 1 gauge per 1000 km^2

These densities are, of course, very tentative in that all networks are also subject to economical considerations and specific data requirements.

Water gauging is not only carried out to measure flow rate or volume. Many other parameters are needed. The spacing and sampling interval will always depend on the purpose of the data. Design of gauging networks varies considerably on what data are to be collected and when, such as

1. Water quality
2. Sediment measurements
3. Groundwater table depth
4. Precipitation
5. Humidity
6. Temperature
7. Solar radiation
8. Atmospheric wind speed and direction
9. Soil moisture
10. Evaporation and evapotranspiration
11. Water temperature
12. Ice on rivers and lakes

Networks for a few gauge types are discussed in the following. Most of these networks are of a special type, and network densities are frequently very dense in areas of extensive abstraction.

4.4 Streamflow Gauging

The objective in developing a stream-gauged network is to improve the data available for water resources evaluation and flood estimation. A practical limit should be set on the number of proposed gauges, bearing in mind the difficulty of obtaining and processing many gauges and the cost of construction of weirs or even the equipment for measuring and recording water levels of flows. Gauge stations should be spaced to obtain a reasonable indication of flows from various types of catchments, and some gauges are established specifically for selected rivers. Gauge stations must be sited bearing in mind access and the suitability of the river for either construction of a weir or for rating the cross section (which should be of reasonably stable). A preference for weirs as opposed to rated natural sections is made where trained hydrologists for rating natural rivers are scarce and because weirs generally give more accurate data, especially at low flows. Access is important for the purpose of retrieving data as well as construction [10].

The decision as to whether or not to construct a weir should be based on the importance of the site but also on whether a weir can be practically satisfactory for rating of the natural river cross section. Owing to the flat gradient of rivers near the coast, tidal effects and backwater rule out accurate gauging. The most important areas with respect to water utilization are the interior region and the coastal escarpment region where there may be hydroelectric sites. In general, the sections should also be straight and uniform upstream, and a pool should be created behind the weir to ensure low velocities over the weir. There should be no bends or obstacles downstream to cause backwater or nonuniform flow profiles.

Where measured data are to be digitized and fed into computers, the software should be available, and in fact, the effort and cost can be significant. Data processing will make users aware of the problems in data collection and make the data more meaningful to the processor when he sees summaries of flows. Some gauging sites can be influenced by downstream conditions especially at high flows, and this and other effects can affect theoretical rating curves. For this reason, it is advisable to calibrate some sections using both water profile computations and current meters.

A national gauging network may be established to assess water resources, to research into runoff, and to study the environment. For these purposes, sites will be selected with spacing of stations commensurate with the budget and norms in the industry. On the other hand, gauges may be established for specific purpose, for example, at a proposed dam site, for a diversion or to monitor discharges from the quantity or quality point of view. Much effort is being made to compile international data banks. These are useful for poorer countries and for global models.

The easiest but not the most accurate way of estimating flow rate is to measure depth of water in a channel, possibly even in a natural river channel, and use a rating curve to convert the flow depth to flow rate. However, channels may not have stable beds, that is, may be liable to erosion or deposition over time. The method of converting flow depths to flow rates is also important. In a long straight channel reach, it may be possible to assume uniform flow and use a simple resistance equation to calculate flow rate. Obviously, cross-sectional surveys will have to be made and the roughness will have to be assessed.

In many situations, the river bed changes in gradient from point to point, and there are obstructions and deviations in the alignment of the river, and therefore a water profile may have to be determined for alternative flow rates using backwater calculations. Then a stage–discharge curve should be prepared using the gradient of the energy line rather than bed of the river channel, which is used for the simple method indicated earlier. Therefore, the slope/area method requires observations of water level over a reach of river. This may be done theoretically using water profile calculations.

The accuracy and permanence of streamflow records can be improved by designing correct measuring structures. These are frequently expensive, and often a natural section may be calibrated or a suitable natural stable gauging site found. When natural streamflow depth is used to estimate flow rates, then there are various ways of obtaining the rating relationship, that is, the relationship between water level at a point and flow rate in m^3/s. These methods range from slope/area methods through to actual stream gauging at different flow rates.

Slope/area methods use a hydraulic equation, such as that of Manning, to relate water depth, gradient of the energy line, or surface or bed slope to flow rate in the river. The measurement of energy gradient is not as easy as that of measuring water surface gradient or bed gradient. If an equation for converting from water levels and gradient to flow rate is used, it may require a variable referred to as a roughness coefficient. Roughness can vary with the water level as there could be rocks at the bottom of the river, sediment deposits on the banks, and vegetation higher up, for example. There may be also different flow regimes, for example, much faster currents in the center of the stream and almost stagnant water on the lateral flood plains. The method is usually to estimate the roughness coefficient from a number of experiments and then to prepare a graph of water level compared to flow rate.

It is also necessary to account for turbulent losses in energy at bends, obstructions and changes in gradient and therefore a water profile model may be built. Existing software is available for these calculations, and the most common models are those based on the US Core of Engineers HEC program.

If there is a specific control point in the stream, then only one water level needs to be measured. This could occur at a waterfall, or at rapids, where there is a rapid drop downstream of the gauging point. This ensures that backwater effects do not affect the water depth at the control point.

Ideally, a regular shape and definite control point can be made by constructing a measuring weir. Such measuring weir-shapes used to be in the form of rectangular notches made of concrete or other materials. Nowadays, v-notches have become more common for keeping accuracy for both low flow and high flow. Alternatively, a series of notches of increasing width is required, so that a narrow width can be used for low flows and a wider one for higher flows. More often than not, a weir has to be of limited height and extreme floods, which will both inundate the structure and backwater to submerge it, have to be estimated by modeling the channel.

4.4.1 Site Selection

The selection of site for a gauging structure is important [3]. The following items need to be checked:

- *Foundation conditions.* This will ensure no under-scouring or leakage and also a stable structure.
- *Stable flanks.* The erodability of the sides of the channel can affect high flow observations, and therefore stabilizing the banks may be necessary, for example, the stone pitching or flow velocities in the vegetation with minimum obtrusion into the flow.
- *The tortuosity of the channel affects turbulence losses and therefore water surface profile.* A straight section is needed downstream, sufficiently far to ensure minimal backwater effect on the gauging structure.
- *Upstream of the gauging structure should be a quiet pool.* This ensures that the velocity head is negligible. The gauging structure itself may be in the form of a pressure transducer, or float, which measures water level, that is usually transposed to energy level over the structure. The actual water level over the structure is less because of the velocity head and indeed the level is usually at critical depth as it flows over the structure. Critical depth is the depth at which energy is a minimum. The pool upstream, therefore, requires a straight inflow to ensure flow rate across the full width and depth of the pool. Sediment deposition and debris-catching have to be minimized. Sediment collection up to the level of the crest can have a serious impact on the velocity head and therefore give incorrect flow rates for any water level measurement.

4.4.2 Current Gauging

An accurate method of measuring flow is using a current meter. A current meter is a propeller that has the revolutions counted electronically as it rotates. A rating equation gives the water velocity in m/s, and

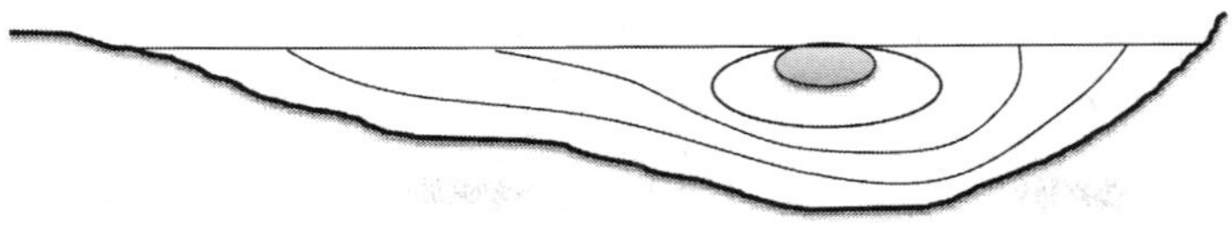

FIGURE 4.6 Velocity distribution across a river section.

flow in cubic meters per second is obtained by multiplying the area. The channel cross section should be divided into vertical strips using a tape measure or knotted line (Figure 4.6). For deep channels, the current gauge can be lowered from a bridge or boat or pulley. In the latter case, a correction is needed for the angle of the suspension tope.

Current gauging using a propeller type meter is a direct method of measuring flow rate instead of flow depth. The cross section is divided into sections and flow velocity measured at each section. The current gauge can be handheld by wading across the river or held over the edge of a boat. If the water is deep or dangerous, this can be done from bridges, but in other cases a cable is required with a bosun chair or mechanically operated gauge, to traverse the section, especially during flood flows. Salt dilution methods are also possible but expensive as are radioactive traces methods. Rotating current gauges can be purchased in various sizes from a few mm diameters to over 200 mm diameter (Figure 4.7). The larger units have heavy torpedo-shaped ballast to align them facing the flow despite hanging from a cable.

Whether a natural stream section or a man-made structure is used, an actual measured relationship between flow rate and water level is desirable for preparing a rating relationship. Current gauging by propeller is an accurate method. Here, for low flows, a wader may carry a gauging propeller across the river and gauge at selected points across the river and at selected depth. Alternatively, for bigger flows, gauging equipment can be lowered from a cableway or a bridge into the water. This is much safer than wading or boating across the river, particularly where untrained assistants do the work.

Distance is measured across the river, and the readings are taken every 1 m or more for wide rivers. When taking readings, first measure the depth of the river. Allow the propeller to run for 40 s or so and then take the reading on the counter. Take the mean of two readings at 0.2 and 0.8 times the depth from the top. For shallow water it may only be possible to take one reading, preferably at 0.6× depth from the top. The average velocity from three readings should be multiplied by depth and width to get flow rate (Figure 4.8). Use waders to keep dry, a life jacket for safety, and a rope tied around observer's chest to shore to shore in case of emergency.

Magnetic and sonar methods are also used, but these are more appropriate to closed conduits where the cross-sectional area is known.

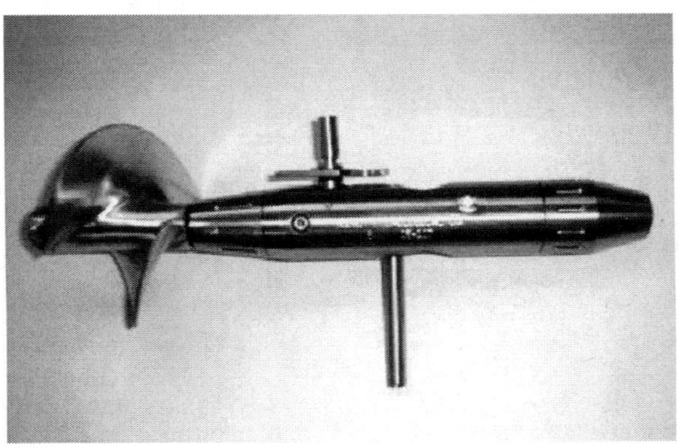

FIGURE 4.7 Propeller current gauge.

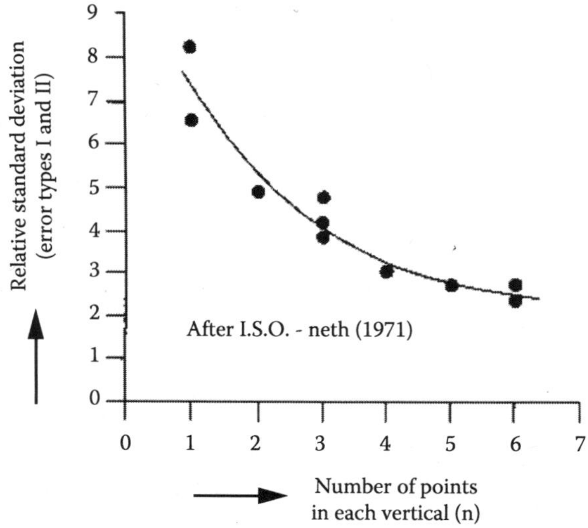

FIGURE 4.8 Relative error for differing number of gaugings on a vertical.

4.4.3 Tracer Measurement

For low flows, an accurate method of gauging the flow is by dye or tracer injection. The concentration of the tracer can be measured downstream of an injection point, and by simple proportion, a flow rate can be calculated. This can be repeated at different flow rates to obtain a rating curve or preferably combined with different methods such as the current gauging and a calculation model.

Continuous injection of a tracer is one method but requires considerable volume of tracer. Another method is the gulp injections whereby the hydrograph of tracer is observed downstream of the injection point.

Continuous injection is usually only useful for low flows because large volumes of tracer would be required for higher flows. In addition, it is necessary to allow a considerable length of river to ensure good mixing before samples are taken. Unsteady flow can also influence the flow rate versus stage relationship. The hydrograph could rise and fall within the time that measurements are made, and the rate of rise and fall would also influence the mechanics, that is, an equilibrium may not have been achieved (Figure 4.9).

For the gulp method, the flow rate in the stream may be computed from the mass of dye injected divided by the downstream integral of dye concentration over time.

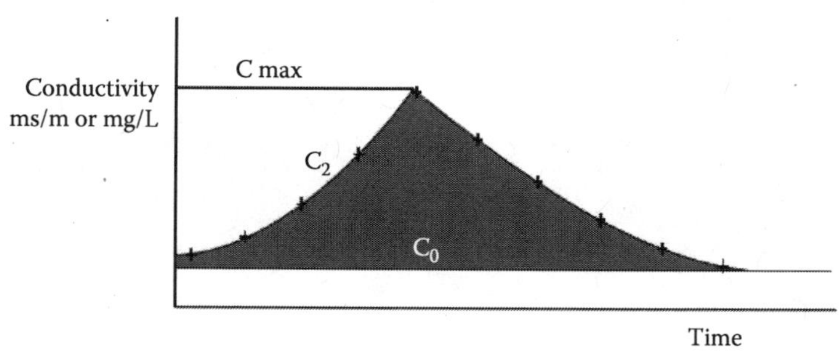

FIGURE 4.9 Tracer concentration over time for gulp injection.

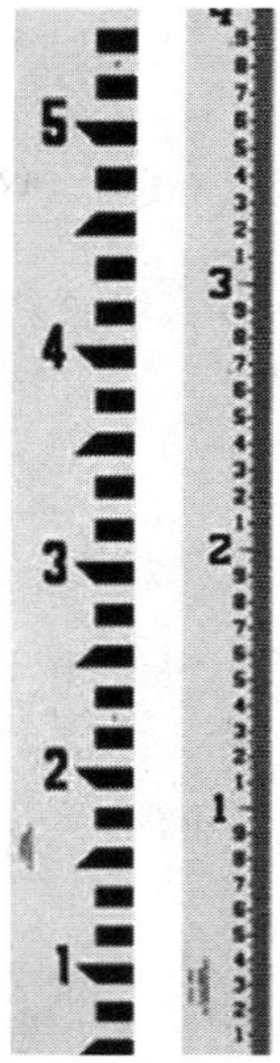

FIGURE 4.10 Different staff gauge markings.

4.4.4 Staff Gauges

The simplest form of a stream gauge is the staff gauge (see Figure 4.10). These are usually of steel, enamelled and bolted to a fixed iron post. They must be read manually. It is subject to the availability of a gauge observer at least once or twice a day and more frequently during flood events. Education of the observer should go hand in hand with the establishment of a gauging site. The staff gauges can also be used to check automatic recorders. And their positioning should be checked to correct for sediment deposits or erosion, which may affect the calibration.

4.5 Recorders

Automated gauges are of several types: float-type gauges requiring stilling wells and those utilizing some other type of device that measures water pressure (e.g., nitrogen bubbler gauges and pressure-measuring strain gauges) or water level, for example, sonar waves. In recent years, many of the automated gauges have been equipped with communication equipment (namely, telephones or radios utilizing relays where necessary or satellites). The previous transducers only measure water stage and rely on a stage–discharge relationship to convert recorded readings to streamflow.

It is advisable to observe water lines along the riverbank and gauge the flow, to obtain calibration curves. Many assumptions must be made for these calculations so it is a good policy to actually measure flow rate at some different water levels.

4.6 Weirs

Weirs offer a reliable and accurate way of measuring streamflow (e.g., [1,9,10,21,22]) the stage–discharge relationship is relatively independent of the downstream water level, and the structure is stable, accurately dimensioned, and easy to construct. Water level measurement is simple and accurate, and the accuracy can be controlled by selecting notch width and shape. Rectangular notches are generally preferred, in

FIGURE 4.11 Crump weir on a stream.

tiers, but triangular notches can be used for low flows or large variations in flow range. The alternative to a sharp-crested weir is a flume or throat wherein water level is dropped by constricting the width of flow.

Older weirs are often sharp-crested rectangular compound weirs. There is a problem of siltation upstream of weirs in some cases. However, due to high sediment loads in many rivers, this is difficult to avoid unless the river reaches are selected carefully or alternative designs are used. The Crump weir (Figure 4.11) is now favored for de-silting, but it requires a good foundation as it is very broad. Most river beds are alluvial and heavy grouting or excavation would be required; also the concrete volume required is considerably greater than for sharp-crested weirs. There is also the problem of peak floods, which will overtop any reasonable weir, and the actual design of the weir is not all that relevant during high allows because of inundation (e.g., Charlton [6]).

In general, weirs are designed to monitor accurately flows up to approximately the 1-year flood without overtopping the flanks (Figures 4.12 and 4.13). Above this, the cross section of the valley will also have to be considered, and rating curves may be required for estimating higher flows. Concrete weir sections are generally grouted and founded on bedrock as seepage under the structure and around the flanks in alluvial material can be a real problem. It is suggested that the bottom crest be a minimum height above bed level to avoid inundation or backwater effects during the 1-year flood flow through the section. Similar criteria should apply to each crest. The maximum rise of each step should be about 400 mm, except in a steep-sided valley; then dividing walls should be used to ensure accuracy of calibration curves. The crest and sides of each level of a sharp-crested weir should be mounted with 100 × 100 × 10 mm angle irons. Weirs should be located in reaches relatively straight upstream to ensure symmetrical flow. Relatively steep sections are preferred [4] because

1. They assist in suspending silt and scouring it out upstream of the weir.
2. They minimize the problems of backwatering and even tidal effects, which inundate the weir.

A correction factor is also necessary for inundation due to downstream flooding, and if the banks can overflow during flood, an estimate of the overbank flow is necessary. Supercritical flow should, however, be avoided upstream of the weir. An upstream pool for stilling flow is also required, although these frequently silt up and require recalibration of the weir. The distance from the weir to the gauge is important, and the water level should be unaffected by the drawdown curve. Bottom and side contractions can occur in the nappe as it flows over the crest, and correction factors are necessary to the calibration equation. A correction factor is also necessary for inundation due to downstream flooding, and if the banks can overflow during flood, an estimate of the overbank flow is necessary.

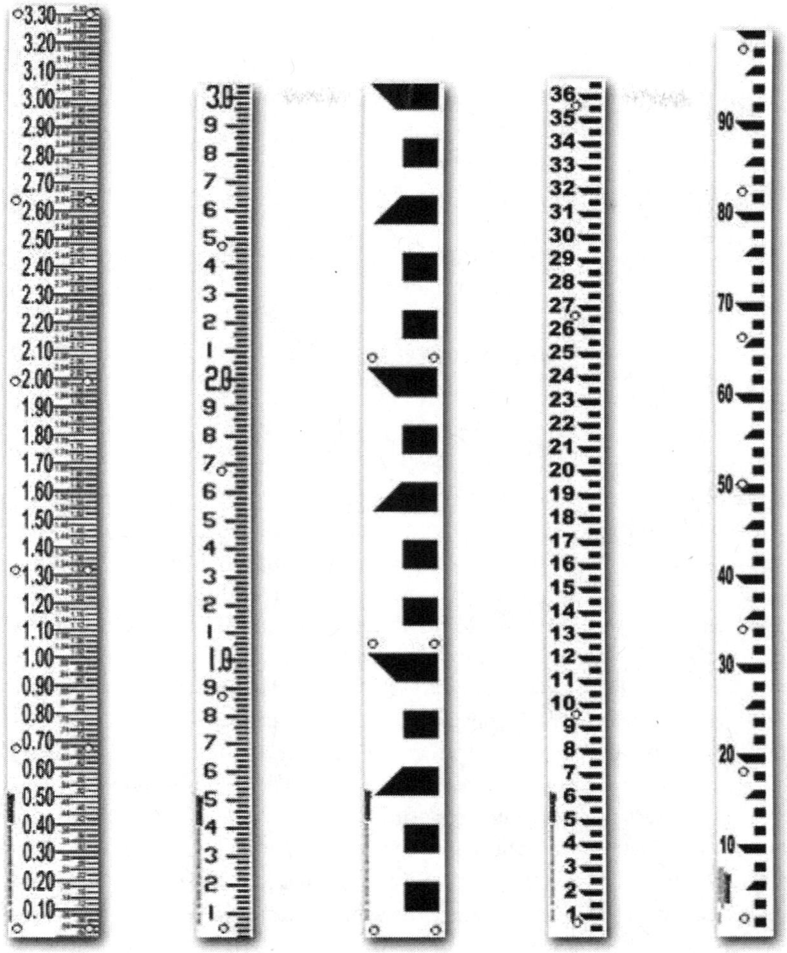

FIGURE 4.12 Different weir designs.

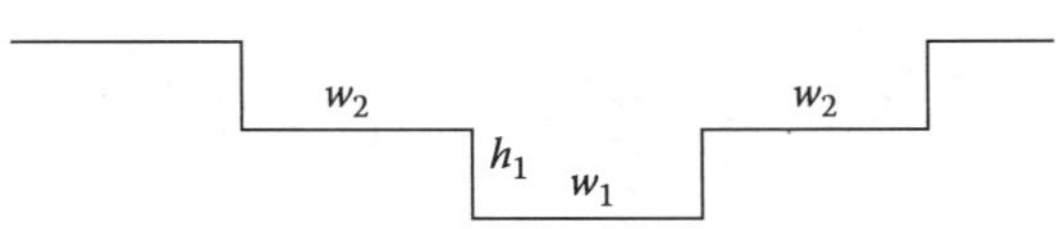

FIGURE 4.13 Front elevation of notched weir.

4.6.1 Optimum Weir Width

Based on a minimum flow of Q_1 for notch 1 $\left(\text{say } 0.1\ \text{m}^3/\text{s}\right)$ and minimum accuracy desired for flow of dQ $\left(\text{m}^3/\text{s}\right)$,

$$Q = kwy^{3/2} \approx 2wy^{3/2}, \quad \therefore y = \left(\frac{Q}{2w}\right)^{2/3}$$

$$dQ = 2w\frac{3}{2}y^{1/2}\,dy$$

$$= 3wy^{1/2}\,dy = 3w\,dy\left(\frac{Q}{2w}\right)^{1/3}$$

$$\text{Solving for } w: \; w = \left[\frac{dQ}{3dy\,Q^{1/3}/2^{1/3}}\right]^{3/2}$$

for example, accuracy of measurement 10 mm and minimum flow of 0.1 m^3/s

$$w = \left(\frac{0.1}{3 \times 0.01 \times 1^{1/3}/2^{1/3}}\right)^{3/2} = 8.6 \text{ m (or less)} \tag{4.2}$$

The older type of float-operated water level gauges is reliable and robust, but piezoelectric or similar electronic water level recorders are much cheaper. These are more economical than float systems also in data collection and processing. It is suggested that provision be made for float chambers: that is, 150 mm diameter pipes should be constructed from upstream of the lowest weir crest into a float chamber on the bank of the river. The chamber shaft should be constructed of masonry or other rigid material to avoid being washed away in floods and to a level that ensures that the housing of the recorder is not washed away with floods of less than, say, the 100 year flood (this will require a risk analysis).

Housings for mechanical recorders need to be above flood level, weatherproof, and secured against vandalism. Provision should be made for floats and chart recorders together with the shaft down to low water level. The housings for electronic data collection systems should be weatherproof secure enclosed boxes away from the stream. The use of batteries and solar panels is open to theft in remote areas.

The rating of weirs can be carried out theoretically or in the field and preferably using a number of ways and at regular intervals if bed changes occur (see Figures 4.14 and 4.15). The rating curve or table, or equation for automatic processing, can be prepared with the assistance of computer programs. Theoretical computations will allow for effects of side contractions, upstream pool depth, and siltation and estimate backwater effect for submergence of the weir. Each crest level can, thus, be selected to avoid submergence, which could affect flow interpretation significantly. Above the top of the weir, the transition to full flood flow will be likewise estimated. In order to minimize error in reading, there is a certain minimum width that can be used for the notches.

4.6.2 Sharp-Crested Weirs

We call weirs short crested when their characteristics in some way look like those of broad-crested or sharp-crested weirs. The streamlines above the crest are curved.

Well-known examples are

- Weir sill with rectangular control section
- V-notch weir sill

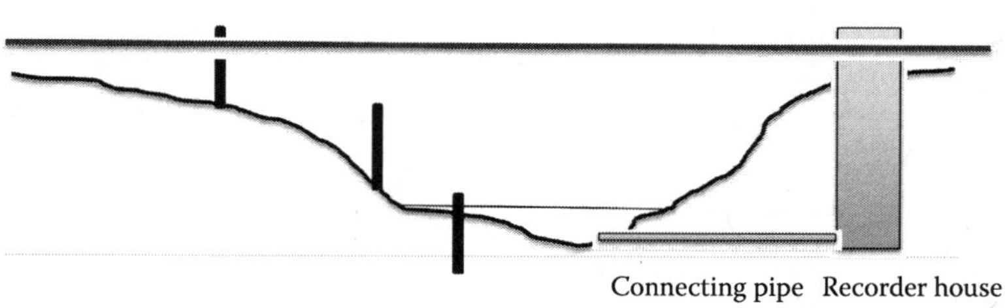

FIGURE 4.14 Stream gauge staff may not be zeroed on bed level, especially in alluvial streams.

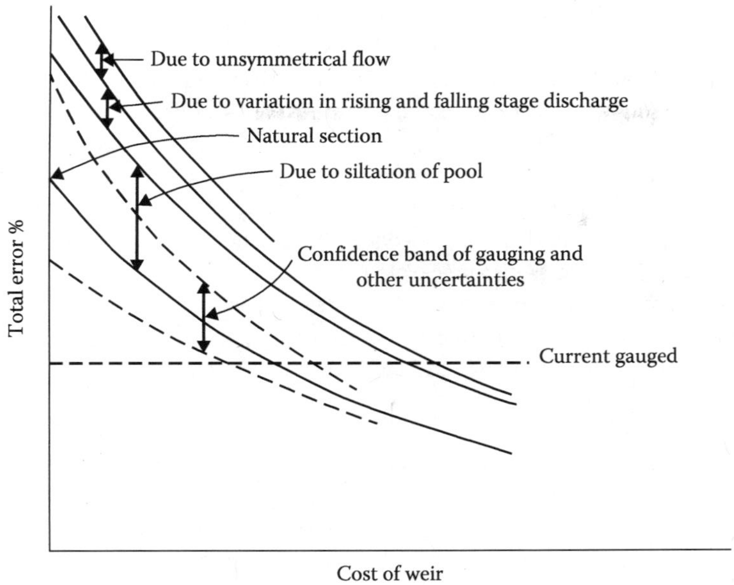

FIGURE 4.15 Cost of weir as a function of accuracy.

- The triangular profile weir (Crump weir)
- The flat weir

4.6.3 End Depth Methods

Where the bottom of the canal drops suddenly, a free overfall is created. The water level is measured exactly above the drop (end depth or brink depth). The discharge is a function of both the end depth and the shape of the cross section. Then, we can identify

- Rectangular channels with a free overfall
- Nonrectangular channels with a free overfall

4.6.4 Flumes

Critical depth flumes and broad-crested weirs have some resemblance. Flumes are less restricted in crest-height, and the downstream section is gradually divergent to regain energy. Distinction is made between long-throated flumes and short-throated flumes. Long-throated flumes are similar to broad-crested weirs (parallel streamlines). Short-throated flumes behave like short-crested weirs (curved streamlines).

4.7 Accuracy in Flow Gauging

Even though there may be flow records, they may be often of dubious value for lack of accuracy. The following factors can affect the accuracy and therefore usefulness of data:

- *The method of measuring flow.* This is usually done from water depth, and an equation or relationship between water depth and flow rate is required.
- *The method of converting from water depth to flow rate* can vary from a rating table to a hydraulic equation and these often are not at the same stage of accuracy. Ratings often change over time due to changing river morphology and they have to be kept up to date.

- *Gauge observer.* The method of changing data and collecting it can range from fitting and removing paper charts to the same for electronic data loggers or receiving data transmitted to a computer. Each of these will have its own accuracy or inaccuracy and requires to be checked. The older method of collecting water levels by manual reading is still of use for cross-checking automatic recordings, and the input of the person doing the gauge recording should be ascertained. Remote areas often require calling on a part-time gauge observer who does not understand his or her responsibility and has many other duties in life so that there may be periods of missing data, synthesized data, or incorrect data. Data outside the gauge calibration should be indicated, for example, overtopping or missing gauge plates.
- *Converting data to desired form.* Although conversion of electronic data is automatic, the digitizing of paper charts or manual readings requires human intervention, and often decisions have to be made as flows or water levels change. There are often complicated reversals in the trace on the paper chart. Missing pieces, because of battery or ink failure, have to be interpreted correctly, that is, not assuming zero flow.
- *Updating rating relationship.* Since some gauges are purely for measuring water depth in a natural channel or over a structure, there are changes in the river morphology, which can affect the relationship between the water level and flow rate. For example, sedimentation upstream of a weir, erosion of the channel downstream, side erosion over a weir, and inundation of a control structure at higher flow rates. There is also the problem of rainfall, or some other source, causing damage to the system, as well as blockage or damage by floating or submerged objects in the stream. These factors need to be assessed and often require visits by a professional rather than relying on a field observer to correct the situation or to notify the responsible department.
- *Methods of digitizing data.* The data from the field can be in tabular form, paper charts, on data loggers, and media, such as tapes or memories, or it can be transmitted by telephone or radio signals to a central computer. The data can then be processed automatically although manual inspection and correction are generally required.
- *Method of converting water levels on paper charts to flow rates* (Figure 4.16). This often uses an electronic digitizer on a tablet, and again human decisions need to be made at many points, such as where suspected incorrect recordings are made, data is missing, recordings go off the side of the paper roll, or there is damage to the system.

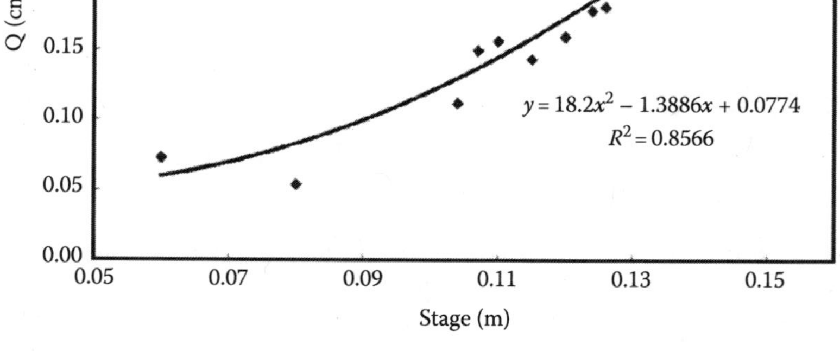

FIGURE 4.16 Stage-flow rating curve.

4.8 Data Processing

The appointment of a team to carry out all the steps, from gauge installation through field observation to data processing and filing in the office, requires training of responsible staff. All too frequently governments who used to be responsible for these activities are reducing their budgets and consequently the amount of accurate data available. Privatization has often entered into the system for data processing and publishing, but the most important is the establishment and accuracy of the field recording equipment and people. It is not possible to go back in time to collect the data. Intensive spot gauging, or installation of gauging as part of the pre-feasibility planning stages, can provide some data. This can be used for calibrating mathematical models. Again, the more data available, the more accurate the results that the model will produce. This is often also necessitated by unsuitable gauges at incorrect locations, and therefore short-term gauging is a very useful way of calibrating any type of transposition model.

The use of stochastic models for extending data into time series has become more and more common. It should always be borne in mind that the data cannot be more accurate than observed data and the statistics of the short data available, that is, mean, variation, skewness, correlation, and other statistical parameters can be made to fit existing data, but again, it should be borne in mind that the actual rainfall runoff process does not follow a mathematical pattern, and there are many more variables that we cannot model. Even global climate models cannot account for local topography or geology.

World weather models are trying to be more and more comprehensive, that is, to use ocean temperature inversions to predict El Niño and other effects on the rainfall patterns. There are also models for predicting climate change and estimating rainfall patterns as the atmosphere changes. Changes in solar radiation and earth temperatures will change rainfall patterns, and therefore again either statistical trends or realistic global climate models are required. This is getting out of the realm of the engineering hydrologist, and therefore statistical projections and curve fitting are more frequently used than climate models. The same, in fact, applies to estimating extremes such as maximum probable flow rate. Estimation of maximum probable precipitation for purposes of estimating maximum probable floods is frequently recommended, but we are dependent on the global climate modeler. Here, the engineer or hydrologist concerned with project design frequently resorts to extrapolations using experienced curves. Curves of maximum-recorded floods versus catchment area are frequently used for estimating, for example, maximum probable floods as a function of catchment area.

4.9 Transposition of Data

In most cases it is necessary to extend records, fill in missing data, correct data, or change one or more parameters in the data series. Therefore, standard methods are required, and because of uncertainty in all methods, it is desirous that more than one method (preferably at least three methods) is used to confirm data. These methods include the following:

- Extrapolation in time. Frequently, short-term records can be extended by correlating with hydrologically similar records or alternative parameters such as rainfall, which may be more extensive than the streamflow record.
- Data may be available at different times at different surrounding gauges, and a multidimensional correlation improves the reliability of the data record to be synthesized.
- Apart from transposition in space, transposition from one medium to another could be used as well, for example, the use of rainfall records to obtain streamflows series. Alternatively, archaeological data, such as geological records of flood levels and debris on flood plains, and vegetation changes can be used for estimating water levels in the past and for extreme events. These should be put into the equations with as many alternative ways of estimating the flows as possible.

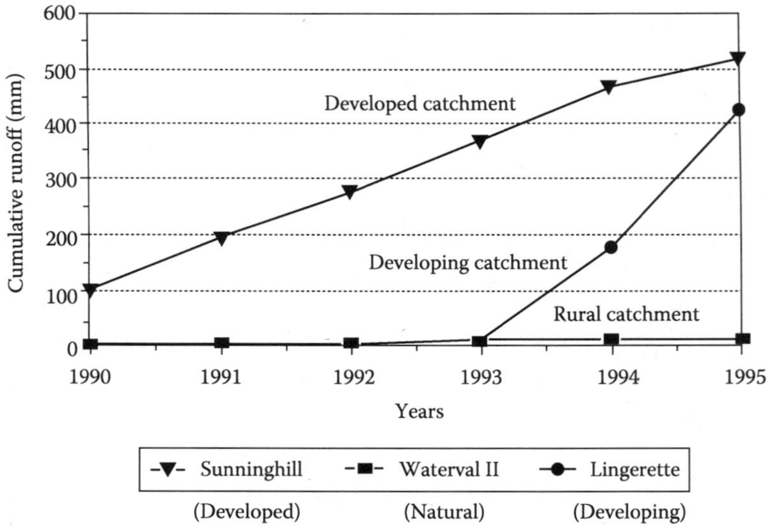

FIGURE 4.17 Mass flow curve for different catchment characteristics.

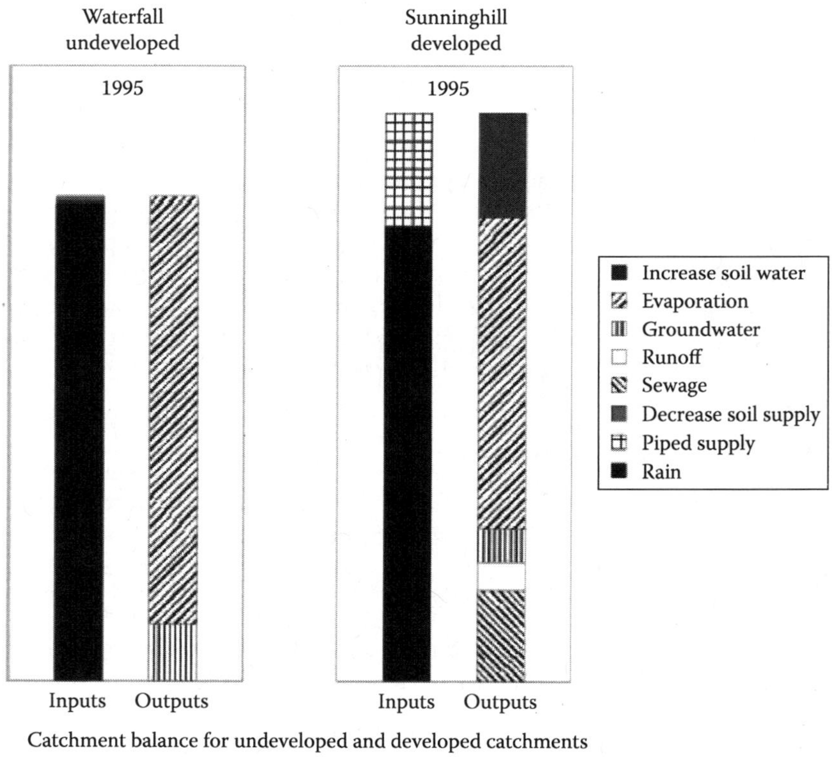

FIGURE 4.18 Comparison of mass balance for different catchments.

Deviations in flow pattern can often be picked up with massed flow curves (Figures 4.17 and 4.18). This may be due to climate change or land use change. Once the data are verified or adjusted, they can be used to extend flow series and to extrapolate to adjacent streams (Figure 4.19). In the case of an ungagged catchment spot, gauging may be made at that site and another gauged site to correlate the two (e.g., Figure 4.20).

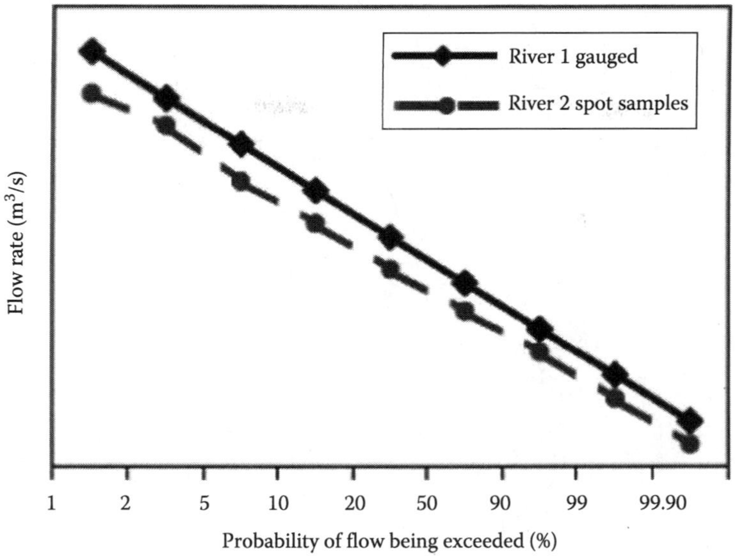

FIGURE 4.19 Transposition of gaugings from one catchment to another.

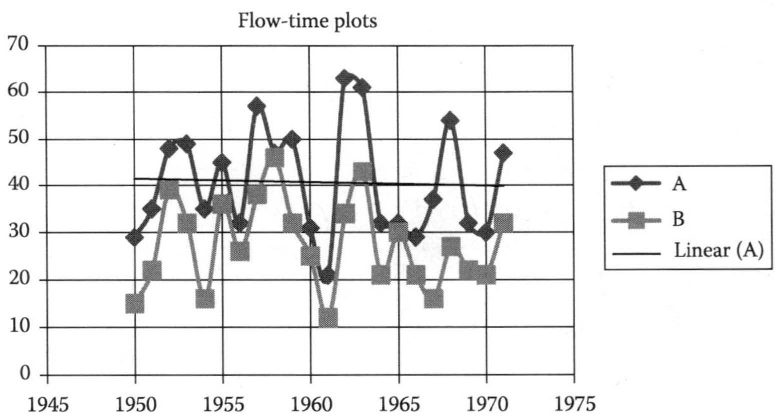

FIGURE 4.20 Comparison of flows at two gauges.

4.10 Trend and Homogeneity

Data can be checked for correctness using plotting. Changes due to land use, climate change, or different observation methods can be picked up by plotting trends (Figure 4.21) or comparison with other gauges. The latter can be isolated with double mass curves (Figure 4.22). Reasons for differences should be considered, for example, catchment area, groundwater contribution, and land use.

4.11 Remote Sensing

Hydrometeorological data are more amenable to remote sensing than hydrological and therefore easier to collect using modern low-cost technology. Radar can measure rainfall accurately as well as storm distribution and movement. It can be done from ground stations or satellites or airplanes. Thermal imaging is also used to obtain much data, for example, ground moisture and cloud characteristics.

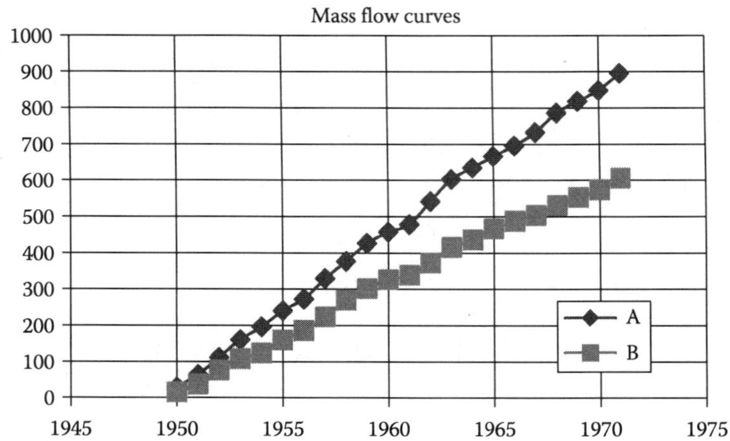

FIGURE 4.21 Mass flow plots for two gauges.

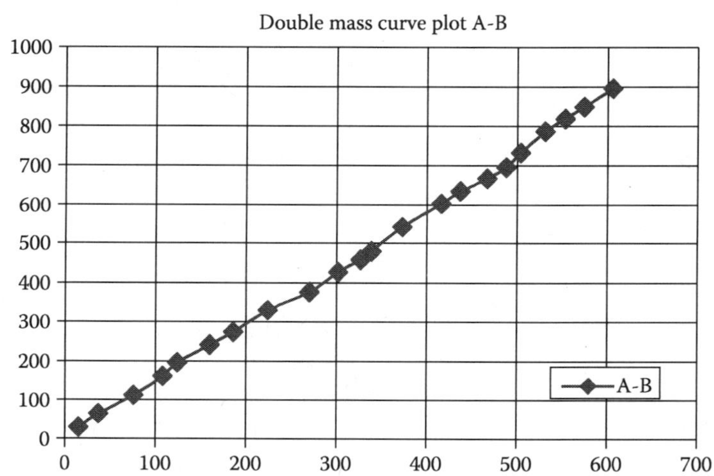

FIGURE 4.22 Double mass flow plot.

Hydrological data are more difficult to estimate by remote sensing. Flood plains may be measured in width, but considerable ground truth is needed. Light detection and ranging (LIDAR) surveys by airplane may offer greater accuracy and even depth measurement.

4.12 Water Quality

Water quality is a complex parameter to measure because there are many variables. It depends largely on the purpose of monitoring. The following need selection:

- Parameters to be detected including suspended and dissolved solids
- Accuracy required and base level in the case of automatic starting instruments
- Time scale and interval
- Spacing of gauges
- Specification of output, for example, total, peak, or range, together with statistical ranges

Laboratory tests for selected parameters are expensive and slow so the number of tests and parameters needs to be selected with care.

Sediment is a special parameter affecting reservoir capacity, fertility, and water quality. It cannot be measured easily because it is suspended and transported at different densities at different depths below the surface of a river. It is generally measured indirectly by sonar surveys of reservoirs that contain sediment.

4.13 Groundwater Measurement

The measurement of groundwater flow rate and total resources is considerably more complicated than surface flow. Water in the ground occupies the pores between grains of soil or fissures in rock and the water can occur in various forms. Below the water table, there are usually fairly high saturation levels, whereas above the water table, water may occur in vapor form or in the form of films around soil grains. There may also be impermeable layers that can then cause successive water tables (perched water tables). We may have the condition of the confined groundwater underneath an impermeable layer, that is, water that is under pressure even at its top level (Figure 4.23). The major problem is the conditions underground cannot be seen [18].

What is usually required is the rate of flow or rate of recharge of an aquifer as well as the rate at which water can be abstracted. This is required for estimating yield as well as estimating the sustainability of an aquifer.

The extent of aquifers can be limited by dykes and surrounding mountains as well as the porosity of the aquifer within the confines. The source of the water needs to be assessed and put into a model of the aquifer: the source could be either from precipitation on the surface, and the infiltration down to water table, or seepage laterally from mountains or other aquifers. All too frequently borehole yields are obtained by short-term pump tests. These do not indicate anything about sustainability, that is, changes in the water level over a considerable period of time. For major abstraction plans, a model of the entire aquifer is desirable.

The two parameters usually estimated for modeling are the transmissivity and the storativity. The transmissivity is a function of the flow rate and the cross-sectional area of the aquifer. Storativity is the volume of water released per unit volume of aquifer. Generally, the objective is to measure a directional flow rate for purposes of estimating recharge rates from the side. The infiltration from the surface is

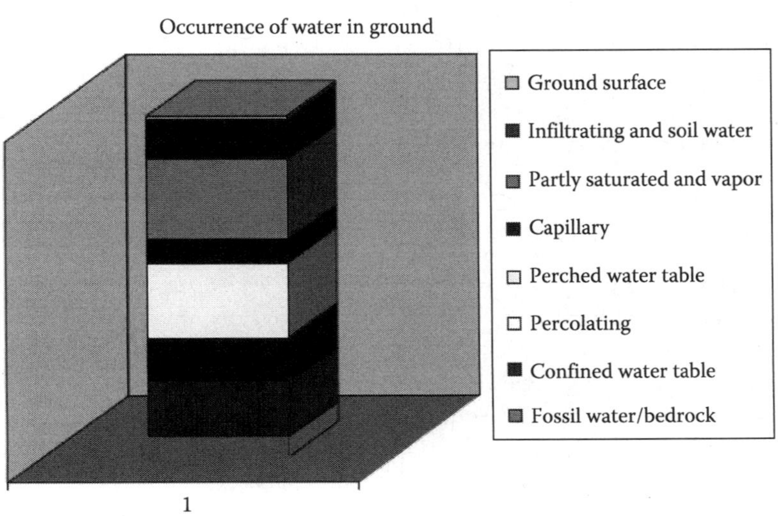

FIGURE 4.23 Distribution of water in the ground.

FIGURE 4.24 Borehole depth gauge.

more difficult to model. Frequently, groundwater ages estimated by tracer tests are thousands of years old. This indicates either a previous wet period or else that the water has infiltrated slowly over thousands of years. It means that the water does not usually seep at a noticeable flow rate but could permeate in vapor form and by osmosis. The actual rate is therefore difficult to measure. The possibility of changes over time in the geology and the permeability cannot be ruled out.

Transmissivity can be measured by permeability tests in the laboratory or on-site. Site tests can be made with tracers and a number of boreholes. Boreholes can be used to measure water levels (Figure 4.24): that is, gradients and velocities can be measured with tracers. Pumping rates can be fitted into a model to obtain permeability and storativity.

Tracer tests in multiple boreholes can yield additional data, that is, dispersivity, homogeneity, and isotropism. A dispersion model is often required with tracers to obtain the parameters. Tracer concentrations, especially if tracer is injected in point form, are measured a number of meters away from the injection points (Figure 4.25).

The extrapolation of laboratory permeability is poor on a field scale, and intensive geological investigation is required in nonhomogeneous aquifers. Geophysical methods, satellite imagery, and Lidar photography are becoming more useful in estimating subsurface conditions.

4.14 Summary and Conclusions

Hydrological conclusions can only be as accurate as the data used. Unless data are realistic and appropriate, incorrect conclusions can be made regarding flood passageways, reservoir sizes, and amount and quality of water available. Data are needed primarily regarding streamflow and runoff, but supplementary data regarding meteorology, groundwater, and other bodies are required to reach conclusions. Data spread over time and space are generally required, for water flow is not constant.

This chapter provides guides for data collection networks, methods of gauging, extrapolation, and compilation of extended time series. Different methods are applicable in different circumstances and to cross-check. The format of data, processing of data, and extrapolations all contribute to the final product before correct design decisions can be made.

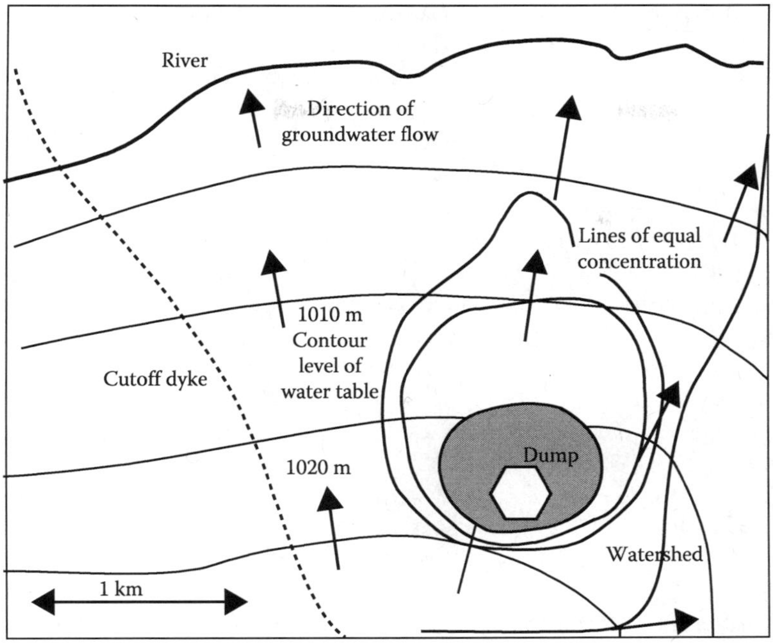

FIGURE 4.25 Distribution of pollution from a buried source.

External factors affecting data now and in the future must be considered. Uncertainty in data and risk of exceedance all require statistical analysis and computer modeling, whether deterministic or stochastic depends heavily on the initial data obtained in the field.

References

1. Ackers, P., White, W.R., Perkins, J.A., and Harrison, A.J.M. 1978. *Weirs and Flumes for Flow Measurement*. John Wiley & Sons, New York.
2. American Geophysical Union, 1979. Chapman Conference on Hydrological Networks. *Water Resources Research,* 15(6): 1673–1871.
3. Boiten, W. 2000. *Hydrometry*. A.A.Balkema, Rotterdam, the Netherlands.
4. British Standard 3680: Part 2B.1986. Integration (gulp injection) method for the measurement of steady flow.
5. Brittan, M.R. 1961. Probability analysis for the development of synthetic hydrology for the Colorado River, In: *Past and Probable Future Variations in Streamflow in the Upper Colorado River, IV*. University of Colorado, Boulder, CO.
6. Charlton, F.G. 1978. *Measuring Flows in Open Channels*. Report 75. CIRIA, London, U.K.
7. Clark, R.A. 1988. *Design of Hydrological Networks*. Course notes University of Witwatersrand, Johannesburg, South Africa.
8. Falkenmark, M. 1982. Water data strategy. *Proceedings of 4th World Congress*. IWRA, Beunos Aires, Argentina.
9. Gunston, H. 1998. *Field Hydrology in Tropical Countries*. Intermediate Technology Publicans. London, U.K.
10. Herschy, R.W. 1985. *Streamflow Measurement*. Elsevier Applied Science, London, U.K.
11. International Association of Hydrological Science (IAHS). 1965. *Design of Hydrological Networks*. IAHS Publications 67 and 68, Gentbrugge, Ghent.

12. International Association of Hydrological Science. 1986. *Integrated Design of Hydrological Networks. Proceedings of a Symposium*, Budapest. IAHS Publication 158, Wallingford, CT.
13. Johansen, R.C. 1971. *Precipitation Network Requirements for Streamflow Estimation*. Department of Civil Engineering, Stanford University, Stanford, CA, Tech. Report 147.
14. Joint Committee of the Meteorological Office, Royal Meteorological Society and the Institution of Water Engineers. 1937. Report on "The determination of rainfall over any area". *Transactions of the Institution of Water Engineers London,* 42: 231.
15. Kovacs, G. 1986. Time and space scales in the design of hydrological networks. *Proceedings of the Budapest Symposium IAHS*, 158: 283–294.
16. Moss, M. and Marshall, E. 1979. Some basic consideration in the design of hydrological data networks. *Water Resources Research*, 15(6): 9.
17. Penman, H.L. 1948. Natural evaporation from open water, bare soil and grass. *Proceedings of the Royal Society of London,* 1032: 193.
18. Rodda, J.C. 1969. *Hydrological Network Design—Needs, Problems and Approaches*. WMO/IHD Report No. 12, WMO, Geneva, Switzerland.
19. Schmugge, T. 1985. Remote sensing of soil moisture, In: *Hydrological Forecasting*, eds. M.G. Anderson and T.P. Burts, John Wiley & Sons, New York.
20. World Meteorological Organisation (WMO). 1972. *Casebook on Hydrological Networks*. WMO Publication, 324, Geneva, Switzerland.
21. World Meteorological Organisation. 1980. *Manual on Stream Gauging*. Vol. 1, Fieldwork, No. 519, Vol. 2: Computation of discharge, Secretariat of the WMO, Geneva, Switzerland.
22. World Meteorological Organisation. 1981. *Guide to Hydrological Practices,* Vol. 1: Data acquisition and processing, No. 168, Vol. 2: Analysis forecasting and other applications, Secretariat of the WMO, Geneva, Switzerland.

5

Ecohydrology for Engineering Harmony in the Changing World

Maciej Zalewski
University of Lodz and European Regional Centre for Ecohydrology Polish Academy of Sciences

AUTHOR

Maciej Zalewski is a professor of Lodz University, Department of Applied Ecology, director of the European Regional Centre for Ecohydrology (EH) under the auspices of the United Nations Educational, Scientific, and Cultural Organization (UNESCO), and founder of the EH concept under the framework of UNESCO–International Hydrological Programme (IHP). His research interests evolved from fish bioenergetics toward river ecosystem ecology, when he proposed the concept of abiotic–biotic hierarchy of factors determining water communities. Based on the scientific experience gained during his scholarship of the Canadian Government and the work for UNESCO Man and Biosphere (MAB) program, he proposed a new paradigm in water-related sciences—EH—based on the requirement for regulation of ecological and hydrological processes to reach sustainability. He is a respected international expert in EH, working closely with the UNESCO–IHP, as well as other United Nations' and European Commission's programs and institutions. He is author and coauthor of numerous scientific articles in prestigious scientific journals, as well as book chapters concerning the EH concept. He is a devoted lecturer in national and international teaching programs, as well as chairman and keynote speaker of more than 70 international scientific conferences and symposia. He is a also the founder and editor in chief of the international journal of *Ecohydrology & Hydrobiology* and member of the editorial board of *Ecological Engineering, Brazilian Journal of Biology, Wetlands Ecology and Management, Fisheries Management*, and other ecology journals. He is a leader of numerous scientific national and international projects. Professor Zalewski is a member of environmental scientific committees of the Polish Academy of Sciences and a member of the advisory board of the Ministry of Environment for Water Resources Management. He served, among others, as a national representative for the Polish Government in the Organisation for economic co-operation and development (OECD) Export Credits and Environment treaty negotiations, member of the scientific council of the Regional Office for Science and Technology for Europe (ROSTE), chairman of the working group "Fish and Land/Inland Water Ecotones" of UNESCO MAB, as chairman of the working group "Physical Habitat Modification and Freshwater Fisheries" of FAO European Inland Fisheries Advisory Commission (EIFAC), member of the Council of the National Centre for Research and Development Poland, representative of Poland at the Thematic Working Group in Environment of the European Strategy Forum on Research Infrastructures (ESFRI ENV TWG), chairman of the steering committee of the program "Ecohydrology" of the UNESCO–IHP, and member of the drafting task force for the European Union (EU) Joint Programming Initiative (JPI) for water and current UNESCO–IHP programs.

PREFACE

The genesis of the ecohydrology (EH) origins from the abiotic–biotic regulation concept (ABRC) and dates back over three decades ago. What is common in science is that it was inspired by contact with many scientists. Using this opportunity I would like to acknowledge all of them, mentioning just some.

The turning point, which gave impetus to my further scientific career, was my postdoctoral Canadian Government fellowship at Guelph University and the Matamek Research Program of the Woods Hole Oceanographic Institution, in the early 1980s, where I had the opportunity to meet many leading American ecologists and limnologists. Nonetheless, the fundamentals to assimilate these inspirations and encouragements I gained in Poland during my studies and research at the University of Łódź where the first important experience was during the intensive field and lab studies on river ecology with Prof. Tadeusz Penczak, my PhD supervisor. I was investigating the role of allochthonous and autochthonous matter in the river trophy, fish production, and reproductive strategies at rivers of different levels of pollution. Then I had an opportunity to meet the leading Polish scientists associated with the International Biological Program (IBP). As head of a hydrobiological students association, I contacted the Department of Ecological Bioenergetics of the Nencki Institute in Warsaw lead by Prof. Romuald Z. Klekowski and Prof. Zofia Fisher, who at that time were finalizing a handbook of the IBP/Oxford series. I spend in their labs several summer months joining the team working on fish bioenergetics. During my PhD studies I had an opportunity to get in close contact with a team of the top Polish hydrobiologists Anna Hilbricht-Ilkowska, Zdzisław Kajak, Ewa Pieczyńska, Ewa Kamler, and Andrzej Prejs, who were involved in a large project on Masuria Lakeland. All that provided me an important background to embrace yet another challenge when I arrived in Canada in 1980.

It was the time when the River Continuum Concept (RCC) was proposed by Vannote and coauthors, and I was fascinated by the new perspectives this change of paradigm opened. It was an especially easy-to-use framework because the RCC considers not only shift in invertebrate community functional groups but also changes of bioenergetic processes along the river course, of which I had experience form previous studies. Nonetheless, after a few months I felt that the new and even more potent tool to deepen the understanding of ecological processes that determine the dynamics of ecosystems was needed. I used to spend hours with my friend Kazimierz Szewczyk, the philosopher, discussing the methodology of science and getting acquainted with the important concepts of Kuhn, Popper, and Feyerabend. This focused my attention on the importance of quantification of ecological processes and reducing them to physics laws, as well as the role of logically constructed hypothesis in formulating of a scientific concept. Thus, in my search I was driven by the tenet that formulation of a complex hypothesis forces highly advanced and sophisticated design of experiments, which further leads to discovering new properties of ecosystems. At the same time, I had long discussions with my friend Józef Robakowski, professor

PREFACE (continued)

of the Łódź Film School and one of the modern art leaders in Poland, which raised my awareness of the importance of creative independent thinking in science.

I believed that there is a need to make a step forward after the RCC. I envisaged it should be more complex and more related to laws of thermodynamics, hydraulics, etc. It was when Robert J. Naiman invited me to participate in his Matamek Summer Program. There I had a chance to see a boreal river for the first time in my life and an appreciated opportunity to read his brand new book about fish of the North American deserts. There I found the key to formulation of the question that should speed up the progress of improving our predictive ability and proactive management of ecosystems—which is regulation. The most important was that Bob Naiman was not only encouraging but even enthusiastic about this idea.

The model of biota regulation across the RCC and temperature gradient was based on fish biology—that is why I submitted an abstract to EIFAC FAO XIII session hosting the Symposium of Habitat Modification and Freshwater Fisheries in Aarhus, Denmark. I was highly surprised when John Alabaster, the convenor of the symposium, made the abiotic–biotic regulation concept (ABRC) the opening lecture. After that, the very next year I got an invitation to present a test of this theory at British Fishery Society's annual meeting, and later I was invited to lead the working groups of FAO EIFAC, UNESCO MAB, and UNESCO IHP programs consequently. Over these years I had a great chance to cooperate with such prominent scientists as Frederic Fournier, John E. Thorpe, Fritz Schiemer, Georg Janauer, William J. Mitsch, András Szöllösi-Nagy, Luís Chícharo, David M. Harper, Richard D. Robarts, Eric Wolanski, José Galizia Tundisi, Zbigniew Kundzewicz, Zdzisław Kaczmarek, Leszek Starkel, Artur Magnuszewski, Ian G. Cowx, Pierre Hubert, Giovanni Bidoglio, Pascal Breil, Michael E. McClain, Azime Tezer, Jim Reynolds, Ângelo A. Agostinho, Shahbaz Khan, Jun Xia, and Demin Zhou, which gave me and still gives further inspiration for my scientific activity. As the result of all the brainstorming we had, the word "regulation" became a key word of the new paradigm and EH process-oriented thinking and has been appearing as a step toward the new integrative environmental science in which the ultimate goal is sustainability. All these international contacts would not have developed if I had not have a great team to do fieldwork and to discuss all sorts of ideas. I am especially obliged to Piotr Frankiewicz, Bogda Brewińska-Zaraś, Malgorzata Lapinska, Malgorzata Tarczynska, Iwona Wagner, Katarzyna Izydorczyk, Agnieszka Bednarek, Joanna Mankiewicz-Boczek, Edyta Kiedrzyńska, Tomasz Jurczak, Magdalena Urbaniak, Kamila Belka, Kinga Krauze, Adrianna Wojtal-Frankiewicz, and Zbigniew Kaczkowski, to mention the most important.

I wanted to highlight so precisely the starting point of the ideas introduced in this chapter having in mind the "iron curtain" times when it was especially difficult to do good science. I must underline that without such support of the scientists from the East and the West, the person like me would never get into the global science network. I use this opportunity to express my appreciation to all those people; many of them were not mentioned here by name. I always feel obliged to do the same and support younger colleagues whenever opportunities arise. I believe that this chapter to a great extent inspired by Prof. Bill Mitsch's Ecological Engineering Concept might open a new phase of dialogue between environmental scientists and hydrology engineers. I believe that synergy obtained by integration of these two fields is the only way to reverse degradation of the global environment and to achieve sustainable use of water, which is the primary resource to persist on the planet.

Finally, I want to express my great appreciation for all forms of support from my wife Anna and my daughter Monika who were always open to share my enthusiasm and to be first expressing inspiring but sometimes critical opinions about my ideas and also the ones who suffered most from my frequent absence at home.

5.1 Introduction

In the Anthropocene era, due to acceleration of the population growth, the relationships between humans, water resources, and ecosystems become highly complex. What is more, over 70% of the Earth's surface is modified, which results with the subsequent modifications of the key ecological processes propelled by water and nutrient circulation. Continuing demographic growth together with climatic changes is expected to amplify those negative trends in many areas, and according to the International panel on climate change (IPCC), water availability may decline by 20% by 2050. For this reason, there is an urgent need to formulate a new strategy for global environmental resources and particularly for water use.

Considering water as a major driver of ecosystem functioning and its sustainability, there are three major threats related to the degradation of the hydrological cycle: (1) *acceleration* of the water and matter circulation due to degradation of natural plant communities, river channeling, and increasing impermeable surfaces in urbanized areas; (2) *emission* of large fluxes of the fossil fuel energy and matter, mineral, organic, and chemical substances to the environment; and (3) *reduction of habitats* for plants and animals in terms of space and connectivity. All of them reduce water resources' availability, biodiversity and a gene pool decline, resilience to various forms of human impact and climate variability, and ecosystem services for society. This in turn amplifies overexploitation of natural resources, which by following the pattern of the "tragedy of the commons" may finally lead to Malthusian catastrophe.

Following the Millennium Ecosystem Assessment conclusions, the first question is: "Should we try to solve the increasing water, food, and environmental problems by applying more and more technology, finally changing the Earth into *techno garden?*"

The answer suggested by ecohydrology (EH) is to apply the understanding of water and biota interplay to use the ecosystem processes as a new tool in water and environment management. Secondly, this new tool has to be harmonized with hydro-engineering solutions. However, a prerequisite to such harmonization should be the awareness of the rudimentary mechanisms of ecological processes and biological evolution (e.g., thermodynamic laws, natural selection, and self-organizing potential of ecosystem that are limiting ecological and biological processes, etc.). An example of the large-scale long-term EH processes is the broadly observed reduction of water retentiveness in the landscape due to the reduction of permanent plant cover [12]. Its consequences, that is, acceleration of water and sediments and nutrient transfer from the landscape to the sea and the atmosphere, have been driving the ecological succession process toward xeroseries (drying ecosystems). This seriously affects the biogeochemistry of these regions, that is, a permanent exchange of carbon, nitrogen, phosphorus, and other biogenic substances in gaseous and sedimentary cycles between abiotic and biotic components of ecosystems. The symptoms are declining food productivity in the catchment scale.

The routine approach to environmental issues, dominating also in engineering sciences, is conservation and, more sophisticatedly, restoration. As a result we usually come to disagreement between pro-conservation and pro-development parties. The EH approach offers in this respect yet another way through *regulation*. To reach this goal, the first step would be to extend the general perception of the environment toward understanding of the key EH processes, that is, water and nutrient circulation in the biogeosphere. With this understanding in mind, all range of the engineering methods should be harmonized with EH regulation methods toward enhancement of ecosystem's carrying capacity, in order to compensate the cumulative human impact on the environment. The result of such combined actions would be the engineering harmony between human and the environment. Lack of understanding of the processes, on the other hand, has already led to overengineering of the environment in many catchments and, as a consequence, resulted in a decline of resilience and ecosystem services, as well as economic loss. Resilience in biological sciences is understood as a capacity of the ecosystem to adapt to the changing "abiotic template" by biotic interactions in such a way that the key ecological processes are maintained within a range of oscillations specific to a given process. This capacity guarantees maintaining current biodiversity and bioproductivity of an ecosystem or allows for transformation into a new sustainable one. Ecosystem services for society are such properties of ecosystems that are used by people, in a deliberate (e.g., wood for timber) or unconscious way (clean air, aesthetic values, etc.). Thus, ecosystem services are understood as the benefits the society receives from the nature, such as provision of water, food, biodiversity, and other natural and cultural values. The increasing human impact on the environment, which modifies the basic EH processes, affects both the resilience and ecosystem services and leads, as a consequence, to economic loss and social and cultural degradation.

As said earlier, the overall goal of EH is *enhancement of ecosystem's carrying capacity*, understood as a simultaneous improvement of water resources, biodiversity, ecosystem services, and resilience to climate and other impacts, by means of regulation of EH processes. This regulation is needed because neither conservation nor restoration prove to be feasible nowadays. However, any interference with the environment, and the processes, must always be cross-verified with the long-term prognoses for socioeconomic and technological development, that is, on the basis of foresight, which includes scenarios and vision for the future [19]. What is more, there is an urgent need to change the socioeconomic approach to the development. The paradigm that we are living in a competitive world has eventually led to the exponentially accelerating exploitation of natural resources, consumption of energy and materials [6], and reducing undisturbed natural heritage. A new approach should embrace a shift where competition for resources is replaced by competition in the resource efficiency use and regulation of ecological processes toward sustainability.

Last but not least, regulation of high complex ecological processes in a basin scale has to be based on an informed action. The key assumption of EH harmonization of society needs with enhanced ecological potential has to be based on accelerated environmental sciences evolution from "information gathering" to "knowledge generation" fed by integration of different disciplines, which will finally generate "wisdom" on how to carry on actions. In this respect, EH tries to combine information on patterns and structures of ecosystems to understand underlying processes to create fundamentals for the formulation of the principles for enhancement of water resources and ecosystem's carrying capacity and other high-complexity problem solving [29].

5.2 Ecohydrology as a Problem-Solving Science

According to the International Council for Science (ICSU), scientific efforts in the twenty-first century must be integrative, problem solving, and policy oriented, rather than solely curiosity driven. Recent integration of environmental sciences toward problem solving was initiated by the concept of ecological engineering formulated by Mitsch [16], which changed the way of thinking about the relationship between the human and the biosphere. The increasing global environmental degradation led to the conclusion that as far as the biosphere has been shaped by evolution, ecosystem processes have to be considered as a management tool to reverse this downward trend [3,10,33]. The future of the biogeosphere will be dependent not only on the development of environment-friendly technologies but also on the harmonization of those technologies with the enhanced potential of the environment, which can be achieved by an integrative science.

EH—the integrative transdisciplinary science—is based on the assumption that the major drivers of the biosphere evolution in a given thermal regime are the interactions between water and biota. Thus, a thorough understanding of the previous mechanisms can be used to regulate the complex interactions between the water cycle, ecosystems, and the societies. The general assumption of EH is that in order to reverse degradation and achieve sustainable water and ecosystems in highly modified river basins, there is a necessity to regulate ecological processes at different levels. The microbial level would include, for example, enhancement of denitrification in the land–water ecotones [2]; the ecosystem level would mean for example, biomanipulation at ecosystem scale [20] and reforestation; and creation of land–water ecotone zones [17] would fall into a type of interventions at a landscape scale.

This regulation, however, should be implemented in parallel to the already recognized actions, such as reduction of erosion, nutrients, and pollutant emission.

5.3 Ecohydrology: Definition

The EH concept was formulated in the framework of the IHP of UNESCO, starting from 1996 and programmed until 2021, as a reply to the need for developing a methodology on how to regulate hydrological cycle at various spatial scales toward sustainability [5,26,32,33,35]. On the other hand, EH, as a subdiscipline of sustainability science, is focused on biological aspects of the hydrological cycle. It provides not only scientific understanding of the hydrology/biota interplay but also a systemic framework on how to use ecosystem processes as a tool complementary to already applied hydro-technical solutions. It is also compliant with the Integrated Water Resources Management (IWRM) guidelines. A new aspect of this approach is that to reverse degradation of the aquatic ecosystems, it is necessary to regulate the ecological performance of the "novel ecosystems" [9] at various scales, from molecular to landscape, towards enhancing carrying capacity.

5.4 Key Elements of Ecohydrology as a Background for System Solutions

For terrestrial ecosystems a fundamental assumption of EH is based on the concept that water availability determines plant yield [23], and an increase in biomass production of certain ecosystems is highly correlated with solar radiation [13,14]. Following these dependencies, the hypothesis introduced by Zalewski [30,33] suggests that in the given geomorphological conditions, water and temperature are the major determinants of *biodiversity*. The water and temperature determine primary productivity and the rate of energy flow through an ecosystem; the temperature, on the other hand, determines the ratio of decomposition of organic matter (thus nutrients cycle dynamics), as well as the ratio of carbon allocation between biomass and organic matter (energy flow dynamics through the ecosystem). Following the van't Hoff law where the rate of chemical reactions increases together with the temperature rise, the growing temperature from boreal to tropical zone, and thus accelerated organic matter decomposition, results in an increased carbon accumulation in the biomass (Figure 5.1). As a consequence, conditions for diversification of microbe, plant, and animal communities, by persistence of favorable mutations and consequently by occurrence of more dynamic natural selection and adaptation processes, become more favorable by an order of magnitude. On the other hand, while moving from tropical to boreal zone, the total amount of dead organic matter accumulated in the soil rises, due to the shortening of the growing period for plants and slower organic matter decomposition rate, indicating the limited flow of energy and nutrients through the biosphere [31].

The EH concept was conceived based on two parallel observations fuelling two simultaneous conclusions, one deductive and the other one inductive in nature. The first observation was derived from the general theory of physics, ecological bioenergetics, hydrology, and evolutionary ecology and led to the formulation of the abiotic–biotic regulatory continuum concept [37]. The other one was inspired by the effect of cascading changes in the trophic structure of a reservoir (a cascading effect) while the water level changes and an ascertainment that biotic interactions induced by abiotic changes may affect the intensity of algal blooms [36].

The concept of the abiotic–biotic regulatory continuum was formulated in the early 1980s [37] and met the rising consciousness of the progressing degradation of the biosphere that implied a need to develop a new way to reach sustainability by ecological process regulation. The concept is based on the understanding of the functional relation between abiotic and biotic factors as a mode of regulating fish communities in different geographical zones and on the responses of fishes to ecological factors, which are the same in any ecosystem, although they may work in different ranges [37–39] (Figure 5.2).

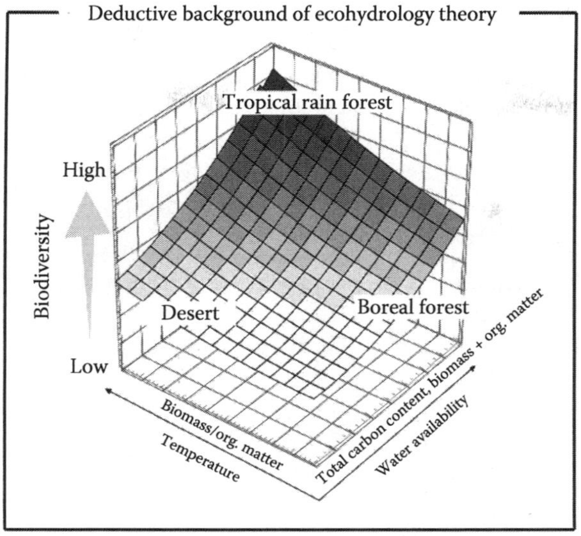

FIGURE 5.1 The amount of water determines the amount of carbon accumulated in the ecosystem, while temperature determines the carbon allocation between biomass and soil organic matter. (From Zalewski, M., *Braz. J. Biol.*, 70(3 Suppl.), 689, 2010.)

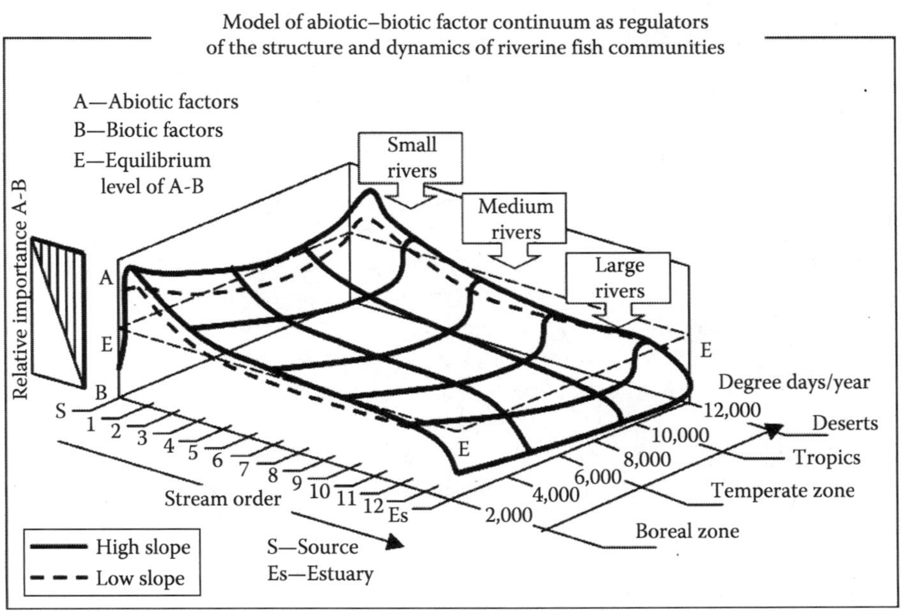

FIGURE 5.2 The model of change of the hierarchy of regulatory factors along the gradients of the stream order, thermal conditions, and slope. The thermal features of a river are presented from the point of view of fish physiological performance by degree-days per year, which is the sum of average daily temperatures. (From Zalewski, M. and R.J. Naiman., The regulation of riverine fish community by a continuum of abiotic–biotic factors, In *Habitat Modification and Freshwater Fisheries*, ed., J. S. Alabaster, pp. 3–9, Butterworths Scientific Ltd, London, U.K., 1985.)

A comparison of riverine fish communities that occur in various geographical areas of the world demonstrates that their structure and dynamics are regulated by a continuum of abiotic and biotic factors. The abiotic factors are usually of primary importance; however, the role of biotic factors gradually increases when the character of the environment approaches the physiological optimum of a fish and when it becomes stable and predictable. The research on fish diversity shows that pertinent physiological adaptations and reproductive strategies of fish species provide an evidence that a certain combination of abiotic–biotic factors regulates a continuum of fish communities in rivers to a large extent. This fish community continuum responds mainly to such factors as

1. Stream order, which expresses a degree of hydrology stabilization and a degree of environmental diversity
2. Temperature regime, which has a strong impact on the physiological performance of the ecosystem and fish productivity
3. Slope of the river, which seems to modify to a huge extent the previous two factors, as well as can modify the quality of abiotic factors, such as the speed of the current, structure of bottom substrate, and dissolved oxygen content (compare Figure 5.2)

Another core element of EH is the concept stating that models of the effects of environmental change on river systems are nonlinear by nature [18], and observed multiple feedbacks and thresholds may be used as a management tool toward a sustainable water management. An example of such correlation is the aforementioned cascading effect of water level manipulation in a reservoir on the whole freshwater ecosystem. The research was conducted on the Sulejow Reservoir, Central Poland, in the early 1980s, and it showed that the reduction of water retention during the spawning period reduced to a great extent the area of spawning grounds and led to the reduction of excessive density of young planktivorous fish from 40 to less than 5 fish per square meter. This created less competitive conditions for young fish to feed and allowed to maintain high density of filtering zooplankton (12–16 mg/L). The latter, thus, was able to reduce the number of phytoplankton and prevented the appearance of algal blooms [36] (Figure 5.3). This case proved a reduction of the algae biomass by 80% and allowed to avoid toxic algal blooms. It serves also as a good example of the other two postulates of EH, that is, (1) harmonization of EH measures with necessary hydrotechnical infrastructure and (2) integration of various regulation actions in a synergistic way to stabilize and improve the quality of water resources [34].

5.5 Ecohydrology: The Terrestrial and Aquatic Dimension

Considering the conceptual division of hydrological cycle into an atmospheric/terrestrial phase (precipitation, evaporation and evapotranspiration, runoff, and storage) and an aquatic phase (river outflow, retention in waters), the EH approach to catchment management distinguished terrestrial (EHT) and aquatic ecohydrology (EHA). This management concept is also in line with the IWRM concept. In both areas, diverse biota appear as moderators of water dynamics. In the terrestrial phase (EHT), where vegetation moderates the water quantity and quality, the major question is how the land cover changes influence the hydrological cycle. In the aquatic phase (EHA) though, where complicated biotic interactions affect water quality, the main concern nowadays is eutrophication and their related symptoms, for example, toxic algal blooms.

In case of pollution, the reduction of point-source pollution is dependent on proper technology development and implementation, monitoring, and law enforcement. Just to the contrary, the reduction of diffuse pollution, which is mainly agriculture dependent, first of all, should be based on a good understanding of the hierarchical complexity of ecological processes in a river basin. Of utmost importance is the interplay between water and biocenosis—both in the terrestrial and aquatic phases of the hydrological mesocycle.

A good example of EHT solutions is creation of the highly efficient land–water transition zones (ecotones) for reduction of pollution fluxes from land to waters from diffused sources. The nonpoint

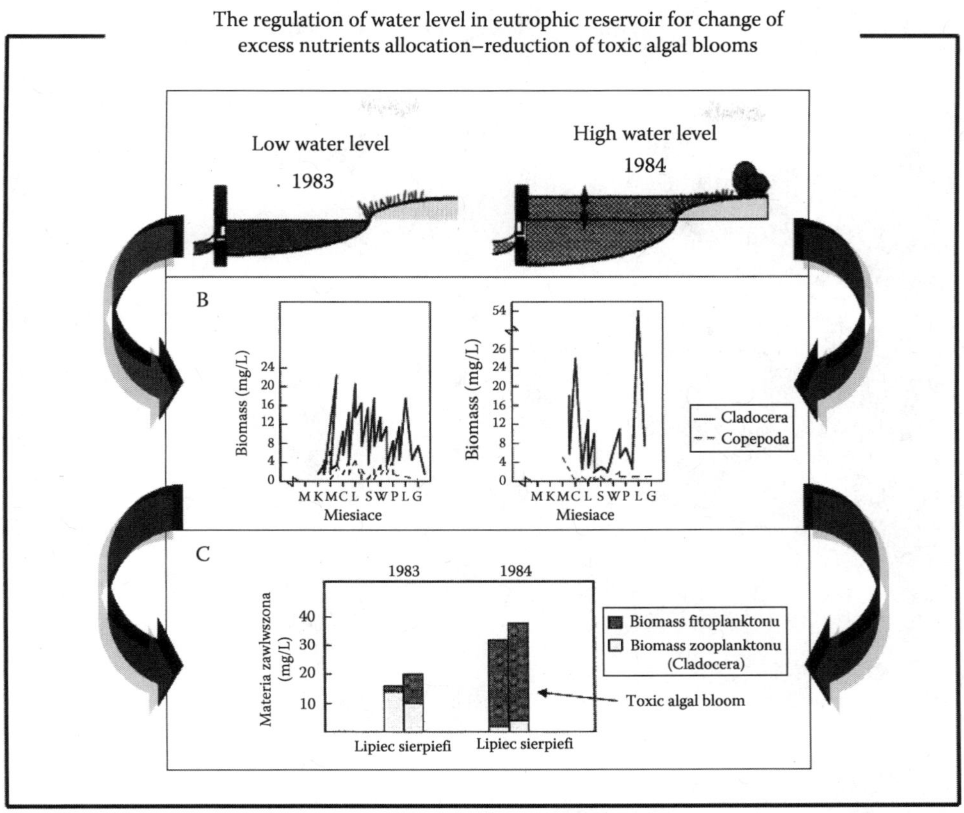

FIGURE 5.3 The cascading effect of water level control on the ecosystem. (From Zalewski, M. et al., *Hydrobiol.*, 200/201, 549, 1990.)

sources of pollution constitute over 60% of the total nitrogen and phosphorus load from the territory of Poland to the Baltic Sea (Figure 5.4). General opinion is that to reduce it by at least 50%, there is a need to use a 15 m wide ecotone plant buffer between agricultural land and water. However, such solution would not be acceptable for farmers, who perceive it as a loss of productive land. In this case, the efficiency of the ecotone belt should be improved. One of the solutions was to create a 2 m wide buried structure that enhanced denitrification processes of the ecotone and allowed for reduction of the width of the transition zone to 6 m and maintaining of 70% efficiency [21,28] (Figure 5.5).

Both EHT and EHA require integration into the water management practice within a basin. The methodology should respect the quantitative side of the hydrological cycle and should be supported by remote sensing, GIS techniques, and mathematical modeling [11].

The aquatic ecosystems are complex entities, and they should be studied in a interdisciplinary approach by considering the hierarchy of various regulatory environmental factors. However, this study needs an additional dimension—the society—that, if combined with natural sciences, will develop into an integrative transdisciplinary science and management.

EH provides three new aspects to environmental sciences and their integration into problem solving: (1) the need to *use ecosystem properties as a management tool* that is complementary to and harmonized with hydro-technical solutions, (2) the necessity to *enhance ecosystem's carrying capacity* toward reaching the United Nations Millennium Development Goals (UN MDG) by using the interplay between hydrology and biota, and (3) a requirement of *regulation of ecological and hydrological process* based on the analysis of dynamic oscillations of the ecosystem, its productivity, and succession

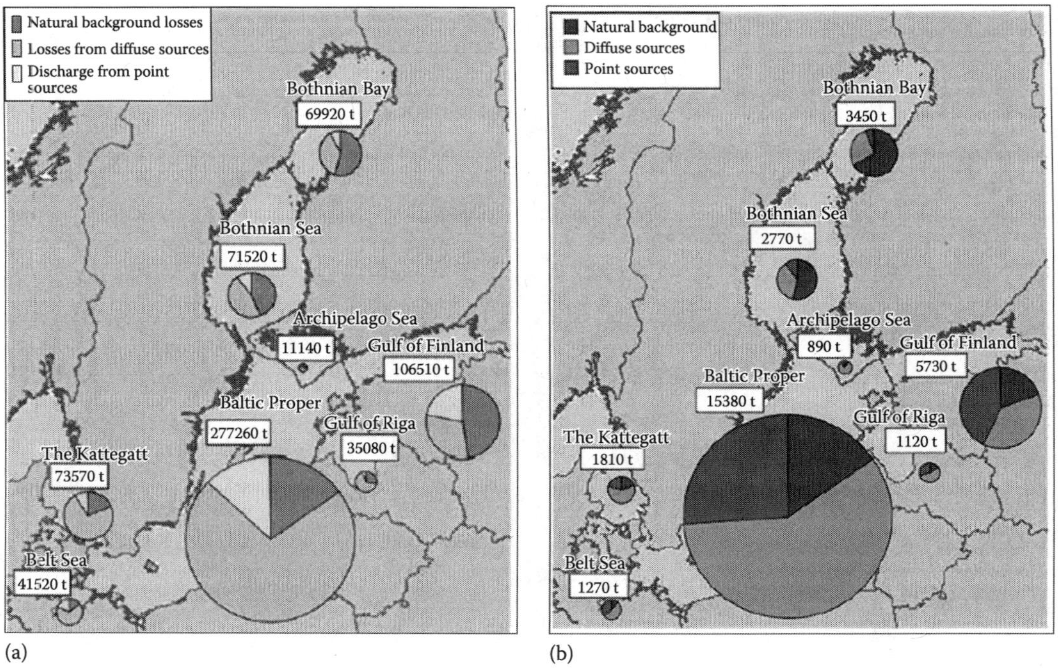

FIGURE 5.4 Proportion of source contribution to nitrogen (a) and phosphorus (b) inputs into the Baltic Sea.

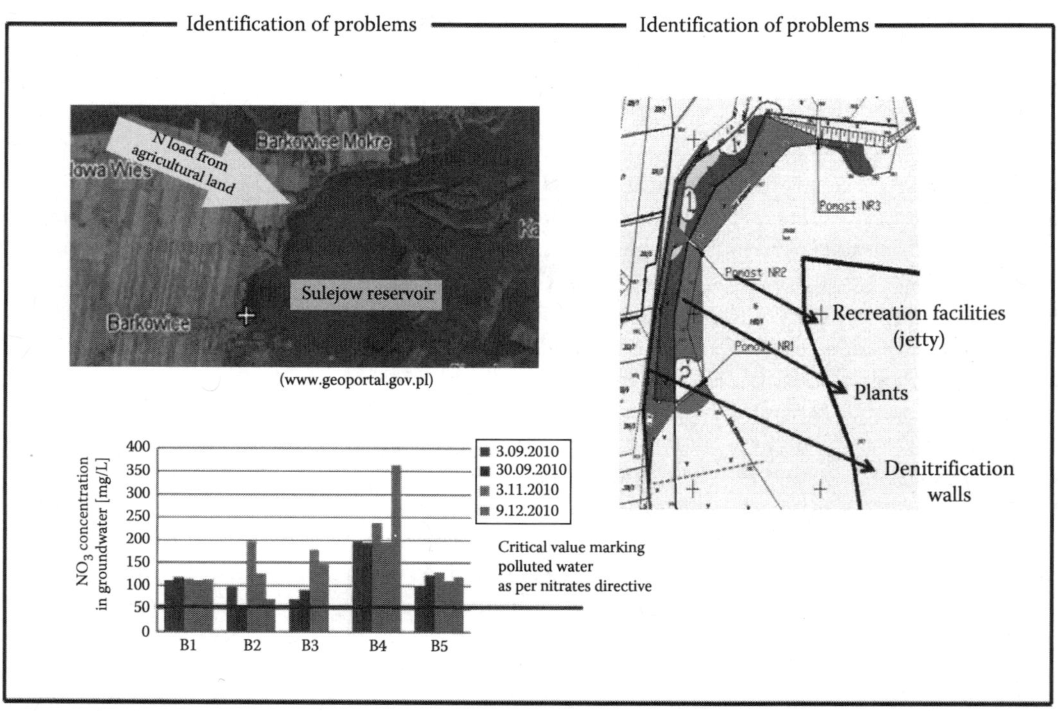

FIGURE 5.5 Reduction of nitrogen pollution from diffuse source by enhancement of plant buffering zones with denitrification walls. (From Izydorczyk, K., et al., 2013. A biogeochemical barrier to enhance a buffer zone for reducing diffuse phosphorous pollution–preliminary results. *Ecohydrology and Hydrobiology* 13(2): 104–112.)

pathways reflected by nutrients'/pollutants' absorbing capacity versus human impacts based on a "dual regulation" approach.

According to Laszlo [15], transformation from industrial to postindustrial global information society characterized by extreme consumption patterns is in reality a choice between further evolution and extinction. In this vein, the EH goals are

1. To slow down the transfer of water from the atmosphere to the sea (considering floods and droughts control as a priority)
2. To reduce input and regulate the allocation of excessive nutrients and pollutants to aquatic ecosystems in order to improve water quality, biodiversity, and human health
3. To enhance ecosystem's carrying capacity (resilience, biodiversity, and ecosystem services for society) by harmonization of the society needs with the ecosystem potential within the framework of EH principles and IWRM

5.6 Principles of Ecohydrology as a Framework for Scientific Investigation and Problem Solving

The hydrological principle—quantification of hydrological processes—includes, for example, patterns of hydrological pulses along the river continuum with special consideration of the role of groundwater and monitoring of point and nonpoint sources of pollution. The pulsing nature of the ecological variables is connected with fluctuations of the hydrological parameters along the river continuum and its catchment. Thus, understanding the seasonal and spatial patterns of those pulses is crucial for developing the strategy of "dual regulation" of hydrological/ecological processes. Implementation of the first principle has to be done with the help of numerical methods supported with geographical information science and technologies (GIS). This is because the complexity of the water–biota interactions and the pulses character is to a great extent determined by geomorphology of the basin, distribution of various types of natural and modified ecosystems, and increasing extent of the urbanized and industrialized areas, which are additionally interconnected with hydro-technical infrastructures. The understanding of the previous processes in the basin scale is fundamental for the regulation of EH processes for the reduction of the impacts of floods and droughts, enhancement of river self-purification processes, and control of reservoirs' eutrophication and degradation of ecosystem services in costal zones.

The ecological principle is the key assumption that, under intensive global changes, it is not enough to protect ecosystems against increasing impacts but is essential to regulate ecosystem structure and processes toward increasing of the ecosystem's carrying capacity, understood as a simultaneous improvement of water quality, restoration of biodiversity, ecosystem services for society, and resilience. Understanding of the role of vegetation in water-cycling processes in pristine ecosystems is of crucial importance [1,4,7,8,24,27]. However, only the modified ("novel") ecosystems should be the subject for regulation by reversing their degradation and restore their potential for maintaining sustainable water and biota processes. The previous actions have to be cross-verified with the understanding of the biological evolution, ecological succession, and paleohydrological processes in order to define the possibility of harmonization of the enhanced ecosystem potential with the society needs.

The ecological engineering principle—integration of the understanding of the hydrological cycle and biological processes in the basin scale—is based on the assumption that abiotic factors geomorphology, climate, and hydrology, are of primary importance, and when they became stable and predictable, then biotic factors start to play a prevailing role [37]. There is a need to use ecosystem processes analyzed under the framework of the first and second principles of EH as a management tool, which relates to the ecological engineering concept. Since natural processes in most of the world's river basins at the beginning of the twenty-first century are to a great extent modified, it is impossible to reverse their degradation without using technical solutions. For example, when biochemical oxygen demand, a common indicator of the organic quality of water, of a large city wastewater runoff is too high, the use of exclusively natural processes (e.g., constructed wetlands) to treat this water would require enormous space. By analogy, if water retentiveness of a catchment was highly reduced and a river outflow significantly accelerated (due to, e.g., deforestation), then, in the face of the expected climate changes, the construction of reservoirs to retain water for human needs is critical. Here, one common problem of designing and constructing of new dams arises, that is, if and to what extent reservoirs can be negative for ecological structure of the environment? The answer is that in highly modified catchments, reservoirs can support the overall biological structure of the catchment, at the same time supplying water for societal and industrial needs and enhancing agricultural productivity, if only they are constructed with consideration of the basic ecological processes (e.g., maintaining organism migration by a well-functioning bypass channel). They may also seriously reduce transfer of nutrients and pollutants downstream [22]. This principle requires that any scientific investigation should have its implementation side. In this respect, the implementation should follow a proposed pattern:

1. *Dual regulation*—we aim at regulating biota by hydrology and, vice versa, hydrology by shaping biota or their controlling interactions.
2. *Synergy of actions*—we aim at integration of various types of biota- or hydrology-based solutions toward achieving synergy in improving water quality, freshwater resources, and biodiversity.
3. *Integration*—we aim at harmonization of the developed EH measures with existing and necessary hydro-technical solutions, such as dams, irrigation systems, and sewage treatment plants.

5.7 Ecohydrology: Implementation

EH implementation for IWRM includes the following four steps: (1) monitoring of threats, (2) assessment of cause–effect relationships, (3) development of EH methods, and (4) development of system solutions (Figure 5.6).

5.8 Summary and Conclusions

In the face of the reducing Earth's global carrying capacity, the key to reverse degradation in line with the UN MDG should be a new global environmental policy and public consciousness. The first is based on the idea of Factor Four [25] where reducing the consumption of materials and emission of pollutants by half would still allow to double the wealth of societies. The latter is based on a profound understanding of the cultural heritage of human civilization and its regional unique values, where the wealth of the cultural diversity should be considered equally important as biodiversity and as a key for new perception of the relation between man and the environment, especially sustainable use of environmental resources (Figure 5.7).

There is an increasing number of evidences that the restrictive nature conservation is not sufficient, because to prevent further global environmental degradation and to reverse it in many areas, it is necessary to expand a new way of thinking among environmental scientists, decision makers, and society. The *conservation* executed by law enforcement is still of primary importance in the case of pristine and not drastically modified ecosystems. These are the important natural and cultural heritage for humanity. On the other hand, *restoration* of ecosystems of local importance is necessary because otherwise the use of the ecosystem services would be too costly. For example, restoration of eutrophic lake into more oligotrophic and elimination of toxic algal blooms could be done by elimination of sewage deposition into the lake and reduction of internal phosphorus loads from bottom sediments. This would provide

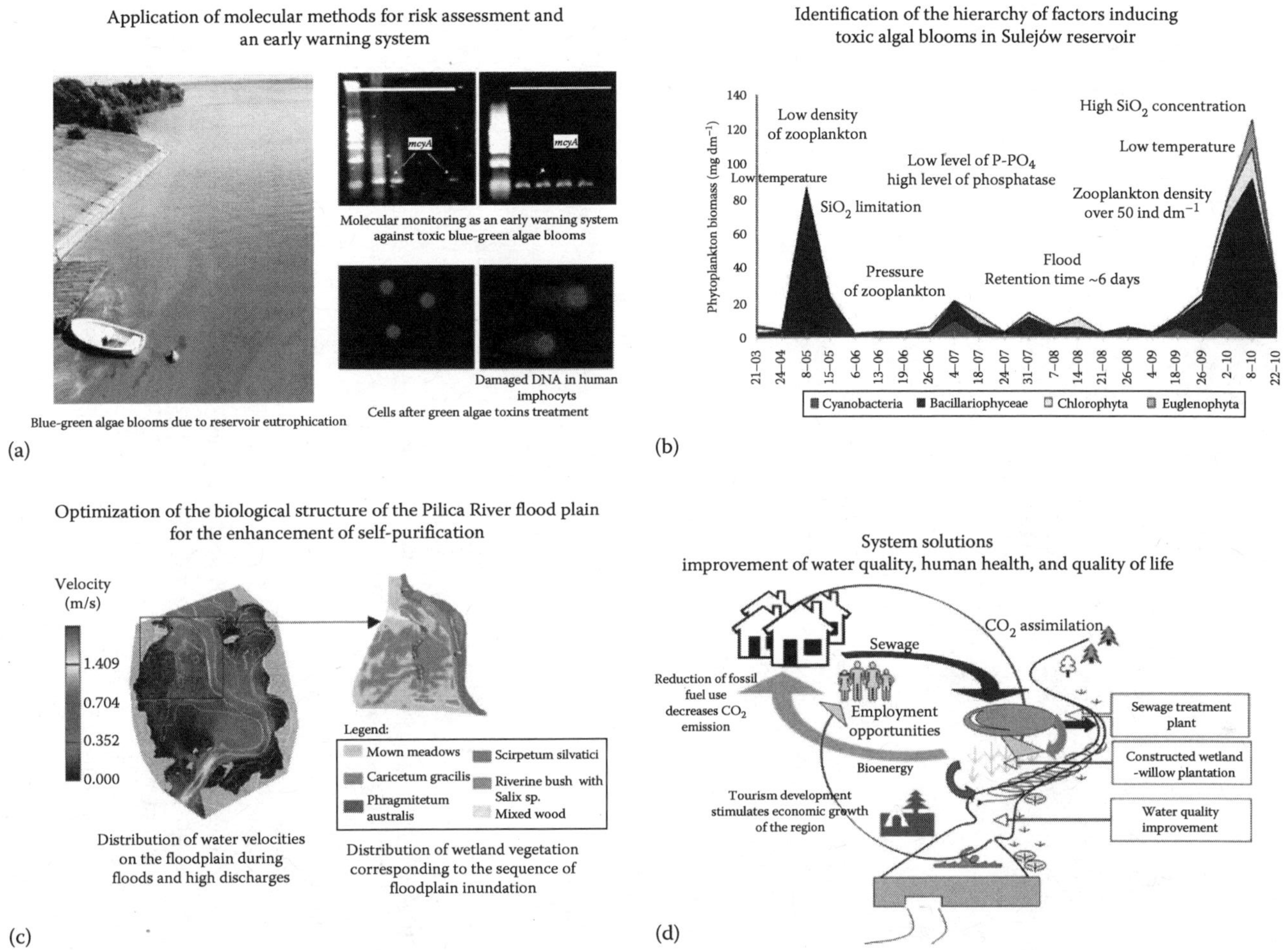

FIGURE 5.6 Four-step cycle of methodology for EH implementation: (a) monitoring of threats. (From Zalewski, 1999; Markiewiez et al., 2005.) (b) identification of cause–effect relationship. (From Zalewski et al., 1999.) (c) elaboration of EH methods. (From Megrusrewski et al., 2006; Kiedrayriska et al., 2009.), and (d) development of system solutions. (From Zalewski, M., *Ecol. Eng.*, 16, 1, 2000.)

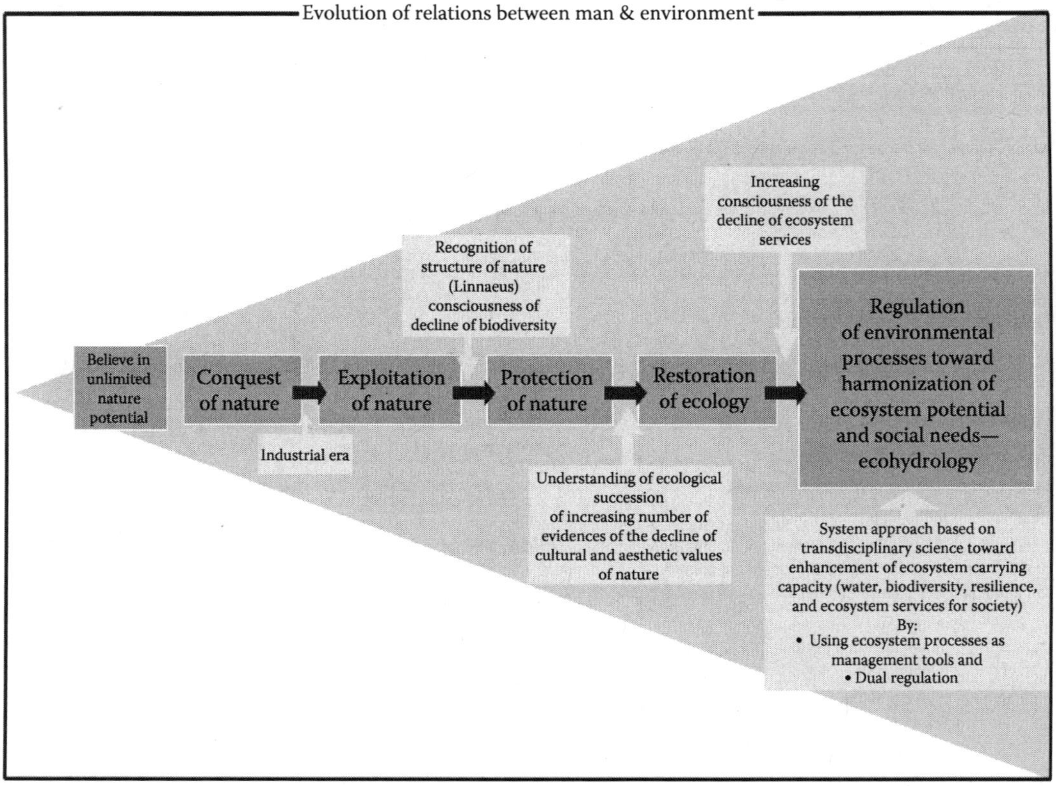

FIGURE 5.7 From the belief in the unlimited potential of nature to the awareness of necessity of regulating ecological processes toward enhancement the ecosystem's carrying capacity—an evolution of the human approach toward use of natural resources.

good water quality and good ecological status of the ecosystem (ecosystem services for society) instead of high economic loss. However, *EH regulation* toward enhancement of carrying capacity of an ecosystem and harmonization of society needs with economic potential has to be done on the basis of the analysis of the entire catchment, even if only improvement of water resources and ecosystem services in one part of the basin is considered. Regulation of EH processes has to be based on the profound understanding of the geomorphology, climate, hydrological dynamics of the basin, ecological structure and processes, its modifications by human impact, and their regulation in the basin scale. The three principles of EH provide the framework towards sustainability.

In the Anthropocene era, we are dealing with socioecological–economical systems, where methodology for reversing negative trends has to be provided by an integrative science.

Scenarios of the proactive strategy for the fulfillment of the UN MDG should be based on the holistic transdisciplinary environmental science, system approach, foresight methodologies, adaptive assessment, and management strategies for the implementation of innovative solutions, education, and social dialogue. Learning outcomes of the new transdisciplinary education have to be focused on the range of practical skills, from the ability to formulate hypotheses related to highly complex processes up to the design of the system solutions and the communication with decision makers and stakeholders. The general goal of the environmental education should be rising of the consciousness among students that we are no more passengers at the spaceship Earth, but we are all members of the crew and we are all responsible (Figure 5.8).

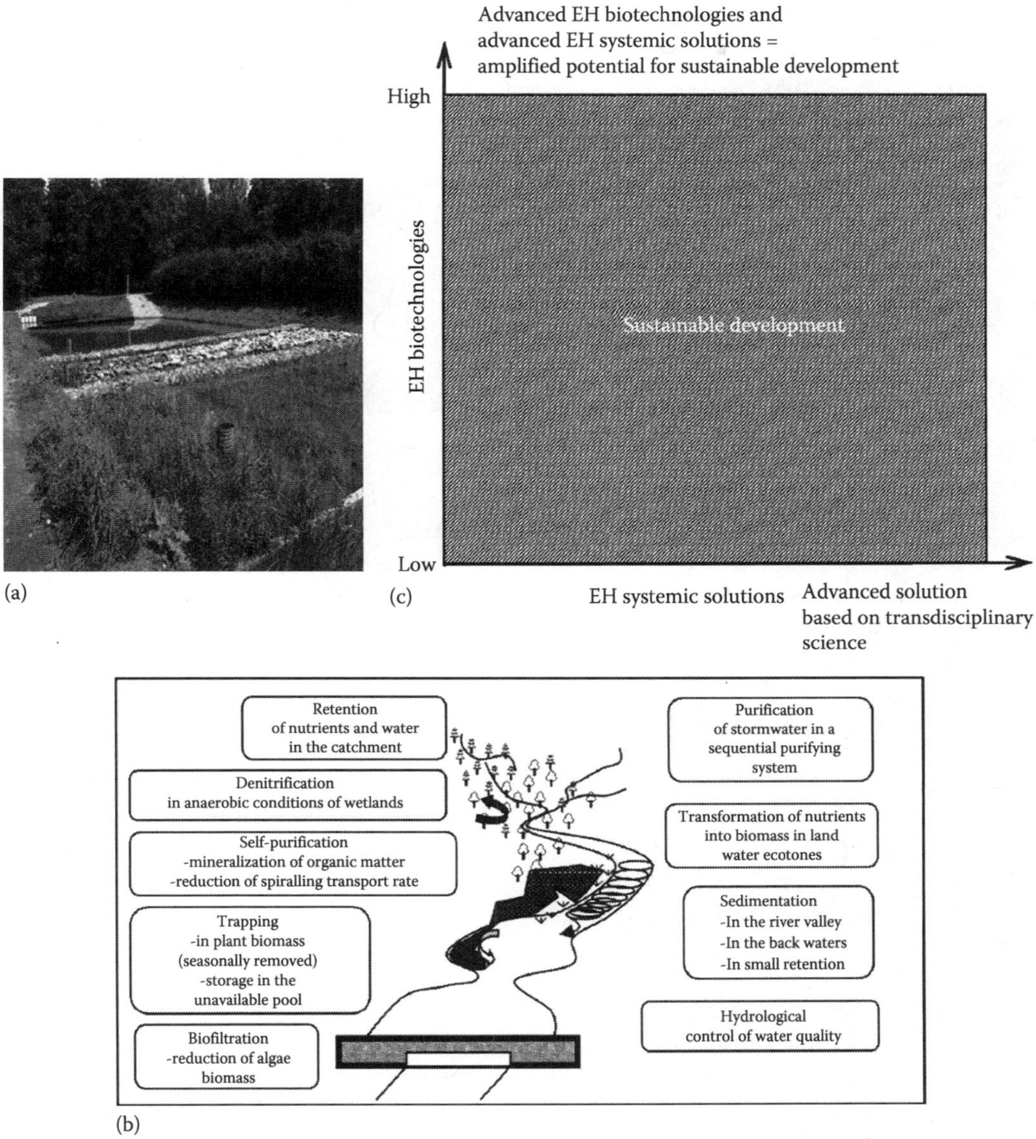

FIGURE 5.8 The integration of EH biotechnologies (a) into systemic framework based on EH principles (b) as a methodology for enhancement of sustainable water, biodiversity, resilience, and ecosystem services for society (c). (a) The sequential biofiltration system for urban storm water purification, which combines enhanced sedimentation zones, biogeochemical barrier, biodegradable geotextiles, constructed wetland, and regeneration system. (b) The example of the systemic solutions for synergy between different EH biotechnologies in river basin. (From Zalewski, M., Ecohydrology for implementation of the EU water framework directive, *Proceedings of the Institution of Civil Engineering Water Management* 164, Issue WM8, 375–385, 2011a.)

References

1. Baird, A.J. and R.L. Wilby. (eds.) 1999. *Eco-Hydrology: Plants and Water in Terrestrial and Aquatic Environments*. London, U.K: Routledge.
2. Bednarek, A., Stolarska, M., Ubraniak, M., and M. Zalewski. 2010. Application of permeable reactive barrier for reduction of nitrogen load in the agricultural areas—Preliminary results. *Ecohydrology and Hydrobiology* 10(2–4): 355–362.
3. Berton, J.P. and M. Bacchi. 1997. Données générales sur le patrimoine écologique ligérien. *Actes du Colloque La gestion patrimoniale de l'Hydrosystème ligérien*: 3–8.
4. Bonacci, O. 2003. Ekohidrologija. Vodnih resursa I otovorenih vodotoka. Zagreb: IGH d. d.
5. Chicharo, L., Chicharo, M.A., Esteves, E., Andrade, P., and P. Morais. 2001. Effects of alterations in fresh water supply on the abundance and distribution of Engraulis encrasicolus in the Guadiana estuary and adjacent coastal areas of south Portugal. *Ecohydrology and Hydrobiology* 1(3): 341–347.
6. Constanza, R., Graumlich, L.J., and W. Steffen. 2007. *Sustainability or Collapse? An Integrated History and Future of People on Earth*. Cambridge, MA: MIT Press.
7. Eagelson, P.S. 1982. Ecological optimality in water limited natural soil-vegetation systems: 1. Theory and hypothesis. *Water Resources Research* 18: 325–340.
8. Eamus, D., Hatton, T., Cook, P., and Ch. Covin. 2006. *Ecohydrology. Vegetation Function, Water and Resource Management*. Collingwood, Victoria, Australia: CSIRO.
9. Hoobs, R.J., Arico, S., Aronson, J. et al. 2006. Novel ecosystems: Theoretical and management aspects of the new ecological world order. *Global Ecology and Biogeography* 15:1–7.
10. Johnston, B.R., Hiwasaki, L., Klaver, I.J., Ramos Castillo, A., and V. Strang. 2012. *Water, Cultural Diversity, and Global Environment Change. Emerging Trends, Sustainable Future*? Paris, France: UNESCO.
11. Jorgensen, S.E. 2002. Explanation of ecological rules and observation by application of ecosystem theory and ecological models. *Ecological Modeling* 158(3): 241–248.
12. Kędziora, A. 2010. Landscape management practice for maintenance and enhancement of ecosystem services in a countryside. *Ecohydrology and Hydrobiology* 10(2–4): 133–152.
13. Kędziora, A. 1996. Hydrological cycle in agricultural landscapes. In *Dynamics of an Agricultural Landscape*, (eds.) L. Ryszkowski, N. French, and A. Kędziora, pp. 65–78. Poznan, Poland: PWiL.
14. Kowalik, P. and H. Eckersten. 1984. Water transfer from soil through plants in the atmosphere in willow energy forest. *Ecological Modeling* 26: 251–284.
15. Laszlo, E. 1994. *VISION 2020. Reordering Chaos for Global Survival*. Amsterdam, the Netherlands: Gordon and Breach.
16. Mitsch, W.J. 1993. Ecological engineering—A co-operative role with planetary life support system. *Environmental Science Technology* 27(3): 438–445.
17. Naiman, R.J. and H. Decamps. 1990. *The Ecology and Management of Aquatic-Terrestrial Ecotones*. Paris, France: Parthenon.
18. Phillips, J.D. 1997. Simplexity and the reinvention of equifinality. *Geographical Analysis* 29(1): 1–15.
19. Rogut, A. and B. Piasecki. 2011. Foresight methodology as a tool for elaboration of plans for sustainable management of water, energy, the environment and society. *Ecohydrology and Hydrobiology* 11(3–4): 261–272.
20. Shapiro, J., Lamarra, V., and M. Lynch. 1975. Biomanipulation: An ecosystem approach to lake restoration. In *Proceedings of a Symposium on Water Quality Management through Biological Control*, (eds.) P.L. Brezonik and J.L. Fox, pp. 85–96. Gainesville, FL: University of Florida.
21. European Regional Centre for Ecohydrology in Lodz and Regional Water Management Authority in Warsaw, 2011. Ecotones for reduction of diffuse pollutions. *The Parliament Magazine*. 328:84.

22. Urbaniak, M. and M. Zalewski. 2010. Large dams as purification systems for toxic PCDD/PCDF and dl-PCB congeners. Hydrocomplexity: New tools for solving wicked water problems. *IAHS Publication* 338: 256–257.
23. Visser, W.C. 1971. Mathematical models in soil productivity studies, exemplified by the response to nitrogen. *Plant Soil* 30(2): 161–182.
24. Vorosmarty, C.J. and D. Sahagian. 2000. Anthropocentric disturbance of the terrestrial water cycle. *Bioscience* 50(9): 753–765.
25. von Weizsacker, E., Lovins, A.B., and L.H. Lovins. 1997. *Factor Four: Doubling Wealth, Halving Resource Use*. London, U.K.: Earthscan.
26. Wolanski, E.J., Chicharo, L., Chicharo, M., and P. Morais. 2006. An ecohydrology model of the Guadiana Estuary (South Portugal). *Estuarine Coastal and Shelf Science* 70: 132–143.
27. Woods, P.J., Hannah, D.M., and J.P. Sadler. 2007. *Hydroecology and Ecohydrology. Past, Present and Future*. Chichester, U.K.: Wiley & Sons Ltd.
28. Zalewski, M., Wagner, I., Fratczak, W., Mankiewicz-Boczek, J., and P. Parniewski. 2012. Blue-Green City for compensating Global Climate Change. The Parliament Magazine, Issue 350.
29. Zalewski, M. 2011a. Ecohydrology for implementation of the EU water framework directive. *Proceedings of the Institution of Civil Engineering Water Management* 164, Issue WM8: 375–385.
30. Zalewski, M. 2011b. Towards engineering harmony between water, ecosystem and society: Editorial. *Ecohydrology and Hydrobiology* 11(3–4): 137–140.
31. Zalewski, M. 2010. Ecohydrology for compensation of global change. *Brazilian Journal of Biology* 70(3 Suppl.): 689–695.
32. Zalewski, M. and I. Wagner-Lotkowska. 2004. *Integrated Watershed Management—Ecohydrology and Phytotechnology—Manual*. Paris, France: UNESCO IHP, UNEP-IETC.
33. Zalewski, M. 2002. Ecohydrology—The use of ecological and hydrological processes for sustainable management of water resources. *Hydrological Sciences—Journal-des Sciences Hydrologiques* 47(5): 825–834.
34. Zalewski, M. 2000. Ecohydrology—The scientific background to use ecosystem properties as management tools toward sustainability of water resources. *Ecological Engineering* 16: 1–8.
35. Zalewski, M., Janauer, G.A., and G. Jolankai. 1997. Ecohydrology: A new paradigm for the sustainable use of aquatic resources. In *Conceptual Background, Working Hypothesis, Rationale and Scientific Guidelines for the Implementation of the IHP-V Projects 2.3/2.4. UNESCO, Paris, Technical Documents in Hydrology No. 7*. Paris, France: UNESCO.
36. Zalewski, M., Brewińska-Zaraś, B., Frankiewicz, P., and S. Kalinowski. 1990. The potential for biomanipulation using fry communities in a lowland reservoir: Concordance between water quality and optimal recruitment. *Hydrobiologia* 200/201: 549–556.
37. Zalewski, M. and R.J. Naiman. 1985. The regulation of riverine fish community by a continuum of abiotic-biotic factors. In *Habitat Modification and Freshwater Fisheries,* (ed.) J.S. Alabaster, pp. 3–9. London, U.K.: Butterworths Scientific Ltd.
38. Zalewski, M. 1982a. An attempt to adapt an electrofishing method for sampling rivers with low conductivity. In: *The Matamek Research Program: Annual Report for 1981,* (ed.) R.J. Naiman, pp. 207–209. Woods Hole, MA: Woods Hole Oceanographic Institution, Technical Report WHOI-82-29.
39. Zalewski, M. 1982b. Growth and biology of Catostomus catostomus (Forster) from different environments near Matamek. In: *The Matamek Research Program: Annual Report for 1981,* (ed.) R.J. Naiman, pp. 210–218. Woods Hole, MA: Woods Hole Oceanographic Institution, Technical Report WHOI-82-29.
40. Izydorczyk, K., Frątczak, W., Drobniewska, A., Cichowicz, E., Michalska-Hejduk, D., Gross, R., and M. Zalewski. 2013. A biogeochemical barrier to enhance a buffer zone for reducing diffuse phosphorous pollution–preliminary results. *Ecohydrology and Hydrobiology* 13(2): 104–112.

41. Zalewski, M. 2002. Ecohydrology - the use of ecological and hydrological processes for sustainable management of water resources. *Hydrological Sciences Journal* 47(5): 825–834.
42. Zalewski, M. 1999. Minimising the risk and amlyfing the opportunities for restoration of shallow reservoirs. In: D.M. Harper, B. Brierley, A.J.D. Ferguson, G. Philips (Eds) The Ecological Bases for Lake and Reservoirs Management. Kluwer Acad. Publ. The Netherlands. *Hydrobiologia* 395/396: 107–114.
43. Mankiewicz-Boczek, J., Urbaniak, M., Romanowska-Duda, Z., Izydorczyk, K. 2006. Toxic Cyanobacteria strains in lowland dam reservoir (Sulejów Res., Central Poland): amplification of mcy genes for detection and identification. *Polish Journal of Ecology* 54(2): 171–189.
44. Altinakar, M. Kiedrzyńska, E., Magnuszewski, A. 2006. Modelling of inundation patterns on the Pilica River flood plain, Poland (Conference Paper) Issue 308, 579–585.
45. Kiedrzeyńska, E. 2001. Comparison of Nutrient Deposition Rate in the Pilica River Floodplain of Different Typology. Msc. Thesis.

6

Ecohydrology Concepts

Neil A. Coles

University of Western Australia

AUTHOR

Neil A. Coles is a professor and the director of the Centre for Excellence in Ecohydrology at the University of Western Australia, Perth, Western Australia, and a guest professor at Zhejiang University, Hangzhou, China. Neil spent his formative years in a coastal country town in the Spencer Gulf of South Australia, where he grew to become enthralled with the coastal and dryland landscapes of this region. He later moved to North Queensland in 1980 to undertake studies at James Cook University in geology and rainforest hydrology. On graduating, he accepted a position in the mining industry, exploring tenements in western and central Queensland, studying not only the geology of these regions, but gaining an understanding of the fragility and extremes experienced in these semi-arid regions. In 1988, he moved to Perth, where he now resides, to undertake postgraduate studies in catchment hydrology and soil physics and was awarded a PhD in 1995. Neil joined the Western Australia (WA) Department of Agriculture in 1998 as a senior research scientist, conducting, directing, and extending research on land and water resource management in the dryland agricultural areas in the southwestern Australia. He continued in this post until 2009, when was appointed as the director for the Centre for Ecohydrology. Under his leadership and direction, this center has become an internationally recognized leader in ecohydrology research with industry applications. Neil has continued to foster the interdisciplinary approach to improving agricultural production and protecting and understanding ecosystems and remains committed to achieving better outcomes for the myriad of environments on this unique and singular planet.

PREFACE

This chapter examines the definition of ecohydrology and its applications in environmental engineering. Ecohydrology in itself is a combination of terms eco- and hydrology and examines the role of water in the ecosystem through its ecology and systems dynamics [44]. Therefore, whether it is defined as hydroecology or ecohydrology has been discussed at length, and to some degree, consensus has been achieved [9,17,68]. While some may argue for a narrower focus for ecohydrology [81,86], a more inclusive approach that has ecosystem cohesiveness at its core, with links to other guiding and governing disciplines as an outer shell, which contributes to the integrated management and transdisciplinary approach to ecohydrology is preferred [69,86,91]. In all cases, a catchment or watershed functionality is an echo of the ecohydrological performance of the system as a whole. Further discussions on the terminology will not be entered into in this chapter; suffice it to say that there is enough evidence to suggest that ecohydrology will be adopted as the most appropriate terminology, and whether the science achieves a broad or narrow focus is beyond the scope of this discussion. Ecohydrology is an emerging science that has the capacity to cross disciplinary boundaries, to become truly multidisciplinary or what can be termed transdisciplinary, as will be discussed later. Ecohydrology has variously been described as "the link between vegetation function (or ecophysiology) and water in landscapes" [70,76] and can be broadly described as the "study of how the movement and storage of water is linked reciprocally to the structure and function of vegetation within an ecosystem" [66] or, for the purposes of this chapter, *a catchment*.

This approach was clarified further by Smettem [116] when he wrote "Ecohydrology seeks to understand the *interaction* between the hydrological cycle and ecosystems, the influence of hydrology on ecosystem patterns, diversity, structure and function coupled with *ecological feedbacks* on elements of the hydrological cycle and processes are central themes" [116]. From an environmental engineering perspective, understanding the interactive links and relationships between vegetation and landscape functionality is paramount in developing and implementing structures that are required to blend into the landscape with minimal disruption of the system in which they are housed. Traditionally, engineered outcomes have assessed the competencies and function of structures based strictly on engineering terms. Modification of the design and intent of these structures, required to accrue improved environmental and sustainability outcomes, is becoming the focus of the engineer, their prospective clients, and the global community. The broader more inclusive approach may require the engineer to move beyond the traditional engineering focus and may entail the engineer engaging with not only the environmental outcomes but also the policy, social, and economic considerations as well. This chapter seeks to provide an overview of ecohydrology, and how it can be harnessed to deliver improved environmentally engineered outcomes, and gives an overview of the status of the discipline and its potential for the future.

6.1 Introduction

Ecohydrology, considered as a broader more inclusive science, as defined by the UNESCO Hydrology, environment, life, program (HELP) is "a transdisciplinary and applied science."* Furthermore, it is stated that ecohydrology

> Uses the understanding of relationships between hydrological and biological processes at the catchment scale to achieve water quality improvement, biodiversity enhancement and sustainable development [145].

* www.erce.unesco.lodz.pl/story/ecohydrology-concept (accessed September 2012).

This concept is based on the assumption that sustainable management of resources can be achieved by restoration and maintenance of ecosystem function, enhancement of the system capacity and flux transference mechanisms, and using the basic properties of resilience as a water management framework [144].

Thus, the study of ecohydrology requires the necessary divergence from single discipline to multidisciplinary approaches, and a movement through the micro- to the meso-scale analytical techniques, to provide a broader understanding of landscape function.

Tress et al. go further to state the following:

> Interdisciplinary studies as projects that involve several unrelated academic disciplines in a way that forces them to cross subject boundaries to create new knowledge and theory and solve a common research goal [126].

And furthermore

> Transdisciplinary studies as projects that both integrate academic researchers from different unrelated disciplines and non-academic participants, such as land managers and the public, to research a common goal and create new knowledge and theory. Transdisciplinarity combines interdisciplinarity with a participatory approach [126].

From their perspective, a broader view of ecohydrology is given, with not only the science being studied, but also the social and economic aspects, combined with on-ground outcomes that involve affected stakeholders and the wider global audience through an understanding of broader implications of the research and the actions undertaken as a result of that research. This more wide-ranging methodology is adopted for purposes of this discussion on ecohydrology concepts.

6.2 Fundamentals of Ecohydrology

Water is critical to all aspects of life as we know it on this planet, and while there is an increasing demand for potable water resources for human consumption, for industry and agricultural production systems, this water comes at a cost to the environment and sustained ecosystem functionality (Figure 6.1). Therefore, there is a fine line between water allocation and ecohydrological functions, such that a trade-off is inevitable and a considered approach must be adopted. The difficulty arises when assigning "value" or "primacy" to a water resource and the potential "value" the water may realize through an alternate use (Figure 6.1). To improve the management of water resources and allocation to appropriate uses for food, industry, and environment on an equitable basis requires knowledge of the dynamic sciences, economics, governance, and regulation thereby harmonizing the transition (Figure 6.2). Ecohydrology by necessity therefore transcends the boundaries between disciplines and provides a more holistic approach to managing landscapes and the water that is transferred and exchanged within them [140,141].

The current perception that the environment or ecosystem has no net present value or is considered of lesser value than alternate uses of water reflects the long-term exclusion of ecosystem function from business models and considerations (Figure 6.1). Ecohydrology seeks to redress this anomaly through improved understanding of hydrological dynamics and ecosystem performance. By understanding the need for trade-offs and assigning primacy to water allocation on an equitable basis through an evidentiary-based system, better management decisions on allocation can be determined (Figure 6.2).

Scale is an important factor in this determination, as the scaling factors that impact on rainfall inputs (or climate) such as adiabatic effects, inversions, storm fronts, thermal systems, and atmospheric circulation are quite different to those impacting on landscapes, such topography, vegetation cover, land use, geology, and soils [9,81]. Thus, capturing global scale data from climatic observations is challenging, but classifying and analyzing shifts using models that are then downscaled to the level of

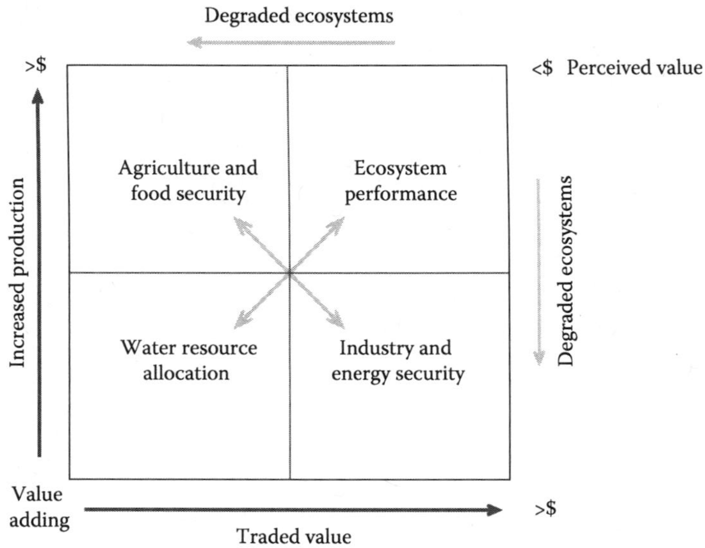

FIGURE 6.1 Determining the underlying "value" of resource allocation in a water-limited environment.

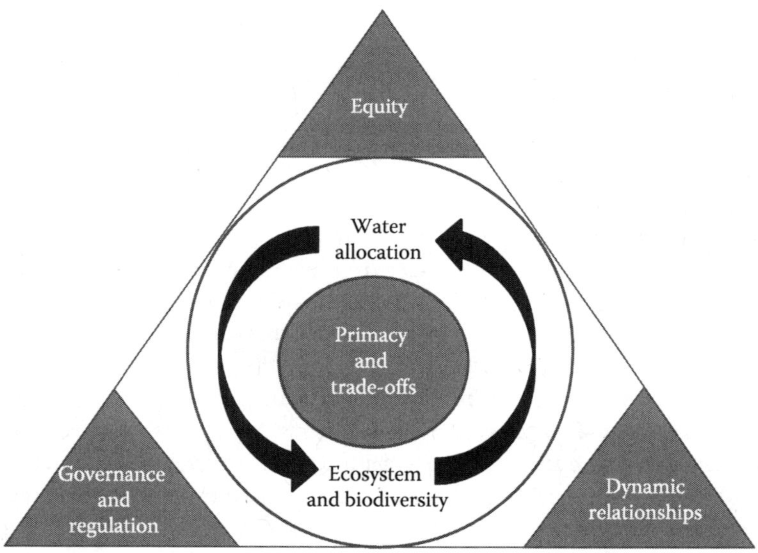

FIGURE 6.2 Drivers and actors that impact on water allocation and ecosystem performance, which link water "primacy" with trade-offs, in a policy and science framework.

data application within a catchment, or single landscape unit, become increasingly difficult, much less relevant at fine scales [9,69,77].

Ecohydrology deals with functioning and process that surround hydrological dynamics in an ecosystem context [9,45,112]. Such knowledge is fundamental to understanding actors and drivers for change in these systems, and at what point pressures exerted on these systems through external forces will cause that system to adapt or fail (Figure 6.2). Available data and the consistent monitoring of system or catchment performance are required to unravel these relationships [20], and this datum is often unavailable, at an inappropriate scale and location, or fails to capture the key component of the dynamics or event processes that will provide the cogent understanding necessary to provide for informed decisions [14].

In part, modeling solutions are often limited by the data available to calibrate and validate appropriately, without undertaking some form of assumption on the part of the modeler, or the events and the landscape being modeled [20,31,32,69]. Such modeled outcomes are often applied, as a best approximation, to evaluate both impacts and to derive solutions. Thus, our understanding of the ecohydrology and the interrelationship network of the ecosystem go hand in glove in clarifying the affect of the human dimension, as both source and solution to ecosystem degradation [69].

6.2.1 Global Perspective

Aquatic ecosystems, both marine and freshwater, are under increasing pressure globally. Increasing water allocations to provide for food security and energy production, access to safe drinking water and sanitation, and industrial uses are contributing to degradation of aquatic ecosystems with inherent loss of biodiversity as well as their use as ecosystem service providers [20,92]. This loss is exacerbated by an increase in intensive urbanization through the creation of megacities and extensive urban sprawl, which is now compounded by climate change phenomena that generate divergence from the "norm" that is causing extremes in either too much or too little water, as highlighted in global forums and recent reports [49,50,130]. Water, energy, and food are inextricably linked (Figure 6.1), and they underpin the current global trade and productivity models that have an expansionary developmental nature at their core [136]. The "forever" nature of this exploitative approach, which relies on physically and organically limited resources, presents a future challenge to either find targeted niche technological innovations, or shift to a near zero growth as alternatives, or run the risk of catastrophic contraction [30].

Cumulatively, climate changes are affecting hydrological cycles and will pose threats to the security of the existing water and food production and allocations systems, and will particularly impact on water quality and quantity available in different regions [50,92]. In some regions, this will take the form of increased rainfall (e.g., humid subtropics), and in others, it will result in declining rainfall (e.g., semiarid regions). Each shift brings its own set of problems and issues to solve, from both environmental and engineering design perspectives. Aquatic ecosystems are very dynamic and can rapidly adapt or collapse depending on the external pressures placed on them. Invasions of nonendemic species pose significant bio-security risks and threaten the biodiversity values of rivers, estuaries, and coastal areas [25,142].

The tendency for human migration and congregation toward coastal regions and river systems increases the potential to induce stress and subsequent degradation in these systems. Forecasting scenarios and adaptive management frameworks as applied in recent times, such as Integrated Catchment Management (ICM) [7,48,137], Integrated Water Resources Management [13,83], and Integrated River Basin Management [1,11,16,94,124], as the naming convention would suggest call for integrated solutions for water quality, water quantity sustainability, and the maintenance of ecosystem performance. Therefore, these solutions, by default, are focused on an in-depth knowledge of ecosystem interrelationships and hydrological process [91].

In the Murray–Darling Basin (Australia), ICM is largely directed toward practical ecological outcomes such as flow and hydroperiod, water quality, biosecurity, and river health [36]. However, the social sciences, if they are considered, are often marginalized, but at the same time, social and natural scientists agree that social and cultural issues should be integrated into land and water management, promoting research collaborations [121]. This is contrary to the current opinion in which ecohydrology is focused on the physical sciences alone; therefore, as discussed previously, it is suggested here that the adoption of a wider and more inclusive approach will deliver better environmental and humanitarian outcomes.

However, there are significant hurdles to overcome if these essential collaborations are to occur. Research policy must move toward favoring projects that integrate disciplinary knowledge and involve nonacademic stakeholders, creating transdisciplinary opportunities in land and water research and planning. In addition, funding providers are required to account for the variable time-frames associated with integrative research, with research outputs delivered from different disciplines within multiple temporal frameworks. For example, The European Union Water Framework Directive (WFD) [46] represents a new approach to the management of water across Europe. Integrated, catchment-scale plans for the protection and restoration of aquatic ecosystems are required to be developed within this framework [124]. By necessity, this will require the collaborative creation of new scientific knowledge, extracting expertise from multiple academic disciplines. Using existing scientific knowledge, to generate a new awareness that is incorporated into policy and practice, scientists and managers need to foster collaborative partnerships that promote both the coproduction and the bidirectional exchange of knowledge [124]. While there has been a fundamental shift in thinking, groups often struggle with the challenges of operational integration, refocusing on more extensive and expansive spatial and temporal scales, defining collective terminology, appropriate methodologies, and managing the expectations of stakeholders [126].

One conceptual approach developed by the UNESCO International Hydrological Programme (IHP) and the Man and Biosphere Programme is based upon the assumption that the sustainable development of water resources is derived from the resistance and resilience of the system to change. This theory is based on an ecosystem's adaptive capacity and its ability to cycle water, nutrients, and energy fluxes at multiple scales, from a farm plot to the basin scale [70,145]. This suggests that for ecohydrology to move beyond monitoring and sensing environmental health, transdisciplinarity is essential, such that boundaries between disciplines dissolve; disciplines such as ecology, biology, hydrology, geography, environmental engineering, and law and socioeconomics transform into a seamless evaluation continuum, achieving new insights into the true nature of the effect that human and natural interventions may have on aquatic ecosystem.

This transcendary process is essential to propose revised remedial solutions and alternate practices to industry, agricultural or rural communities, governments, and other stakeholders that will be impacted by the shift toward sustainable production and environmental systems [48,70,91]. The role of ecosystem engineering, for example, restorative river flows to control eutrophication or regeneration of forest ecosystem to manage erosion and flooding [5], restoration of estuarine habitat (i.e., mangroves) to revive fish nurseries [139], and provision of tree and shelter belts to mange livestock and soil stability [123], demonstrates the integrative nature of the ecohydrology approach [25,47,91] that will be presented.

6.2.2 Ecohydrology Dynamics

Knowledge of the interrelationship dynamics and the impact of scales associated with water allocation is not only one of the fundamental pillars of understanding ecohydrology but forms the basis of process and management frameworks in which to optimize resource allocation. A significant barrier to developing this framework is the range of spatiotemporal scales to be considered and the lack of available data to predetermine local and regional impacts of water redistribution [77]. The water balance at the basin scale based on Budyko principles [21] (Figure 6.3) illustrates that for a given time interval, the surface and groundwater that flows out of a basin, plus any changes in storage, are equivalent to the amount of water entering the basin through atmospheric discharge (e.g., rain, snow, fog), minus that volume returned to the atmosphere through evapotranspiration, from plants, soils, and water surfaces [20,57,146].

The water balance model provides a useful summary of the hydrological processes within a basin; however, it tends to gloss over the finer interrelationships at the water–plants–soils interface [21,123]. The ecohydrology approach levers this dynamic interrelationship between hydrological and ecological factors. Zalewski [143] and others [70,128,145] propose to increase ecosystem's robustness and resilience to external impacts by adopting this approach based on the proposition that water quality and biodiversity are managed through controlling hydrological parameters such as hydroperiods, by influencing residence times and/or discharge volumes, or through the use of biological parameters, such as biofilters in the case of riparian vegetation or filter feeders that can be merged into

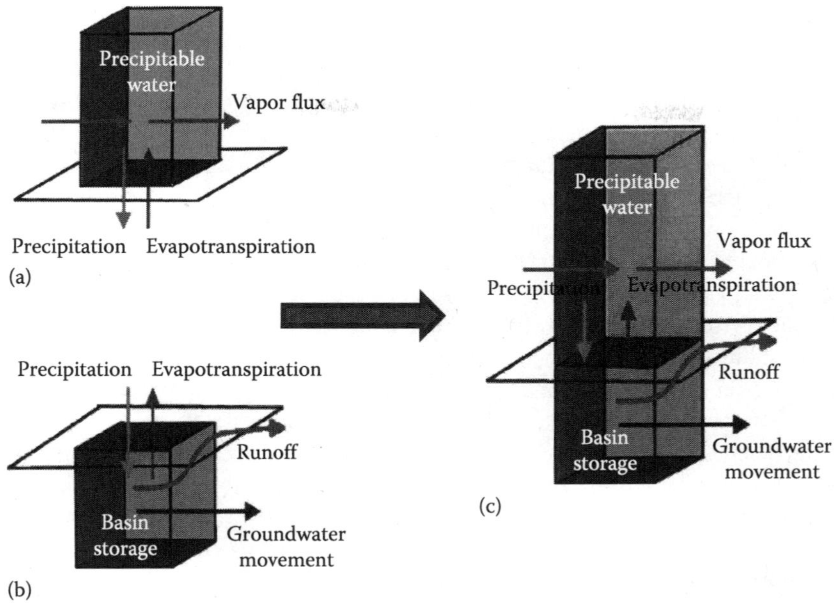

FIGURE 6.3 Linking process and water fluxes at the basin scale: (a) atmospheric storage and fluxes, (b) basin storage-land-based component, and (c) combined atmosphere–land surface water balance.

existing systems and structures. Through these synergies, integrative strategies can be optimized and extended beyond single-use technologies or strategies, thereby encouraging innovative approaches to water sciences and engineering [128].

This ecohydrology approach is based on three principles that are expressed as sequential components:

Hydrologicalal: The quantification of the hydrological cycle of a basin should be a template for functional integration of hydrological and biological processes.

Ecological: The integrated processes at river basin scale can be steered in such a way as to enhance the basin's carrying capacity and its ecosystem services.

Ecological engineering: The regulation of hydrological and ecological processes, based on an integrative system approach, is thus a new tool for Integrated Water Basin Management and Integrated Coastal Management [128,140,145].

From this discussion, we can derive that linkages are important to understanding ecosystem performance, and water and energy fluxes are drivers of transition between components of these systems. Also that it is possible to engineer and enhance the performance of an ecosystem if we understand the fundamental and dynamic nature of these interrelationships. Using these principles to apply ecohydrology to real-world outcomes is the next challenge.

6.3 Ecohydrology and Ecoservices

The use of the collective term "ecosystem services" was assigned in the early 1980s to explain a framework for structuring and synthesizing biophysical relationships within an ecosystem setting that delivered benefits to human well-being [95]. Ecosystems provide hydrological services that are coupled with a range of essential services, including air quality, carbon dioxide exchange, and soil development. These systems and services, as outlined previously, are often interrelated in dynamic and complex ways, and establishing their functional relationships requires approaches spanning diverse fields of inquiry. This section, for the purposes of this discussion, is focused on the terrestrial freshwater hydrological services as an example of what an ecoservice is or what may be achieved through a greater understanding of ecosystems.

6.3.1 Ecoservices: Value and Benefits

Ecosystem services, described as the benefits obtained from ecosystems [19,20], form the basis for the design and development of environmental policy, trading schemes, benchmarking performance, and monetization of natural resources and are a powerful tool in assessing the relationships that people have with their surrounding tangible environment and the more distant intangible global benefits. The inclusion of beneficiaries, as shown in Figure 6.4, links values intrinsic to ecosystem services, to both the fundamental economics of dollar value and the trade-offs required to realize that value. By closing the cycle of service delivered and exploited through benefits derived from reduced ecosystem performance, it provides a direct assessment process that may enable the value proposition to be revised in relation to service provision. Whether or not those values are monetized, this framework should provide for the opportunity to assess theoretical trade-offs based on alternative scenarios of water resource use, land cover, and land-use change through comparative value and degradation of service [20].

The ecosystem services framework (Figure 6.4) highlights the complex feedbacks and trade-offs among service providers and direct and indirect beneficiaries, such that service provided in one portion of an ecosystem is often at the expense of that system through functional degradation [20]. The complex interactive nature of the ecosystem defines the framework into which the ecohydrologist and engineer must plunge, in order to deliver informed decisions to their clients that may be at the expense of consumption by others, elsewhere and in the future. The role of the ecohydrology engineer is to identify that combination of services, support, and ecosystem resilience that will enable the long-term success of the planned activities with minimal immediate cumulative impacts on the ecosystem being evaluated. These terms have gained currency because of the values that they convey and the idea that ecosystems are "socially valuable" and may have "existential value" that is not yet apparent [19,41].

Beyond this simple terminology, there is limited accounting and measurement standardization between economics, ecology, hydrology, and what is defined as an ecosystem service [19]. Assigning and trading values are the key to this process and has been at the forefront of adaptive management frameworks in recent years, employed in assessing biodiversity or natural resource values [34]. An accounting system has been suggested by Boyd et al., which has broad-scale applications that deliver

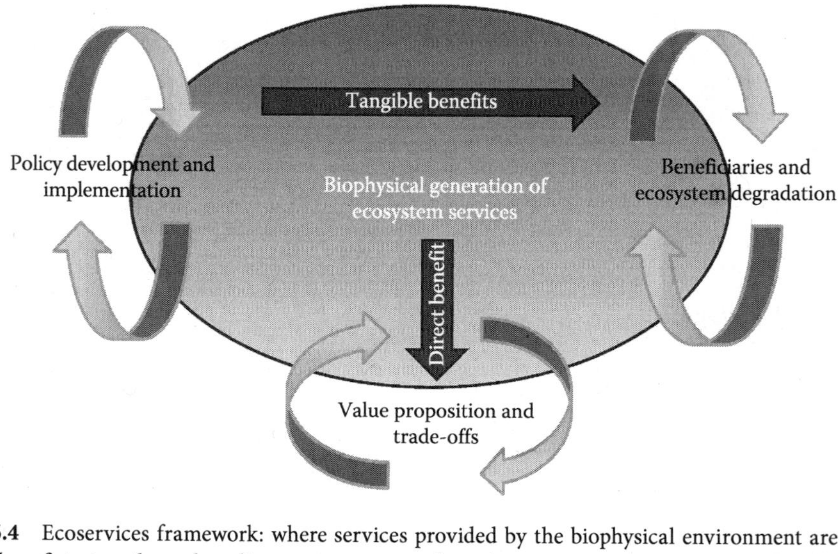

FIGURE 6.4 Ecoservices framework: where services provided by the biophysical environment are valued and traded to beneficiaries, through policy, governance, and market instruments. Equity and benefits in service provision determine the level and value of service degradation in ecosystem performance.

both environmental accounting and performance assessment, which can be allocated to a broad "Green GDP"* [19]. The accounting system is designed to provide more exacting measures of service, quantifying service units, linked with both monetary values and ecological principles, but targeted at a narrower range of measurable parameters to enable the prioritization of monitoring and data collection [19]. De Fries et al. (2004) and others [75,106], following the same theme, indicate that by confining the focus of the parameters (and their interrelationships) used for accounting purposes, it will enable governments to set priorities, monitor the efficiencies of economic policies more precisely, enact targeted environmental regulations and resource management strategies, and design more efficient market instruments [43]. This will provide an opportunity to deliver a socially responsible economic and environmental framework that has been touted for the last decade at both national and international levels [62].

This framework will enable policy documents and statements to be underpinned by complementary statistical information, or via a scientifically affirmed database, creating an evidentiary-based system of accounting and monitoring that can be utilized to set meaningful qualitative and quantitative targets for these policies. By shifting toward an evidentiary-based policy approach to ecoservices creates opportunities on a number of discipline fronts, in aquatic (ecohydrology, biology), terrestrial (geology, pedology), marine (oceanography, coastal dynamics), atmospheric (meteorology) sciences with linkages to engineering, law, politics, economics, and business applications that will provide an understanding of ecosystems from the perspective of people as beneficiaries [41,122]. This approach has tremendous potential for protecting ecosystems and their ecoservices and brings the science of ecohydrology as a broad concept to the fore [20]. Thus, an ecosystem services framework that links conservation and development through science and policy, which has the basic fundamentals of environmental health, water, and food security and human health, at its core, will provide for a more sustainable future. For a candid review of environmental valuation, its history, and context, the reader is referred to [19,42,49,74].

However, while this idealistic framework is being developed, for ecosystem services to move from a conceptual to an operational framework for decision making, significant changes, not only within the evaluation fields of natural, social, economic, and policy science, but within national and international businesses and governments globally are required [20]. Both Brauman et al. [20] and Boyd and Banzhaf [19] have suggested accounting and assessment approaches that seek to define and apply value to ecosystems and the ecoservices they deliver. For the purposes of this chapter, the focus is on identifying the ecohydrological interactions to define processes, relationship, and functions that may deliver key parameters, indicators, or indices that can be assigned values for accounting purpose, resilience measures, or industry targets.

However, deriving these targets, measures, or metrics is not necessarily straightforward—as has been discussed—to then assign a monetary or service value to each of these is also problematic. With the advent of the "green consciousness" and the need to develop better management strategies for ecosystems, biodiversity, water, and food production systems, green companies offering services have also proliferated. A key U.N. update report [38] on biodiversity recommends massive economic changes to manage and conserve species and protect the natural environment. The study argues that the economic case for global action to protect biodiversity is now much stronger than the case for tackling climate change. Research by the economics of ecosystems and biodiversity (TEEB)† group has led to the increased recognition of the economic value of natural assets and the returns that can be generated through investing in natural capital. The EU Biodiversity Action Plan [38] has increased the understanding of the drivers of biodiversity loss (including climate change) and how biodiversity is linked

* Note that Boyd and Banzhaf [19] suggest that the green GDP should be considered as national-scale environmental welfare accounting and performance assessment system such that the ecosystem measure employed here is thought of as a measure of nature's value, not the value itself. Creating a uniform and consistent accounting system that can be utilized across a number of disciplines and represents the GDP of services delivered to people (e.g., as described by Peskin and Delos Angeles [106]).

† http://www.teebweb.org/ Accessed December 2012.

with other sectoral activities. The implementation of the plan in 2006 has highlighted the basic role and fundamental importance of ecosystems and has strengthened the need for better managed outcomes to facilitate mitigation and adaptation to climate change.*

Through economic incentives or tax concessions, companies and countries can be induced to implement best management practices for biodiversity and a range of ecosystem services that can engender an improved understanding of the value of ecosystems and the services they provide, creating both an economic and a societal framework to reduce overexploitation and generate a new market for ecoservices. Ecoservices is thus termed—the new approach to managing shared space [19].

Increasingly, therefore, the use of soft-engineered solutions—ecoservices approach—is being championed. Soft systems such as those that employ sustainable drainage systems [37] and soft flood defenses are more responsive to extreme events [120]. Projects linking reforestation and river reconstruction are prime examples of soft engineering reposes to manage altered ecosystems in an effort to reintegrate and repair these natural systems [5,110,135].

6.3.2 Ecoservices: Accounting and Decision Making

Daily et al. [41] and others [19,20,42,49,74] have emphasized that understanding and providing ecoservices requires not only a scientific basis for this determination but also a valuation and accounting process that enables the value proposition to be defined, commoditized, traded, and regulated. They suggest three fundamental steps to the decision-making process to integrate both ecological and economic terminologies. First, possible alternatives to the proposed land use, system modification, or exploitation should be considered. The alternative framework in which this decision is often made is guided by the narrow focus of service, for example, water treatment, made to be only examined from an engineering and economic perspective, with alternative natural options given scant regard. Therefore, alternative natural treatments that may be implemented are not even considered [42].

Second, all impacts that may arise from this decision need to be identified and measured for each alternative (including labor, capital, biophysical, and social impacts). These decisions are also often undertaken with incomplete knowledge, either due to the lack of data or a limited understanding of the side effects that may occur (or might be long term and cumulative). However, it is suggested that quantification of uncertainties and risks should be undertaken before proceeding [42].

Third, and most significantly, the assignation of value that will link the risk and uncertainty to the return on investment and the capacity to maintain the status quo, evaluating each alternative scenario in comparable units, nominally the change in human well-being, calculated for the short, medium, and long term [42]. These can be defined as resources that are forgone to obtain the goods, services, or other outcomes associated with any alternative. Predominantly, this is measured in monetary terms [19,20].

In using a monetary approach, often the end result is a system that may not suit our requirements but is one that we are willing to pay for. For instance, extensive commodity and trading markets exist for the products of land- and water-based industries, such agriculture, aquaculture, and forestry [85]. However, the benefits of ecoservices (i.e., protection, treatment, mitigation, and provisioning [93]) remain relatively unpriced and therefore have limited market value [85]. As existing markets rarely reflect the true material cost of production, established initially on the basis that natural resources are "inherently free," and can be used at "no-cost," the opportunity to recover degradation and reparation costs is limited. This also leads to inappropriate resource use, incorrect measures of scarcity and vulnerability for some ecoservices, and no measures for the majority [85]. During the last decade, there has been growing support to utilize market and policy fundamentals to increasingly account for the environmental costs of industry, urbanization, and agriculture [8,26,38,64].

* http://europa.eu/rapid/press-release_IP-10-1303_en.htm, 2010, Accessed December 2012.

Markets for trading in carbon emissions (e.g., Chicago Climate Exchange,* EU Carbon Trading Exchange†), Renewable Energy Certificates (REC),‡ and Biodiversity & Forestry (Reducing Emissions from Deforestation and Forest Degradation—REDD) have been developed in one form or another. These schemes have gained some credence through market and political support but are still relatively small in scale and operation as measured on a global scale, with some additional support for environmental trading offered by the development of market-like mechanisms for the payment for ecoservices (PES) schemes [85]. Through these mechanisms, governments and nongovernmental organizations are able to fund environmental recovery and conservation through habitat provision, watershed protection, or carbon sequestration [57,88].

While progress has been made, there are limitations to their effectiveness that include the high degree of uncertainty concerning greenhouse gas (GHG) emissions and climate change; reticence in public policy decision-making process as a result; and the quantification of the economic and social benefits of reducing GHG emissions is generally lacking [57,58]. So while derivative financial instruments and other mechanisms developed to "fill the gap" between the true cost and sustainability can aid the shift toward a green economy, if they are poorly designed, they can have the reverse effect of creating more problems than they solve [85]. To develop effective approaches to manage increased reliance, sustainability, and productivity, an understanding of the complexity of biodiversity, ecological functioning, and service provision is required [26,74,89], with the benefits accruing within the environmental sphere through a public–private partnership.

While monetization of services and goods is one approach, others have been suggested. For example, an Ecological Sustainability Index (ESI), which is comparable to the ecological footprint or the material flow accounting (MFA) used in economics: these indices seek to measure the sustainability of human use of the environment [75]. A complex index, the ESI, attempts to overcome the issues associated with converting all measures to a uniform numéraire, by creating several hierarchical layers grouping the component variables, rather than using a simple sum, that form indicators, grouped in different categories that capture different elements of environmental impact [75].

Four main mechanisms have been developed over the last decade to initiate the recovery and conservation of ecoservices: (1) regulation and penalty, (2) cap and trade, (3) direct payments, and (4) self-regulation [85]. Each has flexibility in application, implementation, compliance, and distribution among industries, communities, countries, and ecoservice provisions, but will not be discussed in detail here. Suffice is to say that the mechanisms exist and are used with varying degrees of success, depending on the level of industry, public and political support for their introduction, and the degree to which the benefits of such schemes can be realized at variable temporal and spatial scales. However, as with other process of accountability, the determination of value is subjective, and the application and assessment of impact can be a limiting factor, either through a lack of understanding or minimal data availability.

6.4 Ecohydrology: Monitoring and Assessment

Ecohydrology, as a science, can provide essential elements (and data) to assist other sciences and disciplines to create environmental assessment frameworks, compliance targets and indicators, policy directives, governance, and regulatory frameworks, and assist with stakeholder dialogues. Through a process of monitoring, analysis, and discovery, the ecohydrologist can provide parameters to develop

- Evidenced-based system for management and policy development
- Targeted management strategies
- Integrated platform to ensure science basis for decision making

* https://www.theice.com/ccx.jhtml

† http://carbontradexchange.com/

‡ http://www.asx.com.au/products/rec_futures_prices.htm

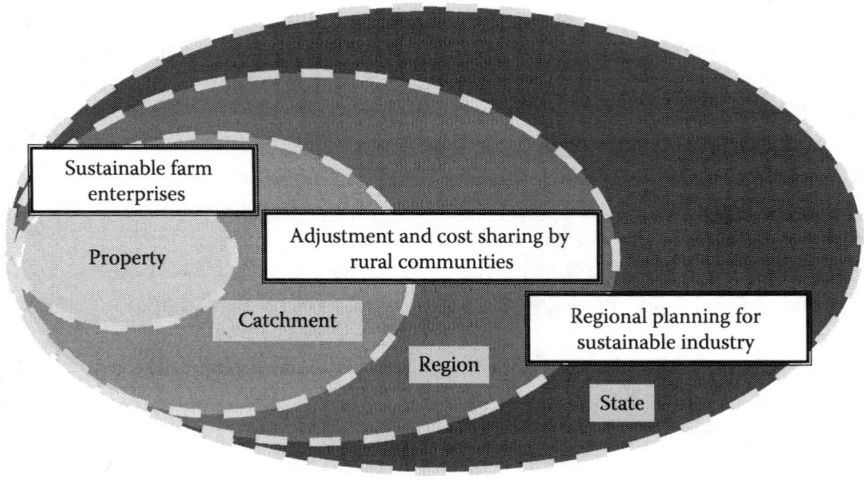

FIGURE 6.5 Scale and process: intervention and support strategies applied at intersecting scale boundaries for effective land and water management from an ecohydrological and policy perspective.

- Models to assess the potential impacts of investment decisions prior to implementation
- Models to assess social and environmental impacts
- A broader basis for an evaluation framework that is more inclusive

In this way, land and water management decisions can be made and evaluated at variable scales and landscapes from on-farm, to catchment and river basin, with each scalar expansion bringing with it added levels of complexity (Figure 6.5).

Some approaches to overcome this complexity have already been discussed, from economic (REC, PES, MFA), to the environmental (ESI, REDD) and the Millennium Ecosystem Assessment principles, all of which have merit in their own right, but essentially are not holistic and inclusive independently. Alternatives for measuring sustainability and use have been suggested, such as ecological and water footprints, via virtual accounting processes [75,89], or broader climate change models to determine cause and effect scenarios, all of which have limitations, in process understanding, data availability, real-time monitoring, and on-ground applications, and in some cases, policy and regulatory vacuums [85,89,107,141].

However, as with most models and accounting principles, there are limitations, and these systems are no different. An ecological footprint is a strong sustainability measure that focuses on natural resources and the environment, whereas the assessment of savings or accounting is based on economics, with measures of health required to integrate them [75]. While having limitations, a footprint (either ecological or water based) can be calculated at any scale and be applied across cities, regions, and countries [132] and provides a simple flag as an indicator of a country's consumption patterns. This model adopted by its proponents attempts to bring more rigor to the calculation of ecological footprints by using the environmental impact of production systems that required to deliver its products irrespective of where that production occurs [75]. How this may be applied to assess the sustainability of industry practices or a country's consumption and productive output is questionable [141].

At the core of these methods is the necessary link to nature and natural systems, which is often overlooked in the current societal and economic models and governance structures [125]. Without understanding these links and the impact of anthropogenic activities either in the short term or thorough cumulative impacts, it is often difficult to determine the true cost and sustainability of such activities as promoted by governments, industry, and market economies. The future focus and that of the engineer is on the green economy and natural solutions that can often be more cost effective and durable than other capital engineered solutions for delivering specific services or policy objectives [125].

A number of cornerstone processes need to be in place to assist in the transition to a more ecohydrologically sensitive green economy, which include

1. Minimizing losses and avoiding inappropriate trade-offs
2. Investing in environmental infrastructure
3. Active management of environmental risks
4. Proactive investment in natural capital
5. Good governance
6. Identification of direct and indirect benefits to various economic sectors and industries
7. Further ecoefficiency for relative decoupling
8. Absolute decoupling of the economy from resource use and its negative impacts [125]

The capacity and ability to implement these fundamentals are dependent on the context of the national and international economies, global environmental considerations (i.e., climate change), conservation frameworks, and the delivery of local and regional benefits that have realizable on-ground returns.

6.4.1 Ecohydrology and Knowledge-Based Systems

As this pressure on the world's water resources increases, it has generated new and increasingly interconnected challenges, resulting in the introduction of new disciplines, new techniques, new language, and new thinking [23,99]. Ecology, economics, and other social sciences have all progressively added to the historic hydrology and engineering base. To continue this evolution in the wake of new challenges, the adoption of "imported" terms from other sciences becomes a precursor to transdisciplinary learning and collaboration [98]. This is where ecohydrology, in the broadest sense of the term, can play a significant role in developing water and ecosystem management tools as proposed by Zalewski [143], Harper et al. [70], McClain et al. [91], and others.

This has led to the use of concepts and terminologies that were originally defined for a specific disciplinary context, being commonly used in a broad and diverse range of disciplines. According to Neto [98], "....this gave rise to a certain conceptual confusion and the construction of a mixed terminology." For example, the broad use of economic efficiency terminology that has been introduced into the water management sphere, breaching contextual constraints such that "....the term 'supply' and 'demand' of water or the 'water market,' restricting the assertion to a service provision and interpreting water as a mere 'commodity' subject to natural monopoly...." [99]. More recently, sociology, anthropology, and social psychology have also become part of solution to the challenges of sustainable water management [98], thus they form part of the revised ecohydrology agenda, supporting the drive toward an expanded theoretical and practical application within this field.

In the last decade, there has been a transition from seeking solutions within individual scientific disciplines to multidisciplinary, interdisciplinary solutions, and latterly transdisciplinary thinking and approaches [98]. *Multidisciplinarity,* in its strictest sense, does not require the researcher to transition to another discipline [110]; it requires that a group of researchers in a range of disciplines use traditional theories and methods to integrate the findings from these disciplines into a single understanding. However, a *transdisciplinary* approach focuses on developing a common conceptual framework that delivers a coherent integral process that allows for the cross-pollination of ideas, policy development, and implementation strategies [10,79].

Part of the processes driving this shift has been the tendency toward the globalization of both problems and solutions, driven in part by climatic and economic factors that have the capacity to impact on water and food resources. This globalization has also increasingly enabled the extension of localized solutions and practical experience to a much larger audience, providing opportunities to build on existing knowledge frameworks to solve interlinked and complex problems [10,98]. History provides a stark realization of the consequences of water-use decisions, made with limited science or awareness of the risks to the environment, society, and the water source itself (e.g., Murray Darling Basin, Australia [8]; Lake Taihu, Jiangsu, China [108]). Communities are now better trained, connected, and equipped to monitor government and industry decisions, and there is an expectation that new developments will not only deliver acceptable environmental impacts, but also generate social and economic benefits [8,23,98]. These actors are placing increasing pressure on decision-makers, engineers, governments, and developers to evaluate the broader view of localized to global-scale impacts and deal with the interrelationship complexities and dynamics as the shared need emerges. To deal with these complex water management issues, a more adaptive and flexible policy approach is required, and when coupled with online technologies, it will deliver interactive learning and best practice to the wider community, stakeholders, and industry partners [103].

There is a growing need for transdisciplinary approaches, such as that proffered for ecohydrology, to develop beyond water and ecology science into a more holistic visionary framework that is able to incorporate nonformal knowledge and effective participatory processes. This framework is aimed at achieving multilevel learning processes that address the hierarchical structure of governments, markets, and networks and provides for both top-down and bottom-up actions for the identification of problems and solutions [98,102]. However, there is tension and competition between specializations within different disciplines, and without specialization, it is difficult to obtain formal and academic recognition. The 2010 Capacity Building Workshop on Water Education in Paris identified the need to build tertiary or vocational-level learning opportunities, which includes participants with different backgrounds in order to facilitate cooperation and exchange of views [98], thus providing broader opportunities for engineers and hydrologists to expand their knowledge base and become more interactive with other disciplines.

However, both personal and institutional barriers can limit the development of these partnerships. Particular issues exist around a willingness to operate outside disciplinary comfort zones and their professional norms and associated the reward and value systems. Scientific knowledge must compete with both personal and institutional values, with issues of moral judgments and social equity involved in the consensus-building processes that occur within decision-making forums [75,83,124]. This integrated approach, when coupled with the evaluation and decision support framework outlined previously, should enable communication between specialists addressing specific problems, as well as other expert practitioners, providing new skills and knowledge competencies promoting a paradigm shift that delivers a new "water practitioner" or ecohydrologist profile [102], thereby creating the freedom and flexibility needed to tackle the increasingly interrelated and complex water challenges the world is now facing.

Pahl-Wostl [103] suggested adopting management practice theory to develop an adaptive evaluation framework using triple-loop learning theory (Figure 6.6) to refine the influence of governing variables that underpin assumptions and values. This approach follows on from double-loop learning [6], which assesses the "cause and effect" relationships within a value-normative framework, and enables the user to test the underlying theoretical assumptions employed within a management or evaluation context [102,103]. By running through a sequence of actions, key criteria, reflections, and transitions delivered from the actionable outcomes, it is possible to provide a theoretical framework: One in which all participants can agree on actions without having to go through the longer-term practical application and evaluation cycles (i.e., trial and error) necessary to deliver agreed outcomes.

A further step from Argyris and Schön's conceptualization, to learn by simply reflecting critically upon the theory-in-action, Finger and Asún [50] suggest that it is now possible to readjust the theory through double-loop learning (Figure 6.6) to avoid experiencing the entire learning circle. Smith [117] suggested that this process will often entail capturing the maximum participation of stakeholders, minimizing the risks of participation, starting at an agreed baseline, and designing methodologies that value rationality and honesty. Pahl-Wostl [103] argues further that in using triple-loop learning that provides avenue to move away from the formative norms, to an informal adaptive process, it is possible to drive innovation in water management, policy development, and water governance.

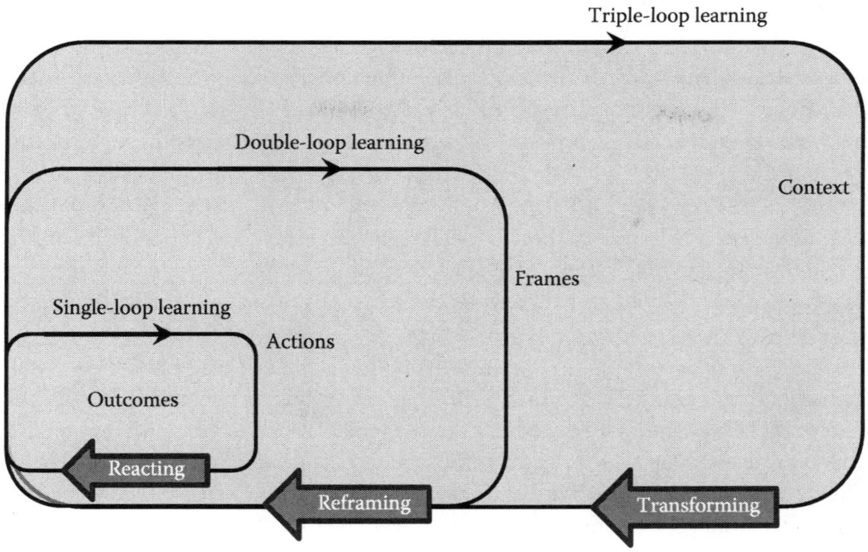

FIGURE 6.6 Learning loops, pathways, outcomes, and dynamics of single, double, and triple learning loops and applications to water management, where the single-loop actions and outcomes provide incremental changes, double-loop learning provides for reflection and reevaluation, and the triple-loop learning provides for process transformation and implementation to deliver broader contextual outcomes.

Through the development of these analytical frameworks, which are developed in an open and honest learning cycle, it will enable stakeholders to move away from "defended positions" and "protectionist" approaches [102]. This triple-loop adaptive approach provides the ecohydrological engineer with a robust framework in which complex issues can be addressed, which will involve a limited resource (i.e., water), multiple stakeholders, vested interests, and both government and communities, which will deliver transparent equitable and reasoned outcomes.

6.5 Applying Ecohydrology Concepts

The services generated by ecosystems are many and varied; increasingly as we are forced to move to a green economy, understanding the functionality, fragility, and resilience of these systems is important in determining appropriate measures for monitoring, exploitation, and service extraction. To do this requires these services to be organized in a coherent framework [19]. The causal interactionary nature of these services requires a robust framework that considers multiple scalar processes and spatiotemporal considerations as cumulative impacts may manifest locally (e.g., salinization) or globally (e.g., carbon accumulation); therefore, classification of what constitutes a benign or degrading activity is proving to be manifestly difficult with current classification systems offering only arbitrary evaluations. The next section examines approaches taken to monitoring and evaluating ecosystems, through better governance frameworks, technological innovation, and assessment tools.

6.5.1 Ecohydrology and Ecoservices as an Evaluation Tool

The Millennium Ecosystem Assessment [93] has suggested that by dividing services into four categories, interactive services and relationships can be more easily defined. The first of these categories is *provisioning services* that provide food, freshwater, timber, and fiber for direct human use and is a significant part of the current economy. The second category and much less widely appreciated is the *regulating services* that maintain the global environment and localized habitats such that it is biophysically possible for people to live and prosper (e.g., pollination of crops, climate stabilization) [111]. The third category is based around *cultural services*, those that provide intellectual and spiritual inspiration and aesthetic values. The last category groups the *supporting services* that underpin ecosystem processes that generate the previously described ecoservices [19,93]. To increase the tractability of the ecosystem services concept, there have been an increasing number of attempts to quantify services through some process of ecosystem accounting; however, the limitations of these efforts become clear once their benefits defined at the local or service delivery scale are removed, to support a more diverse range of products, products in which the immediate ecoservice support, processes, or provisions are not readily self-evident [19]. The Millenium Assistant (MA's) system can be used to emphasize that supporting services are fundamentally intermediate, and not an end product in themselves, which are essential to managing and maintaining the delivery of ecosystem end products [20,93].

Therefore, a system of accounting is required that enables the tracking of base products, support services, transition services, and that which ultimately links the end users with all the services provided and recognizes the true cost of the utilization of that service. To date, we have not developed or created such an accounting system, nor do we currently have the capabilities in place to trace and track all levels of services, let alone place a value (either monetized or service) that will enable the global community to be accountable for the full range of ecosystem services, support service processes, and degradation costs.

As a basis for the foundation of an accounting service, Brauman et al. [20] refined the term of hydrological services, through the division of these services into five broad categories:

1. Extractive water supply services
2. In-stream water (supply) services (both quantity and quality)
3. Water damage mitigation
4. Provision of water-related cultural services
5. Water-associated supporting services [20]

In this way, services can be defined more readily (and visibly), impacts on extraction and management can be quantified, and the beneficiaries identified. By assessing the chain of processes through which hydrological services are delivered, and evaluating the impacts of these activities on the service sustainability, and thus identifying the beneficiaries, appropriate actions can be undertaken. These may include assigning value, calculating the long-term cumulative impacts, determining the costs and benefits of services, developing appropriate regulation and governance structures, or even terminating the activities if they are deemed to be inappropriate.

However, to measure and assess the activities and the longer-term impacts, an effective and efficient monitoring and evaluation framework is also required. Measurement of change, if associated with medium to long-term impacts or cumulative influences, may require a wide range of monitoring techniques that are able to collect data on both rapid and incremental changes in the ecosystem performance. Measuring small incremental changes may be enough to detect the influence of change across the system network before they have a significant impact allowing for adjustments in both practice and target values [42].

In reality, these measurements and conditions are difficult to amass for ecosystem services, as the relevant systems tend toward a high degree of interdependency, where seemingly small changes in one sector can result in significant changes in the overall system [42,85]. The amount of uncertainty in the knowledge of ecological processes points toward either a small-step approach or cessation of activities if the outcome is unsure to avoid irreversible consequences. In addition, there is conundrum on where to place the value of service relative to current versus future costs and benefits [73,85].

There are contending views as to the effectiveness of the commoditization as an instrument of conservation that range from full support of this tool, which provides a pragmatic and transitory valuation with marketable solutions as core strategies, to a completely opposite view in which the market has no place in this process [65]. However, there is some consensus that economic valuation could lead to the commodification of ecosystem services with potentially counterproductive effects in the longer term for biodiversity conservation and the equity of access to ecosystem services and net benefits [65]. Some social discounting is inevitable if future generations are likely to benefit and have greater access to sustainable ecoservices, a situation that if current practices continue is unlikely to prevail [85]. This debate is beyond the scope of this discussion, but has been highlighted to demonstrate that while intentions can be innocent, the outcomes can be as unintended as the consequences of uninformed actions.

6.5.2 Ecohydrology as a Management Framework

As in other case studies, Newman et al. [100] suggest that improvements in the understanding of the linkages between hydrological, biogeochemical, and ecological processes are required to effectively foster integrated, interdisciplinary approaches to developing better resource management frameworks. Be they environmental, social, economic, or political in nature, new methodologies and ways of thinking about complex interdisciplinary approaches will improve the impact forecasting capacity [100]. They state that ecohydrology recognizes that vegetation, water, and nutrients are intimately coupled and that changes in one bring about changes in the others. Furthermore, Newman et al. [100] argue that while individual sciences and disciplines have carried out varied and wide-ranging studies on plant, soils, nutrient transport, and uses, an in-depth complete understanding of the interdependencies and interaction of these three components is yet to emerge.

By applying an ecohydrological approach that merges the concepts and tools from numerous disciplines (i.e., geology, biogeochemistry, plant physiology, soil science, and atmospheric science), advancements in unraveling the complex relationships in the vegetation–water–nutrient nexus that dictate fluxes and transport in the critical soil–root zone can be achieved [100].

By adopting this "merger" perspective, a broadening of the individual disciplines of hydrology and ecology is more likely to produce a general or "universal" understanding about ecohydrological functionality. While arguing this case, in which closer ties between hydrological and ecological disciplines will deliver more unified understanding, this broadening also opens the door to other possibilities that stand outside the circle of the sciences, but are of equal importance if desired sustainability and production goals are to be achieved. Environmental transformational and agronomic adaptation is not new; however, the pace at which it proceeds at global and localized scales has reduced the time that allows for adoption and/or adaptation [36]. A number of examples follow, of different approaches taken in which ecohydrology, and perhaps the broader interpretation of this transdisciplinary science, could be adopted and adapted to deliver better environmental and engineered outcomes.

6.5.2.1 Ecohydrology of Water-Limited Environments

Water-limited environments* [68,105] occupy approximately 50% of the land surface and support some of the fastest growing populations, with nearly two billion people living in these regions [35,100]. Scarcity and/or variability in the distributions of available nutrient and water resources ensure that these environments are highly responsive to changes in land cover–land use and water capture or redistribution [36,77]. Given their importance, in terms of human needs and the competition between environmental considerations, the impacts of increasing demands on water and other natural resources create significant social problems and scientific challenges [100]. There are numerous examples of broad-scale environmental changes that have occurred over vast areas, as previously discussed, that have brought about significant changes in water-limited environments including desertification, woody plant encroachment, groundwater depletion, salinization, and soil erosion [43,81,92].

* Defined as water limited due to annual precipitation (P) being typically less than annual potential evapotranspiration (Ep), such that the ratio of P to Ep ranges from about 0.03 to 0.75, and because extreme temporal variability results in extended periods with little to no precipitation.

To address the future needs of these water-limited environments, Newman et al. [100] describe the significant challenges ahead for the ecohydrologist, and their capacity to influence and manage the changes that are upon us at this time. These scientific challenges are stated as follows:

1. Partitioning of evaporation and transpiration
2. Water and nutrient interactions
3. Vegetation and streamflow
4. Vegetation and groundwater recharge
5. Hydrological change and vegetation
6. Landscape interactions in the paleodominated and human-dominated ages [100]

By incorporating these challenges into a research framework, improved understanding can be delivered by addressing issues associated with scale and complexity (Figure 6.5), thresholds and balance, and feedback and interactive loops (Figure 6.6). These processes drive the biotic and landscape response to anthropogenic intervention within the ecosystem delivered by the expanding requirements of human population and limited or unevenly distributed resources [81,100]. This is particularly the case in semiarid (or dryland) environments that are fragile and prone to collapse under external pressures.

6.5.2.2 Salinization in the Dryland Agricultural Areas of Western Australia

Broadacre land clearing for agriculture, resulting in secondary salinization, is widespread in Australia and exists in other countries as a result of clearing and poor irrigation practices. In Western Australia, successive governments promoted agricultural development in the "wheatbelt" or dryland areas of the state, with the last great land-clearing program, "a million acres a year" started in the 1960s and completed in the 1970s [31]. In all, about 15.7 million hectares of perennial, deep-rooted forests and open woodlands were cleared, removing an estimated 15 billion trees and vast areas of mallee scrub and heath. By 2001, when clearing had effectively ceased, over 19 Mha of land had been converted to annual crops and pastures, thus paving the way and unknowingly creating the right conditions to develop one of the worst examples of dryland salinity in the world [12].

The extent of secondary salinity development has been particularly severe and widespread in the dryland agricultural areas of southwest Western Australia [72], with more than 1.1 Mha impacted and a further 4.4 Mha at risk [58]. The main cause of the salinity problems in the southwest was largely the result of rising groundwater due to the extensive clearing of native vegetation and its replacement by short rooted, annual cropping systems [72,114]. This created "leaky" landscapes, with increased water (i.e., rainfall) drained below the root zone into the water table, thus mobilizing the significant salt store within the regolith. Despite lower than average rainfall over much of the wheatbelt since 2000, there is continued salinization expansion in all regions, especially following episodic floods, such as occurred in 1999–2000, 2001, and 2006 [58]. The irony of this situation is that in a dryland environment, poor management of excess water (i.e., rainfall) in the landscape has produced to one of the most widespread degradation issues in Australia.

In response to national legislation on the natural environment, which tied state funding to natural resources policy, the Western Australian State Government, between 1996 and 2003 released a number of policy documents that were targeted at providing a policy framework for the improved management of the state's natural resources. These included the Western Australian Salinity Action Plan [2], State Salinity Strategy [3], State Water Strategy [65], and State Sustainability Strategy [66], which were significant in integrating the government's approach to natural resource management for a decade between 1996 and 2006. The formative basis of these strategies was that the

1. Fundamental understanding of the environment existed
2. Scientific and economic farming systems and management strategies were available

3. New or revised land and water management practices will be developed
4. Decision support tools would follow that would enhance land managers' (farmers) capacity to deal with change within their industry

During this decade, research science, in the main, concentrated on developing engineering and bioremedial management strategies and farming systems to provide a multipronged approach to tackling the environmental and economic challenges facing rural communities. However, the social capacity of these communities was given limited consideration. This has led to a patchwork delivery and adoption system that has had varied success and did not achieve the stated outcomes of the state's strategies at the level of adoption and participation required, given the scale and impacts of land degradation in the southwest of the state [34].

Part of the problem was associated with the level of funding support to transition farmers and industry from current land management practice to a more sustainable approach. Hydrology, biodiversity conservation, catchment water management, and soil conservation were discussed as providing potential avenues for change [31]. As a whole, neither the ecosystems of the southwest, nor the ecoservices they provided were discussed, in a holistic sense. A major focus of the community was on drainage to manage land degradation, and how this may be funded, implemented, and sustained over decades required to remediate the degradation cause by secondary salinity [24,37,82]. Limited thought was given to the scale and integrative issues, or the complexity of issues surrounding land ownership, access rights, common law, and regulation in place at the time (Figure 6.7).

The outcome of this brief evaluation of the adoption of the state's natural resources management strategies, relative to the wheatbelt (or dryland agricultural) areas of southwestern Australia, highlights issues of scale and integration as shown in Figure 6.7. While the state's focus was on the water, salinity management, and agricultural production issues, the crisis of implementation and adoption was considered to be external to the natural sciences and thus crossed disciplinary boundaries. This is an important conclusion and supports the notion that while ecohydrology is in itself a new "science," it is now poised to link key sciences and may need to look outside this science to develop networks that are more inclusive of other stakeholders, which in turn may ultimately determine the effectiveness of on-ground activities.

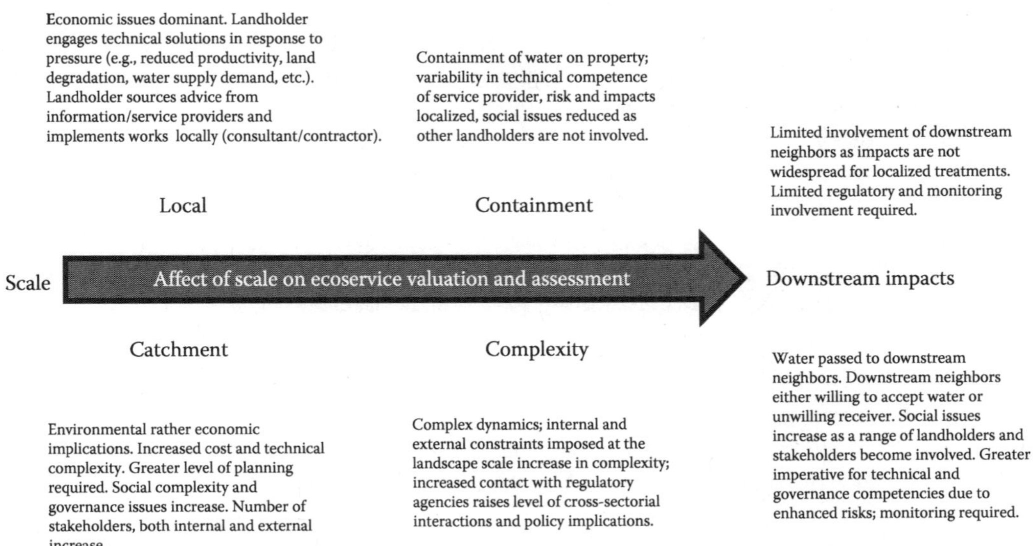

FIGURE 6.7 Increasing complexity of implementing water management through drainage on a farm within a catchment. Evaluations of services delivered and services disrupted become increasingly complex from farm to catchment scales.

6.5.2.3 Humid Zone Ecohydrology

Humid lands are a prime example of a water-controlled environment where the exchange and transfer dynamics are notably different from those of arid and semiarid (dryland) ecosystems previously discussed. Soil moisture, its availability and replenishment frequency, is a key factor in maintaining the health of terrestrial ecosystems [112]. Soils are instrumental in creating and maintaining feedback loops between the Earth's surface and the atmosphere, and its composition, flux transfer capacity (i.e., water and nutrients), is a key determinant in the health and distribution of vegetation cover and/or agricultural productivity. The dynamics of soil water in humid areas, and in particular wetlands, present challenges in relation to understanding the performance of the soil profile (and underlying regolith), including linking the stochastic fluctuations of both the water table and the soil moisture in the unsaturated zone [112].

These fluctuations are in themselves dependent on the climate, soil, and vegetation of the region being studied [131]. Therefore, to build on our understanding of the inherent dynamics of humid land ecosystems will necessitate the construction of a process-based quantitative framework. This contextual model will be used to evaluate the water table and soil moisture dynamics relative to the random nature of localized rainfall patterns, the vegetation (i.e., root profile, water tolerance, root uptake strategies, plant water relations), and the soil physical properties (e.g., storage capacity, hydraulic properties) [112]. In addition, the potential to generate overland flow (either from infiltration or saturation excess runoff) and ponded conditions caused by raised water tables, or perched systems—owing to near-saturated surface conditions—further complicates the classification of water redistribution processes within humid land landscapes [112]. Furthermore, understanding the influence of various localized factors such as topography, vegetation patterns, and soil heterogeneities on the lateral redistribution of soil water is problematic in deriving a simplistic process-based quantitative framework [18,131]. Also, critically important to understanding the ecohydrology of regions with shallow water tables and humid lands is divining the distribution probabilities of soil moisture content for soil layers at different depths in the unsaturated zones, for inclusion in this framework [112]. Rodriguez-Iturbe et al. [111] suggest that this approach will unravel the complex ecohydrology of humid areas, so as to provide a platform for the development of revised scientifically based management practices that are required for the adaptation of food production systems facing a changing climate. Targeted best management practices based on a refined understanding of humid regions are of paramount importance, given they are some of the most productive ecosystems and critical to the continued function of the planet [93].

Not only are there significant ecohydrology challenges within the sphere of humid regions, but it is also incumbent upon us to look to other fields of endeavor for research assistance and support, which can link to the characteristics of soil, climate, and vegetation to provide for a strong platform to underpin adaptive management strategies and policy decisions. As part of understanding the interconnectedness and impact causality of this research, Rodriguez-Iturbe et al. [111] suggest that a broadening of the research network should occur to establish links with related fields, including

1. Environmental fluid mechanics
    a. For the effect of wetland vegetation on surface and subsurface flow (e.g., [90,113])
    b. Sediment deposition/erosion (e.g., [97])
    c. The mixing/transport of dissolved contaminants or nutrients (e.g., [112,138])
2. *Humid land restoration*, to develop quantitative ecohydrological tools for restoration projects (e.g., [54,96,134])
3. *Landscape architecture and urban ecology*, to investigate how in urban humid lands the interactions between hydrological processes and ecosystem function are modified by the built environment (e.g., [43,130])
4. *Environmental health*, to relate ecohydrological environmental conditions to human health (e.g., [119])

While no silver bullet solution is offered in this discussion, nor is it comprehensive, it does highlight the specific challenges related to humid regions in which excess water is a common phenomenon.

6.5.2.4 Ecohydrology and Biodiversity

One approach taken to applying an ecohydrological model to assess a humid ecosystem [61] is that adopted to evaluate the performance of the Serengeti National Park (SNP), in Tanzania. The ecohydrological model is a simplified triple trophic model that assesses the movement of water within the system based on rainfall inputs, river flows, and population dynamics [60]. The model assesses the numbers and migrations of animals (both herbivores and carnivores) within the SNP, in monthly time steps, relative to water availability, grass production, and climatic factors [61]. The model predicts the future population densities of herbivores and carnivores within the SNP over a 50 year period (until 2060) based on known animal behavior, vegetation patterns, and historical climate data.

The water-use scenarios are based on flows from the Mara River, namely, (1) to evaluate performance relative to the implementation of remediation measures in the upper reaches of the river in Kenya, maintaining perennial flows in the Mara River (based on preclearing data prior to 1972); and (2) to evaluate the impact of continued deforestation of the Mau forest and overuse of Mara River water in Kenya (i.e., business as usual) [60]. The model is sufficiently representative to demonstrate that the changes in the hydrological processes during the previous 50 years have occurred as the direct result of land-use change (i.e., deforestation and irrigation in Kenya). These changes are accelerating particularly over the last decade and now threaten the survival of the ecosystem [60]. The authors stress that it will require significant effort on the part of the authorities of both the countries (Kenya and Tanzania) to find ecohydrology-based solutions to prevent the collapse of the Serengeti ecosystem [60].

While this model is relatively simplistic in process capture (and the authors caution against its potential flaws), it was able to identify divergent trends and potential threats to the continued ecosystem functionality of the SNP. This approach also highlights a salient point and one of the major themes of this chapter. That is, while the science has established that there are significant threats, and potential disruption of the ecosystems in the SNP, the capacity to deliver changes across borders and between countries currently lies outside the realm of the ecohydrologist and engineer. Increasingly, there will be a need for these professionals to link with and broaden their understanding of the geopolitical and economic circumstances that have a tendency to override the science and drive decision-making processes.

6.5.3 Ecohydrology Applications and Lessons Learned

As has been demonstrated [31,34,104], a shift in the agricultural industry approach to dryland production systems, so as to include a water management focus that addresses the industries' productivity, profitability, and environmental concerns, is likely to result in more sustainable long-term outcomes. The biophysical, social, and economic complexities all increase at the various spatiotemporal scales, as does the need for information to be able to address these issues [34] at each scalar expansion (Figure 6.7). For landholders to effectively manage their landscapes across these scales, their need for improved, timely performance and system health information is crucial.

The recognition that the broader community, through government, is investing in a partnership with the local communities to protect the natural environment within an agricultural catchment is an important concept. Previous negotiations with landholders indicated that a willingness to proceed with on-ground works is generated where there are financial incentives for implementation and a good understanding of the on-farm and catchment scale benefits. New approaches to implementing water management and engineering solutions involve community capital and empowerment as well as technical advice [34]. The challenge for the ecohydrologist and the environmental engineer in this case is the development of broadly inclusive agricultural systems that provide information on the impacts of these activities on the complex ecological systems they disturb: a system that provides an understanding of the trade-offs and benefits to landholders, and deals with complex scale issues—at local, regional, and global scales—that lead to the adoption of sustainable agricultural practices.

Jackson et al. [81], as with others [69,91,100,140], have also embraced this concept that in order to develop an integrated ecohydrological perspective, there is a need to build a framework that fosters proactive collaboration of ecologists and hydrologists. However, the author supports the case that this integrative approach should be further augmented to include a wider group of sciences and disciplines to create greater opportunities for success. The ideology is commensurate with the notion of basin scale applications and monitoring that would enable an explicit focus on ecohydrological principles, but also provide opportunities for the involvement of a wider network of stakeholders.

6.6 Ecohydrology: Future Directions and Applications

A key part of understanding the ecohydrological effects of human actions is based around landscape connectivity, understanding and observing ecohydrological connections that link surface and groundwater flow and flux exchange capacities [81]. While no panacea is offered in this discussion, nor is it comprehensive, it does highlight the specific challenges. Some would consider the world's ecosystems as capital assets [42] as they are the basis for continued life on this planet. How are they to be managed into the future? There is an incessant call for growth, in all aspects of society, on the premise that this will improve the life expectancy and life styles of the majority. However, this must be set as backdrop against the cost of this "growth" relative to the limitations of the capital assets to continue to provide the raw materials and services necessary to maintain and deliver the aspirations and goals of the world's population.

If sustainably managed, ecosystems are able to yield a flow of vital services, including the production of goods (i.e., water, food, fiber, and timber); life support processes (e.g., soil formation, pollination, water treatment, climate regulation, and genetics); and life-fulfilling conditions (i.e., aesthetics and spiritual fulfillment) [42,92]. Ecosystems, however, as capital assets are poorly understood, rarely monitored, and are often in rapid degenerative decline, with extensive loss of service capability [95]. This shift in service provision is generally undocumented or unreported until the ecosystem collapses. The recent Millennium Ecosystem Assessment [92] report indicated that of the "Approximately 60% of the ecosystems services evaluated…are being degraded or used unsustainably" and the "….human use of all ecosystem services is growing rapidly…."

The relationship between these services is such that declines in production function also reveal critical points and interdependencies in the supply of a combination of services, which may also be in decline, and reflect subtle variabilities in the time scales over which the ecoservices and ecosystems are amenable to repair [40,42]. Invariably, these scales and interrelationships only reveal themselves as they become overexploited and dysfunctional and, often, typically respond nonlinearly to perturbation [19]. For example, the development of secondary salinity in Western Australia largely went unnoticed for decades, due to the yearly incremental changes in landscape hydrology and subsoil salinity mobilization following widespread vegetation clearing [36]. However, as discussed previously, following widespread flooding and redistribution of localized runoff [22,29,36,58] over decades, a tipping point was reached resulting in significant proportions (1.1 Mha) of the agricultural landscape became salt affected with a further 1.7–4.4 Mha at risk [59]. Reparations to manage this impact are costly, long term, and may prove to be irreversible [36]. Furthermore, ecosystems are formed through interactions of local conditions creating distinctive and individualistic relationships, such that restorative actions initiated in one landscape or regional scale may not apply elsewhere [42]. Therefore, the potential to develop a universal single application solution is rare, and remediatory and conservative actions often require localized "tweaking" to deliver the desired outcomes [22,32].

Remediation of ecosystems, landscapes, and reengineering built environments (e.g., CUD*) through rethinking and assessing the performance of the system and thereby extracting true value from the system (Figure 6.1). Thus, while providing services that can be supported for long term, the system is not significantly degraded at the same time, beyond set parameters. By analyzing the cause and effect (i.e., double-loop learning) and applying revised approaches to the way in which activities are designed, orchestrated, and implemented, there exists the potential to create more efficient and productive living environments that have limited negative impacts. These systems, by necessity and design, will strengthen and create sustainable ecoservices.

Through the use of local and global scalable approaches, ecosystems can be protected or restored to more effectively manage natural disasters such as floods, to enhance soil fertility and improve productivity, to treat and recycle waste water, and to provide healthy rivers, wetlands, and habitat that will offer both productive and aesthetic services [74]. Changes in management approaches and system design bring not only environmental benefits, but are being viewed as an increasingly viable, financially sound alternative. Therefore delivering theory into practice will necessitate locally based information. This is the future role for the ecohydrologist or environmental engineering, that is, being able to determine the synergies within an altered "natural" landscape or urban environment that will provide the necessary levers to deliver the most balanced and sustainable outcome in a given locality.

6.6.1 Ecohydrology and Climate Change

The increasing challenges encountered to satisfy the global water demand are well documented (e.g., [51–53,129,136]). Shifting demographics and increasing consumption associated with rising per capita incomes are key drivers of increased demand [129]. As discussed, the tendency of human populations to gather near coastal regions or water bodies, coupled with population dynamics such as growth, gender, and age distribution, generates unmet or exclusive demands on freshwater resources [79]. This often results in loss of the pristine nature of the river system (e.g., Tigris and Euphrates Rivers, Iraq) or complete failure (e.g., Aral Sea, Kazakhstan and Uzbekistan) through increased water demand and pollution. Projected climate variability, resulting in changes to rainfall patterns and distribution, threatens the viability of food production systems, already stressed under the burgeoning demands of population growth and increasing affluence [51].

As agriculture develops and land and water use intensifies, the impact of agriculture on natural ecosystems will become more apparent, damaging the integrity of these ecosystems, undermining the food-producing systems that they support. Effective adaptation to the impacts of climate change on water availability and food security requires an integral knowledge of agronomic science, water management, hydrology, and ecology, intertwined with the resulting environmental interactions and trade-offs [79]. Water, energy, and food are inextricably linked and underpin the basic requirements for life, long-term development, and the current global trade and productivity models [35]. Limited access to these three fundamentals for life is compounded by growing concerns about their future availability and sustainability in the face of climate change, population expansion, and resource limitations [53].

An increasingly urbanized planet will exert significant pressure on the level and complexity of natural resource trade-offs required, trade-offs that at the same time must act to minimize ecosystem degradation [35]. A new approach, thinking in a Nexus perspective [78], is at the core to redefining our understanding of the interrelationships between the water, energy, and food security and is a fundamental tenet to realizing the green economy [125]. This paradigm shift is vital to achieving the sustainability development goals in an environment of global climate and economic change [136]. Changes in water distribution, quality, and availability associated with short- to medium-term regional climate variability will also create challenges for future water, energy, and food security [35]. To assure the

* http://www.connectedurbandevelopment.org/cities/ Accessed December 2012.

broader community that water and land managers are utilizing natural resources sustainably (and are being independently assessed), there is a requirement for both an adequate and flexible eco-accreditation framework supported by a robust real-time monitoring and reporting system [31]. This is where the enhanced role of the engineer, the hydrologist, or the ecohydrologist fits into the equations and will need to deliver better designed, more effective, and sustainable food production systems, water resource management strategies, and harmonious built environments.

6.6.2 Integrated Systems, Monitoring Networks, and Global Linkages

Part of the promise of modernism was that technology could provide services more efficiently and more reliably than natural systems [74]. However, although it may be possible to augment or replace some ecosystem services—often at great cost, on a limited scale, or in constrained locations—the reliance of technology on functioning ecosystems often goes unrecognized. The challenge now facing us to improve or maintain water quality and access is a growing global concern, typified by the creation of the European Commission WFD [46] and the U.S. Clean Water Act [130], among others. A transformation in thinking and approach is necessary with the adoption of new management and development opportunities, which are enabled by innovative technology [35]. By recognizing their reliance on natural systems, people have long attempted to divorce themselves from the vagaries of this dependence.

6.6.2.1 Monitoring Networks

Place-based or catchment-based research is an effective way of promoting collaboration and focusing efforts on the integration of reductionist and holistic approaches [100]. An ideal starting point would be an effective scalable monitoring network, which links the micro- to the mesoscales and provides for an understanding of water fluxes and water quality variations from the plant-root level through to the basin scale. Within this framework, and of global concern, is the increasing storage of agricultural chemicals in soils and various surface and subsurface water bodies arising from the overapplication of fertilizers, herbicides, and pesticides. Chemical species such as nitrates and chlorides impact on crop growth and adversely on the quality of water supply for both communities and commercial activities [33,39,109]. This could be achieved through new technologies that link targeted molecular sensor monitoring with wireless networks that can deliver real-time responses within catchments and regions, and potentially globally, through satellite monitoring technologies.

Wider monitored areas are required, particularly in semiarid regions, and as suggested by Newman et al. [100], such areas are (1) geographically extensive and contain a significant and growing proportion of the human population, (2) extremely sensitive to ecohydrological processes, and (3) composed of well-defined and broad elevational gradients, with numerous, closely spaced ecotonal and hydrological transition zones ideal for comparative studies. To undertake the challenge of explaining the large spatial and temporal variability observed in patterns of soil moisture across multiple space–time scales, understanding the interactive soil–vegetation–atmospheric fluxes and exchanges is critical [15,87]. Access to this information is fundamental to understanding the impacts of changed vegetative covers or in designing intervention or recovery strategies that may need to be implemented or required to influence the rate of change, at multiple scales [142]. External factors like atmospheric forcing, topography, soil properties, and vegetation, which interact in a complex, nonlinear way (e.g., [67]), present measurement and quantification challenges in determining (or predicting) soil moisture spatial variability [87,131].

In addition to the necessity for evaluating ecological and soil–landscape changes as they come under increasing pressures, there is a need to manage water resources and water quality. The main drivers of poor water quality are economics, poor water management, agricultural practices, and urban development [147]. Given that over 90% of the world's population lives in countries that share river basins, of which 40% lives in river and lake basins that comprise two or more countries [138], access to water, water quality, and water allocation become increasingly problematic from the headwaters, to the discharge point [33]. This becomes more complex within river basins, as there are multiple monitoring and compliance requirements that are often undertaken across borders, under differing governance structures and administrative capabilities [33]. Water-quality monitoring (WQM) is currently undertaken through small-scale and single-application sampling and testing with limited techniques available and expensive highly technical instrumentation and selective decision support tools [39]. The amount and quality of data available clearly limit the amount of extractable knowledge gained and thereby inherently limit the capabilities of the scientist, modeler, or land manager to deliver appropriate information on which to base actionable decision [80]. A framework with key attributes for real-time, spatial–temporal, and multilevel catchment-level monitoring is proposed [147]. Therefore, there is a case for evaluating the multilevel impacts that various stakeholders have in a catchment on water resources. This would allow the implementation of a more inclusive and effective monitoring and management framework. This framework is underpinned by a real-time integrated and targeted monitoring system that allows for the assessment of both catchment function and modifications to those functions or services by the stakeholders (i.e., land managers). This linkage and similarities between individual sensing requirements indicate that there is need for an autonomous vegetation–soil–water-quality monitoring framework based on targeted wireless sensor technologies [147].

The suggested new modeling frameworks will need to be validated and tested against field data. To this end, improved field measurement and data collection networks are required to observe variations in ecosystem performance. In addition to in-field measurements, satellite observations provide spatially distributed data of surface soil moisture and water depth that could be used to investigate ecohydrological processes in spatially extended systems [29]. Using remote sensing technologies will provide an intercomparison analysis of average surface conditions from remote-sensed measurements and ground-based measurements, land surface models can be utilized to determine variability in soil moisture distribution patterns. Thus they can provide an indication of relative soil moisture conditions to improve runoff predictions and analyze land surface–atmosphere interactions for regional climate predictions in data-limited areas [28]. These data sets can be utilized to improve predictive models of land–atmosphere interactions, rainfall–runoff processes, and groundwater recharge processes [27] and thus be used to assess changes in function, management, land cover, and land use.

Coupling remote technologies with ground-based monitoring networks creates a system that encourages both independent measures and verification tools for remotely sensed data. A survey of ground-based monitoring techniques suggests that wireless sensor networks are tools that, despite their limitations, are attractive for real-time spatial–temporal data collection for soil and water monitoring applications [39]. These tools have a huge potential for dense data collection for monitoring agricultural activities. For effective implementation of this system, key attributes can already be determined, which include multiscale catchment monitoring, spatiotemporal data collection, long life, and real-time systems. Other aspects of technology for consideration include the development of appropriate targeted networks, including variable frequency range, variable sampling, well-defined sensor interface, lifetime, ease of deployment and configuration for hydrologists, and a network model for broad environment monitoring. These requirements are not well catered for by using off-the-shelf components [39].

6.6.2.2 Indicators and Metrics

To progress this field of research and to ensure the best managed outcomes, placing a greater emphasis on field data, long-term monitoring, scalable data collection, targeted and well-designed performance indicators (KPIs), and means of verification (MOVs), integration of varied data sets from multiple disciplines is required. While there is broad discussion on the need for improved monitoring technologies, there is also a sustained effort required to develop appropriate targets and indicators of change within catchments. Therefore, while effort goes into the technologies required, additional research is required to derive indicators of sustainability that clearly identify and monitor shifts in ecosystem performance. At best, current practice informs us whether we are moving in the wrong direction or that our current activities are not sustainable [75]. More often than not, these indicators simply draw our attention to the existence of problems, but do little to identify their origins and have limited scope in solving these problems [75]. Therefore, the construction and development of appropriate indicators, monitoring networks, and reporting frameworks are required to assess ecosystem performance and deliver sustainable outcomes at multiple levels. But in order to do this, ecohydrology, as a discipline, by necessity must look to collaborate with a wider transdisciplinary network.

Thus, through the combination of the varied monitoring and tools for individual areas of a catchment, within a river basin or within regions, a greater understanding of geo-bio-physical trends and the quantification of the contributing factors are within our grasp. This can be achieved through new technologies combined with a network of like-minded institutions, industry partners, and governments to deliver real-time global observations of the impacts of anthropogenic activities, climate change, and localized ecosystem variability. Here, then, presents the opportunity for the ecohydrologist and engineers in determining what type of sensors are required and how these may be implanted within the plant soil continuum and linked to the wider network of sensors from the plot to the catchment scale.

6.7 Summary and Conclusions

The review of ecohydrology concepts presented here is by no means comprehensive, but does highlight the changing nature of issues surrounding water–soil–vegetation ecosystem management and the water and food security issues in that area associated with the services provided by these systems. The integrative requirement of sciences to deliver a broader focus to understanding interrelationships within ecosystems and the need to bring in both policy and governance as tools to assess and monitor ecosystem performance relative to the demands placed on it has been discussed. The use of technologies to monitor the local to global responses on impacts to ecosystems, in this time of rapid change and increased demands, is viewed as imperative. Appropriate measures, metrics, and indicators are required to be developed and categorized, without which short-, medium-, and long-term goals, policies, and directions cannot be set. By broadening the nature of ecohydrology, it provides numerous opportunities for the researcher, to develop and test new skills and applications that will deliver benefits into the global community.

This review has also highlighted the changing roles of industry and economics and the need to underpin the current economic models with a greening ecological framework. There is a continued requirement to provide significant support from a wide variety of researchers, sponsors, NGOs, and governments to build the necessary evaluation frameworks, integrative networks, global monitoring systems, and subsequently responsive and responsible governance and regulation. As has been discussed through this chapter, the scope and breadth of ecohydrology are still emerging, and while there is clear evidence of interdisciplinary collaboration, it has not yet consolidated into new ways of approaching the science. There has been limited progress over the last 15 years in integrating the methodological strengths of the respective parent disciplines (i.e., ecology and hydrology), with a dominance of observational correlation studies, which are rarely coupled with empirical measurements and almost never coupled with experimental studies [84]. The existing research is still dominated by one-way interactions (i.e., hydrological impact on biota or biota impact on a hydrological process), and there exists the potential for the broadening of the scope of this discipline through the coupling models with more controlled manipulative field experiments [115].

The review has highlighted, first, the need for stronger research in arid and semiarid ecosystems, where water is scarce and a tight coupling exists between hydrology and ecology. However, these challenges are not dissimilar to those experienced by other regions on the planet. Second, a requirement to develop predictive frameworks for understanding the consequences of vegetation change, with landscape connectivity, through recharge and discharge dynamics, with global climate. Third, a proposal is presented to further widen the role of the ecohydrologist in creating more integrated networks of a transdisciplinary nature that will foster improved linkages with other areas important to the delivery of sustainable ecosystems and the services that they provide. Fourth, there is an identified need to define ecoservices in a way that is broadly understood by the engineers, communities, governments, and industry partners and allow this understanding to be introduced into management frameworks that promote resilience and sustainability in the use of the planet's natural capital.

If the types of political, scientific, and sustainability strategies discussed herein are to be successful, then mechanisms for the provision of adequate and targeted funding and the identification of cost–benefit accruals to the individual, community, and environmental returns are required. In addition, the impacts of physical and soft engineering interventions at short-, medium-, and long-term time periods, and the communities' capacity to accept and generate change, require substantiation. There is a marked shift toward the development of evidentiary-based systems in which the science of ecohydrology can be positioned as a major player in ensuring our future.

References

1. Abdullah, D., Ir.Hj. Keizrul, and B. Christensen. 2004. Integrated river basin management. *Buletin Ingenieur* 21: 21–23.
2. Anon. 1996. *Western Australian Salinity Action Plan*. Perth, WA: Government of Western Australia.
3. Anon. 2000. *State Salinity Strategy*. Perth, WA: Government of Western Australia.
4. Alfieri, A., R. Hassan, and G. Lange. 2004. Using environmental accounts to promote sustainable development. In: *Integrated Economic, Environmental, and Social Frameworks*. Paris, France: OECD Publishing.
5. Archer, N.A.L., M. Bonell, N.A. Coles, A.M. MacDonald, R. Stevenson, and P.D. Hallet. 2012. The relationship of landuse and topography on soil water storage and its implication on Natural Flood Management in the Scottish Borders. In: *Proceedings of the 11th British Hydrological Society Symposium-Hydrology for a Changing World*, (ed.) BHS. Dundee BHS. doi:10.75558/bhs.2012.ns04. p. 6
6. Argyris, C. and D. Schön. 1978. *Organizational Learning: A Theory of Action Perspective*. Reading, MA: Addison Wesley.
7. Australia, Land and Water. 2000. *Evaluating Integrated Catchment Management*. Research Project number CTC7. Factsheet. Australian Government, p. 4.
8. Authority, Murray Darling Basin. 2012. *Basin Plan*. Water Act 2007 44(2)(c)(ii). edited by Australian Government.
9. Baird, A.J. and R.J. Wilby. 1999. *Eco-Hydrology: Plants and Water in Terrestrial and Aquatic Environments*: Palgrave Macmillan. Original edition, London, U.K.: Routledge, Taylor & Francis.
10. Bammer, G. 2005. Integration and implementation sciences: Building a new specialization. *Ecology and Society* 10(2). 25 pages. [online] URL: http://www.ecologyandsociety.org/vol10/iss2/art6/
11. Barton, D.N., T. Saloranta, S.J. Moe, H.O. Eggestad, and S. Kuikka. 2008. Bayesian belief networks as a meta-modelling tool in integrated river basin management—Pros and cons in evaluating nutrient abatement decisions under uncertainty in a Norwegian river basin. *Ecological Economics* 66(1):91–104. doi: http://dx.doi.org/10.1016/j.ecolecon.2008.02.012.
12. Beresford, Q., H. Bekle, H. Philips, and J. Mulock. 2001. *The Salinity Crisis: Landscapes, Communities and Politics*. Crawley, Western Australia, Australia: University of Western Australia Press.
13. Biswas, A.K. 2004. Integrated water resources management: A reassessment—A water forum contribution. *IWRA, Water International* 29(2):248–256.
14. Bloschl, G. 2001. Scaling in hydrology. *Hydrological Processes* 15(4):709–711. doi: 10.1002/hyp.432.
15. Bogena, H.R., M. Herbst, J.A. Huisman, U. Rosenbaum, A. Weuthen, and H. Vereecken. 2010. Potential of wireless sensor networks for measuring soil water content variability. *Vadose Zone Journal* 9(4):1002. doi: 10.2136/vzj2009.0173.
16. Bohensky, E., D. Connell, and B. Taylor. 2009. Experiences with integrated river basin management, international and Murray Darling Basin: Lessons for northern Australia. In *Northern Australia Land and Water Science Review*, 33. Highett Victoria, Australia: CSIRO.
17. Bonell, M. 2002. Ecohydrology–A completely new idea? Discussion. *Hydrological Sciences Journal-Journal Des Sciences Hydrologiques* 47(5):809–810. doi: 10.1080/02626660209492984.

18. Borgogno, F., P. D'Odorico, F. Laio, and L. Ridolfi. 2009. Mathematical models of vegetation pattern formation in ecohydrology. *Reviews of Geophysics* 47(1):RG1005. doi: 10.1029/2007rg000256.
19. Boyd, J. and S. Banzhaf. 2007. What are ecosystem services? The need for standardized environmental accounting units. *Ecological Economics* 63(2–3):616–626. doi: 10.1016/j.ecolecon.2007.01.002.
20. Brauman, K.A., G.C. Daily, T.K. Duarte, and H.A. Mooney. 2007. The nature and value of ecosystem services: An overview highlighting hydrological services. *Annual Review of Environment and Resources* 32(1):67–98. doi: 10.1146/annurev.energy.32.031306.102758.
21. Budyko, M.I. 1974. *Climate and Life*. New York: Academic Press.
22. Callow, J.N., N.A. Coles, and T. Hall. 2010. A new ecohydraulic management paradigm for salt affected ecosystems and wetlands in low-gradient semi-arid environments. In: *Proceedings of the 8th International Symposium on Ecohydraulics*. Seoul, Korea: Republic of Korea.
23. Camkin, J.K., K.L. Bristow, C. Petheram, Z. Paydar, F.J. Cook, and J. Story. 2008. Designs for the future: The role of sustainable irrigation in northern Australia. CSIRO, Australia. *WIT Transactions on Ecology and the Environment* 112:293–302.
24. Chandler, K. and N.A. Coles, Western Australia. Department of Agriculture, and Western Australia. Engineering Evaluation Initiative Committee. 2004. *Review of Deep Drains to Manage Salinity in Western Australia*. South Perth, Western Australia, Australia: Department of Agriculture.
25. Chícharo, L. and M.A. Chícharo. 2006. Applying the Ecohydrology approach to the Guadiana estuary and coastal areas: Lessons learned from dam impacted ecosystems. *Estuarine Coastal and Shelf Science* 70(1–2):1–2.
26. Chichilnisky, G. and G. Heal. 1998. Economic returns from the biosphere. *Nature* 391:629.
27. Choi, M. 2006. Soil moisture dynamics from satellite observations, land surface modeling, and field data, PhD Thesis. Durham, NH: University of New Hampshire.
28. Choi, M., J.M. Jacobs, and D. Bosch. 2008. Remote sensing observatory validation of surface soil moisture using Advanced Microwave Scanning Radiometer E, Common Land Model, and ground based data: Case study in SMEX03 Little River Region, Georgia, U.S. *Water Resources Research* 44(8):2617–2627. doi: 10.1029/2006WR005578.
29. Coles, N.A., R.J. George, and A.D. Bathgate. 1999. *An Assessment of the Efficacy of Deep Drains Constructed in the Wheatbelt of Western Australia Part 1: A Discussion Paper on Drainage Implementation in the Wheatbelt*. NRMS Technical Bulletin 4391. ISSN 1326–415X. Perth, Western Australia, Australia: Department of Agriculture Western Australia.
30. Coles, N.A. and P. Hall. 2012. Water, energy and food security: Technology challenges of thinking in a Nexus perspective. Paper read at *IEEE 2012 Conference on Technology & Society in Asia*, 27–29th October, at Singapore.
31. Coles, N.A. and M. Sivapalan. 1991. The role of soils in the heterogenecity of small agricultural catchment hydrology: A modelling exercise. In: *Proceedings of the International Hydrology and Water Resources Symposium*. Perth, Western Australia, Australia: Institution of Engineers, Barton, A.C.T.
32. Coles, N.A., D. Farmer, T.J. Cattlin, and D. Stanton. 2004. Managing water, the key to preserving biodiversity in the dryland agricultural areas of Western Australia. Paper read at *Proceedings of International Conference: Hydrology Science and Practice for the 21st Century. British Hydrological Society. London, July 2004*, at London, U.K.
33. Coles, N.A., T.J. Cattlin, D. Farmer, and D. Stanton. 2004. Water management: What's in a name? Paper read at *Proceedings of 13th International. Soil Conservation Organisation Conference: Conserving Soil and Water for Society: Sharing Solutions ISCO 2004*, at Brisbane, Queensland, Australia.
34. Coles, N.A. and S.M. Ali. 2000. Implications for surface water management on recharge and catchment water balance. Paper read at *Proceedings of Hydro 2000 Conference*. Perth, Western Australia, Australia: Institute of Engineers Australia.

35. Coles, N.A., M. Sivapalan, J.E. Larsen, P.E. Linnet, and C. Fahner. 1998. Modelling runoff generation on small agricultural catchments. In: *Distributed Hydrological Modelling: Applications of the TOPMODEL Concept*, (ed.) K. Beven, pp. 289–314. New York: John Wiley & Sons.
36. Coles, N.A., J. Warren, and J. Roe. 2004. Engineering partnerships for salinity management. Paper read at *Proceedings of 1st National Salinity Engineering Conference*. Perth, Western Australia, Australia: Institute of Engineers Australia.
37. Coles, N.A., J. Camkin, N. Harris, A. Cranny, P. Hall, and H. Zia. 2013. Water-boundaries and borders-the great intangibles in water quality management: Can new technologies enable more effective compliance? In *TWAM 2013 International Conference: Transboundary Water Management Across Borders and Interfaces: Present and Future Challenges*. Aveiro, Portugal.
38. Commission, European. 2010. Report from the commission to the council and the European parliment: The 2010 assessment of the implementing of EU Biodiversity Action Plan. In *SEC(2010) 1165*. Brussels 8.10.2010 COM (2010) 548 final.
39. Cranny, A., N.R. Harris, N.M. White, E. Barrett-Lennard, N.A. Coles, M. Rivers, K. Smettem, and J. Wu. 2012. Screen-printed potentiometric sensors for chloride measurement in soils. *Procedia Engineering* 47:1157–1160. doi: 10.1016/j.proeng.2012.09.357.
40. Daily, G.C., S. Alexander, P. Ehrlich, R., L. Goulder, J. Lubchenco, P. Matson, A., A. Mooney et al. 1997. *Ecosystem Services: Benefits Supplied to Human Societies by Natural Ecosystems*. Washington, DC: Ecological Society of America 2(Issues in Ecology).
41. Daily, G.C., T. Söderqvist, S. Aniyar, K. Arrow, P. Dasgupta, P.R. Ehrlich, C. Folke et al. 2000. The value of nature and the nature of value. *Science* 289(5478):395–396.
42. Daily, G. 1997. *Nature's Services: Societal Dependence on Natural Ecosystems*. Washington, DC: Island Press.
43. De Fries, R., G.P. Asner, and R. Houghton. 2004. Ecosystems and land use change. In *Geophysical Monograph Series*. Washington, DC: AGU.
44. Eagleson, P.S. 2002. *Ecohydrology: Darwinian Expression of Vegetation form and Function*. Cambridge, U.K.: Cambridge University Press.
45. Eamus, D., T. Hatton, P. Cook, and C. Colvin. 2006. *Ecohydrology: Vegetation Function, Water and Resource Management*, (ed.) CSIRO. Collingwood, Victoria, Australia: CSIRO.
46. EC, E.C. 2000. Directive 2000/60/EC establishing a framework for community action in the field of water policy. Official Journal of the European Communities. 22.12.200. I.327/1–72.
47. Ecohydrology ecosystems, land and water process interactions, ecohydrogeomorphology. 2008, Summaries. Chichester, U.K.: John Wiley & Sons.
48. National Committee on Water Engineering. 1995. *Position Paper on Integrated Catchment Management*. Canberra, Australia: The Institution of Engineers, Australia.
49. Farber, S.C., R. Costanza, and M.A. Wilson. 2002. Economic and ecological concepts for valuing ecosystem services. *Ecological Economics* 41(3):375–392. doi: http://dx.doi.org/10.1016/S0921-8009(02)00088-5.
50. Finger, M. and M. Asún. 2000. *Adult Education at the Crossroads. Learning Our Way Out*. London, U.K.: Zed Books.
51. Food and Agricultural Organisation (FAO). 2011. *Climate Change, Water and Food Security*. FAO Water Report 36. (eds.) J. Burke, J.-M. Faurès, and H. Turral. Rome, Italy: Food and Agriculture Organization of the United Nations.
52. Food and Agriculture Organization of the United Nations and Earthscan. 2011. *The State of the World's Land and Water Resources for Food and Agriculture: Managing Systems at Risk*, (1st edn.). Oxford, U.K.: Food and Agriculture Organization of the United Nations and Earthscan.
53. Foresight. 2011. *The Future of Food and Farming: Final Project Report*. (ed.) The Government Office of Science. London, U.K.: The Government Office of Science.

54. García, M., F. Soto, J.M. González, and E. Bécares. 2008. A comparison of bacterial removal efficiencies in constructed wetlands and algae-based systems. *Ecological Engineering* 32(3):238–243. doi: http://dx.doi.org/10.1016/j.ecoleng.2007.11.012.
55. Gatto, M. and G.A. De Leo. 2000. Pricing biodiversity and ecosystem services: The never-ending story. *BioScience* 50(4):347–355.
56. Gatto, M., A. Caizzi, L. Rizzi, and G.A. De Leo. 2002. The Kyoto Protocol is cost-effective. *Conservation Ecology* 6(1): 211. [online] http://www.consecol.org/vol6/iss1/resp11/ (3 pages)
57. Gentine, P., P. D'Odorico, B. Lintner, G. Sivandran, and G.D. Salvucci. 2012. Interdependence of climate, soil, and vegetation as constrained by the Budyko curve. *Geophysical Research Letters* 39(19)39, L19404. doi: 10.1029/2012GL053492.
58. George, R., J. Clarke, and E. English. 2008. Modern and paleographic trends in the salinisation of the Western Australian wheatbelt: A review. *Australian Journal of Soil Research* 46:751–767.
59. George, R.J., R.J. Speed, J.A. Simons, R.H. Smith, R. Ferdowsian, G.P. Raper, and D.L. Bennett. 2008. Long-term groundwater trends and their impact on the future extent of dryland salinity in Western Australia in a variable climate. Paper read at *2nd International Salinity Forum*, 30 March–3 April. Adelaide, South Australia, Australia: Adelaide Convention Centre.
60. Gereta, E., E. Mwangomo, and E. Wolanski. 2009. Ecohydrology as a tool for the survival of the threatened Serengeti ecosystem. *Ecohydrology and Hydrobiology* 9(1):115–124.
61. Gereta, E., E. Wolanski, M. Borner, and S. Serneels. 2002. Use of an ecohydrological model to predict the impact on the Serengeti ecosystem of deforestation, irrigation and the proposed Amala weir water diversion project in Kenya. *Ecohydrology and Hydrobiology* 2:127–134.
62. Giovannini, E. 2004. Accounting frameworks for sustainable development. In *Integrated Economic, Environmental, and Social Frameworks*, pp. 111–128. Paris, France: OECD Publishing.
63. Gómez-Baggethun, E. and M. Ruiz-Pérez. 2011. Economic valuation and the commodification of ecosystem services. *Progress in Physical Geography* 35:613–628. doi: 10.1177/0309133311421708.
64. Gopalakrishnan, M., A.D. Mohile, L.N. Gupta, R. Kuberan, and S.A. Kulkarni. 2006. An integrated water assessment model for supporting India water policy. *Irrigation and Drainage* 55(1):33–50.
65. Government of Western Australia. 2003. Securing our water future. Published by the Department of Water, State Govt. of WA. pp. 64. ISBN 073070209X
66. Government of Western Australia. 2003. Hope for the future. Perth, WA: The Western Australian State Sustainability Strategy, Department of the Premier and Cabinet. Published by the State Govt. of WA. pp. 153. ISBN 073070212X
67. Grayson, R. and G. Blöschl. 2000. *Spatial Patterns in Catchment Hydrology: Observations and Modelling*. Cambridge, U.K.: Cambridge University Press.
68 Guswa, A.J., M.A. Celia, and I. Rodriguez-Iturbe. 2004. Effect of vertical resolution on predictions of transpiration in water-limited ecosystems. *Advances in Water Resources* 27:467–480. doi: 10.1016/j.advwatres.2004.03.001.
69. Hannah, D.M., P.J. Wood, and J.P. Sadler. 2004. Ecohydrology and hydroecology: A "new paradigm" *Hydrological Processes* 18(17):3439–3445. doi: 10.1002/hyp.5761.
70. Harper, D.M., M. Zalewski, and N. Pacini. 2008. *Ecohydrology: Processes, Models and Case Studies: An Approach to the Sustainable Management of Water Resources*. Wallingford, U.K.; Cambridge, MA: CABI Pub.
71. Hassan, J. 1995. The impact of EU environmental policy on water industry reform. *European Environment* 5(2):45–51.
72. Hatton, T., J.K. Ruprecht, and R.J. George. 2003. Preclearing hydrology of the Western Australia wheatbelt: Target for the future? *Plant and Soil* 257:341–356.
73. Heal, G.M. 2000. *Nature and the Marketplace: Capturing the value of Ecosystem Services*. Washington, DC: Island Press.

74. Heal, G.M., Barbier, E.E., Boyle, K.J., Covich, A.P., Gloss, S.P., and Hershner, C.H. 2005. *Valuing Ecosystems Services: Toward Better Environmental Decision-Making*. Washington, DC: National Research Council.
75. Hecht, J.E. 2007. *Can Indicators and Accounts Really Measure Sustainability?* Washington, DC: Considerations for the US Environmental Protection Agency.
76. Heinmiller, B.T. 2003. Harmonization through emulation: Canadian federalism and water export policy. *Canadian Public Administration* 46(4):495–513.
77. Hipsey, M.R., R. Vogwell, D. Farmer, and B. Busch. 2011. A multi-scale ecohydrological model for assessing floodplain wetland response to altered flow regimes. In *19th International Congress on Modelling and Simulation*. Perth, Western Australia, Australia.
78. Hoff, H. 2011. Understanding the Nexus. Background paper for the Bonn 2011 Nexus Conference. In *The Water, Energy and Food Security Nexus*. Bonn, Germany: Stockholm Environment Institute.
79. Hommes, S. 2008. Conquering complexity: Dealing with uncertainty and ambiguity in water management. PhD Thesis. Enschede, Netherlands: University of Twente.
80. Huyen Le, T.T., M. Lorenz, K. Prilop, Q.D. Lam, and G. Meon. 2012. An ecohydrological, ecohydraulic model system for water management of the saigon river system under tide effect. In *9th International Symposium on Ecohydraulics* Vienna, Austria.
81. Jackson, R.B., E.G. Jobbágy, and M.D. Nosetto. 2009. Ecohydrology bearings-invited commentary ecohydrology in a human-dominated landscape. *Ecohydrology* 2:383–389. doi: 10.1002/eco.81.
82. Johns, A., N.A. Coles, and E. Crossley. 2001. Networks of partnerships towards the adoption and best use of engineering water management options. Paper read at *Proceedings of the State Landcare Conference*. Mandurah. 2001, at Mandurah, Western Australia, Australia.
83. Jønch-Clausen, T. 2004. Integrated Water Resources Management (IWRM) and Water Efficiency Plans by 2005: Why, What and How?: Global Water Partnership.
84. King, E.G. and K.K. Caylor. 2011. Ecohydrology in practice: Strengths, conveniences, and opportunities. *Ecohydrology* 4(4):608–612. doi: 10.1002/eco.248.
85. Kinzig, A.P., C. Perrings, F.S. Chapin III, S. Polasky, V.K. Smith, D. Tilman, and B.L. Turner II. 2011. Paying for ecosystem services—Promise and peril. *Science Letters* 33. 10 February 2012. www.sciencemag.org
86. Kundzewiez, Z.W. 2002. Ecohydrology-seeking consensus on interpretation of the notion. *Journal of Hydrological Sciences* 47(5):799–804.
87. Laio, F., S. Tamea, L. Ridolfi, P. D'Odorico, and I. Rodriguez-Iturbe. 2009. Ecohydrology of groundwater-dependent ecosystems: 1. Stochastic water table dynamics. *Water Resources Research* 45(5):W05419. doi: 10.1029/2008wr007292.
88. Ludwig, D. 2001. The Era of management is over. *Ecosystems* 4:758–764. doi: 10.1007/s10021-001-0044-x.
89. Ludwig, D. 2000. Limitations of economic valuation of ecosystems. *Ecosystems* 3:31–35.
90. Luhar, M. and H.M. Nepf. 2011. Flow-induced reconfiguration of buoyant and flexible aquatic vegetation. *Limnology and Oceanography* 56(6):2003–2017. doi: 10.4319/lo.2011.56.6.2003.
91. McClain, M.E., L. Chícharo, N. Fohrer, M. Gaviño Novillo, W. Windhorst, and M. Zalewski. 2012. Training hydrologists to be ecohydrologists and play a leading role in environmental problem solving. *Hydrology and Earth System Sciences Discussions* 9(2):1481–1514. doi: 10.5194/hessd-9-1481-2012.
92. Millennium Ecosystem Assessment. 2005. *Ecosystem and Human Well-Being: Desertification Synthesis*, (eds.) A. Adeel, U. Safriel, D. Niemeijer, and R. White. Washington, DC: World Resources Institute.
93. Millennium Ecosystem Assessment. 2005. *Ecosystems and Human Well-Being: Synthesis*. Washington, DC: World Resources Institute.
94. Miller, F. and P. Hirsch. 2003. *Civil Society and Internationalized River Basin Management*. In Working Paper Series, (ed.) Australian Mekong Resource Centre, Sydney, New South Wales, Australia: University of Sydney.

95. Mooney, H.A. and P.R. Ehrlich. 1997. Ecosystem services: A fragmentary history. In *Nature's Services: Societal Dependence on Natural Ecosystems*, (ed.) G.C. Daily, pp. 392. Washington, DC: Island Press.
96. Naiman, R.J., H. Decamps, and M.E. McClain. 2005. *Riparia: Ecology, Conservation, and Management of Streamside Communities*. Amsterdam, the Netherlands, Boston, MA: Elsevier Academic Press.
97. Nepf, H., M. Ghisalberti, B. White, and E. Murphy. 2007. Retention time and dispersion associated with submerged aquatic canopies. *Water Resources Research* 43:W04422. doi: 10.1029/2006WR005362.
98. Neto, S. 2010. Water, territory and planning. contemporary challenges: Towards a territorial integration of water management. PhD Thesis. Lisbon, Portugal: Technical University of Lisbon.
99. Neto, S. and J.K. Camkin. 2012. New learning foundations for building water knowledge bridges. Paper read at *Proceeding of the Second Symposium on Building Knowledge Bridges for a Sustainable Water Future*, at Panama.
100. Newman, B.D., B.P. Wilcox, S.R. Archer, D.D. Breshears, C.N. Dahm, C.J. Duffy, N.G. McDowell, F.M. Phillips, B.R. Scanlon, and E.R. Vivoni. 2006. Ecohydrology of water-limited environments: A scientific vision. *Water Resources Research* 42(6):W06302. doi: 10.1029/2005wr004141.
101. Nuttle, W.K. 2002. Is ecohydrology one idea or many? *Hydrological Sciences Journal* 45(5):805–807.
102. Pahl-Wostl, C. 2007. Transitions towards adaptive management of water facing climate and global change. *Water Resources Management* 21(1):49–62.
103. Pahl-Wostl, C. 2009. A conceptual framework for analysing adaptive capacity and multi-level learning processes in resource governance regimes. *Global Environmental Change* 19(3):354–365. doi: http://dx.doi.org/10.1016/j.gloenvcha.2009.06.001.
104. Pannell, D.J. 2001. Explaining non-adoption of practices to prevent dryland salinity in Western Australia: Implications for policy. In *Land Degradation*, (ed.) A. Conacher, pp. 335–346. Dordrecht, the Netherlands: Kluwer.
105. Parsons, A.J. and A.D. Abrahams. 1994. Geomorphology of desert environments. In *Geomorphology of Desert Environments*, (eds.) A.D. Abrahams and A.J. Parsons, pp. 1–12. Boca Raton, FL: CRC Press.
106. Peskin, H.M. and M.S. Delos Angeles. 2001. Accounting for environmental services: Contrasting the SESA and the ENRAP approaches. *Review of Income and Wealth* 47(2):203–219.
107. Yale Center for Environmental Law and Policy and Center for International Earth Science Information Network. 2005. *Environmental Sustainability Index: Benchmarking National Environmental Stewardship (2005)*, (ed.) Columbia University. New Haven, CT: Yale University.
108. Qin, B., Z. Liu, and K. Havens. 2007. *Eutrophication of Shallow Lakes with Special Reference to Lake Taihu, China. Developments in Hydrobiology*, Vol. 194. pp. 328. Dordrecht, the Netherlands: Springer.
109. Rivers, M.R., D.M. Weaver, K.R.J. Smettem, and P.M. Davies. 2011. Estimating future scenarios for farm-watershed nutrient fluxes using dynamic simulation modelling. *Physics and Chemistry of the Earth* 36:420–423.
110. Roderick, M.L. and G.D. Farquhar. 2011. A simple framework for relating variations in runoff to variations in climatic conditions and catchment properties. *Water Resources Research* 47(12): W00G07. n/a-n/a. doi: 10.1029/2010WR009826.
111. Rodriguez-Iturbe, I., P. D'Odorico, F. Laio, L. Ridolfi, and S. Tamea. 2007. Challenges in humid land ecohydrology: Interactions of water table and unsaturated zone with climate, soil, and vegetation. *Water Resources Research* 43(9):W09301–W09400. doi: 10.1029/2007wr006073.
112. Rodríguez-Iturbe, I. and A. Porporato. 2004. *Ecohydrology of Water-Controlled Ecosystems: Soil Moisture and Plant Dynamics*. Cambridge, U.K.: Cambridge University Press.
113. Rominger, J., A. Lightbody, and H.M. Nepf. 2010. The effects of vegetation on sand bar stability and stream hydrodynamics. *Journal of Hydraulic Engineering* 136(12):994–1002. doi: 10.1061/(ASCE)HY.1943-7900.0000215.
114. Ruprecht, J.K. and N.J. Schofield. 1991. Effects of partial deforestation on hydrology and salinity in high salt storage landscapes. 1 Extensive block clearing. *Journal of Hydrology* 129:19–38.

115. Smettem, K., D. Breshears, and M. Waddington. 2012. Observations on the Development of Ecohydrology as a Discipline: History and Future Opportunities. In *EGU General Assembly 2012*. Vienna, Austria: EGU.
116. Smettem, K.R.J. 2008. Welcome address for the new 'Ecohydrology' journal. *Ecohydrology* 1(1):1–2. doi: 10.1002/eco.2.
117. Smith, M.K. 2001. Chris Argyris: Theories of action, double-loop learning and organizational learning. *The Encyclopedia of Informal Education*. www.infed.org/thinkers/argyris.htm. Accessed December 2012.
118. Smith, R. 2004. A capital-based sustainability accounting framework for Canada. In *Integrated Economic, Environmental, and Social Frameworks*, pp. 111–128. Paris, France: OECD Publishing.
119. Speldewinde, P.C., A. Cook, P. Davies, and P. Weinstein. 2011. The hidden health burden of environmental degradation: Disease comorbidities and dryland salinity. *EcoHealth* 8(1):82–92. doi: 10.1007/s10393–011–0686-x.
120. Spray, C., T. Ball, and J. Rouillard. 2009. Bridging the water law, policy, science interface: Flood risk management in Scotland. *Water Law* 20(2/3):165–174.
121. Strang, V. 2006. *Social Ecohydrology: Integrating the Social and Natural Sciences in Water Management*. Auckland, New Zealand: University of Auckland.
122. Strang, V. 2009. Integrating the social and natural sciences in environmental research: A discussion paper. *Environmental Development Sustainability* no. 11:1–18. doi: 10.1007/s10668-007-9095-2.
123. Sudmeyer, R.A., D. Bicknell, and N.A. Coles. 2007. *Tree Windbreaks in the Wheatbelt, Bulletin. No 4723*, p. 27. Perth, Western Australia, Australia: Western Australia Department of Agriculture and Food.
124. Surridge, B. and B. Harris. 2007. Science-driven integrated river basin management: A mirage? *Interdisciplinary Science Reviews* 32(3):298–312. doi: 10.1179/030801807X211711.
125. ten Brink, P., L. Mazza, T. Badura, M. Kettunen, and S. Withana. 2012. *Nature and Its Role in the Transition to a Green Economy*, (ed.) IEEP. Brussels, Belgium: IEEP.
126. Tress, B., G. Tress, and G. Fry. 2006. Defining concepts and process of knowledge production in integrative research. In *From Landscape Research to Landscape Planning: Aspects of Integration, Education and Application*, (eds.) B. Tress, G. Tress, G. Fry and P. Opdam, pp. 434. Dordrecht, the Netherlands: Springer.
127. UN Water. 2008. *Transboundary Waters: Sharing Benefits, Sharing Responsibilities*. In *Thematic Paper*, (ed.) Task Force on Transboundary Waters. New York: UNDP.
128. UNESCO-IHP. 2011. *Ecohydrology for Sustainability*. Paris, France. UNESCO-IHP.
129. UNWWAP. 2009. *The United Nations World Water Development Report 3: Water in a Changing World*, (ed.) Earthscan. Paris, France London, U.K.: UNESCO.
130. US C. 1972. An act to amend the Federal Water Pollution Control Act. PUBLIC LAW 92–500-October18.1972.
131. Vereecken, H., T. Kamai, T. Harter, R. Kasteel, J. Hopmans, and J. Vanderborght. 2007. Explaining soil moisture variability as a function of mean soil moisture: A stochastic unsaturated flow perspective. *Geophysical Research Letters* 34(22):L22402. doi: 10.1029/2007GL031813.
132. Wackernagel, M. and R. William. 1996. *Our Ecological Footprint: Reducing Human Impact on the Earth*. Gabriola Island, BC, and Philadelphia, PA: New Society Publishers.
133. Walsh, C.J. 2000. Urban impacts on the ecology of receiving waters: A framework for assessment, conservation and restoration. *Hydrobiologia* 431:107–114.
134. Walsh, C.J., T.D. Fletcher, and A.R. Ladson. 2005. Stream restoration in urban catchments through redesigning stormwater systems: Looking to the catchment to save the stream. *Journal of the North American Benthological Society* 24(3):690–705. doi: 10.1899/04–020.1.
135. Wang, Y., P. Yu, W. Xiong, Z. Shen, M. Guo, Z. Shi, A. Du, and L. Wang. 2008. Water yield reduction after afforestation and related processes in the semiarid Liupan Mountains, Northwest China. *Journal of the American Water Resources Association* 44(5):1086–1097.

136. Waughray, D. (ed.) 2011. *Water Security: The Water-Food-Energy-Climate Nexus: The World Economic Forum Water Initiative*, (ed.) D. Waughray. Washington, DC: Island Press, World Economic Forum.
137. Williams, J. 2010. *Is Intergrated Catchment Management now a possibility in New South Wales*, Canberra, Australian Capitol Territory, Australia: Wentworth Group.
138. Wipfli, M.S., J.S. Richardson, and R.J. Naiman. 2007. Ecological linkages between headwaters and downstream ecosystems: Transport of organic matter, invertebrates, and wood down headwater channels1. *Journal of the American Water Resources Association* 43(1):72–85. doi: 10.1111/j.1752-1688.2007.00007.x.
139. Wolanski, E. 2007. *Estuarine Ecohydrology*. Amsterdam, the Netherlands: Elsevier.
140. Wood, P.J., D.M. Hannah, and J.P. Sadler. 2007. *Hydroecology and Ecohydrology: Past, Present and Future*. Chichester, U.K.: John Wiley & Sons.
141. WWF. 2004. Living Planet Report 2004. Published by World wild-life fund for nature. Gland Switzerland. Ropress. ISBN 288085265X.
142. Zacharias, S., H. Bogena, L. Samaniego, M. Mauder, R. Fuß, T. Pütz, M. Frenzel et al. 2011. A network of terrestrial environmental observatories in Germany. *Vadose Zone Journal* 10(3):955. doi: 10.2136/vzj2010.0139.
143. Zalewski, M. 2000. Ecohydrology-scientific background to use ecosystem properties as management tools toward sustainability of water resources. *Ecological Engineering* 16:1–8.
144. Zalewski, M., G.A. Janauer, and G. Jolankai. 1997. *Ecohydrology: A New Paradigm for the Sustainable Use pf Aquatic Resources*, in IHP-V Projects 2.3/2.4. Paris, France: UNESCO, IHP.
145. Zalewski, M. and I. Wagner-Lotkowska. (eds.) 2004. *Integrated Watershed Management—Ecohydrology & Phytotechnology—Manual*. Japan: UNESCO-UNEP-IHP.
146. Zhang, L., N. Potter, K. Hickel, Y. Zhang, and Q. Shao. 2008. Water balance modeling over variable time scales based on the Budyko framework—Model development and testing. *Journal of Hydrology* 360(1–4):117–131. doi: http://dx.doi.org/10.1016/j.jhydrol.2008.07.021.
147. Zia, H., N.R. Harris, G.V. Merrett, A. Cranny, M. Rivers, and N.A. Coles. 2012. A review on the impact of catchment-scale activities on water quality: A case for collaborative wireless sensor networks. *Journal of Computers and Electronics in Agriculture* (in review).

7

Ecohydrology: Plant Water Use

Lixin Wang
Indiana University-Purdue University
University of New South Wales

Matthew F. McCabe
King Abdullah University of Science and Technology

AUTHORS

Lixin Wang is an assistant professor in the Department of Earth Sciences of Indiana University–Purdue University Indianapolis (IUPUI). His research focuses on dryland ecohydrology and biogeochemical processes. He received his PhD from the University of Virginia in 2008. Before moving to IUPUI, he was a research associate at Princeton University (2008–2011) and vice-chancellor research fellow at the University of New South Wales (2011–2012).

Matthew F. McCabe is an associate professor at the Water Desalination and Reuse Center, King Abdullah University of Science and Technology (KAUST). His research focuses on remote sensing hydrology and global water/energy cycle. He received his PhD from the University of Newcastle in 2003. Before moving to KAUST, he was an associate professor at the University of New South Wales and a research scientist at Princeton University and Los Alamos National Laboratory.

PREFACE

Water is vital to vegetation function and up to 90% of plant weight is water. The amount of water plants use is substantial but varies drastically for different species. Vegetation water use not only affects the ecosystem water cycle through groundwater uptake and transpiration water loss but also affects regional climate through transpiration cooling. Quantification of plant water use has been an active research area for many years, and multiple methods have been developed to estimate vegetation water use at different scales. In this chapter, common modeling and measuring techniques to estimate plant water use from organ to landscape scale are summarized. The methods range from stomata modeling to landscape evapotranspiration partitioning using stable isotopes. The principle of these methods as well as the advantages and limitations of each method is discussed.

7.1 Introduction

Water is vital to plants, with key cellular activities such as photosynthesis, nutrient transport, and the regulation of plant temperature all requiring, or occurring in, the presence of water molecules. Apart from playing an integral role in biophysical processes, water also helps plants maintain turgor pressure ensuring internal structure and morphology. Up to 90% of plant weight is water and the amount of water plants use is substantial. A mature houseplant can transpire its body weight every day. For natural woody vegetation, water use can cover a broad range: from 10 kg day^{-1} for trees in a 32-year-old *Quercus petraea* plantation in eastern France to 1180 kg day^{-1} for an overstory *Euperua purpurea* tree growing in the Amazonian rainforest (Wullschleger et al., 1998), with use often governed by a combination of need and availability.

The major water input to terrestrial ecosystems is precipitation, although vegetation in riparian zones can utilize additional lateral water and many desert vegetations can access deep groundwater. As a consequence, vegetation plays a critical role in controlling the water cycle across various scales. Indeed, vegetation-based transpiration is the major pathway through which precipitation is recycled back into the atmosphere. Globally, evapotranspiration, the combination of transpiration and soil/canopy evaporation, accounts for 65% of the total precipitation input (Trenberth et al., 2007), and the number can be greater than 90% in drylands (Huxman et al., 2005; Wang et al., 2012). The process of evapotranspiration results in significant cooling of the Earth surface. Indeed, without evapotranspiration, it has been projected that the northern hemisphere would be 15°C–25°C warmer (Shukla and Mintz, 1982).

Given the important role vegetation play in water cycles across scales, vegetation water use has been an area of research in plant physiology, hydrology, and meteorology studies. In this chapter, some common monitoring and modeling tools that are used to estimate plant water use are discussed, moving from the organ level (e.g., leaf and root) to the landscape scale. The aim here is to provide a general overview of the various methods used to monitor plant water use and highlight the advantages and limitations of each approach.

7.2 Leaf Transpiration and Root Water Uptake

Transpiration is the process of evaporation of water from plants into the atmosphere. Water enters a plant through its root system and exits via stomata on the leaf surface, with approximately 90% of the water utilized by a plant lost via transpiration. Transpiration is an essential plant function and plays multiple roles to maintain plant function and form, including the following: (1) transporting water from roots to growing tissues and therefore maintain the turgor of plant cells, (2) delivering mineral nutrients from the soil/root system to leaves, and (3) reducing and regulating the leaf temperature.

In practice, leaf transpiration rates can be directly measured using commercial analyzers such as Licor 6400 (LI-COR Biosciences, Lincoln, NE), which measures air flow rate, water mole fractions, and leaf area inside a leaf chamber. Leaf stomatal conductance, the process that directly controls transpiration, is often modeled empirically. The commonly used Ball–Berry model (Ball et al., 1987) and its modification by Leuning (Leuning, 1990, 1995) is expressed as

$$g = g_o + \frac{A_n a_1}{(C_s - \Gamma)\left(1 + \left(\frac{D_s}{D_o}\right)\right)} \tag{7.1}$$

where

g is the stomatal conductance (mol CO_2 m^{-2} s^{-1})

g_o is the value of g at the light compensation point (mol CO_2 m^{-2} s^{-1})

A_n is the rate of net photosynthesis (mol CO_2 m^{-2} s^{-1})

D_s is the vapor pressure deficit (the difference between saturated vapor pressure at the leaf surface and the actual vapor pressure of the surrounding atmosphere) (dimensionless)
C_s is the CO_2 concentration at the leaf surface (mol CO_2 (mol air)$^{-1}$)
Γ is the CO_2 compensation point (dimensionless)
a_1 and D_o are empirical coefficients (dimensionless)

Ultimately, the model seeks to characterize the environmental (e.g., VPD, CO_2 concentration) and physiological (e.g., photosynthesis rate) controls on stomatal conductance. There are several commercial instruments to conduct field measurements of leaf stomatal conductance. For example, it can be measured by Licor 6400, with additional measurements such as leaf temperature and total atmospheric pressure besides air flow rate, water mole fractions, and leaf area inside leaf chamber, plus boundary layer conductance correction. Alternatively, it can be measured by steady-state porometers such as a Decagon SC-1 Leaf Porometer (Decagon Devices, Inc., Pullman, WA), which uses a sensor head with a fixed diffusion path to the leaf and measures the vapor concentration at two different locations in the diffusion path. It can also be measured by a dynamic porometer such as a Delta AP4 Leaf Porometer (Delta-T Devices Ltd., Cambridge, United Kingdom), which measures diffusion conductance by comparing the precise rate of humidification within a small cuvette to readings obtained with a calibration plate.

Roots are the primary organ responsible for plant water uptake, though recent studies show that leaf water absorption can improve vegetation water status in drylands during drought (e.g., Breshears et al., 2008). The root tip of fine roots presents the most active point of plant water uptake. Water moving within the soil matrix is driven by pressure gradients associated with gravity, osmotic pressure, and matric forces, which collectively determine soil water potential. Water moves from the surrounding soil to plant roots by flowing from high to low water potential. The flow equation for a single root inside an infinite, 2D soil medium is

$$\partial\theta/t = \left(\frac{1}{r}\right)\left(\frac{\partial}{\partial r}\right)\left[rD\left(\frac{\partial\theta}{\partial r}\right)\right] \tag{7.2}$$

where
θ is the soil volumetric water content
D is the diffusivity of the soil
t is time
r is the radial distance from the axis of the root (Kirkham, 2005)

The steady-state solution to determine water movement from soil medium to plant root is

$$\Psi_b - \Psi_a = \left(\frac{q}{4\pi k}\right)\left[\ln\left(\frac{b^2}{a^2}\right)\right] \tag{7.3}$$

where
Ψ_b is the matric potential midway between roots (bars)
Ψ_a is the matric potential at the plant root–soil boundary (bars)
q is the rate of root water uptake per unit root length (cm^3 cm^{-1} s^{-1})
k is the unsaturated hydraulic conductivity of the soil (cm^2/s)
a is the radius of the root
$b = 2(Dt)^{1/2}$ (Gardner, 1960)

For more details, please refer to Kirkham (2005).

Transpiration and root water uptake represent the water fluxes in and out of a plant at the organ level. The pool size of water at organ level is also an important parameter to indicate plant water status. Typically, leaf water content and leaf water potential are good indicators of plant water status. Leaf water content can be attained using gravimetric drying method, while leaf water potential can be quantified using a pressure chamber technique. One example of using a pressure chamber measuring leaf water potential is the Scholander pressure bomb (ICT International, Armidale, NSW, Australia). In this approach, a leaf is cut from a plant and placed inside a pressurized chamber, and a reading is made when a water drop first appears at the cut after slowly increasing the chamber pressure. The operational principle is that the amount of pressure required to force water out of the cut end of the petiole equals the leaf water potential. More details about leaf water and leaf water potential measurements can be found in Turner (1981).

7.3 Plant-Scale Water Use

There are a number of methods and techniques that have been developed and applied to estimate the water use of the whole plant. These include methods such as lysimeters (Figure 7.1; e.g., Allen et al., 1991; Fritschen et al., 1973) that quantify the total water loss from both the soil and vegetation continuum; large-tree potometers/transpirometers (e.g., Knight et al., 1981) that measure the rate of water uptake, ventilated chambers (e.g., Greenwood and Beresford, 1979), whole-tree chamber (Ryan et al., 2000), as well as radio isotope (e.g., tritium, Kline et al., 1970; Waring and Roberts, 1979); and stable isotope approaches (e.g., deuterium, Calder et al., 1986) that trace whole-plant transpiration. As expected, each of these methods has their advantages and intrinsic limitations in quantifying plant water use. For example, potometers have limitations for transpiration measurements. The potometer does not measure transpiration rate accurately because not all of the water that is taken by the plant is used for transpiration (water might be used for photosynthesis or by the cells to maintain turgidity). In addition, potometer measurements tend to be labor-intensive and provide relatively poor time resolution. Whole-tree chambers can measure transpiration of the whole plant, but the application of this method is constrained by its portability issue and the possibility of affecting the microclimate of the measuring plants and thereby influencing the intended measurements. Both radioisotope and stable isotope approaches provide nondestructive ways to trace and calculate plant water use. However, partially due to environmental concerns, tritium-based plant water use studies are less frequently seen in the recent literature. Stable isotope applications have been applied more widely,

(a) (b)

FIGURE 7.1 (a) A diagram and (b) a field photo of lysimeter setup for evapotranspiration measurements.

but the number of studies using stable isotopes to directly calculate transpiration rates is somewhat limited (Calder, 1992), compared with studies using stable isotopes to differentiate evaporation and transpiration (Wang et al., 2010; Yepez et al., 2003) (see Section 7.4.2).

Sap flow measuring techniques have become widely used to quantify the whole-plant water use due to their advantage of continuous measurements and automated setup (Smith and Allen, 1996). The sap flow methods have several variations and include stem heat balance, trunk sector heat balance, and heat-pulse and heat dissipation approaches. All of these methods use heat as a tracer for sap movement, whereby the stem is heated electrically and the heat balance is solved for the amount of heat taken up by the moving sap stream. This heat component is then used to estimate the mass flow of sap in the stem. However, these methods have different operating principles and practical considerations. A detailed review of operational theory and practical considerations can be found in Smith and Allen (1996), but a brief summary of the operating principles of these methods is provided in the succeeding text.

In the heat-pulse method, short pulses of heat are applied and the sap mass flow is determined from the velocity of the heat pulses moving along the stem. The obvious advantage of this method is a lower power consumption compared with continuous heating. However, because the heater and sensors (two sensors are typically used) must be installed by drilling holes into the sapwood, the heat-pulse method is only really suitable for a range of woody stems. The diameter of the stems has to be large enough to accommodate the drilling but not too large to access the full depth of the sapwood. Typically, it is used on tree diameters larger than 30 mm (Smith and Allen, 1996). For the heat-pulse method, each set of heat-pulse probes comprises one heater and two sensor probes, which are above (downstream) and below (upstream) the heater probe at unequal distances, with the upstream one typically closer to the heater probe. All of the probes are put radially into the stem. When a heat pulse is released, the upstream sensor is heated up earlier because of the closer distance and the heat carried by the moving sap quickly warm the downstream sensor. The temperature of the two sensors will tend to equalize after a certain period of time. This time is related to the velocity of the heat pulse and can be converted into velocity of sap flow (sap flux). Mass flow of the sap through the stem (F_m) is the integration of sap flux over the cross section of the sapwood, which is calculated as

$$F_m = \rho_s 2\pi \int_h^R r_d \mu'_v(r_d) \mathrm{d}r_d \tag{7.4}$$

where

ρ_s is the wood density
h is the heartwood boundary
R is the cambium boundary
r_d is the radial depth
μ'_v (r_d) is the sap flux density as a function of r_d
μ_v is calculated from the corrected heat-pulse velocity

The stem heat balance and trunk sector heat balance methods use the heat balance principle with a continuous heating. They are based on the principle that sap flow rates can be determined from the heat balance of the heated stem tissue. The difference between the two methods is that heat is applied to the entire circumference for stem heat balance method, but only to a segment of the trunk for the trunk sector method. As a consequence of the heating location difference, the stem heat balance method can be used to measure sap flux in both herbaceous and woody plants. The trunk sector heat balance method on the other hand is intended for trees with a diameter larger than 120 mm (Smith and Allen, 1996). For the stem heat balance method, F_m is calculated as

$$F_m = \frac{2(P - q_v - q_r)}{C_s(\Delta T_a + \Delta T_b)} \tag{7.5}$$

where

P is the total heat provided to the stem
q_v is the rate of vertical heat loss by conduction in the stem
C_s is the specific heat capacity of sap
$(\Delta T_a + \Delta T_b)/2$ is the sap temperature increase across the heater

For the trunk sector heat balance method, F_m is calculated as

$$F_m = \frac{P - q_v - q_r - q_l}{C_s \Delta T} \tag{7.6}$$

where

q_l is the heat loss laterally into neighboring sapwood by conduction
ΔT is the sap temperature increase across the heater

The heat dissipation method is an empirical method, which has an easier setup than the aforementioned heat flux approaches. In the heat dissipation method, two probes are used, with the upper one containing a heater element with constant heating and a thermocouple junction, while the lower probe contains another junction (Granier, 1987). The temperature change is empirically related to the volumetric sap flux, and F_m is calculated as

$$F_m = \rho_s A_{sw} 0.000119 \left(\frac{\Delta T_0 - \Delta T}{\Delta T} \right)^{1.231} \tag{7.7}$$

where

A_{sw} is the area of sapwood cross section
ΔT_0 is the sap temperature change when there is no sap flow

7.4 Large-Scale Vegetation Water Use

To better understand the influence of vegetation on water resources, individual plant-based water use ultimately needs be converted into plant water use at the stand or larger scale. The measurements of large-scale vegetation water use have always been a challenging task. Both bottom-up and top-down methods have been used to achieve this goal. Bottom-up methods involve the scaling up of individual plant measurement (e.g., chamber measurements, sap flow measurements), whereas top-down approaches measure total evapotranspiration (e.g., eddy-covariance method) and seek to partition its contributing components.

7.4.1 Bottom-Up Methods

Bottom-up methods basically upscale the individual tree-based transpiration measurements (e.g., tree sap flux, whole-tree chamber observations, refer to Section 7.3 for more details) to the stand scale using different scaling functions. The volume or mass flow rates (e.g., $g\ H_2O\ m^{-2}\ s^{-1}$) of the individual plants need to be converted into transpiration rate per unit ground area (Oren et al., 1998). Upscaling typically consists of two steps: distributing the individual-scale measurements over the interested area and aggregating the spatial distribution of the measurements into one single value (Blöschl and Sivapalan, 1995). In terms of sap flux scaling, the flux measurements need to be scaled to the tree scale using tree sapwood area, tree circumference, or projected crown area, which can then be scaled up to the stand scale. The scaling issue is relatively simple for uniform stands (e.g., monoculture forest plantations), though variability in the canopy gaps and leaf area is still an important factor. If vegetation structure is more complex (e.g., trees of varying size and structure), significantly more sampling covering the whole range of heterogeneous units are often needed. Alternatively, quantitative relationship between sap flow rate and tree geometry (e.g., stem diameter, leaf area) could be established and used in the scaling (Smith and Allen, 1996).

7.4.2 Top-Down Methods

7.4.2.1 Large-Scale Evapotranspiration Estimation: Measurements and Modeling

There are diverse methods available to quantify large-scale evapotranspiration. While there are a number of detailed physical descriptions that seek to detail the plant physiological constraints on water loss (e.g., soil–vegetation–atmosphere transfer [SVAT] schemes, Arora, 2002), the modeling technique most commonly used to estimate water loss from ecosystems is the Penman–Monteith equation, based largely on simplifications of Monin–Obukhov similarity theory (MOST) (Wang and Dickinson, 2012). MOST provides a basis from which to estimate latent heat flux (*LE*) and sensible heat fluxes (*H*) by relating turbulence fluxes to gradients of mean wind, temperature, and humidity (Monin and Obukhov, 1954). It is valid for horizontally homogeneous and stationary surface layers.

From an observational perspective, the approaches typically applied include the Bowen ratio, eddy-covariance, and scintillometer-based techniques. These approaches can provide both diurnal and long-term evapotranspiration fluxes. Other methods such as remote sensing–related techniques and water balance method can provide larger spatial coverage for evapotranspiration estimates, but sometimes at an expense of temporal scale (e.g., while providing daily and subdaily evapotranspiration, the water balance approach for evapotranspiration estimation is typically at a monthly to yearly timescale).

The Penman–Monteith equation predicts net evapotranspiration, requiring a minimum input of daily mean temperature, wind speed, relative humidity, and solar radiation. The typical form is

$$LE = \frac{\Delta(R_n - G) + \rho c_p g_a \left(e_s(T_a) - e\right)}{\Delta + \gamma\left(1 + \left(\frac{g_a}{g_s}\right)\right)} \tag{7.8}$$

where

R_n is the total radiation input in W m^{-2}
G is the ground heat flux in W m^{-2}
Δ is the rate of change of saturation specific humidity with air temperature (Pa K^{-1})
γ is the psychrometric constant (~66 Pa K^{-1})
ρ is the dry air density (kg m^{-3})
g_a is the atmospheric conductance (m s^{-1})
g_s is the surface conductance (m s^{-1})
c_p is the specific heat capacity of air (J kg^{-1} K^{-1})
$e_s(T_a)$ is the saturated vapor pressure at temperature T_a in Pa
e is the actual vapor pressure in Pa
$e_s(T_a)-e$ is the VPD (Pa)

The sensitivity analyses of Penman–Monteith parameters to different environmental conditions have been investigated in a number of studies. It was found that the parameter sensitivity depends on many factors such as vegetation type and soil water status but in general aerodynamic and canopy resistance present as the most sensitive parameters (Beven, 1979; Rana and Katerji, 1998).

The Bowen ratio method uses simultaneous measurements of vertical gradients of T_a and humidity (q) to partition the surface available energy to sensible heat flux and latent heat flux (Bowen, 1926). The method is based on the surface energy balance, requiring radiation and ground heat flux measurement as well as environmental variables such as temperature and vapor pressure. It is typically quantified using

$$LE = \frac{R_n - G}{1+\beta} \tag{7.9}$$

where

LE is the latent heat flux (evapotranspiration) in W m^{-2}
R_n is the total radiation input in W m^{-2}
G is the ground heat flux in W m^{-2}
β is the Bowen ratio

For a particular system, the Bowen ratio can be estimated using

$$\beta = \gamma \frac{T_2 - T_1 + \Gamma(z_2 - z_1)}{e_2 - e_1} \tag{7.10}$$

where

γ is the psychrometric constant
T_2 and T_1, e_2 and e_1 are the measured air temperature (°C) and vapor pressure (Pa) at heights z_2 and z_1 (m)
Γ is the adiabatic lapse rate in Km^{-1}

The eddy-covariance method quantifies turbulent heat fluxes by considering the high frequency of 3D wind speed and water vapor concentration variations and is calculated as

$$ET = \frac{\overline{w'q'}}{\rho w} \tag{7.11}$$

where

w' and q' are the deviations from the mean vertical wind and absolute humidity in m s^{-1} and g m^{-3}, respectively
ρw is the density of air in kg m^{-3}

Eddy-covariance is by far the most common technique to quantify the evapotranspiration flux at ecosystem to landscape scale and is widely adapted in the global monitoring networks (e.g., FLUXNET) (Baldocchi et al., 2001).

The scintillometer system is based on the principal of scintillation, which is the intensity fluctuation in electromagnetic wave when passing through a turbulent atmosphere (Figure 7.2). While the phenomenon has been known for a long time, its application to evapotranspiration estimation is fairly recent, and it is a newer method compared with the Bowen ratio and eddy-covariance methods. One of the advantages of scintillometers over other approaches is the large spatial coverage (10 km^2). The evapotranspiration can be determined based on C_T (the temperature structure parameter) as

$$LE = \frac{AC_T^{3/2}}{\beta} \tag{7.12}$$

where

A is a constant, depending on temperature and boundary layer parameters
β is the Bowen ratio (Gieske, 2003; Kohsiek, 1982)

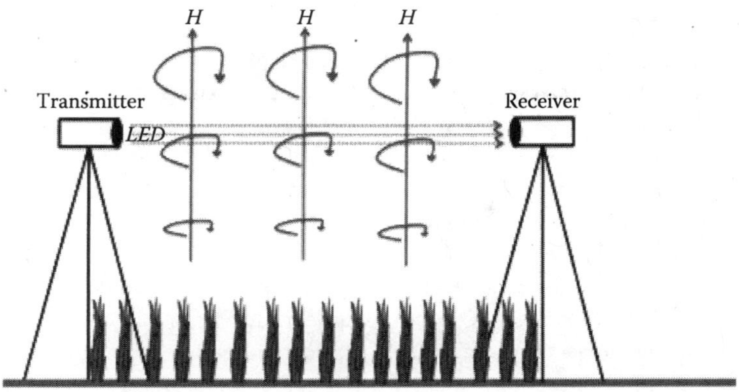

FIGURE 7.2 A diagram of scintillometer measurements of sensible heat flux (*H*). Light from a light-emitting diode (*LED*) is bundled into a parallel beam. The light signal is amplified by a receiver to produce a signal that is representative of changes in the refractive index of the atmosphere. The changes in the refractive index of the atmosphere can be converted into *H*.

C_T can be calculated using

$$C_T = 10^6 \frac{C_n T^2}{0.78p}\left(1 + \frac{0.03}{\beta}\right) \tag{7.13}$$

where

- C_n is the refractive index structure parameter, which is directly calculated from the light signal of the scintillometer
- T is the temperature (K)
- p is the atmospheric pressure (bar)
- β is the Bowen ratio (Gieske, 2003)

For scales larger than watershed, remote sensing and water balance approaches (surface water balance or atmospheric water balance) have been shown to be useful in estimating net ecosystem evaporative flux. A variety of methods have been developed to process remotely sensed data for evapotranspiration retrieval at regional to global scales (Kalma et al., 2008). While satellite remote sensing provides reasonable estimates of land surface variables, they do not measure evapotranspiration directly. Therefore, most satellite evapotranspiration algorithms relate satellite-derived land surface variables (e.g., surface temperature and air temperature) to evapotranspiration using either empirically based approaches or the Penman–Monteith equation (Wang and Dickinson, 2012). Various global-scale products are currently available (e.g., multisatellite global land surface evapotranspiration from the Vrije University of Amsterdam). For a summary and assessment of the global evapotranspiration product and variability in flux retrieval approaches, please refer to Jimenez (2011) and Mueller et al. (2011).

The surface water balance approach provides another means to estimate evapotranspiration, using a simple mass balance of the net water exchange within a system:

$$ET = P - R - \frac{dw}{dt} \tag{7.14}$$

where

- P is the precipitation
- R is the runoff/river discharge
- dw/dt is the water storage change

Evapotranspiration can be estimated at monthly to yearly timescale.

7.4.2.2 Evapotranspiration Partitioning

The evapotranspiration includes several components such as plant transpiration, canopy/soil evaporation and plant interception. Quantifying large-scale vegetation water use requires separating evapotranspiration into its corresponding components, a process referred to as evapotranspiration partitioning. Evapotranspiration partitioning, especially at large spatial scales, remains both an observational and theoretical challenge (Huxman et al., 2005; Wang et al., 2010), due mainly to the lack of methodologies available to quantify large-scale evaporation or transpiration in an easy and reliable way.

Among other methods, stable water isotopes provide a powerful tool to separate evaporation and transpiration, as evaporation and transpiration carry distinct isotopic signatures (Gat, 1996). Both hydrogen (^{2}H and H) and oxygen (^{16}O, ^{17}O, and ^{18}O) have multiple naturally occurring stable isotopes that are of use in tracing elements of the hydrological cycle. The isotopic composition is defined as $\delta = (R/R_{vsmow} - 1)$, where δ is the reported water vapor isotope composition (δ^{18}O or δ^2H), and R and R_{vsmow} are the atomic ratios of heavy/light isotope in samples and the international standard (VSMOW), respectively. The process of evaporation is accompanied by a high degree of isotopic fractionation that leads to evaporated water with an isotopic composition (δ_E) that is depleted in heavy isotopes (Craig and Gordon, 1965). At the same time, the rapid turnover of water in transpiring leaves means that the signature of transpiration is usually very similar to the isotopic composition of plant source water, especially during midday (Ehleringer and Dawson, 1992). While some isotopic enrichment can occur in the leaf due to the same kinetic and diffusive effects that lead to evaporative fractionation in soils (Flanagan et al., 1991), these nonsteady-state leaf-scale effects usually occur only during early morning hours (Flanagan et al., 1991). Therefore, the isotopic composition of transpiration is always much heavier than the isotopic composition of evaporation (Craig and Gordon, 1965), and the distinct isotopic signature of these two fluxes can be used to partition total evapotranspiration into relative rates of evaporation and transpiration in landscapes.

Assuming a simple two-source model of total evapotranspiration, the fractional contribution of transpiration to total evapotranspiration can be quantified as

$$\frac{T}{ET} = \frac{\delta_{ET} - \delta_E}{\delta_T - \delta_E} \tag{7.15}$$

where

T and ET represent transpiration and evapotranspiration, respectively

δ_{ET}, δ_E, and δ_T are the isotope signatures of evapotranspiration, evaporation, and transpiration, respectively

The stable water isotopic compositions are traditionally measured by the isotope ratio mass spectrometry (IRMS), while the vapor-phase measurements are usually based on cryogenic water vapor collection coupled with the IRMS method. Such methods are labor-intensive and time-consuming, as ^{18}O measurements require offline CO_2–H_2O equilibrium. The typical vapor equilibrium takes about 24 h. Over the past decade, a revolution in isotope measurements has been realized through the development of spectroscopy-based isotope instruments capable of making continuous measurements of water vapor isotopic compositions. This new type of instrument does not usually require pretreatment and have precisions similar to traditional cryogenic-based mass spectrometry methods (Griffis et al., 2010; Lee et al., 2005; Wang et al., 2009; Wen et al., 2008). The continuous measurement of water vapor isotope compositions allows for the direct quantification of the water vapor and will expand the capabilities of using stable isotopes for ecohydrological insight.

Recent efforts have focused on developing, refining, and assessing estimation methods of all three isotopic components (δ_{ET}, δ_E, and δ_T) using the newly developed spectroscopy-based monitoring techniques. δ_{ET} is typically measured using the Keeling plot approach coupled with traditional cryogenic methods (e.g., Yepez et al., 2005). In the Keeling plot approach, δ_{ET} is calculated as

$$\delta^2 H_M = c_A\left(\delta^2 H_A - \delta^2 H_s\right)\left(\frac{1}{c_M}\right) + \delta^2 H_s \tag{7.16}$$

where

$\delta^2 H_M$, $\delta^2 H_A$, and $\delta^2 H_s$ are the isotope signatures of observed water vapor (mixed), ambient background water vapor, and evapotranspiration, respectively
c_A is the background water vapor concentration
c_M is the observed water vapor concentration

With the development of new laser techniques, Wang et al. (2013) extended this technique to direct chamber measurements coupling with laser instruments. Good et al. (2012) quantified and compared the uncertainties using different δ_{ET} estimation methods at tower scale and showed that the eddy-covariance method has the largest uncertainties, while the Keeling plot and flux gradient methods have smaller and similar uncertainties.

δ_E is typically calculated by the Craig–Gordon model (Craig and Gordon, 1965), which describes δ_E as a function of humidity, kinetic, and equilibrium isotope fractionation, the isotopic composition of water of the evaporation surface, and atmospheric vapor. Specifically, in the following Craig–Gordon model,

$$\delta_E = \frac{\alpha\delta_L - h\delta_A - (1-\alpha)10^3 - \varepsilon_K}{(1-h) + \varepsilon_K 10^{-3}} \tag{7.17}$$

where

α is the temperature-dependent equilibrium fractionation factor ($\alpha < 1$ for liquid–vapor transformation), which can be calculated based on soil temperature
δ_L is the isotopic composition of liquid water at the evaporating front
δ_A is the isotopic composition of the background atmospheric water vapor
ε^* is calculated as $(1-\alpha) \times 1000$
ε_K is the kinetic fractionation factor for hydrogen
h is the relative humidity normalized to the soil temperature

Typically, δ_T is assessed indirectly using stem water measurements or leaf water measurements with leaf enrichment corrections. The use of stem water measurements is based on the assumption that leaves operate under isotopic steady state, so that δ_T is equal to the isotopic composition of plant source water. The assumption that leaves are in isotopic steady state is generally valid during midday because the magnitude of transpiration relative to the volume of leaf water is large and there is a rapid turnover of water in transpiring leaves. However, non-steady-state isotopic enrichment is also common in many natural systems, especially during the early morning and late afternoon (e.g., Lai et al., 2005), when transpiration fluxes are lower. Taking advantage of the new laser technique, Wang et al. (2010) reported the first direct measurements of δ_T using a customized leaf chamber and off-axis integrated cavity output spectroscopy (OA-ICOS) water vapor isotope analyzer with pure nitrogen as purging gas. This method, however, has two limitations: (1) the limited availability of ultra-purity nitrogen gas makes this method unsuitable for many field applications, and (2) the water-free and CO_2-free inline environment affects stomata openings since humidity and CO_2 have opposite effect on stomata openings, which may alter instantaneous δ_T values. Building on this configuration, Wang et al. (2012) developed a new framework to remove the need for dry air by employing a mass balance approach for both isotopes and water vapor inside the leaf chamber (Figure 7.3). This method can measure both instantaneous and steady-state δ_T. The steady-state solution of the δ_T estimate is

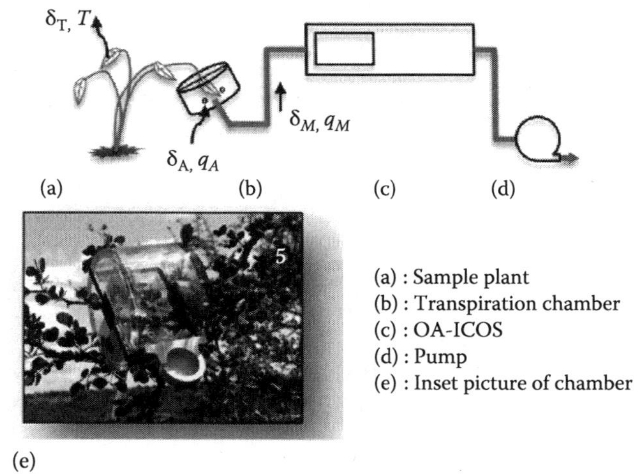

FIGURE 7.3 Schematic of the experimental setup for directly quantifying leaf transpiration isotopic composition. (a) Leaves of the plant to be sampled are placed in (b) the transpiration chamber, which is connected directly to (c) the water vapor isotope analyzer (OA-ICOS) and (d) a vacuum pump. The part (e) shows a photo of the chamber setup. (Modified from Wang, L. et al., *Agr. Forest Meteorol.*, 154, 127, 2012.)

$$\delta^2 H_s = \frac{C_M \delta_M - C_A \delta_A}{C_M - C_A} \tag{7.18}$$

where

$\delta^2 H_s$ is the isotope signatures of evapotranspiration (‰)

C_A and C_M are the water vapor concentrations inside the leaf chamber (mol m^{-3})

δ_A and δ_M are the isotope signatures of water vapor inside the leaf chamber (‰)

This direct and continuous δ_T quantification method has been shown to effectively capture the fast δ_T responses to radiation variations (Wang et al., 2012).

7.5 Summary and Conclusions

Significant advances have been made to understand plant water use and their controls from leaf to landscape scale, with a corresponding range of modeling and monitoring techniques to facilitate such advances. In this chapter, common approaches to model and measure plant water use at various scales are summarized. The strength and limitations of these different methods are discussed. The organ-scale water flux (e.g., leaf transpiration) can be measured accurately using several commercially available instruments (e.g., Licor and Decagon devices) and modeling approaches also adequately reproduce the observed patterns. Plant-scale water use is more challenging but has been routinely measured in many cases using sap flow or whole-tree chamber techniques. The most challenging issue is the large-scale (e.g., landscape-scale) plant water use measurements and how to scale from individual tree measurements to regional and global scales or how to partition large-scale measurements into their respective components, which remain active research areas. In terms of plant water use at scales larger than the watershed, remote sensing–based techniques are one of the most powerful tools available, and new products such as vegetation optical depth (VOD) from several satellite algorithms provide exciting opportunities to monitor the regional- to global-scale evapotranspiration fluxes. Stable isotope–based partitioning techniques, at the same time, provides a powerful tool to partition the large-scale evapotranspiration fluxes into evaporation and transpiration fluxes, thus providing information on large-scale vegetation water use. Not unique to plant water use, but equivalently important in this area, scale and scaling are always confounding and critical issues. We briefly discussed the scaling of individual tree-based sap flux measurements to stand-scale plant water use; more scale and scaling issues can be referred to in other reviews and syntheses (Blöschl and Sivapalan, 1995).

References

Allen, R., Howell, T., Pruitt, W., Walter, I., and Jensen, M. 1991. *Lysimeters for Evapotranspiration and Environmental Measurements,* ASCE, New York.

Arora, V. 2002. Modeling vegetation as a dynamic component in soil-vegetation-atmosphere transfer schemes and hydrological models. *Reviews of Geophysics*, 40(2): 1006.

Baldocchi, D., Falge, E., Gu, L., Olson, R., Hollinger, D., Running, S., Anthoni, P. et al. 2001. FLUXNET: A new tool to study the temporal and spatial variability of ecosystem, ÄìScale carbon dioxide, water vapor, and energy flux densities. *Bulletin of the American Meteorological Society*, 82(11): 2415–2434.

Ball, J., Woodrow, I., and Berry, J. 1987. A model predicting stomatal conductance and its contribution to the control of photosynthesis under different environmental conditions, in *Progress in Photosynthesis Research*, Biggins, J., (ed.), Martinus Nijhoff, Zoetermeer, the Netherlands.

Beven, K. 1979. A sensitivity analysis of the Penman-Monteith actual evapotranspiration estimates. *Journal of Hydrology*, 44(3): 169–190.

Blöschl, G. and Sivapalan, M. 1995. Scale issues in hydrological modelling: A review. *Hydrological Processes*, 9(3–4): 251–290.

Bowen, I. 1926. The ratio of heat losses by conduction and by evaporation from any water surface. *Physical Review*, 27(6): 779–787.

Breshears, D.D., McDowell, N.G., Goddard, K.L., Dayem, K.E., Martens, S.N., Meyer, C.W., and Brown, K.M. 2008. Foliar absorption of intercepted rainfall improves woody plant water status most during drought. *Ecology*, 89(1): 41–47.

Calder, I., Narayanswamy, M.N., Srinivasalu, N.V., Darling, W.G., and Lardner, A.J. 1986. Investigation into the use of deuterium as a tracer for measuring transpiration from eucalypts. *Journal of Hydrology*, 84: 345–351.

Calder, I.R. 1992. Deuterium tracing for the estimation of transpiration from trees Part 2. Estimation of transpiration rates and transpiration parameters using a time-averaged deuterium tracing method. *Journal of Hydrology*, 130(1): 27–35.

Craig, H. and Gordon, L.I. 1965. Deuterium and oxygen-18 variations in the ocean and marine atmosphere, in *Proceedings of the Conference on Stable Isotopes in Oceanographic Studies and Paleotemperatures*, Tongiorgi, E. (ed.), Laboratory of Geology and Nuclear Science, Pisa, Italy.

Ehleringer, J. and Dawson, T. 1992. Water uptake by plants: Perspectives from stable isotope composition. *Plant Cell Environment*, 15: 1073–1082.

Flanagan, L., Comstock, J., and Ehleringer, J. 1991. Comparison of modeled and observed environmental influences on the stable oxygen and hydrogen isotope composition of leaf water in *Phaseolus vulgaris. Plant Physiology*, 96: 588–596.

Fritschen, L., Cox, L., and Kinerson, R. 1973. A 28-meter Douglas-fir in a weighing lysimeter. *Forest Science*, 19: 256–261.

Gardner, W. 1960. Dynamic aspects of water availability to plants. *Soil Science*, 89(2): 63–73.

Gat, J. 1996. Oxygen and hydrogen isotopes in the hydrologic cycle. *Annual Review of Earth and Planetary Sciences*, 24: 225–262.

Gieske, A.S. 2003. Operational solutions of actual evapotranspiration, in *Understanding Water in a Dry Environment-Hydrological Processes in Arid and Semi-Arid Zones*, Simmers, I. (ed.), A.A. Balkema Publishers, Lisse, the Netherlands.

Good, S., Soderberg, K., Wang, L., and Caylor, K. 2012. Uncertainties in the assessment of the isotopic composition of surface fluxes: A direct comparison of techniques using laser-based water vapor isotope analyzers. *Journal of Geophysical Research*, 117: D15301.

Granier, A. 1987. Evaluation of transpiration in a Douglas-fir stand by means of sap flow measurements. *Tree Physiology*, 3: 309–320.

Greenwood, E. and Beresford, J. 1979. Evaporation from vegetation in landscapes developing secondary salinity using the ventilated-chamber technique. I. Comparative transpiration from juvenile *Eucalyptus* above saline ground-water seeps. *Journal of Hydrology*, 42: 369–382.

Griffis, T.J., Sargent, S.D., Lee, X., Baker, J.M., Greene, J., Erickson, M., Zhang, X. et al. 2010. Determining the oxygen isotope composition of evapotranspiration using eddy covariance. *Boundary-Layer Meteorology*, 137(2): 307–326.

Huxman, T., Wilcox, B., Breshears, D.D., Scott, R.L., Snyder, K.A., Small, E.E., Hultine, K., Pittermann, J., and Jackson, R. 2005. Ecohydrological implications of woody plant encroachment. *Ecology*, 86(2): 308–319.

Jimenez, C. 2011. Global inter-comparison of selected 1993–1995 monthly averaged land surface heat fluxes. *Journal of Geophysical Research*, 116.

Kalma, J., McVicar, T., and McCabe, M. 2008. Estimating land surface evaporation: A review of methods using remotely sensed surface temperature data. *Surveys in Geophysics*, 29: 421–469.

Kirkham, M. 2005. *Principles of Soil and Plant Water Relations*, Elsevier Academic Press, Amsterdam, the Netherlands.

Kline, J.R., Martin, J.R., Jordan, C.F., and Koranda, J.J. 1970. Measurement of transpiration in tropical trees with tritiated water. *Ecology*, 5: 1068–1073.

Knight, D.H., Fahey, T.J., Running, S.W., Harrison, A.T., and Wallace, L.L. 1981. Transpiration from 100-yr-old lodgepole pine forests estimated with whole-tree potometers. *Ecology*, 62(3): 717–726.

Kohsiek, W. 1982. Measuring CT2, CQ2, and CTQ in the unstable surface layer, and relations to the vertical fluxes of heat and moisture. *Boundary-Layer Meteorology*, 24(1): 89–107.

Lai, C., Ehleringer, J., Bond, B., and Paw, U.K. 2005. Contributions of evaporation, isotopic non-steady state transpiration and atmospheric mixing on the delta ^{18}O of water vapour in Pacific Northwest coniferous forests. *Plant Cell and Environment*, 29: 77–94.

Lee, X., Sargent, S., Smith, R., and Tanner, B. 2005. In situ measurement of the water vapor $^{18}O/^{16}O$ isotope ratio for atmospheric and ecological applications. *Journal of Atmospheric and Oceanic Technology*, 22: 555–565.

Leuning, R. 1990. Modelling stomatal behaviour and photosynthesis of *Eucalyptus grandis*. *Australian Journal of Plant Physiology*, 17(2): 159–175.

Leuning, R. 1995. A critical appraisal of a combined stomatal-photosynthesis model for C_3 plants. *Plant Cell and Environment*, 18: 339–355.

Monin, A.S. and Obukhov, A.M. 1954. Basic laws of turbulent mixing in the ground layer of the atmosphere (in Russian). *Tr. Akad. Nauk SSSR Geophiz. Inst.*, (Contrib. Geophys. Inst. Acad. Sci. USSR) 24(151): 163–187.

Mueller, B., Seneviratne, S.I., Jimenez, C., Corti, T., Hirschi, M., Balsamo, G., Ciais, P. et al. 2011. Evaluation of global observations-based evapotranspiration datasets and IPCC AR4 simulations. *Geophysical Research Letters*, 38: L06402.

Oren, R., Phillips, N., Katul, G., Ewers, B., and Pataki, D. 1998. Scaling xylem sap flux and soil water balance and calculating variance: A method for partitioning water flux in forests. *Annales Des Science Forestieres*, 55(1–2): 191–216.

Rana, G. and Katerji, N. 1998. A measurement based sensitivity analysis of the Penman-Monteith actual evapotranspiration model for crops of different height and in contrasting water status. *Theoretical and Applied Climatology*, 60(1): 141–149.

Ryan, M.G., Bond, B.J., Law, B.E., Hubbard, R.M., Woodruff, D., Cienciala, E., and Kucera, J. 2000. Transpiration and whole-tree conductance in ponderosa pine trees of different heights. *Oecologia*, 124(4): 553–560.

Shukla, J. and Mintz, Y. 1982. Influence of land-surface evapotranspiration on the Earth's climate. *Science*, 215(4539): 1498–1501.

Smith, D.M. and Allen, S.J. 1996. Measurement of sap flow in plant stems. *Journal of Experimental Botany*, 47(12): 1833–1844.

Trenberth, K.E., Smith, L., Qian, T., Dai, A., and Fasullo, J. 2007. Estimates of the global water budget and its annual cycle using observational and model data. *Journal of Hydrometeorology*, 8: 758–769.

Turner, N.C. 1981. Techniques and experimental approaches for the measurement of plant water status. *Plant and Soil*, 58(1): 339–366.

Wang, K. and Dickinson, R.E. 2012. A review of global terrestrial evapotranspiration: Observation, modeling, climatology, and climatic variability. *Review in Geophysics*, 50(2): RG2005.

Wang, L., Caylor, K., and Dragoni, D. 2009. On the calibration of continuous, high-precision $\delta^{18}O$ and $\delta^{2}H$ measurements using an off-axis integrated cavity output spectrometer *Rapid Communications in Mass Spectrometry*, 23: 530–536.

Wang, L., Caylor, K.K., Villegas, J.C., Barron-Gafford, G.A., Breshears, D.D., and Huxman, T.E. 2010. Evapotranspiration partitioning with woody plant cover: Assessment of a stable isotope technique. *Geophysical Research Letters*, 37: L09401.

Wang, L., D'Odorico, P., Evans, J., Eldridge, D., McCabe, M., Caylor, K., and King, E. 2012. Dryland ecohydrology and climate change: Critical issues and technical advances. *Hydrology and Earth System Sciences*, 16: 2585–2603.

Wang, L., Good, S.P., Caylor, K.K., and Cernusak, L. 2012. Direct quantification of leaf transpiration isotopic composition. *Agricultural and Forest Meteorology*, 154–155: 127–135, doi:10.1016/j.agrformet.2011.10.018.

Wang, L., Niu, S., Good, S., Soderberg, K., Zhou, X., Xia, J., Sherry, R., Luo, Y., Caylor, K., and McCabe, M. 2013. The effect of warming on grassland evapotranspiration partitioning using laser-based isotope monitoring techniques. *Geochimica et Cosmochimica Acta*, 111: 28–38.

Waring, R. and Roberts, J. 1979. Estimating water flux through stems of Scots pine with tritiated water and phosphorus-32. *Journal of Experimental Botany*, 30: 459–471.

Wen, X., Sun, X., Zhang, S., Yu, G., Sargent, S., and Lee, X. 2008. Continuous measurement of water vapor D/H and $^{18}O/^{16}O$ isotope ratios in the atmosphere. *Journal of Hydrology*, 349: 489–500.

Wullschleger, S., Meinzer, F., and Vertessy, R. 1998. A review of whole-plant water use studies in trees. *Tree Physiology*, 18: 499–512.

Yepez, E., Williams, D., Scott, R., and Lin, G. 2003. Partitioning overstory and understory evapotranspiration in a semiarid savanna woodland from the isotopic composition of water vapor. *Agricultural and Forest Meteorology*, 119(1–2): 53–68,10.1016/S0168–1923(03)00116–3.

Yepez, E.A., Huxman, T.E., Ignace, D.D., English, N.B., Weltzin, J.F., Castellanos, A.E., and Williams, D.G. 2005. Dynamics of transpiration and evaporation following a moisture pulse in semiarid grassland: A chamber-based isotope method for partitioning flux components. *Agricultural and Forest Meteorology*, 132: 359–376.

8

Evapotranspiration and Water Consumption

Sadiq Ibrahim Khan
The University of Oklahoma

Yang Hong
The University of Oklahoma

Wenjuan Liu
Ningxia University

AUTHORS

Sadiq Ibrahim Khan received his PhD from the Department of Geography and Environmental Sustainability at the University of Oklahoma (2011). He is currently working as a postdoctoral fellow at the School of Civil Engineering and Environmental Science, University of Oklahoma. His fields of interest include flood hydrology and remote sensing in hydrology, particularly optical and microwave remote sensing for flood hydrology and management. Dr. Sadiq Khan is the recipient of the NASA Earth and Space Science Fellowship and was selected for an outstanding student paper award for the research work he presented at the American Geophysical Union Fall 2009 Meeting in San Francisco, CA. He is a member of the American Geophysical Union, Association of American Geographers, and American Society for Photogrammetry and Remote Sensing.

Yang Hong is currently professor of hydrometeorology–climatology and remote sensing in the School of Civil Engineering and Environmental Sciences and in the School of Meteorology at the University of Oklahoma. Previously, he was a research scientist at NASA's Goddard Space Flight Center and postdoctoral researcher at the University of California, Irvine. Dr. Hong is currently the director of Hydrometeorology and Remote Sensing Laboratory (HyDROS Lab: http://hydro.ou.edu) at the National Weather Center, Norman, OK. Dr. Hong's areas of research span the wide range of hydrology–meteorology–climatology, with particular interest in bridging the gap among the water–weather–climate–human systems across scales in space and time. He is a recipient of the 2008 Group

Achievement Award from the NASA Headquarter. Dr. Hong received a PhD major in hydrology and water resources and PhD minor in remote sensing and spatial analysis from the University of Arizona (2003) and an MS (1999) in environmental sciences and a BS (1996) in geosciences from the Peking (Beijing) University, China.

Wenjuan Liu received her PhD in environmental sciences from the College of Resources and Environment, Northwest A&F University. She is currently working as a lecturer at the school of Agriculture Ningxia University, China. Dr. Liu also worked at the Institute of Earth Environment, Chinese Academy of Sciences. Her field of interest includes remote sensing in hydrology and land use/land cover change detection.

PREFACE

Terrestrial evapotranspiration (ET) is the main process of water circulation in the hydrological cycle. Water is transformed into vapor and transported from the land surface to the atmosphere. This water exchange usually involves a phase change of water from liquid to gas, which absorbs energy and cools the land surface. ET is an essential variable for energy and water balances on the Earth's surface. The process of evaporation (E) from soil and transpiration (T) from vegetation occurs simultaneously for vegetated areas. In terrestrial water balance, ET is the second largest term after precipitation. The three foremost factors controlling ET are availability of water, amount of available radiant energy, and transport mechanism to remove the water vapor from any surface. The previously mentioned elements in turn depend on other variables such as air temperature, wind speed, land surface temperature, vegetation cover, vapor pressure, and soil moisture that vary for different geographical regions, seasonality, and diurnal cycle. In this chapter, an overview of remote sensing principles and sensors on surface energy flux estimation is presented, with additional reviews of energy retrieval models and case studies on a regional and field scale.

8.1 Introduction

Evapotranspiration (ET) is an important variable for energy and water balances on the Earth's surface. The energy fluxes influence the main process of water circulation in the hydrological cycle, as liquid is transformed into vapor and transported from the land surface to the atmosphere. Therefore, estimation of land surface energy fluxes is essential for many environmental monitoring applications including water resources management, agricultural efficiency, global vegetation analysis, climate dynamics, and meteorological and ecological applications. Understanding the distribution of ET is essential for many environmental monitoring applications including water resources management, agricultural efficiency, global vegetation analysis, climate dynamics, and ecological applications. The evaporation plus transpiration from a vegetated surface with unlimited water supply is known as potential evaporation (PE) or potential evapotranspiration (PET), and it constitutes the maximum possible rate due to the prevailing meteorological conditions. Actual evaporation is the amount of water that is evaporated during a normal day that averages that if, for instance, the soil becomes dry, the actual evaporation is the amount of water that has been evaporated and not the amount of water that could have been evaporated if the soil had had an infinite amount of water to vaporize.

Over the course of the last three decades, hydrologists, environmental engineers, and scientists trusted on meterological station-based observations as a primary method for estimating ET. Sometimes these methodologies are insufficient to observe the spatially distributed ET over large regions that only focuses on PET instead of actual ET (AET). Their scales are substantially different from those of other methods, which can estimate ET at higher temporal resolution, for example, 30 min, at finer spatial resolution. However, satellite data with high spatial and sometimes temporal resolution with large areal coverage provide added advantage for ET estimation. Remote sensing–based ET retrieval has been a subject of many studies [3–8,40,42,46].

As demand for water increases, water managers need to know how much water is actually consumed in agriculture, urban, and natural environments. Increased demand for scarce water supplies has shifted water management strategy from increasing water supply to innovatively managing water use at sustainable levels. However, in order to more effectively allocate limited water, water resources managers must understand water consumption patterns over large geographical areas. There is a particular need to understand and measure the ET flux where irrigated agriculture is the primary consumptive use in arid and semiarid regions. In a broader sense, measurement of ET would be useful in all watersheds where streamflow must satisfy demand by current and future sustainable use. Because of the variability of region and seasons, water managers who are responsible for planning and adjudicating the distribution of water resources need to have a thorough understanding of the ET process and knowledge about the spatial and temporal rates of ET.

This chapter is focused on how numerical methods and remote sensing data are utilized to estimate the Earth's energy fluxes within the perspective of water resource monitoring. Section 8.1 outlined the fundamental theories of ET and their assumptions. Section 8.2 examines the leading remote sensing-based ET estimation methods. Section 8.3 focuses on how the basic theory and the understanding of surface energy balance (SEB) can be used to relate satellite-derived land surface variables to ET retrievals. Agricultural applications of ET products are discussed in Section 8.4. Afterward, the commonly used energy balance estimation models and satellite sensors are briefly reviewed, and two case studies are presented on AET retrieval algorithms that integrate satellite remote sensing data. The last section concludes and highlights ongoing developments and topics that require additional research.

8.2 ET Estimation Paradigms

Over the past two decade, accuracy of ET estimation methods has been improved that are attributable to the new and increasingly sophisticated techniques. In general, there are well-established ET retrieval techniques, for example, hydrological methods, direct measurement, micrometeorological methods, and empirical or combination of the previously listed methods [44]. Some of these methods are very accurate but can only offer point measurements of ET, which are insufficient for large-scale assessment. Hydrologic models can estimate ET estimates but demand immense ground-based observations, which is often inaccessible for much of ecosystems and watersheds around the world. Recently, satellite remote sensing-based ET retrieval has emerged as a practical method with the accessibility of shear amount of remote sensing estimates and development of several modeling techniques. The following sections discuss different methods used for ET retrievals.

8.2.1 Meteorological Network–Based ET Estimation

8.2.1.1 Penman–Monteith Equation

The Penman and Penman–Monteith equations were developed to use surface radiation, temperature, and humidity data to estimate ET. The Penman equation describes evaporation from an open water surface or from short vegetation [33]. The reference ET approach is established on the well-known Penman–Monteith reference ET Equation 8.1 adopted by the American Society of Civil Engineers (ASCE). More details on the estimation are outlined in [1,2]. The ASCE standardized reference ET equation is computed as

$$ET_0 = \frac{0.408\Delta(R_n - G) + \gamma(C_n / (T + 273))u_2(e_s - e_a)}{\Delta + \gamma(C_d u_2)} \quad (8.1)$$

where

ET_0 is the reference ET (mm day^{-1})
R_n is the calculated net radiation at the crop surface (MJ m^{-2} day^{-1})
G is the soil heat flux density at the soil surface (MJ m^{-2} day^{-1})
T is the average daily or hourly air temperature at 1.5–2.5 m height (°C)
u_2 is the average daily or hourly wind speed at 2 m height (m s^{-1})
e_s is the saturation vapor pressure at 1.5–2.5 m height, for daily computation
e_a is the average actual vapor pressure at 1.5–2.5 m height (kPa)
Δ is the delta, the slope of the saturation vapor pressure–temperature curve (kPa °C^{-1})
γ is the psychrometric constant (kPa °C^{-1})
C_n is the numerator constant that fluctuates with reference type and calculation time step
C_d is the denominator constant that changes with reference type and calculation time step

8.2.1.2 Reference ET (ET_0) and Crop Coefficients (K_c) to Calculate Actual ET

Meteorological network-based reference ET combined with crop coefficient provides AET at station scale. The crop coefficient approach to estimate ET was developed for agricultural crops but has been extended to natural ecosystems and has been adapted to remote sensing methods for ET retrieval. In agricultural applications, actual crop ET (ET_c) is related to ET_0 through an empirical crop coefficient (K_c) that must be determined experimentally (Allen et al., 1998). The crop water use is directly related to ET, and this can be determined by multiplying the reference ET_0 by a crop coefficient (K_c) as later shown in Equation 8.2:

$$ET_c = ET_0 \times K_c \quad (8.2)$$

where

ET_c is crop ET or crop water use (mm)
ET_0 is calculated reference ET for grass (mm)
K_c is crop coefficient

Crop coefficients (K_c) are used with ET_0 to estimate specific crop ET rates. The crop coefficient is a dimensionless number that is multiplied by the ET_0 value to arrive at a crop ET (ET_c) estimate. The resulting ET_c can be used to help an irrigation manager schedule when an irrigation should occur and how much water should be put back into the soil.

The crop coefficient regulates the calculated reference ET_0 to obtain the crop ET ET_c. Different crops will have a different crop coefficient values resulting in varying water demand. Crop coefficients vary by crop, stage of growth of the crop, and by some agricultural practices. For example, if $ET_0 = 0.25$ in./day (for June 8, 2012) and $K_c = 1.20$ (for apple in June), then, using Equation 8.2, ET_c will be 0.30 in./day.

8.2.1.3 Pan Evaporation

Pan evaporation (E_{pan}) is one of the first methods that has been used for several centuries to show a rough estimate of PET. In simple terms, (E_p) is the volume of water evaporated from the pan certain time, usually in an entire day. Several types of evaporation pans are used for evaporation measurements at meteorological stations, but the most commonly used is the U.S. Weather Bureau Class A pan evaporimeter.

The step-by-step process for pan evaporation measurements described by [10] is as follows: (1) The pan is set up in the field; (2) a known quantity of water is used to fill the pan and the water depth is measured; (3) the water is allowed to evaporate for a 24 h time period and again water depth is measured (simultaneously, the precipitation during that time period is measured); (4) after 24 h, another water depth measurement is taken; (5) the amount of evaporation per time unit, which is the difference between the two measured water depths, plus precipitation amount during the 24 h time period is calculated using Equation 8.3 shown in the following:

$$E_{pan} = P + (h_1 - h_2) \tag{8.3}$$

where

E_{pan} denotes daily pan evaporation (mm/day)

P is the daily precipitation that is simultaneously measured at the same location and with the same unit as pan evaporation

h_1 and h_2 are water surface heights measured in the evaporation pans for the previous and present measurements, respectively

8.2.2 Water Balance–Based ET Calculation

At a watershed scale, runoff is roughly the balance between precipitation received from the atmosphere and ET (green water) lost to the atmosphere. ET consists of evaporation from soil, evaporation from intercepted precipitation by plants, and transpiration via plant tissues.

If we consider a watershed the only water input from precipitation (P), then the only paths that water can take are into the soil as groundwater recharge/flow (D), surface runoff through streamflow (Q), and back to the atmosphere as AET. The precipitation is therefore "balanced" by the sum of these respective processes. A water balance budget is depicted by the typical water balance Equation 8.4 as

$$P = Q + S + \text{AET} \tag{8.4}$$

We may be able to measure P as precipitation (from rain gauges), Q as the river discharge (from stream measurements), and S as the change of terrestrial water storage (from monitoring wells), but we may not be able to measure AET. Because of the large spatial scale on which the water balance equation is often applied, it may be difficult to measure all terms accurately and to "close" the equation, that is, the left-hand side of the equation should balance out the right-hand side of the equation.

8.2.3 Energy Balance and Radiation Balance–Based ET Retrieval

The ET process requires source of heat energy to convert water from the liquid to the vapor state. This is ultimately supplied by net radiation (R_n), the amount of incident solar radiation (R_s) that is absorbed at the Earth's surface; a simplified equation for the *SEB* is

$$\lambda ET = R_n - H - G \tag{8.5}$$

where the available net radiant energy R_n (W m^{-2}) is combined between the soil heat flux G and the atmospheric convective fluxes (sensible heat flux H) and latent heat flux LE, which is readily converted to ET. The R_n and other variables (H and G) of Equation 8.5 can be solved using remotely sensed data of surface characteristics such as vegetation cover, surface temperature albedo, and leaf area index. The basic concept is to calculate the energy coming from the sun less any radiation that gets reflected (or emitted as thermal infrared [TIR] radiation) back to the atmosphere. This net radiation (R_n) is the energy available for ET, some of that R_n can be felt as the sensible heat flux (H), some of it is stored in the soil and other objects such as woody material, and the rest of the energy is absorbed by water that can be converted to water vapor for ET. A certain amount of energy per mass of water is required to vaporize water, and this is called the latent heat of vaporization.

8.3 Remote Sensing–Based Evapotranspiration Models

Notable advancement has been made in AET estimation by assimilating satellite remote sensing data into existing models [27,35]. In comparison to meteorological network-based in situ observations, satellite ET estimation has the benefit of better areal footprint, moderate resolution, and reliable quality [31,32]. Data availability, cost-effectiveness, and easy manipulation of these products are the added advantages. Furthermore, over the last two decades, several ET estimation prototypes have been developed that utilize these remote sensing and supplementary ground-based data products [13,22,41]. These developments in remote sensing ET estimation have been applied in North America, Asia, Africa, and many other countries around the world [36]. Remote sensing ET algorithms mainly solve the SEB of the land surface for latent heat flux (λET) at the time of satellite overpass. However, further research work is needed on high-resolution satellite remote sensing–based ET retrievals for decision-making and operational water management in real time.

In this chapter, the main satellite remote sensing ET estimation algorithms, that is, the Surface Energy Balance Algorithm for Land (SEBAL) [7,8] and later Mapping EvapoTranspiration with high Resolution and Internalized Calibration (METRIC) [3,4], are briefly introduced.

8.3.1 Surface Energy Balance Algorithm for Land

Emerging technologies based on satellite remote sensing are developed and refined for several operational ET algorithms that are now routinely used in hydrological studies at different spatiotemporal resolutions ranging from field to basin and global scales. These applications in ET calculations have advanced the understanding in agricultural water use and in some cases groundwater resources management at different scales and diverse ecosystems. The central and technical basis of SEBAL methods is to calculate the λET as the residual of the energy balance equation. The process is based on a complete energy balance for each satellite-retrieved pixel, where ET is predicted from the residual amount of energy remaining from the classical energy balance. SEBAL computes that the R_n is net radiation flux, G is soil heat flux, and H is sensible heat flux to the atmosphere as explained in the following.

8.3.1.1 Net Radiation (R_n)

The most common method for net radiation (R_n) calculation is by subtracting all outgoing radiant fluxes from all incoming radiant fluxes and includes solar and thermal radiation:

$$R_n = RS\downarrow - \alpha RS\downarrow + RL\downarrow - RL\uparrow - (1-\varepsilon o)RL\downarrow \tag{8.6}$$

where

$RS\downarrow$ is incoming short-wave radiation (W m^{-2})

α is surface albedo (dimensionless)

$RL\downarrow$ is incoming long-wave radiation (W m^{2})

$RL\uparrow$ is outgoing long-wave radiation (W m^{2})

εo is broadband surface thermal emissivity (dimensionless)

The $(1 - \varepsilon o)$ $RL\downarrow$ term represents the fraction of incoming long-wave radiation reflected from the surface

8.3.1.2 Soil Heat Flux (G)

Soil heat flux (G) is the rate of heat storage in the soil and vegetation due to conduction. General applications compute G as a ratio G/R_n using an empirical equation by Bastiaanssen (2000) representing values near midday:

$$G = Rn \times (Ts - 273.16)\ (0.0038 + 0.0074\alpha)\ (1 - 0.98\text{NDVI}^4) \tag{8.7}$$

where
 Ts is surface temperature (K)
 α is surface albedo

The Normalized Difference Vegetation Index (NDVI) is used to predict surface roughness and emissivity.

8.3.1.3 Sensible Heat Flux (*H*)

Sensible heat flux (H) is defined by the bulk aerodynamic resistance equation, which uses aerodynamic temperature (*Taero*) and aerodynamic resistance to heat transfer (*rah*):

$$H = \frac{\rho_{air}\, Cpa(Taero - Tat)}{rah} \tag{8.8}$$

where
 ρ_{air} is air density (kg m^{-3})
 Cpa is specific heat of dry air (1004 J kg^{-1} K^{-1})
 Tat is mean air temperature (K)
 Taero is mean aerodynamic temperature (K)

which is defined for a uniform surface as the temperature at the height of the zero plane displacement (d, m). The earlier Equations 8.5 through 8.7 establish the basic interpretation of SEB.

8.3.2 Mapping Evapotranspiration at High Resolution with Internalized Calibration

METRIC [4] is an extension of SEBAL, through integration with reference ET, which is computed using ground-based weather data as recorded by weather networks. As the name implies, METRIC approach adapts an internalized calibration approach to identify hot and cold pixels. These two varying conditions from within a satellite image are selected to fix boundary conditions for the energy balance and to internally calibrate the sensible heat computation. This internal calibration somewhat eliminates the need for exhaustive atmospheric correction of the data products, that is, temperature or albedo. The satellite image is simplified as a combination of vegetated areas and soil surface. This landscape is differentiated by the fractional canopy coverage (f_c), with the value ranging between 0 and 1, and is related to Moderate-Resolution Imaging Spectroradiometer (MODIS) NDVI:

$$f_c = \frac{\text{NDVI} - \text{NDVI}_{\min}}{\text{NDVI}_{\max} - \text{NDVI}_{\min}} \tag{8.9}$$

The SEB computation is then based on the determination of the relative instantaneous ET fraction (ET_f) given by

$$ET_f = \frac{\lambda E}{H + \lambda E} = \frac{(H_{dry} - H)/(H_{dry} - H_{wet})\lambda E_{wet}}{R_n - G} \tag{8.10}$$

A simplified approach to estimate the ET_f is to assume, according to [4,26,37], that dry-hot pixels exhibit low ET and wet-cold pixels show highest ET over the study domain, and the temperature of cold and hot pixels is used to estimate relative elements of ET on a pixel basis. Therefore, the

ET_f can also be estimated for each pixel by using Equation 8.10 to each of the MODIS land surface temperature (LST) image:

$$ET_f = \frac{T_{hot} - T_{i,j}}{T_{hot} - T_{cold}} \tag{8.11}$$

where

T_{hot} is the mean value of *hot* pixels
T_{cold} is the mean value of the *cold* pixels designated for a given scene
$T_{i,j}$ is the LST value for any MODIS pixel in the scene

Rationally, the ET_f is used in combination with reference *ET* (ET_r) described in Equation 8.1 to calculate the per pixel instantaneous AET (ET_a) estimates in a given scene according to METRIC:

$$ET_a = ET_f \cdot ET_r \tag{8.12}$$

A fundamental hypothesis of this approach is that the ET_f value is nearly constant, which is revealed on other studies [3,11,39]. This permits instantaneous retrieval of ET_f at MODIS overpass times and then to be inferred to daily mean ET. The daily ET can be calculated as

$$ET_{daily} = \sum_{i=1}^{day} \left(ET_f \times ET_r^i \right) \tag{8.13}$$

where

ET_{daily} is the daily AET (mm day^{-1})
i is temporal resolution of calculated reference *ET*

The next section describes several ET estimation techniques and studies water consumption variability by different authors at different scales, that is, local [28], regional [26], and global scale [29,31,32].

8.4 Evapotranspiration and Water Consumption Variability

8.4.1 Climate Variability and ET Estimation

The impact of the human activities on the terrestrial water cycle by unsustainable direct consumption for domestics, industrial and major portion for agricultural use is putting the already scarce freshwater resources at risk of depletion [24,25]. Furthermore, climate change projections show an impact on the global water cycle with higher frequency of extreme events and intensified ET globally [20]. This will ultimately influence the availability of the scarce water resources. Thus, consistent ET estimates are critical for the development of techniques that can be deployed for sustainable water resource use at the required spatiotemporal resolutions, especially in irrigation water management [9,18,19]. Conventionally, in order to estimate the spatial distribution of ET, meteorological observations are needed through a network of instruments that provided data at that particular location. These data are then interpolated to study ET at a regional scale.

8.4.2 Evapotranspiration Trends

To date, long-term changes in ET have been evaluated by studying reference evaporation using measurements of pan evaporation. One of earlier research work published showed that, on average, pan evaporation had decreased over North America, Europe, and Asia from the beginning of 1950 until the 1990s

[19,34]. Recent studies have reiterated this to be an overall trend throughout the northern latitudes. For instance, over the last half century, decreases in pan evaporation have been reported in India [12], China [43], and parts of Europe [30], although some mixed trends have also been reported, for example, East Asia [47], and similar anomaly in the Middle East [14,16]. Another comprehensive study on the decrease of pan evaporation over the conterminous United States for the past half century is presented by [19]. One of the critical points in most of these studies is that mean observations are used over a wide area with some sites showing decreasing trend, while others with increasing anomalies.

8.4.3 Water Consumption Variability at Different Spatial and Temporal Scales

8.4.3.1 Local Case Studies

A meteorological gauged river basin in central Oklahoma, near Lovell, OK, with an area of approximately 1033.59 km^2 was studied. The catchment includes very diverse land use/land cover (LULC) from agriculture (Garfield County) to urban (Oklahoma County). The annual precipitation is around 870 (mm), and average high and low temperatures are 21°C and 8°C, respectively. The study watershed is in a semiarid region where the agriculture activities were mainly sustained by irrigation. This study evaluates the possible closure of the heat balance equation using unique environmental monitoring network and to estimate ET_a and determine the variation with regard to varying types of LULC in urban settings.

The remotely sensed ET_a estimates were compared with calculated ET_a based on water balance over this watershed. The study sites and the LULC of the area are shown in Figure 8.1. In order to study the

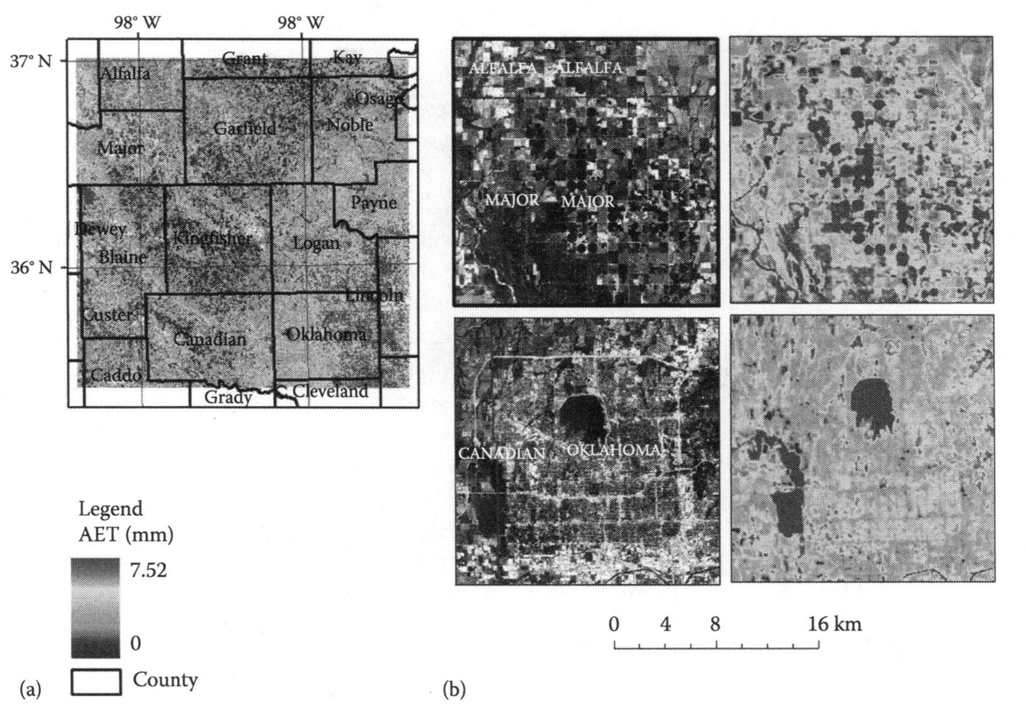

FIGURE 8.1 (a) Spatial variations of the ET_a (mm) over the study area in 2005. (b) Upper left side is Landsat false color image. Part (b) located at the right side shows local spatial variations of the ET_a (mm) at agricultural areas, and part (b) located at the lower right side shows local spatial variations of the ET_a (mm) at urban areas near a water body on July 31, 2005. Note: AET stands for the actual ET. Darker shade shows high ET values. (Cited from Liu, W. et al., *J. Appl. Remote Sens.*, 4(041873), 041873, 2010.)

TABLE 8.1 Annual ET for Two Counties with Different LULCs during 2005

LULC Types	Entire Study Area (mm)	Garfield County (mm)	Oklahoma County (mm)
Grass	774.26	804.12	763.89
Developed area	717.18	767.83	708.69
Open water	1019.37	919.27	990.28
Forest	854.24	858.76	831.59
Wetland	897.90	870.77	835.89
Agriculture	778.18	796.51	732.18
Shrubland	769.95	807.41	743.31

ET_a for distinctive LULC types, seven types of LULC were chosen in the watershed that include agriculture lands, water bodies, forests, grassland, wetlands, urban areas, and shrublands.

This study presents the retrieval of remotely sensed ET_a by SEB approach and examines the spatio-temporal variations of ET_a in terms of four types of LULC in the urban region of Oklahoma. Landsat 5 data and Oklahoma Mesonet (http://www.mesonet.org) data were used for AET estimation.

Statistics were extracted by overlaying the LULC map of the whole study area. Figure 8.1b shows that the ET_a values for irrigated croplands are high during crop growing season, but the yearly ET_a values for agriculture are not necessarily higher than those for grass- and shrublands. This can be attributed mainly to the low ET values for the croplands during nongrowing seasons in comparison to the other two vegetated lands. The top three values of the ET_a associated with designated types of LULC include water bodies, wetlands, and forests, and all ET_a values are over 800 mm per year (Table 8.1). Figure 8.2 shows that for the year 2005 ET_a values in the agriculturally dominant Garfield County are generally higher than those in Oklahoma County except for water bodies and densely vegetated areas. The urbanized areas mostly have lower ET values because of the lower soil moisture accessibility. Thus, the energy transformation is mainly limited in the form of sensible heat exchange.

Using site-specific flux towers and a water budget model at the basin scale, ET estimates are evaluated. Figure 8.3 compares SEB-based ET estimate at the regional level, that is, for different types of LULC that significantly reflects different values in urban and rural regions. Overall, wetlands have the highest ET, wetlands and forests present a higher rate of ET than grass and agricultural lands, and the highly urbanized areas have the lowest ET. With seasonal water use computed for distinctive types of land cover, those estimates of ET_a could help create decision-making tactics to improve water management. The estimates ET_a reveal that the higher the ET, the lower the development levels in urban regions.

However, ET calculated through SEB potentially has systematic errors. The sources of these errors include the characteristic adjustment bias of Landsat LST data. With a single thermal band, obtaining the LST from Landsat data is very difficult and might cause systematic errors. Consequently, the LST derived from Landsat data might have bias due to varying emissivity of infrared radiation. In addition, the ground truth observations might also have some measurement errors. Nevertheless, the estimation of ET using a high-resolution satellite remote sensing technology in urban regions can be still deemed as promising. Such a method complements the usual procedures that exclusively rely on ground- and point-based ET measurements.

8.4.3.2 Regional ET Patterns

A simplified form of the SEB method is employed to retrieve AET while maintaining the major rules of METRIC and SEBAL method. The principal basis of SEB is employed to calculate ET as the residual of the energy balance. This technique computes the AET by integrating the MODIS daily products and Oklahoma Mesonet meteorological network, that is, MOD/METRIC (thereinafter M/M-ET) was developed for southern plains in the United States and evaluation over Oklahoma. The goal of this

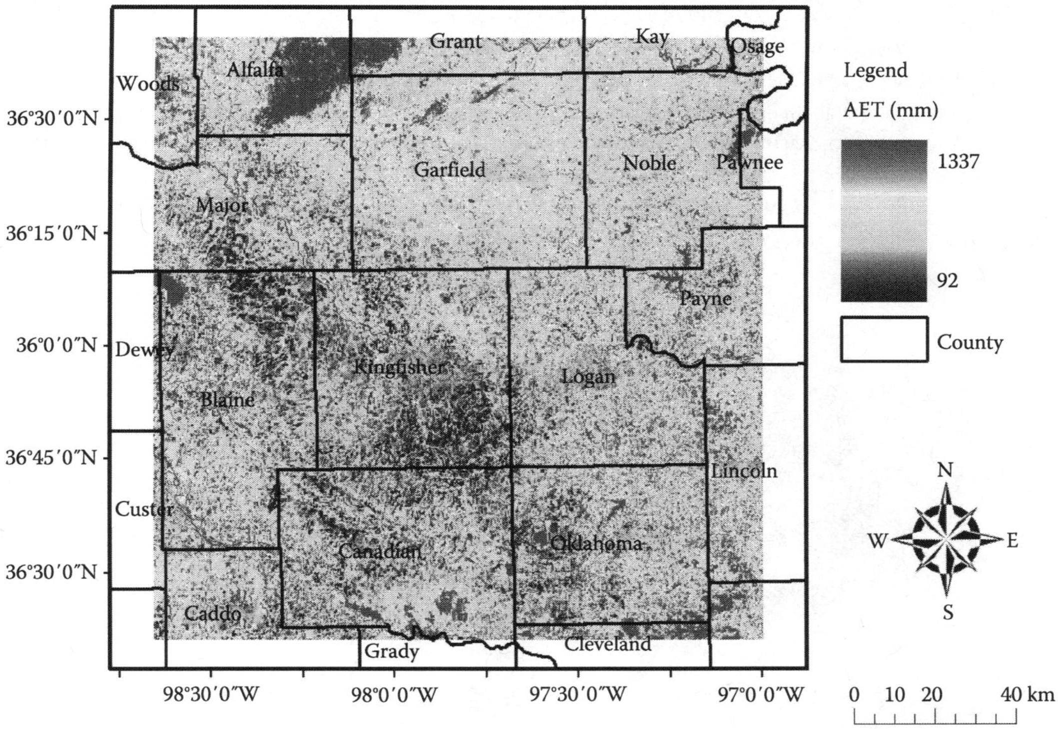

FIGURE 8.2 Spatial variations of the ET_a (mm) in the study area, 2005. Note: Darker shade shows high ET value. (Cited from Liu, W. et al., *J. Appl. Remote Sens.*, 4(041873), 041873, 2010.)

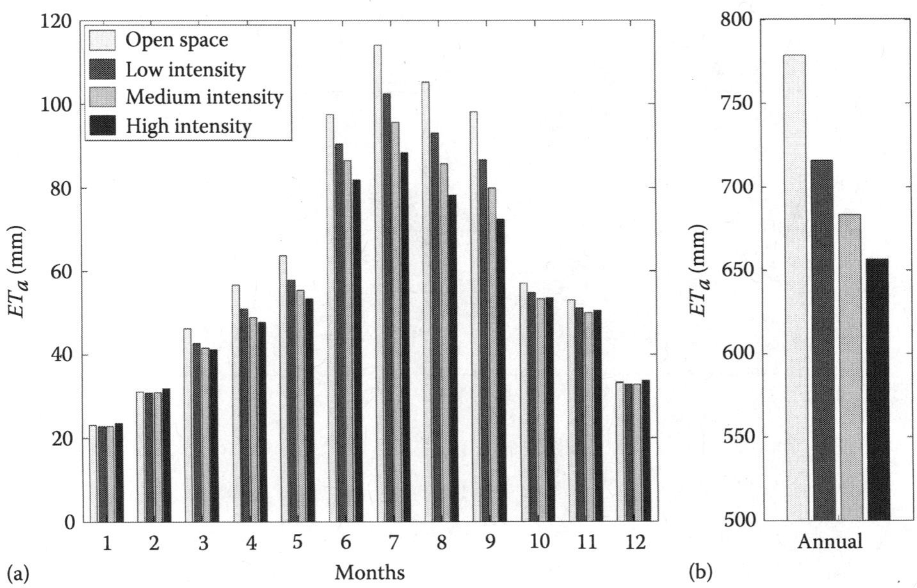

FIGURE 8.3 Temporal variability of ET by land use intensities in Oklahoma County during the 2005: (a) monthly and (b) annual ET. (Cited from Liu, W. et al., *J. Appl. Remote Sens.*, 4(041873), 041873, 2010.)

study was to evaluate the possibility of applying the M/M-ET for a functioning AET retrieval method suitable for regional scales, for example, the scale of irrigation projects, rather than individual fields in near real time.

Assessment of the M/M-ET measurements is on a daily, 8 day, and regular basis at both field and watershed scales. Two distinctive field sources explained in the following texts are used to compare the assessed results: one with metrological towers for latent heat flux observation and the other with Mesonet sites for crop ET. The location of these sites is illustrated in Figure 8.4. The observations of latent heat flux from the two Atmospheric Radiation Measurements (ARM) (http://www.arm.gov)

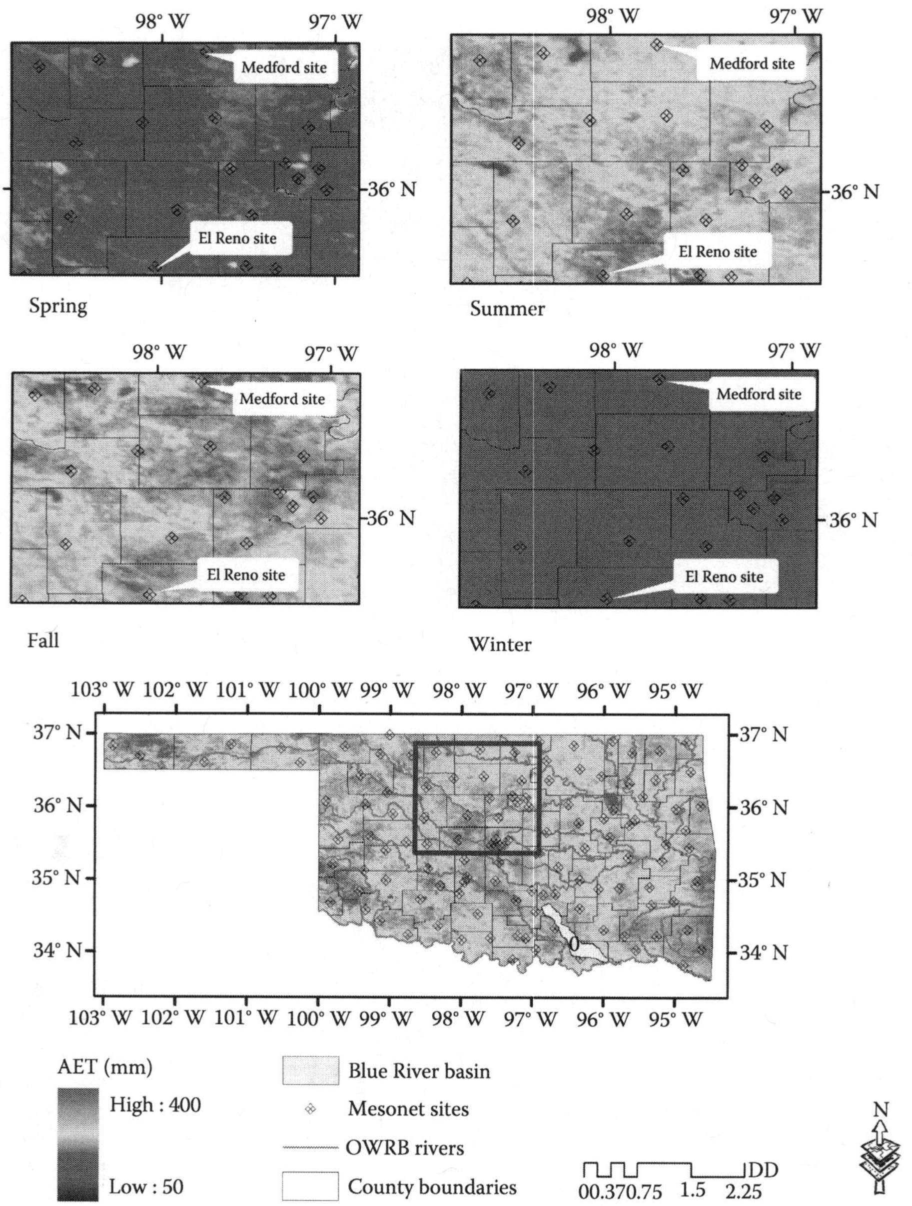

FIGURE 8.4 Seasonal AET based on M/M-ET with Mesonet site locations at Grant and Canadian Counties during 2004. (Cited from Khan, S.I. et al., *Int. J. Remote Sens.*, 31(14), 3799, 2010.)

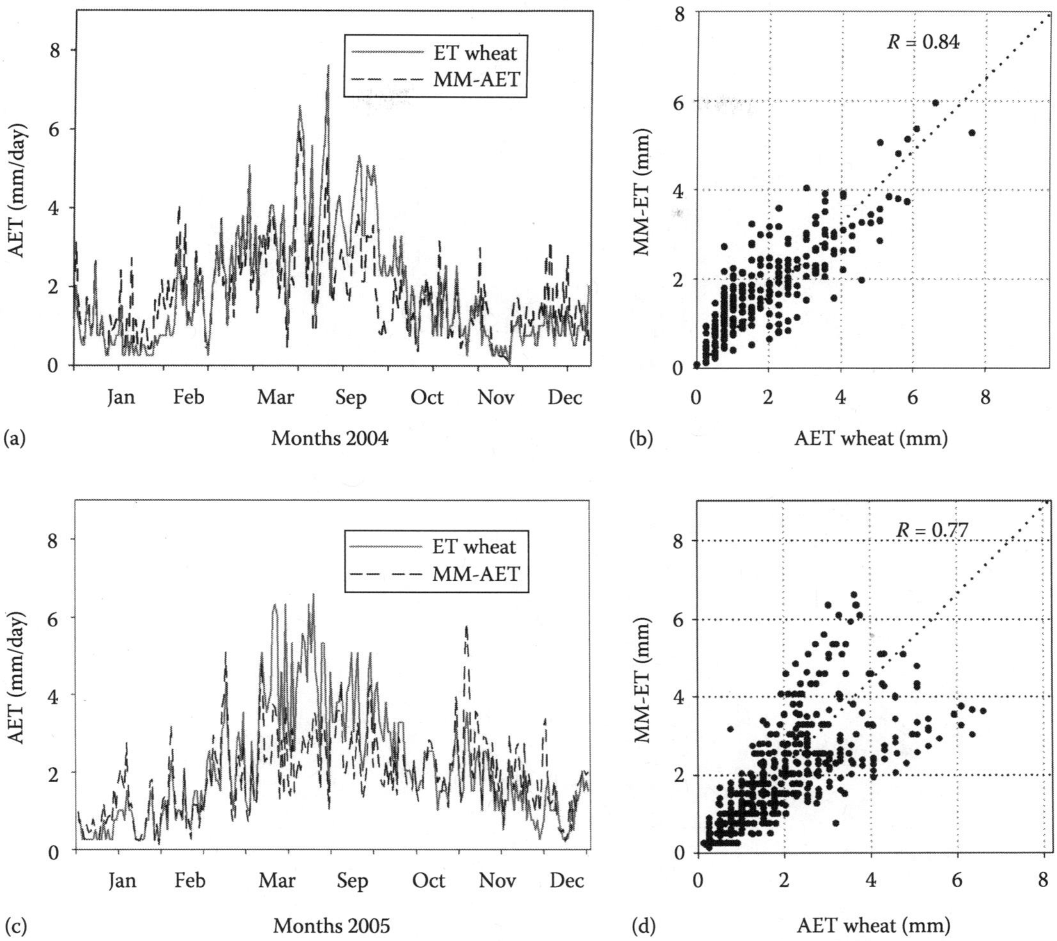

FIGURE 8.5 Comparisons of crop ET (wheat) and SEB-based MM-AET estimates at Mesonet Medford site. Panels (a) and (b) show the 2004 time series and scatter plot and, (c) and (d) are for 2005.

AmeriFlux eddy covariance tower sites were used for the comparison. These sites are located at the ARM Southern Great Plain (SGP) extended facilities in Lamont and El Reno, Oklahoma. The two Mesonet sites at El Reno and Medford with crop ET data are also selected for the evaluation.

The estimated ET is also compared with the crop ET at the selected sites during wheat growing season for multiple years. Figures 8.5 and 8.6 show the daily time series and scatter plots for both Medford and El Reno sites for years 2005 and 2006, respectively. The estimated ET is in good agreement with the observed crop ET. There is an underestimation of –7% and slight overestimation with 3% at Medford site for 2005 and 2006, respectively. Similarly, the ET estimates from the model agreed strongly with the observations at the Medford site with correlation coefficient of 0.84 and 0.80 for 2005 and 2006, respectively. M/M-ET calculations at El Reno show slightly higher biases but overall in agreement with the measurements. The correlation coefficient values also indicate that the ET estimates correlate measurements at the El Reno site relatively well with values of 0.82 and 0.51 for 2005 and 2006, respectively. A detailed analysis of these results is provided in [26].

8.4.3.3 Global ET Variability

Global terrestrial ET retrieval at a fine spatial scale was never achieved before until the satellite-driven estimation of the terrestrial ET, using one of the satellite sensors the MODIS onboard the Aqua satellite

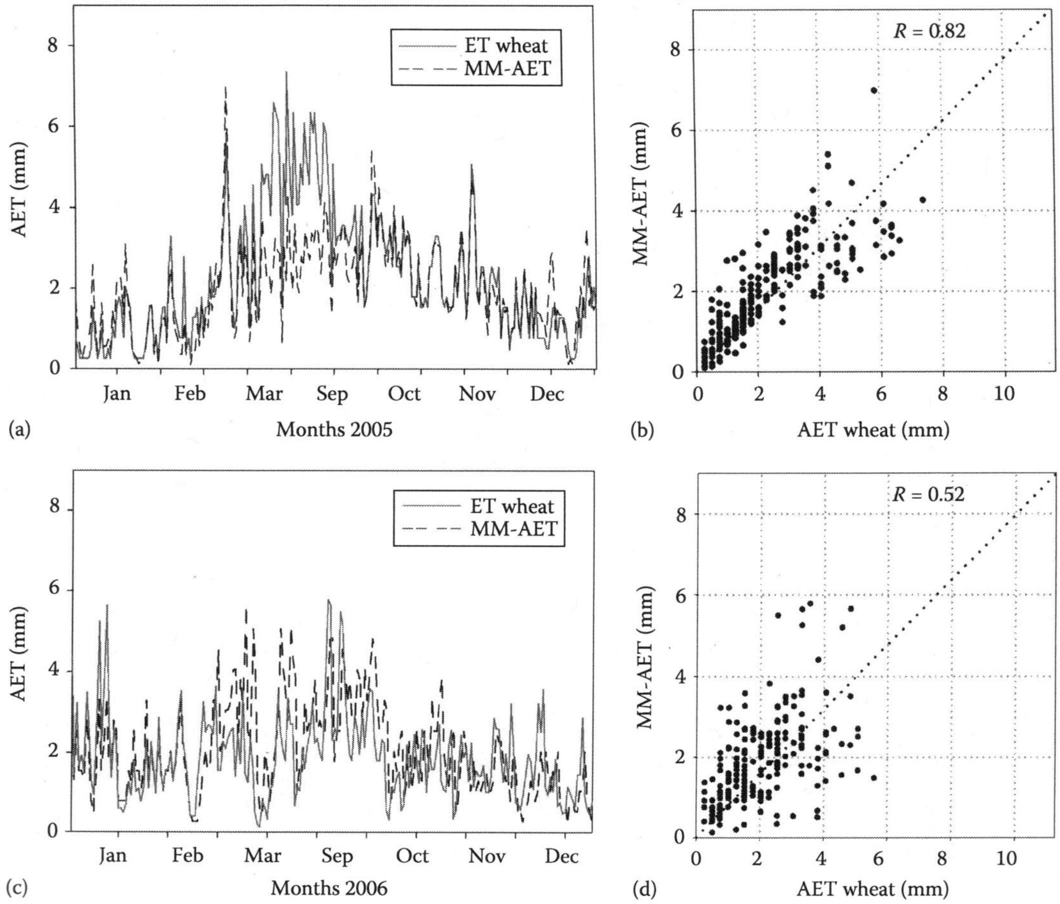

FIGURE 8.6 Comparisons of crop ET (wheat) and SEB-based MM-AET estimates at Mesonet El Reno site. Panels (a) and (b) show the 2005 time series and scatter plot and (c) and (d) are for 2006.

launched on May 4, 2002, with 36 spectral bands, 20 reflective solar bands, and 16 thermal emissive bands. MODIS provides exceptional data observing vegetation and surface energy [23], which is utilized to develop a remotely sensed ET model [31]. The moderated resolution global terrestrial ET algorithm is developed and refined by the Numerical Terradynamic Simulation Group (NTSG), a research laboratory at the University of Montana in Missoula. The derived data product is known as the "MOD16" that is defined as an "evaporation fraction (EF)," the energy budget equivalent of an index actual to PET, over the global land surface with 1 km resolution every 8 days [31] (Figure 8.7).

ET-relevant MODIS data related to ET retrieval are LST and emissivity (MOD11), surface reflectance products (MOD09), vegetation index (MOD13), and albedo (MOD43B3) obtained by assimilating the bihemispherical reflectance data modeled. All these listed NASA land products are quality-controlled data sets that account the atmospheric conditions in terms of cloud cover and aerosol content, algorithm choices, processing failure, and error estimates. These data sets were acquired from the Land Processes Distributed Active Archive Center (LP DAAC) at the United States Geological Survey (USGS) Earth Resources Observation and Science (EROS) Center, with the standard Hierarchical Data Format (http://LPDAAC.usgs.gov). For more information on MODIS, please refer to http://modis.gsfc.nasa.gov.

Another global reference ET product is generated by extracting the meteorological variables from the Global Data Assimilation System (GDAS) analysis fields. The GDAS data are produced at 6 h temporal resolution and coarse spatial resolution of 1° by the National Oceanic and Atmospheric Administration

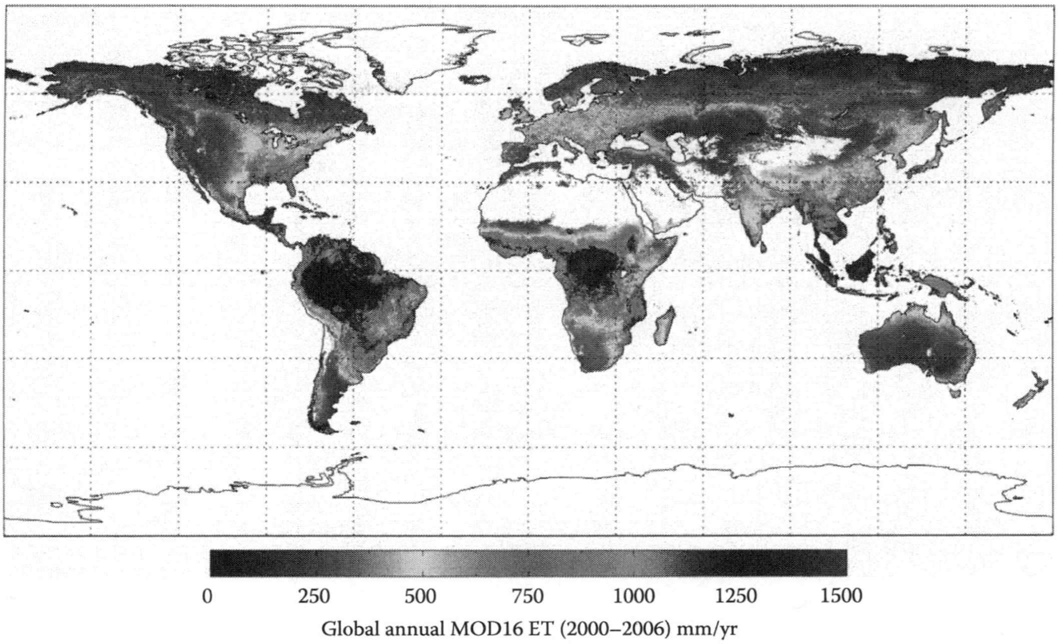

FIGURE 8.7 Global annual MODIS ET (2000–2010) mm/year. (Cited from Mu, Q. et al., *Remote Sens. Environ.*, 115, 1781, 2011.)

(NOAA). The GDAS fields that are used as input to the reference ET calculation include air temperature, atmospheric pressure, wind speed, relative humidity, and solar radiation (incoming, outgoing, long wave, and short wave). Here, we provide the summary of the comparison between Mesonet ET_0 and GDAS ET_0 that was conducted primarily at daily time scale (Figure 8.8).

Figure 8.8 presents a statistical evaluation of daily ET estimates for years 2005 and 2006. The bias values in all groups are within the range of −7.12% to 7.19%. The lowest bias is 1.53%, observed at Mesonet station number five. The lowest absolute bias is 14.69%, observed in GDAS estimates. Correlation coefficient (CC) for all groups are above 0.9. The maximum CC obtained is about 0.94 in Mesonet station number six. The RMSEs from all stations are within 28.67% and Mesonet station six gave the best result among all sites (RMSE = 19.70%). As shown in Figure 8.8, GDAS ET_0 and Mesonet ET_0 have shown strong linear correlations among the 11 Mesonet stations for the years 2005 and 2006. However, the variations among the stations shown in Figure 8.8 can be attributed to different LULC types and microclimate difference within the SGP.

8.5 Applications of ET Estimates in Agriculture

Assimilating meteorological network measurements and satellite remote sensing estimates with best possible spatial and temporal resolution can overcome many of the limitations associated with low spatial coverage of field-scale ET estimation models. Consequent applications of ET estimates have begun the era of precise agricultural water use monitoring and management techniques at diverse scales and various ecosystems. The World Bank estimates that 70% of freshwater use is for agriculture. In arid and semiarid regions, ET is the major source of water removal from the land surface. Particularly in arid regions, approximately 90% of the annual rainfall can be evapotranspired, and therefore, ET determines the freshwater recharge and discharge from aquifers in these environments [21]. The United States irrigates over 50 million acres of agricultural land and 32 million acres of recreational landscapes. Therefore, estimation of spatially distributed ET from agricultural areas is critical as irrigation consumes the main

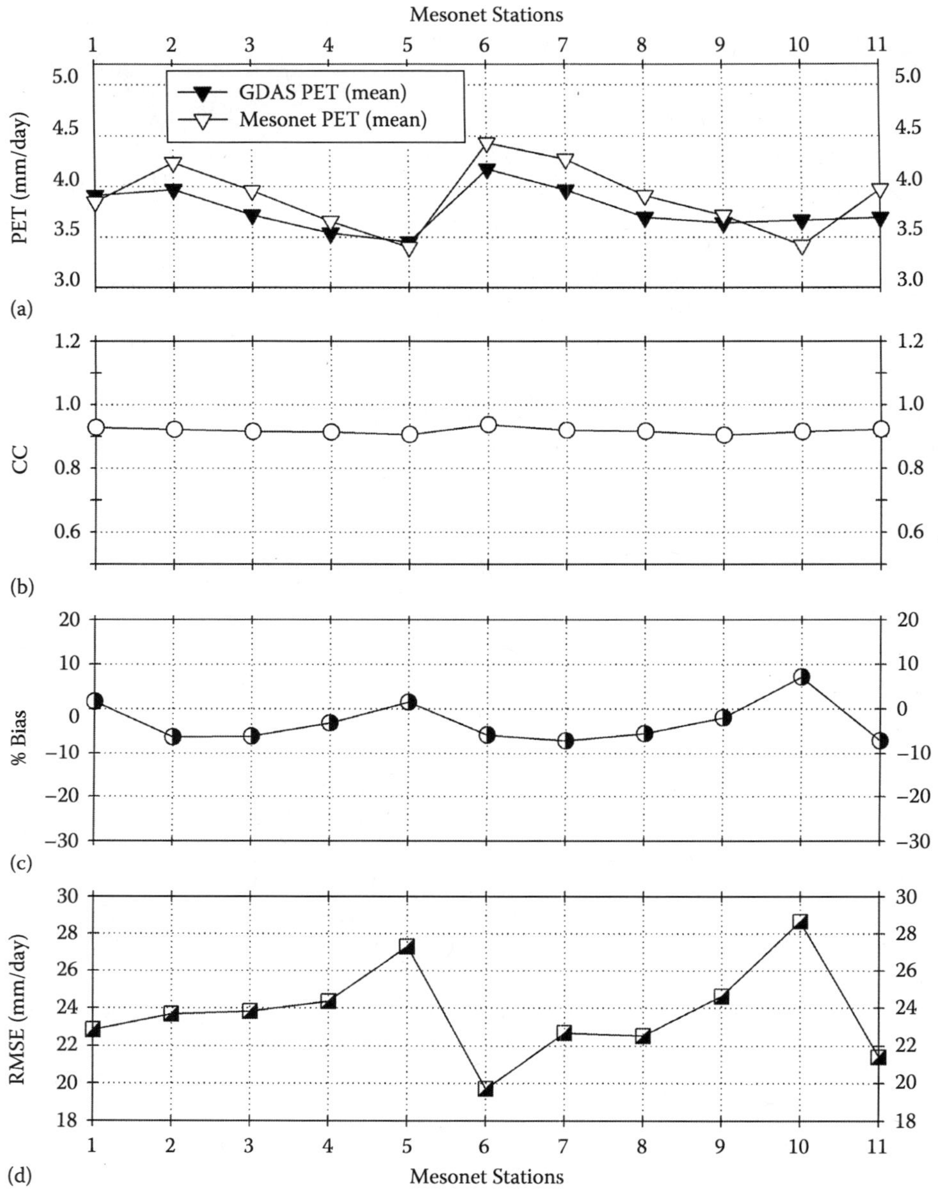

FIGURE 8.8 Statistical evaluation of GDAS daily ET_0 using Mesonet ET_0 for 2005–2006.

portion in water use [17,38]. In brief, AET estimation at various scales is vital to water monitoring and sustainable agricultural management.

8.5.1 Agricultural Water Use Efficiency

Conventionally, water management models are used for irrigation scheduling that are established on water allocations, and not on actual water consumed, in part because irrigation water use is complex to estimate from direct observations. As discussed at the beginning of this chapter, the traditional techniques of estimating ET classically provide potential or reference ET at points, rather than spatial distribution of AET. Furthermore, these techniques require immense ground observations that may be difficult to obtain at large scales. Satellite remote sensing is a promising tool for estimating the spatial distribution of ET at regional (and larger) scales. Different methods have been developed to use satellite remote sensing data in surface flux estimation schemes.

ET is the term used for total water consumption of the crop; it includes water loss due to evaporation and plant transpiration. Agricultural water use efficiency (WUE) is defined as crop yield per accumulated AET for the growing season:

$$\text{WUE} = \frac{\text{Crop yield}}{\text{AET}} \tag{8.14}$$

Accurate computation of WUE is possible only with reliable AET estimates, therefore making ET estimation an important variable in agricultural productivity that can ultimately help water manager to increase the amount of crop per unit of water consumed.

8.5.2 Water Deficit and Water Stress Indices

The association of surface temperature and water stress is based on the theory that as vegetation transpires, the evaporated water cools the leaves in the surrounding and temperature. In a drought condition, the vegetation becomes water stressed leading to low transpiration and thus an increase in the surface temperature of the plants. Water or moisture stress is often quantified in terms of the reduction of ET from the potential rate (PET) expected under non-moisture-limiting conditions. Satellite remote sensing instruments with TIR instruments provide land surface information. In this regard, the Evaporative Stress Index (ESI) proposed by [5,6] quantifies variance in the actual to PET. This method uses fine- and moderate-resolution TIR imagery from polar-orbiting systems to generate daily ET maps at subfield scales.

8.6 Summary and Conclusions

ET is the most vital process in the hydrologic cycle and an essential variable in various disciplines such as hydrology, agriculture, ecology, and climate science. Even in the same climatic and meteorological conditions, AET may exhibit remarkable spatial variability across different vegetated regions, agricultural land use practices, and differing types of built-up urbanized areas. Spatially distributed ET estimation for agricultural areas is a challenging and important task, as irrigation consumes the largest share in water use in arid and semiarid regions. Furthermore, with the projected climate change, these areas might be adversely affected, as it will intensify ET, globally impacting the scarce water resources.

Traditional ET retrieval methods typically present reference ET at point scale and do not contain information on the spatial variation of ET. Recent advances in satellite remote sensing of ET have showed advances in water resource monitoring at local, regional, and global scales. Integrating satellite data with available ground-based measurements by using a simplified surface energy balance (SSEB) method renders opportunities to utilize remote sensing data products for sustainable water management. The ability to map ET and moisture availability has broad applications in monitoring droughts and consumptive water use, planning for irrigation, predicting local and regional water demand, and providing important boundary conditions to hydrological and weather forecast models.

References

1. Allen, R.G., Smith, M., Perrier, A., and Pereira, L.S. (1994a). An update for the definition of reference evapotranspiration, *ICID Bulletin*, 43(2):1–34.
2. Allen, R.G., Smith, M., Perrier, A., and Pereira, L.S. (1994b). An update for the calculation of reference evapotranspiration, *ICID Bulletin*, 43(2):35–92.

3. Allen, R.G., Tasumi, M., Morse, A., and Trezza, R. 2005. A Landsat-based energy balance and evapotranspiration model in Western US water rights regulation and planning, *Irrigation and Drainage Systems*, 19(3):251–268.
4. Allen, R.G., Tasumi, M., and Trezza, R. 2007. Satellite-based energy balance for mapping evapotranspiration with internalized calibration (METRIC)—Model, *Journal of Irrigation and Drainage Engineering*, 133:380.
5. Anderson, M.C., Kustas, W.P., Norman, J.M., Hain, C.R., Mecikalski, J.R., Schultz, L. et al. 2011. Mapping daily evapotranspiration at field to continental scales using geostationary and polar orbiting satellite imagery, *Hydrology and Earth System Sciences*, 15(1):223–239. doi: 10.5194/hess-15-223-2011.
6. Anderson, M.C., Norman, J.M., Mecikalski, J.R., Otkin, J.A., and Kustas, W.P. 2007. A climatological study of evapotranspiration and moisture stress across the continental United States based on thermal remote sensing: 2. Surface moisture climatology, *Geophysical Research: Atmospheres*, 112(D11):D11112. doi: 10.1029/2006jd007507.
7. Bastiaanssen, W., Menenti, M., Feddes, R., and Holtslag, A. 1998. A remote sensing surface energy balance algorithm for land (SEBAL). 1. Formulation, *Journal of Hydrology*, 212:198–212.
8. Bastiaanssen, W., Noordman, E., Pelgrum, H., Davids, G., Thoreson, B., and Allen, R. 2005. SEBAL model with remotely sensed data to improve water-resources management under actual field conditions, *Journal of Irrigation and Drainage Engineering*, 131:85.
9. Bouwer, L.M., Biggs, T.W., and Aerts, J.C.J.H. 2008. Estimates of spatial variation in evaporation using satellite-derived surface temperature and a water balance model, *Hydrological Processes*, 22(5):670–682.
10. Brouwer, C. and Heibloem, M. 1986. Irrigation water management: Irrigation water needs, *Irrigation Water Management Training Manual* (3). Food and Agriculture Organization of the United Nations, Rome, Italy.
11. Brutsaert, W. and Sugita, M. 1992. Application of self-preservation in the diurnal evolution of the surface energy budget to determine daily evaporation, *Journal of Geophysical Research: Atmospheres*, 97(D17):18377–18382.
12. Chattopadhyay, N. and Hulme, M. 1997. Evaporation and potential evapotranspiration in India under conditions of recent and future climate change, *Agricultural and Forest Meteorology*, 87(1):55–73.
13. Choudhury, B.J., Ahmed, N.U., Idso, S.B., Reginato, R.J., and Daughtry, C.S.T. 1994. Relations between evaporation coefficients and vegetation indices studied by model simulations, *Remote Sensing of Environment*, 50(1):1–17.
14. Cohen, S., Ianetz, A., and Stanhill, G. 2002. Evaporative climate changes at Bet Dagan, Israel, 1964–1998, *Agricultural and Forest Meteorology*, 111(2):83–91.
15. Crago, R.D. 1996. Comparison of the evaporative fraction and the Priestley-Taylor α for parameterizing daytime evaporation, *Water Resources Research*, 32(5):1403–1409.
16. Eslamian, S., Khordadi, M.J., and Abedi-Koupai, J. 2011. Effects of variations in climatic parameters on evapotranspiration in the arid and semi-arid regions, *Global and Planetary Change*, 78(3–4):188–194. doi: 10.1016/j.gloplacha.2011.07.001.
17. Glenn, E.P., Huete, A.R., Nagler, P.L., Hirschboeck, K.K., and Brown, P. 2007. Integrating remote sensing and ground methods to estimate evapotranspiration, *Critical Reviews in Plant Sciences*, 26(3):139–168.
18. Gowda, P.H., Chavez, J.L., Colaizzi, P.D., Evett, S.R., Howell, T.A., and Tolk, J.A. 2008. ET mapping for agricultural water management: Present status and challenges, *Irrigation Science*, 26(3):223–237.
19. Hobbins, M.T., Ramírez, J.A., and Brown, T.C. 2004. Trends in pan evaporation and actual evapotranspiration across the conterminous US: Paradoxical or complementary, *Geophysical Research: Atmospheres*, 31(13):1–5.

20. Huntington, T.G. 2006. Evidence for intensification of the global water cycle: Review and synthesis, *Journal of Hydrology*, 319(1):83–95.
21. Huxman, T.E., Smith, M.D., Fay, P.A., Knapp, A.K., Shaw, M.R., Loik, M.E. et al. 2004. Convergence across biomes to a common rain-use efficiency, *Nature*, 429(6992):651–654.
22. Jiang, L. and Islam, S. 2001. Estimation of surface evaporation map over southern Great Plains using remote sensing data, *Water Resources Research*, 37(2):329–340.
23. Justice, C., Townshend, J., Vermote, E., Masuoka, E., Wolfe, R., Saleous, N. et al. 2002. An overview of MODIS Land data processing and product status, *Remote Sensing of Environment*, 83(1–2):3–15.
24. Kabat, P., Claussen, M., Dirmeyer, P.A., Gash, J.H., de Guenni, L.B., Meybeck, M., Pielke Sr, R.A., Vörösmarty, C.I., Hutjes, R.W., Lütkemeier, S. 2004. *Vegetation, Water, Humans and the Climate: A New Perspective on an Interactive System*. Global Change: The IGBP Series, p. 24, Springer Verlag, Heidelberg, Germany.
25. Kalma, J.D., McVicar, T.R., and McCabe, M.F. 2008. Estimating land surface evaporation: A review of methods using remotely sensed surface temperature data, *Surveys in Geophysics*, 29(4):421–469.
26. Khan, S.I., Hong, Y., Vieux, B., and Liu, W. 2010. Development evaluation of an actual evapotranspiration estimation algorithm using satellite remote sensing meteorological observational network in Oklahoma, *International Journal of Remote Sensing*, 31(14):3799–3819.
27. Kustas, W. and Norman, J. 1996. Use of remote sensing for evapotranspiration monitoring over land surfaces, *Hydrological Sciences Journal*, 41(4):495–516. doi: 10.1080/02626669609491522.
28. Liu, W., Hong, Y., Khan, S.I., Huang, M., Vieux, B., Caliskan, S. et al. 2010. Actual evapotranspiration estimation for different land use and land cover in urban regions using Landsat 5 data, *Journal of Applied Remote Sensing*, 4(041873):041873.
29. Liu, W., Hong, Y., Khan, S., Huang, M., Grout, T., and Adhikari, P. 2011. Evaluation of global daily reference ET using Oklahoma's environmental monitoring network—MESONET, *Water Resources Management*, 25(6):1601–1613.
30. Moonen, A., Ercoli, L., Mariotti, M., and Masoni, A. 2002. Climate change in Italy indicated by agrometeorological indices over 122 years, *Agricultural and Forest Meteorology*, 111(1):13–27.
31. Mu, Q., Heinsch, F.A., Zhao, M., and Running, S.W. 2007. Development of a global evapotranspiration algorithm based on MODIS and global meteorology data, *Remote Sensing of Environment*, 111(4):519–536.
32. Mu, Q., Zhao, M., and Running, S.W. 2011. Improvements to a MODIS global terrestrial evapotranspiration algorithm, *Remote Sensing of Environment*, 115:1781–1800.
33. Penman, H.L. 1948. Natural evaporation from open water, bare soil and grass. *Proceedings of the Royal Society of London. Series A. Mathematical and Physical Sciences*, 193(1032):120–145. doi: 10.1098/rspa.1948.0037.
34. Peterson, T.C., Golubev, V.S., and Groisman, P.Y. 1995. Evaporation losing its strength, *Nature*, 377(6551):687–688.
35. Rango, A. and Ahlam, I.S. 1998. Operational applications of remote sensing in hydrology: Success, prospects and problems, *Hydrological Sciences Journal*, 43(6):947–968.
36. Seguin, B., Courault, D., and Guerif, M. 1994. Surface temperature and evapotranspiration: Application of local scale methods to regional scales using satellite data, *Remote Sensing of Environment*, 49(3):287–295.
37. Senay, G.B., Budde, M., Verdin, J.P., and Melesse, A.M. 2007. A coupled remote sensing and simplified surface energy balance approach to estimate actual evapotranspiration from irrigated fields, *Sensors*, 7(6):979–1000.
38. Shiklomanov, I.A. 2000. Appraisal and assessment of world water resources, *Water International*, 25(1):11–32.
39. Shuttleworth, W.J. 1991. Insight from large-scale observational studies of land/atmosphere interactions, *Surveys in Geophysics*, 12(1):3–30.

40. Sobrino, J., Gómez, M., Jiménez-Muñoz, J., and Olioso, A. 2007. Application of a simple algorithm to estimate daily evapotranspiration from NOAA-AVHRR images for the Iberian Peninsula, *Remote Sensing of Environment*, 110(2):139–148.
41. Sugita, M. and Brutsaert, W. 1991. Daily evaporation over a region from lower boundary layer profiles measured with radiosondes, *Water Resources Research*, 27(5):747–752.
42. Tang, Q., Peterson, S., Cuenca, R.H., Hagimoto, Y., and Lettenmaier, D.P. 2009. Satellite-based near-real-time estimation of irrigated crop water consumption, *Geophysical Research: Atmospheres*, 114:D05114.
43. Thomas, A. 2000. Spatial and temporal characteristics of potential evapotranspiration trends over China, *International Journal of Climatology*, 20(4):381–396.
44. Thornthwaite, C. and Mather, J. 1955 The water balance. Centerton, Drexel Institute of Technology-Laboratory of Climatology, *Publications in Climatology*, 8(1):104.
45. Vörösmarty, C.J., McIntyre, P., Gessner, M.O., Dudgeon, D., Prusevich, A., Green, P, Glidden, S., Bunn, S.E., Sullivan, C.A., Liermann, C.R. 2010. Global threats to human water security and river biodiversity, *Nature*, 467(7315):555–561.
46. Wang, K., Wang, P., Li, Z., Cribb, M., and Sparrow, M. 2007. A simple method to estimate actual evapotranspiration from a combination of net radiation, vegetation index, and temperature, *Geophysical Research: Atmospheres*, 112:1984–2012.
47. Xu, J. 2001. An analysis of the climatic changes in eastern Asia using the potential evaporation, *Journal of Japan Society of Hydrology and Water Resources*, 14:151–170.

9

Fundamentals of Hydrodynamic Modeling in Porous Media

Shaul Sorek
Ben-Gurion University of the Negev

AUTHOR

Shaul Sorek is a professor, which affiliated with the Department of Environmental Hydrology and Microbiology (EHM), Zuckerberg Institute for Water Research, Blaustein Institutes for Desert Research at the Ben-Gurion University (BGU) of the Negev, Israel. His research interests are transport phenomena through heterogeneous media, numerical methods, shock waves through porous media, and decision support systems for water resources.

PREFACE

Using mathematical formulations for quantitative expressions of physical concepts is the essence of modeling, aiming at when and where predictions of primary state variables being subject to temporal and spatial conditions. Models do not portray exactness, as being addressed by assumptions, uncertainty, lack of information, and correctness of observations, yet they enable trackable outcome to cause–effect situations. Hence, the usefulness of models as working tools is the ability to reveal a quantitative trend. In such regard, specifically for heterogeneous media, the chapter sweeps through a general framework of three modeling approaches: (1) the lumped parameter model using ordinary differential equations for the temporal dependent state prime variables (this can be considered worth implementing to media without homogeneous spatial distribution [e.g., fractured domain with preferential flow paths, dead ends, and/or not interconnected flow paths]), (2) the continuum microscopic approach using partial differential equations for the temporal and spatial dependent state prime variables (this can be considered worth implementing for relative simple tortuous interconnected flow pathways through the solid matrix and interactions at common phase interfaces such that it can be addressed by phenomenological relations), and (3) the continuum macroscopic approach with spatially averaged microscopic partial differential equations, for the temporal and spatial dependent state prime variables. The averaging is implemented over a representative elementary volume, which for the macroscopic domain represents a point at which spatial averaging is applied to at least two phases also interacting at their common microscopic interface. Such an approach is worthy of implementing for geometrical features that are impractical to account for all details at the microscopic level.

Examples for these approaches are considered. Special emphasis is devoted to the question of concurrent macroscopic balance equations for the phase extensive quantity at adjacent significant different spatial scales. The modeling of interacting multiphase and multicomponents is also addressed.

9.1 Introduction

The vadose zone (VZ) comprised from the unsaturated zone (UZ) and the saturated/unsaturated interface region (SUIR) is an important link between the saturated zone and land surface. The VZ provides storage capacity for both water as one possible fluid and contaminants; a reactor medium for physical, chemical, and biological processes; and a delay time between the release of multicontaminants into the UZ and their transfer by flow of multifluids into the saturated zone. The quantitative relationship between the amount of contaminants released at the soil surface by anthropogenic activities external and/or internal to the VZ, for example, agricultural, industrial, and urban activity, and the resulting concentrations in groundwater is uncertain [6]. The UZ is believed to be with "natural attenuation" potential, which led to consider soil aquifer treatment as an efficient tertiary process for sewage effluents after secondary treatment, a technique that seemed to be a psychological projection of the granular filter methodology [7]. Qualitative decision driven from water scarcity conditions triggered in Israel the directing of rainwater falling on the roofs into the subsurface. Yet, simulations prove that in an urban environment characterized by patchworks of permeable and impervious surfaces, the underground migration of contaminants may be enhanced due to the highly nonuniform recharge of rainwater [1]. In what follows we will specifically refer to the need of accounting for nonhabitual approaches to address the modeling of hydrodynamics in porous media.

9.2 Modeling Approaches for Characterization of Physical Processes

9.2.1 Mathematical Modeling

Let temporal and spatial measurable observations (information) of prime (state) variables made in reference to a tangible system B (tsB). One wishes to establish a virtual system A (vsA) that predicts quantifiable assessments of state variables at a prescribed spatial domain and/or temporal extent of the tsB. If the assessments consistently do not deviate beyond a desired accuracy from the observations, then one follows the premise that the vsA mathematically models the tsB. Thus, further hypotheses can be investigated via the vsA as it is assumed to represent the hydrodynamic essentials of the tsB.

The entire spectrum of temporal and spatial information is always not available. Hence, the vsA structure consists of mathematically formulated physical assumptions referred to as a generic model. Such are, for example, balance equations (conservation laws) of the phase's extensive quantities (e.g., its volume, mass, linear momentum, angular momentum, and energy), constitutive relations (i.e., state functions), and definitions. The adaptation of an obtained generic model to a subset of prescribed observations (i.e., the inverse problem) is done via estimation of the model sought parameters and the choice of appropriate temporal and spatial conditions (i.e., initial and boundary conditions). The input of the model parameters, initial and boundary conditions, leads to simulation predicting state variables (i.e., the direct problem) that can also be compared against the residual set of the available information. Verification of the calibrated model enables further simulations for predictions resolved from different operation scenarios. The general process of establishing a mathematical model is described in Figure 9.1.

9.2.2 Lumped Parameter Approach

The premise of lumping is that the behavior of a distributed system can be conformed into an interacting set of information characterized by typical, not spatially dependent, entities that approximate the behavior of the original system subject to physical assumptions. As the spatial distribution within each constituent of the set is meaningless, its configuration is irrelevant and can thus be described by a node, or a compartment representing in reality, for example, a cluster of similar information. Implementation of the lumped parameter (LP) approach can be found in different network sets. Figure 9.2 presents two of such LP examples: a mechanical set of mass–spring–damper and an electric set of resistor–capacitor–conductance.

In what follows let us consider a compartment as a region of identifiable physical space characterized by lumped (i.e., not a function of space) quantities. A network section of virtual compartments addressing the hydrodynamic interactions to be modeled by the LP approach for the displacement of fluids and components through a solid matrix is described in Figure 9.3.

We note that as the LP approach is not accounting for spatial dependency, the balance equations conform to ordinary differential equation (ODE) forms. Let us first consider some definitions:

- Phase refers to an identifiable γ body (subject to one's definition) composed of a sum of α (= 1, 2, …., $N\alpha$) components (i.e., species of the γ phase such as ions, molecules, particles). Hence, we write $(\,)^{\gamma} \equiv \sum_{\alpha=1}^{N\alpha} (\,)^{\gamma}_{\alpha}$.
- A phase (or component) extensive quantity addresses an additive quantity like mass, volume, momentum, and energy that is bounded by the phase volume.

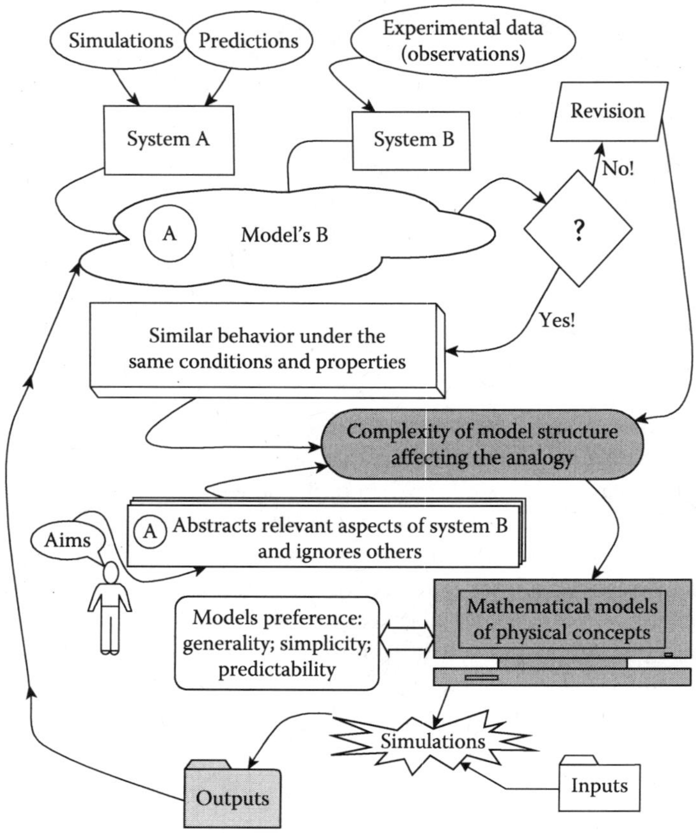

FIGURE 9.1 Processes in enabling a vsA to model a tsB.

Assuming that flow directions through the compartmental boundaries are a priori known, we write the mass balance of a γ fluid in any compartment N of the compartmental network in the form

$$\frac{d}{dt}(\theta_n^\gamma \rho_n^\gamma V_n) = \sum_i^{In} \theta_i^\gamma \rho_i^\gamma q_{in}^\gamma - \sum_j^{Jn} \theta_n^\gamma \rho_n^\gamma q_{nj}^\gamma + \sum_d^{Dn} \theta_n^\gamma J_{dn}^\gamma + \sum_s^{Sn} \tilde{\rho}_{sn}^\gamma Q_{sn} - \theta_n^\gamma \rho_n^\gamma W_n \tag{9.1}$$

where

- ρ_n^γ denotes the γ fluid density in compartment N
- ρ_i^γ denotes the incoming γ fluid density from compartment I to N
- q_{in} denotes the advective flux entering N from the I compartment, through their common boundary
- q_{nj} denotes the crossing boundary advective flux leaving N into the J compartment
- In and Jn denote, respectively, the number of incoming and outgoing fluxes across the boundaries of the N compartment
- J_{dn}^γ denotes the diffusive mass flux of the γ fluid crossing the boundary between the N and the D compartment with a total of Dn such compartments
- $\tilde{\rho}_{sn}^\gamma$ denotes the γ fluid source density injected into N with one of its s ($s \in Sn$) source flux intensity Q_{sn}
- W_n denotes all of its sink pumping intensity
- θ_n^γ denotes the volume fraction (fluid content) of the γ fluid $\left(\sum_\gamma \theta_n^\gamma = 1\right)$ occupying the V_n volume of the N compartment

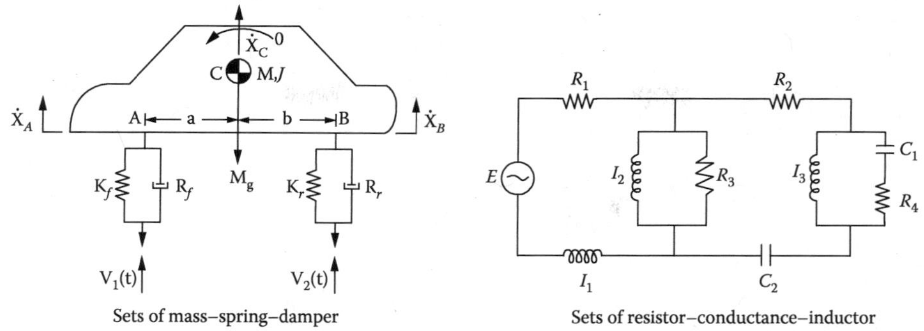

FIGURE 9.2 Application examples of modeling using the LP approach.

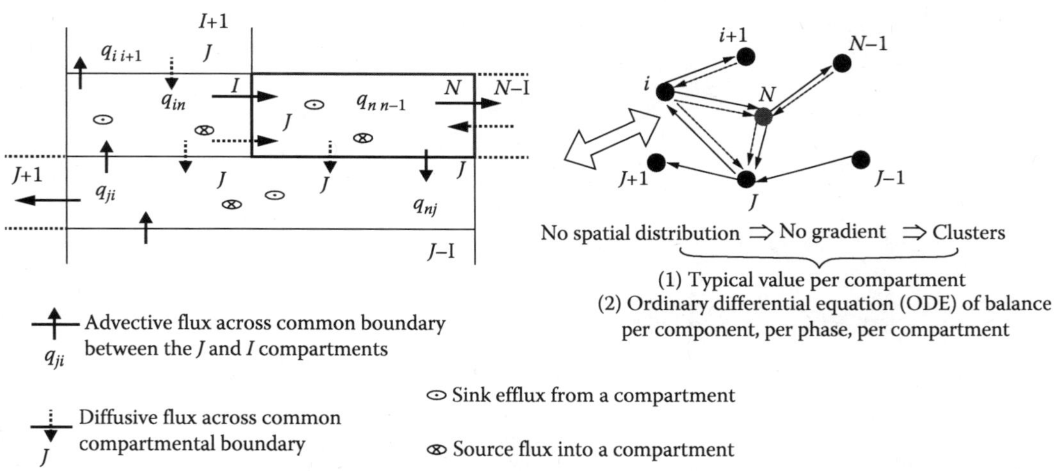

FIGURE 9.3 Compartmental (or nodal) network for hydrodynamic LP modeling.

Considering the advective flux, say crossing the boundary from the I compartment to the N one, the momentum balance equation for the γ fluid that accounts for inertia (rate and quadratic terms of the flux, the latter of which gives rise to vertical motion) and drag (linear flux term) reads

$$\eta_{in}\frac{dq^{\gamma}_{in}}{dt}+\lambda_{in}(q^{\gamma}_{in})^2+\Psi_{in}q^{\gamma}_{in}=P^{\gamma}_{i,n}+\rho^{\gamma}_{i}gZ_{i,n} \tag{9.2}$$

where

- $()_{i,n}\equiv()_i-()_n$
- P^{γ} denotes the γ fluid partial pressure
- g denotes gravity acceleration
- Z denotes altitude along gravity
- η_{in}, λ_{in}, and Ψ_{in} denote, respectively, inductance, transfer, and resistance coefficients across the compartment's boundary

If the flux rate is negligible, then (9.2) conforms to the LP form of Forchheimer's law. If inertia terms are assumed negligible, then (9.2) conforms to the LP form equivalent to Darcy's law.

A flow model addresses (9.1) for all network compartments with (9.2), which is assigned to the common compartmental boundaries. Removing the assumption of knowing a priori the flux direction, we rewrite (9.1), for the fluid mass balance equation, in the form

$$\frac{d}{dt}(\theta_n^\gamma \rho_n^\gamma V_n) = \sum_i^{In}\left(\theta_i^\gamma \rho_i^\gamma \frac{q_{in}^\gamma + |q_{in}^\gamma|}{2} + \theta_n^\gamma \rho_n^\gamma \frac{q_{in}^\gamma - |q_{in}^\gamma|}{2}\right) + \sum_d^{Dn} \theta_n^\gamma J_{dn}^\gamma + \sum_s^{Sn} \tilde{\rho}_{sn}^\gamma Q_{sn} - \theta_n^\gamma \rho_n^\gamma W_n \tag{9.3}$$

where In denotes the total number of compartment boundaries through which flux q_{in} is entering or leaving compartment N. Instead of (9.2), we now rewrite the fluid momentum balance equation assigned to each of the In boundaries of compartment N to read

$$\eta_{in}\left[\frac{d}{dt}\left(\frac{q_{in}^\gamma + |q_{in}^\gamma|}{2}\right) + \frac{d}{dt}\left(\frac{q_{in}^\gamma - |q_{in}^\gamma|}{2}\right)\right] + \lambda_{in}\left[\left(\frac{q_{in}^\gamma + |q_{in}^\gamma|}{2}\right)^2 - \left(\frac{q_{in}^\gamma - |q_{in}^\gamma|}{2}\right)^2\right] + \Psi_{in}\left[\left(\frac{q_{in}^\gamma + |q_{in}^\gamma|}{2}\right) + \left(\frac{q_{in}^\gamma - |q_{in}^\gamma|}{2}\right)\right] = P_{i,n}^\gamma + \rho_i^\gamma g Z_{i,n} \tag{9.4}$$

We note that (9.3) and (9.4) represent the LP flow model for the compartmental network when the flow direction is not assumed to be known a priori.

Referring to the γ fluid diffusive mass flux J_{dn}^γ across the (membrane) boundary between the N and D compartments in (9.1), we write

$$J_{dn}^\gamma = D_{dn} P_{d,n}^\gamma - \chi_{dn} \Pi_{d,n}^\gamma \tag{9.5}$$

where at the dn boundary of all Dn diffusive boundaries ($dn \in Dn$) of the N compartment, D_{dn} denotes the diffusion coefficient due to the partial pressure difference $P_{d,n}^\gamma$ across it, and χ_{dn} denotes the diffusion coefficient due to the osmotic pressure difference $\Pi_{d,n}^\gamma$ across it.

Note that the diffusive flux associated with the osmotic pressure Π_{dn}^γ is in contrary direction to the one generated by the hydrodynamic pressure difference. The osmotic pressure depends linearly on the concentration difference (i.e., follows the notion of a potential flow law) of all α components associated with the γ fluid within the D $\left(C_{\alpha D}^\gamma\right)$ and N $\left(C_{\alpha N}^\gamma\right)$ compartments:

$$\Pi_{dn}^\gamma = \sum_\alpha \sigma_{\alpha\, dn}\left(C_{\alpha D}^\gamma - \xi_\alpha C_{\alpha N}^\gamma\right) \tag{9.6}$$

in which ξ_α denotes a chemical equilibrium (partition) coefficient and $\sigma_{\alpha\, dn}$ denotes a reflection coefficient across a $dn(\in Dn)$ boundary so that (1) $\sigma_{\alpha\, dn} = 1$, when the boundary is permeable to the solvent (fluid) and not to the solute, and (2) $\sigma_{\alpha\,dn} = 0$, when the boundary is permeable to both the fluid and the solute.

We assume that the storage capacity addressed by the occupied volume can change, resulting from deflection of the compartment boundaries. This is similar to the change in a specific yield due to fluctuation of the unsaturated–saturated interface. In reference to (9.3), we associate changes in V_n of the N compartment with the sum of all γ fluid partial pressure difference change $\left(P^{\gamma}_{n,e}\right)$ across its ne ($\in En$) boundary:

$$dV_n = \sum_{e}^{En} \varepsilon_{ne} \sum_{\gamma} dP^{\gamma}_{n,e} \tag{9.7}$$

in which ε_{ne} denotes a compliance factor of the ne wall. We note that (9.7) is similar to Hook's constitutive law for an elastic deformation with V_n/ε_{ne} as its measure of stiffness (i.e., Young's modulus) and dV_n/V_n associated with dilatation (i.e., skeleton strain).

Accounting for the assumption of not a priori determined flow direction, following (9.3), we write the mass balance equation for an α solute being a species of a γ fluid occupying an N compartment:

$$\begin{aligned}\frac{d}{dt}\left(\theta^{\gamma}_n C^{\gamma}_{\alpha n} V_n\right) &= \sum_{i}^{In}\left(\theta^{\gamma}_n C^{\gamma}_{\alpha_i}\frac{q^{\gamma}_{in} + \left|q^{\gamma}_{in}\right|}{2} + \theta^{\gamma}_n C^{\gamma}_{\alpha n}\frac{q^{\gamma}_{in} - \left|q^{\gamma}_{in}\right|}{2}\right)\\ &+ \sum_{d}^{Dn}\theta^{\gamma}_n J^{\gamma}_{\alpha\,dn} + R^{\gamma}_{\alpha n} + \sum_{s}^{sn}\tilde{C}^{\gamma}_{\alpha\,sn} Q_{sn} - \theta^{\gamma}_n C^{\gamma}_{\alpha n} W_n\end{aligned} \tag{9.8}$$

in which $\tilde{C}^{\gamma}_{\alpha\,sn}$ denotes the α solute external sn ($\in Sn$) source concentration injected with the γ fluid and $R^{\gamma}_{\alpha n} = R^{\gamma}_{\alpha n}\left(C^{\gamma}_{n\alpha=1}, C^{\gamma}_{n\alpha=2}, \ldots, C^{\gamma}_{n\alpha=N\alpha}\right)$ denotes the generation rate of the α solute in the γ fluid, occupying the N compartment, by chemical reactions with other solutes, with their corresponding concentrations, of that fluid. We note that summation of (9.8) over the α solutes, based on $\sum_{\alpha} C^{\gamma}_{\alpha n} = \rho^{\gamma}_n, \sum_{\alpha} J^{\gamma}_{\alpha\,dn} = J^{\gamma}_{dn}, \sum_{\alpha} R^{\gamma}_{\alpha n} = 0$, yields the γ fluid mass balance equation in the form of (9.3). For concentration-dependent fluid density, we address the mass fraction $f^{\gamma}_{\alpha n}$ of the α solute in the γ fluid (i.e., $\sum_{\alpha} f^{\gamma}_{\alpha n} = 1$) and assign $C^{\gamma}_{\alpha n} \equiv \rho^{\gamma}_n f^{\gamma}_{\alpha n}$. Further, by mathematical manipulation between (9.3) and (9.8), we can rewrite the latter without terms of mass efflux through the N compartmental boundaries and via a sink term. In (9.8), we refer to the diffusive flux $J^{\gamma}_{\alpha\,dn} = J^{\gamma}_{\alpha\,dn}\left(C^{\gamma}_{\alpha\,D,N}\right)$ of the α solute associated with the γ fluid across the dn boundary as a function of the concentration difference of that solute in the D and N compartments. This mass diffusion of the α solute assuming to be uncharged and with no account for chemical reactions reads

$$J^{\gamma}_{\alpha\,dn} = \omega_{\alpha\,dn}\left(C^{\gamma}_{\alpha D} - \xi_{\alpha} C^{\gamma}_{\alpha N}\right) + J^{\gamma}_{dn}(1 - \sigma_{\alpha\,dn}) C^{*\gamma}_{\alpha\,dn} \tag{9.9}$$

in which $\omega_{\alpha\,dn}$ denotes the α solute diffusion coefficient and $C^{*\gamma}_{\alpha\,dn}$ denotes a concentration subject to conditions associated with the dn boundary:

$$C^{*\gamma}_{\alpha\,dn} \equiv \begin{cases} \dfrac{C^{\gamma}_{\alpha N} - C^{\gamma}_{\alpha D}}{\ell n\left(C^{\gamma}_{\alpha N}/C^{\gamma}_{\alpha D}\right)} & \text{When concentration varies on membrane surface} \\ C^{\gamma}_{\alpha D} & \text{When } J^{\gamma}_{\alpha\,dn} \text{ from } D \text{ to } N \\ C^{\gamma}_{\alpha N} & \text{When } J^{\gamma}_{\alpha\,dn} \text{ from } N \text{ to } D \end{cases} \tag{9.10}$$

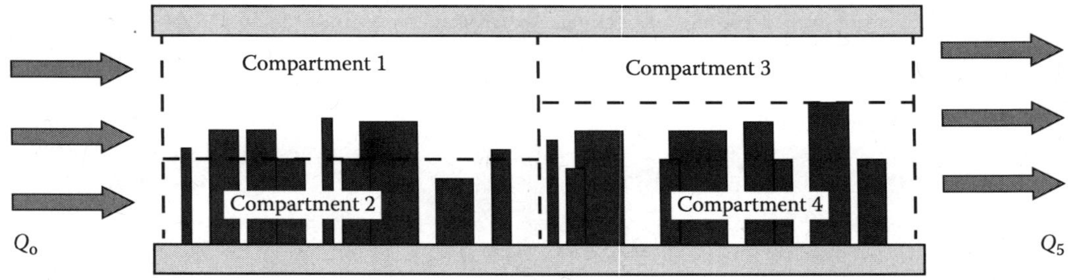

FIGURE 9.4 Inflow (Q_0) and outflow (Q_5) through a "Pin Forest" configuration, for which a subdivision into four virtual compartments is chosen.

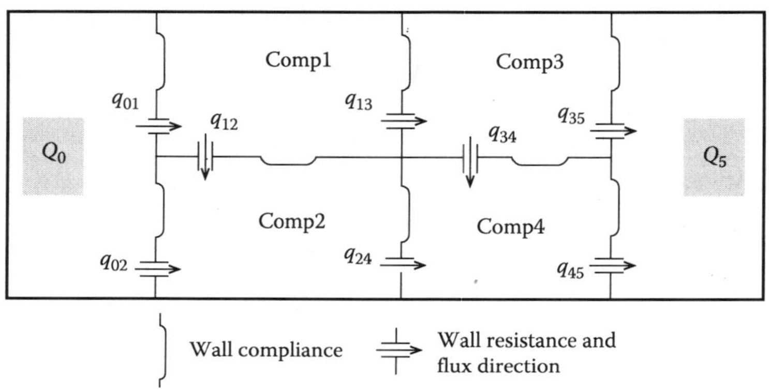

FIGURE 9.5 Compartmental setup chosen to represent the "Pin Forest" system depicted in Figure 9.4. Compartmental cross boundary flow directions and mutual interactions are ascribed a priori.

As a concluding example, let us consider modeling the problem of a one fluid flow through a "Pin Forest" (e.g., urban landscape) carrying one solute, as described in Figure 9.4.

We note that the layout of virtual compartments is identifiable throughout the investigated domain, and moreover, these are chosen to cover certain geometrical characterizations that will be reflected by a volume ratio associated with each subdomain. The actual "Pin Forest" reality (Figure 9.4) is replaced by a virtual compartmental layout (Figure 9.5), which will be our modeling premises.

In view of Figure 9.5 we note that compartments 0 and 5 represent the environment surrounding the assigned 1–4 compartments. Thus, we consider P_0, P_5, ρ_0, ρ_5, C_0, and C_5 as known constant quantities, V_0 and V_5 are constant volumes, and $\theta_0 = \theta_5 = 1$. For the single fluid for which we prescribed its flow direction through the "Pin Forest," let us assume that its compressible density state function ($\rho = \rho(P_{i,n})$) is given for which the compressibility β ($\equiv(1/\rho)(d\rho/dP_{i,n})$) is constant, assign $\theta_2 < \theta_4 < \theta_1 < \theta_3 = 1$ as the fluid volume ratio distribution, and neglect its diffusive flux through any boundary. By virtue of (9.1), we can now write the fluid mass balance equations. For the surrounding environment, we obtain $Q_0 = q_{01} + q_{02}$ for compartment 0 and $\rho_5\, Q_5 = \rho_3\, q_{35} + \rho_4 q_{45}$ for compartment 5. Four additional fluid mass balance equations will be written for compartments 1–4. This, for example, for the fluid-occupying compartment 2, by virtue of (9.1) accounting for (9.7), will read

$$\theta_2\rho_2\left[(\beta V_2 + \varepsilon_{20})^{dP_2}\!\big/_{dt} + (\beta V_2 + \varepsilon_{21})^{dP_{2,1}}\!\big/_{dt} + (\beta V_2 + \varepsilon_{24})^{dP_{2,4}}\!\big/_{dt}\right] = \rho_0 q_{02} + \theta_1\rho_1 q_{12} - \theta_2\rho_2 q_{24}$$

We note that the left-hand side (LHS) of the fluid mass balance equation in compartment 2 is associated with storage factors due to its compressibility and the fluctuation of some of its boundaries. To complete the flow model, we need to write the fluid momentum balance equation for any common compartmental boundary with a crossing flux. For the eight such common boundaries (Figure 9.5), let us assume that inertia due to temporal and spatial changes together with friction between the fluid and the solid interface are contributors to the momentum fluxes; hence, in view of (9.2), for example, for boundary 12 we write

$$\eta_{12}\frac{dq_{12}}{dt}+\lambda_{12}q_{12}^{2}+\Psi_{12}q_{12}=P_{1,2}+\rho_{1}gZ_{1,2}$$

The flow model is well posed as for the prescribed values of P_0, P_5, ρ_0, ρ_5, η_{ij}, λ_{ij}, ψ_{ij}, and the 14 unknowns Q_0, q_{01}, q_{02}, q_{12}, q_{13}, q_{24}, q_{34}, q_{35}, q_{45}, Q_5, P_1, P_2, P_3, P_4 can be solved by eight momentum balance equations and six mass balance equations.

Considering the one solute carried by the fluid, we suppose that mass generation by chemical reactions is negligible, and in addition to the previous assumptions, we propose that its diffusive mass fluxes cannot be neglected in comparison to its advective fluxes. In view of (9.8), when addressing flow direction and accounting for (9.9), we write four solute mass balance equations addressing compartments 1–4. Hence, this, for example, for compartment 2, reads

$$\frac{d}{dt}(\theta_2 V_2 C_2)=C_0 q_{02}+\theta_1 C_1 q_{12}-\theta_2 C_2 q_{24}+\omega_{02}(C_0-\xi C_2)\theta_2+\omega_{12}(C_1-\xi C_2)\theta_2+\omega_{24}(C_4-\xi C_2)\theta_2$$

We note that after solving for the flow problem, the various across-boundary fluxes were resolved and the compartmental volumes can be evaluated by virtue of (9.7). Hence, with prescribed values for C_0, C_5, ω_{ij}, ξ, the four unknowns C_1, C_2, C_3, C_4 can be solved using the four solute mass balance equations.

9.2.3 Microscopic Continuum Approach

The continuum approach applies if average, measurable manifestations associated with the smallest typical length of the problem structure are significantly greater than a typical lattice constant (e.g., mean free path of the molecules). Mathematically, this represents a phase body as a region bounded by a surface (Figure 9.6), assuming that process relevant state quantities and properties are homogeneous within that region and can be assigned to every point within it. This spatial domain is thus referred to behave as a continuum, and the level of description of phenomena within the phase is addressed as description at the microscopic level.

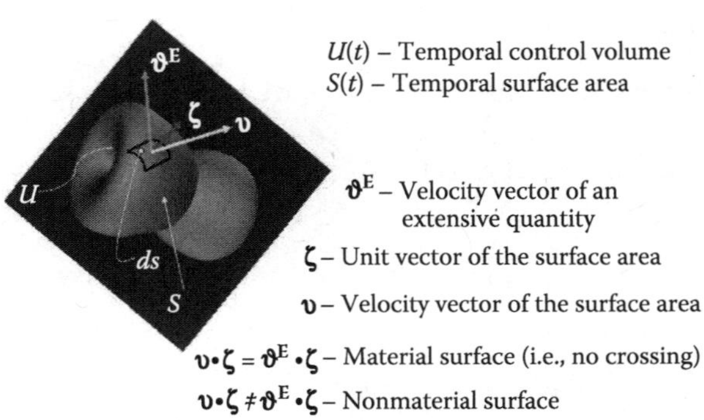

FIGURE 9.6 A phase extensive quantity bound within some continuum domain.

With reference to the continuum approach, let E denote an extensive quantity of any γ phase (m mass, V volume, M momentum vector, ε energy) bounded by U, the phase volume; e ($\equiv dE/dU$) denotes the intensive quantity (i.e., specific E value); $\mathbf{X}^E$ denotes a position vector of a particle moving along a trajectory (Lagrangian concept) carrying E; and $\vartheta^E \equiv \partial \mathbf{X}^E/\partial t\big|_{\xi^E = Const.}$ denotes a velocity vector of E with reference to a ξ^E location at an initial time relative to a fixed frame of reference. We note that this velocity can be measurable only for $E \equiv V$, $\mathbf{J}^E \equiv e\boldsymbol{\vartheta}^E$ denotes the flux vector of E, Γ^E denotes the rate of generating E (source of E) per unit mass of the phase, and $D_{\boldsymbol{\upsilon}}/Dt \equiv (\partial/\partial t) + \boldsymbol{\upsilon} \cdot \nabla$ denotes the material (hydrodynamic or Lagrangian) time derivative that addresses the rate at a fixed frame of reference and the rate due to travel with a velocity $\boldsymbol{\upsilon}$ tangent to a track of spatial change.

In view of the Reynolds transport theorem, we write

$$\frac{D_\upsilon}{Dt}\left(\int_U e dU\right) = \int_U \frac{\partial e}{\partial t} dU + \int_S e\upsilon \cdot \zeta ds \tag{9.11}$$

where on the left hand side (LHS) of (9.11), we account for the material derivative over the total extensive quantity within the phase volume. As we may want to account for this derivative addressing $\boldsymbol{\vartheta}^E$, we add and subtract $\int_S e\vartheta^E \cdot \zeta ds$, and thus, (9.11) conforms to

$$\begin{aligned}\frac{D_\upsilon}{Dt}\left(\int_U e dU\right) &= \int_U \frac{\partial e}{\partial t} dU + \int_S e\vartheta^E \cdot \zeta ds - \int_S e(\vartheta^E - \upsilon) \cdot \zeta ds \\ &= \frac{D_{\vartheta^E}}{Dt}\left(\int_U e dU\right) - \int_S e(\vartheta^E - \upsilon) \cdot \zeta ds \end{aligned} \tag{9.12}$$

The conservation of E in U expressing the balance equation for E reads

$$\frac{D_{\vartheta^E}}{Dt}\left(\int_U e dU\right) = \int_U \rho \Gamma^E dU \tag{9.13}$$

By virtue of (9.12) and (9.13), the integral form for the balance equation for E becomes

$$\frac{D_\upsilon}{Dt}\left(\int_U e dU\right) + \int_S e(\vartheta^E - \upsilon) \cdot \zeta ds = \int_U \rho \Gamma^E dU \tag{9.14}$$

We note that (9.14) is mostly suitable when the investigated domain is of a known spatial configuration. Using the relation of (9.11), the divergence theorem, and in view of (9.13), for infinitesimal phase volume (i.e., $U \rightarrow 0$), the partial differential equation (PDE) form of the balance equation for E, per unit volume of the phase, reads

$$\frac{\partial e}{\partial t} + \nabla \cdot (e\vartheta^E) - \rho \Gamma^E = 0 \tag{9.15}$$

As $\boldsymbol{\vartheta}^E$ in (9.15) is not a measurable quantity (unless for $E \equiv U$), we replace it for the γ phase by an E-weighted velocity, using the first moment notion (i.e., balancing $E_\gamma \vartheta_\gamma = \sum_\alpha E^\alpha \vartheta^\alpha, E_\gamma = \sum_\alpha E^\alpha$ between E fluxes). Hence, the mass-weighted velocity $\boldsymbol{\vartheta}$ for the γ phase will be obtained from the summation over its α components $m\vartheta = \sum_{\alpha} m^{\alpha}\vartheta^{\alpha}$ and its mass $m = \sum_{\alpha} m^{\alpha}$. To circumvent the inability to measure the E flux $\mathbf{J}^E$ ($\equiv e\boldsymbol{\vartheta}^E$), we add and subtract the $e\vartheta$ flux and obtain $\mathbf{J}^E \equiv e\boldsymbol{\vartheta} + e(\boldsymbol{\vartheta}^E - \boldsymbol{\vartheta})$ in which $e\boldsymbol{\vartheta}$ refers to the advective flux of E with respect to $\boldsymbol{\vartheta}$ the phase mass-weighted velocity, and $\mathbf{J}^{Em} \equiv e(\boldsymbol{\vartheta}^E - \boldsymbol{\vartheta})$ refers to the diffusive flux of E relative to that velocity. We note that $\mathbf{J}^{Em}$ vanishes for the phase mass-weighted velocity and, as it refers to the nonmeasurable $\boldsymbol{\vartheta}^E$, its form is habitually proposed to follow a potential flow pattern and is empirically (state function or constitutive law) adopted.

The *Eulerian* form of the *balance equation for E* reads

$$\frac{\partial e}{\partial t} + \nabla \cdot (e\vartheta + \mathbf{J}^{\mathrm{Em}}) - \rho\Gamma^{\mathrm{E}} = 0 \tag{9.16}$$

The *Lagrangian* form of the *balance equation for E* reads

$$\frac{D_{\vartheta}}{Dt}(e) = -e\nabla \cdot \vartheta - \nabla \cdot \mathbf{J}^{\mathrm{Em}} + \rho\Gamma^{\mathrm{E}} \tag{9.17}$$

We note that both (9.16) and (9.17) are PDEs constructed of specific flux terms (i.e., per unit volume of the phase). The divergent terms in (9.16) and (9.17) address fluxes perpendicular crossing the enclosing boundaries for which boundary conditions need to be assigned. The general form of conditions that balance between ascribed fluxes at an Ω boundary segment reads

$$(e\vartheta + \mathbf{J}^{\mathrm{Em}}) \cdot \zeta + \alpha_{\Omega}(e - e_{\Omega}) = \rho\Gamma^{E}\Big|_{\Omega} \tag{9.18}$$

in which the fluxes in (9.18) are per plan square length units for a 3D problem, per curve length units for a 2D problem, and at a point for a 1D problem; ζ denotes an outward-directed unit vector perpendicular to the Ω boundary segment; and $\alpha_\Omega, e_\Omega, \rho\Gamma^E\big|_\Omega$ are prescribed at the Ω boundary segment. The dimensional parameter α_Ω serves as a variable controlling the type of boundary conditions resulting from (9.18). These are (1) Dirichlet's (1st kind) condition $e = e_\Omega$ for $\alpha_\Omega \to \infty$, (2) Neumann's (2nd kind) condition $\left(e\vartheta + \mathbf{J}^{\mathrm{Em}}\right) \cdot \zeta = \rho\Gamma^E\big|_\Omega$ for $\alpha_\Omega = 0$, and (3) the mixed Cauchy's (3rd kind) condition $(e\vartheta + \mathbf{J}^{\mathrm{Em}}) \cdot \zeta + B(e - e_\Omega) = \rho\Gamma^E\big|_\Omega$ for $\infty > \alpha_\Omega = \mathrm{B} > 0$.

We note that (9.16) or (9.17) describes a general "cookbook" formula to which one assigns the specific ingredients to construct the balance equation of the phase (or component) extensive quantity. Let us, for example, construct the mass balance equation of component α within any phase. We assign $E \equiv m^\alpha \Rightarrow e \equiv c^\alpha$, $\mathbf{J}^{m^\alpha m} \equiv c^\alpha\left(\vartheta^{m^\alpha} - \vartheta\right) = -\mathbf{D}^\alpha \cdot \nabla c^\alpha$, using Fick's constitutive law for which $\mathbf{D}^\alpha$ denotes the diffusion tensor. By virtue of (9.16) and the aforementioned ingredients, *Eulerian* form of the *mass balance equation of component* α reads

$$\frac{\partial C^\alpha}{\partial t} = -\nabla \cdot \left(C^\alpha \vartheta - \mathbf{D}^\alpha \cdot \nabla C^\alpha\right) + \rho\Gamma^{m^\alpha}$$

Upon summing over all α components and noting that $\sum_\alpha C^\alpha = \rho$, $\sum_\alpha \mathbf{J}^{m^\alpha m} = 0$, $\sum_\alpha \Gamma^{m^\alpha} = 0$, we obtain the phase mass balance that reads

$$\frac{\partial \rho}{\partial t} = -\nabla \cdot (\rho\vartheta)$$

For constructing a phase momentum equation, we assign $\mathbf{E} = \mathbf{M} \Rightarrow \mathbf{e} = \rho\boldsymbol{\vartheta}$; $\boldsymbol{\Gamma}^{\mathrm{M}} = \boldsymbol{F}$ where $\boldsymbol{F}$ denotes a specific (i.e., per phase mass, e.g., $-g\nabla Z$) body force being the momentum source flux, and the diffusive momentum flux is given by $\mathbf{J}^{Mm} \equiv \rho\boldsymbol{\vartheta}(\boldsymbol{\vartheta}^{M} - \boldsymbol{\vartheta}) = -\boldsymbol{\sigma}$ where $\boldsymbol{\sigma}$ denotes a stress tensor that follows a constitutive relation (e.g., $\tau = \mu\left[\nabla\vartheta + (\nabla\vartheta)^{T}\right] + \lambda''\nabla\cdot\vartheta\mathbf{I}$, the Newtonian fluid shear stress with μ and λ'' denoting, respectively, the first and second viscosities and I denoting the unit tensor). Moreover, in case of a fluid phase, we have $\boldsymbol{\sigma} = \boldsymbol{\tau} - p\mathbf{I}$. By virtue of (9.16) and the aforementioned ingredients, *Eulerian* form of the *momentum balance equation of any phase* reads $\frac{\partial}{\partial t}(\rho\vartheta) = -\nabla\cdot(\rho\vartheta\vartheta - \sigma) + \rho\mathbf{F}$, from which after subtracting the phase mass balance equation, we can write the *Lagrangian* form of the *momentum balance equation of any phase*:

$$\frac{D_{\vartheta}}{Dt}(\vartheta) = \nabla\cdot\sigma + \rho\mathbf{F}$$

To conclude let us note that any continuum-defined microscopic balance PDE can be transformed into its LP counterpart ODE. As an example, let us consider the 2D, Cartesian coordinates, mass balance PDE for a component carried by a fluid, which reads

$$\frac{\partial c}{\partial t} = \bar{\nabla}\cdot\left(\bar{\mathbf{D}}\cdot\bar{\nabla}c - \bar{\vartheta}c\right);\ \bar{\boldsymbol{D}} \equiv \begin{bmatrix} D_{xx} & D_{xy} \\ D_{yx} & D_{yy} \end{bmatrix};\ \bar{\nabla} \equiv \begin{bmatrix} \dfrac{\partial}{\partial x} \\ \dfrac{\partial}{\partial y} \end{bmatrix};\ \bar{\vartheta} \equiv \begin{bmatrix} u \\ v \end{bmatrix}$$

This PDE is to be solved addressing a 2D plan, discretized into a map of virtual compartments, each of which with a prescribed pole point where we seek the estimation of the state variable C (Figure 9.7).

Let S_I denote the area of the I compartment and $L_{I,J}$ denote the distance between two poles at the I and J compartments, perpendicular to their common boundary, which its length is denoted by $\Lambda_{I,J}$. Following the notion of lumping the aforementioned PDE, we write $\int_S \frac{\partial c}{\partial t} dS = \int_S \bar{\nabla}\cdot\left(\bar{\mathbf{D}}\cdot\bar{\nabla}c - \bar{\vartheta}c\right)dS$ and conform it to $\frac{\partial c}{\partial t}\int_S dS = \oint_{\Lambda}\left(\bar{\mathbf{D}}\cdot\bar{\nabla}c - \bar{\vartheta}c\right)\cdot\zeta d\Lambda$ as $\frac{\partial c}{\partial t}$ is assumed to be not space dependent. Using a first-order

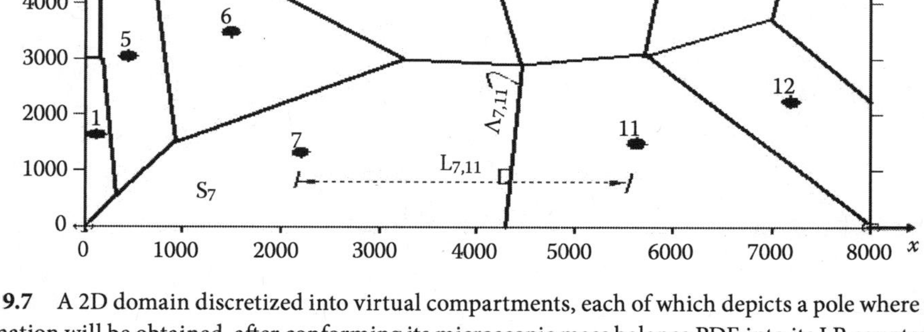

FIGURE 9.7 A 2D domain discretized into virtual compartments, each of which depicts a pole where concentration estimation will be obtained, after conforming its microscopic mass balance PDE into its LP counterpart ODE.

difference, the aforementioned relation is approximated to $\frac{\partial c_I}{\partial t} S_I = \sum_\Lambda \Lambda_{IJ}\left(D_{IJ}\frac{C_J - C_I}{L_{IJ}} - C_I \frac{\vartheta_{IJ} + |\vartheta_{IJ}|}{2} - C_J \frac{\vartheta_{IJ} - |\vartheta_{IJ}|}{2}\right)$ for any I compartment with no a priori presumption of the flux direction. Finally, the mass balance PDE conforms to its ODE equivalent in the form

$$\frac{dc_I}{dt} = \sum_\Lambda \omega_{IJ}(c_J - c_I) - \lambda_{IJ}\left(C_I \frac{q_{IJ} + |q_{IJ}|}{2} + C_J \frac{q_{IJ} - |q_{IJ}|}{2}\right); \; \omega_{IJ} \equiv \frac{\Lambda_{IJ} D_{IJ}}{S_I L_{IJ}}; \; \lambda_{IJ} \equiv \frac{\Lambda_{IJ}}{(S_I)^2}$$

which can be referred as an elaboration of the previously introduced LP approach, as the location of the sought state variable inside each compartment directly dictates the apparent diffusion coefficient ω_{IJ}.

A similar procedure is applicable to any E balance PDE, using, for example (9.16).

9.2.4 Macroscopic, Volume-Averaged, and Continuum Approach

The basic notions of obtaining the macroscopic continuum based on spatial averaging over a representative elementary volume (REV), and from its macroscopic balance equations of phase (or components) extensive quantities, can be found in [2]. Figure 9.8 describes some of the essences addressing the REV concept, intensive quantities (e_α, e_β,...) associated with interacting phases (α, β,....) in it, and their spatial averaging over their respective volumes within the REV. On top of the fundamental definitions appearing in Figure 9.8, we also find in [2] the relations

$$\overline{e_1 + e_2}^\alpha = \overline{e_1}^\alpha + \overline{e_2}^\alpha; \quad \overline{e}^\alpha \Leftrightarrow \overline{e_1 e_2}^\alpha = \overline{e_1}^\alpha\,\overline{e_2}^\alpha + \overline{\overset{\circ}{e}_1 \overset{\circ}{e}_2}^\alpha; \quad \overset{\circ}{e}_i = e_i - \overline{e_i}^\alpha \tag{9.19}$$

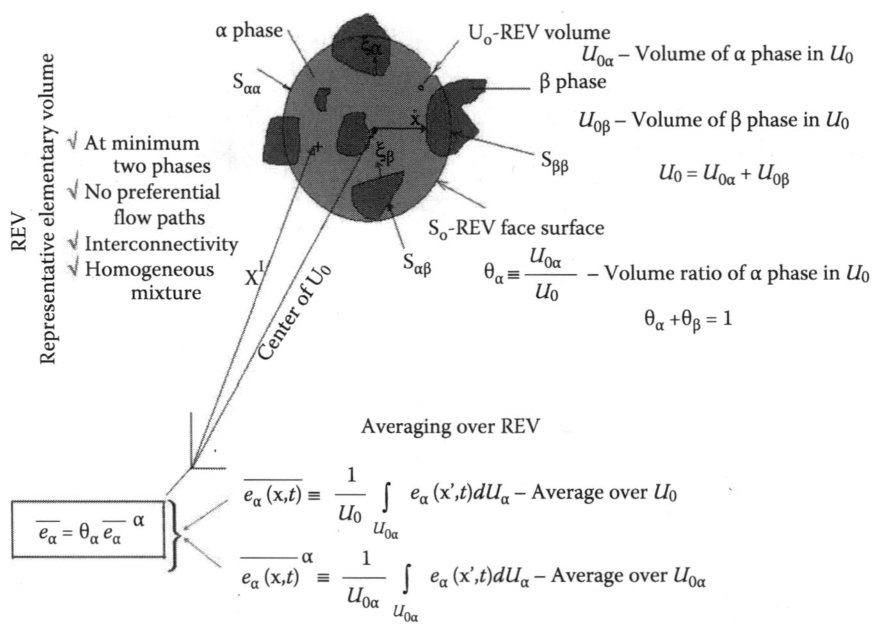

FIGURE 9.8 Fundamental notions concerning the REV and spatial fundamental notions concerning the representative elementary volume and spatial averaging over intensive quantities associated with interacting phases in it.

where $\overline{e_i}^{\alpha}\left(\equiv \int_{U_{0\alpha}} e_i \, dU \Big/ U_{0\alpha}\right)$ denotes the i intensive macroscopic quantity associated with the α phase spatially averaged over the volume it occupies within the REV. Actually the notion of $\overline{e}^{\alpha} \Leftrightarrow \overline{e_1 e_2}^{\alpha}$ in (9.19) can, for example, refer to the microscopic phase extensive quantity being the momentum ($\mathbf{E} \equiv \mathbf{M}$) vector for which $\boldsymbol{e} = \rho\boldsymbol{\vartheta}$; $e_1 \equiv \rho$; $\boldsymbol{e}_2 \equiv \boldsymbol{\vartheta}$ (see second example in Section 9.3). Following [8], we note that phase motion is not of a Brownian type, as the vanishing of its macroscopic intensive quantity will cause the disappearance of the deviation from that intensive quantity (i.e., $e = 0 \Leftrightarrow \overline{e}^{\alpha} = 0 \Rightarrow \overset{\circ}{e} = 0$). Hence, adding to the definition in (9.19), we write

$$\overset{\circ}{e} = \Lambda_E e \tag{9.20}$$

in which Λ_E denotes a factor (e.g., obtained empirically) that can be a scalar or an element of a vector or of a tensor. Following [2], we write the averaging over the REV a spatial derivative that represents *the macroscopic spatial derivative* of $G_{\alpha_{jk}}$ element (scalar, vector, or tensor associated with α) of e_α:

$$\overline{\frac{\partial}{\partial x_i} G_{\alpha_{jk}}} = \frac{\partial}{\partial x_i}\overline{G_{\alpha_{jk}}} + \frac{1}{U_0}\int_{S_{\alpha\beta}} G_{\alpha_{jk}} \zeta_i \, ds \tag{9.21}$$

The modified rule for *the macroscopic spatial derivative* reads

$$\overline{\frac{\partial e}{\partial x_i}}^{\alpha} \cong \frac{\partial \overline{e}^{\alpha}}{\partial x_j} T^*_{\alpha ij} + \frac{1}{\theta_\alpha U_0}\int_{S_{\alpha\beta}} \overset{\circ}{x}_i \frac{\partial e}{\partial x_j} \xi_j \, dS; \quad T^*_{\alpha ij} \equiv \frac{1}{\theta_\alpha U_0}\int_{S_{\alpha\alpha}} \overset{\circ}{x}_i \, \xi_j \, dS \tag{9.22}$$

where T^*_α denotes the α phase tortuosity tensor summing over the α–α part of the REV enclosing interface of the tortuous passage length along the interface of two adjacent phases.

The REV average time derivative, representing *the macroscopic time derivative*, reads

$$\frac{D_\upsilon}{Dt}(\overline{e_\alpha}) = \overline{\frac{\partial e_\alpha}{\partial t}} + \frac{1}{U_0}\int_{S_{\alpha\beta}} e_\alpha \upsilon \cdot \zeta \, ds \tag{9.23}$$

We note in (9.23) that $1 \ll St^e \Leftrightarrow \frac{D_\upsilon}{Dt}(\overline{e_\alpha}) \cong \frac{\partial}{\partial t}\overline{e_\alpha}$ where $St^e \equiv \frac{L^e_c / t^e_c}{\vartheta_c}$ denotes the Strouhal number, for which L^e_c, t^e_c, ϑ_c denote characteristic values of length, time, and velocity, respectively, of the phase intensive quantity.

In view of (9.16), (9.19), (9.21), and (9.23), the macroscopic form of the general balance equation for the α phase E quantity reads

$$\begin{aligned} \frac{D_\upsilon}{Dt}\left(\theta_\alpha \overline{e_\alpha}^{\alpha}\right) = {} & -\nabla \cdot \left[\theta_\alpha \left(\overline{e_\alpha}^{\alpha}\,\overline{\vartheta_\alpha}^{\alpha} + \overline{\overset{\circ}{e}_\alpha \overset{\circ}{\vartheta}_\alpha}^{\alpha} + \overline{\mathbf{J}^{Em}_\alpha}^{\alpha}\right)\right] + \theta_\alpha \overline{\rho_\alpha \Gamma^E_\alpha}^{\alpha} \\ & - \frac{1}{U_0}\int_{S_{\alpha\beta}} \left[e_\alpha(\vartheta_\alpha - \upsilon) + \mathbf{J}^{Em}_\alpha\right] \cdot \zeta \, ds \end{aligned} \tag{9.24}$$

in which $\overline{\overset{\circ}{e}_\alpha \overset{\circ}{\vartheta}_\alpha}^{\alpha}$ is referred to as the macroscopic dispersive flux, which is an artifact of the spatial averaging procedure. The combined $\overline{\overset{\circ}{e}_\alpha \overset{\circ}{\vartheta}_\alpha}^{\alpha} + \overline{\mathbf{J}^{Em}_\alpha}^{\alpha}$ macroscopic dispersive and diffusive flux in (9.24) is referred as the macroscopic hydrodynamic dispersive flux. Assuming that (1) a material α–β interface, that is, $\upsilon\cdot\zeta = \vartheta_\alpha^E\cdot\zeta \Leftrightarrow \left[e_\alpha\left(\vartheta_\alpha - \upsilon\right) + \mathbf{J}_\alpha^{Em}\right]\cdot\zeta = 0$, and (2) $1 \ll St^e \Leftrightarrow \frac{D_\upsilon}{Dt}\overline{(e_\alpha)} \cong \frac{\partial}{\partial t}\overline{e_\alpha}$, we rewrite (9.24) the general α *phase balance equation for its E quantity* to read

$$\frac{\partial}{\partial t}\left(\theta_\alpha \overline{e_\alpha}^\alpha\right) = -\nabla\cdot\left[\theta_\alpha \overline{e_\alpha}^\alpha \overline{\vartheta_\alpha}^\alpha + \theta_\alpha\left(\overline{\overset{\circ}{e_\alpha}\,\overset{\circ}{\vartheta_\alpha}}^\alpha + \overline{\mathbf{J}_\alpha^{Em}}^\alpha\right)\right] + \theta_\alpha \overline{\rho_\alpha \Gamma_\alpha^E}^\alpha \tag{9.25}$$

As an example let us consider the macroscopic mass balance of a δ component in α phase, considering ϑ^U its volume-weighted velocity. Hence, considering $E = m^\delta \Rightarrow e = \rho_\alpha^\delta\ (\equiv C^\delta)$; $\mathbf{J}^{m^\delta U}\ (\equiv \mathbf{J}^\delta)$; $\Gamma^{m^\delta}\ (\equiv \Gamma^\delta)$ and assigning Fick's law as the potential flow pattern defining the hydrodynamic dispersion flux $\left(\overline{\overset{\circ}{e_\alpha}\,\overset{\circ}{\vartheta_\alpha^U}}^\alpha + \overline{\mathbf{J}_\alpha^{Em}}^\alpha \equiv -\mathbf{D}_h^\delta\cdot\nabla\overline{C^\delta}^\alpha\right)$, then in view of (9.25) we write

$$\frac{\partial}{\partial t}\left(\theta_\alpha \overline{C^\delta}^\alpha\right) = -\nabla\cdot\left(\theta_\alpha \overline{C^\delta}^\alpha \overline{\vartheta^U}^\alpha - \theta_\alpha \mathbf{D}_h^\delta\cdot\nabla\overline{C^\delta}^\alpha\right) + \theta_\alpha \overline{\rho\Gamma^\delta}^\alpha$$

in which $\mathbf{q} \equiv \theta_\alpha \overline{\vartheta^U}^\alpha$ can be referred to as Darcy's specific flux, for negligible solid phase velocity.

9.2.5 Spatial Scale–Dependent Macroscopic Balance Equations

Accounting for the possibility that $\overline{e}^\alpha \Leftrightarrow \overline{e_1 e_2}^\alpha$ and by virtue of (9.19), we note that the advective flux can become $\overline{e\vartheta}^\alpha = \overline{e}^\alpha\,\overline{\vartheta}^\alpha + \overline{\overset{\circ}{e}\,\overset{\circ}{\vartheta}}^\alpha = \left(\overline{e_1}^\alpha\,\overline{e_2}^\alpha + \overline{\overset{\circ}{e_1}\,\overset{\circ}{e_2}}^\alpha\right)\overline{\vartheta}^\alpha + \overline{\overset{\circ}{e}\,\overset{\circ}{\vartheta}}^\alpha$ in which $d^E \equiv \overline{\overset{\circ}{e_1}\,\overset{\circ}{e_2}}^\alpha$ accounts for macroscopic deviations from the intensive macroscopic quantity $\overline{e}^\alpha$ of α phase and $\overline{\overset{\circ}{e_1}\,\overset{\circ}{e_2}}^\alpha\,\overline{\vartheta}^\alpha \equiv d^E\overline{\vartheta}^\alpha$ presents deviation from the advective flux. When substituting this into (9.25), we obtain the α phase balance equation of its *E* quantity that reads

$$\frac{\partial}{\partial t}\left[\theta_\alpha\left(\overline{e_1}^\alpha\,\overline{e_2}^\alpha + d^E\right)\right] = -\nabla\cdot\left[\theta_\alpha\left(\overline{e_1}^\alpha\,\overline{e_2}^\alpha + d^E\right)\overline{\vartheta}^\alpha + \theta_\alpha\left(\overline{\overset{\circ}{e}\,\overset{\circ}{\vartheta}}^\alpha + \overline{\mathbf{J}^E}^\alpha\right)\right] + \theta_\alpha \overline{\rho\Gamma^E}^\alpha \tag{9.26}$$

Hereinafter, for the sake of reducing the overload of notations, we will refer to macroscopic quantities without the $\overline{()}^\alpha$ designator. By virtue of (9.20) we can relate the dispersive flux in (9.26) to the advective flux and the deviation from the intensive quantity, reading

$$\overset{\circ}{e}\,\overset{\circ}{\vartheta} = \mathbf{A}\cdot e\vartheta,\ \mathbf{A} \equiv \Lambda_E\Lambda_\vartheta,\ \overset{\circ}{\vartheta} = \Lambda_\vartheta\cdot\vartheta;\ d^E \equiv \overset{\circ}{e_1}\,\overset{\circ}{e_2} = \Lambda_{e_{12}} e_1 e_2,\ \Lambda_{e_{12}} \equiv \Lambda_{e_1}\Lambda_{e_2} \tag{9.27}$$

where Λ_ϑ denotes a tensor quantity. Referring to (9.26) and in view of (9.27), we chose to make the following assumptions:

[A.1] $|e_1 e_2| \gg |d^E| = |\Lambda_{e_{12}} e_1 e_2| \Leftrightarrow \varepsilon_\alpha \equiv |\Lambda_{e_{12}}| \ll 1 \Rightarrow \Lambda_{e_{12}} = \varepsilon_\alpha B_e \Leftrightarrow |B_e| = 1$;

[A.2] $|e\vartheta| \gg |\overset{\circ}{e}\,\overset{\circ}{\vartheta}| = |\mathbf{A}\cdot e\vartheta| \Leftrightarrow \varepsilon \equiv \|\mathbf{A}\| \ll 1 \Rightarrow \mathbf{A} = \varepsilon\mathbf{B} \Leftrightarrow \|\mathbf{B}\| = 1$.

By virtue of [A.1] we rewrite (9.26) to read

$$\frac{\partial}{\partial t}\left(\theta e_1 e_2\right)+\varepsilon_\alpha \frac{\partial}{\partial t}\left(\theta B_e e_1 e_2\right)=-\nabla\bullet\left[\theta\left(e_1 e_2 \vartheta+\mathring{e}\,\mathring{\vartheta}+\mathbf{J}^E\right)\right]+\theta\rho\Gamma^E-\varepsilon_\alpha\nabla\bullet\left(\theta B_e e_1 e_2\vartheta\right) \tag{9.28}$$

which accounts for the advective and the hydrodynamic dispersive fluxes. The latter is habitually adopting a potential flow pattern fitting an empirical relation.

By virtue of [A.1] and [A.2], we rewrite (9.26) to read

$$\frac{\partial}{\partial t}\left(\theta e_1 e_2\right)+\varepsilon_\alpha \frac{\partial}{\partial t}\left(\theta B_e e_1 e_2\right)=-\nabla\bullet\left[\theta\left(e_1 e_2 \vartheta+\mathbf{J}^E\right)\right]+\theta\rho\Gamma^E-\nabla\bullet\left[\theta\left(\varepsilon_\alpha B_e \mathbf{I}+\varepsilon\mathbf{B}\right)\bullet e_1 e_2\vartheta\right] \tag{9.29}$$

which addresses the advective and diffusive fluxes.

As both solutions of (9.28) and (9.29), when $\varepsilon \approx 0$ (ε_α) are functions of $\varepsilon \ll 1$, we can expand their dependent variables using terms associated with a polynomial of ε powers. Approximate solutions to (9.28) and (9.29) can thus be obtained by solving an infinite set of the corresponding balance equations of descending order magnitudes. While the first (i.e., when $\varepsilon = 0$) balance equation may be nonlinear and of the highest order of magnitude, the others are linear balance equations as they are combined with variables obtained from the solution of equations that are of higher-order magnitude. By virtue of [A.1] and in view of (9.28), we obtain the set of *general macroscopic balance equations for any α phase of its E extensive quantity:*

The *primary (dominant)* one corresponding to the REV (larger) length scale reads

$$\frac{\partial}{\partial t}\left(\theta e_1 e_2\right)=-\nabla\bullet\left[\theta e_1 e_2 \vartheta+\theta\left(\mathring{e}\,\mathring{\vartheta}+\mathbf{J}^E\right)\right]+\theta\rho\Gamma^E \tag{9.30}$$

Concurrently, the *secondary* one that corresponds to a length span much smaller than that of (9.30) reads

$$\frac{\partial}{\partial t}\left(\theta d^E\right)=-\nabla\bullet\left(\theta d^E \vartheta\right) \tag{9.31}$$

We note that (9.30) and (9.31) are based on a different premise and thus can represent different characteristics. At the REV (larger) spatial scale and subject to different assumptions, the governing balance equation (9.30) can accommodate pure advection (hyperbolic PDE), advection–dispersion (hyperbolic or parabolic PDE, respectively), or pure dispersion (elliptic PDE) fluxes. The secondary governing balance equation (9.31) describes the concurrent pure advection (i.e., hyperbolic PDE characterized by a sharp front migration of the state variable) of the deviation (d^E) from the intensive quantity, at a spatial scale much smaller than that of the REV. The solution for d^E of (9.31) also on ϑ obtained from the solution of (9.30). Moreover, one can seek further resolution modes of d^E and ϑ, via an iterative process between (9.31) and the combined form of (9.30) and (9.31), which is the form expressed in (9.28).

By virtue of [A.1] and [A.2] and in view of (9.29), another possible set of *general macroscopic balance equations of the E extensive quantity for any $\propto$ phase:*

The *primary (dominant)* one valid at the REV (larger) length scale reads

$$\frac{\partial}{\partial t}\left(\theta e_1 e_2\right)=-\nabla\bullet\left[\theta\left(e_1 e_2 \vartheta+\mathbf{J}^E\right)\right]+\theta\rho\Gamma^E \tag{9.32}$$

Concurrently, the *secondary* one that corresponds to a length span much smaller than that of (9.32) reads

$$\frac{\partial}{\partial t}\left(\theta d^E\right)=-\nabla\bullet\left\{\theta\left[\mathbf{I}+\left(\Lambda_{e_{12}}\right)^{-1}\mathbf{A}\right]\bullet d^E\vartheta\right\} \tag{9.33}$$

in which the dispersive flux $\left(\overset{\circ}{e}\overset{\circ}{\vartheta}\right)$ is also addressed. The discussion and procedure referring to (9.28), (9.30), and (9.31) correspond, respectively, also to (9.29), (9.32), and (9.33).

On the premises of sets (9.30) and (9.31), let us exemplify the construction of a spatial-scaled component mass balance equations in a fluid. In [5], we consider a γ component for which m^{γ} denotes its mass so its extensive quantity is with the intensive quantity $\left(e^{m^{\gamma}} \equiv C = \rho\omega\right)$ given by its concentration $C\left(= e_1^{m^{\gamma}} e_2^{m^{\gamma}};\ e_1^{m^{\gamma}} \equiv \omega, e_2^{m^{\gamma}} \equiv \rho\right)$ related to its mass fraction (ω), and let us account for the specific component source $\left(\Gamma^{m^{\gamma}}\right)$. Let us consider a saturated domain, and following Fick's law based on a potential flow pattern, we combine the component diffusive flux vector $\left(\mathbf{J}^{m^{\gamma}}\right)$ and its dispersive flux vector $\overset{\circ}{C}\overset{\circ}{\mathbf{V}}\left[=\left(\overset{\circ}{\rho\omega}\right)\overset{\circ}{\mathbf{V}}\right]$ into $-\mathbf{D}_h\boldsymbol{\cdot}\nabla C\left(=\overset{\circ}{C}\overset{\circ}{\mathbf{V}}+\mathbf{J}^{m^{\gamma}}\right)$ for which $\mathbf{D}_h$ denotes the component constant hydrodynamic dispersion tensor. In view of (9.30), for the REV length scale, we obtain the *primary component mass balance equation* in the form

$$\frac{\partial}{\partial t}(\phi C) = -\nabla\boldsymbol{\cdot}\left[\phi\left(C\boldsymbol{\vartheta} - \mathbf{D}_h\boldsymbol{\cdot}\nabla C\right)\right] + \phi\rho\Gamma^{m^{\gamma}}$$

that describes the transport of a component subject to advective and dispersive flux mechanisms, interpreted by hyperbolic and parabolic PDE characteristics, respectively.

In view of (9.31), at the smaller scale, the *secondary component mass balance equation* becomes

$$\frac{\partial}{\partial t}\left(\phi d^{m^{\gamma}}\right) = -\nabla\boldsymbol{\cdot}\left(\phi d^{m^{\gamma}}\boldsymbol{\vartheta}\right);\ d^{m^{\gamma}} \equiv \overset{\circ}{\omega}\overset{\circ}{\rho}$$

that governs transport at the pore scale by pure advection (hyperbolic PDE) and in which $d^{m^{\gamma}}$ denotes the deviation from the component concentration.

In reference to sets (9.32) and (9.33), let us exemplify the construction of the spatial-scaled momentum balance equations of a phase. Following [4] and [5], here the phase extensive quantity is its momentum (i.e., $\boldsymbol{E}\equiv\boldsymbol{M}$), for which its macroscopic quantities are $\mathbf{e}^{M}\equiv\rho\boldsymbol{\vartheta}$ (i.e., $e_1^{M}\equiv\rho$, $\mathbf{e}_2^{M}\equiv\boldsymbol{\vartheta}$, $\boldsymbol{d}^{M}\equiv\overset{\circ}{\rho}\overset{\circ}{\boldsymbol{\vartheta}}$ denote the deviation from the intensive quantity, and $\left(\overset{\circ}{\rho\boldsymbol{\vartheta}}\right)\overset{\circ}{\boldsymbol{\vartheta}}$ represents the momentum dispersive flux tensor), the momentum diffusive flux $\mathbf{J}^{M}=-\boldsymbol{\sigma}$ ($\boldsymbol{\sigma}$-phase stress tensor), and $\boldsymbol{\Gamma}^{M}=\mathbf{F}$ ($\boldsymbol{F}$-specific body force acting on the phase) refers to the momentum source flux vector. In view of (9.32) for the REV length scale, we obtain the *primary phase momentum balance equation* in the form

$$\frac{\partial}{\partial t}(\theta\rho\boldsymbol{\vartheta}) = -\nabla\boldsymbol{\cdot}\left[\theta\rho\boldsymbol{\vartheta}\boldsymbol{\vartheta} - \theta\boldsymbol{\sigma}\right] + \theta\rho\mathbf{F}$$

In view of (9.33), at a scale much smaller than the REV length scale, the *secondary phase momentum balance equation* becomes

$$\frac{\partial}{\partial t}\left(\theta\boldsymbol{d}^{M}\right) = -\nabla\boldsymbol{\cdot}\left\{\theta\left[\left(\mathbf{I}+\boldsymbol{\Omega}\right)\boldsymbol{\cdot}\boldsymbol{d}^{M}\right]\boldsymbol{\vartheta}\right\};\quad \boldsymbol{\Omega}\equiv\left(\Lambda_{\rho}\Lambda_{V}\right)^{-1}\Lambda_{\rho V}\Lambda_{V}$$

Letting the α phase represent a Newtonian fluid, then the macroscopic Navier–Stokes equation resulting from the aforementioned primary momentum balance equation can govern the propagation of shock waves through porous media [3] and conform to Forchheimer's law accounting for the transfer of fluid inertia through the microscopic solid–fluid interface or to Darcy's law when friction at that interface is dominant [4]. At the much smaller scale, the hyperbolic PDE addressing the secondary phase momentum balance equation describes an inertia momentum balance equation in the form of a wave equation that governs the concurrent propagation of the $\boldsymbol{d}^M$ quantity.

9.2.6 Modeling of Multiphase Multicomponent Transport Phenomena

We maintain that the modeling of the temporal and spatial distribution of contaminants in the unsaturated and saturated zones is best demonstrated via an example [1]. This will thus be developed through the case of three $l \equiv q, n, g$ interacting mobile phases: q denoting the aqueous fluid, n denoting the nonaqueous phase liquids (NAPL), and g denoting the gas fluid. The described model accounts for $\mathrm{p} \equiv w, a, c_m$ three components, w denoting water, a denoting air, and c_m denoting a contaminant (e.g., trichloroethylene [TCE]), undergoing equilibrium or nonequilibrium partitioning between the three mobile phases as well as sorption into the solid matrix. Schematization of the interactions between the phases and the components is depicted in Figure 9.9.

Let us thus consider the case of equilibrium partitioning of components between phases under isothermal conditions; we then write the following:

The *mass balance equation of the contaminant, c_m, in the three phases, q, g, n,* reads

$$\frac{\partial}{\partial t}\left[\phi\left(S_q M_q X_{c_m q} + S_g M_g X_{c_m g} + S_n M_n X_{c_m n}\right) + \rho_b K_d M_q X_{c_m q}\right] = \nabla\cdot\left(\phi S_q \mathbf{D}_{hq}^{c_m}\cdot M_q \nabla X_{c_m q} - M_q X_{c_m q}\vartheta_q\right)$$
$$+\nabla\cdot\left(\phi S_g \mathbf{D}_{hg}^{c_m}\cdot M_g \nabla X_{c_m g} - M_g X_{c_m g}\vartheta_g\right) + \nabla\cdot\left(\phi S_n \mathbf{D}_{hn}^{c_m}\cdot M_n \nabla X_{c_m n} - M_n X_{c_m n}\vartheta_n\right) + q_{c_m} \quad (9.34)$$

The *mass balance equation of the water component, w, in the two phases, q, g,* reads

$$\frac{\partial}{\partial t}\left[\phi\left(S_q M_q X_{wq} + S_g M_g X_{wg}\right)\right] = \nabla\cdot\left(\phi S_q \mathbf{D}_{hq}^{w}\cdot M_q \nabla X_{wq} - M_q X_{wq}\vartheta_q\right)$$
$$+\nabla\cdot\left(\phi S_g \mathbf{D}_{hg}^{w}\cdot M_g \nabla X_{wg} - M_g X_{wg}\vartheta_g\right) + q_w \quad (9.35)$$

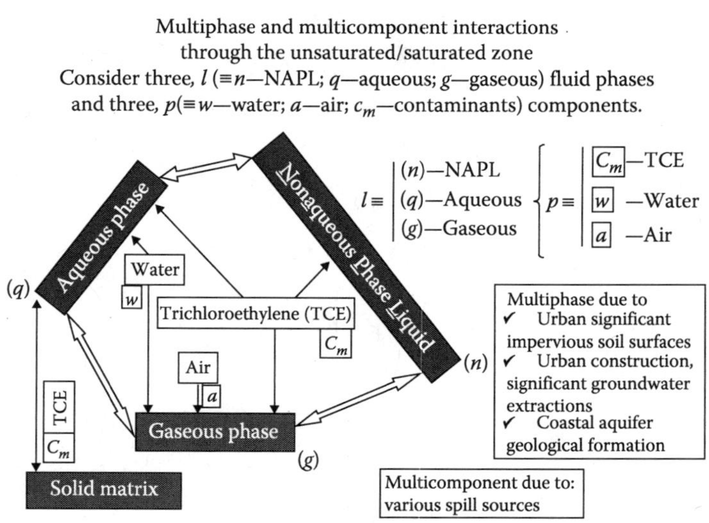

FIGURE 9.9 Schematization of four interacting phases (three mobile fluids) with three components. A possible transport phenomenon in an urban area aquifer.

The *mass balance equation of the air component, a, in the g phase* reads

$$\frac{\partial}{\partial t}\left[\phi S_g M_g X_{ag}\right] = \nabla\bullet\left(\phi S_g \mathbf{D}^a_{hg}\bullet M_g \nabla X_{ag} - M_g X_{ag}\boldsymbol{\vartheta}_g\right) + q_a \tag{9.36}$$

In (9.34) through (9.36), $X_{pl} = n_{pl}/n_l, \left(\sum_p X_{pl} = 1,\ \forall\, l\right)$ denotes the mole fraction of component p in fluid l; porosity is denoted by ϕ, fluid content by ϕS_l in which $S_l\left(\sum_l S_l = 1\right)$ denotes saturation of the l fluid; the relation between the l fluid mass density ρ_l and its molar density M_l is defined by

$$\rho_l/M_l = m_l/n_l \Rightarrow \rho_l = M_l \sum_p m_{pl} \Big/ \sum_p n_{pl} = M_l \sum_p W_p \frac{n_{pl}}{n_l} \Big/ \sum_p \frac{n_{pl}}{n_l} = M_l \sum_p W_p X_{pl}$$

where

$m_l\left(= \sum_p m_{pl}\right)$ denotes the l fluid mass in [mole/volume]

$n_l\left(\approx \sum_p n_{pl}\right)$ denotes a quantity of substance of phase l in [mole]

$W_p = m_p/n_p$ denotes the molecular weight of the p component
ρ_b denotes the solid mass density
K_d denotes the sorption coefficient onto the solid phase
q_p denotes a source/sink mass flux of the p component

Using the fluid linear momentum balance equations given by Darcy's law (friction dominant between fluid and solid phases, at their common microscopic interface), we obtain $\boldsymbol{\vartheta}_l$ the l fluid velocity vectors (Darcy's specific flux) in the form

$$\vartheta_l = -\mathbf{K}\bullet\frac{k_{rl}}{\mu_l}\left(\nabla P_l + \rho_l g \nabla Z\right);\ \ \forall\, l \tag{9.37}$$

where

$\boldsymbol{K}$ is the absolute permeability tensor
k_{rl} denotes the relative permeability of the l fluid
μ_l denotes its viscosity

A typical graph for k_{rn} is depicted in Figure 9.10.

The l fluid partial pressure P_l in (9.37) is given by constitutive relations in the form

$$\begin{matrix} P_g = P_n + \hat{\alpha} P_{c,\,gn}(S_g) + (1-\hat{\alpha})\left[P_{c,\,gq}(S_g) - P_{c,\,nq}(S_q = 1)\right] \\ P_n = P_q + \hat{\alpha} P_{c,\,nq}(S_q) + (1-\hat{\alpha}) P_{c,\,nq}(S_q = 1) \end{matrix};\ \ \hat{\alpha} \equiv \min\left(1, S_n/S_n^*\right) \tag{9.38}$$

where

$P_{C,l_i l_j}$ denotes the capillary pressure between the l_i and l_j fluids
S_n^* denotes a blending parameter

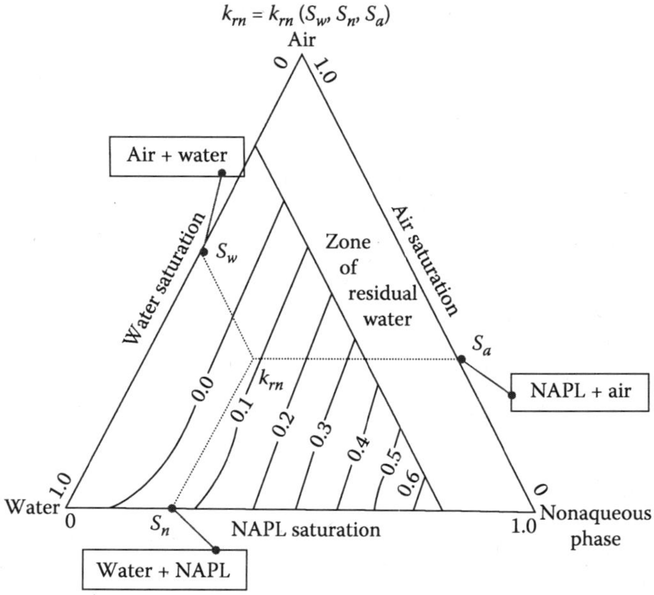

FIGURE 9.10 NAPL relative permeability k_{rn} as a function of saturation S_l in a three fluid system—$l \equiv g, q, n$ system.

The ij element of the hydrodynamic dispersion tensor $D^p_{h,lij}$ for component p in the l fluid is defined by

$$\phi S_l D^p_{h,lij} = \left(\phi S_l D^p_l + \alpha_{lT}\left|\vartheta_l\right|\right)\delta_{ij} + \left(\alpha_{lL} - \alpha_{lT}\right)\frac{\vartheta_{li}\vartheta_{lj}}{\left|\vartheta_l\right|}; \quad \forall l \tag{9.39}$$

where

α_{lT} and α_{lL} denote, respectively, the transverse and longitudinal dispersivities of the l fluid
D^p_l denotes the diffusion coefficient of component p in the l fluid
δ denotes the Kronecker delta tensor

Equilibrium partitioning coefficients between the gas, g, and the nonaqueous fluids, n, as well as the gas, g, and the aqueous fluids, q. In these cases we have state functions in the form

$$X_{c_m g} = \frac{X^*_{c_m g} P^*_g}{P_g} X_{c_m n}; \quad X_{c_m q} = \frac{X^*_{c_m q} P_g}{X^*_{c_m g} P^*_g} X_{c_m g}; \quad X_{wg} = \frac{\beta_{wgq}}{P_g} X_{wq} \tag{9.40}$$

where

$X^*_{c_m g}, X^*_{c_m q}$ denote the mole fractions of contaminant in equilibrium with nonaqueous fluids
P^*_l denotes a reference pressure for the l fluid
β_{wgq} denotes the partial pressure of water in gas fluid in equilibrium with the aqueous fluid

The state functions associated with the fluid molar densities M_q, M_n, M_g are expressed in the form

$$M_l = \frac{1 + \beta_l\left(P_l - P^*_l\right)}{\sum_p X_{pl} M^*_l}, \quad l \equiv q, n; \quad M_g = \frac{P_g}{RT^*} \tag{9.41}$$

where

M_l^* denotes the l fluid molar density at $P_l = P_l^*$
the l fluid compressibility is denoted by β_l
the ideal gas constant is denoted by R
T^* denotes a reference temperature

Finally, for the case of need to accommodate the point-to-point variability in a heterogeneous permeability field, the capillary pressure P_{c,l_il_j} curves associated with the l_i and l_j fluids are scaled using P^D_{c,l_il_j} the dimensionless capillary pressure between these fluids in the form

$$P^D_{c,l_il_j} = \frac{P_{c,l_il_j}}{\sigma_{l_il_j}}\left(\frac{|\mathbf{K}|}{\phi}\right)^{\tilde{\alpha}}; \ \tilde{\alpha}\approx 0.5; \quad l_il_j \equiv gn, gq, nq \tag{9.42}$$

where $\sigma_{l_il_j}$ denotes the interfacial tension between the l_i and l_j fluids.

9.3 Summary and Conclusions

Fundamental formulation was addressed for modeling the transport of multiphase and multicomponent through heterogeneous media. Although beyond the scope of the current interest, in the case of nonequilibrium dissolution, separate mass balance equations for the contaminant will need to be prescribed. Moreover, as simulations may be consuming significant computational time, seeking for the possibility of studying small-scale (spatial span and evolution periods) examples can yield qualitative and quantitative assessments on trends for larger-scale examples. This scale-up process can be achieved by using nondimensional analysis of the involved balance equations.

References

1. Sorek, S., Kuznetsov, M., Yakirevitch, A., and Ronen, D., 2005, Multiphase and multi-component interactions through the unsaturated saturated zone: Field and model study. In *Reactive Transport in Soils and Groundwater*, (eds. G. Nutzmann, P. Viotti, and P. Aagaard), Springer Verlag, New York, pp. 171–185.
2. Bear, J. and Bachmat, Y., 1990, *Introduction to Modeling of Transport Phenomena in Porous Media*, Kluwer Academic Publishers, Dordrecht, the Netherlands.
3. Levy, A., Sorek, S., Ben-Dor, G., and Bear, J., 1995, Evolution of the balance equations in saturated thermoelastic porous media following abrupt simultaneous changes in pressure and temperature, *Transport in Porous Media*, 21, 241–268.
4. Sorek, S., Levi-Hevroni, D., Levy, A., and Ben-Dor, G., 2005, Extensions to the Macroscopic Navier-Stokes Equation, *Transport in Porous Media*, 61, 215–233.
5. Sorek, S. and Ronen, D., 2006, Scale dependent fluid momentum and solute mass macroscopic balance equations: Theory and observations, CMWR XVI, Copenhagen, Denmark, http://proceedings.cmwr-xvi.org/getFile.py/access?contribId=80\&sessionId=11&resId=0&materialId=paper&confId=a051
6. Ronen, D. and Sorek, S., 2005, The unsaturated zone–a neglected component of nature, *Reactive Transport in Soils and Groundwater* (eds. G. Nutzmann, P. Viotti, and P. Aagaard), Springer Verlag, Berlin, Germany, pp. 3–15.
7. Ronen, D., Sorek, S., and Gilron, J., 2012, Rationales behind irrationality of decision-making in groundwater quality management, *Journal of Ground Water*, 50, 1, 27–36. doi: 10.1111/j.1745–6584.2011.00823.x.
8. Sorek, S., Ronen, D., and Gitis, V., 2010, Scale-dependent macroscopic balance equations governing transport through porous media: Theory and observations, *Transport in Porous Media*, 81, 1, 61–72.

10

Green Infrastructure: Hydrological and Hydraulic Design

Sandeep Joshi
Shrishti Eco-Research Institute

AUTHOR

Sandeep Joshi has professional experience of 25 years in environmental monitoring and pollution elimination technologies and has done his masters in environmental sciences at the University of Pune, Maharashtra, India, in 1989, and doctoral work I in river sediments and doctoral work II in water and wastewater technologies. He is well versed in environmental instrumentation, impact analysis, conventional and nonconventional pollution treatment, ecological engineering, and ecological restoration of rivers and lakes. He innovated ecotechnological treatment systems for point and nonpoint sources of pollution through extensive research for 20 years and field applications for the waste flows ranging from 1 to 600,000 cu m/day. He has also designed more than 400 sewage and effluent treatment systems ranging from 1 to 14,000,000 cu m/day flows. Till date, he has worked on designing the ecorestoration of 42 rivers (flows 10–600 MLD) and 62 lakes (area ranging from 2 to 20,000 ha), conducted extensive studies on catchment of Ujjani Reservoir's (Pune District, Maharashtra, India) pollution load assessment, and conducted research on the enrichment of nutrients and their impacts on ecological health of the water body for the last 12 years.

PREFACE

This chapter has been written for engineers, hydrologists, ecologists, and ecotechnologists who wish to understand the vagaries of river dynamics, hydraulics, and corresponding ecological processes and attune themselves to rejuvenating self-purification capacities based on green infrastructure units. It also provides information about rectilinear engineering river models and physical–biological processes with the concept of river continuum.

However, the quality of river water is of major concern, which is subject to change due to contaminants and pollutants from point sources and nonpoint sources. So it also is of wider interest to ecologists, students, and members of the general public who increasingly question the hydromodifications of the rivers and their impacts on physical–biological processes and entities. Environmentally aware individuals or groups are now assessing the impacts of infrastructure and pollution on water bodies. This chapter will assist them in learning the usefulness of green infrastructure in reviving the self-purification capacity of the river.

In this chapter, an attempt has been made to move systematically from artificial mechanized infrastructure to natural structure to regulate and restore the quality of water using aquatic ecological processes to improve the performance. There are many ways of dealing with river issues. Inevitably, there are no unique solutions for them. In this chapter, an attempt has been made to emphasize ecosystem approaches with river engineering and supportive structures to revive the lotic water bodies.

10.1 Introduction

River basins are very significant with their open character in lentic–lotic systems of a catchment. They accomplish a number of services such as water supply for domestic, industrial, and agricultural sectors; means of transport such as navigation; means of occupation such as fishing and aquafarming; and additionally hydropower generation, recreation, and as a "living habitat." Economic and social development and even life itself cannot be sustained without sufficient water at the right time and place and of the right quality. Though it serves as a religious entity, it is considered as the most suitable medium to clean, disperse, transport, and dispose of wastes.

Occupying less than 1% of the world's surface, freshwater ecosystems contribute hugely to human welfare and survival. People have recognized the provisioning services of rivers for water, food, and energy from thousands of years. Persuasion of insatiable aspirations to maximize the profits has led to extensive modifications of lentic–lotic water systems for modern irrigation, water supply, and energy procurements. These services have "economic value." Regulating services of freshwater systems encompass flood mitigation and control, sediment transport and deposition, and water purification through self-purification processes, etc. Cultural services take into account the recreation, aesthetic, and spiritual provisions. Regulating services can be translated into economic valuation, but cultural services are invaluable and transferable from one generation to the next generation. Supporting services of freshwater ecosystems are nutrient cycling and primary production in appropriate habitats with biodiversity. These services are provided by both free-flowing and dammed rivers. Benefits obtained by water infrastructure are often at the cost of environmental and ecological processes, which need to be quantified and subsequently to be taken into account.

Streams, rivers, and groundwater flow pathways are the connection between reservoirs, lakes, and oceans—the earth's water cycle. Precipitated water flows downhill in surface–subsurface channels in geological time and spatial scales that interact with the bio-chemo-physical form of landscapes that forms its catchment basins. A landscape view of a river catchment comprises the networks of streams, tributaries, subtributaries, and interconnections with groundwater flows, embedding in its highest terrestrial setting to the point of confluence with another catchment system or hydrostatic (lentic) water body or ocean.

All river flows, irrespective of their size, mold landscape of their catchment basins over a period of time, eroding at some places and building (alluviating) at other places. Rivers transport sediments and nutrients from the highlands to the lowlands. Constant modification of catchment features is directly related to its geography, climate, geology, geomorphology, and biotic factors coupled with spatial and temporal variations. The riverscape (landscape of river) at any point within the stream network is four dimensional—longitudinal (from headwaters to ocean, i.e., upstream–downstream direction); lateral (from river channel into terrestrial environment, i.e., aquatic to terrestrial); vertical (penetration of water deep substrata of river bottom, i.e., surface–subsurface flows); and change in size over time due to floods and droughts altering hydrology, sediment treatment, and distribution of flora and fauna [30].

Within a catchment, stream channels normally grow in size and complexity in a downstream direction. First-order stream is the smallest, beginning as surface runoff or spring below porous substrata forming ridges dividing one microcatchment from another. Two first-order streams form a second-order channel and so on continue to create the big network of streams [27]. A large river (seventh order), like the Ganga, Brahmaputra, Nile, Amazon, and Mississippi, has numerous tributaries and subtributaries that are fed by numerous small first-order and second-order streams. Thus, the catchment of large river has abundant medium or small catchments. The ecological setup of river has distinct features commensurate with biophysical gradients expressed by the four-dimensional nature of stream network within the catchments. Species residing in high mountains (upland rhithron environment) cannot be found in warmer, sandy, turbid water in lowland near ocean (potamon environment) [29]. Therefore, biological species of riverine ecosystem are distinctively distributed at different locations with longitudinal (upstream–downstream), lateral (aquatic–terrestrial), vertical (surface–groundwater), and temporal (timescale) gradients. These observations are substantiated by decade-long observations of rivers in the Ujjani Basin of Western Maharashtra in India [13,16]. Environment conditions and stresses in the urban tributaries of Bhima River on which the Ujjani Dam is constructed have resulted in distribution and abundance of riverine species as a function of its landscape setting and land use.

10.2 River System: Hydrodynamic System

The complexity of natural systems and sediments have been tried to define on the basis of water movement, its chemical balance [18] indicating that only abiotic factors are not considered in various uniform steady-flow rectilinear channel models and biotic factors are not given due significance. These models are based on the analysis of observations of physical and chemical processes of the river system. The different processes and mechanisms operating in river drainage basins and channels act together to shape and govern the functioning of the whole hydrodynamic environment (Figure 10.1). It was believed that rivers can be defined in terms of mathematical models based on physical and chemical principles supplemented by experiments under controlled and simplified conditions. Two approaches—empirical and classical—are adapted to describe hydrology and the hydraulic nature of the river system [2,19,25].

The characteristics and behavioral patterns of the river system are largely dependent on the climatic variables operating over the river drainage basin such as rainfall, temperature, local elevation, and some unquantifiable variables such as lithology and geological structure. All these physical–chemical variables contribute to determine the basin-wide hydrological regime. Important status parameters of hydrologic regime are elevation (H), runoff (R), sediment (S) and flow (Q).

10.2.1 Flows in the River and Mass Movement

Local hydraulic conditions can be described by the local mean flow velocity and local dry mass sediment transport rate, Q_{st}. These are functions of water surface slope, S; local channel width, w;

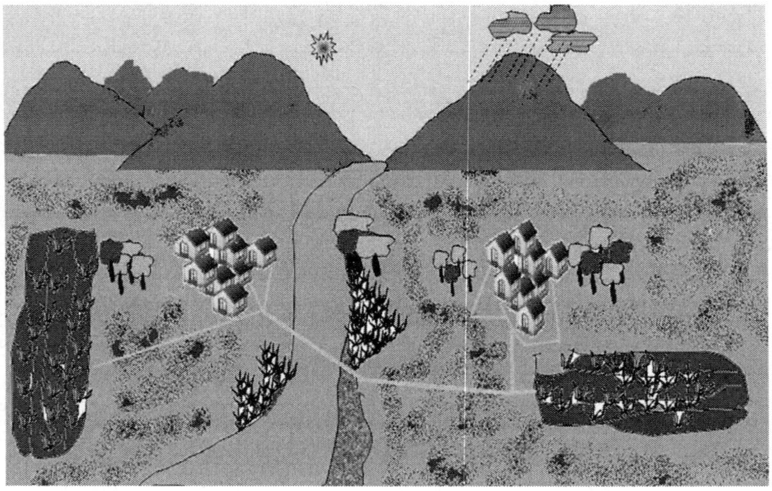

FIGURE 10.1 The river system and its components.

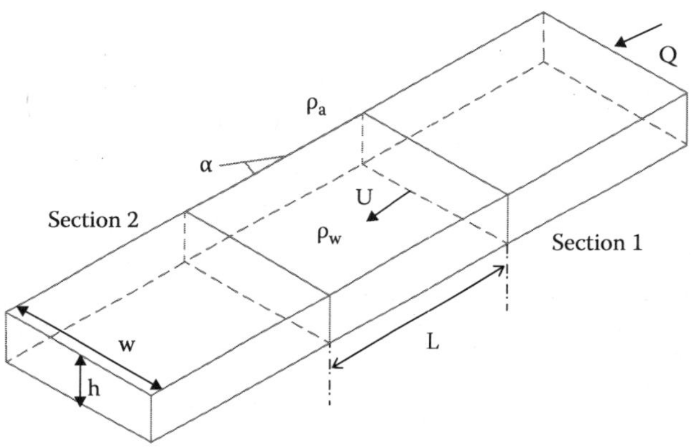

FIGURE 10.2 Uniform steady flow in a straight open channel.

local water depth, h; and the channel pattern by the sinuosity. Additionally, an index of braiding needs to be considered.

A simple mathematical model for uniform steady flow in a rectilinear channel can be developed on the basis of nonecological, physiographic variables (Figure 10.2). It assumes that water flows in a uniform, steady, and rectilinear manner in a straight, open, constant-depth channel. There is no change in the discharge with time and depth in the downstream direction [23,24]. According to a uniform steady flow, model discharge (Q) is directly proportional to the flow depth, width, and slope of the terrain. At the same time, the flow velocity increases with increasing discharge and reduces with increasing flow resistance. The value of weight flow rate (QWF) coefficient of significant size of river is generally 0.05. This value increases proportionately with the increase in boundary's relative roughness and rooted vegetation in the river bed and on the banks.

Using f, the overall Darcy–Weisbach friction (DWF) coefficient (dimensionless), and U, the mean velocity, the flow is calculated as

$$Q=h_wU,\quad U=\frac{Q}{h_w}=\sqrt{\frac{8gQS}{f}} \tag{10.1}$$

Where

f = Darcy–Weisbach friction coefficient
Q = discharge
U = mean flow velocity
H = height
w = width
S = slope
g = gravity

and

$$U=\sqrt{\frac{8gQs}{f}} \quad \text{and} \quad \left(\frac{Q}{h_w}\right)^2=\frac{8gQs}{f}=\frac{(h_w)^2\cdot 8gQs}{f}$$

$$Q=\frac{(h_w)^2\cdot 8gs}{f} \quad \text{and} \quad h_wU=\frac{(h_w)^2\cdot 8gs}{f}$$

Therefore,

$$Q=\frac{(h_w)^2\cdot 8gs}{f} \tag{10.2}$$

Natural rivers vary in discharge with time, and they are nonuniform. Whether they flow straight or meander, the thalweg (line joining the deepest point of the channel in successive downstream cross sections) alternates along its water course with respect to the rising and falling of the valley slope between the deep pools, shallow riffles, or crossings. Nonuniformity of natural rivers giving rise to complexity of behavior and a number of uncertainties is yet to be resolved using mathematical models. River channel width and depth at a point varying with time due to pattern of discharge change become unsteady, leading to variations in hydraulic geometry.

Mass movement is largely affected by events of modest magnitude and high frequency rather than infrequent large discharges. Channel-forming discharges on meandering streams having floodplains are mostly of modest magnitude though catastrophic events lead to great erosion with exceptionally high flows. It is also noted that the relationship of peak flow with the delivery of water and sediment is independent of basin area and of recurrence interval [7].

10.2.2 Sediment Transport and Bed Forms

Most rivers meander through channels curving alternatively to the left and right. A helically spiraling motion is generated as a result of gradual curving of near-bed inward toward the convex bank and traveling outward toward the concave shore. The channel near the apex of meander loop is observed to be relatively deeper and steeper in transverse profile with increasing sinuosity. It is most vulnerable to change when the bend radius depth ratio is minimum [1]. Similarly, the complexity of development and changeability of river morphology is the function of weathering deposition of sediments, episodic cutoffs, and transport bed load and suspended load. Finer grains are transported in suspension and are widely dispersed.

Common bed forms of river are current ripples, dunes, antidunes, plane with lineations, etc. These forms are shown in Figure 10.3. Ripples are composed of finer sand arranged transversely to the flow steeper on the down current side. The height of ripples is normally less than 0.04 m and the wavelength less than 0.6 m. Dunes are much steeper transversely arranged sand or gravels with a height of more than 0.04 m and a wavelength of more than 0.6 m. A plane bed is very gently curved or plain subhorizontal surface underlain by finer grades of sand. Depositions are shown in the photograph of the Ganga River bed in Figure 10.4.

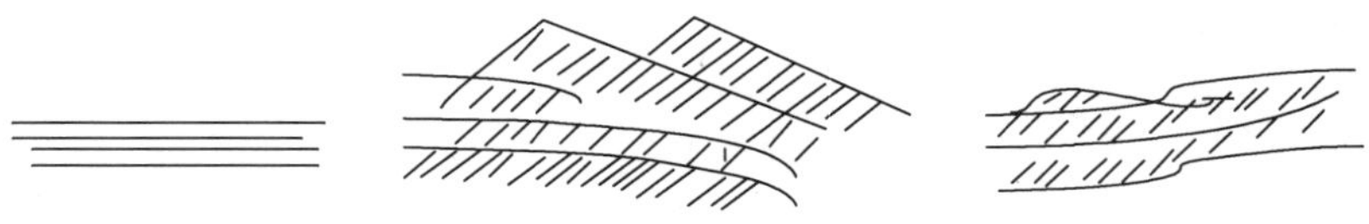

FIGURE 10.3 River bed forms.

FIGURE 10.4 Depositions on the Ganga River bed, Uttaranchal State, India. (Photo by Sandeep Joshi.)

In the river system with abiotic factors, biological factors also contribute to structures and functions of the system. Different segments of the river, valleys, reaches, and channels within a watershed influence the distribution and abundance of aquatic plants and animals depending on the water flow characteristics, transport of sediments, and transformation of organic material [20]. Within the hierarchy of spatial scales in riverscape, all geometric units represent physical features (zones) that are directly or indirectly altered by human activities. This is shown in Figure 10.5 [26,31].

Stream ecologists or ecohydrologists can study river geomorphology and hydrometery of hills, valleys, reaches, and channels in the first place to describe

- Physical changes over time
- Natural disturbances
- Human impacts
- Suitability of a stream for hydromodification
- Habitat restoration with chronology of environmental degradation

10.2.3 Physical Processes of the River System

Hill slopes and valleys are the principal topographic features of any watershed of the river. Water converges and products of erosion sediments and organic debris concentrate in the valleys. Valleys are the

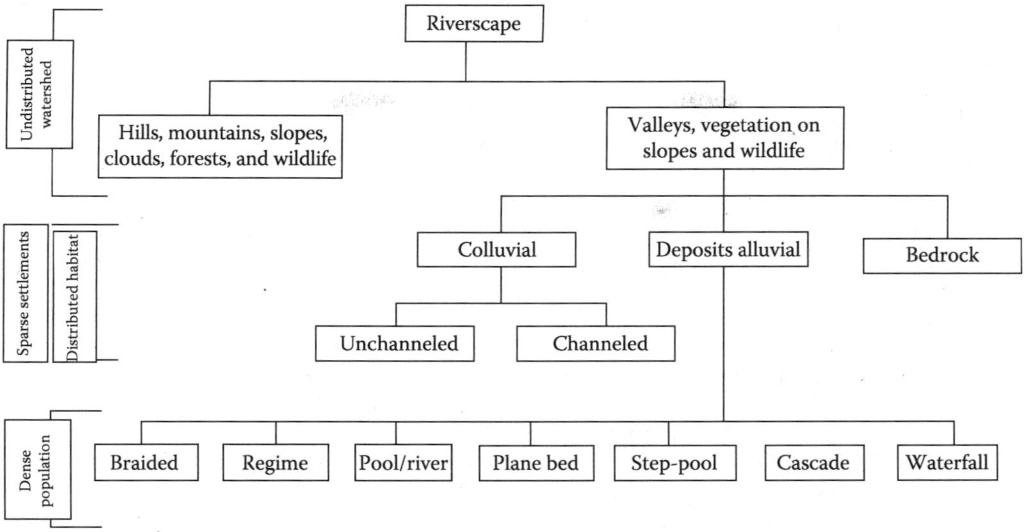

FIGURE 10.5 Physical features of riverscape (aquatic + terrestrial ecosystems) with ecotones.

sections of stream network that have distinct geomorphic and hydrological transport properties. They can be of three types—colluvial (temporary reposting of the sediments and organic matter), alluvial (moving and sorting of the sediments at intervals in confined hill slopes, narrowing valley floor, and unconfined developed floodplain terrains), and bedrock (confined channels lacking alluvial bed).

In colluvial valleys, after the removal of sediment by substantial hydrological disturbances, refilling processes occur again and again [6]. These cyclic processes of emptying and refilling occur at different rates in different geoclimatic regions depending on patterns of precipitation, geolithological conditions, and hill slope vegetation.

10.2.4 Hydraulic Geometry

Hydraulic geometry is a branch of knowledge of functional relationship between discharge and width, depth, velocity, slope, and roughness, riffle–pool spacing, meander wavelength, and sinuosity. Hydraulic geometry has immense importance in channel change studies based on assumptions of dynamic equilibrium; reversible, continuous, and linear relationships; and responses between discharges and variables. In the green infrastructure design for the restoration of ecological quality, hydraulic geometry is significant for in the modeling of the self-purification water body using techniques like green bridges.

Essentials of hydraulic geometry are the following:

1. Relation with climate
2. Dynamic, reversible, continuous, and linear
3. Gradual response between discharge, input changes, and variable changes in time
4. Complexity between area and discharge
5. Topological hydraulic geometry (of macroscale correlations), microscale processes, steady-state geometry of channels, and hydraulic autogeometry (truly downstream hydraulic geometry)
6. Transient behavior due to nonuniform unsteady causal mechanisms
7. Kinematic models based on short-term and long-term responses of channel system

Hydraulic geometry, an empirical model [17], describes stream behavior along the course of water or at some cross section in the water path. Three relationships between the dependent variables and the discharge can be expressed as

$$w = aQbd = cQfv = kQm \tag{10.3}$$

where

w is the width
d is the mean depth
v is the mean velocity
Q is the discharge

Discharge is a product of width, depth, and velocity ($Q = w \times d \times v$), where cross-section area is $w \times d$; therefore, $b + f + m = 1$.

10.2.5 Patterns of Changes

Channel changes result mostly due to major flood events. Factors like slope and width of valleys and plains; frequency, intensity, and volume of flood; physiography; morphology; and geochemistry determine the extent of channel changes. Pattern changes can be autogeneic (inherent in the stream regime involving channel cutoffs), migration, and allogeneic (climatic fluctuation, human-induced flow, or sediment changes, etc.).

Channel changes can be classified into six classes from no changes to the formation to completely new channels as described in the following:

Class 1
No physical change; changes associated with increase or decrease in stage height; extended long base flow recession leading to gradual subsidence of stages

Class 2
Minor change; usually undetectable, prolonged low flows may cause changes

Class 3
Accelerated change; limited collapse of unstable banks

Class 4
Moderately accelerated change; bank scour widespread; movement of course material

Class 5
Highly accelerated change; extensive scouring; appearance of new small channels

Class 6
Formation of new channel

Sometimes channel changes are reversible as in the case of the Ganga River in plains. The Ganga River's course swings between two extremes of banks over a period of time; thus, a no-water zone becomes a floodplain.

10.2.6 Change in Ephemeral Perennial Streams and Rivers

Two-thirds of world's land is occupied by basins of arid and semiarid climates or seasonally occurring concentrated flow patterns. These streams or rivers are severely affected by the abstraction of water for domestic, industrial, and agricultural use and/or discharge of used water. These activities change the solids of the natural water and make the water body unfit for further applications due to change in siltation pattern and chemical and biological species composition.

Hydrology and hydraulics of ephemeral streams or rivers are very complex as compared to perennial rivers because of the uncertainty of flow forecasting, quantum, and quality of sedimentation (natural, urbane, agricultural, etc.). Sporadic precipitation in the catchment and varying infiltration rates due to modern development add to the complexity of hydrograph of ephemeral water bodies. Nowadays, rivers flowing through the cities have become perennial sewage canals because every day they receive a fixed

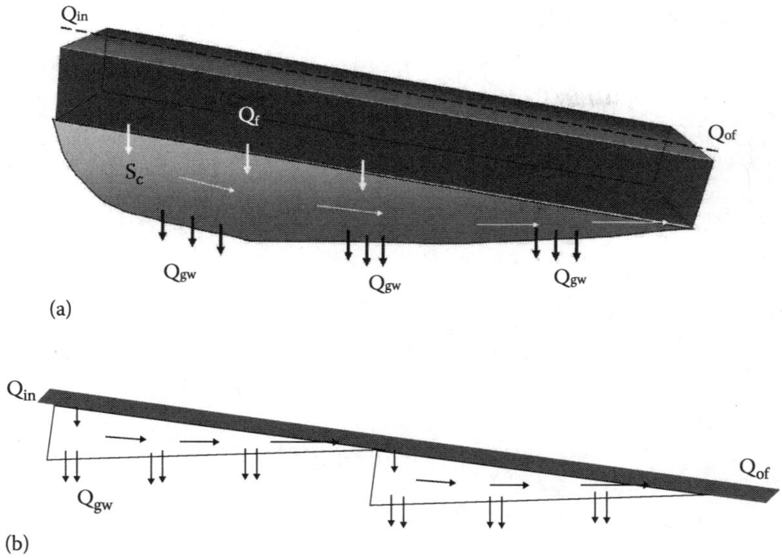

FIGURE 10.6 Surface–groundwater linkages in hydrological regime: (a) schematic of water flow in the stream channel and (b) line diagram of water flow in the stream channel.

amount of used water flow from its urban catchment. They have lost their floodplains to the encroachment of development and have lost their natural meander as well as recharging by surface–groundwater linkages (Figure 10.6a and b).

10.2.7 Short-Term Change

Small-scale spatial variations in hydraulic geometry are temporary owing to sediment depositions, especially in the case of ephemeral rivers. Velocity gradient is also responsible for short-term changes in the course of the river. These changes are random and systemic and need to be considered while designing green infrastructure for the ecological restoration of polluted streams and rivers.

10.3 River: Lotic System

10.3.1 Aquatic Ecology

Aquatic ecology is a branch of ecology that studies the structures and functions of various freshwater, coastal, estuarine, and marine ecosystems. It involves all the interactions and interrelationships of biotic and abiotic components of the ecosystem encompassing various hydrological and hydraulic characteristics. It may be termed as ecohydrology—a combination of ecology and hydrology. Components of ecohydrological studies are water chemistry, geomorphology, shaping of river channels, wetlands, and lakes by vegetation in combination with physical processes and other responses by biological communities, etc.

Principles of ecohydrology can be articulated in three components—qualitative and quantitative estimation of hydrological cycle in a basin, basin's carrying capacity, and ecosystem services and integrative ecosystem approach to regulate hydrological processes to improve sustainability of water resources [32]. Hydrological and hydraulic natural designs of rivers generally regulate biota, but biological systems can be used as a tool to control hydrological processes with hydrotechnical infrastructure to achieve sustainable water resources management and ecosystem services without external energy and chemical inputs [15]. Ecotechnological solutions for sewage- and effluent-receiving rivers can be restored by meticulously using such ecosystem approach to remediate the degraded status of the ecological health.

10.3.2 Flows and Habitat Structure

In addition to physical, structural, and morphological features in the stream corridor (virtual box of water channel, banks, floodplains), there are observable characteristic features of stream ecosystems from the headwaters to the mouth—a longitudinal aquatic–terrestrial zone of ecosystem. Diversity of biological communities varies in different reaches of a river system. Gradual or quick changes in species in the stretches of river–stream zones are associated with corresponding changes in stream size, water zone width, gradient, flow, and temperature. River stretch is divided into three distinct zones—headwater zone (steep, high-velocity, narrow water zone width, and more erosion), transfer zone (moderate velocity, widening of stream channel, lesser gradient), and depositional zone (flat gradient, wide banks, silt deposition) (Figure 10.7). Every zone has its characteristic resident and migratory biological communities and species. This means that the river and its riparian is a longitudinal, dynamic ecotone, which has its own equilibrium of inputs–outputs, interactions, and interrelationships of abiotic and biotic factors.

A longitudinal model for rivers, the river continuum concept (RCC) [28], has been proposed to identify the relationships between gradual changes in stream structure—channel size, stream flow, and type and distribution of species and communities. Characteristics of river stretches are dependent on discrete factors such as water temperature, chemical composition, and positions along the river. RCC is at the basin and catchment scale, and its limitation is that it is applicable only to perennial streams and rivers.

Some rivers that frequently spill over their banks and inundate the adjoining floodplain areas are defined by a "Flood Pulse" model. It describes the habitat's flood and wetland characteristics and biotic communities along a temporal continuum of water–land interface. Inclusion of temporal dimension in the RCC with a time-based river model (e.g., duration of inundation) contemplates predictable longitudinal variation. The RCC model deals with ideal conditions without any natural and man-induced disturbances and their impacts on the river continuum. Many a times, severe disturbances may upset the connections between the watershed, streams, and river continuum.

Another model accepted by ecologists is the "patch dynamics"—collage of landscapes—as stream habitat and species distribution often show signs of unevenness or inconsistency. The patch dynamics concept reflects the unpredictable nature and distribution of biotic communities on stream reach. It complements the RCC because most running waters that show local patch effects exhibit predictable longitudinal patterns over larger scales and longer time periods [5].

(a) (b) (c)

FIGURE 10.7 Zoning of river: (a) headwater zone, (b) transfer zone, and (c) deposition zone. (Photo by Sandeep Joshi.)

10.3.3 Communities of Flowing Water Body

Physical and morphometric features of the stream are responsible for the suitability of habitat and the diversity of biological communities. Unidirectional stream flow of water characterized by slope of the terrain is one of the most important factors controlling the survival of biological life in rivers. Organisms adapted to varying velocities of flow guard themselves from being detached from their niche and washed away. Factors deciding the livability of habitat of river ecotone other than hydraulic characteristics are physicochemical environmental factors such as temperature, sunlight, water quality, and oxygen, as well as biological factors such as food in every trophic level and protection from predators. Therefore, the ecological quality index (EQI) denoting the ecological health of the water body is an important parameter for people at large to judge the appropriateness for consumptive and nonconsumptive uses. Any selective component of EQI can be used for adjudicating the effectiveness of pollution control measures to maintain the purity of the water body with ecologically corrected water.

Organisms in an ecosystem are interrelated and form a complex food chain and food web. The basic principle in the ecosystem is "one's waste is another's food." Food relationships commonly observed among stream biota are shown in Figure 10.8. In the riverine ecosystem, decomposers provide nutrients to the primary producers (aquatic plants—macrophytes and algae) in utilizable form that compose the base of the food chain and web. Periphyton, some types of algae attaching to surfaces (such as stones, bank soil, and branches of macrophytes) in the stream channel, can cope with the varying flow velocity. Phytoplankton, suspended forms of algae growing in relatively still water (minimal velocity), may get washed away in floods. Macrophytes (rooted, floating, or submersed) and planktons are the primary producers, which develop food using energy from sunlight to turn dissolved inorganic nutrients (nitrogen, phosphorus, carbon) through inimitable photosynthesis process. Plant tissues get converted into animal tissues by grazing animals and by animals that feed on grazers. Then, nutrients get recirculated through the ecosystem by the decomposition of organic materials of dead plant and animal tissue and wastes.

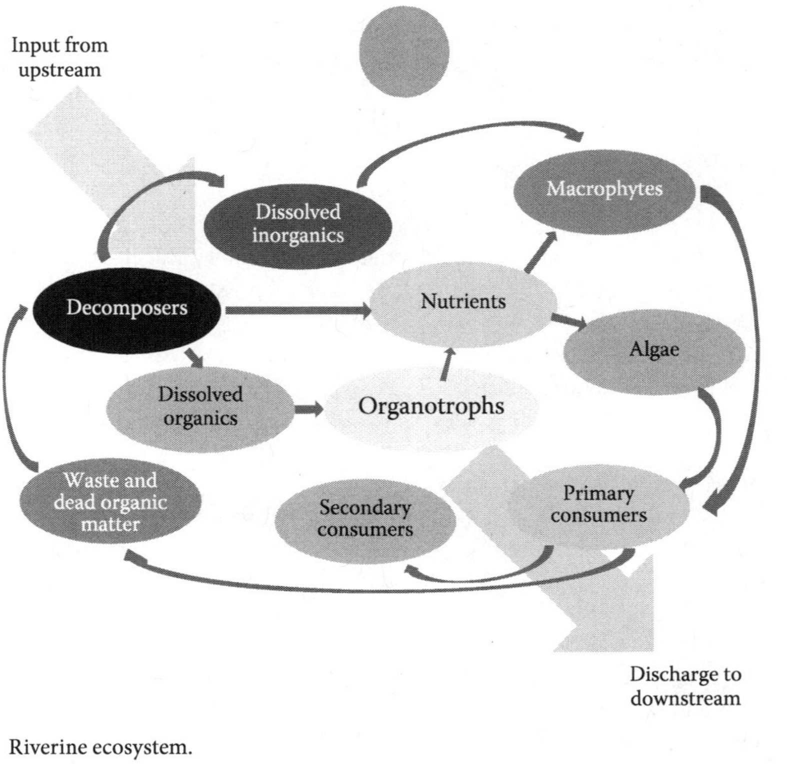

FIGURE 10.8 Riverine ecosystem.

The rate of decomposition and subsequent primary production moderate the suitability of habitat for the growth, sustenance, and diversity of stream biota. If primary production is more than decomposable matter, the stream reach can be termed as autotrophic; in other terms, the rate of decomposition exceeds the rate of primary production. In some stream stretches, the rate of solid accumulation due to higher rate of primary production and exterior upstream inputs exceeds the rate of decomposition, as it occurs in most of streams that receive waste from urban, industrial, and agricultural discharges, and the aquatic ecosystem becomes highly eutrophic to moderately eutrophic. This condition may be termed as heterotrophic.

10.3.4 Population Dynamics and Movement

Microorganisms such as bacteria, fungi, and microinvertebrates feed on dead organic matter (detritus) and convert it into nutrients. They are the recyclers of nature, keeping the ecosystems free from piling on waste. Most of these organisms inhabit headwater streams where there is ample influx of organic matter from dead leaves and other plant wastes and where the substratum is available for attachment. Algae and other small plant species that find attachments in high-velocity upland streams become food for small scraper and grazer animals. Invertebrates (collectors–accumulators) feeding on fine particulate organic material grow profusely in lowland reaches in association with wetland plant species accumulating solids discharged by upstream activities. Fish and other invertebrate–vertebrate predators devour on all these collectors, shredders, converters, and assimilators of the waste. Upland high-speed streams serve as habitat for spawning and fingerlings of some fish species, while lowland stream is home for adults. In the floodplains of streams and rivers, intermittent floods make available a full range of habitats such as high flowing to still water through gradual decrease in flow velocity. Fish prefer shallow, seasonally flooded pools for reproduction and maturation. These are numerous eventualities of life in the river, and its occurrence is solely dependent on hydrological and hydraulic characteristics of the water body.

10.4 Hydrological and Hydraulic Green Design

Ecology of free-flowing rivers—hydrodynamic water bodies—as well as the specific contributions made by free-flowing rivers to human society including water supply, absorption of wastes, and self-purification is the basis of ecosystem approach and ecotechnological solutions for sustainable maintenance of lentic–lotic water systems from pollution and unwanted contamination. On the basis of flow characteristics, the significance of rivers can be enlisted as follows:

Low (base) flows

- Enough habitat space and water for aquatic and terrestrial biological communities to thrive
- Conducive environmental conditions such as water chemistry, water temperatures, dissolved oxygen
- Retain water table in the floodplain and soil moisture for vegetative growth
- Accumulate more waste and pollution

High rhythmic flows

- Morphometric changes in river channel
- Determination and deposition of stream bed substrates (formation of pool, riffles)
- Restoration of water quality after prolonged low flows with the accumulation of waste and pollutants

Bulk floods

- Contouring of physical habitats in the floodplains and formation of new habitats such as secondary channels and/or oxbow lakes
- Deposition of nutrients, gravel, and cobbles in floodplain and spawning areas

- Recharging of water table
- Maintaining balance of species in aquatic and riparian communities
- Flushing of organic materials (food) and debris (waste) downstream
- Cleansing of invasive species from aquatic and riparian communities

Considering all these hydrological, hydraulic, and ecological variations of the river receiving wastewaters from human establishments, the ecotechnological—use of ecological processes—approach for the restoration of highly polluted streams and rivers was evolved and implemented from the last two decades to validate the acceptance of in-stream zero-electricity remediation of water quality of lentic-lotic systems using green bridge technologies [14].

10.4.1 Ecological Health of Hydrodynamic Water Body

It is very difficult to evolve a common yardstick for expressing the ecological health of the river as discussed earlier for physical and ecological features. But certain parameters such as multidiversity ecosystem, water balance, energy cycle, and biogeochemical cycle of microhabitat can be useful in defining the healthiness of the water bodies. These yardsticks are much more elaborate than physicochemical and engineering norms for the qualitative and quantitative assessment of usefulness of water body for human applications. Physicochemical norms of water quality such as chemical oxygen demand (COD), biochemical oxygen demand (BOD), and heavy metals are not enough to describe the ecological correctness of the water body.

Therefore, nutrient concentration in association with conversion into primary, secondary, and tertiary productivity, and ecological toxicity are some of the indicator parameters that define the inhabitable status of the water body. One yardstick can be termed as comprehensive, that is, multidiverse populations and communities in the water body indicate low or zero ecotoxicity of nonbiological species.

10.4.2 Velocity Profiles

The velocity of a river is defined as the rate of water movement in the channel expressed as meters per second. The average velocity of river flow increases gradually with distance from the origin. Velocity appears to be higher in upland streams than in the lowlands. River velocity is determined by the efficiency of the river in overcoming friction with the in-course materials, bed, and banks. Much of a river's kinetic energy is lost to friction (up to 90%–95%). In case of the deeper, wider river with higher discharge, relatively less water is in contact with the wetted perimeter, so friction from the bed and the banks is reduced, and that is why there is uniformity of velocity in the water body. The presence of pebbles, stones, and boulders on the beds and banks increases the roughness of the channel. Channel roughness is higher in the upland stretch of the river than further downstream. This also adds to the uniformity of flow velocity in the lowland river stretch. Hydraulic radius (ratio of cross-sectional area to wetted perimeter) is a measure of the efficiency of the river channel. The higher the hydraulic radius, the more efficient the river channel. Velocity in a river cross section actually varies from bank to bank and from river bed to free water surface and, hence, can be called a two-dimensional variation in a vertical plane. There are various models like the Schumm model that describe the mean velocity of the river in different zones (Figure 10.9). Ecological biodiversity is found to vary in different velocity zones [8].

10.4.3 Habitat Mapping

Habitat studies are undertaken to understand the detailed terrain and interrelationship between landscape and habitat characteristics. Current habitat uses several methods relying mainly on the description of habitat units based on depth, velocity, substrate, and water surface patterns. Water surface patterns are controlled by local geomorphology and hydraulics. They can be mapped considering distinctive physical and biological features.

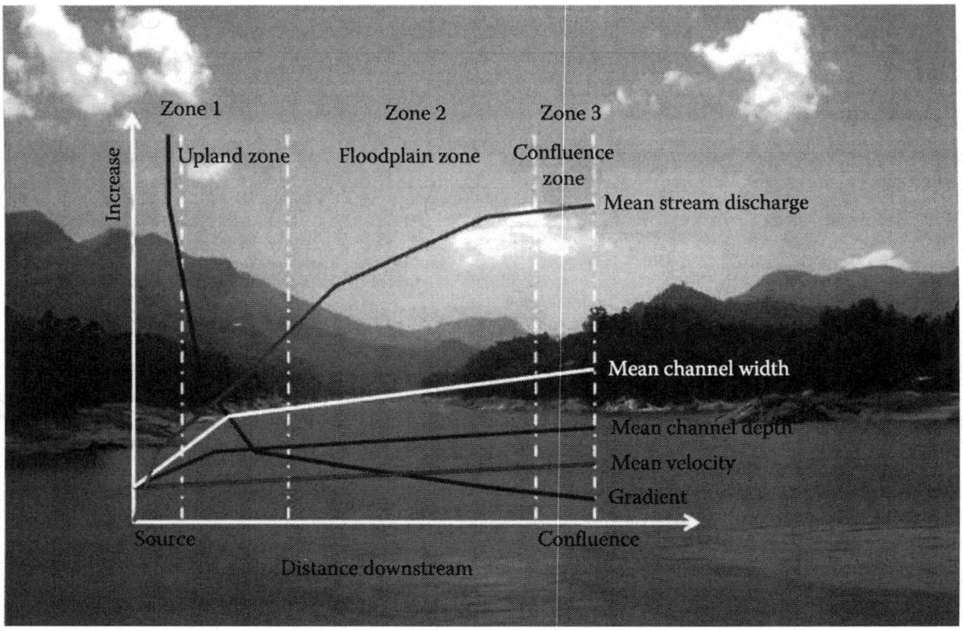

FIGURE 10.9 Different velocity zones of river.

FIGURE 10.10 Microhabitat of Manyame River, Harare, Zimbabwe.

River channels contain an infinitely variable mosaic of morphohydraulic habitat cells (microhabitats) that form the in-stream characterization of biological communities, ranging from microbes to benthic to floating organisms having different nutritional requirements and different environmental conditions (Figure 10.10). All these microhabitats of different conditions form mesohabitat, and many such mesohabitats form one macrohabitat of a river stretch [8]. Habitat maps serve as a key to restoration of streams and rivers.

10.4.4 Environmental Factors

Environmental factors of the microhabitat are stream hydraulic and morphological parameters in addition to water quality and biodiversity, which influence the restoration of the water body from highly polluted condition to oligotrophic condition. Measurement and monitoring of these parameters are essential to design an operable restoration green mechanism (zero electricity and no machinery), that is, green infrastructure grafted on polluted stretches of streams and rivers to revive the self-purification capacity of water body.

Measurement of depth, width, wetted perimeter, velocity, discharge, long profile, bed load, and suspended load will help establishing the strength of the banks. Erodability and energy of flow will help in designing the infrastructure that supports the growth of ecosystem components having inexorable pollution–conversion–assimilation capabilities. The emphasis is on how to use the river's own natural physiography, morphology, and hydraulics without much alteration in association with ecological factors to enhance the efficiency of the green infrastructure to revive the river's ecosystem by reducing toxicities of pollutants.

10.4.5 Trophic Relations

In the riverine ecosystem, when the river starts receiving contaminants, pollutants from different sources, its aqua–bio–chem equilibrium gets disturbed. The dispersion or dilution of these pollutants is dependent on the river's volumetric flow, velocity, and turbulence. Heterogeneity or homogeneity due to mixing in the river water decides the extent of the distribution of pollutants, their assimilation in the food chain, and their toxicity to various individuals in populations and communities. There is a temporary shift in the ecological equilibrium of lotic water body due to external materials, and if this ingress continues, it leads to permanent change in the composition of biological communities.

Ingress of sewage into the river depletes dissolved oxygen concentration temporarily till the organic matter is mineralized. Then, there is temporary change in the populations of aerobic organisms, ranging from bacteria, to single-celled animals to microinvertebrates. There may not be a major impact on aquatic macroinvertebrates and vertebrates. After restoration of the condition by diverting or stopping sewage ingress, the populations of bacterial and dependent microinvertebrates restore over a period of time, and once again the ecological equilibrium is attained. This is a very complex phenomenon where the restoration is a result of attaining ecological equilibrium of oligotrophic conditions as a result of stress removal or destruction. But when the influx of stress-generating pollutants is continued, the aquatic system of river tries to attain a new ecological equilibrium while coping with the hydraulic properties of the water body.

10.4.6 Stream Food Webs

Food webs are different for different quantitative and qualitative stream types, geoclimatic conditions, morphological, geological features, and human impacts [22]. They differ in structure and function for different streams. Actually, it starts from the processing of organic matter in the water body and subsequently goes up to tertiary level with complex interactions and interrelationships of biological populations and communities at every trophic level. Detritus-feeding and primary producers support the populations of dependent animals, ranging from detritivores, grazers, carnivores, omnivores, and predators.

There are grossly two types of measurement of food webs—qualitative and quantitative. A qualitative approach identifies and describes the components and their linkages within the complex food web with no focus on interactions. But it is rational to measure the strength of the food web is to study and enumerate the population dynamics, species diversity at various trophic levels to evolve ecological pyramids of the water body under study. Lotic food webs are more complicated than lentic systems as far as high species diversities, high trophic overlaps, and difficulties in taxonomic resolution are concerned. But the quantitative estimations of food webs are based on carbon or energy budgets and tracer studies. Overall, food webs have a very important role in the ecological restoration of water bodies. For example, the restoration of Ahar River at Udaipur, Rajasthan, India, is thought to be complete with the return of resident freshwater turtles and crocodiles, which ran away a decade before because of pollution [12].

10.4.7 River as an Ecohydrological Unit

Hydrological, hydraulic, morphological, and ecological processes occur in natural river systems autonomously. River catchments with their natural boundaries and hierarchical structures with water-related interactions characterize suitable scale for ecohydrological regime. An ecohydrological process-based model for river catchment inevitably contains different modules such as hydrological, vegetation, nutrient transport with water, biogeochemical cycles coupled with interactions and feedbacks, and water and nutrient drivers for plant growth.

Since long, engineering activities in river basins are centered on water supply for human settlements, agriculture, and industries. This approach is based on quantitative applications of river water, while the ecological quality of water is given second preference again from the point of view of its economized use. Considering the river as an ecohydrological unit, multiple uses and functions of river systems can be optimized by integrating different hydrological and ecological processes into robust and sustainable management plans for the catchment and lentic–lotic systems in the basin.

River ecohydrological management involves a spatial framework of landscape–riverscape interactions and interrelationships with respect to human development. River basins and watersheds are the appropriate spatial units for resource management, contributing quality and quantity of water at suitable points on streams and rivers to human activities and growth. They integrate the surface and subsurface flow of water. Watersheds allow drainage basin-specific hydrological and ecological accountings factoring point and nonpoint sources of pollutants, being transported with the movement of water. An ecosystem approach for the rivers involves the prime objectives of quality, integrity, and healthiness of aquatic ecosystems and their components considering mixture of biotic and abiotic components in both riverscape and adjacent landscape (Figure 10.11).

10.4.8 Quality and Ecological Health of Streams and Rivers

Ecohydrological processes of the river are a tool for upgrading the quality of water in an efficient way to reduce costs of restoration and protection. Low-quality water discharge from conventional mechanistic treatment systems has higher costs of water purification often responsible for unsatisfactory ecological disturbances in receiving streams and rivers thereby reducing the recreational potential. This affects the health and income of local people, and the need emerges to reinvest in upgrading the quality of polluted streams and rivers. Many examples can be cited such as many cities along the shoreline of Lake Victoria, settlements in the Lake Laguna Basin in the Philippines, and fast-growing cities in India discharging wastewaters into the urban stretches of rivers despite of having installed well-engineered sewage and effluent treatment plants. In case of Indian cities in the

FIGURE 10.11 Integration of biotic–abiotic factors of riverscape–landscape in ecological restoration of Ahar River, Udaipur, India.

Ganga Basin, studies cited the inadequacy of electricity for the failure of state-of-the-art wastewater treatment technologies [9–11,21].

A further positive socioeconomic outcome from the efficient restoration of quality and ecological health of the aquatic ecosystems is the improvement of water quality at moderate cost [3,4] without compromising the recreational use of resources balanced with the carrying capacity of the environment, which is concurrent with the idea of sustainable development.

10.5 Summary and Conclusions

Ecohydrological studies, ecological engineering, and ecotechnology are the basis of green infrastructure of the restoration of polluted streams and rivers. Understanding of river's morphometry with ecological health is most essential to design the restoration process on the basis of hydrological, hydraulic, and ecological parameters to attain maximum efficiency in terms of reduction in pollution levels, improvement of self-purification capacity, and elimination of ecological toxicity of pollutants.

References

1. Allen, J.R.L. 1977. Changeable rivers: Some aspects of their mechanics and sedimentation. In: *River Channel Changes*, ed. Gregory, K.J. John Wiley & Sons Ltd, Chichester, England.
2. Anderson, A.G. 1967. On the development of stream meanders. In: *Proceedings of the 12th Congress, International Association of Hydraulic Research,* Fort Collins, CO. Vol. 1, pp. 370–378.
3. Anonymous. 2012a. Chronology of the ecological restoration of Ahar River and Udaisagar Lake. *SERI News* 6(5), 3–8. Available on www.seriecotech.com
4. Anonymous. 2012b. Highlights of the ecological restoration of Rasoolabad stream complex project at Allahabad, Uttar Pradesh. *SERI News* 6(6), 3–8. Available on www.seriecotech.com
5. Brezonik, P.L. 1996. Organizing paradigms for the study of inland aquatic ecosystems. In: *Freshwater Ecosystems: Revitalizing Educational Programs in Limnology*, National Academy Press, Washington, DC. p. 364.
6. Dietrich, W.E., Wilson, C.J., and Reneau, S.L. 1986. Hollows colluviums and landslides in soil mantled landscapes. In: *Hill Slope Processes*, ed. Abrahams, A.D. Allen and Unwin, Boston, MA. pp. 361–388.
7. Dury, G.H. 1977. Peak flows, low flows and aspects of geomorphic dominance. In: *River Channel Changes*, ed. Gregory, K.J. John Wiley & Sons, Chichester, England, pp. 61–74.
8. Hawkins, C.P., Kershner, J.L., Bisson, P.A., Bryant, M.D., Decker, L.M., Gregory, S.V., and McCullogh, D.A. Geography fieldwork. http://www.geography-fieldwork.org/Stage 1: Before you start: Background information. http://www.geography-fieldwork.org/riverfieldwork/downstream_changes/stage1.htm (accessed on August 22, 2012).
9. Controller and Auditor General (CAG) of India Ganga Action Plan. http://www.cag.gov.in/reports/scientific/2000_book2/gangaactionplan.htm (accessed on August 22, 2012).
10. SWOT analysis of Ganga Action Plan. Report Code: 006_GBP_IIT_GEN_ANL_01_Ver 1_Dec 2011. http://gangapedia.iitk.ac.in/sites/default/files/Second%20Set%20of%20Report/006_GEN_SWOT%20of%20GAP.pdf (accessed on August 22, 2012).
11. Clean Ganga and Yamuna mission—A failure. http://www.downtoearth.org.in/content/clean-ganga-and-yamuna-mission-failure (accessed on August 22, 2012).
12. Kodarkar, M.S. and Joshi, S.S. 2010. ILBM impact story—Ecological restoration of highly polluted stretch of Ahar river, Udaipur and ecological improvement of Udaipsagar lake, Rajasthan, India. In: Final review meeting and international symposium of a project entitled *Integrated Lake Basin Management (ILBM), Basin Governance, Challenges and Prospects,* November 2–7, 2010. International Lake Environment Committee (ILEC) Foundation, Headquarters, Kusatsu, Japan.
13. Joshi, S.S. 2008. Use of ecotechnology for sustainable management of the lakes. In: *Souvenir of ILEC-IAAB Workshop on Integrated Lake Basin Management (ILBM)*, Hyderabad, India, pp. 55–60.

14. Joshi, S.S. 2010. Indian experiences of integrated lentic & lotic basin management (*IL*[2]BM). In: Final review meeting and international symposium of a project entitled *Integrated Lake Basin Management (ILBM), Basin Governance Challenges and Prospects*, November 2–7, 2010. International Lake Environment Committee (ILEC) Foundation, Headquarters, Kusatsu, Japan.
15. Joshi, S.S. and Joshi, S.S. 2008. Ecotechnological application for the control of lake pollution. In: *Proceedings of TAAL 2007: The 12th World Lake Conference*. Ministry of Environment and Forests, Government of India and International Lake Environment Committee, Kusatsu, Japan, pp. 864–867.
16. Joshi, S.S., Joshi, S.S., and Kodarkar, M.S. 2010. Ujjani Reservoir in Pune District, Maharashtra, India: A world lake vision candidate waiting for ecological restoration. In: Final review meeting and international symposium of a project entitled *Integrated Lake Basin Management (ILBM), Basin Governance, Challenges and Prospects*, November 2–7, 2010. International Lake Environment Committee (ILEC) Foundation, Headquarters, Kusatsu, Japan.
17. Leopold, L.B. and Maddock, T., Jr. 1953. The hydraulic geometry of stream channels and some physiographic implications. U.S. Geological Survey Professional Paper 252, pp. 1–57.
18. Leopold, L.B. and Maddock, T., Jr. 1953. The hydraulic geometry of stream channels and some physiographic implications. U.S. Geological Survey Professional Paper 282-A.
19. Leopold, L.B. and Wolman, M.G. 1957, River channel patterns: Braided, meandering and straight. U.S. Geological Survey Professional Paper 282-B, pp. 39–85.
20. O'Neill, R.V., DeAngelis, D.L., Waide, J.B., and Allen, T.F.H. 1986. *A Hierarchical Concept of Ecosystems*. Princeton University Press, Princeton, NJ.
21. Ong'ang'a, O. 2010. *Lake Victoria (Kenya) and Its Environs: Resources, Opportunities, and Challenges*. OSIENALA, Kisumu, Kenya.
22. Overton, C.K., Reeves, G.H., Steedman, R.J., and Young, M.K. 1993. A hierarchical approach to classifying stream habitat features. *Fisheries* 18, 3–12.
23. Rouse, H. 1938. *Fluid Mechanics for Hydraulic Engineers*. McGraw Hill, New York; Chow, V.T. 1959. *Open Channel Hydraulics*. McGraw Hill, New York.
24. Sellin, R.H.J. 1969. *Flow in Channels*. Macmillan, London, U.K.
25. Simons, D.B. and Richardson, E.V. 1966. Resistance to flow in alluvial channels. U.S. Geological Survey Professional Paper 222-J.
26. Statzner, B., Gore, J.A., and Resh, V.H. 1988. Hydraulic stream ecology: Observed patterns and potential applications. *Journal of the North American Benthological Society* 7, 307–360.
27. Strahler, A.N. 1963. *The Earth Sciences*. Harper & Row, New York.
28. Vannote, R.L., Minshall, G.W., Cummins, K.W., Sedell, J.R., and Cushing, C.E. 1980. The river continuum concept. *Canadian Journal of Fisheries and Aquatic Sciences* 37(1), 130–137.
29. Ward, J.V. 1986. Altitudinal zonation in a rocky mountain stream. *Archivfiir Hydrobiologic* 2(Suppl. 74), 133–199.
30. Ward, J.V. 1989. The four-dimension nature of lotic ecosystems. *Journal of the North American Benthological Society* 8, 2–8.
31. Report TFW-SH-10-93-002. Timber/fish/wildlife agreement. Department of Natural Resources, Olympia. Washington State. V. A.
32. Zalewski, M., Janauer, G.A., and Jolankai, G. 1997. Ecohydrology. A new paradigm for the sustainable use of aquatic resources UNESCO IHP Technical Document in Hydrology No. 7. IHP-V Projects 2.3/2.4, UNESCO IHP Technical Document in Hydrology No. 7.; IHP - V Projects 2.3/2.4, UNESCO Paris, 60 pp.

11

Groundwater Exploration: Geophysical, Remote Sensing, and GIS Techniques

Satyanarayan Shashtri
Jawaharlal Nehru University

Amit Singh
Jawaharlal Nehru University

Saumitra Mukherjee
Jawaharlal Nehru University

Saeid Eslamian
Isfahan University of Technology

Chander Kumar Singh
Jawaharlal Nehru University
TERI University

AUTHORS

Satyanarayan Shashtri received his MSc in organic chemistry from Banaras Hindu University and his PhD in remote sensing applications in water resources from Jawaharlal Nehru University (JNU), New Delhi, in 2009. He has worked with several renowned institutions of the country, such as the Defence Research and Development Organisation, National Informatics Centre, and Department of Science and Technology. He worked as a scientist from 2009 to 2010 and is currently working as a research associate on geomorphology and lineament mapping at the School of Environmental Sciences, JNU, for the Indian Space Research Organisation (ISRO). His research interests are in remote sensing, geographic information system (GIS), and geophysics applications for groundwater exploration.

Amit Singh received his MSc in environmental sciences and PhD in remote sensing applications for morphotectonic influences on groundwater, both from Jawaharlal Nehru University, India, in 2007 and 2013, respectively. Currently, he is working on assessment of tectonic implications on groundwater in the vicinity of Faridabad and Ghaziabad faults across Yamuna River. He has assisted Professor S. Mukherjee in establishing SEVAN cosmic ray detector (funded by AOARD, NASA, USA) and in subsequently studying the influence of the Sun and other cosmic factors on environment of the space around Earth. His research interests are in hyperspectral remote sensing, tectonics, and groundwater exploration.

Saumitra Mukherjee received his PhD from Banaras Hindu University in 1989. He is currently Professor of Geology and Remote Sensing in School of Environmental Sciences, Jawaharlal Nehru University, India; with a distinguished career of more than 27 years in remote sensing applications in geosciences and space sciences. He has published six books and 86 papers with international publishers and peer-reviewed journals. He was awarded as Commonwealth Fellow in 2004–2005 to extend his research in the United Kingdom, where he was appointed as a visiting professor in the Department of Earth and Ocean Sciences, University of Liverpool, United Kingdom. He has promoted the scientific concept of climate change by extraterrestrial phenomena. His correlation of heliophysical and cosmic variability with global climate change is acclaimed as the Mukherjee correlation of climate change.

Saeid Eslamian received his PhD from the University of New South Wales, Australia, with Professor David Pilgrim. He was a visiting professor in Princeton University, United States, and ETH Zurich, Switzerland. He is currently an associate professor of hydrology in Isfahan University of Technology. He is the founder and chief editor of the *Journal of Flood Engineering* and the *International Journal of Hydrology Science and Technology*. He has published more than 200 publications, mainly in statistical and environmental hydrology and hydrometeorology.

Chander Kumar Singh received his MSc in analytical chemistry from Banaras Hindu University and his MPhil and PhD in remote sensing, GIS, and geophysical applications in hydrology from JNU, New Delhi, in 2007 and 2011, respectively. He has worked for many premier organizations of India. He is currently working as an assistant professor in TERI University. He has published one book and 23 international scientific papers in peer-reviewed journals. His research interests are in LIDAR, hyperspectral remote sensing, and GIS applications for natural resources management, geochemistry, hydrology, urban sprawl, urban heat islands, and public health. He has been using a very integrated use of remote sensing, GIS, and geophysical techniques for hydrogeological studies.

PREFACE

Groundwater plays a major role in the livelihood of mankind by providing water for drinking, irrigation, and industrial purposes. Groundwater, a renewable and finite natural resource, vital for man's life and social and economic development, and a valuable component of the ecosystem, is vulnerable to natural and human impacts. There is a great need for the assessment and monitoring of quality and quantity of groundwater resource required at local level to develop an exact scenario of watershed. The rapid population growth in the last three decades all over the globe resulted in exploiting more groundwater. In many countries, the decline of water level indicates that the resources are depleted very fast. It is, therefore, necessary to assess the available subsurface resource in a more judicious scientific manner and then apply it for evolving optimal utilization purposes.

The chapter introduces the concept of integrated approach of various cutting edge techniques for groundwater exploration. The motivation for integrating a number of techniques stems from the realization that from conceiving the plan to explore new groundwater potential sites, groundwater modeling, quantification, and interpretation of large quantities of geohydrological data becomes a cumbersome approach. The integration of techniques like remote sensing, GIS, and geophysical methods helps us to identify, explore, model, and manage the water resources in a very scientific manner.

11.1 Introduction

Groundwater resource is a multidimensional concept—defined by its location, its occurrence over time, its size, properties, conditions of accessibility, and the effort required to mobilize it, all of which are to be considered in the context of demand [11]. The use of groundwater for drinking and irrigation depends on environmental factors affecting long-term sustainability and the costs of extraction that affect the economic viability. The cost of extraction depends upon the depth of drilling required and the rates of groundwater extraction that can be achieved [10]. In arid and semiarid areas, surface water resources are generally scarce and highly unreliable; therefore, groundwater is often the primary water resource in these areas [9]. Thus, groundwater monitoring is particularly important for sustainable groundwater resource management in arid and semiarid areas. Geophysical exploration is the scientific measurement of physical properties of the earth crust for investigation of mineral deposits or geological structures. Electrical and electromagnetic geophysical methods have been widely used in groundwater investigations because of good correlation between electrical properties (electrical resistivity), geology, and fluid content. Presence of groundwater largely depends upon the availability of porosity in the rocks, on degree of weathering, and intensity/nature of fractures and lineaments [5–7]. The various surface features are normally captured through remote sensing data, which serve as guides for groundwater investigation, and could be grouped into two categories: first-order or direct indicators, that is, features directly related to groundwater occurrence and movement, and second-order or indirect indicators, that is, surface features indirectly linked with groundwater regime [1,2,12].

Subsurface geological features play an important role in groundwater replenishment. The access to surface data such as landforms, drainage density, and water bodies is easier than the subsurface data. Subsurface data characterization such as lithology, geological structure, weathered/fractured thickness, and stratigraphy, relying on geologic framework, is difficult due to the variability of geologic environments and the sparseness of geologic data. In addition, a variety of data types must be combined or synthesized. Through conventional methods alone, it is not an easy task to study all the surface and subsurface parameters of a large area and to identify a satisfactory hydrogeological framework. Since many controlling parameters must be independently derived and integrated, this involves additional cost, time, and man power. Consequently, a centralized database capable of manipulation and analysis of a great variety of data is required such as GIS techniques. These technologies have many advantages over older conventional methods, due to their facility of integration and analysis of large volumes of data, and have proved to be useful for studying geological, structural, and geomorphologic, hydrodynamic conditions together with conventional surveys. Each of these steps can be accomplished utilizing a GIS for (1) data management, analysis, and visualization; (2) integration of diverse data sources; and (3) rapid development, visualization, and testing of alternative hypotheses. The integration of GIS has proven to be an efficient tool in groundwater studies [4,8]. GIS allows the hydrogeologist to interact with the database in ways that retain the spatial relationships in order to visualize the subsurface in two or three dimensions [3,13]. The development of 3-D spatial models is the result of surface and subsurface characterization and field conceptualization. Surface and subsurface characterization is accomplished using careful data gathering and preparation techniques. GIS tools have proved their usefulness in hydrogeology over the years, but standard multilayered systems are quite limited for modeling, visualizing and editing subsurface data, and geologic objects and their attributes. Hence, the main purposes of this chapter are to present and define the following:

- The suitable locations based on remotely sensed data and GIS and then going for selected resistivity and magnetic survey in these locations
- Performing drilling at these selected sites using DTH rigs
- Constructing hydrogeological framework and 3-D geological model of a multilayered aquifer system based on GIS approach
- Constructing hydrogeological framework (solid litholog and aquifer model) that thereof can be used to expedite geological and hydrogeological categorization and conceptualization

To conceptualize these objectives, we decided to opt for a multilayered micro-watershed in the university campus, which actually had problems with groundwater availability and required a suitable plan of action for sustainable hydrogeosciences.

11.2 Study Area

The study area lies 77.15°E and 28.52°N in national capital of India, Delhi. Geologically, the area can be divided into three categories (Figure 11.1):

1. Delhi (quartzitic) Ridge
2. Older alluvium on both sides of the Delhi Ridge
3. Alluvium deposits of Chattarpur enclosed basin

The quartzite ridge enters the area from the southeastern part and passes through the eastern part extending up to the eastern bank of river Yamuna and Wazirabad. The rocky ridge has a length of about 35 km and shows trends in a NNE–SSW direction. Isolated exposures of the quartzite are also found in the western part of the area. The elevation of the crest of the ridge varies from 213 to 314 m above mean sea level with an average elevation of 40 m from the surrounding plain. The alluvial plain in the area is almost flat and is interrupted by a cluster of sand dunes and quartzite ridges. The alluvial plain in the area is almost flat and is interrupted by a cluster of sand dunes and quartzite ridges. Geologically, the area is occupied by the alluvium/buried pediment plain at an elevation ranging from 198 to 270 m above mean sea level. It is occupied by the quartzite interbedded with mica schist belonging to Delhi Supergroup. The rock type exposed in the area belongs to Delhi Supergroup of Lower Proterozoic age and consists of quartzite of the Alwar Group, phyllite, and slate of the Ajabgarh Group. The quartzite is massive, thickly bedded, hard, compact, highly jointed, and intercalated with thin beds of phyllite and slates. The strike of the beds is NNE–SSW and dip westerly at moderate angles. These rocks are mostly covered by quaternary sediments and are exposed in isolated residual and structural hills and pediments. The hills are exposed in south and southwest of Delhi.

11.3 Methodology

Surface manifestations of subsurface geological features can be inferred using satellite data. We used satellite imageries of the year 2000 and 2005 of the month of October to investigate the changes in the

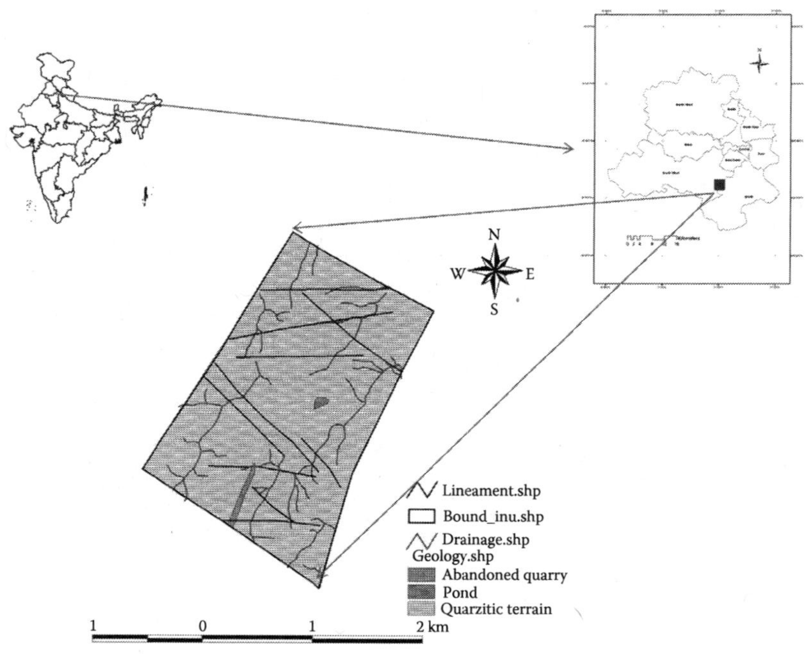

FIGURE 11.1 Location and geology of study area.

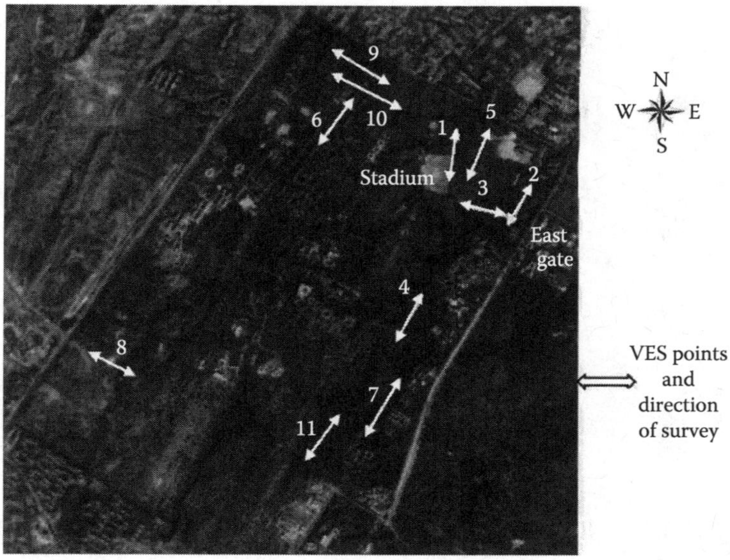

FIGURE 11.2 Vertical electrical sounding locations.

land cover patterns. Normalized difference vegetation index (NDVI) was used to study the changes in vegetation pattern in these 2 years.

Geoelectrical survey was carried out in and around JNU campus, using DC resistivity meter (IGIS, DDR3). Geological investigation based on electrical measurements help in understanding the subsurface geological conditions so that the exploratory drilling can be restricted to only favorable areas. In VES, the goal is to observe the variation of resistivity with depth. The Schlumberger configuration is most commonly used for VES investigation. The midpoints of array are kept fixed, while the distance between current electrodes is progressively increased.

During VES, the alignments were restricted to particular orientation so that it should encounter minimum lateral inhomogeneities (Figure 11.2). The interpretation of sounding data is complicated and more of quantitative nature. The apparent resistivity for a given electrode spacing of one area was compared with that of the adjoining area to evaluate the relative geological and hydrogeological condition. The data was used to identify zones of low resistivity of weathered and fractured rock, which are generally favorable location for groundwater storage.

11.3.1 Gridding Methods

The program RockWorks® reads the depth intervals for the selected water level data and internally translates depths to elevations based on the boreholes' surface and downhole surveys. There are various gridding methods available in RockWorks®—closest point, cumulative, directional weighting, distance to point, inverse distance, kriging, multiple linear regression, sample density, trend surface polynomial, trend surface residuals, triangulation (grid based), and hybrid (Figure 11.3). Since the dataset is not very huge, thus, inverse-distance gridding methods was selected for application due to its simplicity.

11.3.1.1 Inverse Distance

It is a common method using a weighted average approach to compute node values.

11.3.1.2 Triangulation (Grid Based)

It uses a network of triangles to determine grid node values.

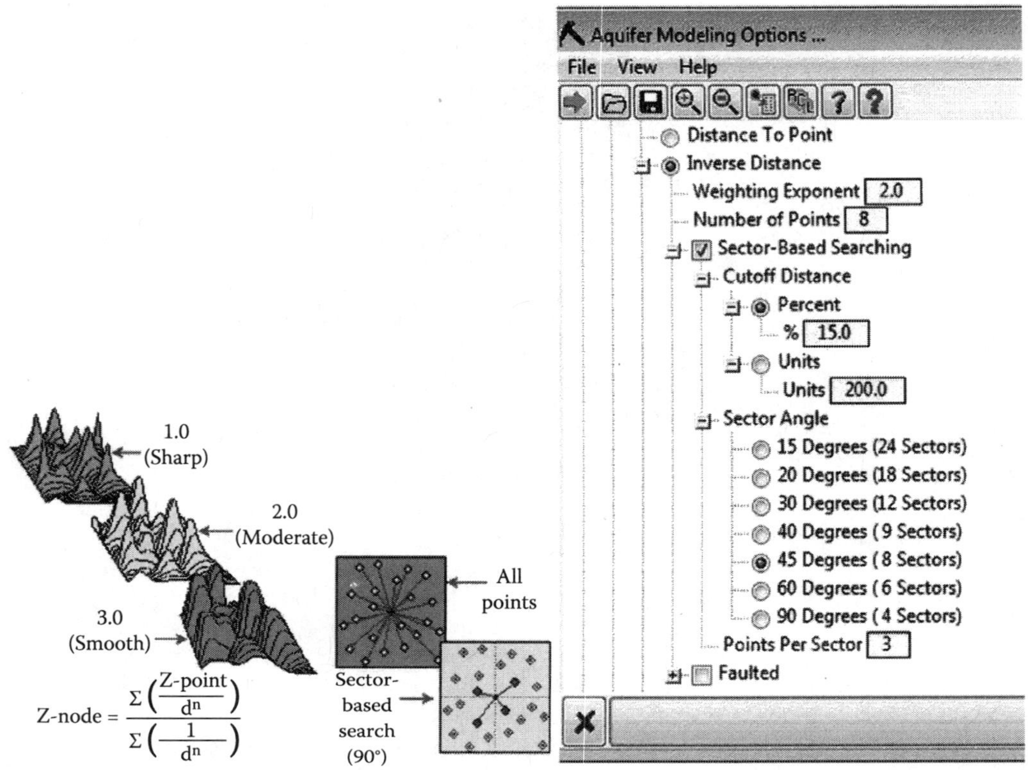

FIGURE 11.3 The weighting exponent was chosen as 2.0 and sector angle as 45° (8 sectors) with 3 points per sector.

Both gridding methods produced quite similar results, but finally, the output from inverse-distance method was chosen because it produces smooth and continuous grid and will not exaggerate its extrapolations beyond the given data points (range of grid values will be smaller than the data point range: The highest grid value will be less than the maximum data point, and the lowest grid value will be greater than the minimum data point). Inverse-distance is one of the more common gridding methods. With this method, the value assigned to a grid node is a weighted average of either all of the data points or a number of directionally distributed neighbors. The value of each of the data points is weighted according to the inverse of its distance from the grid node, taken to a user-selected power. The greater the value of the exponent specified, the more localized the gridding since distant points will have less influence on the value assigned to each grid node.

Weighting exponent: This value determines how "local" or "global" the gridding process will be in assigning node values; the value of each data point is weighted according to the inverse of its distance (d) from the grid node, taken to the nth power, as shown in the following diagram.

Sector-based searching: Tells the program that instead of simply finding the nearest neighbors for the inverse-distance gridding, regardless of where they lie, it should look for points in each x-degree sector around the node. This kind of directional search can improve the interpolation of grid values that lie between data point clusters. It can also increase processing time.

More refined gridding was not required simply because the area of the region covered by seven drilling sites is not much and shows very less variability in terms of slope, strike, and aspect that was again reconfirmed from the statistical report of the generated grid. Refining the grid parameters even more would have resulted in unnecessary increase in processing time and varied outputs. Thus, a grid model of the layer's upper elevations or thickness was created using the selected gridding method, and from this model a 2-D map with the selected layers and 3-D model was created. The maps illustrate the surface elevations (superface) of the aquifer or the aquifer's thickness (isopach).

11.4 Normalized Difference Vegetation Index

The NDVI map generated showed that the NDVI values were quite higher in year 2000 than in 2004 and 2005 (Figure 11.4). The decrease in NDVI values can be attributed to cutting of plantations for new built-ups, but what was more interesting to observe is that the NDVI increased in the area from year 2004 to 2005. The drilling site 1 showed that the NDVI value that was 0.395 in year 2000 decreased to mere 0.175 in year 2004 and then increased to 0.246 in year 2005. Similar trends were observed for drilling sites 3, 4, 5, 6, and 7. Only site 2 accounted for a good amount of increase in NDVI values even in comparison to year 2000, that is, the NDVI increased from 0.338 in year 2000 to 0.393 in year 2005. The NDVI values were not very high but it was observed that the selected sites showed anomalous vegetation growth, and thus, these sites were selected for drilling as this suggested that the vegetation growth was not hampered as it regained in one year (Figure 11.5).

11.5 Geophysical Survey

However, the resistivity survey was carried out in several locations, but to summarize the results of these sites, we will be dealing with interpretation of only a few locations (Figure 11.6). The soil resistivity on site 2 shows more than 100 Ω-m initially that is suggestive of arenaceous soil and low moisture content that is supported by low NDVI value (0.228873) and the litholog also. Further down to 6 m,

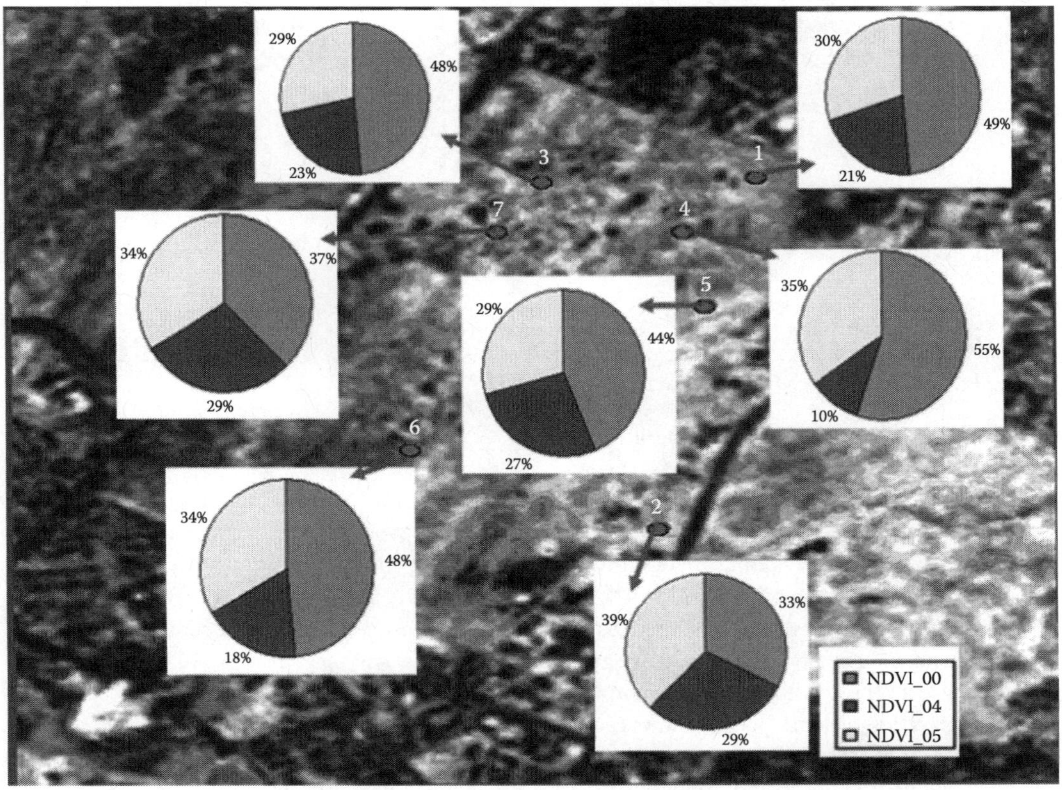

FIGURE 11.4 The NDVI variation at selected locations on temporal scale.

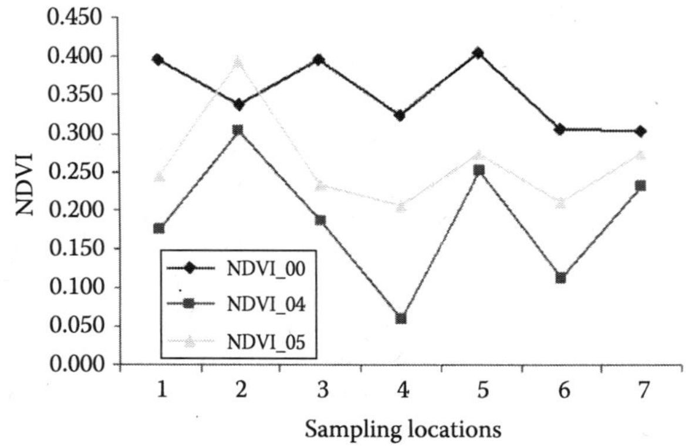

FIGURE 11.5 NDVI values at different inferred drilling sites in JNU in the year 2000, 2004, and 2005.

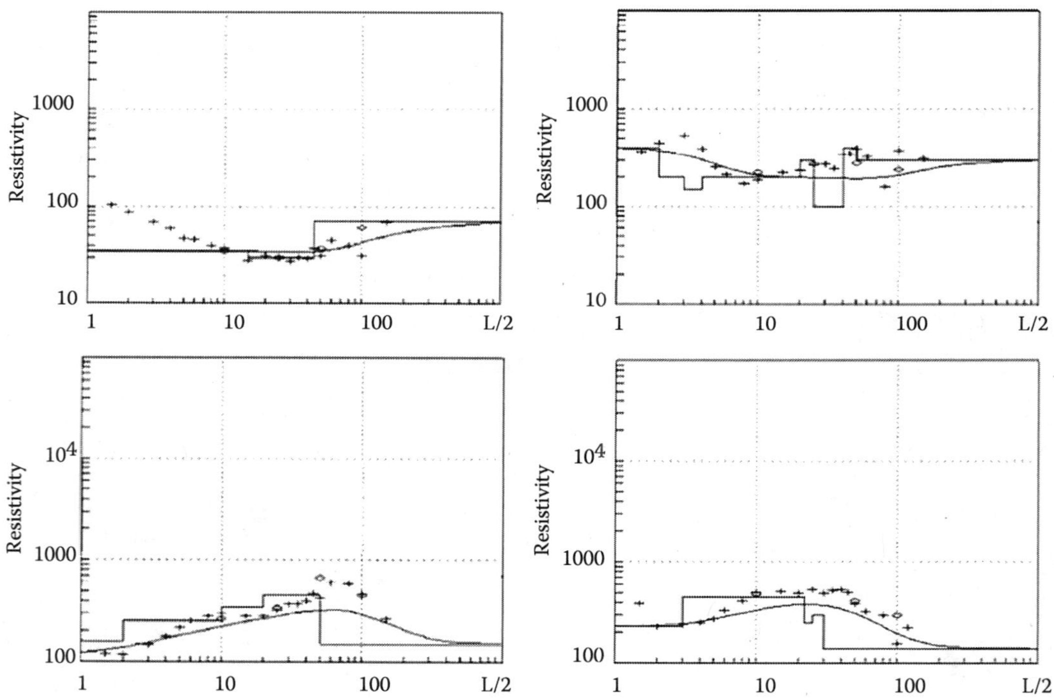

FIGURE 11.6 VES curve at some of the drilling locations.

the resistivity value sharply decreases to 35 Ω-m, which is supported by silt and clay in the litholog record. Further down to 10 m, the ferrogenous intercalation overlying ferrogenous quartzite shows the possibility of shallow aquifer material that is further supported by the resistivity value, ranging in between 25 and 40 Ω-m. Further down to 50 m, the resistivity value remains low ranging between 40 and 60 Ω-m, which is further supported by the presence of weathered mica schist, quartz vein, and ferrogenous quartzite. Magnetic value recorded at this site was very low (29,000 gamma), which further suggests the possibility of interconnected fractures with good aquifer zones. The resistivity data interpretation of site 3 (JNU3) shows that initially top soil mixed with weathered rock shows very high resistivity of 400 Ω-m. Down to the depth of 4 m, the resistivity value shows further rise up to 600 Ω-m, which is suggestive of top layer of hard ferrogenous quartzite, and from 10 m, the resistivity value started decreasing and it was ranging in between 200 and 300 Ω-m down to 20 m. It shows the fractured quartzite with potential groundwater zone possibility at a shallow depth. The resistivity values were found to be fluctuating in between 150 and 400 Ω-m in between 12 and 40 m that comprises groundwater-bearing fracture zone. From 42 to 45 m, the resistivity value has shown a rising trend that is suggestive of mica schist and quartzite intercalations. From 46 to 80 m, the resistivity data show a sharp downfall; it was ranging in between 300 and 150 Ω-m. This zone comprises of two sets of groundwater-bearing fracture in quartz mica schist and gneissose granite. Further down to 100 m, the resistivity value shows a rise up to 300 Ω-m, which is suggestive of granite and quartzite with less fracture. From 100 m down to 150 m, a sharp decrease in the resistivity value infers groundwater-bearing fractured quartzite. Resistivity value at site 4 shows compact overburden below the ferrogenous quartzite. At a depth of 27 m, ferrogenous quartzite was found overlying on a quartzite and aplite veins, thus, beginning of potential groundwater-bearing zone. Further down to 50 m, fractured quartzite with high discharge of groundwater encountered. The resistivity curve shows an initial rise due to presence of ferrogenous quartzite. When the curve was matched with standard curve, it was found that the angle in between X–Y plot shows less than 45°, which further suggests the possibility of groundwater-bearing zone at 27 m. The litholog also supports the interpreted resistivity curve. The magnetic value shows very high anomaly from its standard value in Delhi area that is suggestive of multiple fractures below ground level. Beyond 100 m, the presence of mica schist intercalated with thin quartzite veins further lowers the resistivity value that suggests that wherever there is change in lithofacies, it is a marker of potential fracture zone. Resistivity survey carried out at site 5 shows high top soil resistivity down to 2 m that falls abruptly from 120 to 100 Ω-m at 2.5 m, which is suggestive of very shallow weathered quartzite with some moisture content. Resistivity value further rises up to 300 Ω-m from 5 to 35 m below ground depth that suggests presence of massive quartzite. From 35 to 45 m, the resistivity shows the presence of groundwater in the fractured metamorphics. The striplogs for the four sites discussed are shown in Figure 11.7. Quantitative interpretation of VES data in corroboration with borehole data suggests with precision the availability of groundwater in-depth. Based on the resistivity and magnetic surveys carried out in the study area, seven sites were selected within JNU campus area for primary drilling (Table 11.1).

11.6 Aquifer Thickness and Solid Lithology Model

The pumping results at the selected drilling sites are shown in Table 11.2. The interconnected fracture in Aravalli quartzite has been found to have potential for groundwater exploration. Within Aravalli quartzite, the ferrogenous variety was found more fracture prone. Pegmatites, aplite, and quartz vein intercalated with schistose rocks have multiple fracture system. Aquifer thickness was modeled for the area covered by seven primary drilling locations in JNU based on the water level data from each of the borewells, total depth, and elevation contours (Figure 11.8). As per the results, thickest layers of water are found toward northeastern part of the sampled area. Southwestern part of the sampled area also shows moderate aquifer thickness, but for this part, the accuracy of predicting aquifers with ample water is lesser due to lesser number of primary drilling sites. But for the southeastern and eastern parts, it was found that aquifer thickness decreases to 5–10 m. Aquifer thickness alone can't be criteria for groundwater exploration or management. It should be considered coupled with the topography as per the elevation contours (or rather flow direction of rainwater) as well as the aquifer depth model and the geology of the terrain. As water follows the natural depressions over the terrain, the chances of water getting accumulated in these areas increase. If such pockets are having high primary porosity or interconnected fractures (particularly in hard-rock terrain such as JNU), then the chances of finding water increase. Confined aquifers in such cases may show lesser aquifer thickness. These conditions can be studied remotely using satellite data but that has to be again reconfirmed through ground-based geophysical surveys (resistivity and magnetic). Even in this case, the change in aquifer thickness toward northeastern and southwestern parts coincides to a larger extent with the changes in elevation contours (rapid decrease in elevation).

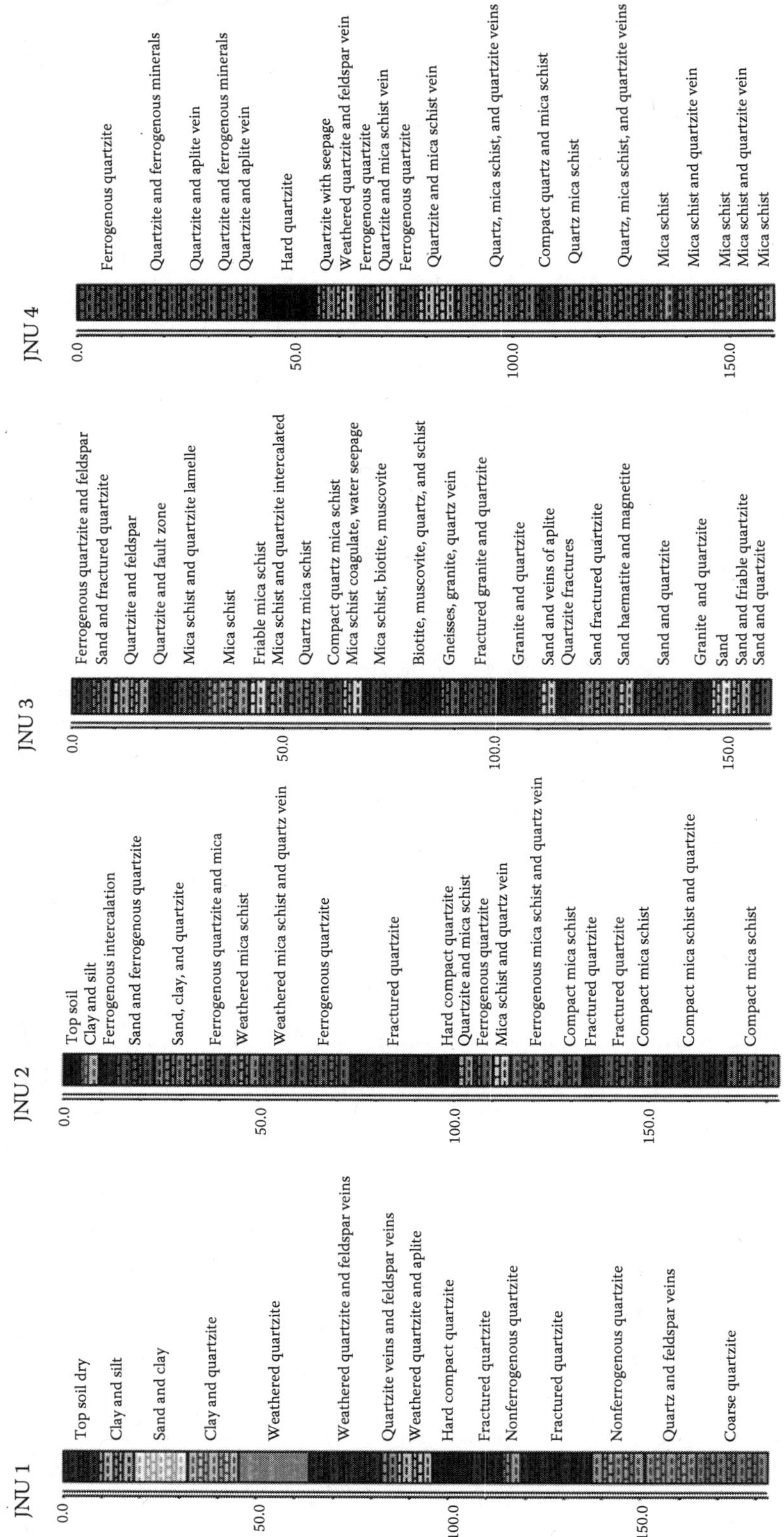

FIGURE 11.7 Striplogs constructed based on drill cuts collected during drilling.

TABLE 11.1 Location of the Seven Drilling Sites with Total Depth Drilled

Bore	Longitude	Latitude	Easting	Northing	Elevation (m)	Total Depth Drilled (m)
JNU1	77.17186	28.54092	712494.9	3159050	248	182.92
JNU2	77.17406	28.54642	712698.5	3159663	236	182.92
JNU3	77.17019	28.54608	712321.4	3159619	250	164.63
JNU4	77.17078	28.54283	712385	3159260	253	160.06
JNU5	77.1563	28.5381	710977.6	3158710	263	166.15
JNU6	77.1619	28.544	711513.9	3159374	250	198.17
JNU7	77.17251	28.5376	712565.1	3158683	256	146.34

TABLE 11.2 Individual Cumulative Water Yields of Seven Borewells

Drilling Sites	JNU 1	JNU 2	JNU 3	JNU 4	JNU 5	JNU 6	JNU 7
Head of water in tube above the center of orifice in in.	9	18	25	10	20	9	20
4 in. pipe with 2.5 in. opening in L/h	20,475	28,938	34,125	21,840	30,576	20,475	30,576

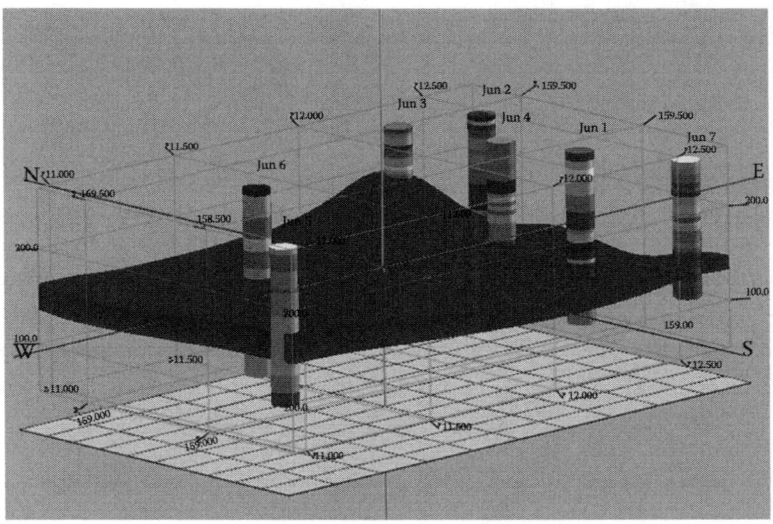

FIGURE 11.8 Aquifer model generated for seven primary drilling sites.

The six lithologic categories of the quaternary system (clay, clay and sand, fine sand, coarse sand, sand and gravel, and gravel) are represented as spatially repeated sequences that have significant spatial changes in terms of their occurrence, thickness of individual categories, and elevation of top and bottom of each layer. Interfingering and presence of lenses is a main characteristic of the sedimentary basin represented in the study area. Due to these characteristics, heterogeneity of the aquifer system is represented by a spatial variation in hydraulic conductivity ranging between that of clay and gravel. Results of the lithologic models and the generated 3-D fence diagrams revealed that there is a wide range of hydraulic conductivities in the modeled area, which vary spatially and control the groundwater flow regime.

Subsurface lithologic information was used to generate a lithology solid model for drilling sites predicting the possible subsurface lithologic connections and extrapolating the information to the whole region apart from primary drilling sites. The software (RockWorks®) determines the lithology types along each borehole in the project and assigns certain values to those nodes along the wells. It then uses the "lithoblending" method to assign lithology to nodes lying between wells. Finally, it will reset those nodes above the ground surface to a value of 0.

11.7 Summary and Conclusions

The land use of National Capital Region (NCR) in general and Delhi in particular has gone under drastic change in the last few decades. The rapid change in land use/land cover, population growth, and seismic instability, all have contributed in changing the hydrogeoenvironment of NCR. The present work aims at hard rocks and colluvial and alluvial aggregates for exploration, exploitation, and management of water resources in one of the most water-vulnerable regions of India. Although it is difficult to accurately estimate subsurface water available in a region, nevertheless, the present work attempts to estimate and give a clear-cut estimation of groundwater in hard-rock area of Delhi. To achieve the specific objective, a systematic approach of understanding the terrain characteristic using satellite data and then a detailed geological mapping, geophysical surveys (resistivity and magnetic), DTH drilling, and analysis of drill cuttings have been adopted. The information from satellite data, resistivity, and magnetic anomaly values were used to locate seven locations for deep drilling at identified favorable sites for groundwater exploration. Deep drilling was conducted at selected locations using DTH rig up to maximum depth of 198.17 m, and it was found that the discharge at all the locations was more than 20,475 lph. However, it was conceptualized that these wells need some time to recover as the recharge in this area takes place through laterally interconnected fractures that have high transmissivity.

It was found that wherever the lineament density were high, there were also resistivity and magnetic anomaly with lower values; at all such places, groundwater availability is very high. If NDVI is high, the vegetation is thick due to high moisture-laden lineament that is suggestive of high mineral availability and hence the groundwater availability. Resistivity survey threw light on the different levels of water availability. Further, it was found that wherever there has been change in the land use, the natural recharge potential too has declined. Digital elevation model could tell us about the course of water runoff and it will help in recharging the aquifer. The study also came to the conclusion that wells may not be directly recharged. There can be indirect method of recharging them. Recharging the lateral dry wells can be done by the lateral homogeneity. In view of the total area of study in JNU that is approximately 5 sq. km, the number of tube wells was restricted to seven based on the delineation of micro-watershed. As per the National Water Policy, there should not be more than one tube well in one micro-watershed. The discharge of the tube wells ranged between 24,475 and 34,125 L/h with less than 10 m drawdown in 72 h of pumping. In this area, most of the drilling site fractures are interconnected with high transmissivity; it has been observed that 80% recovery of drawdown takes place within 1 h, if surrounding tube wells are also stopped. Remaining 20% recovery takes 4 h due to elastic nature of the aquifer.

References

1. Ellyett, C.D. and Pratt, D.A. 1975. *A Review of the Potential Applications of Remote Sensing Techniques to Hydrogeological Studies in Australia*. Australian Water Resource Council Technical Paper 13. Australian Government Publishing Service: Canberra, Australia.
2. Gupta, R.P. 1991. *Remote Sensing Geology*. Springer Verlag: Berlin, Germany.
3. Kolm, K.E. 1996. Conceptualization and characterization of groundwater systems using geographic information systems. *Engineering Geology* 42:111–118.
4. Krishnamurthy, J., Kumar Venkatesa, N., Jayaraman, V., and Manivel, M. 1996. An approach to demarcate ground water potential zones through remote sensing and a geographical information system. *International Journal of Remote Sensing* 17(10):1867–1884.

5. Mukherjee, S. 2006. Integrated water resources management in Aravalli Quartzite of Delhi, India by remote sensing and geophysical techniques. *Proceedings of the International Workshop on Impacts of Reforestation of Degraded Land on Landscape Hydrology in Asian Region*, Roorkee, India, March 6–10.
6. Mukherjee, S. 2008. Role of satellite sensors in groundwater exploration. *Sensors* 8:2006–2016.
7. Mukherjee, S. 1998. Eco-conservation of a part of J.N.U. campus, by GIS analysis. *Proceedings of the National Seminar on Artificial Recharge of Ground Water*, New Delhi, India, pp. 103–119, December 15–16.
8. Saraf, A.K. and Choudhary, P.R. 1998. Integrated remote sensing and GIS for groundwater exploration and identification of artificial recharge sites. *International Journal of Remote Sensing* 19(10):1825–1841.
9. Scanlon, B.R., Keese, K.E., Flint, A.L., Flint, L.E., Gaye, C.B., Edmunds, W.M., and Simmers, I. 2006. Global synthesis of groundwater recharge in semiarid and arid regions. *Hydrological Processes* 20:3335–3370.
10. Sikandar, P. and Christen, E.W. 2011. Geoelectrical sounding for the estimation of hydraulic conductivity of alluvial aquifers. *Water Resource Management* 26:1201–1215.
11. Singh, C.K., Shashtri, S., Mukherjee, S., Kumari, R., Avatar, R., Singh, A., and Singh, R.P. 2011. Application of GWQI to assess effect of land use change on groundwater quality in lower Shiwaliks of Punjab: Remote sensing & GIS based approach. *Water Resources Management* 25(7):1881–1898.
12. Singhal, B.B.S. and Gupta, R.P. 2001. *Applied Hydrogeology of Fractured Rocks*. Kluwer Academic Publishers: Dordrecht, the Netherlands.
13. Turner, A.K. 1989. The role of three-dimensional geographic information systems in subsurface characterization for hydrogeological applications. In: Raper, J.F. (ed.) *Three Dimensional Applications in Geographic Information Systems*. Taylor & Francis: London, U.K., pp. 115–127.

12

Groundwater Hydrology: Saturation Zone

Giovanni Barrocu
University of Cagliari

AUTHOR

Giovanni Barrocu is a professor of engineering geology, Faculty of Engineering of the University of Cagliari, Italy, and the chairman of the International Association of Hydrogeologists (IAH) network on "Coastal aquifer dynamics and coastal zone management," and scientific adviser of the UNESCO-IHP MED-MAP project on the management of coastal aquifers, groundwater and land involving all Mediterranean extra-EC countries.

He has carried out an intensive research activity as responsible for the Italian party in a number of bilateral and multilateral projects with European, North African, and Middle East research institutions for groundwater applied studies in fractured and porous media. He is the author or coauthor of more than 110 papers, mainly on engineering geology, hydrogeology of hard rocks, coastal aquifers, and integrated water resources management.

PREFACE

This chapter aims at providing the basic physical principles of groundwater flow in the saturated zone to be considered in well designing and installation, groundwater sampling methods for hydrochemical analysis and geotechnical tests, procedures for groundwater drainage, soil consolidation, and land reclamation.

Groundwater is an important component of the natural environment and plays an important role in the development and management of water resources to be developed through springs and wells and the calculation of aquifer yields in terms of integrated watershed resource management.

In many cases, the solving of groundwater flow problems may offer a medium for the prevention and remediation of water contamination and land degradation. Among the groundwater flow problems encountered in civil, environmental, mining, and geomechanical engineering practice are those connected with the assessment of water resources, seepage through earth dams and underneath hydraulic structures, and water intrusion into underground constructions.

Groundwater flow is a key to understanding a wide variety of geological processes, such as the generation of earthquake, the migration and accumulation of petroleum, the genesis of certain types of ore deposits, soil types, land forms, and geotechnical problems as slope stability and land subsidence.

Groundwater hydrology is a discipline of broad interest, basic for engineers, hydrologists, geologists, soil scientists, agronomists, foresters, geographers, ecologists, and petroleum reservoir analysts.

12.1 Introduction

Groundwater hydrology, or *geohydrology*, can be defined as the discipline that deals with the occurrence, distribution, and movement of water beneath the surface of the earth. Hydrogeology differs from it as it gives greater emphasis to study of the interactions between waters and geology. *Groundwater catchment* or *watershed* is the area of land where all surface waters infiltrating into the ground go into the same groundwater body along vertical and horizontal flow lines from *recharge areas* to *discharge areas*. *Aquifers*, also named *groundwater reservoirs*, or *water-bearing formations*, are geological bodies that yield important water quantities in their discontinuities. *Saturated zone* is the part of the water-bearing material in which all voids, large and small, are filled with water. A *recharge* (*intake* or replenishment) *area* is the region of a hydrogeological catchment that contributes water to an aquifer, either by direct infiltration or by runoff and subsequent infiltration of precipitations, and surface waters from streams, lakes, and wetlands. As shown in Figure 12.1, in confined aquifers, the recharge area is open to the atmosphere so that the saturated zone is overlain by the unsaturated zone, but water may also percolate into and leak through a confining bed. The *discharge area* is the region of the hydrogeological catchment where groundwater may (1) naturally issue from in springs; (2) outflow in rivers, lakes, and wetlands, contributing to their baseflow; (3) escape by evaporation and transpiration; or (4) is nearest to the surface, and it is artificially withdrawn by wells. If groundwater infiltrates in an area partly or totally outside the surface water divide, the *hydrogeological catchment* may not coincide with the river basin collecting surface waters. In fact, whereas surface water flow strictly depends on basin geomorphology, groundwater flow is generally controlled by aquifer structures, consisting of intergranular spaces and fractures. A drop of water infiltrated outside of the boundary will drain to another watershed. *Saturated zone* is the lower part of a geological formation, in which all interconnected openings are filled up with water.

Aquifers may be interbedded between two confining beds, consisting of low-permeability materials, classified in the literature as follows [30]:

- *Aquicludes* are layers of very low permeability materials such as clay, which do not yield appreciable quantities of water to wells.

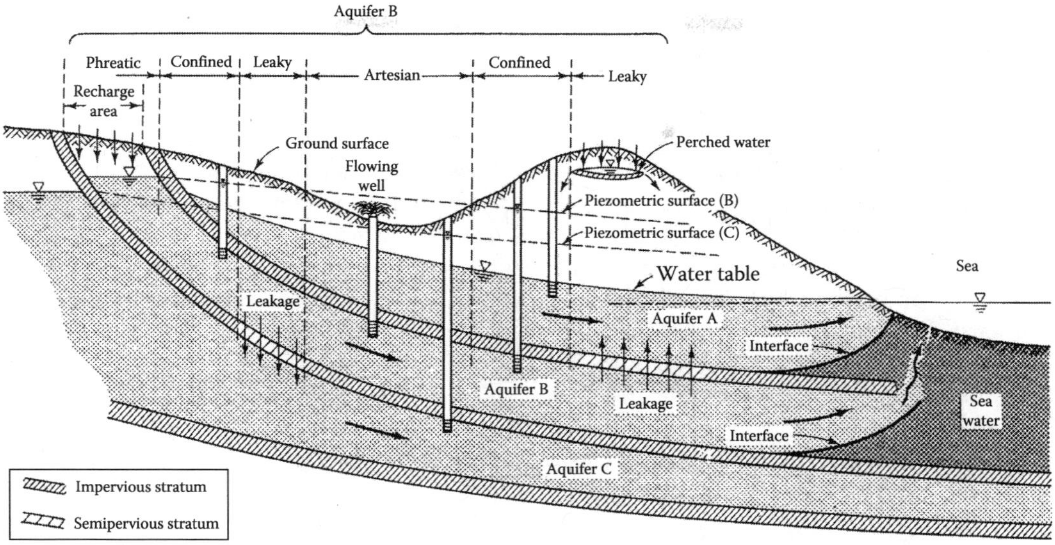

FIGURE 12.1 Types of aquifer system. (After Bear, J., *Hydraulics of Groundwater*, McGraw-Hill Book, Inc., New York, 1979.)

- *Aquitards* are saturated but poorly permeable and semiconfining layers made of clays, shales, loam, and silt where groundwater flow, predominantly vertical, is very slow but not insignificant toward overlain and underlain adjacent aquifers. Where sufficiently thick, they may store important groundwater quantities.
- *Aquifuges* are solid rock formations, practically tight as they have no interconnected openings and hence cannot absorb or transmit water.

An aquifer is called *phreatic* (or *unconfined*) if its saturated zone is in direct contact with the atmosphere; its surface, overlain by the capillary zone and the unsaturated zone, is named *water table*. The saturated zone extends downward from the capillary fringe to the depth where voids are no longer interconnected or absent, so that the rock is practically tight. The vertical flow component is generally lower than the horizontal component so that it can be neglected. A *perched aquifer* consists of a saturated lens included in the unsaturated zone. Where sufficiently thick and extended, perched aquifers may yield significant water quantities of local interest. An aquifer is *confined* when it is interbedded between two aquitards and its piezometric surface given by its water level is higher than the overlain confining bed. If the water level in a well rises above the ground surface, the aquifer is *under artesian conditions* and the well is defined as a flowing artesian well. Common aquifers are geological formations consisting of unconsolidated sand and gravel, poorly cemented sandstones and conglomerates, fractured and karstic limestones, and fractured volcanic and crystalline rocks. The distinction between the water table and the piezometric surface is given by the difference between the naturally occurring surface of water in an unconfined aquifer and the water surface in a monitoring well in a confined aquifer. *Leaky* or *semiconfined aquifers* are interbedded between an underlain aquiclude and an overlain leaky aquitard, from which they receive a significant flow. They are fully saturated and when the groundwater level is withdrawn by pumping from a leaky aquifer, water is removed by horizontal flow within the aquifer and by vertical flow through the overlaying aquitards. The *leakage coefficient* or *leakeance* is the quantity of water that flows across a unit area of the boundary between the main aquifer and its overlying or underlying semiconfining layer per unit head difference across this semiconfining layer. A *multilayered aquifer* consists of a succession of leaky aquifers sandwiched between aquitards. *Aquifer systems* made of a complex of different water-bearing strata, alternated with aquicludes, aquitards, and/or aquifuges, are very common in deep sedimentary basins.

12.2 Total and Kinematic or Effective Porosity

Sedimentary, igneous, and metamorphic rocks consist of solid mineral matter and empty spaces known as *interstices*, *voids*, *pores*, or *pore spaces*, which can serve as water conduits and reservoirs, characterized by their size, shape, irregularity, and distribution. A distinction is made between *primary porosity*, made of the original interstices formed with the rock, and *secondary porosity*, consisting of discontinuities developed after the rock was formed, such as joints, fractures, solution openings, and openings created by plants and animals. With respect to size, interstices may be classed as *capillary*, *supercapillary*, and *subcapillary*. Capillary pores are sufficiently small that surface tension forces will retain water within them, supercapillary interstices are larger than capillary ones, and subcapillary interstices are so small that water is held primarily by adhesive forces. They may be classed as interconnected or communicating, or isolated. The porosity (n) is the ratio of void volumes V_v to the total volume V_t of a *porous medium*, irregardless of whether the voids are interconnected or not:

$$n = \frac{V_v}{V_t} = \frac{V_t - V_s}{V_t} \tag{12.1}$$

where

V_v is the void volume
V_s is the solid volume
V_t is the total volume (*bulk volume*)

Porosity may also be expressed by

$$n = \frac{\rho_m - \rho_d}{\rho_m} = 1 - \frac{\rho_d}{\rho_m} \tag{12.2}$$

where

ρ_m is the density of the solid rock, varying with the type of their mineral components
ρ_d is the *bulk density*

The porosity is strictly related to the *void ratio* e, commonly used in soil mechanics, defined as $e = V_v/V_s$, related to n by

$$e = \frac{n}{1-n} \quad \text{or} \quad n = \frac{e}{1+e} \tag{12.3}$$

The different types of rock porosity are generally classified considering the types of rock interstices, as follows: (1) well-sorted sedimentary deposit having high porosity; (2) poorly sorted sedimentary deposit having low porosity; (3) well-sorted sedimentary deposit consisting of pebbles that are themselves porous, so that the deposit as a whole has a very high porosity; (4) well-sorted sedimentary deposit whose porosity has been diminished by deposition of mineral matter in the interstices; (5) rock rendered porous by solution; and (6) rock rendered porous by fracturing [30].

The *kinematic* or *effective porosity* considers *only* interconnected void space. Practically, it is given by the water contained in a saturated porous medium. In some rocks, such as pumice and pumiceous tuffs, total porosity, due to gas expansion in very viscous magma, is generally very high, but effective porosity may be from very low to almost insignificant as voids are mostly isolated. However, for unconsolidated and many consolidated rocks, the two porosities are of the same order. Porosity may also be expressed by multiplying the right-end side of Equations 12.1 and 12.2 by 100. Porosity may increase in time and space owing to weathering, chemical solution, gas expansion, clay shrinkage and flocculation, openings formed by vegetables and animals, or decrease for sediment compaction, clay deflocculation, and cementation, mainly with $CaCO_3$, SiO_2, and clays.

12.3 Darcy's Law

Darcy's law applies to the flow of groundwater and states that the velocity of the flow is linearly proportional to the fall in pressure and soil permeability [10]. In his experiment through a sand column, as schematized in Figure 12.2, Darcy found that the *volumetric discharge Q* (or the *specific discharge* $v = Q/A$) is directly proportional to the head difference $\Delta h = h_1 - h_2$ when the separation distance $\Delta l = l_1 - l_2$ between the observation points 1 and 2 is kept fixed and inversely it is proportional to Δl when Δh *is* kept constant, that is, $v \propto \Delta h$ and $v \propto 1/\Delta l$:

$$v = -K\frac{\Delta h}{\Delta l} \tag{12.4}$$

or in differential equation form

$$v = -K\frac{dh}{dl} \tag{12.5}$$

where

- v is the *specific discharge* or *Darcy's velocity*, or *Darcy's flux* as it has the dimensions of a velocity [L/T = (L³/T)/L²]
- K is the *hydraulic conductivity* [L/T]

Equation 12.4 is often given in the compact form

$$Q = -KiA \tag{12.6}$$

where $i = dh/dl$ is the *hydraulic gradient.*

The reciprocal $1/K$ is the *hydraulic resistivity*, which measures the frictional forces opposing groundwater flow in a porous medium. Darcy's law is an experimental law, valid in any space direction. If i and K are constant, v is independent of the inclination θ of the sand column from the vertical. In reality, groundwater flows through complex paths within the porous medium so that A is only the total area of the pores intersected in a given section, but in practice, Q is assessed in the whole sample section, encompassing also the total surface of the grains.

Practically, flow is regarded on a *macroscopic* rather than a *microscopic* scale, considering the porosity of a hypothetical porous medium sample of volume V_1, V_2,... at a point P, larger than a single pore [25]. The volume V_3 in Figure 12.3 is the *representative elementary volume* (*REV*), which must enclose

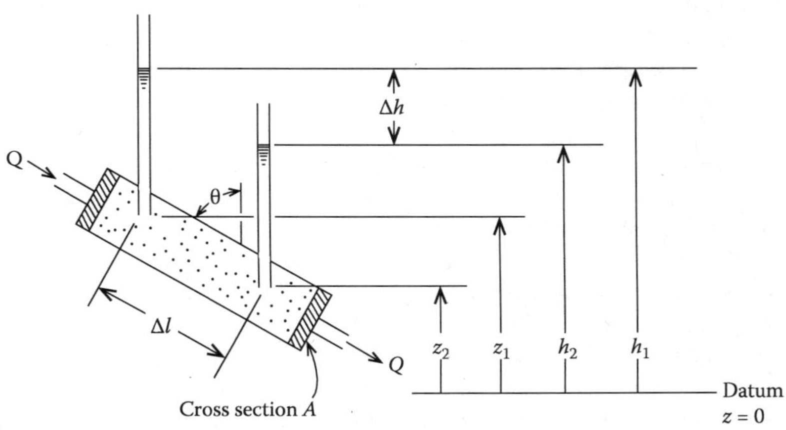

FIGURE 12.2 Pressure distribution and head loss in flow through a sand column.

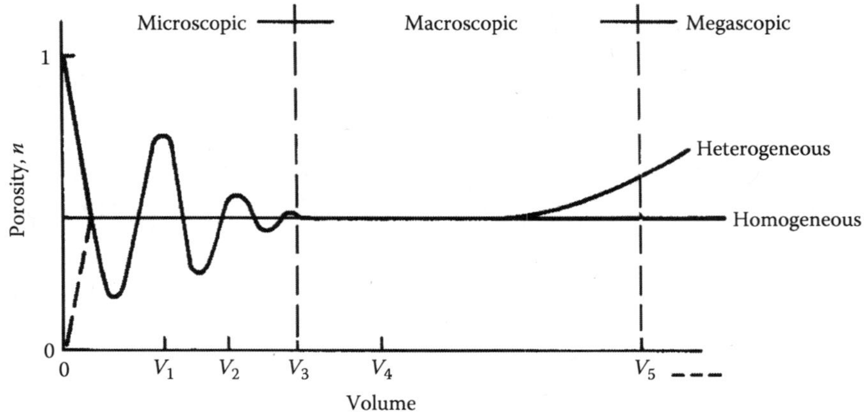

FIGURE 12.3 Microscopic and macroscopic domains and the REV V_3. (After Bear, J., *Hydraulics of Groundwater*, McGraw-Hill Book, Inc., New York, 1979.)

a sufficient number of voids so that they can be statistically treated as needed to represent the porosity of the whole medium in the continuum approach [3,4]. *REV* dimensions and scale of representation depend on the type of medium taken into consideration. In fractured and karstic rock formations, *REV*s should encompass the pattern of all their discontinuities. Porosity, hydraulic conductivity, and compressibility are generally assessed on samples larger than the *REV*, on the usual sizes of cored soil samples. The scale is called *megascopic* when volumes are larger than V_5 in Figure 12.3, so that they may include more than one layer in heterogeneous media.

Laminar flow occurs where viscous forces are dominant and is characterized by smooth, constant fluid motion without lateral mixing; *turbulent* flow is dominated by inertial forces, which tend to produce chaotic eddies, vortices, and other flow instabilities. A measure of the ratio of inertial forces to viscous forces during flow is given by *Reynolds' number* R_e, a dimensionless number that quantifies the relative importance of these two types of forces for given flow conditions. It is widely used in fluid mechanics to distinguish between laminar flow and turbulent flow at high velocities:

$$R_e = \frac{\rho v d}{\mu} \tag{12.7}$$

where

- ρ and μ are the fluid density and (dynamic) viscosity
- v is the specific discharge
- d is a representative length dimension for the porous medium, variously taken as a mean pore dimension, a mean grain diameter, or some function of the square root of the permeability k

According to Bear, "Darcy's law is valid as long as the Reynolds number based on average grain diameter does not exceed some value between 1 and 10" [3]. The transition from laminar to turbulent flow and vice versa is gradual, as flow tends to keep its original characteristics by inertia so that the limit is represented by a range of Reynolds number values. As velocity increases, turbulence begins in the largest pore spaces and proceeds into the smallest pores. In turbulent flow, the head loss decreases with the second power of the velocity rather than linearly. Darcy's law may not be valid for very slow water flow through consolidated clay, because of nonlinearities between hydraulic gradient and flow rate in small pores, as effects of electrically charged clay particles on water. Deviation from Darcy's law can occur near pumped wells, where hydraulic gradients are steep, and in the large underground openings of karstic limestone and basalt lavas. Bear rigorously recognizes the difference between the *volumetric*

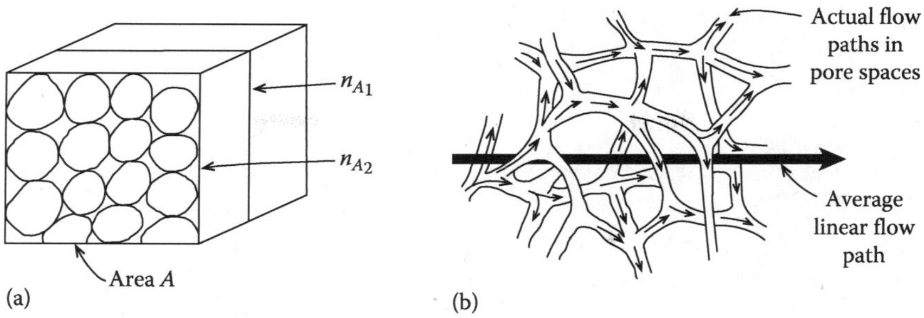

FIGURE 12.4 (a) Areal porosity and (b) average linear porosity. (After Freeze, R.A. and Cherry, J.A., *Groundwater*, Prentice Hall, Englewood Cliffs, NJ, 1979.)

porosity, n, expressed by (12.2), and *areal porosity* $n_a = A_v/A_T$, where A_v is the area occupied by the pores and A_T the total area. The *volumetric porosity, n*, is an average of the various areal porosities nA_i as represented in Figure 12.4. The specific discharge $\bar{v}_i = Q/n_{A_i}A$ indicates the volumetric flux divided by the actual cross section area through which flow occurs. The *average linear velocity* for the various sections $A_1, A_2, \ldots$ is

$$\bar{v}_i = \frac{Q}{nA} = \frac{v}{n} = \frac{-K}{n}\frac{\partial h}{\partial l} = \frac{Q}{n_{Ai}A} \tag{12.8}$$

The microscopic velocities through the irregular and longer paths given by interstices are generally higher than $\bar{v}$ represented by a straight line in Figure 12.4. However, Equation 12.8 is commonly adopted in groundwater tracers, groundwater contamination studies, and soil grouting actions.

12.4 Hydraulic Conductivity and Permeability

The constant of proportionality K in Darcy's law is a function of both the porous medium and the fluid [7]. Column experiments, carried out with various liquids of density ρ and dynamic viscosity μ flowing under a constant hydraulic gradient in columns through an ideal porous medium consisting of uniform class beads of diameter d, may put in evidence the following proportionalities: $v \propto d^2$, $v \propto \rho g$, $v \propto \mu^{-1}$, so that Darcy's law may be reformulated as follows:

$$v = -\frac{Cd^2 \rho g dh}{\mu dl} \tag{12.9}$$

which compared with (12.5) can be expressed as

$$K = -\frac{Cd^2 \rho g}{\mu} \tag{12.10}$$

where ρ and μ are functions of the fluid alone. If we put

$$k = Cd^2 \tag{12.11}$$

$$K = \frac{k\rho g}{\mu} \tag{12.12}$$

The parameter k is the *specific* or *intrinsic permeability*, or simply the *permeability*, which is a function of the medium and has dimensions [L²], while K is always the *hydraulic conductivity*.

By substituting (12.12) in (12.5), Darcy's law becomes

$$v = -\frac{k\rho g dh}{\mu dl} \tag{12.13}$$

which represents the *unit of intrinsic permeability* defined as the permeability of a medium in which a liquid of dynamic viscosity of 1 centipoise (a·1 cP = 1 mPa·s) discharges 1 $cm^3\ s^{-1}$ through a cross section of 1 cm^2, under a hydraulic gradient normal to the section of 1 atm cm^{-1}. A millidarcy (md) is equal to 10^{-3} darcy and a microdarcy (μd) equals 10^{-6} darcys. Typical values of permeability range as high as 100,000 darcys for gravel to less than 0.01 microdarcy for granite (Figure 12.5). Sand has an intrinsic permeability of nearly 1 darcy.

In an anisotropic medium, flow is 3D so that in Darcy's law, the velocity v is a vector with components v_x, v_y, and v_z, and the simplest generalization is

$$v_x = -K_x \frac{\partial h}{\partial x} \quad v_y = -K_y \frac{\partial h}{\partial y} \quad v_z = -K_z \frac{\partial h}{\partial z} \tag{12.14}$$

where K_x, K_y, and K_z are the hydraulic conductivity values in x, y, and z directions of anisotropy. A more generalized set of equations can be written in the form

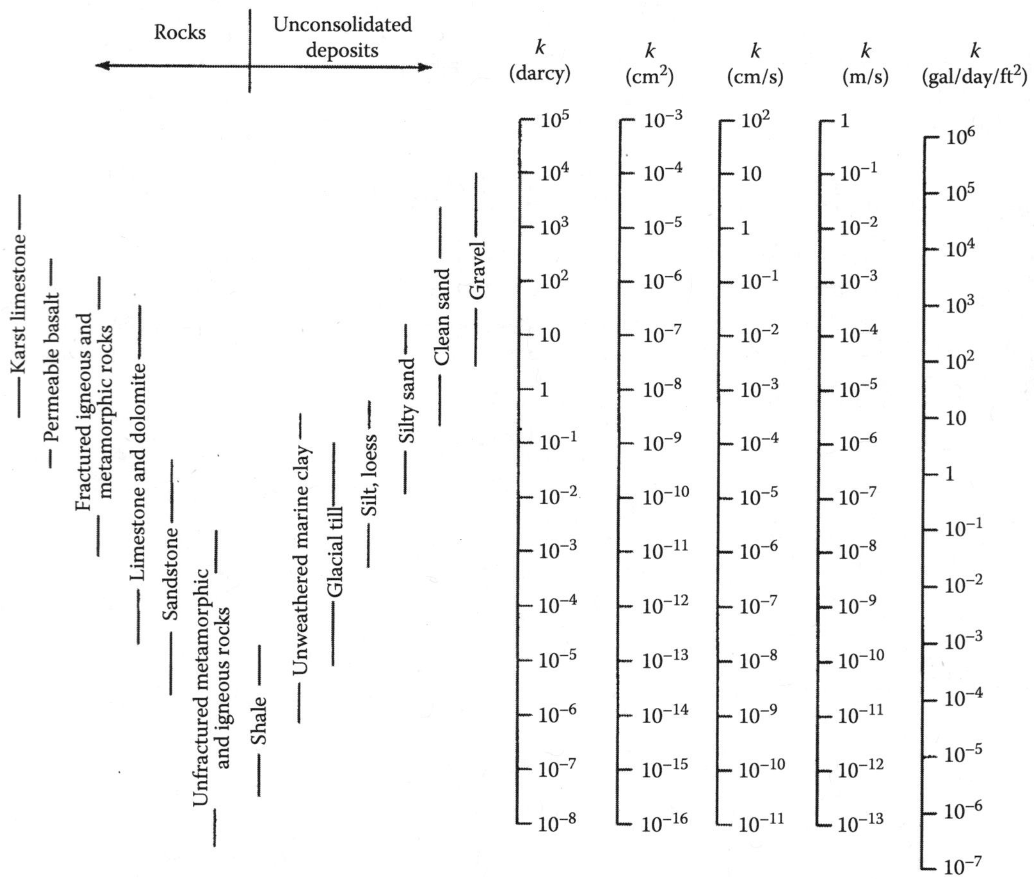

FIGURE 12.5 Range of hydraulic conductivity and permeability values. (After Freeze, R.A. and Cherry, J.A., *Groundwater*, Prentice Hall, Englewood Cliffs, NJ, 1979.)

$$v_x = -K_{xx}\frac{\partial h}{\partial x} - K_{xy}\frac{\partial h}{\partial y} - K_{xz}\frac{\partial h}{\partial z}$$

$$v_y = -K_{yx}\frac{\partial h}{\partial x} - K_{yy}\frac{\partial h}{\partial y} - K_{yz}\frac{\partial h}{\partial z} \quad (12.15)$$

$$v_z = -K_{zx}\frac{\partial h}{\partial x} - K_{zy}\frac{\partial h}{\partial y} - K_{zz}\frac{\partial h}{\partial z}$$

with the vector notation $v = -K\nabla h$, which is a second-rank symmetric tensor known as the *hydraulic conductivity tensor* [3]. If $K_{xy} = K_{xz} = K_{yx} = K_{yz} = K_{zx} = K_{zy} = 0$, the nine hydraulic conductivity components reduce to three, and (12.14) is a convenient generalization of Darcy's law, which may be adopted in practice considering that the principal anisotropy directions vary from one formation to another.

12.5 Fluid Potential

Groundwater flow through porous media is a mechanical process implying an irreversible transformation of mechanical energy into thermal energy due to the frictional resistance setup between the moving fluid and porous medium grains [3,16,26,27,32,42,45]. Hubbert introduced the concept of fluid potential defined as the mechanical energy per unit mass of fluid [25]. With reference to Figure 12.6, three components are to be considered when calculating the work required to lift a unit mass of fluid from the standard state, at elevation $z = 0$, to some point P in the flow system where the elevation is z, the fluid pressure is p, the velocity is v, the density is ρ, and the volume per unit mass is $V_0 = 1/\rho_0$. The fluid potential Φ (the mechanical energy per unit mass) is the sum of $w_1 = mgz$, the work required to lift the mass at the standard state from elevation $z = 0$, representing the loss in potential energy; $w_2 = mv^2$, the loss in kinetic energy, given by the work required to accelerate the fluid from velocity $v = 0$ to velocity v; and $w_3 = m\int_{p_0}^{p}(V/m)\,dp = m\int_{p_0}^{p} dp/\rho$, the loss in elastic energy, the work carried out on the fluid in raising the fluid pressure from $p = p_0$ to p. For a unit mass of fluid, $m = 1$, we have

$$\Phi = gz + \frac{v^2}{2} + \int_{p_0}^{p}\frac{dp}{\rho} \quad (12.16)$$

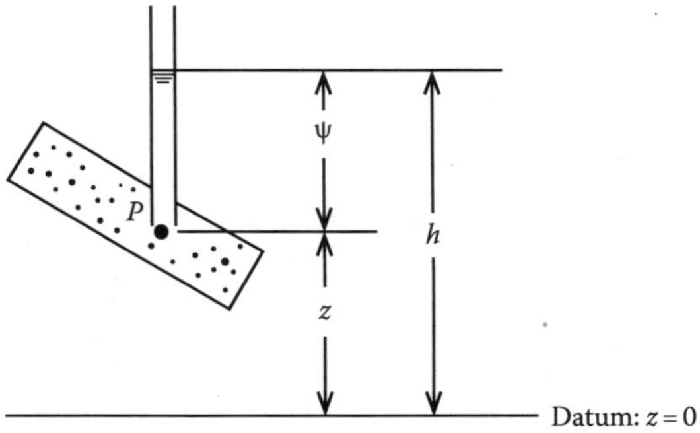

FIGURE 12.6 Hydraulic head h, pressure ψ, and elevation head z at P.

well known as *Bernoulli's equation*, which expresses the loss of energy during fluid flow. By simplifying it, we obtain

$$\Phi = gz + \frac{p - p_0}{\rho} \tag{12.17}$$

which involves the elevation z and the fluid pressure p.

In Figure 12.6, the fluid pressure at P is given by

$$p = \rho g \psi + p_0 \tag{12.18}$$

where

ψ is the height of the liquid column above P

p_0 is the atmospheric pressures at the standard state

From Figure 12.6 and Equation 12.19, we obtain

$$p = \rho g(h - z) + p_0 \tag{12.19}$$

and from (12.19) and (12.17), it issues that

$$\Phi = gz + \frac{\left[\rho g\left(h - z\right) + p_0\right] - p_0}{\rho} \tag{12.20}$$

so that simplifying

$$\Phi = gh \tag{12.21}$$

The fluid potential Φ at any point P in a porous medium is strictly dependent on the hydraulic head, h, as the gravity acceleration, g, is practically constant in the vicinity of earth's surface. In practice, we may assume that the atmospheric pressure $p_0 = 0$ and $h_0 - h_w = (Q/2\pi Kb)\ ln\ (r_0/r_w)$ work in gage pressures (i.e., pressures above atmospheric). From (12.18) and (12.22),

$$\Phi = gz + \frac{p}{\rho} = gh \tag{12.22}$$

which, dividing by g, gives

$$h = z + \frac{p}{\rho g} \tag{12.23}$$

From (12.19) in terms of gage pressures we obtain

$$p = \rho g \psi \tag{12.24}$$

so that

$$h = z + \psi \tag{12.25}$$

The hydraulic head h is the sum of two components: the elevation head, z, and the pressure head, ψ, as in Figure 12.6. Noting that the elevation head $h_z = z$, the pressure head $\psi = h_p$, and the velocity head $h_v = v_2/2g$, the total head is

$$h_T = h_z + h_p + h_v \tag{12.26}$$

12.6 Piezometers and Piezometer Nests

The hydraulic head is measured in a simple tube, as in Figure 12.6, sealed along its length, open to water flow at the bottom and to the atmosphere at the top. The water level in the tube, named *manometer* in the laboratory and *piezometer* in the field, indicates the water pressure at its bottom. Recently, in some applications, more complex instrumentation consisting of pressure transducers, pneumatic devices, and electronic components are used instead of the simple standpipe piezometer. If the hydraulic gradient *dh/dl* between two piezometers and the hydraulic conductivity *K* of the geological formation are known, Darcy's law can be used to calculate the *specific discharge*, representing the flow volume rate through any cross-sectional area perpendicular to the flow direction. *Piezometer nests* consist of two or more piezometers installed side by side at the same location or possibly in the same hole, each bottoming at a different depth and possibly in a different geological formation, so as to measure the vertical potential gradient. The positions of the equal hydraulic heads may be drawn in two dimensions by contours, similar to topographic map contours, which trace lines of equal altitude. Thus, head contours are *equipotential lines* that represent the *piezometric* or *potentiometric* surface of the water body that can be used to calculate the flow direction perpendicular to them in both confined and unconfined aquifers. In practice, if we have a minimum of three observation wells, groundwater levels may be plotted on a scale map at the vertex of a triangle and related to a common datum to draw contours, by joining points of equal height on each of the lines, as shown in Figure 12.7, so that they represent equipotential lines. If the pattern of hydraulic heads is known in a crossed section, flow lines, everywhere tangent to the macroscopic velocity vector, can be constructed perpendicular to the equipotential lines in the direction of the maximum potential gradient. This procedure may be applied to a much larger number of water level values measured within an aquifer basin to construct its contour map. The resulting set of intersecting equipotential and flow lines is named flownet, as in Figure 12.8. Flownets are commonly used to represent the steady-state flow described by Darcy's law [43]. With reference to Figure 12.8, *streamflows* or *streamtubes*, of a width *dm* defined by two adjacent flow lines, let pass through an equivalent amount of flow, concentrated in the smallest squares of a flownet. The hydraulic gradient *i* is given by

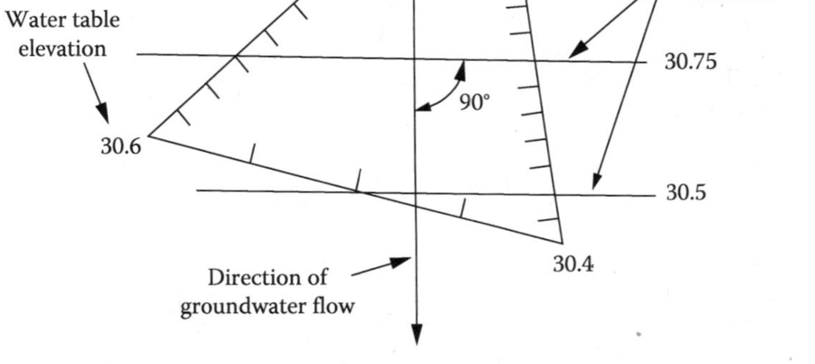

FIGURE 12.7 Groundwater contours and flow directions, based on water levels in three observation boreholes.

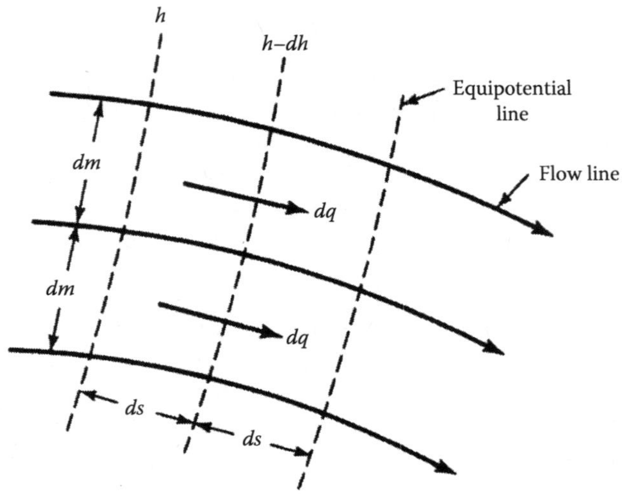

FIGURE 12.8 Portion of an orthogonal flownet formed by flow and equipotential lines.

$$i = \frac{dh}{ds} \tag{12.27}$$

where *dh* and *ds* are, respectively, the difference in head and distance between two adjacent flow lines, so that the constant flow *q* between them for unit thickness is

$$q = K\frac{dh}{ds}dm \tag{12.28}$$

For $ds \cong dm$, (12.28) reduces to

$$q = Kdh \tag{12.29}$$

The total head loss *h* may be divided into *n* squares between any two adjacent flow lines, so that $dh = h/n$, and the total flow divided into *m* channels by flow lines is

$$Q = mq = \frac{Kmh}{n} \tag{12.30}$$

Thus, the geometry of the flownet, together with the hydraulic conductivity and head loss, enables to define the total flow in the section.

Groundwater recharge areas are indicated by divergent flow lines and groundwater discharge areas by convergent flow lines. Furthermore, narrow contour spacings are associated with areas of higher hydraulic conductivity. Local, intermediate, and regional systems of groundwater flow may be differentiated in depth [4,16,43,44].

Contour and flow lines are useful for locating new wells. High flow due to high pore pressure near the surface of a porous loose material may produce *soil piping*, which generally begins with the formation of a sand boil with land surface erosion and collapse. Soil piping may cause landslides and the failure of levees and earth dams.

12.7 Coupled Flow

Water can flow through porous media under the influence of *temperature*, *electrical*, and *chemical* gradients even in the absence of hydraulic gradients [7,18,21,23,31]. An electrical gradient with earth currents can entrain a water flow from high voltage to low, owing to an interaction between charged ions in the water and the electrical charge associated with clay minerals in the soil. The principle is used in soil mechanics and engineering geology to drain clayeous soils [37]. Chemical gradients can also produce flow of groundwater and especially transport of its chemical components from zones with higher water salinity to zones with lower salinity. This process is to be taken in due considerations in the assessment of groundwater contamination. If each of these gradients plays a role in producing a flow, the general flow law may be written in the form

$$v = -L_1 \frac{dh}{dl} - L_2 \frac{dT}{dl} - L_3 \frac{dc}{dl} \tag{12.31}$$

where

- h is the hydraulic head
- T is the temperature
- c is the chemical concentration
- L_1, L_2, and L_3 are constants of proportionality

If $dc/dl = 0$, the fluid flow is governed by both a hydraulic gradient and a temperature gradient:

$$v = -L_1 \frac{dh}{dl} - L_2 \frac{dT}{dl} \tag{12.32}$$

12.8 Aquifer Compressibility

In the volume V_w of a given mass of water of density ρ, an increase in pressure dp produces a decrease, given by the *compressibility*

$$\beta = \frac{-dV_w/V_w}{dp} \tag{12.33}$$

which can be rewritten as

$$\beta = \frac{d\rho/\rho}{dp} \tag{12.34}$$

By integrating (12.35), we obtain the *equation of state* for water:

$$\rho = \rho_0 e^{[\beta(p - p_0)]} \tag{12.35}$$

which for atmospheric p_0 can be written in terms of gage pressure as

$$\rho = \rho_0 e^{\beta p} \tag{12.36}$$

For incompressible fluids, $\beta = 0$ and $\rho = \rho_0 = \text{constant}$.

The total stress, due to the weight of overlying rock and water, acting downward on the plane of a saturated geological formation at depth is

$$\sigma_T = \sigma_e + p \tag{12.37}$$

where σ_e is the effective stress representing the actual stress, not due to water, applied to the grains of the porous medium and p the fluid pressure of the water in the pores [37]. Rock and water weight overlying each point in the system often remains essentially constant through time, so that $d\sigma_e = 0$ and

$$d\sigma_e = -dp \tag{12.38}$$

The compressibility of a porous medium is defined as

$$\alpha = \frac{-dV_T / V_T}{d\sigma_e} \tag{12.39}$$

The total stress applied over time on a saturated soil sample produces water drainage and *soil Consolidation*. If the soil sample has an original height b and an original void ratio e_0 (where $e = V_v/V_s$), and assuming $dV_T = dV_v$, (12.39) can be written as

$$\alpha = \frac{-db/b}{d\sigma_e} = \frac{-de/(1+e_0)}{d\sigma_e} \tag{12.40}$$

The soil compressibility α is a function of the applied stress and it is dependent on the previous loading history. Unlike the fluid compressibility β, α is not a constant. If in a compressible aquifer of thickness b the weight of overlying material remains constant, and the hydraulic head in the aquifer is decreased by an amount $-dh$, the increase in effective stress is given by

$$d\sigma_e = \rho g dh \tag{12.41}$$

and the aquifer compaction by

$$db = -\alpha b d\sigma_e = -\alpha b \rho g dh \tag{12.42}$$

The reduction in the thickness b of the aquifer, due to a head decrease, is indicated by the minus sign. Horizontal hydraulic gradients toward the well, induced by pumping, cause hydraulic head decreases at each point near the well, and, as a result, aquifer compaction due to effective stress increases. Conversely, pumping water into an aquifer increases hydraulic heads, decreases effective stresses, and produces aquifer expansion. If the compaction of an aquifer–aquitard system due to groundwater pumping is propagated to the ground surface, *land subsidence*, *sinkholes*, and differential settlements occur, which may endanger structure and infrastructure stability.

12.9 Specific Storage, Transmissivity, and Storativity

Saturated groundwater flow is characterized by water density ρ, viscosity μ, and compressibility β, and media porosity n (or void ratio e), permeability k, and compressibility α. All the other parameters used to describe the hydrogeological properties of geologic formations can be derived from these six physical properties.

The *specific storage* S_s of a saturated aquifer is defined as the volume of water that a unit volume of aquifer releases, or absorbs, from storage under a unit decline, or increases, in hydraulic head.

The water released from storage under conditions of decreasing h is produced by the compaction of the aquifer caused by increasing σ_e, and the expansion of the water caused by decreasing p. The first of these mechanisms is controlled by the aquifer compressibility α and the second by the fluid compressibility β. The volume of water expelled from the unit of volume of aquifer during compaction will be equal to the reduction in volume per unit of volume of aquifer. The volumetric reduction dV_T will be negative, but the amount of water produced dV_w will be positive, so that from (12.42),

$$dV_w = -dV_T = \alpha dV_T d\sigma_e \tag{12.43}$$

Considering a unit volume, $V_T = 1$, and (12.42), $d\sigma_e = \rho g dh$, for a unit decrease in hydraulic head, $dh = -1$, we obtain

$$dW_w = \alpha \rho g \tag{12.44}$$

From (12.34), the water volume produced by expansion is

$$dW_w = -\beta V_w dp \tag{12.45}$$

Since $V_w = nV_T$, where n is the porosity, and $dp = \rho g d\psi = \rho g d(h - z) = \rho g dh$, (12.45) becomes, for $dh = -1$,

$$dV_w = \beta n \rho g \tag{12.46}$$

The specific storage S_s is given by the sum of two terms of (12.44) and (12.46):

$$S_s = \rho g(\alpha + n\beta) \tag{12.47}$$

Its dimensions are $[L^{-1}]$.

For a confined aquifer of thickness b, the *transmissivity* (or *transmissibility*) T is defined as

$$T = Kb \tag{12.48}$$

As K dimensions are $[LT^{-1}]$, T dimensions are $[L^2T^{-1}]$. The range of T values can be assessed by multiplying the K values from Figure 12.5 by the range of aquifer thickness. Transmissivities greater than 0.015 m^2/s represent good aquifer for water well development adequate for industrial, municipal, or irrigation purposes. An aquifer with a transmissivity lower than 1.4 m^2/day can supply only enough water for wells for low-yield uses.

The *storativity* S (or *storage coefficient* S) is defined as

$$S = S_s b \tag{12.49(a)}$$

By substituting (12.47) in (12.49 (a,b)), we have

$$S = \rho g b(\alpha + n\beta) \tag{12.49(b)}$$

The *storage coefficient* (or *storativity*) is the water volume that an aquifer releases from or takes into storage per unit surface area per unit change in the component of head normal to surface. For a vertical column of unit area extending through a confined aquifer, as in Figure 12.9a, the *storage coefficient* S equals the volume of water released from the aquifer when the piezometric surface declines a unit distance. In confined aquifers, S is the result of compression of the aquifer and expansion of the confined water when the hydraulic head is reduced during well pumping [12,13,43]. The coefficient is a dimensionless quantity involving a volume of water per volume of aquifer $[(L^3/L^2)/L = -]$. Storativity can best be determined from pumping tests or from groundwater fluctuations due to atmospheric pressure or ocean tide variations. Values of S for confined aquifers range from 10^{-5} to 10^{-3}. As S generally varies with confined aquifer thickness, the rule-of-the-thumb relationship

$$S = 3 \times 10^{-6} b \tag{12.50}$$

may be adopted, where b is the saturated aquifer thickness. Transmissivities and storativities can be assessed also for aquitards, although the vertical hydraulic conductivity of an aquitard is more significant than its transmissivity. In clay aquitards, $\alpha \gg \beta$, and the $n\beta$ term in the definition of specific storage

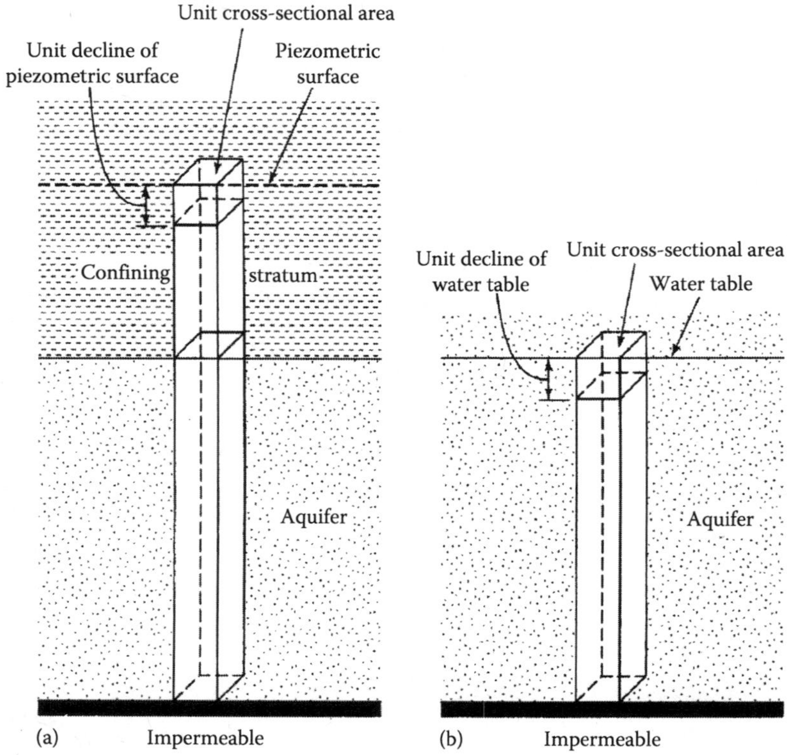

FIGURE 12.9 Representation of the storativity in (a) confined and (b) unconfined aquifers. (After Todd, D.K. and Mays, L.W., *Groundwater Hydrology*, 3rd edn., John Wiley & Sons, Inc., Hoboken, NJ, 2005.)

and storativity, as in (12.47) and (12.50), is negligible. The concept of aquifer storage inherent in the storativity term also implies an instantaneous release of water from any elemental volume in the system as the head drops in that element.

The *hydraulic diffusivity D*, defined as

$$D = \frac{T}{S} = \frac{K}{S_s} \tag{12.52}$$

is a parameter commonly used as it couples the transmission properties T and K and the storage properties S or S_s. Equation 12.49 (a,b) is generally used also in unconfined aquifer, considering b as the saturated thickness of the aquifer between the water table and the top of the underlain aquitards. The *specific yield Sy*, sometimes called *unconfined storativity*, is defined as the water volume that an unconfined aquifer releases from storage per unit area of aquifer per unit decline in the water table, as shown in Figure 12.9b. Its values for unconfined aquifers range from 0.01 to 0.3.

12.10 Groundwater Flow Equations

Groundwater flow equations, well known to mathematicians and mathematical engineers, are good examples of a boundary-value problem of which the solution depends on governing equation, flow region size and shape, material special properties (conductivity), initial conditions, boundary conditions, and mathematical method of solution. For solving problems of groundwater flow, Darcy's law alone is not enough, as it only gives three relations between four unknown quantities: the three components of the specific discharge vector and the head. A fourth equation is given by the *equation of continuity*, which satisfies the fundamental physical principle of conservation of mass for the flow for steady incompressible flow in x, y, z directions when the density ρ is a constant:

$$\frac{\partial v_x}{\partial x}+\frac{\partial v_y}{\partial y}+\frac{\partial v_z}{\partial z}=0 \tag{12.53}$$

This is the so-called *equation of continuity*, which holds in general [3,16,42,45]. By substituting Darcy's law for v_x, v_y, and v_z in (12.53), we obtain the equation of flow for steady-state flow through an anisotropic saturated porous medium:

$$-\frac{\partial}{\partial x}\left(K_x\frac{\partial h}{\partial x}\right)+\frac{\partial}{\partial y}\left(K_y\frac{\partial h}{\partial y}\right)+\left(K_z\frac{\partial h}{\partial z}\right)=0 \tag{12.54}$$

For an isotropic homogeneous medium, $K_x=K_y=K_z$, so that $K(x,y,z)=\text{constant}$, and (12.55) reduces to the differential equation

$$\frac{\partial^2 h}{\partial x^2}+\frac{\partial^2 h}{\partial y^2}+\frac{\partial^2 h}{\partial z^2}=0 \tag{12.55}$$

often written in the abbreviated form

$$\nabla^2 h=0 \tag{12.56}$$

known in mathematical physics as *Laplace's equation*, which represents the steady incompressible groundwater flow through a homogeneous, isotropic medium. Its solution is a function $h(x,y,z)$ that gives the hydraulic head value h at any point in a 3D flow field. By solving (12.55), a contoured equipotential map of h, and with the addition of flow lines, a flownet may be obtained. For steady-state, saturated flow in 2D flow field, say in the xz plane, the central term of (12.55) would drop out and the solution would be a function $h(x,z)$. The operator ∇^2 (*nabla squared*, or *del square*), called *Laplace's operator*, denotes the operation of taking the second derivatives with respect to the coordinates x, y, z, and adding the results

$$\nabla^2=\frac{\partial^2}{\partial x^2}+\frac{\partial^2}{\partial y^2}+\frac{\partial^2}{\partial z^2} \tag{12.57}$$

The basic problem of groundwater movement is to find solutions to (12.55). The equation of flow for *transient flow* through a saturated anisotropic porous medium is

$$\frac{\partial}{\partial x}\left(K_x\frac{\partial h}{\partial x}\right)+\frac{\partial}{\partial y}\left(K_y\frac{\partial h}{\partial y}\right)+\frac{\partial}{\partial}\left(K_z\frac{\partial h}{\partial z}\right)=S_s\frac{\partial h}{\partial t} \tag{12.58}$$

If the medium is homogeneous and isotropic, (12.58) reduces to

$$\frac{\partial^2 h}{\partial x^2}+\frac{\partial^2 h}{\partial y^2}+\frac{\partial^2 h}{\partial z^2}=\frac{S_s}{K}\frac{\partial}{\partial t} \tag{12.59}$$

or expanding S_s

$$\frac{\partial^2 h}{\partial x^2}+\frac{\partial^2 h}{\partial y^2}+\frac{\partial^2 h}{\partial z^2}=\frac{\rho g(\alpha+n\beta)}{K}\frac{\partial h}{\partial t} \tag{12.60}$$

Equation 12.61 is known as the *diffusion equation*. For the special case of a horizontal confined aquifer of thickness b, $S=S_s b$ and $T=Kb$, and the 2D form of (12.59) becomes

$$\frac{\partial^2 h}{\partial x^2}+\frac{\partial^2 h}{\partial y^2}=\frac{S}{T}\frac{\partial h}{\partial t} \tag{12.61}$$

The solution $h(x,y,z,t)$ describes the hydraulic head field at any point on a horizontal plane through the horizontal aquifer at any time. Solution requires knowledge of the aquifer parameters, the storativity S and transmissivity T, used primarily for the analysis of well hydraulics in confined aquifers. For 2D, horizontal flow toward a well in a confined aquifer of thickness b, the terms are well defined, but they lose their meaning in other groundwater applications. If the problem is to be seen in three dimensions, it is better to revert to the use of hydraulic conductivity K and specific storage S_s, or perhaps, even better, to the fundamental parameters *permeability k*, *porosity n*, density ρ, and *compressibility* α.

12.10.1 Confined Aquifer

Let groundwater flow with a velocity v in the x-direction of a confined aquifer of uniform thickness. For 1D, steady flow (12.55) reduces to

$$\frac{\partial^2 h}{\partial x^2}=0 \tag{12.62}$$

Assuming $h=0$ when $x=0$ and $\partial h/\partial x=v/K$ from Darcy's law, by integration, we have

$$h=-\frac{vx}{K} \tag{12.63}$$

This relation states that the head decreases linearly, as sketched in Figure 12.10 [43].

12.10.2 Unconfined Aquifer

In an unconfined aquifer, direct analytic solution of the Laplace equation is not possible as the water table in two dimensions represents a flow line, and the shape of the water table, which determines flow

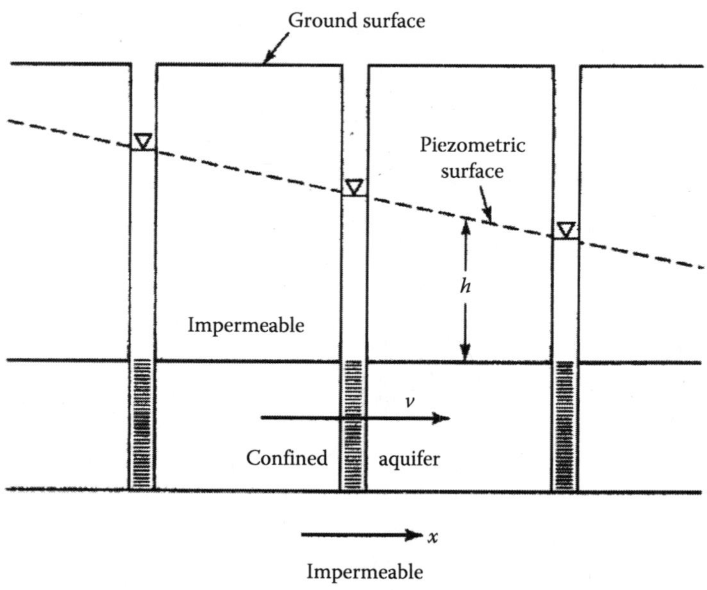

FIGURE 12.10 Steady unidirectional flow in a confined aquifer of uniform thickness. (After Todd, D.K. and Mays, L.W., *Groundwater Hydrology*, 3rd edn., John Wiley & Sons, Inc., Hoboken, NJ, 2005.)

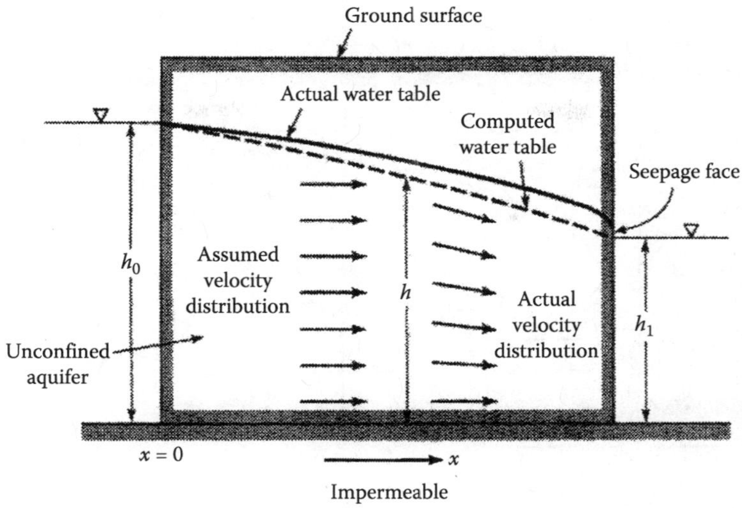

FIGURE 12.11 Steady unidirectional flow in an unconfined aquifer. (After Todd, D.K. and Mays, L.W., *Groundwater Hydrology*, 3rd edn., John Wiley & Sons, Inc., Hoboken, NJ, 2005.)

distribution, is strictly affected by surface morphology. To obtain a solution, Dupuit assumed (1) the velocity of the flow to be proportional to the tangent of the hydraulic gradient instead of the sine as defined in Darcy's law, and (2) the flow to be horizontal and uniform everywhere in a vertical section. Thus, a solution can be obtained, but the application of the results is limited.

For unidirectional flow, as shown in Figure 12.11, the discharge per unit width q at any vertical section is

$$q = -Kh\frac{dh}{dx} \tag{12.64}$$

where

- K is the hydraulic conductivity
- h is the height of the water table above an impervious bed
- x is the direction of flow

By integration, it issues

$$qx = -\frac{K}{2}h^2 + C \tag{12.65}$$

and for $h = h_0$ where $x = 0$, we obtain *Dupuit's equation*

$$q = -\frac{K}{2x}\left(h_0^2 - h^2\right) \tag{12.66}$$

which shows that the head of the water table varies with parabolic law [43]. The actual water table is above the computed one and deviates more and more in the direction of flow, as a greater saturated thickness is needed for the same discharge. The piezometric surface tangentially approaches the downstream boundary above the water table and emerges at atmospheric pressure with a *seepage face*.

12.11 Steady Radial Flow to a Well

When a well is pumped, water is removed from the aquifer around the well, and the water table or piezometric surface, depending on the type of aquifer, is lowered. A *drawdown curve* shows the variation of drawdown with distance from a well (Figures 12.12 and 12.13). Its outer limit (zero drawdown) defines

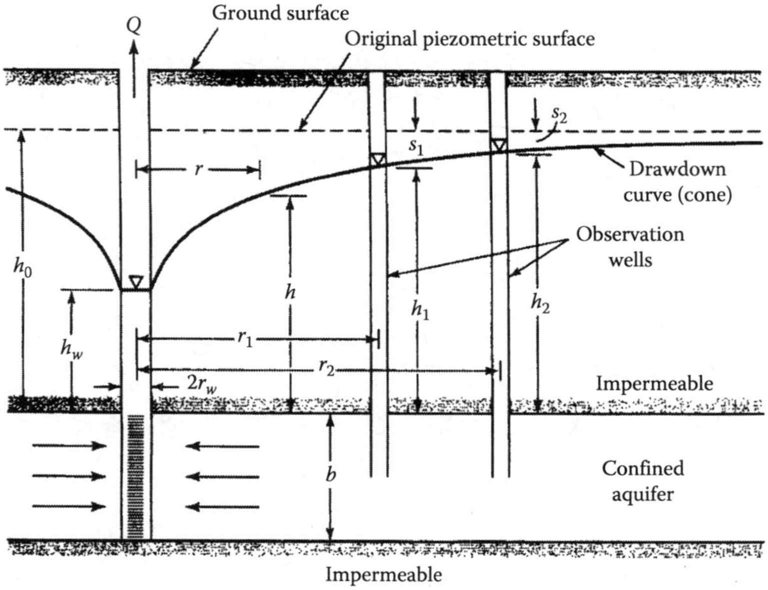

FIGURE 12.12 Steady radial flow to a well completely penetrating a confined aquifer. (After Todd, D.K. and Mays, L.W., *Groundwater Hydrology*, 3rd edn., John Wiley & Sons, Inc., Hoboken, NJ, 2005.)

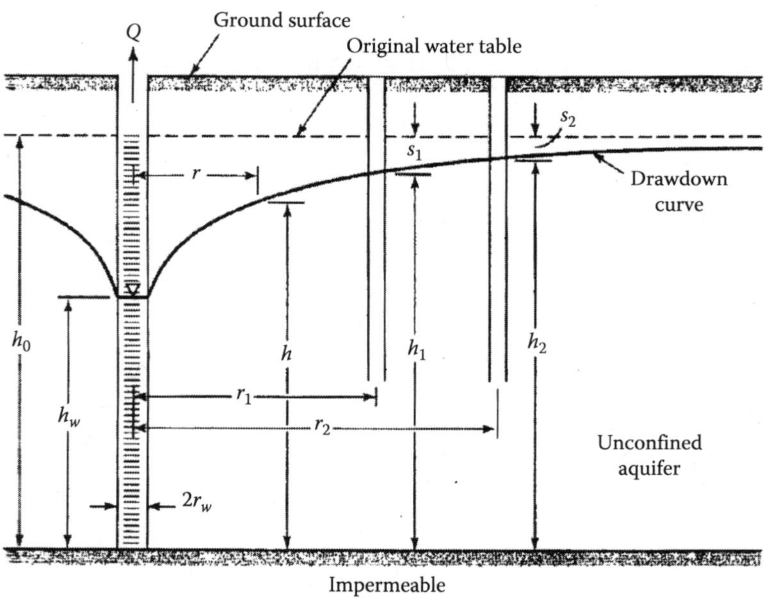

FIGURE 12.13 Radial flow to a well penetrating an extensive unconfined aquifer. (After Todd, D.K. and Mays, L.W., *Groundwater Hydrology*, 3rd edn., John Wiley & Sons, Inc., Hoboken, NJ, 2005.)

the area of influence of the well. In three dimensions, the drawdown curve describes a *cone of depression*. By introducing the relation $r=\sqrt{x^2+y^2}$ in (12.60), the *diffusion equation* for the radial flow into a well can be represented in radial coordinates as

$$\frac{1}{r}\frac{\partial}{\partial}\left(r\frac{\partial h}{\partial r}\right)=\frac{\partial^2 h}{\partial r^2}+\frac{1}{r}\frac{\partial h}{\partial r}=\frac{S}{T}\frac{\partial h}{\partial t} \tag{12.67}$$

where r is the radial distance from a pumped well and t is the time since the beginning of pumping. For steady-state conditions, ∂h/∂t reduces to

$$\frac{1}{r}\frac{\partial}{\partial r}\left(r\frac{\partial h}{\partial r}\right)=0 \tag{12.68}$$

To derive the radial flow equation (which relates the well discharge to drawdown) for a well completely penetrating a confined aquifer, the flow is assumed 2D to a well centered on a circular island and penetrating a homogeneous and isotropic aquifer. Because the flow is horizontal everywhere, Dupuit's assumptions apply without error. Using plane polar coordinates with the well as the origin, we obtain the well discharge Q at any distance r as

$$Q = Av = -2\pi rbK\frac{dh}{dr} \tag{12.69}$$

By integration between the boundary conditions at edge of the island, $h=h_0$ and $r=r_0$, and at the well, $h=h_w$ and $r=r_w$, we obtain

$$h_0 - h_w = \frac{Q}{2\pi Kb}\ln\frac{r_0}{r_w} \tag{12.70}$$

or

$$Q = 2\pi Kb\frac{h_0 - h_w}{\ln(r_0/r_w)} \tag{12.71}$$

with the negative sign neglected. In the more general case of a well penetrating an extensive confined aquifer, as in Figure 12.13, (12.71) may be rewritten as

$$Q = 2\pi Kb\frac{h - h_w}{\ln(r/r_w)} \tag{12.72}$$

which indicates that the maximum h approaches h_0 with distance from the well, and the drawdown varies with the logarithm of the distance from the well. Equation 12.72, known as the *equilibrium* or *Thiem's equation*, which represents the steady-state flow to a pumping well fully penetrating an aquifer with no nearby boundaries [39]. The equation is an approximation to actual conditions generally used in practice to assess the value of the hydraulic conductivity or transmissivity of a confined aquifer by measuring the drawdowns in two observation wells at different distances from a well pumped at a constant rate.

The transmissivity is given by

$$T = Kb = \frac{Q}{2\pi(h_2 - h_1)}\ln\frac{r_2}{r_1} \tag{12.73}$$

where

r_1 and r_2 are the distances
h_1 and h_2 are the heads of the respective observation wells

From a practical standpoint, the drawdown s rather than the head h is measured so that (12.73) can be rewritten as

$$T = \frac{Q}{2\pi(s_1 - s_2)}\ln\frac{r_2}{r_1} \tag{12.74}$$

An equation for the steady radial flow to a well in an unconfined aquifer can also be derived on the basis of Dupuit's assumptions. As shown in Figure 12.13, the well completely penetrates the aquifer to the horizontal base and a concentric boundary of constant head surrounds the well. The well discharge is

$$Q = 2\pi r K h \frac{dh}{dr} \tag{12.75}$$

which, when integrated between the limits $h = h_w$ at $r = r_w$ and $h = h_0$ at $r = r_0$ yields.

$$Q = \pi K \frac{h_0^2 - h_w^2}{\ln(r_0 / r_w)} \tag{12.76}$$

The cone of depression shows a variation in drawdown with distance from the pumped well:

$$Q = 2\pi r K h \frac{dh}{dr}$$

Converting the heads and radii at two observation wells (Figure 12.13),

$$Q = \pi K \frac{h_2^2 - h_1^2}{\ln r_2 / r_2} \tag{12.77}$$

and rearranging to solve for the hydraulic conductivity

$$K = \frac{Q}{\pi\left(h_2^2 - h_1^2\right)} \ln \frac{r_2}{r_1} \tag{12.78}$$

Although (12.78) does not properly describe the drawdown curve near the well as the large vertical flow components contradict Dupuit's assumptions, it is conveniently used to approximate the transmissivity by

$$T \cong K \frac{h_1 + h_2}{2} \tag{12.79}$$

Where drawdowns are appreciable, the heads h_1 and h_2 can be replaced by $(h_0 - s_1)$ and $(h_0 - s_2)$, respectively, as shown in Figure 12.14, so that the transmissivity becomes

$$T = Kh_0 = \frac{Q}{2\pi\left[\left(s_1 - \frac{s_1^2}{2h_0}\right) - \left(s_2 - \frac{s_2^2}{2h_0}\right)\right]} \ln \frac{r_2}{r_1} \tag{12.80}$$

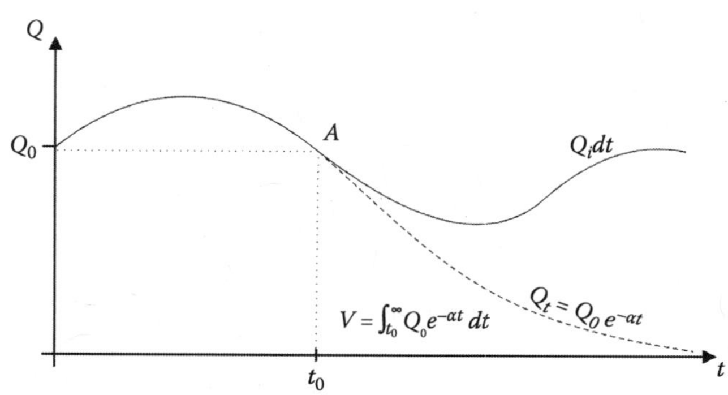

FIGURE 12.14 Groundwater flow recession curve.

12.12 Nonequilibrium Well Pumping Equation

When a confined aquifer is pumped at a constant rate, the influence of the discharge extends outward with time from a cylinder of initial radius R and height H to a coaxial cylinder of radius r and height h given by the filtering section of the well. The rate of decline times the storage coefficient summed over the area of influence equals the discharge. Because the water must come from a reduction of storage within the aquifer, the head will continue to decline as long as the aquifer is effectively infinite; therefore, unsteady, or transient, flow exists. The rate of decline, however, decreases as the area of well influence expands. The applicable differential equation (12.68) in plane polar coordinates is

$$\frac{\partial^2 h}{\partial r^2}+\frac{1}{r}\frac{\partial h}{\partial r}=\frac{S}{T}\frac{\partial h}{\partial t} \tag{12.81}$$

where

- h is the head
- r is radial distance from the pumped well
- S is the storage coefficient
- T is the transmissivity
- t is the time since the beginning of pumping

Theis obtained a solution for (12.81) based on the analogy between groundwater flow and heat conduction [38]. By assuming that a well is replaced by a mathematical sink of constant strength and imposing the boundary conditions $h-h_0$ for $t=0$, and $h\to h_0$ as $r\to\infty$ for $t\geq 0$, the solution

$$s=\frac{Q}{4\pi T}\int_u^{\infty}\frac{e^{-u}du}{u}=\frac{Q}{4\pi T}W(u)=\frac{Q}{4\pi T}\left[-0.5772-\ln u+u-\frac{u^2}{2\cdot 2!}+\frac{u^3}{3\cdot 3!}+\frac{u^4}{4\cdot 4!}+\cdots\right] \tag{12.82}$$

is obtained, where s is the drawdown, Q is the constant well discharge, and

$$u=\frac{r^2 S}{4Tt} \tag{12.83}$$

is a function of the distance r, in m, from the center of a pumped well to a point where the drawdown is measured, S is the coefficient of storage (dimensionless), T is the transmissivity (in m^2/day), and t is the time since pumping started, in days. Equation 12.82 is known as *nonequilibrium*, or *Theis'*, equation. The integral is a function of the lower limit u and is known as an exponential integral. It can be expanded as a convergent series as shown in (12.82) and is termed the well function, $W(u)$. The equation is widely applied in practice and is preferred over the equilibrium equation because (1) a value for S can be determined, (2) only one observation well is required, (3) a shorter period of pumping is generally necessary, and (4) no assumption of steady-state flow conditions is required. Theis' equation gives good results if (1) the aquifer is homogeneous, isotropic, nonleaky, confined, of uniform thickness, and of infinite area extent; (2) before pumping, the piezometric surface is horizontal; (3) the well is pumped at a constant discharge rate; (4) the pumped well penetrates the entire aquifer, and flow is everywhere horizontal within the aquifer to the well; (5) the well diameter is infinitesimal so that storage within the well can be neglected; and (6) water removed from storage is discharged instantaneously with decline of head [43]. Theis solved (12.82) simplifying it to

$$s=\left(\frac{Q}{4\pi T}\right)W(u) \tag{12.84}$$

where $W(u)$, called the *well function*, is a convenient symbolic form of the exponential integral. Rewriting (12.83) as

$$\frac{r^2}{t} = \left(\frac{4T}{S}\right)u \tag{12.85}$$

average values of S and T can be obtained in the vicinity of a pumped well by measuring in one or more observation wells the change in drawdown with time under the influence of a constant pumping rate. Because of the mathematical difficulties, several investigators have developed simpler approximate solutions that can be readily applied for field purposes. Theis' solution to obtain S and T is based on a superposition graphic method, consisting of matching the Theis' type curve to transient (nonequilibrium) drawdown data collected during an aquifer pumping test (APT). Hantush extended Theis' method for use with partially penetrating wells [19,20]. For small values of u, Cooper and Jacob proposed to determine the transmissivity and storativity of nonleaky confined aquifers by matching a straight line to drawdown data as a function of the logarithm time since the start of pumping [8]. In the method suggested by Chow, drawdown data, s, measured in a piezometer near a pumped well, are plotted on a semilogarithmic paper, and Δs is calculated as a drawdown increment for one log time cycle [9]. APTs with different methods are widely used, and well described in literature with useful practical suggestions [4,16,43]. A number of commercial and free software products, mainly based on Excel spreadsheet, are available to calculate S, T, and K as function of drawdown in time with different methods, so that the solutions obtained in well and APTs may be easily compared. It is fundamental that drawdown data measured in pumping test are accurately recorded, and pumping rates must be periodically checked and recorded, so as to avoid misleading result interpretations.

12.13 Base Flow Recession Curve as an Indicator of Groundwater Resource and Reserve Assessment

The recession curve or falling limb of a hydrograph represents the diminishing discharge from storage in the absence of further replenishment, when the effective infiltration is practically ended and the aquifer discharges only its reserves, given by the water volume stored above its minimal piezometric surface. The quantitative analysis of hydrograph recession derives from the work of Boussinesq and Maillet, who proposed that the discharge of a spring is a function of the volume of water held in storage [6,28]. Maillet analyzed hydrographs of springs near Paris and described their flow recession using the simple exponential relation

$$Q_t = Q_0 e^{-\alpha t} \tag{12.86}$$

as shown in Figure 12.14, where Q_t is the discharge (m^3 s^{-1}) at time t, e is the base of the Naperian logarithms, and α is termed the recession coefficient of dimension [T^{-1}]. Maillet equation can be represented in logarithmic form as $\log Q_t = \log Q_0 - 0.43429\alpha t$, from which α may be evaluated as

$$\alpha = \frac{\log Q_0 - \log Q_t}{0.443t} \tag{12.87}$$

The storage capacity of the reservoir at t_0 may be assessed considering that

$$V = \int_{t_0}^{\infty} Q_0 e^{-\alpha t} dt \tag{12.88}$$

By integrating between $t = 0$ and $t = \infty$,

$$V = \frac{Q_0}{\alpha} \tag{12.89}$$

and for $t \neq 0$,

$$V = \frac{Q_0 - Q_t}{\alpha} \tag{12.90}$$

representing the *dynamic resource* of the water volume V stored at t above the minimal piezometric surface of the aquifer. It was widely demonstrated that part or all of the hydrograph recession can be fitted empirically by Maillet's equation [2,24,46].

Boussinesq expressed recession in quadratic form as

$$Q_t = \frac{Q_0}{(1+\alpha t)^2} \tag{12.91}$$

where

$$Q_0 = \frac{1.724 K h_m^2 l}{(1+\alpha t)^2} \tag{12.92}$$

and

$$\alpha = \frac{1.115 K h_m}{n_e L^2} \tag{12.93}$$

in which

- K is the aquifer hydraulic conductivity
- n_e is the effective porosity
- l is the aquifer width perpendicular to its length L
- h_m is the initial hydraulic head at distance L
- Kh_m is the transmissivity

Both Boussinesq and Maillet assumed the aquifer to be porous, homogeneous, isotropic, and unconfined. The recession curve appears to be closer to exponential when flow has a very important vertical component and closer to quadratic when horizontal flow is dominant. As a consequence, aquifer permeability anisotropy also changes the recession form. The combined use of the two fitting methods allows one to quantify the thickness of the aquifer under the outlet.

In French literature, the equation of Boussinesq is often quoted as equation of Tison [41], and elsewhere as equation of Werner and Sundquist [46]. Maillet's exponential equation gives an approximate analytical solution for the diffusion equation that describes flow in a porous medium, whereas the depiction of baseflow recession by Boussinesq's quadratic equation provides an exact analytical solution capable of yielding quantitative data on aquifer characteristics [12,14].

12.14 Groundwater Intrusion: Damages and Counteractions

In urban areas, where natural aquifer recharge due to direct rainfall infiltration may decrease with open space, artificial recharge increases owing to park irrigation and especially to inadvertent leakage from water supply and sewer distribution systems, septic systems, and cesspools. Groundwater level rise in urban areas has been reported from cities all over the world [1,5,15,35]. Groundwater intrusion occurs where the water table raises up to footing-foundation levels, so that the hydrostatic pressure, exerted underneath the floors and against the walls of the basement, causes water encroachment through openings and cracks. Thus, groundwater interacts with concrete cement and aggregates and eventually corrodes its steel reinforcing bars (*rebar*). The expansion of the corrosion products (iron oxides) of carbon steel induces mechanical stresses that cause cracks and disrupt concrete structures. If rebars have been poorly installed and are located too close to the concrete surface in contact with the air, spalling can easily occur. Flat fragments of concrete are detached from the concrete mass by the rebar corrosion and may fall down. The result is loss of concrete tensile strength.

Nonhydraulic cements (e.g., gypsum plaster) must be kept dry in order to retain their strength.

Portland cement, essentially consisting of tricalcium silicate (Ca_3SiO_5), dicalcium silicate (Ca_2SiO_4), tricalcium aluminate ($Ca_3Al_2O_5$), and calcium aluminoferrite ($Ca_4Al_nFe_{2-n}O_7$), is a hydraulic cement widely used to make concrete. Groundwater may be particularly aggressive for its acidity due to transported solutions of sulfates, sulfides, and chlorides [17]. Aqueous solutions may derive from acid rainfall rich in sulfur dioxide (H_2S) dissolved in the airshed and from decomposition of sulfates and iron pyrite (FeS_2) often present in concrete aggregates. Iron pyrite is unstable in the natural environment: when exposed to air and water decomposes into iron oxide and sulfate. This process is hastened by the action of *Acidithiobacillus* bacteria that consumes sulfur and produces sulfuric acid (H_2SO_4). Though first isolated from the soil, it has also been observed corroding concrete sewer pipes, altering hydrogen sulfide sewage gas into sulfuric [33,34]. When aggressive water seeps through cracks and pores present in concrete, it may dissolve various minerals present in the hardened cement paste or in the aggregates. The sulfuric acid dissolves the carbonates in the cured cement and produces sulfates that are harmful to concrete. The saturation of foundations with sulfate-rich groundwater deteriorates the structural capacity of cement [36]. Concrete floors lying on ground that contains pyrite are also at risk. Sodium chloride (NaCl) leaches calcium hydroxide ($Ca(OH)_2$) and causes chemical changes in Portland cement. Dissolved ions, such as calcium (Ca^{2+}) are leached out and transported in solution some distance away. If the physicochemical conditions prevailing in the seeping water evolve with distance along with water path, and water becomes supersaturated with respect to certain minerals, they can further precipitate, making deposits or efflorescences inside the cracks or at the concrete outer surface. Corrosive effects due to seawater are more pronounced on the concrete above the tidal zone than in the submerged zone, where magnesium and hydrogen carbonate (H_2CO_3) ions precipitate a layer of brucite ($Mg(OH)_2$) about 30 μm thick, on which a slower deposition of calcium carbonate as aragonite ($CaCO_3$) occurs. These layers somewhat protect the concrete from other processes, which include attack by magnesium, chloride and sulfate ions and carbonation. Above the water surface, mechanical damage may occur by erosion by waves carrying sand and gravel and by crystallization of salts from water soaking into the concrete pores and then drying up. Pozzolanic cements and cements using more than 60% of slag as aggregate are more resistant to sea water than pure Portland cement. The first known pozzolan was *pozzolana*, a volcanic vitreous siliceous ash used since the Roman times, which, when combined with calcium hydroxide ($Ca(OH)_2$), forms calcium silicates (Ca_2SiO_4), a material with cementitious properties.

Groundwater intrusion generally also produces basement dampness, which can ruin expensive electrical and mechanical equipment exposed in basement space, can increase maintenance requirements through frequent repainting or cleaning to combat mold growth, and can make affected areas uninhabitable or even unusable due to poor air quality.

In selective problem areas, the usual approach to the treatment of water intrusion problems is to *trench and drain*, consisting in excavating and exposing the wall area down to the base of the foundation, to replace waterproofing on the wall surface, and to install a drain tile system and a sump pump around the building or affected area, so as to reduce the hydrostatic pressure and move groundwater away from the building. Of course, all floor drains and sump pump cavities are to be kept clear so as to avoid water stagnation. Elastomeric membranes, able to expand and contract without failure, fixed against the outside face of the concrete, prevent lateral groundwater infiltration into the concrete but are practically ineffective against capillary rise from below the foundations. In fact, foundation structures may be endangered up to the maximum level of the capillarity zone above the water table. Conventional methods are unsuitable to treat basement floors. The water table and the above standing capillary zone may be lowered by pumping but incoherent foundation soil dewatering should be accompanied by consolidation obtained by resin grouting injections. A method called electroosmotic pulse (EOP) is starting to be used, where possible, for both new construction and repair work [22]. An EOP system is realized by inserting anodes (positive electrodes) into the concrete wall or floor on the inside of the structure and by placing cathodes (negative electrodes) in the soil directly outside the structure. The pulse sequence consists of a pulse of positive voltage, a pulse of negative voltage, and a period of rest when no voltage is applied. The positive-voltage pulse has the longest interval and the negative pulse has the shortest interval. The amplitude of the signal is typically between 20 and 40 volts DC (VDC). The positive electrical pulse causes cations (e.g., Ca^{2+}) and associated water molecules to move from the dry side toward the wet side, against the direction of flow induced by the hydraulic gradient, thus preventing water penetration through the below-grade concrete structure. The objective is to achieve a certain current density and thus create an electric field strength in the concrete sufficient to overcome the force exerted on the water molecules by the hydraulic gradient [29,40].

Groundwater intrusion may be a serious problem, in tunnels especially if they are excavated in the saturation zone. Remedial actions to waterproof tunnels include sealing of the liner with either chemical or cementations with various types of grout, directly injected into the crack or joint of the liner through drilled holes so as to seal off the leak by creating a tight grout jacket outside of the tunnel. Grout selection depends on the head, amount of inflow, turbidity, pH, hardness, and chemical composition of the groundwater entering the tunnel.

12.15 Summary and Conclusions

The general aspects concerning the presence and flow of groundwater in porous media were taken into consideration, focusing on the physical characteristics of geological formations acting as hydrostructures. Aquifers are not only a source but also water reservoirs, natural conduits, and filter plants. Engineers and water planners should be well aware that surface and groundwater divides may not coincide, so as to manage groundwater in the best way, in terms of integrated water resources considering their quantity and quality for different uses. The knowledge of hydrogeological catchment physical boundaries is needed to zone protection areas for springs and pumping stations, define the capacity and operational yield of installations for pumping and artificial recharge, and foresee interactions with surface water on the basis of water level field measurements and modeling.

Laws governing groundwater steady and unsteady flow were reviewed keeping in mind their validity and applicability for the solution of problems of major interest in engineering practice. Laminar flow may produce soil piping consisting in the erosion of the finest part of loose materials, with consequent compaction and subsidence. In soluble rocks like calcareous and gypsum formations, turbulent flow in structural discontinuities enlarged by solution generally causes conspicuous hydromechanical erosion in depth, followed by rock collapse and sinkhole formations, typical of karstic areas characterized by rough surface and underground hydrographies, so that surface runoff is mostly ephemeral. Groundwater flow equations are well known and widely used to determine water velocity and pressure in the saturated zone of a soil mass with given boundaries, under certain imposed conditions along them. Mathematically speaking, the problem is a typical case of boundary-value problems; for the solution of which, several more or less effective methods exist, mainly based on pumping test data processing, aiming at defining confined and unconfined aquifer characteristics such as transmissivity, storativity, hydraulic conductivity, specific storage, permeability, porosity, and compressibility. Flow equations are the starting point to develop the differential equations to describe the transport of contaminant solutes into and out of a fixed elemental volume within the flow domain.

The method based on the study of the base flow recession curve, widely adopted to assess the dynamic resource of the water volume stored above the minimal piezometric surface of the aquifer, was also reviewed.

Hydraulic engineers, geomechanical engineers, and engineering geologists are well aware of the relevance that groundwater head and flow may have for dam site stability. Groundwater head may produce failures such as piping, leakage, landslides, and simple embankment slipouts. The flownet approach can be utilized to examine uplift pressures on the base of the dam to calculate the fluid pressure considering the hydraulic head from the flownet. Grouting and drainage to improve foundation rock quality are based on the understanding of the hydrogeological properties and flow characteristics. Drainage systems serve to reduce uplift pressures and gradients and lower water table to counteract groundwater intrusion in building foundations.

References

1. Abu-Rizaiza, O. 1999. Threats from groundwater table rise in urban areas in developing countries, *Water Int.*, 24(1), 46–52.
2. Barnes, B.S. 1939. The structure of discharge recession curves, *Trans. Am. Geophys. Union*, 20, 721–725.
3. Bear, J. 1972. *Dynamics of Fluids in Porous Media*. New York: American Elsevier.
4. Bear, J. 1979. *Hydraulics of Groundwater*. New York: McGraw-Hill Book, Inc.
5. Bianci, H. and Lopardo, R. 2003. Diagnosis and mitigation of groundwater level rise in a highly populated urban system, *Proceedings for the Congress of International Association of Hydraulic Research (IAHR)*, 30, Theme B, pp. 629–636.
6. Boussinesq, J. 1904. Recherches Théoriques sur l'Ecoulement de Nappes d'Eau Infiltrées dans le Sol et sur le Débit des Sources, *J. Math. Pure Appl.*, 10, 5–78.
7. Casagrande, L. 1952. Electro-osmotic stabilization of soils, *J. Boston Soc. Civil Eng., Contribution to Soil Mechanics*. 1941–1953, 285–317.
8. Cooper, H.H., Jr. and Jacob, C.E. 1946. A generalized graphical method for evaluating formation constants and summarizing well-field history, *Trans. Am. Geophys. Union*, 27, 526–534.
9. Chow, V.T. 1952. On the determination of transmissibility and storage coefficients from pumping test data, *Trans. Amer. Geophys. Union*, 33, 397–404.
10. Darcy, H. 1856. *Les fontaines Publiques de la Ville de Dijon*. Paris, France: Victor Dalmont.
11. Dewandel, B., Lachassagne, P., Bakalowitz, M., Weng, Ph., and Al-Malki, A. 2003. Evaluation of aquifer thickness by analysing recession hydrographs. Application to the Oman ophiolite hard-rock aquifer, *J. Hydrol.*, 274, 248–269.
12. Domenico, P.A. 1972. *Concepts and Models in Groundwater Hydrology*. New York: McGraw-Hill.
13. Ferris, J.C., Knowles, D.B., Browne, R.H., and Stallman, R.W. 1962. Theory of aquifer tests, U.S. *Geological Survey Water-Supply Paper*, Washington, DC, 1536-E, 69–174.
14. Ford, D. and Williams, P. 2007. *Karst Hydrogeology and Geomorphology*. Hoboken, NJ: John Wiley & Sons, Inc.
15. Foster, S., Lawrence, A., and Morris, B. 1998. Groundwater in urban development: Assessing management needs and formulation policy strategies, *World Bank Technical Paper 390*, Washington, DC, 55.
16. Freeze, R.A. and Cherry, J.A. 1979. *Groundwater*. Englewood Cliffs, NJ: Prentice Hall.
17. Glasstone, S. 1946. *Textbook of Physical Chemistry*, 2nd edn. Princeton, NJ: D. Van and Company, Inc.
18. Gurr, C.G., Marshall, T.J., and Hutton, J.T. 1952. Water movement in soil due to a temperature gradient, *Soil Sci.*, 24, 335–344.
19. Hantush, M.S. 1961a. Drawdown around a partially penetrating well, *J. Hyd. Div., Proc. Am. Soc. Civil Eng.*, 87(HY4), 83–98.
20. Hantush, M.S. 1961b. Aquifer tests on partially penetrating wells, *J. Hyd. Div., Proc. Am. Soc. Civil Eng.*, 87(HY5), 171–194.
21. Harlan, R.L. 1973. Analysis of coupled heat-fluid transport in partially frozen soil, *Water Resour. Res.*, 9, 1314–1323.
22. Hock, V.F., McInerney, M.K., and Kirstein, E. 1998. Demonstration of electro-osmotic pulse technology for groundwater intrusion control in concrete structures, Technical Report, U.S. Army, Corps of Engineers, CERL, Champaign, IL, Report No. 98/68, ADA No. 354112. http://www.cecer.army.mil/td/tips/pub/details.cfm?PUBID=702&TOP=1

23. Hoekstra, P. 1966. Moisture movement in soils under temperature gradients with the cold-side temperature below freezing, *Water Resour. Res.*, 2, 241–250.
24. Horton, R.E. 1933. The role of infiltration in the hydrologic cycle, *Trans. Am. Geophys. Union,* 14, 446–460.
25. Hubbert, M.K. 1940. The theory of groundwater motion, *J. Geol.*, 48, 785–944.
26. Hubbert, M.K. 1956. Darcy's law and the field equation of the flow of underground fluids, *Trans. Am. Inst. Min. Met. Eng.*, 207, 222–239.
27. Irmay, S. 1958. On the theoretical derivation of darcy and forchheimer formulas, *Trans. Am. Geophys. Union,* 39, 702–707.
28. Maillet, E. 1905. *Mécanique et Physique du Globe. Essai d'Hydraulique Souterraine et Fluviale.* Paris, France: Hermann.
29. Marshall, O.S. Jr., Morefield, S., McIrney, M., and Hock, V. 2007. Electro-osmotic pulse technology for corrosion prevention and control of water intrusion in below grade concrete structures, *Proceedings of the Tri-Service Corrosion Conference 2007*, Department of Defence, Washington, DC.
30. Meinzer, O.E. 1923. The Occurrence of Groundwater in the United States, with a Discussion of Principles, *U.S. Geological Survey, Water Supply Paper* 489.
31. Philip, J.R. and de Vries, D.A. 1957. Moisture movement in porous materials under temperature gradients, *Trans. Am. Geophys. Union,* 38, 222–232.
32. Polubarinova-Kochina, P.Ya. 1962. *Theory of Groundwater Movement.* Princeton, NJ: Princeton University Press.
33. Sand, W. and Bock, E. 1987. Biotest system for rapid evaluation of concrete resistance to sulfur-oxidizing bacteria, *Mater. Performance,* 26 (3), 14–17.
34. Selman, A., Waksman, S.A., and Joffe, J.S. 1922. Microorganisms concerned in the oxidation of sulfur in the soil II. *Thiobacillus thiooxidans*, a new sulfur-oxidizing organism isolated from the soil, *J. Bacteriol.*, 7(2), 239–256. PMC 378965. PMID 16558952. http://www.pubmedcentral.nih.gov/articlerender.fcgi?tool = pmcentrez&artid = 378965
35. Shamsi, A. and Ardeshir, A. 1999. *Pollution of Tehran Groundwater*. Southampton, U.K.: WIT Press, pp. 484–492.
36. Shanin, M. 1988. Impacts of urbanization of the Greater Cairo area on the groundwater in the underlying aquifer, *Proceedings of the International Symposium on Hydrological Processes and Water Management, Duisburg, Germany,* April 24–29, pp. 517–524.
37. Terzaghi, K. and Peck, R.B. 1967. *Soil Mechanics in Engineering Practice,* 2nd edn. New York: John Wiley & Sons, Inc.
38. Theis, C.V. 1935. The relation between the lowering of the piezometric surface and the rate and duration of discharge of a well using groundwater storage, *Trans. Am. Geophys. Union*, 16, 519–524.
39. Thiem, G. 1906. *Hydrogeologische Methoden*. Leipzig, Germany: J. M. Gebhardt, p. 56.
40. Tikhomolova, K.P. 1993. *Electro-Osmosis*. Chichester, West Sussex, England: Ellis Horwood Limited.
41. Tison, G. 1960. Courbe de tarissement, coefficient d'écoulement et perméabilité du bassin, *Proc. AIHS Ass. Gen. Helsinki*, 1, 229–243.
42. Todd, D.K. 1959. *Groundwater Hydrology*. New York: John Wiley & Sons, Inc.
43. Todd, D.K. and Mays, L.W. 2005. *Groundwater Hydrology*, 3rd edn. Hoboken, NJ: John Wiley & Sons, Inc.
44. Tóth, J. 1963. A theoretical analysis of groundwater flow in small drainage basins, *J. Geophys. Res.*, 68, 4795–4812.
45. Verruijt, A. 1970. *Theory of Groundwater Flow*. New York: McMillan.
46. Werner, P.W. and Sundquist, K.J. 1951. On the ground water recession curve for large watersheds, *Int. Assoc. Sci. Hydrol.*, 33, 202–212.

13

Groundwater–Surface Water Interactions

Saeid Eslamian
Isfahan University of Technology

Saeid Okhravi
Isfahan University of Technology

Faezeh Eslamian
Isfahan University of Technology

AUTHORS

Saeid Eslamian received his PhD from the University of New South Wales, Australia, with Professor David Pilgrim. He was a visiting professor in Princeton University, United States, and ETH Zurich, Switzerland. He is currently an associate professor of hydrology in Isfahan University of Technology (IUT). He is the founder and chief editor of the *Journal of Flood Engineering* and *International Journal of Hydrology Science and Technology*. He has published more than 200 publications, mainly in statistical and environmental hydrology and hydrometeorology.

Saeid Okhravi received his degree in water engineering from IUT, Iran, and currently is a distinguished graduate student in water structure engineering in IUT. He has contributed in more than 10 scientific publications, mainly in surface and groundwater interaction, rainwater harvesting systems, and wastewater treatment. He also has experiences in the construction of artificial wetlands and rainwater harvesting systems.

Faezeh Eslamian received her BS in civil engineering and her master's degree in environmental engineering from IUT, Iran, and currently serves as an expert engineer in environmental engineering. In addition to the English language, she has also an official degree in French from the University of Neuchatel, Switzerland. Faezeh has contributed in more than five publications in natural waste solid management, remote sensing, and groundwater–surface water interaction.

PREFACE

Groundwater and surface water have been studied and managed as the separate resources in ancient. However, in recent years, the interactions between the two water resources have been acknowledged worldwide.

Only a small number of surface water management plans consider groundwater as part of their resource management process. The majority does not take groundwater into consideration directly and relies on groundwater management plans to consider surface water allocations and environmental water requirements. There are some exceptions, with more recent area and regional management plans across the country taking a more inclusive approach to surface groundwater–surface water interactions (GSWIs).

Identifying the GSWI is essential to water managers and scientists. Groundwater and surface water interact throughout the landscape, which explains groundwater interactions with all types of surface water, such as streams, lakes, and wetlands, in many different terrains, from the mountains to the oceans.

Historically, regulatory agencies have dealt with these issues through separate groundwater and surface water programs. Increased awareness of GSWI can lead to improved water quality in the bay region. The integration of groundwater and surface water programs can help to avoid problems that arise from managing one resource at the expense of the other particularly as solutions for better stormwater management.

Numerical models developed to date have generally not accounted well for GSWI. There has been limited attention given to considering the groundwater and surface water systems together. Information about catchment water balance modeling with GSWI and increasing our conceptual understanding are the key objectives of this chapter.

13.1 Introduction

Groundwater is a fundamental natural resource for life. Understanding the groundwater–surface water interaction (GSWI) is essential to water managers and scientists. The management of one component of the hydrologic system, such as a stream or an aquifer, is only partly effective because each hydrologic component is in continuing interaction with other components. Interactions between groundwater and surface water play a fundamental role in water resource management. The movement of water between groundwater and surface water should be investigated in view of both quantity and quality aspects of water resources. The movement of water in the atmosphere and on the land surface is relatively easy to visualize, but the movement of groundwater is not. Groundwater moves along flow paths of varying lengths, from areas of recharge to areas of discharge. The generalized flow paths start at the water table, continue through the groundwater system, and terminate at the stream or at the pumped well. The source of water to the water table (groundwater recharge) is the infiltration of precipitation through the unsaturated zone. In the uppermost, unconfined aquifer, flow paths near the stream can be tens to hundreds of feet in length and have corresponding travel times of days to a few years. The longest and deepest flow paths may be thousands of feet to tens of miles in length, and travel times may range from decades to millennia. In general, shallow groundwater is more susceptible to contamination from human sources and activities because of the close proximity to the land surface. Therefore, shallow, local patterns of groundwater flow near surface water are emphasized in this chapter. The following are some concepts of common water resource issues where understanding the interconnections of groundwater and surface water is fundamental to the development of effective water resource management and policy.

The GSWI is complex. To understand these interactions in relation to climate, landform, geology, and biotic factors, a sound hydrogeoecological framework is needed. In addition, the mechanisms of interactions between groundwater and surface water as they affect recharge–discharge processes should be comprehensively outlined, and the ecological significance and human impacts of such interactions should be emphasized as well. Surface water and groundwater ecosystems are viewed as linked components of a hydrologic continuum leading to related sustainability issues. The conceptual landscape in this chapter shows, in a simplified way, groundwater interaction with all types of surface water, such as streams, lakes, and wetlands and in many different terrains, from mountains to oceans. The interaction of groundwater and surface water has been shown to be a significant concern in many of these cases. For example, contaminated aquifers that discharge to streams can result in long-term contamination of surface water; conversely, streams can be a major source of contamination to aquifers. Surface water commonly is hydraulically connected to groundwater, but the interactions are difficult to observe and measure and have been occasionally ignored in water management considerations and policies. Many natural processes and human activities affect the GSWI. The purpose of this chapter is to present our current understanding of these processes and activities as well as the limitations in our knowledge and the capability to characterize them.

Modeling of GSWI requires not only knowledge of groundwater modeling but also special understanding of the exchange processes that occur between surface water and groundwater. In some cases, it becomes necessary to simulate the dynamics of both surface and groundwater flows, using techniques and softwares that are appropriate to the timescales of all flow processes.

Modeling of GSWI calls for all the same stages of development as the modeling of the groundwater flow: conceptualization, design and construction, calibration and sensitivity analysis, and prediction and uncertainty analysis. Each of these is discussed in this chapter with a focus on the specific requirements of GSWI, beyond those of groundwater flow models.

13.2 Federal Statutory/Regulatory/Policy with respect to GSWIs

The Victorian Government of Australia has been committed to manage water resources to support a thriving economy, healthy environment, and growing communities [7]. The government, policy makers, scientists, and land managers should be informed on the interactions between surface and groundwater systems, enabling improved water resource allocation as well as river and floodplain management.

The GSWI zone is important because 75% of the Superfund and the Resource Conservation and Recovery Act (RCRA) sites are located within a half mile of a surface water body. Forty-seven percent of Superfund sites have recorded impacts to surface water. Most of the RCRA sites are located adjacent to or near surface water (presumably for ease of transportation and manufacturing). Within the last 25 years, the Clean Water Act has succeeded in cleaning up point sources in the United States, and Environmental Protection Agency (EPA) now needs to consider nonpoint sources [29].

The Superfund National Contingency Plan offers greater detail; the RCRA relies more on program guidance. The Superfund's goal is to return usable groundwater to beneficial uses (current and future) where practical. When this is not applied, the Superfund strives to prevent further migration and exposure and to evaluate opportunities for further risk reduction. Groundwater generally is considered "potable" if it is so designated by the state or considered so under federal drinking water guidelines.

Final cleanup levels should be attained throughout the plume and beyond the edge of any wastes left in place. The "point of compliance" for a surface water body is where the release enters the surface water [29]. Alternate concentration limits (ACLs) may be considered where contaminated groundwater discharges to surface water, in which contaminated groundwater does not lead to an increase in contaminants of surface water, where enforceable measures are available to prevent exposure to groundwater, or where restoring groundwater is "not practicable" [36]. The Environmental Protection Agency of the United States (USEPA) expects to use treatment to address "principal threats" posed by site where practical.

This guidance document describes key principles and expectations, interspersed with "best practices" based on program experience that should be consulted during the Superfund remedy selection process. These remedy selection "rules of thumb" are organized into three major policy areas:

1. Risk assessment and management
2. Developing remedial alternatives
3. Groundwater response actions

The purpose of this guide was to briefly summarize key elements of various remedy selection guidance documents and policies in one publication. The USEPA believes that consistent application of national policy and guidance is an important means by which we ensure the reasonability, predictability, and cost-effectiveness of our decisions [32]. The RCRA has similar requirements to the Superfund with respect to returning usable groundwater to beneficial uses, points of compliance for groundwater and surface water, protection of surface water from contaminated groundwater, provisions for ACLs (but without an explicit link to "practicability"), and treatment of principal threats. If current human exposures are under control and no further migration of contaminated groundwater is expected, primary near-term goals are established using two environmental indicators [35]. Surface water becomes the boundary if the discharge of contaminated groundwater is within "protective" limits.

In summary, the majority of contaminated sites have serious potential to affect surface waters. The federal framework allows for risk-based decision making (RBDM) with respect to GSWI, but we must still achieve the expectation of restoring groundwater to beneficial use and ensure discharges of groundwater to surface water are protective. Key policy issues to ponder and to pass to senior managers include [7] the following:

- How to achieve short- and long-term protection
- Where, how, and how often to measure compliance
- Whether to restore groundwater, even if it has no impact to surface water
- The diversity of surface bodies
- The relation of cleanup goals to the Clean Water Act's National Pollutant Discharge Elimination System (NPDES) approach
- How to account for, track, and communicate total loads in watersheds

These regulatory approaches provided an improved basis for targeted and prioritized action on prevention and cleaning up of the environmental pollution. Developing these approaches, the most recent European directives and North American laws are increasingly risk based. They stress the need to assess the likelihood and magnitude of adverse effects on the wider environment and to develop more sustainable risk-management approaches that balance costs and benefits.

The Water Framework Directive (WFD) is the most significant piece of new legislation on water issues in Europe in recent decades [12]. It seeks to integrate environmental management of the different environmental compartments, such as groundwater, rivers, estuaries, and wetlands, and to set risk-based objectives to protect and improve water quality and resources. The WFD contains a timetable for implementation that includes plans to replace existing but superseded directives by 2013.

13.3 Hydrologic Cycle and GSWI

The hydrologic cycle describes the continuous movement of water above, on, and below the surface of the Earth [41]. Surface water occurs as streams, lakes, and wetlands, as well as bays and oceans. Surface water also includes the solid forms of water, snow, and ice. The water below the surface of the Earth primarily is groundwater, but it also includes soil water.

Surface water bodies are hydraulically connected to groundwater in most types of landscapes; as a result, surface water bodies are integral parts of groundwater flow systems. Even if a surface water body is separated from the groundwater system by an unsaturated zone, seepage from the surface water may recharge groundwater. Because of the interchange of water between these two components of the hydrologic system, the development or contamination of one commonly affects the other. The movement of surface water and groundwater is controlled to a large extent by the physiography of an area [40]. In addition, climate change is expected to affect the hydrologic cycle. The recognition of climate change, through the effects of precipitation, air temperature, and evapotranspiration, affects the distribution of water to landscapes. Groundwater recharge, discharge, storage, saltwater intrusion, biogeochemical reactions, and chemical fate and transport may be modified by climate change. Another thing that is now known is that pollution in groundwater is a more serious situation because (1) groundwater is a major source of drinking water and (2) remediation of groundwater is much harder and much more expensive compared to that of surface water [9].

Therefore, it is necessary to understand the effects of physiography and climate on surface water and groundwater flow systems in order to understand the GSWI.

The hydrologic cycle is commonly shown by a very simplified diagram that shows only major transfers of water between continents and oceans, as in Figure 13.1.

Groundwater moves along flow paths of varying lengths from areas of recharge to areas of discharge [41]. The infiltration of precipitation through the unsaturated zone is the source of groundwater recharge. Groundwater flow systems can be of greatly different sizes and depths, and they can overlie one another. Local flow systems are the most dynamic and shallowest flow systems; therefore, they have the greatest interchange with surface water. When local flow systems are recharged, water table increases and discharges to adjacent lowlands or surface water. Local flow systems can be underlined by intermediate and regional flow systems. Water in these deeper flow systems has the longer flow paths, but they also eventually discharge to surface water. Local flow system discharges in the nearest shore, and larger-magnitude flow systems discharge to surface water further offshore [39]. Because of the different lengths and travel times of water within flow paths, the chemistry of water discharging into the surface water from different flow paths can be substantially different.

In some landscapes, surface water bodies perch at mediocre altitudes between major recharge and discharge areas. Surface water bodies commonly receive groundwater inflow on the upgradient side and have seepage to groundwater on the downgradient side. Furthermore, depending on the distribution and magnitude of recharge in the uplands, the hinge line between groundwater inflow and outflow can move back and forth across the part of the surface water bed.

The distribution of seepage to and from surface water is controlled by (1) the slope of the water table with respect to the slope of the surface water, (2) small-scale geologic features in the beds of surface water, and (3) climate [39].

Small-scale geologic features in beds of surface water bodies affect seepage patterns at scales too small to be shown in Figure 13.2.

Furthermore, seepage patterns can be influenced by the size, shape, and orientation of the sediment grains in surface water beds. In a surface water bed consisting of one sediment type, such as sand, inflow seepage is greatest at the shoreline, and it decreases in a nonlinear pattern away from the shoreline (Figure 13.3) [41].

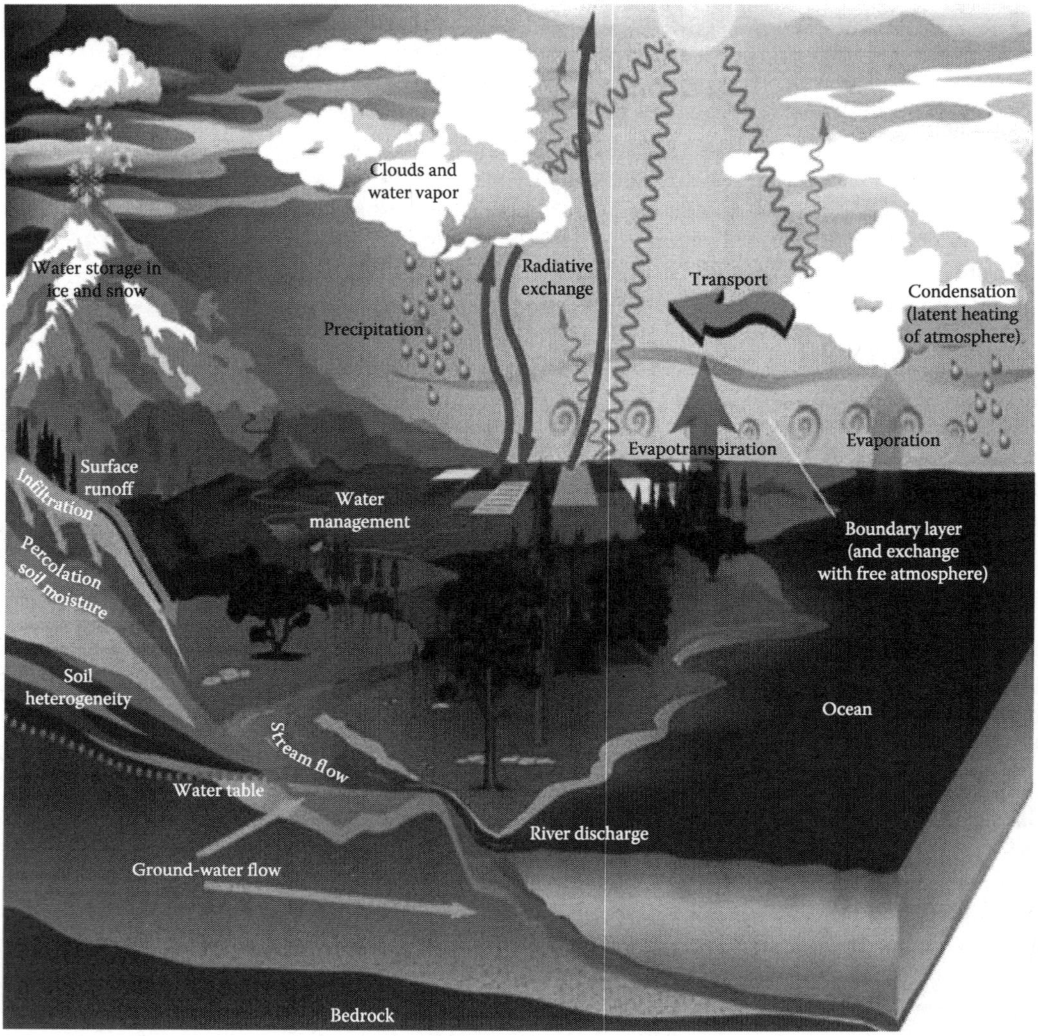

FIGURE 13.1 Hydrologic cycle diagram. (From United States Climate Change Global Research Program, *The Global Water Cycle*, Water Cycle Study Group, Washington, DC, 2001.)

Geologic units having different permeabilities also affect seepage distribution in surface water beds. For example, a highly permeable sand layer within a surface water bed consisting largely of silt will transmit water preferentially into the surface water as a spring (Figure 13.4) [41].

There is a two-way GSWI. One of the most commonly used forms of groundwater comes from unconfined shallow water table aquifers. These aquifers are major sources of drinking and irrigation water. They also interact closely with streams, sometimes water flowing (discharging) into a stream or lake and sometimes receiving water from the stream or lake. An unconfined aquifer that feeds streams is said to provide the stream's baseflow (this is called a gaining stream). In fact, groundwater can be responsible for maintaining the hydrologic balance of surface streams, springs, lakes, wetlands, and marshes [11]. This is why successful watershed partnerships with a special interest in a particular stream, lake, or other surface water body always have a special interest in the unconfined aquifer, adjacent to the water body.

The source of groundwater (recharge) is through precipitation of surface water that percolates downward. Approximately 5%–50% (depending on climate, land use, soil type, geology, and many other factors) of annual precipitation results in groundwater recharges [11].

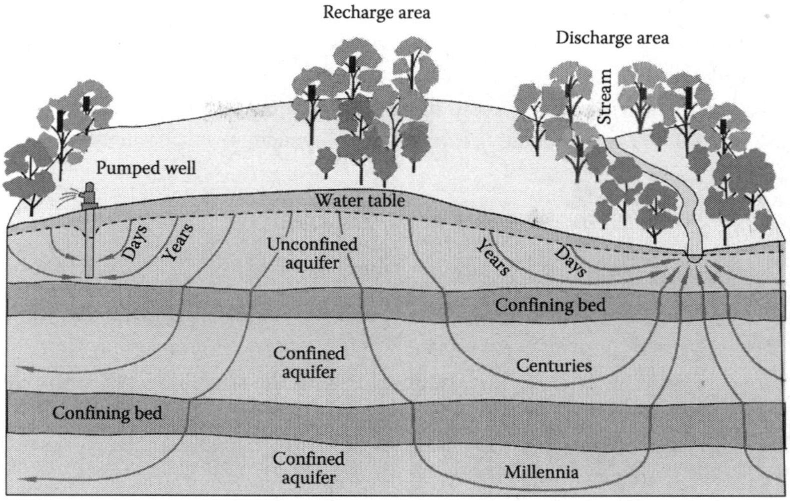

FIGURE 13.2 Groundwater flow paths vary greatly in length, depth, and travel time from points of recharge to points of discharge in the groundwater system. (From Winter, T.C. et al., Groundwater and surface water: A single resource, U.S. Geological Survey Circular 1139, p. 79, 1998.)

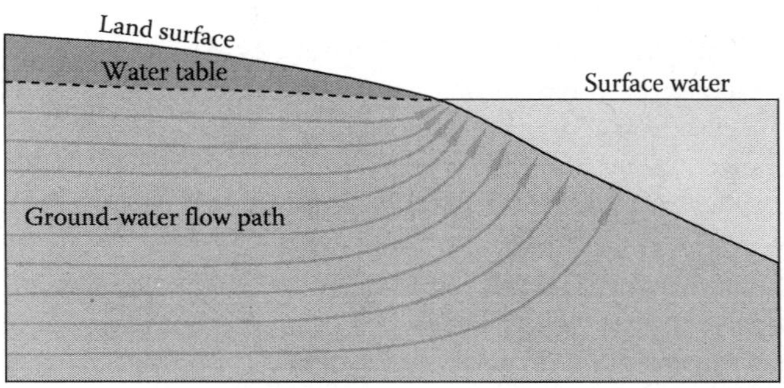

FIGURE 13.3 Groundwater seepage into surface water usually is greatest near shore. In flow diagrams such as that shown here, the quantity of discharge is equal between any two flow lines; therefore, the closer flow lines indicate greater discharge per unit of bottom area. (From Winter, T.C. et al., Groundwater and surface water: A single resource, U.S. Geological Survey Circular 1139, p. 79, 1998.)

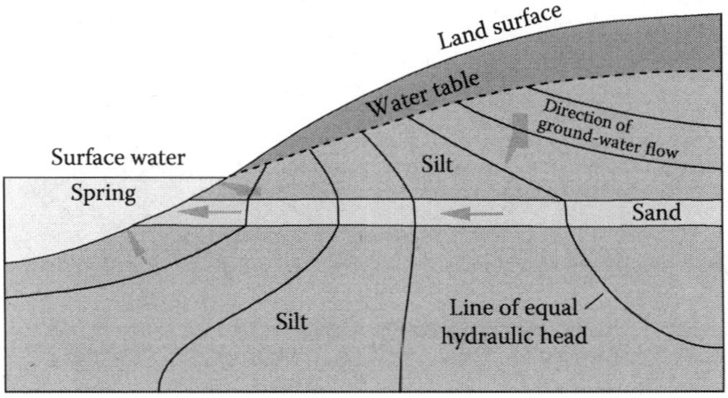

FIGURE 13.4 Subaqueous springs can result from preferred paths of groundwater flow through highly permeable sediments. (From Winter, T.C. et al., Groundwater and surface water: A single resource, U.S. Geological Survey Circular 1139, p. 79, 1998.)

In some areas, streams literally recharge the aquifer through streambed infiltration, called losing streams. Left untouched, groundwater naturally arrives at a balance, discharging and recharging depending on hydrologic conditions.

Integrative studies to fully understand the dynamics and importance of the groundwater/surface water interface are required that combine hydrogeology, biogeochemistry, and aquatic ecology.

13.3.1 Interaction of Groundwater and Streams

The interaction between streams and groundwater takes place in three basic ways:

1. Streams gain water from groundwater through the streambed when the elevation of the water table adjacent to the streambed is greater than that of the stream.
2. Streams lose water to groundwater by outflow through the streambed when the elevation of the water table is lower than that of the stream.
3. A third possibility is that streams may have no flow across their streambed if the stream water and groundwater levels are exactly equal [11].

However, this is a relatively rare coincidence that is unlikely to occur over long reaches for prolonged periods. This variability in streamflow caused by groundwater seepage is often most pronounced around the headwaters of groundwater-fed streams or in stream reaches that go periodically dry due to seepage losses. At times of low groundwater levels, surface flow may only occur due to surface flows in the upper catchment (e.g., from snowmelt) or from stormwater runoff. This flow will contribute seepage through the streambed to the underlying water table. As the water table rises due to general recharge (e.g., rainfall infiltration) to the aquifer, the losing reach may become a gaining reach as the water table rises above the streambed level. Such a change means that the point where groundwater first contributes water to the stream shifts laterally in an upstream direction.

A type of interaction between groundwater and streams takes place in nearly all streams at once, or another is a rapid rise in stream stage that causes water to move from the stream into the stream banks. This process, termed bank storage (Figure 13.5), usually is caused by storm precipitation, rapid snowmelt, or release of water from a reservoir upstream [41].

In addition to bank storage, other processes like pumping may affect the local exchange of water between streams and adjacent shallow aquifers. Changes in streamflow between gaining and losing

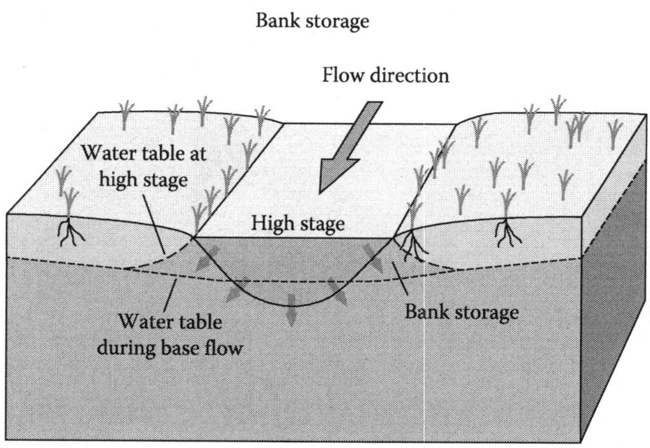

FIGURE 13.5 If stream levels rise higher than adjacent groundwater levels, stream water moves into the stream banks as bank storage. (From Winter, T.C. et al., Groundwater and surface water: A single resource, U.S. Geological Survey Circular 1139, p. 79, 1998.)

conditions can also be caused by pumping groundwater near streams. Pumping can intercept groundwater that would have otherwise been discharged to a gaining stream or at higher pumping rates could have induced flow from the stream to the aquifer.

13.3.2 Interaction of Groundwater and Lakes

Lakes interact with groundwater in three basic methods. Lakes can receive groundwater inflow throughout their entire bed, have outflow throughout their entire bed, or possess both inflow and outflow at different localities (Figure 13.6) [41].

Although these basic interactions are the same for lakes and streams, the interactions differ in several ways [11].

The water level of natural lakes generally does not change as rapidly as the water level of streams; therefore, bank storage is of lesser importance in lakes than it is in streams [41].

Evaporation generally has a greater effect on lake levels than on stream levels because the surface area of lakes is generally larger and less shaded than many reaches of streams and because lake water is not replenished as readily as a reach of a stream. Lakes can be present in many different parts of the landscape and can have complex groundwater flow systems associated with them. This is especially true for lakes in glacial and dune terrain, as it is discussed later in an upcoming section of this chapter. Furthermore, lake sediments commonly have greater volumes of organic deposits than streams. These poorly permeable organic deposits can affect the distribution of seepage and biogeochemical exchanges of water and solutes more in lakes than in streams.

13.3.3 Interaction of Groundwater and Wetlands

GSWIs constitute an important link between wetlands and the surrounding catchment. Wetlands may develop in topographic lows where groundwater discharges. This water has its functions for ecological processes within the wetland, while surface water outflow from the wetland may provide water downstream. Wetlands may also receive inflowing surface water, which may become relatively stagnant giving rise to groundwater recharge. This transition of surface water to groundwater provides groundwater resources for human and ecological purposes further down the groundwater basin. Wetlands are present in climates and landscapes that cause groundwater to discharge to land surface or that prevent rapid drainage of water from the land surface. Similar to streams and lakes, wetlands can receive groundwater inflow, recharge groundwater, or do both. Those wetlands that get depressions in the land surface have interactions with groundwater similar to lakes and streams. Unlike streams and lakes, however, wetlands do not always get low points and depressions in the landscape; they also can be made on slopes or even on drainage divides [41].

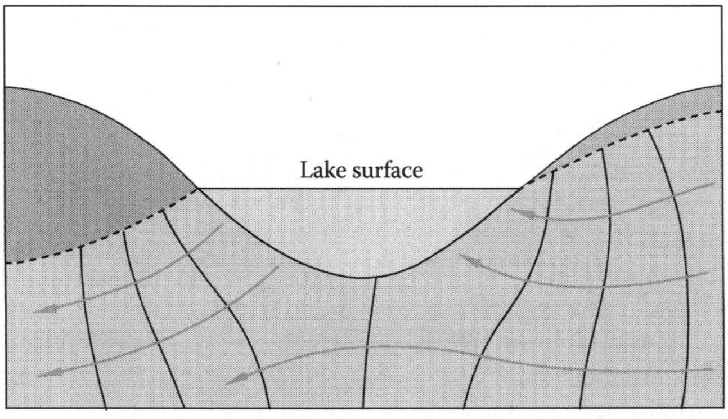

FIGURE 13.6 Lakes can receive groundwater inflow, lose water as seepage to groundwater, or both. (From Winter, T.C. et al., Groundwater and surface water: A single resource, U.S. Geological Survey Circular 1139, p. 79, 1998.)

A major difference between lakes and wetlands, with respect to their interaction with groundwater, is the ease with which water moves through their beds. Lakes commonly are shallow around their perimeter where waves can remove fine-grained sediments, permitting the surface water and groundwater to interact freely. In wetlands, on the other hand, if fine-grained and highly decomposed organic sediments are present near the wetland edge, the transfer of water and solutes between groundwater and surface water tends to be much slower. Another difference in the GSWI for wetlands compared to lakes is determined by rooted vegetation in wetlands [41]. One of the significant interchanges between surface water and pore water can be made with water uptake by roots of emergent plants that results in wetland sediments because fibrous root mat in wetland soils is highly conductive to water flow.

13.4 Chemical Interactions of Groundwater and Surface Water

Information on the same parameters for predicting the geochemical fate of contaminants in both groundwater and surface water bodies needs to be collected. Furthermore, we also need to collect the chemical and physical information commonly used in ecological risk assessments and natural attenuation assessments to determine the dominant biological processes and the potential confounding factors in bioassays. Finally, there is a need to collect chemical information that helps locate zones where a groundwater plume or hyporheic flow is entering a surface water body. The hyporheic zone (HZ) is a region beneath and alongside a streambed, where there is mixing of shallow groundwater and surface water. The flow dynamics and behavior in this zone termed hyporheic flow is recognized to be important for GSWI. There is an overlap among these parameters, but it should be noted that the three different uses of chemical information are basically contaminant chemistry and fate, biological processes, and identification of flow paths [34].

The transport of dissolved contaminants from surface water into the subsurface through hyporheic flow or groundwater recharge from a losing stream is included in our discussion in this chapter.

13.4.1 Chemical Evolution of Groundwater in Drainage Basins

The characterization, interpretation, and understanding of groundwater geochemistry are essential not only to identify groundwater quality but also to comprehend and characterize the water–rock interactions controlling the basic hydrochemistry [23].

Two of the fundamental controls on water chemistry in drainage basins are the type of geologic materials that are present and the length of time that water is in contact with these materials. Chemical reactions that affect the biological and geochemical characteristics of a basin include acid–base reactions, precipitation and dissolution of minerals, sorption and ion exchange, oxidation–reduction reactions, biodegradation, and dissolution (specifically gases) [41]. When water first infiltrates the land surface, microorganisms in the soil have a significant effect on the evolution of water chemistry. Organic matters in soils are degraded by microbes, producing high concentrations of dissolved carbon dioxide (CO_2). This process lowers the pH by increasing the carbonic acid (H_2CO_3) concentration in the soil water. The production of carbonic acid starts a number of mineral-weathering reactions, which result in bicarbonate (HCO_3^-) commonly, being the most abundant anion in the water [41]. Where the contact times between water and minerals in shallow groundwater flow paths are short, the dissolved solid concentration in the water generally is low. In such settings, limited chemical changes take place before groundwater is discharged to surface water.

In deeper groundwater flow systems, the contact time between water and minerals is much longer than of shallow flow systems [41]. As a result, the initial importance of reactions relating to microbes in the soil zone may be superseded over time by chemical reactions between minerals and water.

Hydrogeochemical modeling of the water draining an aquifer not only can characterize their spatial and temporal variations but can also be useful in understanding the interactions between groundwater and the aquifer matrix.

13.4.2 Chemical Interactions of Groundwater with Other Water Resources

Groundwater and surface water chemistry cannot be dealt with separately where surface and subsurface flow systems interact. The movement of water between groundwater and surface water provides a major pathway for chemical transfer between terrestrial and aquatic systems [11]. The transfer of chemical's affects the supply of carbon, oxygen, nutrients such as nitrogen and phosphorus, and other chemical constituents that enrich biogeochemical processes on both sides of the interface. This transfer can ultimately affect the biological and chemical characteristics of aquatic systems downstream [41].

Many streams are contaminated. Therefore, the need to determine the extent of the chemical reactions that take place in the HZ is widespread because of the concern that the contaminated stream water will contaminate shallow groundwater. Streams offer good examples of how interconnections between groundwater and surface water affect chemical processes [34]. Rough channel bottoms cause stream water to enter the streambed and to mix with groundwater in the HZ. This mixing establishes sharp changes in chemical concentrations in the HZ.

A zone of enhanced biogeochemical activity usually develops in shallow groundwater as a result of the flow of oxygen-rich surface water into the subsurface environment, where bacteria and geochemically active sediment coatings are abundant (Figure 13.7). This input of oxygen to the streambed stimulates a high level of activity by aerobic microorganisms if dissolved oxygen is readily available [41]. It is not uncommon for dissolved oxygen to be completely used up in hyporheic flow paths at some distance into the streambed, where anaerobic microorganisms dominate microbial activity. Anaerobic bacteria can use nitrates, sulfates, or other solutes in place of oxygen for metabolism. The result of these processes is that many solutes are highly reactive in shallow groundwater in the vicinity of streambeds.

Biogeochemical processes in the HZ affect the movement of nutrients and other chemical constituents, including contaminants, between groundwater and surface water. For example, the rate at which organic contaminants biodegrade in the HZ can exceed rates in stream water or in groundwater away from the stream. Another example is the removal of dissolved metals in the HZ [41]. As water passes through the HZ, the dissolved metals are removed by precipitation of metal oxide coatings on the sediments.

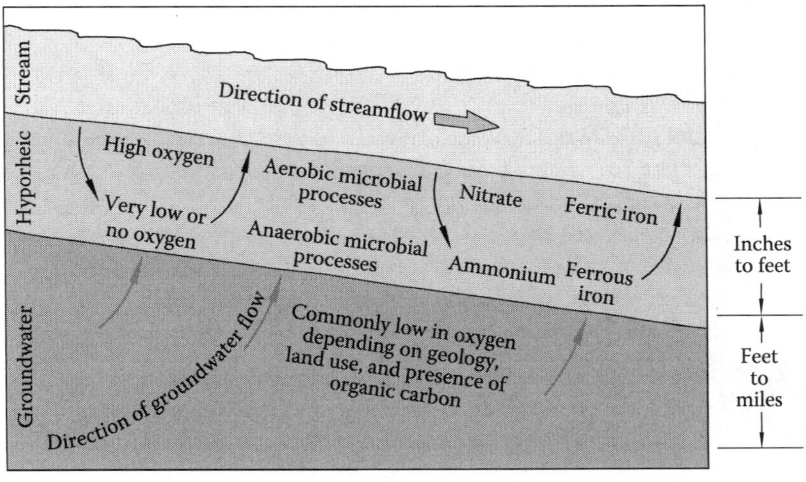

FIGURE 13.7 Microbial activity and chemical transformations commonly are enhanced in the HZ compared to those that take place in groundwater and surface water. This diagram illustrates some of the processes and chemical transformations that may take place in the HZ. Actual chemical interactions depend on numerous factors including aquifer mineralogy, shape of the aquifer, types of organic matter in surface water and groundwater, and nearby land use. (From Winter, T.C. et al., Groundwater and surface water: A single resource, U.S. Geological Survey Circular 1139, p. 79, 1998.)

Lakes and wetlands have also the distinctive biogeochemical characteristics with respect to their interaction with groundwater. The chemistry of groundwater and the direction and magnitude of exchange with surface water significantly affect the input of dissolved chemicals to lakes and wetlands. In general, if lakes and wetlands have little interaction with streams or with groundwater, the input of dissolved chemicals is mostly from precipitation; thus, the input of chemicals is at least. Lakes and wetlands that have a considerable amount of groundwater inflow generally have large inputs of chemicals is minimal because in this case, single source for entering dissolved chemicals. In cases where the input of dissolved nutrients such as phosphorus and nitrogen exceeds the output, primary production by algae and wetland plants is large. When this large amount of plant material dies, oxygen is used in the process of decomposition [11]. In some cases, the loss of oxygen from lake water can be large enough to kill fish and other aquatic organisms.

The magnitude of surface water inflow and outflow also affects the retention of nutrients in wetlands. If lakes or wetlands have no stream outflow, the retention of chemicals is high. The tendency to retain nutrients usually is less in wetlands that are flushed substantially by throughflow of surface water. In general, as surface water inputs increase, wetlands vary from those that strongly retain nutrients to those that both import and export large amounts of nutrients [11]. Furthermore, wetlands commonly have a significant role in altering the chemical form of dissolved constituents.

13.5 GSWI in Different Landscapes

Groundwater and surface water interact throughout the landscape. Groundwater interacts with all types of surface water, such as streams, lakes, and wetlands, in many different terrains, from mountains to oceans. Some common features of the interaction for various parts of the conceptual landscape are described in the succeeding text. The five general types of terrain discussed are mountainous, riverine, coastal, glacial and dune, and karst.

13.5.1 Mountainous Terrain

An overview of GSWI in mountainous terrain is presented by Silar [26] and extended by Bencala et al. [4].

The hydrology of mountainous terrain is characterized by highly variable precipitation and water movement over and through the steep land slopes.

A general concept of water flow in mountainous terrain includes several pathways by which precipitation moves through the hillside to a stream. The water from precipitation moves to mountainous streams along several pathways. Between storms and snowmelt periods, most inflow to streams is commonly from groundwater. During storms and snowmelt periods, much of the water inflow to the streams is from shallow flow in saturated macropores in the soil zone. If infiltration to the water table is large enough, the water table will rise to the land surface and the flow to the stream is from groundwater, soil water, and overland runoff. In arid areas where soils are very dry and plants are sparse, infiltration is impeded and runoff from precipitation can occur as overland flow.

Near the base of some mountain sides, the water table intersects the steep valley wall some distance up from the base of the slope. These activities result in perennial discharge of groundwater and, in many cases, the presence of wetlands (Figure 13.8) [41].

Another aspect of GSWI in mountain settings is caused by the marked longitudinal component of flow in mountain valleys. The high gradient of mountain streams, coupled with the coarse texture of streambed sediments, results in a strong downvalley component of flow accompanied by frequent exchange of stream water with water in the HZ.

Favorable new methods of estimating groundwater recharge, along mountain fronts, are being developed. These methods include the use of environmental tracers, measuring vertical temperature profiles in streambeds, measuring hydraulic characteristics of streambeds, and measuring the difference in hydraulic head between the stream and underlying aquifer.

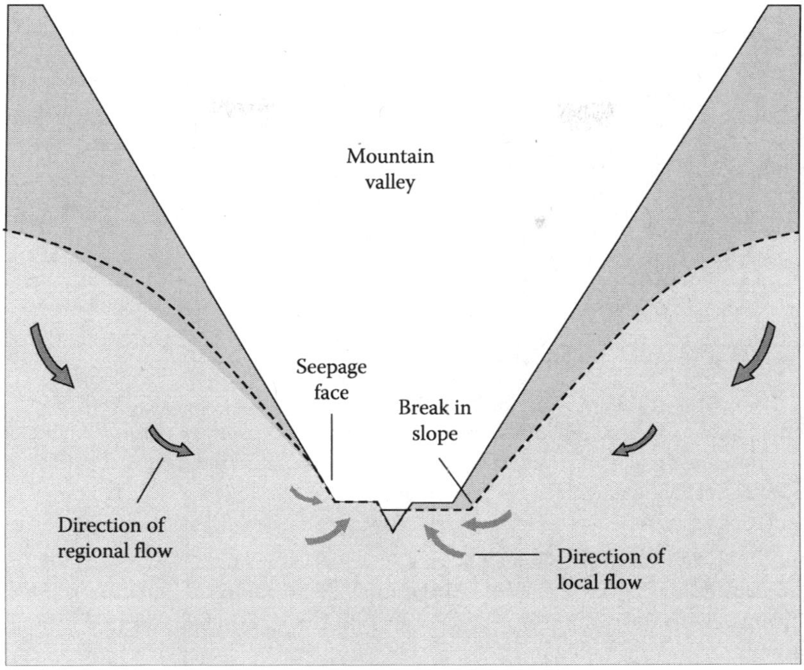

FIGURE 13.8 In mountainous terrain, groundwater can discharge at the base of steep slopes (left side of valley), at the edges of flood plains (right side of valley), and to the stream. (From Winter, T.C. et al., Groundwater and surface water: A single resource, U.S. Geological Survey Circular 1139, p. 79, 1998.)

Water extraction facilities have to be protected against erosion and are often placed near the stream or even in the streambed as not much space is available [15]. Therefore, low pumping rates and short residence times of the bank filtrate are typical for mountainous regions.

The geochemical environment of mountains is quite diverse owing to the effects of highly variable climate and many different rocks and soil types on the evolution of water chemistry. Geologic materials can include crystalline, volcanic, and sedimentary rocks and glacial deposits [41]. During heavy precipitation, much water flows through shallow flow paths, where it interacts with microbes and soil gases. In the deeper flow through fractured bedrock, longer-term geochemical interactions of groundwater with minerals determine the chemistry of water that eventually discharges to streams. The baseflow of streams in mountainous terrain is derived by drainage from saturated alluvium in valley bottoms and from drainage of bedrock fractures.

13.5.2 Riverine Terrain

In some landscapes, stream valleys are small and they usually do not have well-developed flood plains. However, the major rivers have valleys that often become increasingly wider at the downstream. Terraces, natural levees, and abandoned river meanders are common landscape features in major river valleys, and wetlands and lakes are normally associated with these features [15]. Riverine alluvial deposits range in size from clay to boulders, but in many alluvial valleys, sand and gravel are the predominant deposits, offering high yields of well fields adjacent to the river.

For larger rivers that flow in alluvial valleys, GSWI usually is more spatially diverse than it is for smaller streams. Groundwater from regional flow systems discharges to the river as well as at various places across the flood plain (Figure 13.9). If terraces are present in the alluvial valley, local groundwater flow systems may be associated with each terrace, and lakes and wetlands may be formed because of

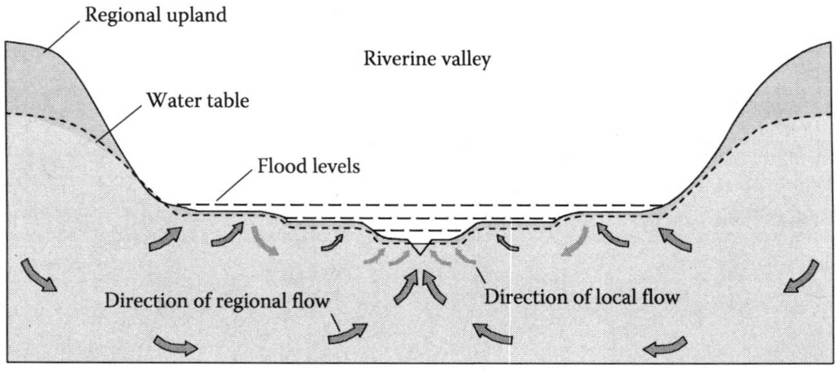

FIGURE 13.9 In broad river valleys, small local groundwater flow systems associated with terraces overlie more regional groundwater flow systems. Recharge from flood waters superimposed on these groundwater flow systems further complicates the hydrology of river. (From Winter, T.C. et al., Groundwater and surface water: A single resource, U.S. Geological Survey Circular 1139, p. 79, 1998.)

this source of groundwater [41]. At some locations, such as at the valley wall and at the river, local and regional groundwater flow systems may discharge in close proximity. Furthermore, in large alluvial valleys, significant downvalley components of the flow in the streambed and in the shallow alluvium also may be present.

For a river, GSWI is often described as of gaining or losing conditions. Rivers can exhibit both gaining and losing conditions in different reaches or at different times of the year within the same reach. Long losing stretches cause the extensive redox zones to develop and have a higher microbial activity within this zone compared to that of a zone under gaining conditions, since the nutrient-poor groundwater enters the river in the gaining stretches [15].

Chemical reactions including dissolution or precipitation of minerals commonly do not have a significant effect on water chemistry in sand and gravel alluvial aquifers because the rate of water movement is relatively fast compared to weathering rates. On the other hand, sorption and desorption reactions as well as oxidation/reduction reactions related to the activity of microorganisms probably have a greater effect on the water chemistry in these systems. As in small streams, biogeochemical processes in the HZ may have a significant effect on the chemistry of groundwater and surface water in larger riverine systems [41]. The movement of oxygen-rich surface water into the subsurface, where chemically reactive sediment coatings are abundant, causes increased chemical reactions relevant to the microorganisms' activity.

13.5.3 Coastal Terrain

Coastal terrain extends from inland scarps and terraces to the ocean. This terrain is characterized by low scarps and terraces that were formed when the ocean was higher than present time such as streams, estuaries, and lagoons that are affected by tides; ponds that are commonly associated with coastal sand dunes; and barrier islands. Wetlands cover extensive areas in some coastal terrains.

GSWI in coastal terrain is affected by discharge of groundwater from regional flow systems and from local flow systems associated with scarps and terraces (Figure 13.10), evapotranspiration, and tidal flooding. The local flow systems associated with scarps and terraces are caused by the configuration of the water table near these features. Where the water table has a downward break in slope near the top of scarps and terraces, downward components of groundwater flow are present, and where the water table has an upward break in slope near the base of these features, upward components of groundwater flow are present [41].

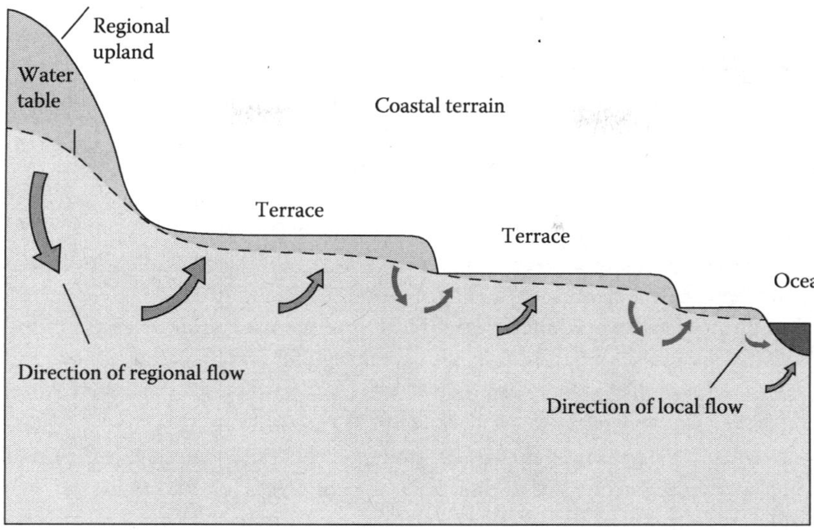

FIGURE 13.10 In coastal terrain, small local groundwater flow cells associated with terraces overlie more regional groundwater flow systems. In the tidal zone, saline and brackish surface water mixes with fresh groundwater from local and regional flow systems. (From Winter, T.C. et al., Groundwater and surface water: A single resource, U.S. Geological Survey Circular 1139, p. 79, 1998.)

Evapotranspiration directly from groundwater is widespread in coastal terrain. The land surface is flat and the water table generally is close to land surface. Therefore, many plants have root systems deep enough to transpire groundwater at nearly the maximum potential rate. The result is that evapotranspiration causes a significant water loss, which affects the configuration of groundwater flow systems as well as how groundwater interacts with surface water. In the parts of coastal landscapes that are affected by tidal flooding, GSWI is similar to that of alluvial valleys affected by flooding. The principal difference between the two is that tidal flooding is more predictable in both timing and magnitude than river flooding [41]. The other significant difference is in water chemistry. The water that moves into bank storage from rivers is generally fresh, but the water that moves into bank storage from tides generally is brackish. Estuaries are a highly dynamic interface between the continents and the ocean, where discharge of freshwater from large rivers mixes with saline water from the ocean. In addition, groundwater discharges to estuaries and the ocean, delivering nutrients and contaminants directly to coastal waters. However, few estimates of the location and magnitude of groundwater discharge to coasts have been made.

Many wetlands exist along streams, especially slow-moving streams. Although these riverine wetlands commonly receive groundwater discharge, they are dependent primarily on the stream for their water supply.

Wetlands in riverine and coastal areas have especially complex hydrologic interactions because they are subject to periodic water-level changes. Some wetlands in coastal areas are affected by very predictable tidal cycles. Other coastal and riverine wetlands are more affected by seasonal water-level changes and by flooding. The combined effects of precipitation, evapotranspiration, and GSWI result in a pattern of water depths in wetlands that is distinctive.

Hydroperiod is a term commonly used in wetland science that refers to the amplitude and frequency of water-level fluctuations. Hydroperiod affects all wetland characteristics, including the type of vegetation, nutrient cycling, and the types of invertebrates, fish, and bird species present [41].

The driving force of these interactions is the yearly flood season giving rise to extensive groundwater recharge. The infiltrated flood waters are transported as groundwater to wetland islands bordering the river delta channels driven by the evaporative force of island plants [25].

13.5.4 Glacial and Dune Terrain

Glacial and dune terrain is characterized by a landscape of hills and depressions. Although stream networks drain parts of these landscapes, many areas of glacial and dune terrain do not supply runoff to an integrated surface drainage network [15].

A common conception is that the lakes and wetlands that are present in topographically high areas recharge groundwater and that the lakes and wetlands that are present in low areas receive discharge from groundwater.

However, the lakes and wetlands underlain by deposits having low permeability can receive discharge from local groundwater flow systems even if they are located in a regional groundwater recharge area. As an overall result, in glacial and dune terrain, local, intermediate, and regional groundwater flow systems interact with lakes and wetlands. It is not uncommon for wetlands that receive recharge from local groundwater flow systems to be present in lowlands and for wetlands that receive discharge from local groundwater flow systems to be present in uplands (Figure 13.11).

Lakes and wetlands in glacial and dune terrain underlain by highly permeable deposits commonly have groundwater to seepage on one side and seepage to groundwater on the other side.

Transpiration directly from groundwater has a significant effect on the interaction of lakes and wetlands with groundwater in glacial and dune terrain [41]. Transpiration from groundwater has perhaps a greater effect on lakes and wetlands underlain by low-permeability deposits than in any other landscape.

The hydrologic and chemical characteristics of lakes and wetlands in glacial and dune terrain are determined to a large extent by their position with respect to local and regional groundwater flow systems [41].

13.5.5 Karst Terrain

A type of landscape that needs special attention is the areas underlain by limestone and dolomite. These landscapes, referred to as karst terrains, normally have fractures and solution openings that become larger with time because of dissolution of the rocks.

Karst terrains are characterized by closed surface depressions of various sizes and shapes known as sinkholes [41].

Water moves at different rates through karst aquifers; it moves slowly through fine fractures and pores and rapidly through solution-enlarged fractures and conduits.

Typically, karst terrain forms where water has dissolved and eroded soluble bedrock particularly carbonate rocks such as limestone and dolomite and gypsum, by enlarging cracks into underground conduits that can enlarge into caverns and sometimes collapse to form sinkholes.

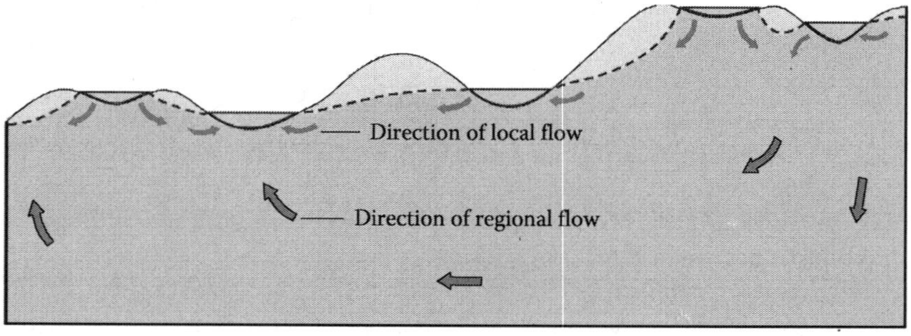

FIGURE 13.11 In glacial and dune terrain, local, intermediate, and regional groundwater flow systems interact with lakes and wetlands. It is not uncommon for wetlands that receive recharge from local groundwater flow systems to be present in lowlands and for wetlands that receive discharge from local groundwater flow systems to be present in uplands. (From Winter, T.C. et al., Groundwater and surface water: A single resource, U.S. Geological Survey Circular 1139, p. 79, 1998.)

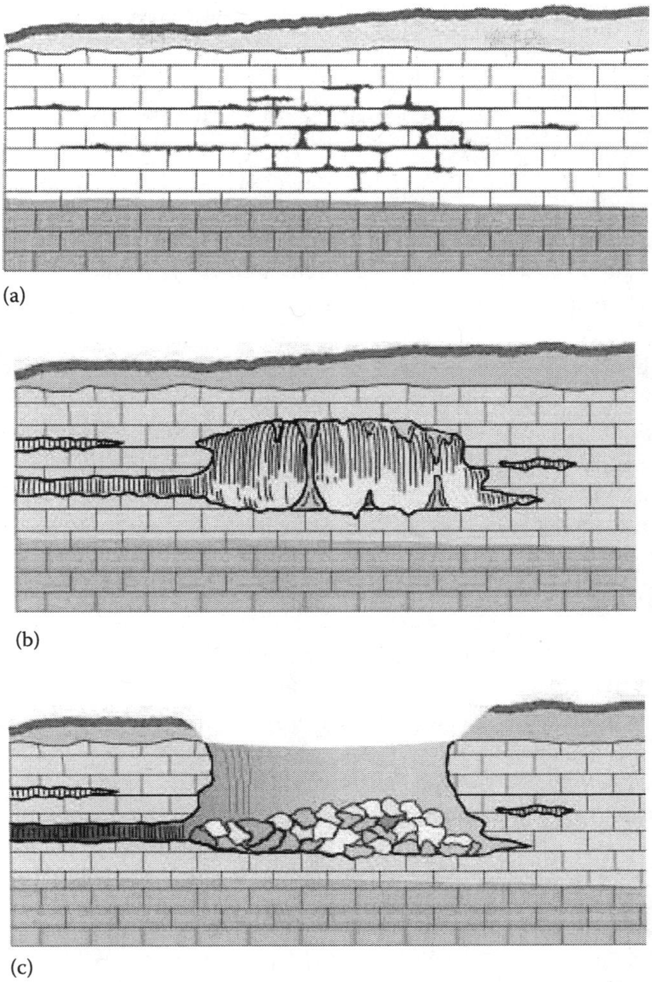

FIGURE 13.12 Process of collapse sinkholes. (a) Water works its way through the "vadose" zone above the water table through joints and cracks in the limestone bedrock. The slightly acidic water erodes the carbonate-rich rock. (b) Small passages grow into caverns as water continues to dissolve and carry away rock. (c) If the cavern ceiling grows thin enough, or it otherwise cannot support the weight of the rock and soil above, it collapses into a sinkhole. (From Karst and Coastal Landscape, Forms and processes of collapse sinkholes, www.caves.org/conservancy/mkc/about_karst)

Collapse sinkholes usually occur progressively. For example, the process of collapse has been shown in Figure 13.12.

The dissolution of limestone and dolomite guides the initial development of fractures into solution holes that are diagnostic of karst terrain. Perhaps nowhere else is the complex interplay between hydrology and chemistry so important to changes in landform. Limestone and dolomite weather quickly, producing calcium and magnesium carbonate waters that are relatively high in ionic strength. The increasing size of solution holes allows higher groundwater flow rates across a greater surface area of exposed minerals, which stimulates the dissolution process further, eventually leading to the development of caves. The development of karst terrain also involves biological processes [41].

Microbial production of carbon dioxide in the soil affects the carbonate equilibrium of water as it recharges groundwater, which then affects how much mineral dissolution will take place before solute equilibrium is reached.

Chemical process for calcium–carbonic acid equilibrium is included in the following phases:
Finally, carbonic acid is produced, which dissolves calcium carbonate [28]:

$$H_2O + CO_2 = H_2CO_3 = H^{+1} + HCO_3^{-1}$$

$$CaCO_3 + H_2CO_3 = Ca^{+2} + 2\left(HCO_3\right)^{-1}$$

Water movement in karst terrain is especially unpredictable because of the many paths groundwater takes through the maze of fractures and solution openings in the rock. Because of the large size of interconnected openings in well-developed karst systems, karst terrain can have true underground streams.

Water chemistry is widely used for studying the hydrology of karst aquifers. Extensive tracer studies and field mapping to locate points of recharge and discharge have been used to estimate the recharge areas of springs, rates of groundwater movement, and the water balance of aquifers [41]. Variations in parameters such as temperature, hardness, calcium/magnesium ratios, and other chemical characteristics have been used to identify areas of groundwater recharge, differentiate rapid- and slow-moving groundwater flow paths, and compare spring flow characteristics in different regions.

13.6 Delineation and Field Technology of the Hyporheic Zone

The term hyporheic is derived from Greek roots *hypo*, meaning under or beneath, and *rheos*, meaning a stream (*rheo* means to flow). It is a middle zone bordered by the surface water of the stream or river above and by the true groundwater below. Although it receives water from both of these sources, the relative strengths of input depend on the configuration of the bed materials and interstitial flow paths and on the prevailing hydraulic heads. These heads vary spatially and seasonally to alter hyporheic habitat volume and to produce ragged-edged boundaries to the zone [37].

Some common themes in the definitions of the HZ in the literature are as follows:

- It is the zone below and adjacent to a streambed in which water from the open channel exchanges with interstitial water in the bed sediments.
- It is the zone around a stream in which fauna characteristic of the HZ (the *hyporheos*) is distributed and alive.
- It is the zone in which groundwater and surface water mix [27].

HZs in sediments adjacent to streams are usually defined as one of the three ways [22]. In the geochemical definition, an HZ must contain at least 10% surface water. This surface water delivers nutrients, including dissolved organic carbon (DOC), to hyporheic sediments. The influx of nutrients from surface waters is reflected in the biological definition, in which the presence of macroinvertebrate riverine fauna delineates the HZ. This faunal population is distinct from typically subterranean species. The hydraulic definition includes a return of hyporheic water (and solutes) to surface water, encompassing hydrologic flow paths that begin and end in the stream [14].

The majority of authors in the freshwater ecology literature emphasize the importance of the HZ as a dynamic ecotone. For example, Boulton and Foster [5] describe the HZ as "an active ecotone between surface and groundwater," in which water, nutrients, and organic matter exchange occurs as a result of hydraulic and chemical gradients, topography, and sediment lithology.

To the hydrogeologist, the HZ is a part of the groundwater system. The conventional definition of groundwater is "water beneath the surface of the ground in the saturation zone and in direct contact with the ground or subsoil" (Council of the European Community [27]).

The HZ is a 3D aquatic interstitial ecotone formed within the mixed substrate particles that comprise the bed of a natural, running water channel (Figure 13.13).

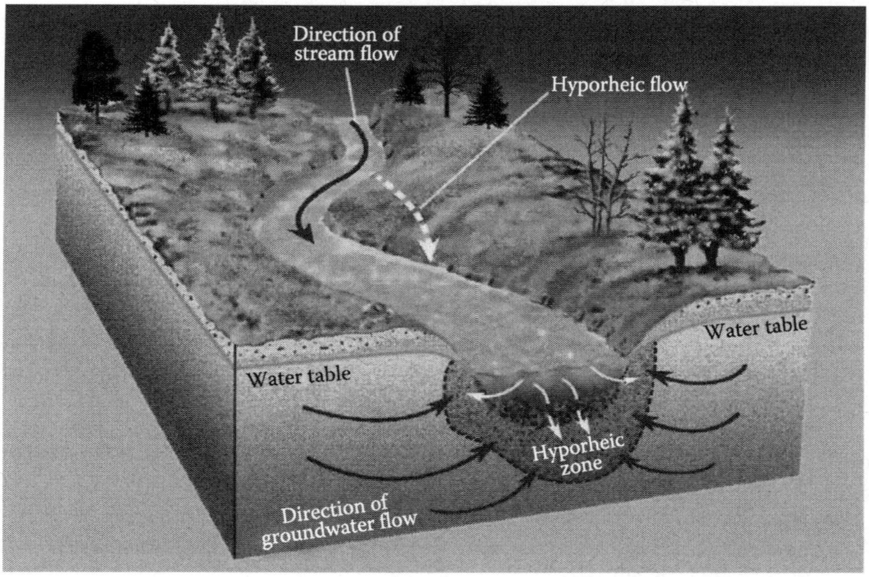

FIGURE 13.13 Diagrammatic section through a stream channel showing the approximate position of the HZ during winter, low flow conditions. Groundwater system involving the HZs. (From Alley, W.M. et al., *Science*, 296, 1985, 2002.)

This chapter adopts a conceptual definition of the Environment Agency's concentration on the attenuation of pollutants at the groundwater–surface water interface [31].

Water-saturated transitional zone between surface water and groundwater is an ecotone with hydraulic and biogeochemical gradients in which there is often mixing of the respective waters, and as a habitat for living organisms, it can play an important and protective role for both surface and groundwater [27]. It is often characterized by the presence of dense microbial communities (relative to a typical aquifer population), variably high organic matter concentrations, and a faunal assemblage distinct from both stream channel and true groundwater.

The most important ecological parameters and services that cause groundwater–surface water interface are as follows:

- Controlling the flux and location of water exchange between stream and subsurface
- Providing a habitat for benthic and interstitial organisms
- Providing a spawning ground and refuge for certain species of fish
- Providing a rooting zone for aquatic plants
- Providing an important zone for the cycling of carbon, energy, and nutrients
- Providing a natural attenuation zone for certain pollutants by biodegradation, sorption, and mixing
- Moderating river water temperature
- Providing a sink/source of sediment within a river channel [32].

Hyporheic sampling techniques roughly fall into four categories (Table 13.1). Unfortunately, virtually all of these samplers have limitations. For example, well digging cannot be used in midstream and is not very quantitative, freeze cores may drive organisms away as they form, mechanical corers may have depth or substrate particle size limitations, and artificial substrates may fail to reestablish natural sediment profiles and/or detrital components [34]. Further, many of these samplers have neither been evaluated in more than one location nor evaluated against each other.

TABLE 13.1 Four Main Categories of Hyporheic Samplers

Main Categories	Descriptions
Digging of small wells	In the exposed (above water) areas of gravel bars and stream margins to reach the water table and then straining the interstitial water so exposed through a fine-mesh net
Freeze cores	That use chemicals such as liquid nitrogen, liquid carbon dioxide, or a mixture of "dry ice" (crushed solid carbon dioxide) and acetone or alcohol to freeze the substrate around a standpipe driven into the bed
Mechanical corers	That, when driven into the bed, either isolate a sample of the surrounding substratum and its fauna for subsequent removal or suck up interstitial water and organisms from a desired depth
Artificial substrate samplers	That involve placing a sterilized portion of natural streambed into perforated containers that are sunk into the bed and then removed after a desired period of colonization

13.6.1 Hyporheic Exchange

Major pathways of water movement through riparian areas are as follows: (1) groundwater flow, (2) overland flow and shallow subsurface flow from adjacent uplands, and (3) in stream water sources such as overbank flow, bank storage, and hyporheic exchange. Streambeds and banks are unique environments because they are where groundwater that drains much of the subsurface of landscapes interacts with surface water. "Hyporheic exchange" is the term given to the process of water and solute exchange in both directions across a streambed [38]. The direction of seepage through the bed of streams commonly is related to abrupt changes in the slope of the streambed or to meanders in the stream channel (Figure 13.14). This process creates subsurface environments that have variable proportions of water

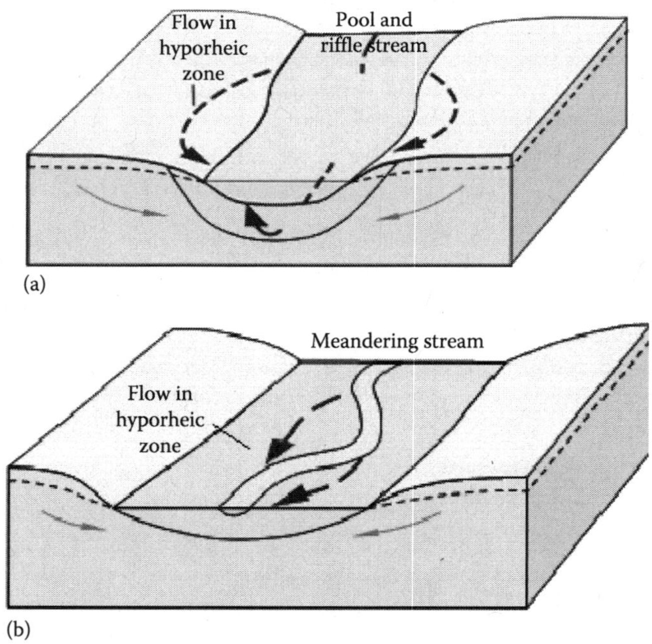

FIGURE 13.14 Surface water exchange with groundwater in the HZ is associated with abrupt changes in streambed slope (a) and with stream meanders (b). (From Winter, T.C. et al., Groundwater and surface water: A single resource, U.S. Geological Survey Circular 1139, p. 79, 1998.)

from groundwater and surface water. Depending on the type of sediment in the streambed and banks, the variability in slope of the streambed, and the hydraulic gradients in the adjacent groundwater system, the HZ can be as much as several feet in depth and hundreds of feet in width. The dimensions of the HZ generally increase with increasing width of the stream and permeability of streambed sediments.

Because of this mixing between groundwater and surface water in the HZ, the chemical and biological characters of the HZ may differ markedly from adjacent surface water and groundwater [41]. Although most works related to hyporheic-exchange processes have been done on streams, processes similar to hyporheic exchange also can take place in the beds of some lakes and wetlands because of the reversals in flow caused by focused recharge and transpiration from groundwater near surface water, as discussed earlier. Therefore, it is not enough to know only the relationship of surface water to groundwater flow systems and to small-scale seepage patterns in surface water beds, because hyporheic-exchange processes also can be important in some types of landscapes [38].

13.6.2 Chemical Processes in the Hyporheic Zone

Chemical and microbial substances in the environment are subject to a range of physical, chemical, and biological processes that often act to decline concentration and mass. Within the subsurface environment, processes that act to degrade pollutants or retard their movement through soils or aquifers are named "natural attenuation" processes [30]. The U.K. Environment Agency [31] has defined natural attenuation within groundwater systems as:

The effect of naturally occurring physical, chemical and biological processes, or any combination of those processes to reduce the load, concentration, flux, or toxicity of polluting substances in groundwater.

For natural attenuation to be effective as a remedial action, the rate at which those processes occur must be sufficient to prevent polluting substances entering identified receptors and to minimize the expansion of pollutant plumes into currently uncontaminated groundwater. Dilution within a receptor, such as in a river or borehole, is not natural attenuation [31].

Natural attenuation processes are commonly considered as part of the assessment of risks associated with contaminated groundwater or soils.

Metals and other nondegradable cations are normally attenuated as a result of sorption onto clay minerals and oxides and/or hydroxides, complexation, and precipitation of insoluble metal compounds.

The conditions present in the HZ, such as steep chemical (e.g., redox) gradients; dynamic exchange of oxygen, carbon, and nutrients from the stream channel; a dense microbial and invertebrate community; and potential for elevated water temperatures (both relative to aquifer conditions), mean that attenuation processes may be more rapid in hyporheic sediments than in aquifer sediments [27]. However, the residence time in the HZ is likely to be small relative to that in an aquifer.

Cycling of diffuse pollutants plays a significant role on natural attenuation in hyporheic region. Nitrogen, along with carbon, oxygen, hydrogen, and phosphorus, is the most important element in living matter.

13.6.2.1 Dissolved Organic Nitrogen

Nitrogen is a vital component of proteins and nucleic acids and contains around 10% of the dry-weight mass of bacteria [8]. Bacteria, fungi, algae, and plants assimilate nitrogen as nitrate, NO_3^-, or ammonium, NH_4^+.

Dissolved organic nitrogen (DON) is usually rare in pristine aquatic environments, and in consequence nitrogen is often the limiting variable in biological productivity. Most nitrogen is bound to organic matter and is not bioavailable. Nitrogen assimilation is generally via mineralization of NH_4^+, which can be taken up or converted into nitrate by bacteria.

Nitrates are recognized to be the most important pollutants of groundwater being produced mainly from agriculture activities, because the excessive levels of nitrate are dangerous for human life and environment. Treatment of contaminated water is also hard or sometimes impossible [10]. Therefore, investigating nitrate pollution in groundwater has been found very necessary.

13.6.2.2 Dissolved Organic/Inorganic Phosphorus

There is an extensive literature on phosphorus dynamics in lakes, wetlands, streams, and marine environments but relatively little on the HZ or aquifer environments [22].

Phosphorus, P, is often a limiting nutrient in aquatic systems and is subject to complex processes that control its speciation, sorption, and fate. In aquatic systems, P may be present in a number of forms. It is usual to classify P into dissolved and particulate forms, which can then be further subdivided [27]. Dissolved P is either dissolved inorganic P (DIP), which is generally bioavailable, or dissolved organic P (DOP), which includes colloidal organic P (generally unavailable to biota until it is converted into DIP).

13.6.2.3 Dissolved Organic Carbons

The quality, or chemical and physical character, of DOC affects the degree to which it is transformed by biological and chemical processes [14]. In turn, the DOC quality affects microbial ecology, metal mobilization in streams, and anthropogenic organic contaminant solubility.

HZs play a significant role in stream nutrient and DOC processing.

As a source of organic carbon, DOC is both consumed and produced by biota in streams and HZs and can be a limiting factor in ecosystem metabolism [14]. DOC transformative processes, considerably microbial processing and photodegradation, can occur in the stream channel, but HZ nutrient processing is so significant in riverine DOC metabolism that HZs have been dubbed "the river's liver" [13] and are generally considered a sink of DOC. GSWI serves as a "control point" for nutrient fluxes with a variety of terminal electron accepting processes active in HZs such as denitrification, iron and sulfate reduction, and methanogenesis [42]. In semiarid catchments with highly variable hydrology, DOC export and biogeochemical activity can be controlled by the degree of connectivity between groundwater and surface water [14].

13.7 Field Methods for Determining GSWI

Measuring GSWI is an important component for integrated river basin management. Numerous methods exist to measure these interactions, which are applied either in the aquifer, in the surface water, or in the transition zone itself. The methods differ in resolution, sampled volume, and the timescales they represent. Often, the choice of methods constitutes a trade-off between resolutions of heterogeneities and sampled subsurface volume. Furthermore, the measurement scale on which a selected technique operates may have a significant influence on the results, leading to differences between estimates obtained from a grid of point measurements and estimates obtained from large-scale techniques. Therefore, a better representation of the local conditions including the effects of scale on measurement results can be achieved by conducting measurements at multiple scales at a single study site. Attention should be paid to distinguish between groundwater discharge and hyporheic-exchange flow. Small-scale flow measurements in the shallow streambed may not suffice to make this distinction, so that additional measurements to identify the water source are necessary.

Approaches that should be considered while developing a conceptual model are as follows [2]:

- Hydrologic approaches are used at a regional scale, over periods of years, and perhaps with a focus on long-term yield and water supply. Rainfall runoff is simulated at the land surface and discharge is computed in networks of rivers and streams. The level in a river or stream is estimated from discharge using a stage discharge or rating curve, and this level is used to compute exchange flows between surface water and groundwater.
- Hydraulic approaches are used at a more local scale, over periods of days and weeks, and perhaps with a focus on flood management. These approaches are based not only on conservation of volume (mass) but also on conservation of energy or momentum, either in 2D in plan or in 1D. They assume a single layer of surface water, with constant head and velocity throughout the water

column in 2D or throughout the cross-sectional area in 1D. Hydraulic approaches are often used to simulate flow in river and stream channels and also on flood plains.

- Hydrodynamic approaches are used in deep or density-stratified water bodies, like mine-pit lakes or tidal estuaries. These approaches are also based on conservation of mass, energy, and momentum. They are applied in 3D or 2D in vertical section, and take into account vertical gradients in head and velocity.

13.7.1 Observable Qualitative Indicators of Groundwater Discharge to Surface Water

Various approaches and techniques to measure the GSWI have been studied.

The most common indicators are seeps and springs, infrared mapping, aquatic plants, phreatophytes, unique sediment zones such as mineral precipitates, water color, odor from contaminants, and mapping of lineaments in fractured-rock settings [34].

New techniques to measure regional groundwater discharge such as thermal remote sensing and using noble gases have been developed [16].

Observation of seeps and springs is relatively straight forward if the flow rates are high. In fractured-rock landscapes, mapping of lineaments can be useful if the fractures are open. Groundwater flow concentrated in the fractures enters surface water bodies as springs. In settings where seepage rates are low, it is easier to observe seeps during colder times of the year when groundwater and air temperatures are considerably different, because the water vapor above seeps is visible [34]. The difference in temperature between groundwater and surface water also makes infrared mapping a useful reconnaissance tool, especially in midsummer when the difference in temperatures of groundwater and surface water is at a maximum.

Some chemical constituents dissolved in anoxic groundwater precipitate upon contacting oxygenated surface water. For example, iron and manganese oxides are common indicators of seep areas. Contaminated groundwater commonly has color and odor. Water color and odor from contaminants can be used as an indicator of groundwater inflow, especially if the inflow consists of the contaminated water [34].

Aquatic plants can be indicators of groundwater discharge. In addition to aquatic plants, upland phreatophytic plants near a surface water body are indicators of the presence of groundwater at shallow depths.

Benthic organisms can be indicators of groundwater discharge to surface water. Numerous examples of the relationship of organisms to water flow and chemistry are provided by studies of the HZ beneath streams.

Thermal remote sensing is based on the principle that in shallow groundwater systems, upward fluxes of deep groundwater discharge reduce seasonal temperature variation in layers close to and at the surface. The analysis of satellite-acquired surface thermal data has the potential for estimation of groundwater discharge [3].

This section explores opportunities to apply widely available remotely sensed thermal data to delineate groundwater discharge zones. The approach is based on two fundamental principles:

1. Temperature is a reliable tracer of GSWI.
2. Remotely sensed thermal data may be used to delineate shallow objects with contrasting thermal properties.

The approach does not intend to inspect an absolute temperature at the ground surface. Rather, it is expected that the thermal signature within groundwater discharge zone is likely to be weak, potentially masked by stronger thermal anomalies associated with vegetation, exposure related to topography (mainly slope and aspect), soil type, and meteorological conditions preceding and at the time of remotely sensed data acquisition. Therefore, the approach was focused on defining [3]

- Criteria for image selection when the thermal signatures associated with groundwater discharge zones are most evident
- Adequate image processing targeting to reduce the effect of the aforementioned factors
- Validation of the methods in a data-limited environment

13.7.2 Direct Measurement and Calculated Flow of Water between Groundwater and Surface Water Using Physical Data

Methods for directly measuring the flux of water between groundwater and surface water include the use of seepage meters, minipiezometers, temperature profiles in the sediments, heat-flow meters, hydraulic properties of sediments determined from cores, and direct contact resistivity probes [34].

Large-scale field assessment methods such as hydrograph separation, temperature modeling, and water and solute mass balance also exist where hydrograph separation will overestimate regional groundwater discharge. Thermal modeling of temperature to determine groundwater inflows to streams has potential. Water and solute mass balances work well but can be difficult to apply in complex catchments.

Seepage meters are chambers that are set on the bed of a surface water body. Direct measurements of water flux across the groundwater–surface water interface can be realized by the use of seepage meters. Bag-type seepage meters as proposed by Lee [21] consist of a bottomless cylinder vented to a deflated plastic bag. The cylinder is turned into the sediment, and as water flows from the groundwater to the surface water, it is collected in the plastic bag. To measure the flux, the valve is opened, and the change in water volume in the bag over a given period of time is a measure of flux per that period of time. Seepage meters are perhaps the most commonly used devices to determine the collected volume, the cross-sectional area of the cylinder, and the collection period from which the seepage flux can be calculated. In case of surface water seeping into the sediment, a known water volume is filled into the plastic bag [17]. These bag-type seepage meters have been used extensively in lakes, estuaries, reservoirs, and streams (Figure 13.15) [19].

Minipiezometers are widely used to determine the hydraulic gradient from hydraulic head measurements. Installed in the streambed, piezometers deliver information whether a stream reach is gaining or losing by a comparison of piezometer and stream water level [17]. If the water level in the piezometer is higher than that of the stream, this indicates groundwater flow into the stream, and vice versa. To determine water flux, the hydraulic conductivity of the sediments needs to be determined as well as the cross-sectional area of the flux.

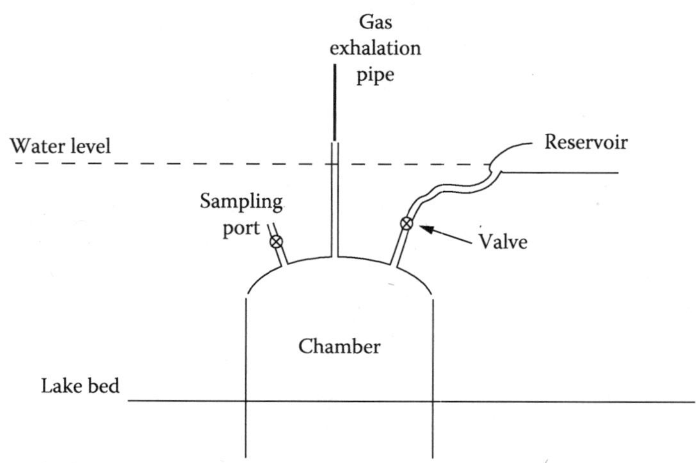

FIGURE 13.15 Figure of seepage meters. (From Landon, M.K., Rus, D.L., and Harvey, F.E., 2001. Comparison of instream methods for measuring hydraulic conductivity in sandy streambeds. *Ground Water*, 39(6): 870–885.)

Temperature profiles measured in the transition zone carry information on whether groundwater is discharging to the surface water or surface water is infiltrating to the groundwater [6]. Heat transport in groundwater is a combination of advective heat transport (i.e., heat transport by the flowing water) and conductive heat transport (i.e., heat transport by heat conduction through the solid and fluid phase of the sediment) [17].

Lapham [20] used sediment temperature data to determine flow rates and hydraulic conductivity of the sediments based on fundamental properties of heat transport. Heat-flow meters, consisting of a heating element and a ring of temperature sensors, placed at a distance from the heater, have been used to measure the rate and direction of water movement through sediments [20].

13.7.2.1 Calculated from Streamflow Data and Analysis

By stream discharge data or groundwater flow nets, the quantity of water moving between groundwater and surface water can be determined.

The most direct method for determining groundwater inflow or stream losses to groundwater is to make stream discharge measurements at different locations along a stream.

The flow net approach is probably the most common method used for determining the GSWI. The term flow net is used broadly herein as any calculation of groundwater flux, including simulation models, that makes use of a network of wells for determining hydraulic gradients, estimates of hydraulic conductivity of the geologic units and sediments, and cross-sectional area of the interface of groundwater and surface water [17]. The accuracy of the values is related to the quantity and quality of the hydrogeologic data and the grid spacing that is justified by these data.

13.7.3 Indicators of Flow between Groundwater and Surface Water Using Chemical Data

Tracer-based hydrograph separation using isotopic and geochemical tracers provides information on the temporal and spatial origin of streamflow components [17]. Stable isotopic tracers, such as stable oxygen and hydrogen isotopes, are used to distinguish rainfall event flow from pre-event flow.

Devices for collection of water samples for determination of the chemical characteristics of water passing through sediments consist of two basic types: (1) collection at the sediment–water interface and (2) collection at various depths in the sediment by inserting a device into the sediments [34]. By calculating mass balances of the constituents, the flux of water can be quantified. Isotopes of some elements, such as nitrogen and radon, are particularly useful because in some cases a specific contaminant source can be identified. Isotopes of water are among the most useful because they are part of the water molecule itself and are not subject to modification by chemical reactions [34]. The age of groundwater can be determined by analyzing for tritium and chlorofluorocarbons, which are useful for identifying groundwater flow paths.

The analytical results of the chemical analysis and the statistical parameters such as minimum, maximum, and mean of the geochemical data and hydrochemical data can be compared to current conditions [24].

A list of required parameters and tools for determining GSWI is as follows:

- Semipermeable membrane devices
- Drag probes for temperature, conductivity, and gamma anomalies, useful in lakes, estuaries, and large rivers to determine zones of groundwater discharge
- Piezometers and minipiezometers
- Dye tracers of groundwater and streamflow
- Bead pipes (ceramic beads)
- Walk river bed with a hand auger
- During low flow, note odor and visual observation
- Photoionization detector (PID)

- Passive diffusion samplers
- Analyze bubbles of gas (marsh or lake setting)
- Multilevel samplers
- Seepage meters
- Cores (solid analysis and visual)
- Field chemistry with spectrometer (nitrate, ammonia)
- Chemetrics for sulfides
- Differential global positioning system (GPS)
- Velocity meter
- Tidal stage
- Multilevel wells
- Everything on screening tools list [38]

13.7.3.1 Sampling at the Sediment–Water Interface

Devices that have been developed for sampling water at the sediment–water interface include drag probes, seepage meters, diffusion bags, bubble collectors, and biosensors [17]. Of these devices, seepage meters are the only ones that actually collect a water sample large enough to be analyzed in the laboratory for many constituents.

13.7.3.2 Sampling at Depth in Sediments

Devices that have been developed for sampling or measuring water chemistry at depth in sediments consist of (1) multilevel samplers that are driven into the sediments and (2) probes through which individual samples can be drawn from any depth or a constituent measured but can then be driven deeper to collect samples at other specific depths [17].

13.8 Summary and Conclusions

Recently, due to rapid increase in population and shortage of suitable water, assessing groundwater quality has become an important issue in groundwater studies of the area. For these reasons, a qualitative evaluation of groundwater has been proposed as a basic prerequisite for efficient groundwater resource management. It is important to note that increased knowledge of geochemical evolution of groundwater in the regions could lead to improved understanding of hydrochemical systems, leading to sustainable development of water resources and effective management of groundwater resources.

The overall goal of this chapter was to provide an opportunity for statement scientific and technical backgrounds to discuss the importance of the groundwater/surface water transition zone and help regulators better understand environmental issues relating to the connections between groundwater and surface water.

This chapter provides an overview of the way that GSWI is conceptualized and the methods to design and construction of models that include GSWI. Each of the development stages such as modeling, calibration, and sensitivity analysis was discussed in this chapter with a focus on the specific requirements of GSWI, beyond those of groundwater flow models.

Attention should be given in order to distinguish between groundwater discharge and hyporheic-exchange flows. Small-scale flow measurements in the shallow streambed may not suffice to make this possible; therefore, additional measurements to identify the water source are strongly recommended. Long-term air entrapment affecting runoff and water table observations should be investigated. The application of thermal remote sensing to delineate groundwater discharge zones and coupling the models to optimize the flooding system is necessary to deal with current needs. Groundwater recharge during spring thaw in view of quality and quantity and bank filtration as a managing tool for GSWIs could be also the other research topics of interest.

References

1. Alley, W.M., Healy, R.W., Labaugh, J.W., and Reilly, T.E., 2002. Flow and storage in groundwater systems. *Science*, 296: 1985–1991.
2. Barnett, B., Townley, L.R., Post, V., Evans, R.E., Hunt, R.J., Peeters, L., Richardson, S., Werner, A.D., Knapton, A., and Boronkay, A., 2012. Australian groundwater modeling guidelines. Sinclair Knight Merz and National Centre for Groundwater Research and Training, Waterlines Report Series, No. 82. National Water Commission, Canberra.
3. Barron, O. and Van Niel, T.G., 2009. Application of thermal remote sensing to delineate groundwater discharge zones. *International Journal of Water*, 5(2): 109–125.
4. Bencala, K.E., Duff, J.H., Harvey, J.W., Jackman, A.P., and Triska, F.J., 1993. Modelling within the stream-catchment continuum. In: *Modelling Change in Environmental Systems*, Jakeman, A.J., Beck, M.B., and McAleer, M.J. (eds.), John Wiley & Sons, New York, pp. 163–187.
5. Boulton, A.J. and Foster, J.G., 1998. Effects of buried leaf litter and vertical hydrologic exchange on hyporheic water chemistry and fauna in a gravel-bed river in northern New South Wales, Australia. *Freshwater Biology*, 40: 229–243.
6. Constantz, J. and Stonestrom, D., 2003. Heat as a tracer of water movement near streams. In: *Heat as a Tool for Studying the Movement of Groundwater Near Streams*, Stonestrom, D. and Constantz, J. (eds.), U.S. Geological Survey Circular 1260, Reston, VA.
7. Department of Primary Industries, Victoria, Australia, 2009. *Groundwater and Surface Water Interactions*. Landscape and Water Sciences Rutherglen, State Government of Victoria, Victoria, Australia.
8. Duff, J.H. and Triska, F.J., 2000. Nitrogen biogeochemistry and surface—Subsurface exchange in streams. In: *Streams and Ground Waters*, Jones, J.B. and Mulholland, P.J. (eds.), Academic Press, London, U.K., pp. 197–220.
9. Eslamian, S., Amiri, M.J., Abedi-Koupai, J., and Shaeri-Karimi, S., 2013. Reclamation of unconventional water using nano zero-valent iron particles: An application for groundwater. *International Journal of Water*, Special Issue on Groundwater and Surface Water Interaction (GSWI): Unconventional Groundwater, 7(1/2): 1–13.
10. Eslamian, S. and Lavaei, N., 2009. Modeling nitrate pollution of groundwater using artificial neural network and genetic algorithm in an arid zone. *International Journal of Water*, 5(2): 194–204.
11. Eslamian, S. and Nekoueineghad, B., 2009. A review on interaction of groundwater and surface water. *International Journal of Water*, 5(2): 89–100.
12. European Commission, 2008. *Groundwater Protection in Europe*. Office for Official Publications of the European Communities, Luxembourg, Germany.
13. Fischer, H., Kloep, F., Wilzcek, S., and Pusch, M.T., 2005. A river's liver—Microbial processes within the hyporheic zone of a large lowland river. *Biogeochemistry*, 76: 349–371.
14. Gabrielsen, P.J., 2012. *Hyporheic Zone Process Controls on Dissolved Organic Carbon Quality*, Department of Earth and Environmental Science, New Mexico Institute of Mining and Technology, Socorro, NM.
15. Grischek, T. and Ray, C., 2009. Bank filtration as managed surface—Groundwater interaction. *International Journal of Water*, 5(2): 125–140.
16. Harrington, G., 2012. New technique to measure regional groundwater discharge in rivers, Examples from northern Australia, Water for Healthy Country National Research Flagship.
17. Kalbus, E., Reinstorf, F., and Schirmer, M., 2006. Measuring methods for groundwater, surface water and their interactions: A review, Department of Hydrogeology, UFZ-Centre for Environmental Research Leipzig-Halle in the Helmholtz Association, Leipzig, Germany.
18. Karst and Coastal Landscape, Forms and processes of collapse sinkholes, www.caves.org/conservancy/mkc/about_karst

19. Landon, M.K., Rus, D.L., and Harvey, F.E., 2001. Comparison of instream methods for measuring hydraulic conductivity in sandy streambeds. *Ground Water*, 39(6): 870–885.
20. Lapham, W.W., 1989. Use of temperature profiles beneath streams to determine rates of vertical groundwater flow and vertical hydraulic conductivity. U.S. Geological Survey Water-Supply Paper 2337, 35pp.
21. Lee, D.R., 1977. Device for measuring seepage flux in lakes and estuaries. *Limnology and Oceanography*, 22(1): 140–147.
22. Reddy, K.R., Kadlec, R.H., Flaig, E., and Gale, P.M., 1999. Phosphorus retention in streams and wetlands: A review. *Critical Reviews in Environmental Science and Technology*, 29: 83–146.
23. Rezaei, M., 2013. The chemical evolution of groundwater in the Kerman plain aquifer, Iran. *International Journal of Water*, Special Issue on Groundwater and Surface Water Interaction (GSWI): Unconventional Groundwater, 7(1/2): 29–43.
24. Saatsaz, M., Azmin Sulaiman, W.N., Eslamian, S., and Mohammadi, K., 2013. Hydrogeochemistry and groundwater quality assessment of Astaneh-Kouchesfahan Plain, Northern Iran. *International Journal of Water*, Special Issue on Groundwater and Surface Water Interaction (GSWI): Unconventional Groundwater, 7(1/2): 44–65.
25. Schot, P. and Winter, T., 2006. Groundwater–surface water interactions in wetlands for integrated water resources management. *Journal of Hydrology*, 320: 261–263.
26. Silar, J., 1990. *Surface Water and Groundwater Interactions in Mountainous Areas*. International Association of Hydrological Sciences. Publication, 190, pp. 21–28.
27. Smith, J.W.N., 2005. Groundwater–surface water interactions in the hyporheic zone. Environment Agency, Science Report SC030155/SR1.
28. Tihansky, A.B. and Knochenmus, L.A., 2000. *U.S. Geological Survey Karst Interest Group Proceedings*. U.S. Geological Survey Water Resources Investigations Report 01-4011, pp. 198–211, St. Petersburg, FL.
29. Tomassoni, G., 2000. A federal statutory/regulatory/policy perspective on remedial decision-making with respect to ground-water/surface-water interaction. *Proceedings of the Ground-Water/Surface-Water Interactions Workshop*. U.S. EPA, Office of Solid Waste, Washington, DC, p. 13.
30. Tonina, D. and Buffington, J.M., 2009. Hyporheic exchange in Mountain Rivers: Mechanics and environmental effects. *Geography Compass*, 3(3): 1063–1086. doi: 10.1111/j.1749–8198.2009.00226.x.
31. U.K. Environment Agency, 2000. *Guidance on the Assessment and Monitoring of Natural Attenuation of Contaminants in Groundwater*. Environment Agency R&D Publication 95, Environment Agency, Bristol, U.K.
32. U.K. Environment Agency, 2009. *The Hyporheic Handbook. A Handbook on the Groundwater–Surface Water Interface and Hyporheic Zone for Environment Managers*. Integrated Catchment Science Programme Science Report: SC050070, Environment Agency, Bristol, U.K.
33. United States Climate Change Global Research Program, 2001. *The Global Water Cycle*. Water Cycle Study Group, Washington, DC.
34. United States Environmental Protection Agency, 2000. *Proceedings of the Ground-Water/Surface-Water Interactions Workshop*. Solid Waste and Emergency Response, Washington, DC. 20460 EPA/542/R-00/007.
35. United States Environmental Protection Agency, 1997. *Rules of Thumb for Superfund Remedy Selection*. US EPA, Washington, DC.
36. United States Environmental Protection Agency, 2005. *Use of Alternate Concentration Limits (CLs) in Superfund Cleanups*. US EPA, Washington, DC.
37. Williams, D.D., 1993. Nutrient and flow vector dynamics at the hyporheic/groundwater interface and their effects on the interstitial fauna. *Hydrobiology*, 251: 185–198.
38. Winter, T.C., 2000. Interaction of ground water and surface water. *Proceedings of the Ground-Water/Surface-Water Interactions Workshop*. Solid Waste and Emergency Response, Washington, DC, 20460 EPA/542/R-00/007, p. 15.

39. Winter, T.C., 2000. Interaction of groundwater and surface water. *Proceedings of the Ground-Water/Surface-Water Interactions Workshop*. U.S. EPA, Office of Solid Waste, Washington, DC, p. 17.
40. Winter, T.C., 2000. The concept of hydrologic landscapes. *Journal of the American Water Resources Association*, 37(2): 335–349.
41. Winter, T.C., Harvey, J.W., Franke, O.L., and Alley, W.M., 1998. Groundwater and surface water: A single resource. U.S. Geological Survey Circular 1139, p. 79.
42. Zarnetske, J.P., Haggerty, R., Wondzell, S.M., and Baker, M.A., 2011. Dynamics of nitrate production and removal as a function of residence time in the hyporheic zone. *Journal of Geophysical Research*, 116: G01025. doi: 10.1029/2010JG001356.

14

Hydrogeology of Hard Rock Aquifers

Patrick Lachassagne
Danone Waters, Evian Volvic World Sources

Benoît Dewandel
BRGM

Robert Wyns
BRGM

AUTHORS

Patrick Lachassagne is presently the environment and water resources manager for Evian Volvic World Sources (Danone Waters France). Patrick is in charge of research projects dealing with improving the knowledge about complex natural mineral water (NMW) aquifers and developing specific management tools and strategies about groundwater protection policies. He is also in charge of the operational management and protection of the Evian, Volvic, Badoit, and La Salvetat NMW aquifers. He previously worked for 20 years at BRGM, the French Geological Survey, as a researcher, project manager, and the head of a research team dealing with the hydrogeology of complex aquifers, hard rocks (HRs), and volcanic, geothermal, thermal, and mineral aquifers, with case studies in various areas of the world (France, West Indies, Comoros, Guiana, Reunion Island, India, South Korea, Burkina Faso, Oman, etc.). He still teaches for students pursing their courses about HR aquifer (HRA) structure and functioning in several French universities for students pursuing their master's degrees.

Benoît Dewandel has been since May 2002 a full-time researcher at BRGM. Having defended his PhD in HRAs hydrogeology in 2002 (Montpellier II University, France), he joined, and then leaded, from 2002 to 2007, the Indo-French Centre for Groundwater Research, a joint research lab from BRGM and National Geophysical Research Institute (Hyderabad, Andhra Pradesh, India) specialized in research projects dealing with the hydrogeology of HRs. Back to France since 2007, he is leading several R&D projects on complex aquifers. Dewandel specializes in the hydrogeology of crystalline massifs in mountainous and basement regions (ophiolitic and granitic aquifer) and focuses his research activity on the links between geology and hydrodynamic parameters. He has published several papers on the conceptualization of HRAs and developed decision support tools for improving the groundwater sustainability in such aquifers and methods for regionalizing hydrogeological parameters in HR areas.

Robert Wyns is a geologist expert in rock weathering at BRGM. After 15 years of working in various geological contexts (sedimentary covers, metamorphic and plutonic rocks, alpine belts) in France and in the Middle East, he developed since 1995 several research projects in order to understand and model the physical properties of weathered crystalline rocks. These projects led him in 1998 to link to the weathering process itself the presence of a fissured layer (FL) below the saprolite cover of lateritic profiles in crystalline rocks, leading, among others, to a new concept of aquifer in HRs. From 2000 up to now, he developed with BRGM's hydrogeologists numerous applications of this new concept in France and in other countries and continents (India, South Korea, Africa, French Guyana, Madagascar, New Caledonia, etc.), for water prospecting and management and recently for petroleum reservoirs characterization in crystalline rocks. Dr. Wyns teaches courses on weathering and physical properties of weathered rocks in several French universities.

PREFACE

HRs make up the basement of the continents and consequently outcrop over large areas throughout the world, particularly in arid and semiarid regions lacking in surface water resources and where consequently groundwater is of paramount importance for the economic development and the well-being of the population, particularly in several developing countries.

The hydrogeological engineering of HRAs derives from recent findings (1990s–2000s) that allow now to precisely describe the conceptual model of HRAs (origin, geometry, hydrodynamic properties, etc.) as a consequence of superficial weathering. These findings are considered as a breakthrough in this branch of the hydrological sciences.

It is the aim of this chapter to introduce this new hydrogeological conceptual model and to cover the most relevant aspects of the practical engineering applications that were developed on this basis, particularly for assessing, developing, protecting, modeling, and efficiently managing HRAs water resources.

14.1 Introduction

14.1.1 Hard Rocks: Hydrogeological Definition

According to [50] HRs or crystalline rocks are plutonic and metamorphic rocks, from which we can exclude marbles because of their potential karstification and consequently the fact that they can exhibit hydrogeological properties of karstic aquifers. Obviously, limestone and nonmetamorphic volcanic rocks (see, for instance, [8]), even if they are often mechanically hard, particularly to drill, are not considered as HR as their hydrogeological properties are very different from those of HR. HRs make up the basement of the continents and, as such, outcrop over large areas throughout the world

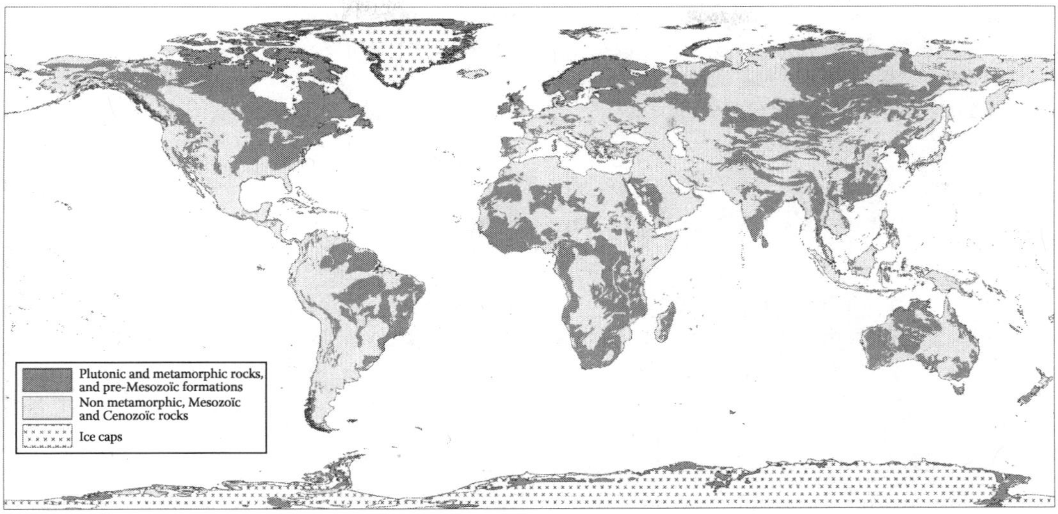

FIGURE 14.1 Main areas of the world with outcropping HRs.

(namely, a large part of Africa, South and North America, India, Australia) mostly in tectonically stable regions such as ancient cratons (Figure 14.1).

It has been shown since a long time (see, for instance, the review proposed by [39]) that, even if plutonic (various kind of granites) and metamorphic (meta-volcanites, meta-sediments, meta-granites, etc., resulting in schists, micaschists, quartzites, gneisses, etc.) rocks are rather different in terms of mineralogy, petrofabrics, or texture, they exhibit similar hydrogeological properties (e.g., [17] for various types of granites and gneisses). It will be shown later (see Section 14.2) that these similarities are mainly related to a unique origin for the hydraulic conductivity: the weathering processes.

14.1.2 Importance of Hard Rock Aquifers in the World

The groundwater resource of HRAs is rather low in terms of sustainable discharge per productive well (from several 100 L/h to a few tens of m^3/h; [13]) as compared to that from porous, karstic, and volcanic aquifers. Additionally, "dry" wells are also very common in HRA [13]. HRAs are, however, well suited for water supply to scattered populations and small- to medium-size cities or the suburbs of the larger cities. Their groundwater resource contributes to the well-being of the population and to the economic development, especially in areas exposed to arid and semiarid climatic conditions where the surface water resource is limited in time (during the year) and space (to the main large rivers only). For instance, in India, HRA water resource largely contributed to the "Green Revolution" that allowed the food self-sufficiency of the country, nevertheless not without any drawbacks such as overexploitation and groundwater quality degradation [47].

14.2 Structure and Hydrodynamic Properties of HRAs

14.2.1 Fissure/Fracture Permeability of Hard Rock Aquifers

HR intrinsically presents a very low (below 10^{-8} ms^{-1}) matrix hydraulic conductivity (e.g., [75]). That explains why, among other arguments, these rock masses are good candidates for nuclear waste storage at least at great depth (several hundreds of meters, e.g., [27]). Their hydraulic conductivity exclusively relies on secondary fissure/fracture permeability (e.g., [5]). Then, a well providing a significant yield in HR (i.e., a few m^3/h; [16,17,48]) crosscuts the low-permeability nonproductive rock matrix alternating with a few centimeters to decimeter thick permeable fractures/fissures ("water strikes") that provide the pumped water. Thus, a pumping well must imperatively tap this secondary "fracture/fissure" permeability.

14.2.2 Origin of HRA Hydrodynamic Properties: The Previous Concepts

The groundwater resource of HR was mainly discovered in the sixties in tropical countries (mostly Africa) by practitioner hydrogeologists (e.g., [57]). The development of a new drilling method (the down-to-the-hole hammer; [58]), very suitable for HR, largely revealed their permeability and their exploitability by small-diameter (a few decimeters) deep wells (a few tens of meters) and thus made it possible to describe their geological structure. Before the development of that drilling technique, only their superficial (meter to decameter thick) unconsolidated (weathered) fringe was prospected and exploited (where saturated and permeable enough) by large-diameter wells [42]. Since that period, 10 of millions of "deep" wells have been drilled in HR throughout the world, particularly in the Sahelian countries of Africa, with the main aim of providing water, and in some cases safer water, to the population [43,77] and particularly in India, mostly for irrigation [25].

These drillings revealed the existence of a permeable fissured/fractured zone located in the first meters or few tens of meters of the "unweathered" HR underlying the unconsolidated weathered layers (regolith or saprolite) [1,68,69,77]. In isotropic granites (s.l.), this permeable zone is composed from (sub)horizontal joints (or sheeting fractures), with a depth-related increasing spacing [10,29,33,39,44, 46,53,65,67,74,77] (Figure 14.2). This permeable horizon has been named the "fissured layer" (FL) or the "stratiform fissured layer" (SFL) [39]. It is noteworthy that the descriptions of this FL are mostly provided by hydrogeologists, surely as a consequence of its remarkable hydrodynamic properties (see all the previous references and also [14,32]). The above-lying unconsolidated weathered profile was described in details by several authors [1,45,66, for instance].

During decades, tectonics, unloading processes, and emplacement and cooling of plutonic rocks were the three most common genetic hypotheses formulated to explain the development of such fissures/fractures, most of the time without any attempt of scientific demonstration (see, for instance, for unloading processes [10,22,30,31,44,68,69,74,77] and for tectonic processes [34,51,60,77]).

Jahns [33] clearly demonstrated that the joints of plutonic rocks have no genetic relationships with the emplacement and cooling of these rocks.

More recently, [39], synthesizing the previous works, demonstrated that:

I. The FL cannot be explained by natural unloading processes. With the exception of some specific conditions (artificial unloading, deep wells, mine galleries, cliffs, deep sedimentary basins), no natural phenomenon able to induce a strong compression parallel to the surface (the only strain state that may produce subhorizontal fractures) appears to exist. In addition, gravity structures can only be invoked in very specific contexts such as high reliefs, for instance, in Alpine valleys. Such process does not exist in the shallow subsurface. Moreover, such process, if it exists, would also be observed for other types of rocks than HR.
II. At the exception of tectonically active areas (mostly plate boundaries, i.e., actively spreading ridges and transform faults, rifts, subduction zones, and convergent or collisional plate boundaries that only occupy a few percents of the earth's emerged surface and that concentration of strain by definition makes them anomalous in regard to the average properties of the crust), the FL and their associated hydrodynamical properties cannot be explained by tectonic fracturing. As a matter of fact, tectonic fracturing faces several inconsistencies. In tectonically stable areas such as most HR areas in the world, it requires the following:

(a)

(b)

FIGURE 14.2 The fissured layer in granites, Lozère, France: (a) outcropping (near the village of Serverette; the ~5 m high phone pile gives the scale) and (b) in a former quarry (near the village of Rieutord de Randon).

- A tectonic process to create the fracture. It is clear that the occurrence of such fractures is very rare both in time and in space.
- That the resulting fracture be permeable. It is obvious that a tectonic fracture is a complex structure that is far from being systematically permeable (see, for instance, [6]).
- That the resulting tectonic fracture reaches the subsurface (i.e., the depth of an HR water well), whereas most tectonic fractures do not reach the last kilometers below-ground surface.
- That a permeable fracture be not fast sealed (at the geological timescale), when in fact rejuvenation is overbalanced by sealing. Outcropping tectonic fractures are very often old and thus sealed.
- The ubiquity of such fractures. Tectonic fractures are spatially unevenly distributed and thus cannot account for the thousands of evenly distributed productive water wells in HR areas.

In addition to that, other arguments can be put forward: (i) the absence of thermal springs in such tectonically stable areas (whereas the main active faults quite systematically exhibit several of these thermal springs) and (ii) the fact that most tectonic fractures have a strong dipping and should thus hardly ever be tapped by standard vertical wells. Infact, the subhorizontal jointing of granitoids and the variously dipping fractures of the other HR FL is easily and systematically cut by vertical wells.

14.2.3 Origin of HRA Hydrodynamic Properties: Fracture Permeability Derives from Weathering

The origin of HRA FL permeability has been demonstrated during the last dozen of years from observations performed all over the world: France [38,81], South Korea [11], Burkina Faso [13], India [16–18,48], France [63], and also from several other observations performed by the authors in various regions of France (Brittany, Massif Central, Vosges, Corsica) and other African countries, French Guiana, etc. (mostly reported within BRGM reports, see www.brgm.fr). The basic concept is that the SFL belongs to the weathering profile as well as the above-lying unconsolidated layers (saprolite) (Figure 14.3).

The development of a weathering profile thick enough to be efficient from a hydrogeological standpoint is first of all a matter of time: millions to tens of millions years are required to create a few tens of meters thick weathering profile [80,87]. In addition, such a development requires the emersion of the rocks but also liquid water—hot or cold deserts or areas with permafrost being not favorable to weathering [1,87]. The weathering profiles are of lateritic type, characterized by leaching, with the evacuation of dissolved chemical elements from the rock. These conditions require a favorable geodynamical context: regional uplift associated with a long wavelength lithospheric deformation leads to the formation of peneplains [26,80,87] with gentle slopes. More precisely, it requires a topographical surface verging on the equilibrium, that is, with an erosion rate not exceeding the weathering one [87], that consequently allows the weathering profile to thicken. The temperature, linked to the climatic conditions, is a second-order factor acting only on the kinetics of the weathering process, with a factor of about 1.7 between a tropical (28°C) and a temperate climate (15°C) [52]. Moreover, the characteristic period of the climate (10^2–10^3 years) cannot explain the functioning of the weathering profiles that is much longer [87]. When the geodynamical context changes, that is, passing from an absence of tectonic activity to an active one, the weathering profile can be (partially) eroded. Conditions again favorable to the development of a new weathering profile lead to polyphased structures with superimposed and/or

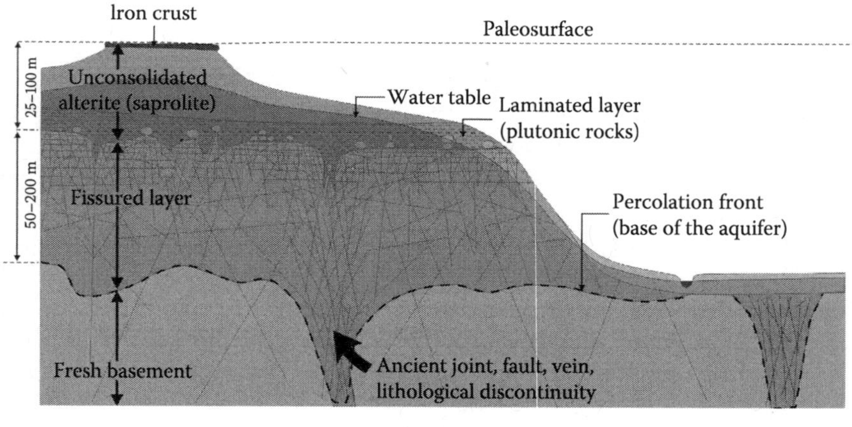

FIGURE 14.3 Conceptual model of a partly eroded paleo-weathering profile on HRs.

totally or partly eroded weathering profiles [17]. Thus, most HR areas of the world exhibit relicts of one or several generations of weathering profiles [9,16,17,33,70,71,72]. In Europe, since the end of the Primary era (during which at least Carboniferous, infra-Permian, infra-Triassic, and infra-Liassic weathering profiles are also recognized), the Early Cretaceous and the Early and Middle Eocene (having lasted 45 and 25 My, respectively) have constituted the most recent periods during which such processes lasted long enough to develop 70–100 m thick weathering profiles (50–70 m thick FL underlying a 20–30 m layer of unconsolidated saprolite; [87]).

14.2.3.1 Stratiform Fissured Layer

The result of the weathering process is a SFL parallel to the paleotopography contemporaneous with the weathering process (Figure 14.3). The development of its fractures network is the response to the stress induced by the swelling of some minerals, particularly the biotite, being progressively hydrated and turning into hydro-biotite, then vermiculite, then mixed clay layers, etc. [28,45,87]. The weathering of biotite is clearly an early process particularly in granitic-type formations [73]. The increase in volume of the weathered mineral can reach 30% [3] and that of the total rock 50% [24]. In granular rocks like granites with a quasi-random orientation of swelling minerals, the potential expansion tensor is isotropic, but expansion is more difficult in the horizontal plane as the medium is infinite in this direction. Consequently, the horizontal stress component accumulates during the weathering. In the vertical axis, however, the stress increases until the lithostatic component is offset, then lets place to vertical expansion, while horizontal stress continues to increase. Consequently, the resulting stress tensor is characterized by a minor vertical component (σ_3) and two major ones (σ_1 and σ_2) that are horizontal. When the stress deviator reaches the elastic limit of the rock, tension cracks appear [81]. For granitic rocks, in accordance with rocks mechanics [46,59], the resulting fractures are perpendicular to the minor stress (subvertical) and consequently are subhorizontal (parallel to the gentle topography contemporaneous with the weathering) and lead to the formation of the subhorizontal jointing of granites (see Section 14.2.2). In foliated and folded rocks, the variability of the orientation of the minerals able to swell as well as the ones of the weaker surfaces of the rock (foliation, schistosity) induce an anarchic fracturing (without any preferential orientation). It is clearly the early stage of biotite weathering that induces fracturing. In fact, the later stages produce clays that exhibit a tendency to fill newly formed pores round about the fractured and weathered minerals [4]. Nevertheless, even if the early weathering of biotites leads to a local clogging of the biotite mineral and its surroundings at a millimeter scale, the density of biotites in the parent rocks is not high enough to clog the aperture of all the fractures of the FL at a meter to decameter scale.

The thickness of the FL is approximately two times that of the saprolite [17] (Figure 14.3). The density of the fractures decreases with depth toward the basis of the weathering profile; thus, the decrease in the hydraulic conductivity of HRA is not a consequence of a lower permeability of the fractures or their "closure" but a consequence of their disappearance in depth [17,48]. Along a vertical view of the profile, the geological observations make it possible to identify several tens of such fractures; however, due to various processes (further healing of the fracture, its weathering, channeling within the fracture plane, etc.), only a maximum of 4–5 of these fractures are permeable enough to be detected as "water strikes" in a water well. This explains, in turn, the large variability of the wells discharge even for nearby wells or a "dry" well close to a productive one [5,17,48]. On top of that, ancient profiles may have been completely sealed particularly as the result of diagenesis due to their burying below marine sediments.

In granitoids, the dimension of each subhorizontal fracture constituting the SFL is a few tens of meters in diameter [48]. Such horizontal permeable fracture set is on average 10 times more permeable, and more numerous, than the subvertical joint set. The hydraulic conductivity of these permeable fractures does not show a very large variability [17,48], which constitutes another argument for a unique origin (weathering).

14.2.3.2 Structure and Hydrodynamic Properties of the Stratiform Fissured Layer

A typical weathering profile (Figure 14.3) comprises the following layers that have specific hydrodynamic properties. All together (where and when saturated with groundwater), these various layers constitute a composite aquifer. From the top to bottom, the layers are the following [17]:

- The laterite (or iron or bauxitic crust) that can be absent, due to erosion or rehydratation of hematite in a latosol (for iron crusts) or resilicification of gibbsite/boehmite into kaolinite (for bauxitic crusts). Where preserved from later erosion and recharged by heavy rainfall (particularly in tropical humid regions, for instance, in French Guiana), the iron or bauxitic crust can give rise to small perched aquifers with locally some springs that exhibit quite an epikarst-type functioning.
- The saprolite or alterite, or regolith, a clay-rich material, derived from prolonged in situ decomposition of bedrock, a few tens of meters thick (where this layer has not been eroded). The saprolite layer can be divided into two subunits (Figure 14.3; [84]): the alloterite and the isalterite. The alloterite is mostly a clayey horizon where, due to the volume reduction related to mineralogical weathering processes, the structure of the mother rock is lost. In the underlying isalterite, the weathering processes only induce slight or no change in volume and preserve the original rock structure; in most of the cases, this layer takes up half to two-thirds of the entire saprolite layer. In plutonic rocks, such as granites, the base of the isalterite is frequently laminated and named "laminated layer." This layer is constituted by a relatively consolidated highly weathered parent rock with coarse sand-size clasts texture and a millimeter-scale dense horizontal lamination cross-cutting the biggest minerals (e.g., porphyritic feldspars) but still greatly preserving the original structure of the rock. Sometimes this particular layer is called "micro-sheeting layer." This clayey layer is generally of quite low hydraulic conductivity, about 10^{-6} m/s. Nevertheless, because of its clayey–sandy composition, in coarse granites the saprolite layer can reach a quite high porosity, which depends on the lithology of the parent rock (bulk porosity is mainly between 5% and 30%). Where this layer is saturated, it mainly constitutes the capacitive function of the global composite aquifer. In fine-grained or low silica content rocks (schists, for instance), this layer is mostly clayey and of very low hydraulic conductivity.
- As described earlier, the FL is characterized by a dense horizontal fissuring in the first few meters and a depth-decreasing density of the fractures, mostly subhorizontal and subvertical in granite-type rocks and anarchic in orientation and shape in metamorphic rocks and foliated granites. The intensification of this fissuring at the top of the layer constitutes, in granites, the overlying laminated layer. This SFL mainly assumes the transmissive function of the global composite aquifer and is drawn from most of the wells drilled in HR areas. However, the covering saprolite layer may have been partially or totally eroded or may be unsaturated. In these cases, the FL assumes also the capacitive function of the composite aquifer; for example, in French Brittany 80%–90% of the groundwater resource is in the FL [82].
- The fresh basement is permeable only locally, where "deep fractures" (see Section 14.2.3.3) are present. Even if these fractures can be as permeable as the fissures belonging to the SFL, their density both with depth and laterally is much more lower. At catchment scale, and for water resources applications, the fresh basement can then be considered as impermeable and of very low storativity [48].

More complex weathering profiles can result from multiphase weathering and erosion processes. In South Korea, for instance, at least two main phases of weathering have been identified ([11]; Figure 14.4):

- An "ancient" one that deeply affects HRs (up to more than 50 m) and results in a typical weathering profile, similar to those observed in other regions of the world (Africa, South America, Europe, etc.).

FIGURE 14.4 Recent weathering (red when seen on the field), related to the present topography, intersecting an old weathering profile (Namwon area, South Korea).

- A more recent one that only affects the rocks (and the ancient weathering profile) within their first meters below the topographic surface. It is clearly posterior to the previous weathering phase(s) as it crosscuts all the horizons of the previously described weathering profile. It is probably quite recent as it clearly follows the actual topographic surface. This recent weathering phase didn't develop a thick FL. It is thus of low hydrogeological interest.

In southern India [17] (Figure 14.5), the nonlaminated saprolite is very thin (1–3 m) and is almost constant in thickness, while it should be thicker in the plateau areas than in the valleys if the weathering had occurred earlier than the erosion that shaped the present topography. The thickness of the laminated layer (10–15 m) is disproportionate compared to that of the nonlaminated saprolite, while in the classical weathering profile (Figure 14.3), the laminated layer occupies only one-third to half of the entire saprolite layer. The thickness of the FL is small compared to that of the saprolite (ratio of ~1 instead of ~2). Moreover, the laminated layer presents preserved fissures, which are usually not observed in the classical model.

In fact, the weathered zone appears to be composed of an old, probably Mesozoic, weathering profile (Figure 14.5a), where only a part of the FL has been preserved. An erosion phase, due to regional uplift, caused the erosion of the entire saprolite layer and a part of the FL (Figure 14.5b). Then, at least one more recent weathering phase, the latest being probably still active, is responsible for the saprolitization of this truncated profile and explains the development of 1–3 m nonlaminated saprolite locally capped by an iron crust and the lamination of a large part of the ancient FL (Figure 14.5c). Consequently, the profile structure is controlled by a multiphase weathering process that was induced by the geodynamic history of the Indian Peninsula. This (or these) more recent weathering phase(s) was efficient enough to hydrogeologically rejuvenate the old weathering profile through the apparition of new permeable weathering induced fractures.

In the Cacao village area, in French Guiana (Figure 14.6; [12], differential weathering between metamorphic rocks and a granodiorite and also a polyphased weathering, uplift, and stripping history explain the complex structure of the weathering profile, with at least four successive paleosurfaces and four "cut-and-filled" weathering profiles.

14.2.3.3 "Deep" "Vertical" Discontinuities

As described by several authors (see, for instance, [1,9,16,54,64]), preexisting heterogeneities within the HR such as veins (quartz veins, pegmatite, aplite, etc.), dykes, ancient fractures or joints, or even contacts

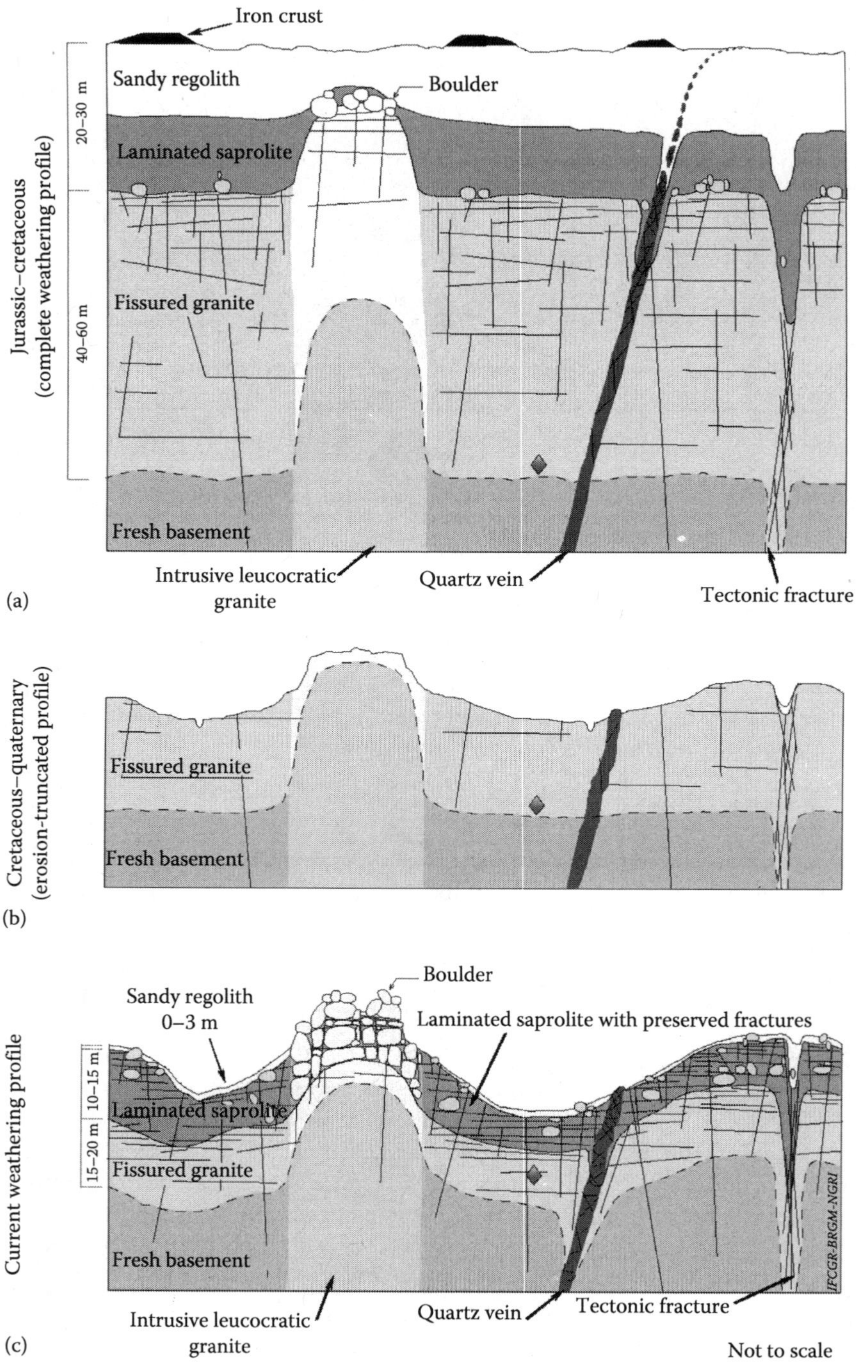

FIGURE 14.5 Idealized multiphase weathering conceptual model, Maheswaram, Andhra Pradesh, India (vertical scale is deliberately exaggerated). (a) Complete Mesozoic Era (Jurassic–Cretaceous) in age-weathering profile, (b) truncated by erosion profile due to Cretaceous to Quaternary uplifts, and (c) current weathering profile. Diamond = benchmark.

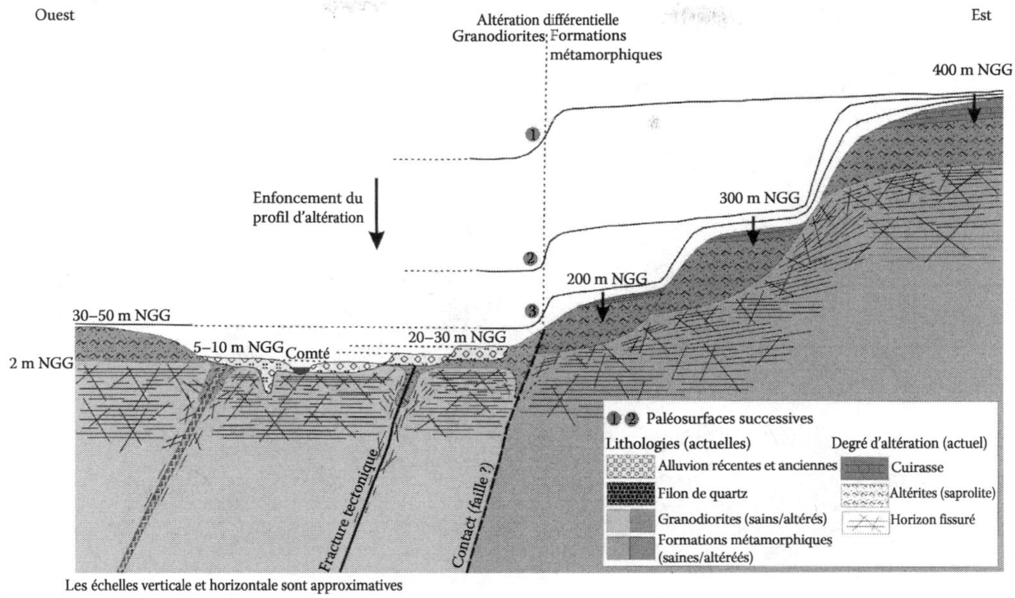

FIGURE 14.6 Differential and multiphase uplift, erosion, and weathering in the Cacao area (French Guiana). (From Courtois, N. et al., Détermination de secteurs potentiellement favorables pour la recherche d'eau souterraine à Cacao (Guyane) Rapport BRGM/RM-52758-FR, Guiana, France, 2003.) Legend (from top to bottom and left to right): west, east, differential weathering, granodiorite, metamorphic formations, sinking of the weathering profile, elevations in m amsl, Comté river, ancient tectonic fracture, contact (fault?), successive paleosurfaces. Lithologies: recent and ancient alluviums, quartz vein, granodiorites (unweathered, weathered), metamorphic formations (unweathered, weathered). Weathering rate: iron crust, saprolite, fissured layer. Vertical and horizontal scales are approximative.

between different geological units (see, for instance, [21,40,41,51]) locally favor the weathering process. It results in a local deepening of the weathering profile that can reach, in the vicinity of such heterogeneities, several hundreds of meters below the surface topography [63], with locally good hydraulic relationships between the subsurface aquifers and these deep structures. In this case, the weathering process turns out to be at the origin of the fissures development and of an enhanced local hydraulic conductivity in the surrounding HR at the periphery of these heterogeneities and also within decameter-wide discontinuities or veins—quartz vein, pegmatite, for instance—as well [16] (Figures 14.3 and 14.7).

Near these discontinuities, the HR-weathering profile (saprolite + FL) is characterized by a sharp deepening of the weathered layers in the HR close to the contact. Because in this case the stress tensor is deviated, σ_3 becomes sub-orthogonal to the heterogeneity; consequently, the FL is "verticalized" as it develops parallel to the vein (Figure 14.7). Consequently, the FL and the saprolite reach much greater depths than within the classical previously described profile (up to several hundred meters—see, for instance, [16,61,63]). Such geological structures are of primary hydrogeological interest in areas where the SFL has been eroded (or is unsaturated, lying above the piezometric level); they constitute there the only available targets for water supply wells.

14.2.4 Influence of Mineralogy, Texture, and Structure on the Development of the Fissured Layer

14.2.4.1 Mineralogy

Biotite is the most common mineral that is responsible for creating fissure network in HRs during weathering: it can be found in many acidic plutonic rocks (granitoids) as in metamorphic rocks

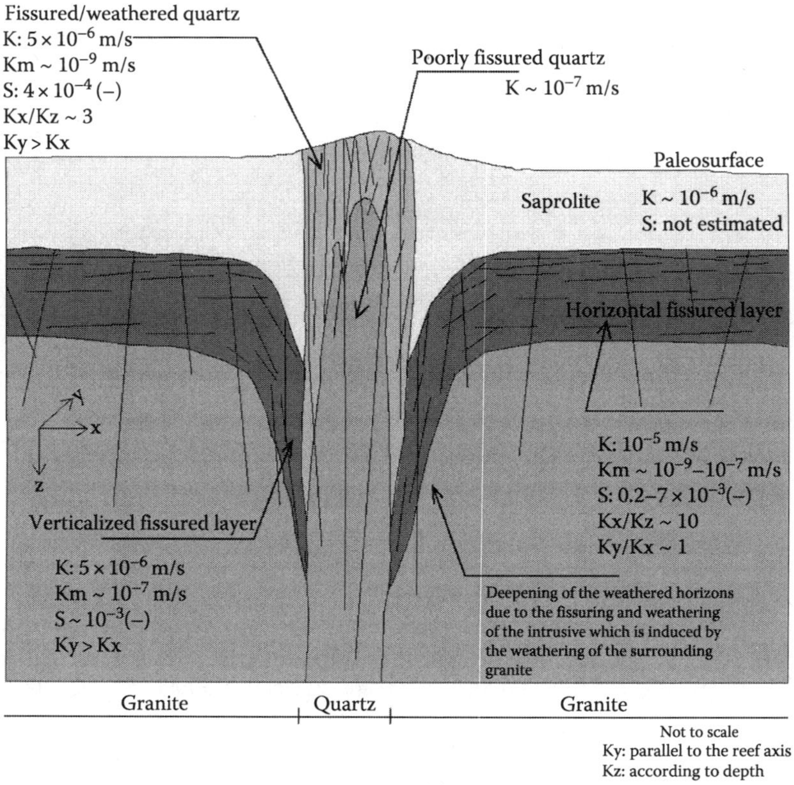

FIGURE 14.7 Conceptual hydrodynamic model of a vertical discontinuity in HR. (From Dewandel, B. et al., *J. Hydrol.*, 405, 474, 2011.)

(micaschists, paragneisses, and orthogneisses). White micas are also able to inflate during weathering, but their transformation into clay minerals starts later than biotite, because they need more activation energy. In 2-mica granites, the weathering of muscovite begins generally when the rock has been transformed into sandy saprolite; in this case, muscovite plays no part in the fissuring of HRs, most of the fissures being due to weathering of biotites. However, in some white mica-bearing rocks as sericitoschists, weathering of sericite can contribute in creating a fissure network, but these cases are poorly documented. Feldspars are not known to inflate during weathering, even if the transformation of plagioclase in clay minerals begins as early as the biotite one.

In basic and ultrabasic rocks, numerous observations show that the existence of a FL is dependent on the presence of only two minerals: pyroxene and/or olivine. Basalts containing phenocrysts of pyroxene and/or olivine generally suffer fissuring during early stage of weathering. Other highly weatherable minerals of basic rocks as plagioclase and amphibole do not induce fissuring, so that when basic lavas undergo slight metamorphism or propylitization, they cannot be fissured later by weathering process; that is particularly the case for palaeovolcanics or some doleritic dykes, where pyroxenes and olivines have been completely retromorphosed.

In peridotites, weathering conduces to inflation of olivine and consequently to intense fissuring of rock. Secondarily, karstification develops in such HRs, below the saprolite cover mainly made up of limonite.

14.2.4.2 Texture

During weathering, the inflation pressure of swelling minerals is dependent upon the mineral size. Therefore, HR fissuring is easier in granular rocks than in microgranular ones: for instance, microgranites, even containing biotite, are less fissured than granites.

14.2.4.3 Structure

During weathering, the inflation of biotite occurs perpendicular to cleavages, the layer spacing changing from 10 Å to 14 Å as biotite turns into chlorite or vermiculite. Consequently, inflation pressure exerts perpendicular to cleavage, and the resulting stress ellipsoid shape depends on the orientation of the minerals. Thus, the permeability tensor is dependent on mineral orientation. Figures 14.8 through 14.11 show, respectively (from base to top and from left to right), the mica orientation, the potential deformation ellipsoid, the resulting stress (and permeability) ellipsoid, and the geometry of induced fissure network.

When inflating minerals are randomly oriented (Figure 14.8), as in the general case of plutonic rocks, the theoretical deformation ellipsoid is isotropic; vertical stress increases until the lithostatic charge is offset, whereas horizontal stresses continue to increase. Horizontal tension joints are created when the difference between minor and major stress components reaches the elastic limit. The resulting reservoir properties (permeability, porosity) are good, the major permeability component being horizontal.

When the inflating minerals are vertically oriented (Figure 14.9), the deformation ellipsoid is anisotropic, with major axis orthogonal to mineral orientation, that is, horizontal; the maximum stress component has the same orientation, and the resulting tension joints are horizontal and generally largely open. The resulting reservoir properties (permeability, porosity) are very good, the major permeability component being horizontal.

When the inflating minerals are horizontally oriented (Figure 14.10), the whole inflating potential is vertical, the horizontal stress being very slight; in this case there is no open tension joints, but the rock acquires a horizontal laminated structure. The resulting reservoir properties are poor.

In the case of metasedimentary folded rocks (schists, micaschists, gneisses) (Figure 14.11), the orientation of inflating minerals varies laterally, and the bedding creates lithological preferential brittleness planes. Breaking joints are created in random directions.

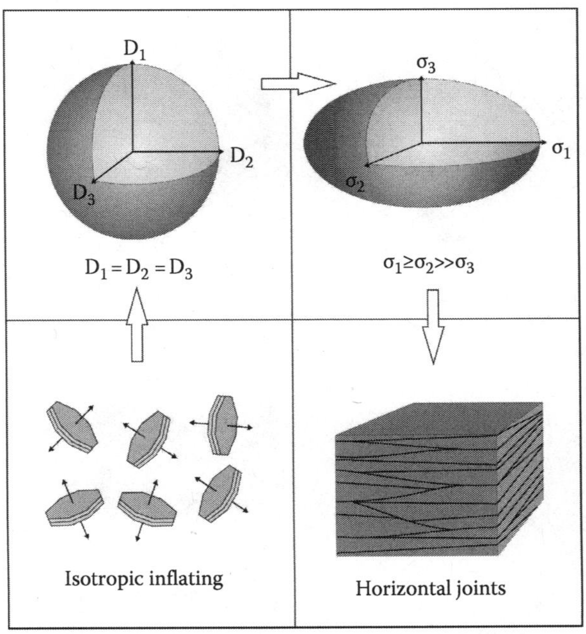

FIGURE 14.8 Parameters of deformation, stress, and fissuration for randomly—oriented inflating minerals (plutonic rocks). From base to top and from left to right: the mica orientation, the potential deformation ellipsoid, the resulting stress (and permeability) ellipsoid, and the geometry of induced fissure network.

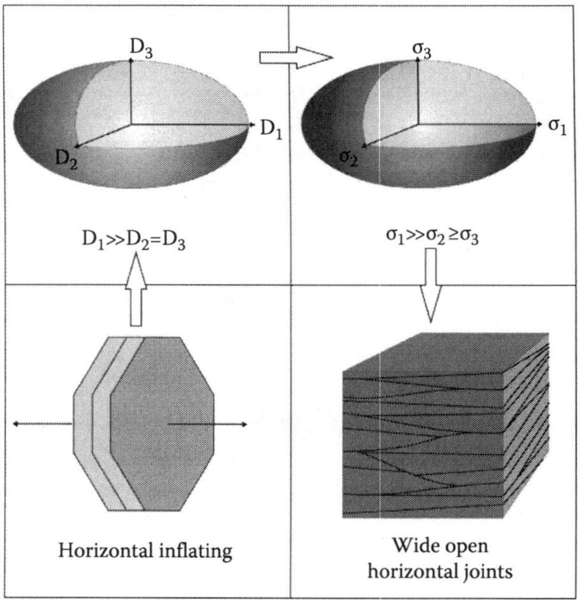

FIGURE 14.9 Parameters of deformation, stress, and fissuration for vertically—oriented inflating minerals (gneisses, orthogneisses, micaschists). From base to top and from left to right: the mica orientation, the potential deformation ellipsoid, the resulting stress (and permeability) ellipsoid, and the geometry of induced fissure network.

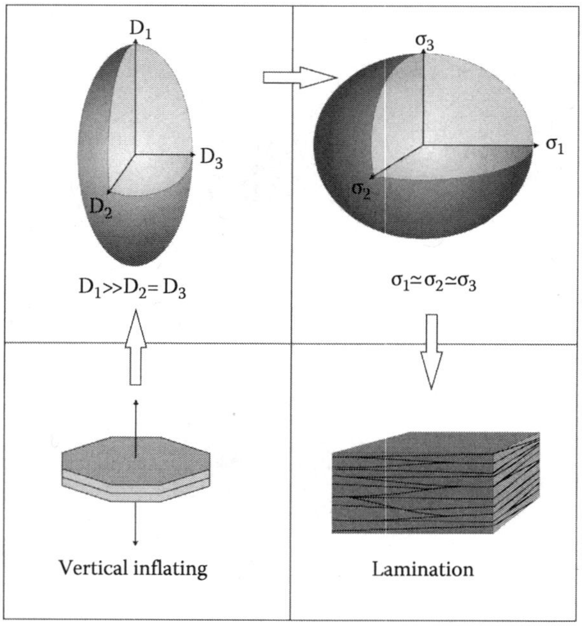

FIGURE 14.10 Parameters of deformation, stress, and fissuration for horizontally—oriented inflating minerals (gneisses, orthogneisses, micaschists). From base to top and from left to right: the mica orientation, the potential deformation ellipsoid, the resulting stress (and permeability) ellipsoid, and the geometry of induced fissure network.

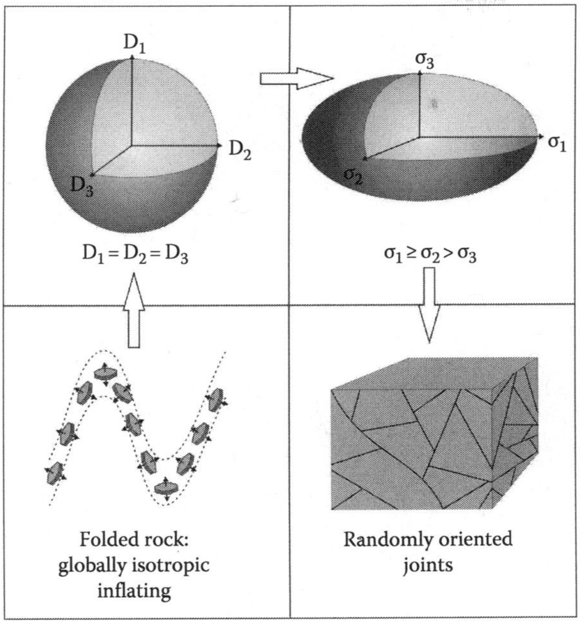

FIGURE 14.11 Parameters of deformation, stress, and fissuration for folded rocks (gneisses, micaschists, schists). From base to top and from left to right: the mica orientation, the potential deformation ellipsoid, the resulting stress (and permeability) ellipsoid, and the geometry of induced fissure network.

14.2.4.4 Conclusion: The Most Favorable Criteria for Good Aquifer Properties

For plutonic rocks, granular, biotite-bearing granites, granodiorites, and diorites along with unmetamorphosed gabbros are generally good aquifers. For metamorphic rocks, orthogneisses, paragneisses, and micaschists can be good aquifers if their foliation dipping is near vertical and are poor ones when foliation dipping is near horizontal. Consequently, biotite-rich rocks (like biotite granites or biotite gneisses, or micaschists, for instance) exhibit a well-developed SFL, with a high hydraulic conductivity (as well as a proportionally thick saprolite), whereas biotite-poor rocks (such as leucogranites, for instance) are poorly weathered/fissured. This often results in a granitic landscape where leucogranite intrusions in the biotite granite now constitute unweathered hills gently emerging (a few tens of meters) from the weathered (often partly eroded) biotite-granite flatty weathering cover. A similar landscape results from differential weathering in metamorphic rock context where unweathered sandstone ridges steeply emerge from flatty weathered schists or micaschists.

Similarly, in granite-type rocks, high hydraulic conductivity SFL develops in coarse-grained rocks (such as porphyritic granite), whereas fine-grained rocks (like aplitic granites) only develop lower hydraulic conductivity SFL.

In foliated rocks, the fracturing intensity is maximum where the foliation is subvertical, whereas it decreases and reaches a minimum where the foliation is subhorizontal.

Basic volcanic rocks can be medium-quality aquifer, only if they contain pyroxene and/or olivine. Paleovolcanics and amphibolites are generally poor aquifers, as are acidic Paleovolcanics.

Peridotites can be good aquifers, with dual aquifer behavior, karstic and fissured.

14.2.5 Hydrogeological Functioning of Hard Rock Aquifers

Due to their globally low hydraulic conductivity, HRAs generally exhibit quite high hydraulic gradients and piezometric heads that well mimic the topographic surface shape (Figure 14.12). At the exception of

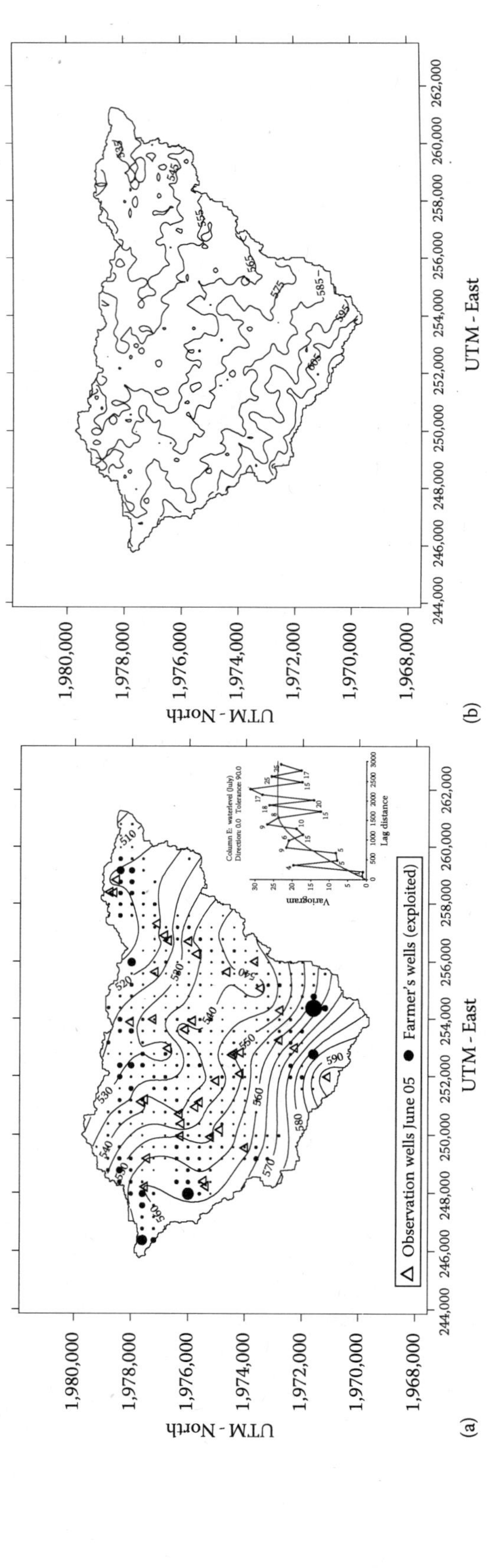

FIGURE 14.12 Piezometric map (a) and topographic map (b) of the 83 km^2 Gajwel watershed (India) characterized by a total groundwater abstraction of about 20×10^6 m^3/year.

areas with a very low recharge, and of course of overexploited areas, the piezometric surface is shallow. Consequently, the streams drain the aquifer, and in wet regions each stream is flowing and most of the topographic depressions give rise to a spring, of course generally of low discharge.

In "natural" conditions, most of the groundwater flow resulting from the recharge occurs in the upper layers of the aquifer, from the topographic surface to the narrowest perennial stream or spring, mostly through the saprolite. Only a few flow lines and a small proportion of the recharge reach the deepest levels. The resulting picture is an aquifer in which groundwater age and mineral content increase with depth.

Pumping in the SFL completely modifies the groundwater flow lines in the aquifer as groundwater from the upper horizon is forced to vertically flow down toward the pumped SFL. Consequently, the groundwater chemistry is also largely modified with, in some cases denitrification processes.

14.3 Characterizing, Valorizing, and Managing Hard Rock Aquifers

This new genetic conceptual model of HRA leads us to revise the methodologies used for characterizing the structure and functioning of such aquifers and thus to optimize the efficiency of these methodologies. Moreover, it enables us to propose a wide range of practical applications in hydrogeology as well as in other domains in applied geosciences.

For instance, the strategy for siting a water well will be different in the general case (looking for the subhorizontal fractures of the SFL that will efficiently be tapped by vertical boreholes; see Figure 14.3) or where the SFL has been completely removed by erosion (searching for a vertical FL located along the walls of a subvertical discontinuity; see Figure 14.7). In the former case, the survey will focus on the SFL preserved from erosion, not clogged by further weathering or diagenesis, and saturated with water (below the piezometric level). In the latter case, standard techniques for well siting such as lineament analysis or radon surveys [37] can be applied. In that case inclined boreholes will be more efficient than vertical ones to increase the probability to crosscut such a vertical structure and also to cut it entirely (not only the low hydraulic conductivity saprolitized core of the structure but also its permeable surroundings).

14.3.1 Mapping the Layers Constituting the Hard Rock Aquifers: From the Local to the Country-Size Scale

The main consequence of this new conceptual scheme is that HRA can now be considered as relatively "continuous aquifers." As a way of consequence, it is now possible to map (or model in 3D) the geometry (elevation, thickness) and the physical properties of their various constituting layers: saprolite, with mostly capacitive hydrodynamic properties, and SFL, with mostly transmissive but also capacitive properties [17,38,81]. This mapping methodology is based on geomorphological (topographic maps, digital elevation model (DEM) processing), geological (outcrop observations, well logs), geophysical (both field and aerial; [7]), and hydrogeological approaches. It comprises a twofold methodology:

- First, characterizing the local (at the scale of the studied area) weathering history and consequently the vertical structure of the weathering profile, for the main rock types of the area (see, for instance, Figures 14.3 through 14.6).
- Second, mapping the elevation of the interfaces separating the main horizons (namely, the saprolite, the SFL, and the fresh substratum). The resulting maps can either show the intersection of these horizons with the topography, as a classical geological map (Figure 14.13a), or isohypses of these interfaces (Figure 14.13b), or thickness of these horizons (Figure 14.13c).

In most cases, the geometry of the base of the saprolite is determined at first, through the combined use of well data, geophysical data if any, and specific field surveys. Where the surface of the base of the saprolite is cut by the present-day ground surface (for instance, near valleys), it is relatively simple, in

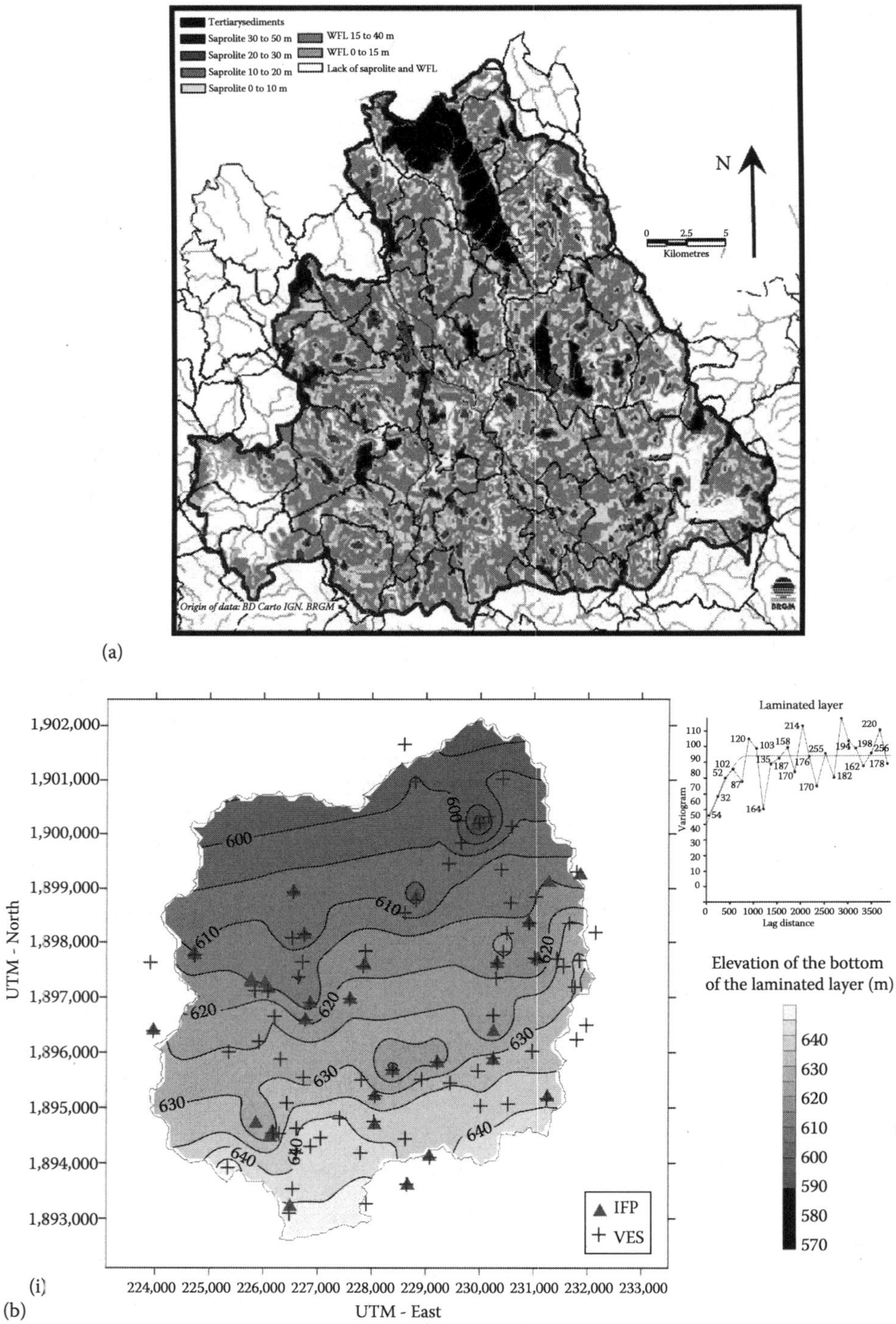

FIGURE 14.13 Examples of maps of the various layers constituting HRAs. (a) Geological map of the weathering cover on the Truyère, Lozère (France) watershed: thickness of the saprolite (increasing thickness from clear to dark and black shading) and the SFL (increasing thickness from dark to clear); white: weathering profile totally eroded. (From Lachassagne, P. et al., *Ground Water* 39(4), 568, 2001.) (b) isohypse map of the elevation (in m above sea level) of the bases (i) of the laminated layer and (ii) of the SFL (= top of fresh and unfissured basement), Maheswaram (India). (From Dewandel, B. et al., *J. Hydrol.*, 330(1–2), 260, 2006.)

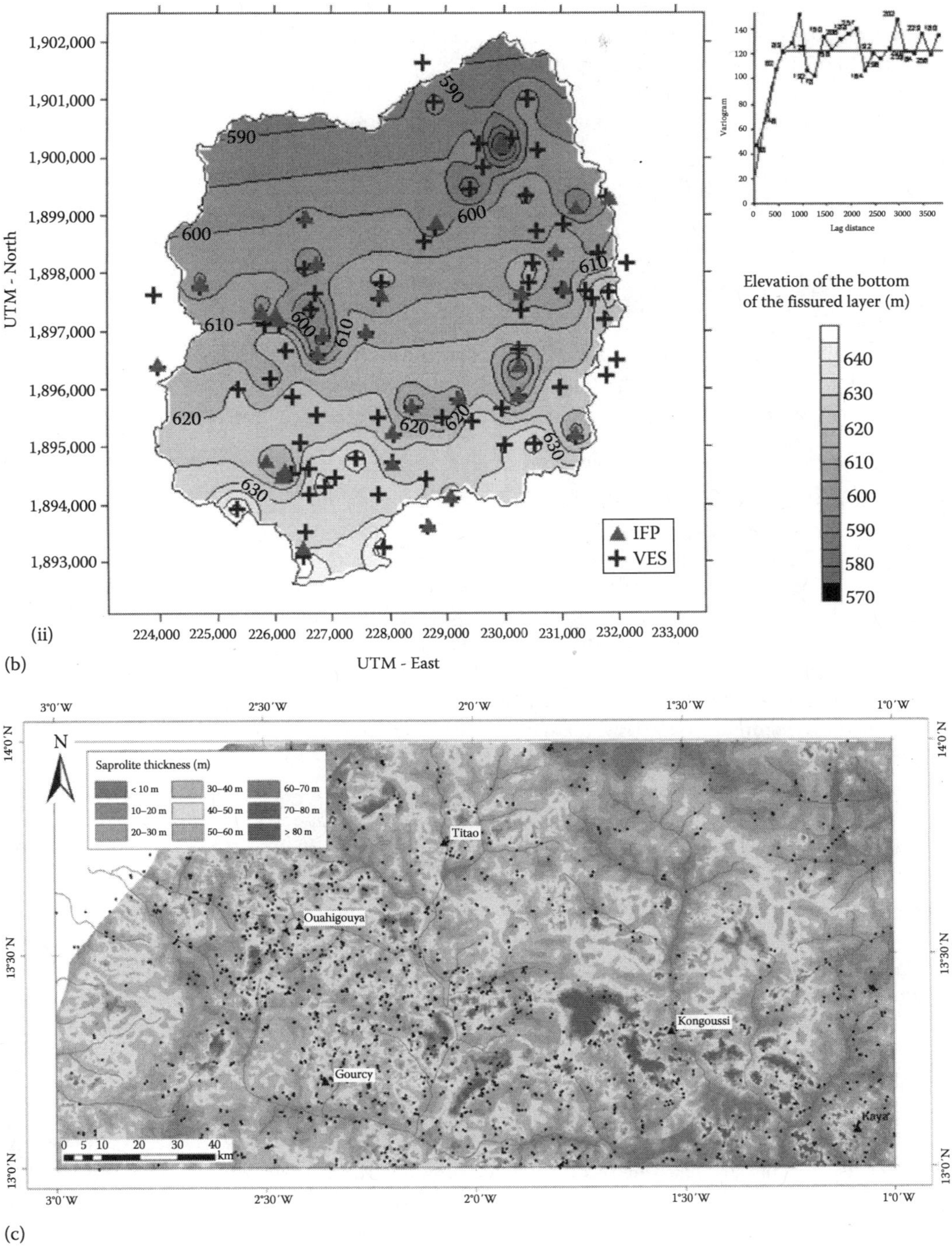

FIGURE 14.13 (continued) Examples of maps of the various layers constituting HRAs. (c) Saprolite thickness over a ~250 × 100 km area in Burkina Faso. (From Courtois, N. et al., *Ground Water*, 48(2), 269, 2009.)

the field, to determine the position of the saprolite/fissured layer interface. On the basis of observations of the thickness of the SFL (observations on outcrops, statistical treatment of existing well data), the altitude of its base can also be directly computed from the one of the alterite base. The altitudes of (1) the saprolite/SFL interface and (2) the elevation of the SFL/fresh rock interface can therefore be subtracted from that of the present ground surface, for instance, as inferred from DEM data, in order to compute the residual thickness of, respectively, the saprolite and the SFL. In regions where the paleosurface(s) has (have) partly been preserved from erosion, slope analysis from DEM data can also be used for the elaboration of such maps in order to reconstruct paleosurfaces.

Several hydrogeological applications have been derived from the concepts and the mapping methodology presented earlier.

14.3.2 Water Well Siting in Hard Rock Aquifers and Mapping Hydrogeological Potentialities

Water well siting (either for the water supply of the population or for agricultural or industrial purposes) answers to a very important hydrogeological need. A complete methodological toolkit, especially devoted to HRAs, has been developed on the basis of the previously presented concepts and results.

14.3.2.1 Regional to Country-Scale Approach

It is known for long that in HRs a basic hydrogeological parameter, the blowing discharge of a well measured at the end of its drilling, is a very robust indicator of the rocks local hydraulic conductivity/transmissivity. It is also a good indicator of the future exploitable discharge of the well, particularly for village hydraulics water supply where quite low discharges (often only >0.5 m^3/h) are required (see, for instance, [43]). The depth to the base of the saprolite is also a parameter measured with a good accuracy during drilling, either by a geologist or by a hydrogeologist but also and more often by the driller himself. These data are very often reported and stored in data bases (see, for instance, [13]). Statistics computed from these data, particularly those gathered in Africa during large village hydraulics programs funded by the international community, are used for a long time to infer, at the regional scale, the hydrodynamic properties of the various lithologies constituting a study area (see, for instance, [43]). These statistical methodologies have been refined on the basis of the previously presented conceptual model [13]. In fact, the hydraulic conductivity of a well (the blowing discharge) logically appears to be strongly correlated to the depth of the well below the base of the saprolite (i.e., to the "linear discharge," the blowing discharge/depth of the well below the base of the saprolite): briefly speaking, the shallowest wells, crosscutting the more densely fissured part of the SFL, provide a higher linear discharge than the deeper ones. The statistical processing of this parameter allows computing, for each lithotype, two important parameters [13]: the thickness of the most productive part of the SFL (in Burkina Faso, for instance, this thickness is comprised between 35 and 40 m according to the lithotype) and its hydrodynamical parameters, namely, basic statistics about the potential discharge of a well (mean, median, standard deviation, number of "dried" wells, i.e., below a discharge threshold).

Such kind of results have direct and concrete uses: In association with the thickness of the saprolite (see Figure 14.13c), they provide an estimation of optimal well length, that is, the length beyond which it is not necessary to drill (little gain in discharge that does not justify the increased drilling cost). Consequently, it is of particular interest for the planning (duration) and financial evaluation of drilling campaigns (cumulative length amount and depth of the wells to be drilled, including the number of "dry" wells), preliminary geophysical surveys included, as they can also be precisely dimensioned. It provides an assessed minimum length for water wells in order to cross the saprolite and reach the SFL and as such can be used to evaluate the related drilling constraints. For example, for a weathered cover between 0 and 30 m thick, drilling can be done with air, whereas if it is between 30 and 60 m thick, the drilling has to be done with a drilling mud. Other applications can also be inferred from the results, such as mapping the mean discharge of wells (at regional/country scale) for different depths and coupling these results with economic parameters such as drilling costs and so on. Applications are also found in fields other than hydrogeology, such as town and country planning, geotechnics, and the search for areas prone to quarrying.

14.3.2.2 Local Scale

Such results are also of high interest at the local scale, the one of a local hydrogeological project. However, the available data from existing wells might not be numerous enough to compute robust statistics. In that case, a high-resolution mapping (see Figures 14.6 and 14.14, and also the beginning of Section 14.2.4) is essential for the preliminary stages of water well siting (see also the beginning of Section 14.3 that describes various strategies depending on the characteristics of the weathering profile). Such preliminary stages enable to locate, in combination with other parameters such as land use, land ownership, and conditions of access, favorable zones to implement geophysical surveys, from which a precise siting of the exploratory drillings can then be inferred.

At such a local scale, geophysical methods, and particularly 2D electric geophysical methods, generally cannot delineate the contours between different lithologies (Figure 14.15). Nevertheless, they are very efficient to characterize the structure of the weathering profile (Figure 14.15). The typical response is, from top to bottom, (1) high resistivity (unsaturated saprolite and/or, where preserved, iron or bauxitic crust); (2) much lower resistivity, corresponding to the clayey saprolite; (3) regularly increasing resistivities within the sandy saprolite and within the SFL; and (4) the highest resistivity (up to more than 10,000 Ωm) being reached within the unweathered (and of course unfractured) rock.

Such a high-resolution mapping (Figure 14.6) is also very suitable for the delineation of groundwater protection zones (Figure 14.14) but also for other applications such as siting a landfill (preferably on the upper part of the saprolite of a rock leading to the formation of a very clayey saprolite) and locating a rock quarry (with a minimum cover cap, preferably on a place where the weathering profile is eroded or at the boundary between two successive weathering profiles).

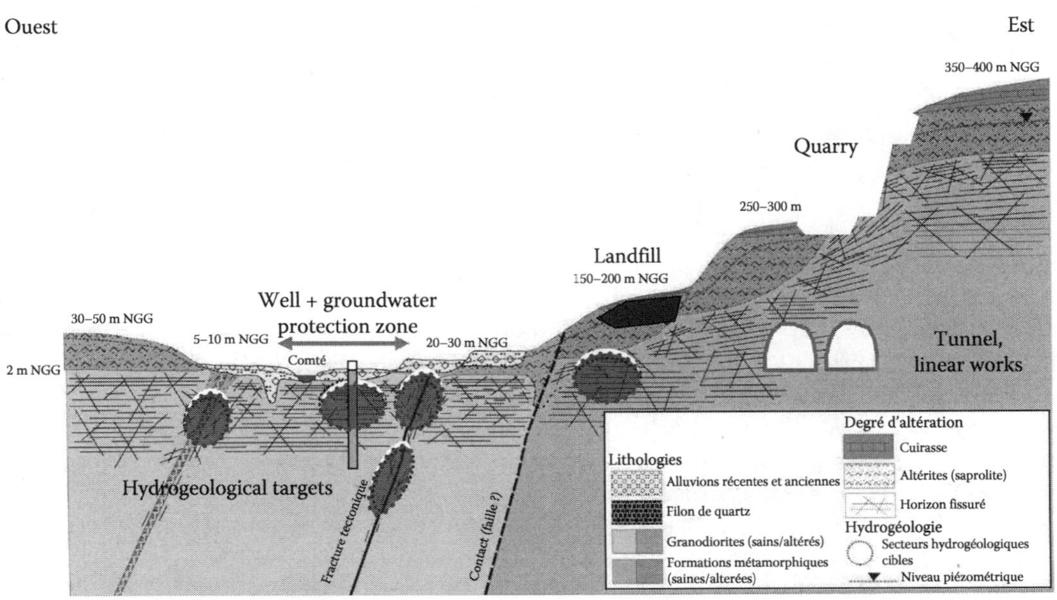

FIGURE 14.14 Various uses of a high-resolution mapping of the weathering cover at the local scale (see Figure 14.6 for the legends and Section 14.2.3.2). (From Courtois, N. et al., Détermination de secteurs potentiellement favorables pour la recherche d'eau souterraine à Cacao (Guyane) Rapport BRGM/RM-52758-FR, Guiana, France, 2003.)

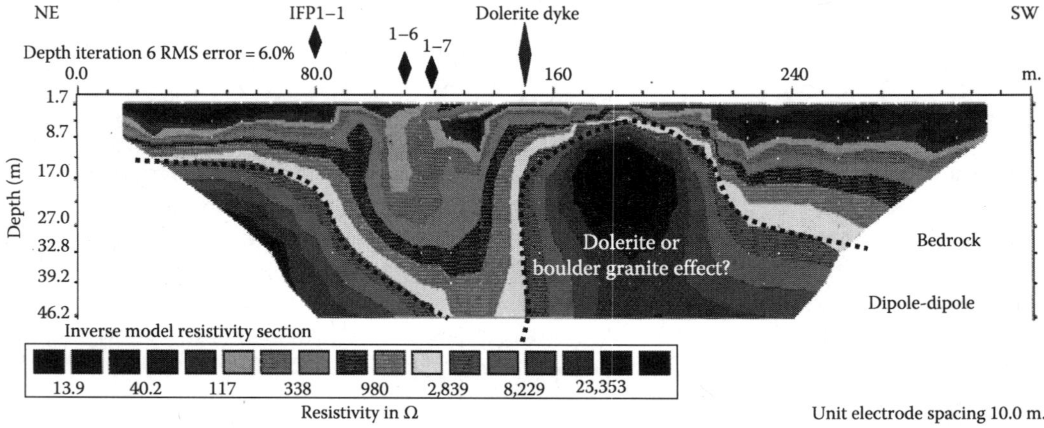

FIGURE 14.15 Examples of geophysical profiles in granitic (Maheswaram, India) weathered HR.

14.3.3 Quantity Management and Modeling of Hard Rock Aquifers Groundwater Resource at the Watershed Scale

On the basis of this concept of stratiform layers and of the relative homogeneity of HRA hydrodynamic properties at a scale of few 100 m [19], the use of water-table fluctuation and groundwater budget methods at the watershed scale is particularly justified [49]. These techniques are very relevant methodologies for the groundwater management of such aquifers and have already been extensively used in the context of the Indian overexploited HRAs [15,20,23,35,49,55,56]. Decision support tools have been developed from these methodologies [15], and strategies built, some of them integrating climate change impacts [23] (Figure 14.16).

14.3.4 Assessing Hard Rock Aquifers Groundwater Reserves from the Watershed to the Regional Scale

These new concepts also allow assessing HRA groundwater reserves from the watershed to the regional scale on the basis of the combined use of the 3D geometry of the weathering profile and the vertical efficient porosity profile determined by protonic resonance soundings (PRS) [2,76,82,84,85].

PRS allows determining the mean effective porosity of each part of the aquifer, which is by convenience divided in three parts: saprolite, upper SFL, and lower SFL. For a given lithology, several PRS are used to determine the mean porosity for each layer, weighted by quality index of PRS. Then, the saturated thickness of each layer, multiplied by the porosity, gives for each cell the water content expressed in water height (Figure 14.17). The total water reserve is given by the addition of the water height of the three layers of the aquifer Figure 14.18.

It also allows upscaling and regionalizing hydrodynamic properties [18]. On this basis, and taking into account the recharge of such aquifers, it makes possible the evaluation of the durability of nonpoint source pollutions (diffuse nitrates contamination, for instance).

14.3.5 Managing and Protecting Hard Rock Aquifers Groundwater Resources

As already discussed earlier, the delineation of groundwater protection zones for water wells has to rely on such a good knowledge of the geometry and hydrodynamic properties of the exploited HRAs. Methodologies for the optimization of piezometric networks have also been developed [88].

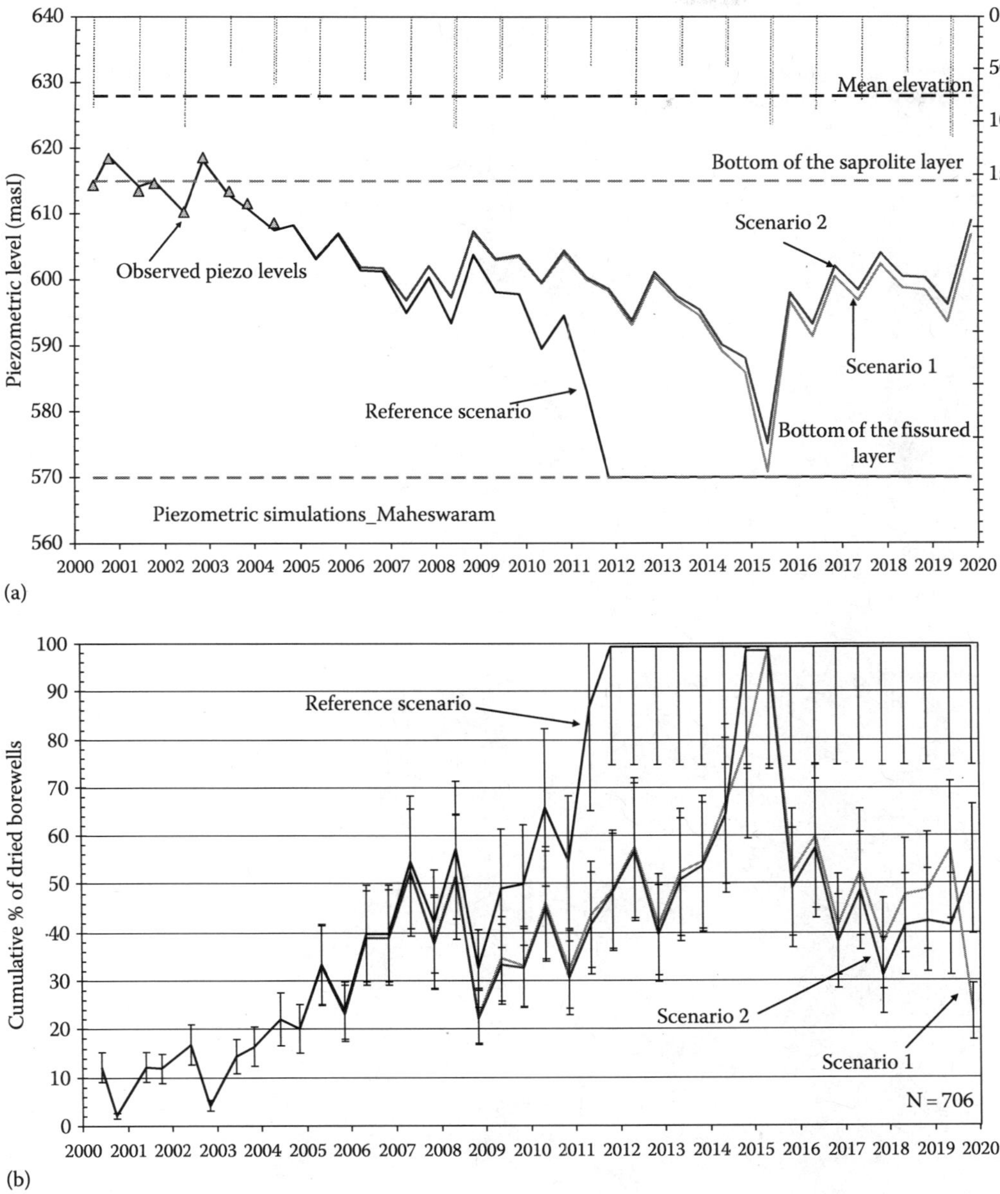

FIGURE 14.16 Example of long-term resource management simulation at the watershed scale (Maheswaram, India): (a) mean piezometric level; (b) number of dried wells.

14.3.6 Drainage Discharge and Hydrological Impacts of Tunnels Drilled in HR

The drainage discharge and the surface hydrogeological (piezometry in wells) and hydrological (discharge of streams) impacts of shallow tunnels in HR (with a depth below ground level between 0 and 300 m) have been forecasted on the basis of (1) the location of the tunnel within or below the various layers constituting the weathering profile, (2) steady-state groundwater discharge measured in existing tunnels, and (3) analytical solutions for tunnel inflow [36]. The actual discharge of the now drilled tunnels validates the efficiency of the methodology.

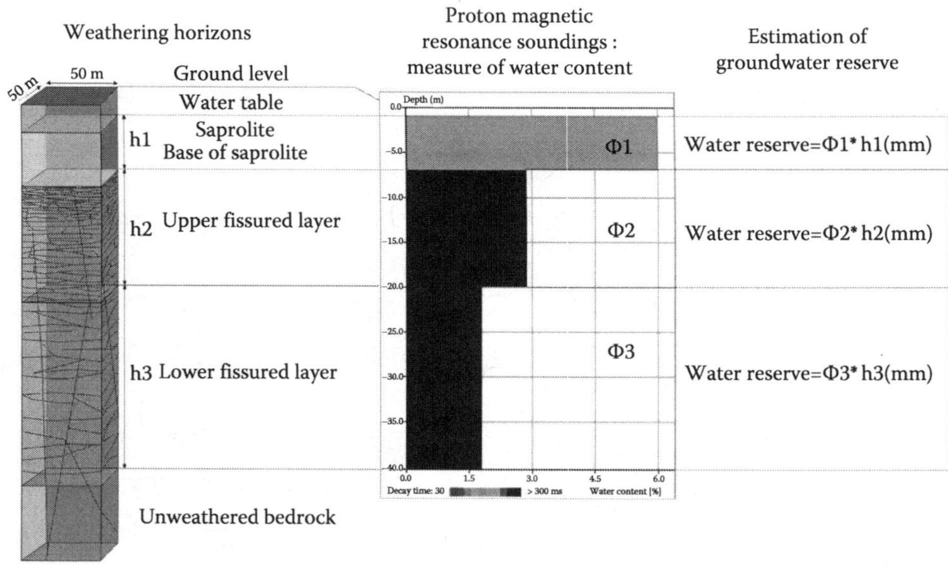

FIGURE 14.17 Principle of water reserve content mapping at watershed to regional scale.

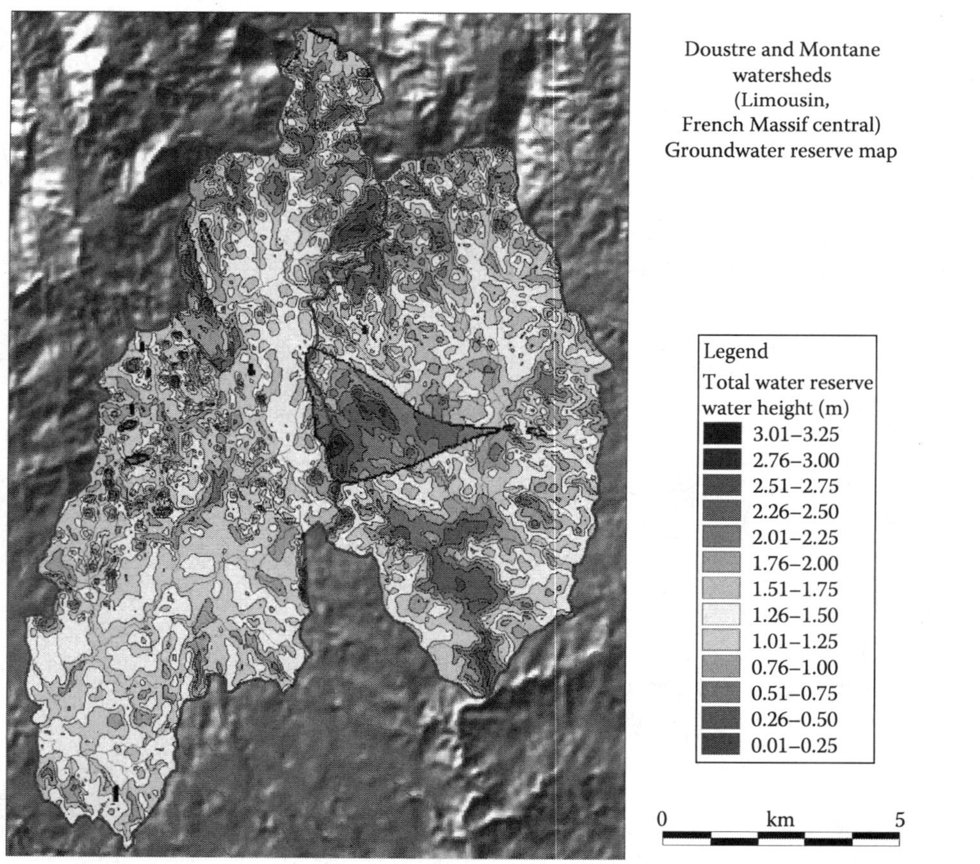

FIGURE 14.18 Example of map of water reserve on two watersheds totalizing 115 km^2 in mainly granitic context (French Massif Central).

14.4 Other Applications in Applied Geosciences

Before being largely improved and developed for hydrogeological applications, these concepts were first used as markers of geodynamic evolutions [79] with geological mapping objectives (see, for instance, [83]) and radioactive waste storage safety analysis [79,80]. Applications have also been developed in the field of soil and rock mechanics, notably for the forecast at cartographic scale of the near subsoil mechanical properties for applications such as the burying of (electrical, telephonic) networks [86] and also for quarrying in HR ([62]; see also Figure 14.14): evaluation of the thickness of the cover cap, determination of optimal geo-mechanic quality of the materials, etc. In the near future, more applications are expected to crop up such as preliminary surveys for linear works (highways, railways).

14.5 Summary and Conclusions

Most HRs are or were exposed to deep-weathering processes, as in large shield of Africa, India, North and South America, Australia, Europe, etc. It turns out that the hydraulic conductivity of HR is inherited from these weathering processes, within their SFL located immediately below the unconsolidated saprolite and, to a much lesser extent, within the vertical FL at the periphery of or within preexisting geological discontinuities (joints, dykes, veins, etc.). The recognition of this conceptual model opens up large perspectives in terms of applied hydrogeology and geology: mapping hydrogeological potentialities, water well siting, quantitative management and modeling of the groundwater resources, computing the drainage discharge of tunnels, quarrying in HR, etc.

References

1. Acworth, R.I., 1987. The development of crystalline basement aquifers in a tropical environment. *Quaternary Journal of Engineering Geology* 20: 265–272.
2. Baltassat, J.-M., A. Legtchenko, B. Ambroise, F. Mathieu, P. Lachassagne, R. Wyns, J.L. Mercier, and J.-J. Schott, 2005. Magnetic resonance sounding (MRS) and resistivity characterization of a mountain hard rock aquifer: The Ringelbach catchment, Vosges Massif, France. *Near Surface Geophysics* 3: 267–274.
3. Banfield, J.F. and R.A. Eggleton, 1990. Analytical transmission electron microscope studies of plagioclase, muscovite, and K-feldspar weathering. *Clays Clay Minerals* 38(I): 71–89.
4. Bisdom, E.B.A., G. Stoops, J. Delvigne et al., 1982. Micromorphology of weathering biotite and its secondary products. Contribution no.3 of the Advisory panel on weathering phenomena and neoformations of the Sub-commission on soil micromorphology of the I.S.S.S. *Pedologie, XXXII* 2: 225–252.
5. Boutt, D.F., P. Diggins, and S. Mabee, 2010. A field study (Massachusetts, USA) of the factors controlling the depth of groundwater flow systems in crystalline fractured-rock terrain. *Hydrogeology Journal* 18: 1839–1854.
6. Caine, J.S., J.P. Evans, and C.B. Forster, 1996. Fault zone architecture and permeability structure. *Geology* 24(11): 1025–1028.
7. Carrier, F., B. Bourdon, E. Pili, C. Truffert, and R. Wyns, 2006. Airborne gamma-ray spectrometry to quantify chemical erosion processes. *Journal of Geochemical Exploration* 8: 226–270.
8. Charlier, J.B., P. Lachassagne, B. Ladouche et al., 2011. Structure and hydrogeological functioning of an insular tropical humid andesitic volcanic watershed: A multi-disciplinary experimental approach. *Journal of Hydrology* 398: 155–170.
9. Chilton, P.J. and A.K. Smith-Carington, 1984. Characteristics of the weathered basement aquifer in Malawi in relation to rural water supplies. *Challenges in African Hydrology and Water Resources (Proceedings of the Harare Symposium,* July 1984). IAHS Publ. no. 144, Malawi, Lilongwe.

10. Chilton, P.J. and S.S.D. Foster, 1995. Hydrogeological characterisation and water-supply potential of basement aquifers in tropical Africa. *Hydrogeology Journal* 3(1): 36.
11. Cho, M., Y. Choi, K. Ha, W. Kee, P. Lachassagne, and R. Wyns, 2003. Relationship between the permeability of hard rock aquifers and their weathering, from geological and hydrogeological observations in South Korea. *International Association of Hydrogeologists IAH Conference on "Groundwater in Fractured Rocks,"* September 15–19, 2003, Prague, Czech Republic.
12. Courtois, N., P. Lachassagne, P. Weng, H. Theveniaut, R. Wyns, B. Joseph, and P. Laporte, 2003. Détermination de secteurs potentiellement favorables pour la recherche d'eau souterraine à Cacao (Guyane). *Rapport BRGM/RM-52758-FR,* Guiana, France.
13. Courtois, N., P. Lachassagne, R. Wyns, R. Blanchin, F.D. Bougaïre, S. Some, and A. Tapsoba, 2009. Large-scale mapping of hard-rock aquifer properties applied to Burkina Faso. *Ground Water* 48(2): 269–283.
14. Day-Lewis, F.D., P.A. Hsieh, and S.M. Gorelick, 2000. Identifying fracture-zone geometry using simulated annealing and hydraulic-connection data. *Water Resources Research* 36(7): 1707–1722.
15. Dewandel, B., J.M. Gandolfi, F.K. Zaidi, S. Ahmed, and K. Subrahmanyam, 2007. A decision support tool with variable agro-climatic scenarios for sustainable groundwater management in semi-arid hard-rock areas. *Current Science* 92(8): 1093–1102.
16. Dewandel, B., P. Lachassagne, S. Chandra, and F.K. Zaidi, 2011. Conceptual Hydrodynamic model of a geological discontinuity in hard rock aquifers: Example of quartz reef in granitic terrain in South India. *Journal of Hydrology* 405: 474–487.
17. Dewandel, B., P. Lachassagne, R. Wyns, J.C. Marechal, and N.S. Krishnamurthy, 2006. A generalized 3-D geological and hydrogeological conceptual model of granite aquifers controlled by single or multiphase weathering. *Journal of Hydrology* 330(1–2): 260–284.
18. Dewandel, B., J.-C. Maréchal, O. Bour, B. Ladouche, S. Ahmed, S. Chandra, and H. Pauwels, 2012. Upscaling and regionalizing hydraulic conductivity and efficient porosity at watershed scale in crystalline aquifers. *Journal of Hydrology* 416–417: 83–97.
19. Dewandel, B., J.-C. Maréchal, O. Bour, B. Ladouche, S. Ahmed, S. Chandra, H. Pauwels, J. Perrin, and S. Aulong, 2012. Upscaling and regionalizing hydraulic conductivity and efficient porosity at watershed scale in crystalline aquifers. *Proceedings of the International Conference on Groundwater in Fractured Rocks,* May 21–24, 2012, Prague, Czech Republic, p. 20.
20. Dewandel, B., J. Perrin, S. Ahmed, S. Aulong, Z. Hrkal, P. Lachassagne, M. Samad, and S. Massuel, 2010. Development of a tool for managing groundwater resources in semi-arid hard rock regions. Application to a rural watershed in South India. *Hydrological Processes* 24: 2784–2797.
21. Durand, V., B. Deffontaines, V. Léonardi, R. Guérin, R. Wyns, G. Marsily de, and J.-L. Bonjour, 2006. A multidisciplinary approach to determine the structural geometry of hard-rock aquifers. Application to the Plancoët migmatitic aquifer (NE Brittany, W France). *Bulletin de la Société Géologique de France* 177: 227–237.
22. Farmin, R., 1937. Hypogene exfoliation in rock masses. *The Journal of Geology* 45(6): 625–635.
23. Ferrant, S., Y. Caballero, J. Perrin, B. Dewandel, S. Aulong, S. Ahmed, and J.C. Maréchal (Submitted). The likely impacts of climate change on farmer groundwater extraction from crystalline aquifer of south-India. Submitted to *Global Environmental Change.*
24. Folk, R.L. and E.B. Patton, 1982. Buttressed expansion of granite and development of grus in central Texas. *Zeitschrift fuer Geomorphologie* 26(1): 17–32.
25. Foster, S., 2011. Hard-rock aquifers in tropical regions: Using science to inform development and management policy. *Hydrogeology Journal* 20: 659–672.
26. Guillou-Frottier, L., E. Burov, P. Nehlig, and R. Wyns, 2007. Deciphering plume-lithosphere interactions beneath Europe from topographic signatures. *Global and Planetary Change* 58(1–4): 119.
27. Gustafson, G., B. Gylling, and J.-O. Selroos, 2009. The Aspö Task Force on groundwater flow and transport of solutes: Bridging the gap between site characterization and performance assessment for radioactive waste disposal in fractured rocks. *Hydrogeology Journal* 17: 1031–1033.

28. Hill, S.M., 1996. The differential weathering of granitic rocks in Victoria, Australia. *Journal of Australian Geology* 16: 271–276.
29. Hill, S.M., C.D. Ollier, and E.B. Joyce, 1995. Mesozoic deep weathering and erosion: An example from Wilson's Promontory, Australia. *Zeitschrift für Geomorphologie* NF 39: 331–339.
30. Holzhausen, G.R., 1989. Origin of sheet structure, 1. Morphology and boundary conditions. *Engineering Geology* 27(1–4): 225–278.
31. Houston, J.F.T. and R.T. Lewis, 1988. The Victoria Province drought relief project, II. Borehole yield relationships. *Ground Water* 26(4): 418–426.
32. Hsieh, P.A. and A.M. Shapiro, 1996. Hydraulic characteristics of fractured bedrock underlying the FSE well field at the Mirror Lake Site, Grafton County, New Hampshire. Technical meeting, Colorado Springs, Colorado, U.S. Geological Survey Water Resources Investigations Report 94-4015.
33. Jahns, R.H., 1943. Sheet structure in granites, its origin and use as a measure of glacial erosion in New England. *Journal of Geology* 51(2): 71–98.
34. Key, R.M., 1992. An introduction to the crystalline basement of Africa. In *Hydrogeology of Crystalline Basement Aquifers in Africa*. E.P. Wright and W.G. Burgess (eds.), Geological Society Special Publication, London, U.K. Vol. 66, pp. 29–57.
35. Lachassagne, P., S. Ahmed, C. Golaz, J.-C. Maréchal, D. Thiery, F. Touchard, and R. Wyns, 2001. A methodology for the mathematical modelling of hard-rock aquifers at catchment scale based on the geological structure and the hydrogeological functioning of the aquifer. *XXXI IAH Congress "New Approaches Characterizing Groundwater Flow." "Hard Rock Hydrogeology" Session*, Munich, Germany, AIHS.
36. Lachassagne, P., F. Lacquement, and C. Lamotte, 2008. Assistance hydrogéologique dans le cadre des études préliminaires afférentes aux fonçages des tunnels de Violay, Bussière et Chalosset, sur le tracé de l'A89, section Balbigny-La Tour de Salvagny. Rapport de phase 3. R. BRGM/RC-56085-FR. Montpellier, France, BRGM.
37. Lachassagne, P. and J.-L. Pinault, 2001. Radon-222 emanometry: A relevant methodology for water well siting in hard rock aquifer. *Water Resources Research* 37(12): 3131–3148.
38. Lachassagne, P., R. Wyns, P. Bérard, T. Bruel, L. Chéry, T. Coutand, J.-F. Desprats, and P. Le Strat, 2001. Exploitation of high-yield in hard-rock aquifers: Downscaling methodology combining GIS and multicriteria analysis to delineate field prospecting zones. *Ground Water* 39(4): 568–581.
39. Lachassagne, P., R. Wyns, and B. Dewandel, 2011. The fracture permeability of hard rock aquifers is due neither to tectonics, nor to unloading, but to weathering processes. *Terra Nova* 23: 145–161.
40. Le Borgne, T., O. Bour, J.-R. de Dreuzy, P. Davy, and F. Touchard, 2004. Equivalent mean flow models for fractured aquifers: Insights from a pumping tests scaling interpretation. *Water Resources Research* 40: 12.
41. Le Borgne, T., O. Bour, F.L. Paillet, and J.-P. Caudal, 2006. Assessment of preferential flow path connectivity and hydraulic properties at single-borehole and cross-borehole scale in a fractured aquifer. *Journal of Hydrology* 328: 347–359.
42. Lelong, F. and J. Lemoine, 1968. Les nappes phréatiques des arènes et des altérations argileuses; leur importance en zone intertropicale; les difficultés de leur exploitation. *Hydrogéologie Bulletin du Bureau de Recherches Géologiques et Minières. Deuxième série. Section III*(2): 41–52.
43. Lenck, P.-P., 1977. Données nouvelles sur l'hydrogéologie des régions à substratum métamorphique ou éruptif. Enseignements tirés de la réalisation de 900 forages en Côte d'Ivoire. *Comptes Rendus de l'Academie des Sciences Paris* 285: 497–500.
44. Mabee, S.B., P.J. Curry, and K.C. Hardcastle, 2002. Correlation of lineaments to ground water inflows in a bedrock tunnel. *Ground Water* 40(1): 37–43.
45. Mac Farlane, M.J., 1992. Groundwater movement and water chemistry associated with weathering profiles of the African surface in Malawi. In *Hydrogeology of Crystalline Basement Aquifers in Africa*. E.P. Wright and W.G. Burgess (eds.), Geological Society Special Publication, London, U.K. Vol. 66, pp. 101–129.

46. Mandl, G., 2005. *Rock Joints. The Mechanical Genesis*. Berlin, Germany, Springer.
47. Maréchal, J.-C. and S. Ahmed, 2003. Dark zones are Human-made. *Down to Earth* 12(4): 54.
48. Maréchal, J.-C., B. Dewandel, and K. Subrahmanyam, 2004. Use of hydraulic tests at different scales to characterize fracture network properties in the weathered-fractured layer of a hard rock aquifer. *Water Resources Research* 40(11): 17.
49. Marechal, J.C., B. Dewandel, S. Ahmed, L. Galeazzi, and F.K. Zaidi, 2006. Combined estimation of specific yield and natural recharge in a semi-arid groundwater basin with irrigated agriculture. *Journal of Hydrology* 329(1–2): 281–293.
50. Michel, J.P. and R.W. Fairbridge, 1992. *Dictionary of Earth Sciences/Dictionnaire des Sciences de la Terre*. Anglais-Français/Français-Anglais. Wiley, New York.
51. Neves, M. and N. Morales, 2007. Well productivity controlling factors in crystalline terrains of southeastern Brazil. *Hydrogeology Journal* 15(3): 471.
52. Oliva, P., J. Viers, and B. Dupre, 2003. Chemical weathering in granitic environments. *Chemical Geology* 202(3–4): 225–256.
53. Ollier, C.D., 1988. The regolith in Australia. *Earth-Sciences Reviews* 25: 355–361.
54. Owen, R., A. Maziti, and T. Dahlin, 2007. The relationship between regional stress field, fracture orientation and depth of weathering and implications for groundwater prospecting in crystalline rocks. *Hydrogeology Journal* 15(7): 1231.
55. Perrin, J., C. Mascré, H. Pauwels, and S. Ahmed, 2011. Solute recycling: An emerging threat to groundwater quality in southern India. *Journal of Hydrology* 398(1–2): 144–154.
56. Perrin, J., S. Ferrant, S. Massuel, B. Dewandel, J.C. Maréchal, S. Aulong, and S. Ahmed, 2012. Assessing water availability in a semi-arid hard-rock regions watershed of southern India using a semi-distributed model. *Journal of Hydrology* 460–461: 143–155.
57. Plote, H., 1968. Les recherches d'eau souterraine dans les régions arides à substratum cristallin et métamorphique de l'Afrique occidentale. *Hydrogéologie. Bulletin du Bureau de Recherches Géologiques et Minières. Deuxième série*. Section III(3): 97–111.
58. Plote, H., 1985. Sondage de reconnaissance hydrogéologique. Méthode du marteau fond-de-trou. Exécution et maintenance. *Manuels et Méthodes* 12. Orléans, France, BRGM.
59. Pollard, D.D. and A. Aydin, 1988. Progress in understanding jointing over the past one hundred years. *Geological Society of America Bulletin* 100: 1181–1204.
60. Razack, M. and T. Lasm, 2006. Geostatistical estimation of the transmissivity in a highly fractured metamorphic and crystalline aquifer (Man-Danane Region, Western Ivory Coast). *Journal of Hydrology* 325(1–4): 164.
61. Rhen, I., H. Thunehed, C.-A. Triumf, S. Follin, L. Hartley, J. Hermansson, and C.-H. Wahlgren, 2007. Development of a hydrogeological model description of intrusive rock at different investigation scales: An example from south-eastern Sweden. *Hydrogeology Journal* 15(1): 47.
62. Rocher, P., A. Mishellany, R. Wyns, F. Lacquement, N. Gobron, G. Greffier, and Y. Germain, 2003. Identification et caractérisation des ressources en matériaux de substitution aux granulats alluvionnaires dans le département du Puy-de-Dôme: La zone des Combrailles. Auvergne, France, R. BRGM/RP52706-FR, BRGM: 97.
63. Roques, C., L. Aquilina, O. Bour et al., 2012. Hydrogeological and geochemical characterization of a deep hard-rock aquifer (Saint-Brice, French Britany). *Proceedings of the International Conference on Groundwater in Fractured Rocks*, Prague, Czech Republic, 21–24 May 2012, p. 39.
64. Sander, P., 1997. Water-well siting in hard-rock areas: Identifying promising targets using a probabilistic approach. *Hydrogeology Journal* 5(3): 32.
65. Shaw, R., 1997. Variations in sub-tropical deep weathering profiles over the Kowloon Granite, Hong Kong. *Journal of the Geological Society of London* 154(6): 1077–1085.
66. Tardy, Y. and C. Roquin, 1998. *Dérive des Continents, Paléoclimats et Altérations Tropicales*. Orléans, France, Editions BRGM.

67. Taylor, G. and R.A. Eggleton, 2001. *Regolith Geology and Geomorphology*. Chichester, U.K., John Wiley & Sons, Ltd.
68. Taylor, R. and K. Howard, 2000. A tectono-geomorphic model of the hydrogeology of deeply weathered crystalline rock: Evidence from Uganda. *Hydrogeology Journal* 8(3): 279–294.
69. Taylor, R.G. and K.W.F. Howard, 1999. Lithological evidence for the evolution of weathered mantels in Uganda by tectonically controlled cycles of deep weathering and stripping. *Catena* 35(1): 65–94.
70. Theveniaut, H. and P. Freyssinet, 1999. Paleomagnetism applied to lateritic profiles to assess saprolite and duricrust formation processes: The example of Mont Baduel profile (French Guiana). *Palaeogeography, Palaeoclimatology, Palaeoecology* 148(4): 209.
71. Theveniaut, H. and P. Freyssinet, 2002. Timing of lateritization on the Guiana Shield: Synthesis of paleomagnetic results from French Guiana and Suriname. *Palaeogeography, Palaeoclimatology, Palaeoecology* 178(1–2): 91.
72. Thiry, M., R. Simon-Coinçon, F. Quesnel, and R. Wyns, 2005. Altération bauxitique associée aux argiles à chailles sur la bordure sud-est du bassin de Paris. *Bulletin de la Société Géologique de France* 176(2): 199–214.
73. Tieh, T.T., E.B. Ledger, and M.W. Rowe, 1980. Release of uranium from granitic rocks during in situ weathering and initial erosion (central Texas). *Chemical Geology* 29(1–4): 227.
74. Twidale, C.R., 1982. *Granite Landforms*. Amsterdam, the Netherlands (NLD), Elsevier Sci. Publ. Co.
75. Uhl, V.W. and G.K. Sharma, 1978. Results of pumping tests in crystalline-rock aquifers. *Ground Water* 16(3): 192–203.
76. Vouillamoz, J.-M., A. Legtchenko, G. Toe, and A. Legtchenko, 2005. Characterization of crystalline basement aquifers with MRS: Comparison with boreholes and pumping test data in Burkina Faso. *Near Surface Geophysics* (3): 205–213.
77. Wright, E.P., 1992. The hydrogeology of crystalline basement aquifers in Africa. In *Hydrogeology of Crystalline Basement Aquifers in Africa*. E.P. Wright and W.G. Burgess (eds.), Geological Society Special Publication, London, U.K. Vol. 66, pp. 1–27.
78. Wright, E.P. and W.G. Burgess (eds.), 1992. *The Hydrogeology of Crystalline Basement Aquifers in Africa*. Geological Society Special Publication No. 66, London, U.K.
79. Wyns, R., 1991. Evolution tectonique du bâti armoricain oriental au Cénozoïque d'après l'analyse des paléosurfaces continentales et des déformations géologiques associées (in French: Structural evolution of the Armorican basement during the Cenozoic deduced from analysis of continental paleosurfaces and associated deposits). *Géologie de la France* (3): 11–42.
80. Wyns, R., 1997. Essai de quantification de la composante verticale de la déformation finie cénozoïque d'après l'analyse des paléosurfaces et des sédiments associés. *Proceedings des Journées Scientifiques CNRS-ANDRA. ANDRA,* Poitiers, France, pp. 36–38.
81. Wyns, R., 2002. Climat, eustatisme, tectonique: Quels contrôles pour l'altération continentale? Exemple des séquences d'altération cénozoïques en France. *Bull. Inf. Géol. Bass. Paris,* 39(2): 5–16.
82. Wyns, R., J.M. Baltassat, P. Lachassagne, A. Legtchenko, and J. Vairon, 2004. Application of proton magnetic resonance soundings to groundwater reserves mapping in weathered basement rocks (Brittany, France). *Bulletin de la Société Géologique de France* 175(1): 21–34.
83. Wyns, R., J.P. Clément, H. Lardeux, M. Gruet, G. Moguedet, R. Biagi, and M. Ballèvre, 1998. *Carte Géologique de la France* (1/50.000), feuille Chemillé (483). Orléans, France.
84. Wyns, R., J.-C. Gourry, J.-M. Baltassat, and F. Lebert, 1999. Caractérisation multiparamètres des horizons de subsurface (0–100 m) en contexte de socle altéré. *PANGEA* 31/32: 51–54.
85. Wyns, R., F. Lacquement, P. Corbier, and J. Vairon, 2002. Cartographie de la réserve en eau souterraine du massif granitique de la Roche sur Yon. Orléans, France, *Rapport BRGM*/RP-51633/FR: 26.

86. Wyns, R., F. Quesnel, F. Lacquement, B. Bourgine, F. Mathieu, F. Lebert, J.M. Baltassat, A. Bitri, and D. Mathon, 2005. Cartographie quantitative des propriétés du sol et du sous-sol dans la région des Pays de la Loire. *R. BRGM*/RP-53676-FR, BRGM: 135.
87. Wyns, R., F. Quesnel, R. Simon-Coinçon, F. Guillocheau, and F. Lacquement, 2003. *Géologie de la France* (1): 79–87.
88. Zaidi, F., S. Ahmed, B. Dewandel, and J.-C. Maréchal, 2007. Optimizing a piezometric network in the estimation of the groundwater budget: A case study from a crystalline-rock watershed in southern India. *Hydrogeology Journal* 15(6): 1131.

15

Hydrograph Analysis and Baseflow Separation

Hafzullah Aksoy
Istanbul Technical University

Hartmut Wittenberg
Leuphana Universität Lüneburg

Ebru Eris
Ege University

AUTHORS

Hafzullah Aksoy studied civil engineering with hydraulics and water resources as his field of expertise at Istanbul Technical University. He is a professor of hydrology and water resources at the same university. His research topics range from statistical hydrology to process-based modeling techniques for rainfall–runoff–sediment transport. He has been awarded with the 2006 Best Paper and 2009 Best Discussion Awards of the *ASCE Journal of Hydrologic Engineering* and the 2011 Engineering Incentive Award of the Scientific and Technical Council of Turkey (TUBITAK). He currently serves as the president of the International Commission on Surface Water (ICSW), International Association of Hydrological Sciences (IAHS).

Hartmut Wittenberg studied and practiced civil and hydraulic engineering. He is emeritus professor of hydrology and water resources management at Leuphana University of Luneburg, Germany. His research in hydrology is documented in 80 scientific papers, many of them on the interaction of ground and surface waters. He also serves as a consultant in international water projects in Europe, Asia, Africa, and Latin America.

Ebru Eris received her bachelor's and master's degrees from Ege University and her PhD from Istanbul Technical University. She is currently working as a research assistant on hydrology, hydraulics, and water resources. She is the vice-president of the International Commission on Statistical Hydrology (ICSH), IAHS.

PREFACE

The shape of a flow hydrograph caused by a major rainfall event and its rise and descent related to time reveals the flow-generating characteristics of a catchment. By analysis of flood events, the typical response of the system to a unit of rainfall in a time interval can be derived that allows computing synthetic design hydrographs for design rainfalls. Typically, direct or quick flow is separated from the slow component for these procedures.

However, hydrograph analysis and separation of flow components is not limited to flood hydrology. The various interactions between ground- and surface waters are strongly dominating the hydrological processes, water balances, movements of water, and transport of dissolved substances in watersheds and landscapes. A high portion of the discharge in most rivers is outflow from the adjacent shallow aquifers exfiltrating through the banks and the bottom not only in rainless periods but also during rainstorms when percolating water quickly raises the levels of groundwater reservoirs. After recharge, the hydrograph recession reflects the storage–outflow relationship of the aquifer. Based on recession analysis, groundwater-fed baseflow can be separated from time series of total flow observed at gauging stations of rivers. Respective computations allow deriving time series of groundwater storage and groundwater recharge in the watershed.

This chapter aims to introduce the basics of hydrograph analysis and baseflow separation. Different approaches are described, accompanied by practical examples.

15.1 Introduction

While for such engineering design problems as dykes, river training, bridges, and stormwater drainage systems peak discharge rates are sufficient, they are inadequate for the design of hydraulic structures where storage is involved. A spillway of a dam, for example, is hit not by the natural hydrograph with its peak discharge of the catchment but by a design discharge significantly attenuated by the retention effect of the reservoir: flood volume counts, and therefore, the design work is usually based on the hydrograph rather than the peak discharge rate.

The streamflow hydrograph is the graphical representation of the variation of discharge through the cross section of a water course over time. It is the superposition of responses of the contributing watershed to precipitation and evapotranspiration according to the properties such as area, topography, morphology, geology, storage capacity, soil moisture, vegetation, and land use.

A typical hydrograph has three parts (Figure 15.1). The rising of the hydrograph from B to C (ascension curve, rising limb, or concentration curve) is due to rainfall generating mainly direct runoff (quick flow). Peak time (B to C in Figure 15.1) shows the time when the maximum flow resulting from the input occurs. Depending on the concentration time of the watershed and the rainfall duration, this can happen as a crest segment [55]. After peaking, direct flow decreases along the recession curve also called depletion curve from C to E corresponding to the discharge of water from the basin after the rainfall ceases. A recession curve can be separated into two parts: the falling limb from C to D and the baseflow recession from D to E in Figure 15.1. The shape of the recession curve does not depend on the input, while that of ascension curve does. When an intermittent stream is concerned, zero-flow portions from H to I arise too.

A flow or flood hydrograph that is caused by a single heavy rainfall event can be represented by three functions: input function, transfer function, and output function. Normally, a rainfall event (hyetograph) is the input, while the total runoff discharge hydrograph is considered as the output function. The watershed itself represents the transfer function, converting parts of the rainfall hyetograph into direct runoff, while the rest is "lost," caught by vegetation (interception), as soil moisture and infiltration, and leaves the catchment later by evapotranspiration or as groundwater outflow (baseflow).

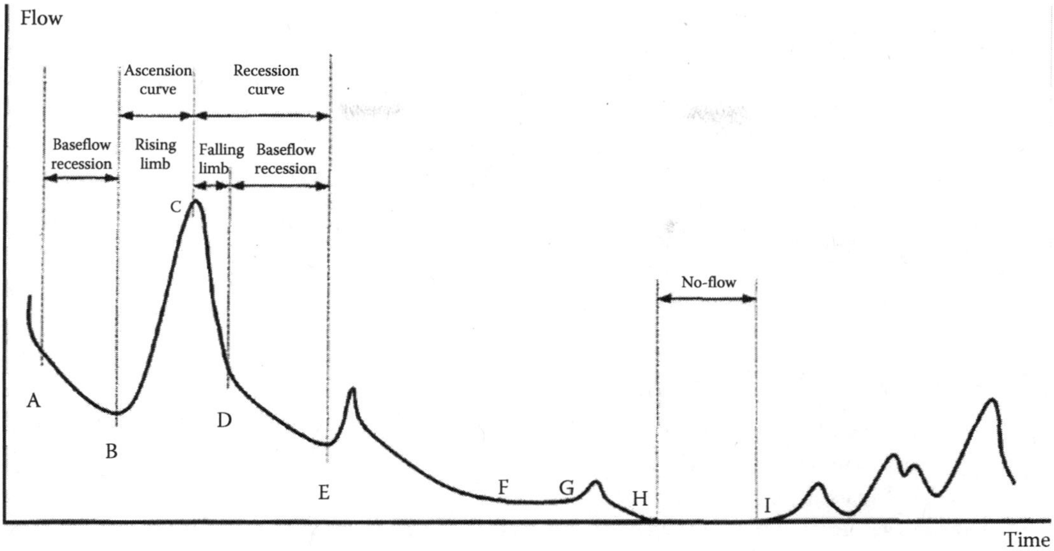

FIGURE 15.1 Schematic representation of a daily flow series and components of a typical hydrograph.

This event-based philosophy is mainly applied in flood analysis and the simulation of design flood hydrographs. A classical and basic transfer or response function is the unit hydrograph method [57]. The unit hydrograph of a catchment is its typical reaction on 1 mm (one unit) of effective rainfall occurring in a defined time step. Once the unit hydrograph is derived, a flood hydrograph caused by different rainfall heights in consecutive time intervals (design hyetograph) is computed by convolution (multiplication and superposition).

The unit hydrograph as the transfer function or response function of a watershed can be determined from measured rainfall and runoff during a flood event. Before processing the observed data, the slow-flow "baseflow" component is separated from the total hydrograph. Likewise, the "losses," that is, the part of rainfall that does not become direct flow, must be separated from the hyetograph so that volume continuity is achieved. The method is only applicable for smaller- or middle-sized watersheds where a fairly equal or typical rainfall distribution during heavy rainfall events can be expected. For larger catchments or with different spatial rainfall patterns, distributed hydrological models are needed.

Runoff is generated either by rainfall or by snowmelt. Glaciers can also contribute to the surface runoff of watersheds. However, it should be kept in mind that not all rainfall contributes runoff due to the absorption rate of the soil changing with water demand of soil storage. Surface flow interacts with atmosphere and subsurface water systems. At the initial stage of precipitation, a large portion of it is devoted to the surface storage to fill the retention and detention capacities of the soil. Retention is depleted by evaporation in a long period of time, while detention has a short period of time to be swept out by surface runoff or infiltration. Subsurface flow takes place beneath the soil surface. It originates from infiltrated precipitation that is diverted from its vertical movement by impermeable layers in the unsaturated zone (interflow). Groundwater is located in the saturated zone under the surface within spaces in gravel, sand, and soil pores and in fractures and fissures of rock formations. Groundwater can be mobile or immobile depending on the soil formation and the gradient. Flow velocity is proportional to the gradient and the permeability of the soil formation (Darcy's law). Under low-permeability conditions, groundwater flow becomes slow or stays immobile within the soil formation.

It was indicated previously that the flow can be separated into two parts, direct runoff born from surface flow and shallow subsurface drainage and baseflow consisting mainly of exfiltrating groundwater. Groundwater-fed baseflow is the largest component of the discharge of most rivers. Groundwater recharge by precipitation is thus an important process. During the initial stage of a rainfall event, the surface layers of the soil are moisturized and little depressions filled. Soon infiltration starts and high portions of rain find their way to the saturated zone, the upper aquifer. Besides the permeability of the unsaturated (vadose) zone, macropores play an important role in vertical water transport and fast groundwater recharge. Layers of lower permeability in the vadose zone may divert the vertical infiltrating water and cause subsurface flow, interflow. When the rate of rainfall exceeds the infiltration rate, infiltration excess overland flow occurs, which is also called Hortonian overland flow or saturated overland flow. Such flows are usual in arid and semiarid regions, where rainfall intensities are high and soil infiltration capacity is low due to soil surface characteristics. In the occurrence of this type of flow, the antecedent soil moisture plays an important role whether or not saturation excess overland flow is observed.

15.2 Recession Curve

Many shortcomings of recession analysis and the large number of methods that exist are due to the high variation in recession behavior, both within and between catchments [74]. Baseflow literature reveals that many problems still remain in recession analysis. However, modern computers have allowed the development of automatic and more objective methods, which eliminate some of the subjective elements of recession analysis [74].

The recession curve can be expressed analytically. However, there is no unique expression due to the high variability encountered in the recession behavior of individual segments. Even when the expression is set, there is no unique parameter set to use; hence, calibration becomes another issue to deal with.

During dry weather, water stored in the catchment is removed by soil and groundwater drainage and by evapotranspiration. The gradual depletion of discharge during periods with little or no precipitation constitutes the drainage or recession rate, graphically presented as the recession curve. The recession curve tells, in a general way, about the natural storages feeding the stream. The recession curve of the streamflow hydrograph contains valuable information related to the storage and aquifer characteristics of a hydrological basin. Information on low-flow characteristics is required for such water resource management issues as water supply, irrigation, hydropower, and water quality and quantity estimation. The understanding of the flow process from groundwater or other delayed sources is also essential in studies of water budgets and catchment response [74].

The recession curve analysis techniques are basic flow equations, linear reservoir models, autoregressive processes, and empirical relations [74]. Also basic flow equations were used to define the recession curve of the hydrograph [32,60]. The recession curve is widely characterized by an exponential equation, which corresponds to the linear reservoir concept in which the discharge is directly proportional to the storage in the reservoir. The reservoir here points the groundwater system. Among others, studies in [34,35] using the linear reservoir model in modeling daily intermittent hydrological processes should be mentioned. The Tank model [45,66,67] is another example of this approach.

Hall [26] and Tallaksen [74] report that the mathematical background of this kind of analysis goes back to the work by Barnes [11] who found that flow is scattered as a straight line on a semilogarithmic paper, even to that by Maillet [42], and originates from the Boussinesq nonlinear differential equation given for aquifers [30,68,70]. Based upon this analysis, the well-known exponential recession equation is obtained as

$$Q_t = Q_0 e^{-\alpha t} \tag{15.1}$$

where

α is the recession coefficient

t is the time

Q_0 is the start or initial value

Q_t flow t days after the peak

During the period when the hydrograph recesses, Q_t can also be expressed as the product of a value K, smaller than one, and flow on the previous day, Q_{t-1}, as

$$Q_t = KQ_{t-1} \tag{15.2}$$

Due to inflows originating from various storage systems, that is, flow components, it should be kept in mind that K is not constant throughout the recession curve but is smaller at the beginning than at the end part [7], which mainly consists of baseflow. The K values vary seasonally too [73]. They are therefore of random structure and have their own probability distribution functions. As an alternative to K, the half-life ($t_{0.5}$), the time required for streamflow to fall down its half value, was introduced [17,18]. The half-life has the advantage of having the dimension of time and hence a physical meaning.

Equation 15.2 can be written as

$$Q_t = KQ_{t-1} + \varepsilon_t \tag{15.3}$$

where ε is a normally distributed independent random variable with zero mean and constant variance. In that case, the recession part of the hydrograph is considered a first-order autoregressive model [76]. Relationships between the recession curve parameters and the ARMA model parameters were developed as another example for this approach [63].

Empirical relations are among the most commonly used approaches. For example, in [15,52], the exponential recession function in the form of

$$Q_t = \left(Q_0 - Q_b\right)e^{-\alpha t} + Q_b \tag{15.4}$$

was used for data from Poland and Denmark, respectively. The same technique was used in [36]. Here, Q_b is the minimum flow in the river that changes from year to year [88]. In [37], Equation 15.1 was used in the double exponential form of

$$Q_t = Q_0 \exp\left(-mt^n\right) \tag{15.5}$$

in which m and n are constants. An approach was given in [53], splitting the complete recession curve into two stages, the upper and lower recessions, to take, respectively, the forms of

$$Q_t = Q_0 t^{-a} \tag{15.6}$$

$$Q_t = Q_0^* e^{b+ct^*} \tag{15.7}$$

where

- t is the time measured from the start of the upper recession, the peak value, Q_0, of the total hydrograph
- t^* is the time from the start of the lower recession
- Q_0^* is the initial flow discharge at the beginning of the lower portion of the curve
- a, b, and c are parameters

An experimental/arbitrary procedure was adopted. Surface runoff is assumed to occur and Equation 15.6 is therefore used as long as $Q_t/Q_{t-1} \leq 0.9$, while the lower recession curve (Equation 15.7) is used in case that the ratio takes values greater than the river basin-specific value, 0.9 in [53]. A similar approach splitting the recession curve was given in [1–4]. The observed monthly mean streamflow discharge was used in splitting the curve into the upper and lower parts. Both parts were calculated by using Equation 15.1 with different parameters determined at monthly time resolution from the observed recession curves of the daily streamflow hydrographs. Such an algorithm increases the number of parameters dramatically. Using the exponential equation to define the point on the recession where the baseflow starts was another attempt [49]. By this approach, a series of recession constants are calculated by regressing flow on time after the flow nearest to the peak value of the hydrograph is successively deleted. Progressively higher values of recession constants are obtained as the number of time units deleted increases, and the constant stabilizes thereafter. The day on which the constant stabilizes is taken as the beginning of the baseflow component. A review on the exponential decay function was supplied [5].

To computerize the recession curve analysis, a number of softwares were developed. For instance, DIFGA [56] and RCA [65] both use the exponential recession curve equation. In the United States, a computer program, HYSEP, was developed [61] for the purpose of baseflow separation in which three methods of hydrograph separation that are referred to in the literature as the fixed-interval, sliding-interval, and local minimum methods were employed. BFI, another computer program implementing the deterministic procedure of the Institute of Hydrology [31], was released at http://www.usbr.gov/pmts/hydraulics_lab/*twahl/bfi*. Using digital filter-based separation modules, a Web-based Hydrograph Analysis Tool (WHAT) system was developed [40]. The user-friendly software, AdUKIH [8], is the most recent. Instead of the linear reservoir theory, the nonlinear reservoir theory can also be applied. The nonlinear concept is detailed later.

15.3 Master Recession Curve

Given a number of hydrographs, they can be combined to give a master recession curve. A master recession curve can be obtained by simply averaging many recession curves. Different techniques are used to obtain the master recession curve of the basin.

The correlation method is based on the plotted logarithms of flow values [log Q_t] against [log Q_{t+N}] some fixed time N later. The slope of the straight line indicated by the data corresponds to the recession constant of the basin.

The matching strip method is based on the simple exponential model. The plot of logarithms of the flows against time result is a straight line. The slope of that line gives the recession constant. Observed individual recession curves are plotted on the same graph, and the recession is then superimposed and adjusted horizontally. A mean line through the set of individual lines represents the master recession curve [48]. A composite recession curve can be similarly developed by piecing together sections of individual recessions or a modified method can be proposed for development of the master recession curve of the basin [17,25].

Data used for the development of a master recession curve should be carefully selected; otherwise, the points will scatter widely and will be difficult to construct a master recession curve [41]. Since the development of master recession curve of the basin is of significant subjectivity, it should be careful in the development of the curve, although it was concluded that the correlation method was less subjective than the matching strip method [48].

The graphical methods of matching strip, correlation, and tabulation, which are all arbitrary methods, are traditionally used in constructing the master recession curve. In the nowadays rarely used tabulating method [33], the recession periods are tabulated and shifted until the discharges agree approximately, and mean discharge is calculated for each time step in the period. In the matching spring method [75], individual recession segments are plotted and adjusted horizontally until they overlap in the main parts. The master recession curve is then constructed as the mean line by eye fitting. In the correlation method [39], the discharge at one time (Q_t) is plotted against the discharge one time interval later (Q_{t+dt}) in the recession period, and a curve is fitted. The time step is determined in such a way that the fitted curve is exponential, for example, a straight line in a semilogarithmic paper (Figure 15.2).

Recession characteristics can be found either from the master recession curve or from the individual recession segments separately. When the latter case is used, a mean value can be considered representative for the catchment.

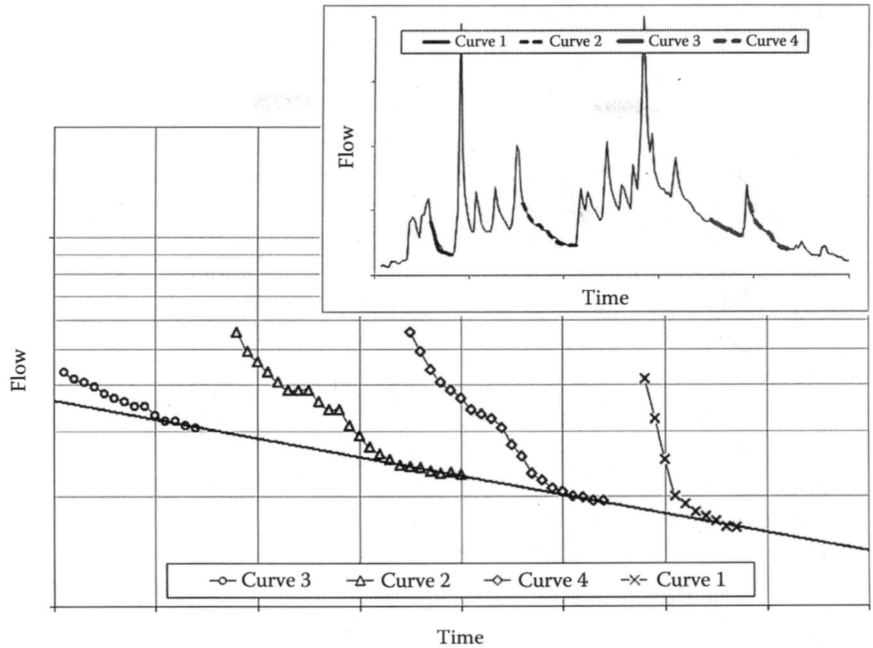

FIGURE 15.2 Individual flow recessions and construction of a master recession (depletion) curve from daily streamflow data.

15.4 Baseflow Separation

An engineering use of the recession curve allows users to separate the flow into its three main components, surface flow, interflow (subsurface flow), and baseflow. Of these, the baseflow is that component of the flow in a river that is not the direct consequence of the rainfall event but is considered as the outflow of the groundwater reservoir feeding the river entirely during rainless periods [21]. It is defined as the net flow from the groundwater storage to the river under the effect of diverse geological, climatologic, and morphological factors, resulting in considerable variations both in time and space [58]. Land use and land cover characteristics such as forestation in the hydrological basin affect the groundwater storage and hence the baseflow [85]. A better understanding of baseflow can help in controlling irrigation withdrawals during low-flow periods, making water supply estimates and forecasts, and determining storage requirements for maintenance of adequate flow for waste dilution [59]. Especially, however, separated baseflow allows the quantification of groundwater recharge in the river basin.

In the planning and management studies of water resources, total flow is separated into baseflow and direct flow where direct flow is considered the sum of surface flow and interflow. As the baseflow and direct flow hydrographs are not measured separately but instead only the total flow is measured, the following question arises: How can the measured hydrograph be separated? In practice, graphical separation techniques to be detailed later are commonly employed for this purpose.

Analytical methods are available as well [30,68,70]. With the help of these methods, subjective points associated with baseflow separation can be reduced by using the analytical solution of the Boussinesq equation.

A certain kind of filter disintegrating the daily streamflow into quick flow and baseflow components was made available by many researchers. A well-known one is the UK smoothed minima approach [31]. This is a widely used method [43,77,78] recently revised [51] and adopted to intermittent streams [8,50]. Another approach is the so-called recursive digital filter algorithm used, among others, in [10,13,19,29,48,62,69,71,72]. The recursive digital filter technique was coupled to the smoothed minima approach [9,38]. Physically based filters were made available [22,23,46] for separating baseflow from the total streamflow. Spectral analysis of baseflow separation [16,64] should also be mentioned.

Baseflow Index (BFI) is calculated as the ratio of baseflow volume to total streamflow volume. Baseflow usually occurs in times of sustained and regionally extensive periods of low precipitation and snowmelt, when streamflow is fed from groundwater. A number of studies highlighted the importance of BFI and automatical calculation procedures have been developed. A procedure for calculating the BFI is described in detail in [25]. The BFI is originally related to the geology of the catchment area and is applied for soil classification and geological regionalization.

Having in mind the definitions mentioned earlier, the BFI over some time period $dt = t_2 - t_1$ can be presented as the ratio of baseflow volume to the total streamflow volume:

$$BFI = \frac{\int_{t_1}^{t_2} Q_{baseflow}(t)\,dt}{\int_{t_1}^{t_2} Q_{total}(t)\,dt} \tag{15.8}$$

where dt could be n consequent days.

15.5 Baseflow Separation Techniques

In applied hydrology, the term "baseflow" has two different meanings that are often confused. By neglecting these differences, unsuited approaches might be applied with unrealistic results. The first and more classical definition of baseflow comes from flood hydrology and aims simply to separate the slow component "baseflow" from the quick "direct flow." This concept of quick flow eases the determination of flood hydrographs for the design of hydraulic structures, for example, spillways of dams. It does not describe physical processes. The second concept aims to separate groundwater flow from total flow. This outflow from the shallow aquifers of the catchments is the main flow component and makes up to more than 80% in the average rivers of the world. Its proper separation enables a reliable determination of groundwater recharge.

Separation of flow into two components, direct flow and baseflow, is sufficient in most of the engineering practices. In this context, direct flow is understood as the total of surface flow and subsurface flow (interflow) reaching the stream with little delay, whereas baseflow is groundwater outflow together with the delayed subsurface flow. For this aim, graphical baseflow separation techniques are commonly used to be detailed as follows.

15.5.1 Graphical Techniques

Graphical separation techniques in Figure 15.3, adapted from [14,44], are used extensively only to separate quick flow from slow flow. These methods detailed later are not apt for the separation of groundwater baseflow and they do not consider flow recession. Instead, these approaches were developed for flood hydrology and do not separate the entire groundwater baseflow but only slow flow from quick flow. They are called the straight-line constant-discharge method (line 1), constant-slope method (line 2), the fixed-base/concave method (line 3), and the variable slope (line 4) (Figure 15.3). Aiming at separating quick flow from the slow flow for the purposes of flood analysis and flood forecasting, these approaches are all based on empirical background yet seem reasonable for using in the unit hydrograph computation to route the portion of precipitation that goes through the quick flow reservoir to the stream. In reality, however, groundwater outflow is considerably higher than slow flow. In many cases, baseflow is the dominant discharge component even during flood events [28]. In this sense, Figure 15.3 appears to be problematic as baseflow is very low and does not follow the recession. The nonlinear baseflow separation technique (line 5) in Figure 15.3 aims to consider the groundwater-fed baseflow and therefore results in much more baseflow than aforementioned graphical baseflow separation techniques.

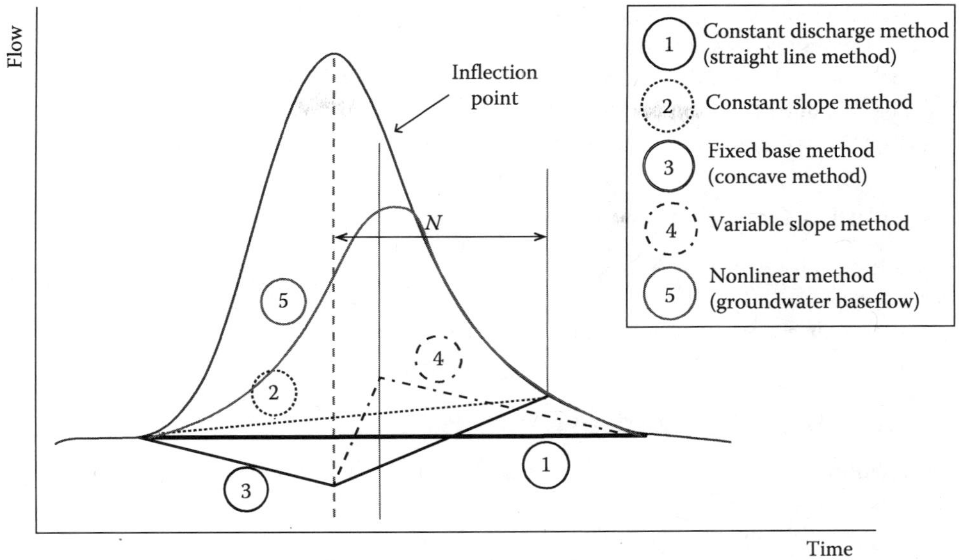

FIGURE 15.3 Graphical methods (1–4) for the separation of slow-flow "baseflow" for flood analysis and synthesis. (Adapted from Chow et al., *Applied Hydrology,* 1988.) The groundwater baseflow hydrograph ⑤ could be as high as to follow the recession.

a. *Constant-Discharge/Straight-Line Method (line 1 in Figure 15.3)*
   In the straight-line method, a horizontal line is drawn from the beginning of the ascension curve of the hydrograph to the intersection with the recession limb. This is applicable to ephemeral streams.
   This is the easiest method commonly used for practical purposes. It is simply the horizontal line A–D in Figure 15.3 starting with the lowest point of the hydrograph before the ascension curve starts. The line is extended until it intersects the recession curve of the hydrograph. Flow under that line is considered baseflow, while flow above it is direct flow. This method is applicable to ephemeral streams. It can easily be argued that the baseflow will continue to decrease beyond the start of the flood runoff, possibly to the time when the flood runoff is maximum. It could also be argued that flow from groundwater storage begins before the time when the total runoff equals the baseflow rate prior to the start of flood runoff. Such arguments have led to proposals of other methods.
b. *Constant-Slope Method (line 2 in Figure 15.3)*
   In the fixed-base method (line 2 in Figure 15.3), the surface runoff can be assumed to end a fixed time *N* days after the hydrograph peak. A line is obtained by projecting the baseflow before the beginning of surface runoff ahead to the time of the peak. This line is connected to the point on the recession limb at the time *N* days after the peak.
c. *Fixed-Base/Concave Method (line 3 in Figure 15.3)*
   For the concave method, the starting and ending points for the line separating baseflow and direct runoff are the same as for the constant-slope method. However, for the concave method, baseflow continues to decrease until the time of the peak discharge of the storm hydrograph. At that time, the separation line is straight between that point and the inflection point on the recession. This is shown in Figure 15.3. While the concave method may require a little more effort to define than the other two methods, it is probably a more realistic representation of the actual separation of flow as determined by the physical processes that control flow during storm events. The distribution of direct runoff equals the difference between the total discharge and the baseflow.

d. *Variable Slope Method (line 4 in Figure 15.3)*
A line is obtained by extrapolating the baseflow curve of the previous hydrograph forward to the time of the peak discharge. A second line is obtained by extrapolating the baseflow of the current hydrograph backward to the inflection point on the recession curve. By connecting the end points of the lines, direct flow is separated from the baseflow.

15.5.2 Manual/Empirical Separation

In hydrological engineering practice, baseflow is often separated manually by intuition and experience, however, following strictly the recessions of the hydrographs. Though subjectivity has an impact, the results are certainly closer to physical reality than the straight-line approaches.

15.5.3 Analytical (Parametric) Techniques

In order to reduce subjectivity and to ease separation, also analytical methods were developed to separate baseflow from the total flow hydrograph. For instance, two automated methods (the smoothed minima and recursive digital filter) were compared to separate baseflow [48]. Use of the digital filter with the parameter set to 0.925 was found to be a fast and objective method of continuous baseflow separation. Another attempt based on exponential recession equation was made [49]. A series of recession constants are calculated by regressing the logarithms of the flow on time by successively deleting the flow nearest the peak of the hydrograph. Progressively higher values of recession constants are obtained as the number of time units deleted increases, and the constant stabilizes thereafter. They day on which the constant stabilizes is taken as the beginning of baseflow. A baseflow separation technique based on analytical solutions of the Boussinesq equation was given in [70]. The technique reduces some of the subjectivities associated with the baseflow separation techniques.

15.5.4 Linear Conceptual Techniques

In these techniques, baseflow is determined as the outflow of one or more linear reservoirs starting at the last recession of the time series and going backward on the time axis. At each recharge phase, normally coincident with direct runoff, the recession curve restarts at a lower point. A transition curve links both recession curves. Separated baseflow allows the determination of groundwater storage and recharge that is computed as inflow.

15.5.5 Separation of Flow into Three Components

The procedure adapted from [12] and illustrated in Figure 15.4 where the hydrograph was plotted on a semilogarithmic paper is used to separate the flow into three components, surface flow, subsurface flow, and baseflow. The baseflow recession is approximated by a straight line and is extended backward. The difference between the total flow and the value on the extended line represents the combined surface and subsurface flow. The combined flow is plotted on the semilogarithmic paper and a straight line extended backward is fitted to the subsurface flow recession.

15.5.6 Separation of a Complex Hydrograph

It is relatively easy to separate baseflow from a simple (one-peak) hydrograph. In most cases, however, baseflow shall be separated from long time series of flows, for example, from a hydrograph of a year with 365 daily values and many peaks. Automated computer-supported methods are used in this case.

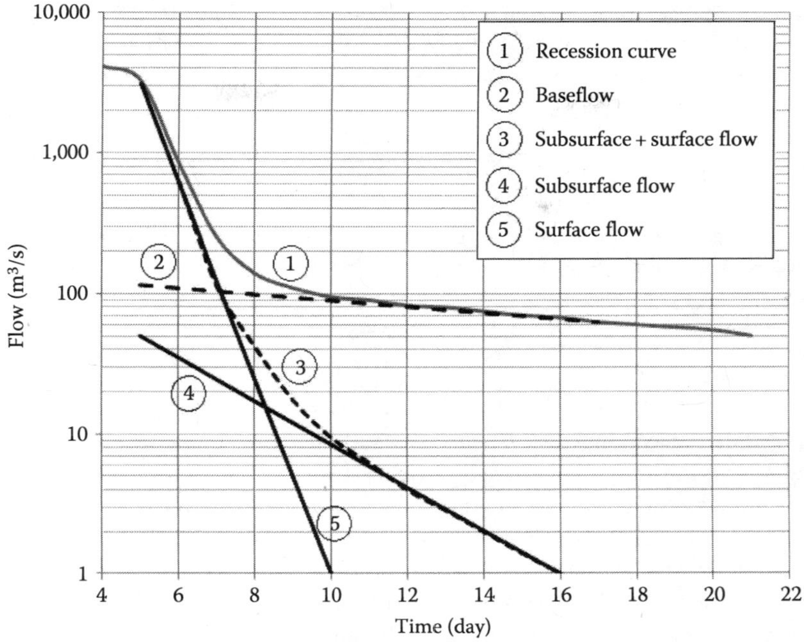

FIGURE 15.4 Separation of flow into three components. (Adapted from Bayazit et al., *Hidroloji Uygulamalari*, 1997.)

15.6 Nonlinear Method

The exponential decay function implying that the groundwater aquifer reacts like a single linear reservoir has, as detailed earlier, been used widely in hydrological literature. Here, the nonlinear reservoir approximation is introduced in order to separate groundwater flow from total flow. Studies [24,47,80–83,86] are all based on the nonlinear theory although there still are such studies as [20] insisting in the use of the linear theory. Later, the nonlinear theory was applied on the data from Australia [87]. A case study for an intermittent streamflow data set in the Thrace region in the European part of Turkey was studied by using the nonlinear algorithm [6]. Another application was performed on the streamflow data from Texas, United States [84]. A recent study [27] tried to understand how applicable the theory is in separating flow into its components during storm events.

In this theory, the storage–discharge relationship was modified by adding a dimensionless exponent b, such that

$$S = aQ^b \tag{15.9}$$

where

S, the storage, is in m^3

Q, the discharge, in m^3/s

the factor a in $m^{3(1-b)}s^b$, whereas the exponent b has no unit

If volumes are expressed as heights over unit area and the time step is a day (S in mm, Q in mm/d), then a is in $mm^{1-b}d^b$. It is seen that the linear reservoir algorithm is reached for $b = 1$.

$$\frac{dS}{dt} = -Q \tag{15.10}$$

can be written for periods with no flow into the reservoir (no groundwater recharge), during the recession curve of the hydrograph. Combining Equations 15.9 and 15.10 for a recession period starting at an initial discharge Q_0 yields, after mathematical transformation, in

$$Q_t = Q_0\left[1+\frac{(1-b)Q_0^{1-b}}{ab}t\right]^{1/b-1} \tag{15.11}$$

The parameter values a and b are calibrated using observed recession discharges on the base of an iterative least-squares method. From Equations 15.9 and 15.10,

$$a = \frac{\sum(Q_{i-1}+Q_i)\Delta t}{2\sum\left(Q_{i-1}^b - Q_i^b\right)} \tag{15.12}$$

is obtained to calibrate the parameter a for the systematically varying parameter b such that the computed outflow volume during the considered time period is equal to that in the given recession curve. The couple of (a,b) with the minimum sum of squared deviation from the observation represents the properties of the aquifer. Analysis of observed flow recession of numerous rivers in different hydrological regimes [6,80–83,87] yielded values $b < 1$ with a typical value of $b \approx 0.5$, which is a finding confirmed by theoretical derivations [54,79] for unconfined aquifers. For practical purposes, such as regionalization, it is more practical to fix the exponent b and calibrate the parameter a. A value of $b = 0.5$ as found typical previously can be suggested for average conditions.

The procedure for computation of the baseflow recession starts at the last value of the recession and proceeds backward along the time axis of the hydrograph. Baseflow at the time $t - \Delta t$ is determined by

$$Q_{t-\Delta t} = \left[Q_t^{b-1} + \frac{t(b-1)}{ab}\right]^{1/b-1} \tag{15.13}$$

that can be derived easily by inverting Equation 15.11. The procedure is applicable to daily time series and Δt therefore corresponds to one day. The computation of baseflow recession is adopted from [82] as follows (Figure 15.5).

When the backward computed baseflow recession curve intersects the rising limb of the hydrograph of the total flow, a one-step transition point is adopted as the peak of baseflow hydrograph. The rising limb of the baseflow hydrograph is then computed from the recession curve computed one step forward for each given total flow value.

The effective groundwater recharge (percolation to the aquifer) can be computed by using the water balance of

$$\begin{aligned}&\text{Storage at time}(i-1) + \text{Input to the storage at time } i(GWR_i)\\ &= \text{Storage at time } i + \text{Outputs from the storage (baseflow, } ET\text{, abstraction) at time } i\end{aligned} \tag{15.14}$$

that mathematically can be written as

$$GWR_i = S_i - S_{i-1} + \int_{t_{i-1}}^{t_i}(Q+A+ET)\,dt \tag{15.15}$$

where

GWR is the groundwater recharge
A is the abstraction
ET is the evapotranspiration, while others stand the same as previously defined.

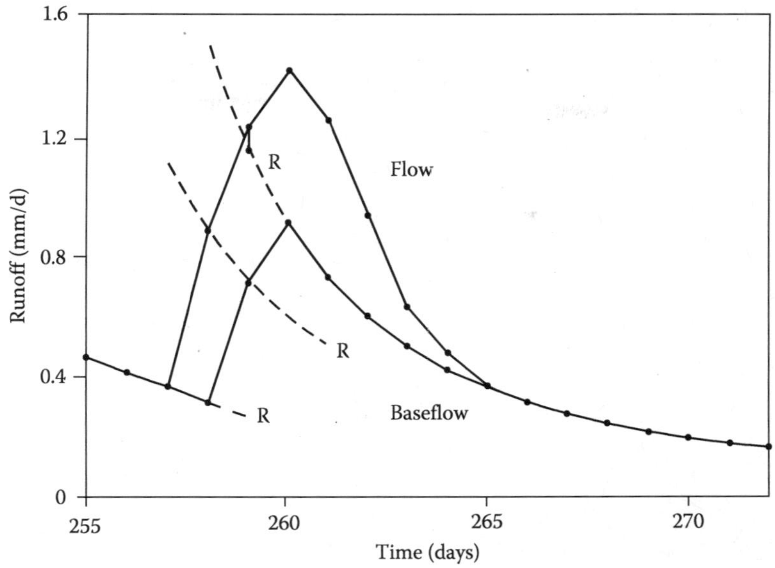

FIGURE 15.5 Construction of transition curve between two recessions. R: recession curves computed by Equation 15.13. (From Wittenberg, H., *Hydrol. Process.*, 13, 715, 1999.)

15.7 Application of Nonlinear Method to Real Data

Figure 15.6 shows baseflow separation for a 1-year hydrograph of a little river in East Germany. Two automated methods were applied, the model DIFGA, describing flow recession as the outflow of two parallel reservoirs [56], and BNL [82] using the nonlinear reservoir algorithm. Results are similar though the nonlinear recession fits slightly better during higher flows.

As lined out earlier, groundwater storage and groundwater recharge can be computed from separated baseflow. Figure 15.7 shows separation by the nonlinear reservoir algorithm for the little river

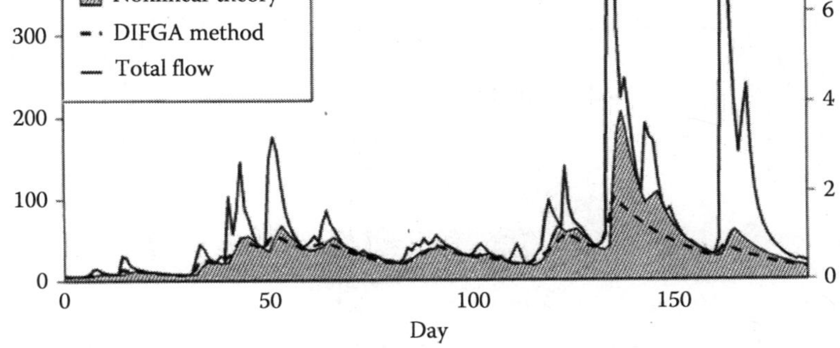

FIGURE 15.6 Baseflow separation: comparison of nonlinear reservoir with two linear reservoirs, Wernersbach creek, Germany. (From Wittenberg, H., *Hydrol. Process.*, 13, 715, 1999.)

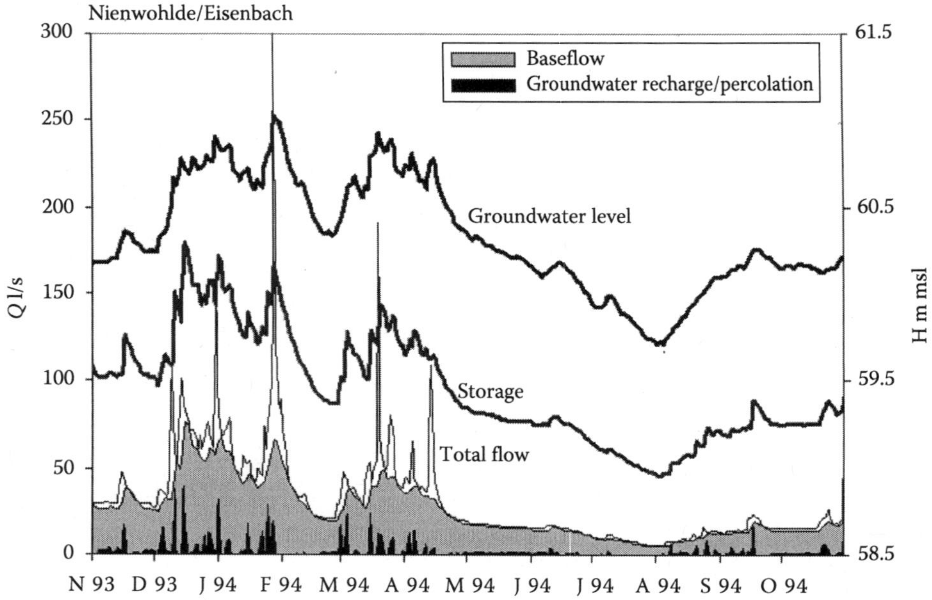

FIGURE 15.7 Baseflow separation by nonlinear reservoir algorithm for Eisenbach creek at Nienwolde, Germany.

Eisenbach in North Germany. The computed storage shows a high correlation and similarity to the observed groundwater level and its variation, confirming thus the aptitude of the method.

15.8 Summary and Conclusions

Baseflow separation is used in hydrology for two different purposes that should not be confused. The older approach is applied in flood hydrology to separate the slow-flow component from quick flow during a flood event. It is a practical but hardly physically based method. The separated flood hydrograph is used in rainstorm-flood modeling, ranging from unit hydrograph up to more sophisticated methods.

The second method aims to separate the groundwater component from total flow not only during floods but for the entire time series. The separated baseflow hydrograph allows computing of groundwater storage and recharge in watersheds, which are important components of water balances. Separation methods take into account the shapes of flow recessions, which reflect indeed the geohydrological conditions of the basin. Recession analysis is therefore an important prerequisite for baseflow separation. Different approaches are applied to model flow recessions, such as digital filters and reservoir algorithms, that is, relationships between storage and outflow of the aquifers and parallel linear and nonlinear reservoirs. Extensive studies of observed recessions at many rivers suggest that the nonlinear reservoir is the most physically based and best-fitting equation.

References

1. Aksoy, H., 1998, Determination of recession curve parameters. *Low Flows Expert Meeting of FRIEND AMHY Group*. Belgrade, Yugoslavia, June 10–12, 1998, pp. 89–96.
2. Aksoy, H., 2003, Markov chain-based modeling techniques for stochastic generation of daily intermittent streamflows. *Adv. Water Resour.*, 26, 663–671.
3. Aksoy, H. and Bayazit, M., 2000, A model for daily flows of intermittent streams. *Hydrol. Processes*, 14, 1725–1744.

4. Aksoy, H. and Bayazit, M., 2000, A daily intermittent streamflow simulator. *Turk. J. Eng. Environ. Sci.*, 24, 265–276.
5. Aksoy, H. and Dakova, S., 1998, Recession curve of the hydrograph. *Seminaire International Annuel du Groupe AMHY de FRIEND*. Istanbul, Turquie, October 1998, pp. 43–53.
6. Aksoy, H. and Wittenberg, H., 2001, Jahreszeitliche Veränderung der Trockenwetterganglinie—Fallstudie für einen Fluß im europäischen Teil der Türkei. *Wasserwirtschaft*, 90(1), 38–41.
7. Aksoy, H., Bayazit, M., and Wittenberg, H., 2001, Probabilistic approach to modelling of recession curves. *Hydrol. Sci. J.*, 46(2), 269–285.
8. Aksoy, H., Unal, N.E., and Pektas, A.O., 2008, Smoothed minima baseflow separation tool for perennial and intermittent streams. *Hydrol. Processes*, 22, 4467–4476.
9. Aksoy, H., Kurt, I., and Eris, E., 2009, Filtered smoothed minima baseflow separation method. *J. Hydrol.*, 372(1–4), 94–101.
10. Arnold, J.G. and Allen, P.M., 1999, Automated methods for estimating baseflow and ground water recharge from streamflow records. *J. Am. Water Resour. Assoc.*, 35(2), 411–424.
11. Barnes, B.S., 1939, The structure of discharge-recession curves. *Trans. Am. Geophys. Union*, 20, 721–725.
12. Bayazit, M., Şen, Z., and Avcı, I., 1997, *Hidroloji Uygulamaları*. Istanbul Technical University Press, İstanbul, Turkey.
13. Chapman, T., 1999, A comparison of algorithms for stream flow recession and baseflow separation. *Hydrol. Process.*, 13, 701–714.
14. Chow, V.T., Maidment, D.R., and Mays, L.W., 1988, *Applied Hydrology*. McGraw-Hill, New York.
15. Clausen, B., 1992, Modelling streamflow recession in two Danish streams. *Nordic Hydrol.*, 23, 73–88.
16. Collado, J., Aldama, A.A., and Acosta-Rodriguez, J.L., 1990, Optimal spectral base-flow estimation. *ASCE J. Hydr. Eng.*, 116(12), 1540–1546.
17. Demuth, S., 1989, The application of the West German IHP recommendations for the analysis of data from small research basins. *IAHS Publ.*, 187, 47–60.
18. Demuth, S. and Schreiber, P., 1994, Studying storage behaviour using an operational recession method. *IAHS Publ.*, 221, 51–59.
19. Eckhardt, K., 2005, How to construct recursive digital filters for baseflow separation. *Hydrol. Process.*, 19, 507–515.
20. Fenicia, F., Savenije, H.H.G., Matgen, P., and Pfister, L., 2006, Is the groundwater reservoir linear? Learning from data in hydrological modelling. *Hydrol. Earth Syst. Sci.*, 10, 139–150.
21. Fröhlich, K., Fröhlich, W., and Wittenberg, H., 1994, Determination of groundwater recharge by baseflow separation: Regional analysis in northeast China. *IAHS Publ.*, 221, 69–75.
22. Furey, P.R. and Gupta, V.K., 2001, A physically based filter for separating base flow from streamflow time series. *Water Resour. Res.*, 37(11), 2709–2722.
23. Furey, P.R. and Gupta, V.K., 2003, Tests of two physically based filters for base flow separation. *Water Resour. Res.*, 39(10), 1297, doi: 10.1029/2002WR001621.
24. Gottschalk, L., Tallaksen, L.M., and Perzyna, G., 1997, Derivation of low flow distribution functions using recession curves. *J. Hydrol.*, 194, 239–262.
25. Gustard, A., Roald, L., Demuth, S., Lumadjeng, H., and Gross, R., 1989, *Flow Regimes from Experimental and Network Data, (FREND)*. 2 Vols. Institute of Hydrology, Wallingford, U.K.
26. Hall, F.R., 1968, Base flow recessions—A review. *Water Resour. Res.*, 4(5), 973–983.
27. Hammond, M. and Han, D., 2006, Recession curve estimation for storm event separations. *J. Hydrol.*, 330(3–4), 573–585.
28. Herrmann, A., 1997, *Global Review of Isotope Hydrological Investigations*. Oberlin, G. and Desbos, E. (eds.), FRIEND 3rd Report, UNESCO, Paris, France.
29. Hughes, D.A., Hannart, P., and Watkins, D., 2003, Continuous baseflow separation from time series of daily and monthly streamflow data. *Water SA*, 29(1), 43–48.

30. Huyck, A.A.O., Pauwels, V.R.N., and Verhoest, N.E.C., 2005, A base flow separation algorithm based on the linearized Boussinesq equation for complex hillslopes. *Water Resour. Res.*, 41, W08415, doi: 10.1029/2004WR003789.
31. Institute of Hydrology, 1980, Low flow studies. Research Report, Institute of Hydrology, Wallingford, U.K.
32. James, L.D. and Thompson, W.O., 1970, Least squares estimation of constants in a linear recession model. *Water Resour. Res.*, 6(4), 1062–1069.
33. Johnson, E.A. and Dils, R.E., 1956, *Outline for Compiling Precipitation, Runoff, and Groundwater Data from Small Watersheds*. Station Paper No. 68, U.S. Department of Agriculture, Southeastern Forest Experiment Station, Asheville, NC, 40pp.
34. Kelman, J., 1977, *Stochastic Modeling of Hydrologic, Intermittent Daily Processes*. Hydrology Paper No. 89, Colorado State University, Fort Collins, CO.
35. Kelman, J., 1980, A stochastic model for daily streamflow. *J. Hydrol.*, 47, 235–249.
36. Kupczyk, E. and Kasprzyk, A., 1997, Research catchments sensitivity to atmospheric drought. *LOC Proceedings of Poster Presentations, 3rd International Conference on FRIEND, October 1–4 1997*, Postojna, Slovenia.
37. Kupczyk, E., Kasprzyk, A., Sztechman, J., and Strzalkowski, K., 1998, Application of a spline function method to streamflow recession curve analysis. *Low Flows Expert Meeting of FRIEND AMHY Group*, Belgrade, Yugoslavia, June 10–12 1998, pp. 79–88.
38. Kurt, I., 2007, Filtered smoothed minima baseflow separation method. MSc thesis, Institute of Science and Technology, Istanbul Technical University, Istanbul, Turkey.
39. Langbein, W.B., 1938, Some channel storage studies and their application to the determination of infiltration, *Trans. Am. Geophys. Union*, 19, 435–445.
40. Lim, K.J., Engel, B.A., Tang, Z.X., Choi, J., Kim, K.S., Muthukrishnan, S., and Tripathy, D., 2005, Automated Web GIS based hydrograph analysis tool, what. *J. Am. Water Resour. Assoc.*, 41(6), 1407–1416.
41. Linsley, R.K., Kohler, M.A., and Paulhus, J.L.H., 1949, *Applied Hydrology*. McGraw-Hill, New York.
42. Maillet, E., 1905, Essai d'hydraulique souterraine et fluviale. *Librairie Scientifique*. A. Hermann, Paris, France.
43. Mazvimavi, D., Meijerink, A.M.J., and Stein, A., 2004, Prediction of base flows from basin characteristics: A case study from Zimbabwe. *Hydrol. Sci. J.*, 49(4), 703–715.
44. McCuen, R.H., 1998, *Hydrological Analysis and Design*. Prentice Hall, Upper Saddle River, NJ.
45. Mizumura, K., 1995, Runoff prediction by simple tank model using recession curves. *ASCE J. Hydraul. Eng.*, 121(11), 812–818.
46. Mizumura, K., 2005, Analyses of flow mechanism based on master recession curves. *ASCE J. Hydrol. Eng.*, 10(6), 468–476.
47. Moore, R.D., 1997, Storage-outflow modelling of streamflow recession, with application to a shallow-soil forested catchment. *J. Hydrol.*, 198, 260–270.
48. Nathan, R.J. and McMahon, T.A., 1990, Evaluation of automated techniques for base flow and recession analyses. *Water Resour. Res.*, 26(7), 1465–1473.
49. Owoade, A. and Bako, M.D., 1989, Riverflow prediction during baseflow recession on some British hard rock catchments. *IAHS Publ.*, 187, 61–65.
50. Pektas, A.O., 2007, Baseflow separation in intermittent streams. MSc thesis, Institute of Science and Technology, Istanbul Technical University, Istanbul, Turkey.
51. Piggott, A.R., Moin, S., and Southam, C., 2005, A revised approach to the UKIH method for the calculation of baseflow. *Hydrol. Sci. J.*, 50(5), 911–920.
52. Radczuk, L. and Szarka, O., 1989, Use of the flow recession curve for the estimation of conditions of river supply by underground water. *IAHS Publ.*, 187, 67–74.
53. Sargent, D.M., 1979, A simplified model for the generation of daily streamflows. *Hydrol. Sci. Bull.*, 24, 509–527.
54. Schoeller, H., 1962, *Les Eaux Souterraines*. Masson, Paris, France.

55. Schulz, E.F., 1989, *Problems in Applied Hydrology*, 7th edn. Water Resources Publications, Littleton, CO.
56. Schwarze, R., Grünewald, U., Becker, A., and Fröhlich, W., 1989, Computer-aided analyses of flow recessions and coupled basin water balance investigations. *IAHS Publ.*, 187, 75–83.
57. Sherman, L.K., 1932, Streamflow from rainfall by the unit graph method. *Eng. News Rec.*, 108, 501–505.
58. Singh, K.P., 1968, Some factors affecting baseflow. *Water Resour. Res.*, 4(5), 985–999.
59. Singh, K.P., 1969, Theoretical baseflow curves. *ASCE J. Hydraul. Div.*, 95(HY6), 2029–2048.
60. Singh, K.P. and Stall, J.P., 1971, Derivation of base flow recession curves and parameters. *Water Resour. Res.*, 7(2), 292–303.
61. Sloto, R.A. and Crouse, M.Y., 1996, *HYSEP: A Computer Program for Streamflow Hydrograph Separation and Analysis*, U.S Geological Survey, Water-Resources Investigations Report 96–4040, Lemoyne, PA, 46pp.
62. Smakhtin, V.U., 2001, Estimating continuous monthly baseflow time series and their possible applications in the context of the ecological reserve. *Water SA*, 27(2), 213–217.
63. Spolia, S.K. and Chanders, S., 1974, Modelling of surface runoff systems by an ARMA model. *J. Hydrol.*, 22, 317–332.
64. Spongberg, M.E., 2000, Spectral analysis of base flow separation with digital filters. *Water Resour. Res.*, 36(3), 745–752.
65. Strzalkowski, K. and Sztechman, J., 1996, *Recession Curve Analyses in Sections. The RCS Program-User's Guide*. Technical University in Kielce, Kielce, Poland.
66. Sugawara, M., 1995, Tank model. *Computer Models of Watershed Hydrology*. Singh, V.P. (ed.), Water Resources Publications, Littleton, CO.
67. Sugawara, M., 1995, The development of a hydrological model—Tank. *Time and the River*. Kite, G.W. (ed.), Water Resources Publications, Littleton, CO.
68. Szilagyi, J., 1999, On the use of semi-logarithmic plots for baseflow separation. *Groundwater*, 37(5), 660–663.
69. Szilagyi, J., 2004, Heurictic continuous base flow separation. *ASCE J. Hydrol. Eng.*, 9(4), 311–318.
70. Szilagyi, J. and Parlange, M.B., 1998, Baseflow separation based on analytical solutions of the Boussinesq equation. *J. Hydrol.*, 204, 251–260.
71. Szilagyi, J., Harvey, F.E., and Ayers, J.F., 2003, Regional estimation of base recharge to ground water using water balance and base-flow index. *Groundwater*, 41(4), 504–513.
72. Szilagyi, J., Harvey, F.E., and Ayers, J.F., 2005, Regional estimation of total recharge to ground water in Nebraska. *Groundwater*, 43(1), 63–69.
73. Tallaksen, L.M., 1989, Analysis of time variability in recessions. *IAHS Publ.*, 187, 85–96.
74. Tallaksen, L.M., 1995, A review of baseflow recession analysis. *J. Hydrol.*, 165, 349–370.
75. Toebes, C. and Strang, C.C., 1964, On recession curves, 1—Recession equations. *J. Hydrol. (New Zealand)*, 3(2), 2–15.
76. Vogel, R.M. and Kroll, C.N., 1996, Estimation of baseflow recession constants. *Water Resour. Manage.*, 10, 303–320.
77. Wahl, K.L. and Wahl, T.L., 1988, *Determining the flow of Comal Springs at New Braunfels, Texas*, pp. 77–86. Texas Water 95, ASCE, August 16–17, 1995, San Antonio, TX.
78. Wahl, K.L. and Wahl, T.L., 1995, Effects of regional ground-water declines on streamflows in the Oklahoma Panhandle. *Symposium on Water-Use Data for Water Resources Management*, pp. 239–249, AWRA, Tucson, AZ.
79. Werner, P.W. and Sundquist, K.J., 1951, On the groundwater recession curve for large water-sheds. *IAHS Publ.*, 33, 202–212.
80. Wittenberg, H., 1994, Nonlinear analysis of flow recession curves. *IAHS Publ.*, 221, 61–67.
81. Wittenberg, H., 1997, Regionalization of nonlinear storage, discharge and recharge of groundwater. *Landschaftsökologie und Umweltforschung*, 25, 343–346.

82. Wittenberg, H., 1999, Baseflow recession and recharge as nonlinear storage processes. *Hydrol. Process.*, 13, 715–726.
83. Wittenberg, H., 2003, Effects of season and man-made changes on baseflow and flow recession: Case studies. *Hydrol. Process.*, 17, 2113–2123.
84. Wittenberg, H., 2004, Recession analysis and baseflow separation—Nonlinear reservoir algorithm—Application on daily flows, Hod Creek, TX. Short Report, September 2004.
85. Wittenberg, H., 2006, Impacts of forestation and land use on water balance and low flows in Northwest Germany. *BALWOIS Conference*, May 23–26, 2006, Ohrid, Macedonia.
86. Wittenberg, H. and Rhode, C., 1997, Impact of groundwater abstractions on storage and baseflow in the Ilmenau basin, Lueneburg Heath, Germany. *Landschaftsökologie und Umweltforschung*, 25, 347–350.
87. Wittenberg, H. and Sivapalan, M., 1999, Watershed groundwater balance estimation using streamflow recession analysis and baseflow separation. *J. Hydrol.*, 219, 20–33.
88. WMO, World Meteorological Organization, 1994, *Guide to Hydrological Practices*, 5th edn. WMO-No. 168, Geneva, Switzerland.

16

Hydrology–Ecology Interactions

Tadanobu Nakayama
National Institute for Environmental Studies (NIES)

AUTHOR

Tadanobu Nakayama is a senior researcher at the National Institute for Environmental Studies (NIES), Japan. He was a visiting scientist at Centre for Ecology and Hydrology, United Kingdom. He has developed a process-based national integrated catchment-based ecohydrology (NICE) model, which includes complex interactions between canopy, surface water, unsaturated zone, aquifers, lakes, and rivers. He is the author of many books and papers on ecohydrological systems, including wetland ecosystem and nature restoration, global ecohydrology, and urban environment and water resources. He is currently conducting research to reevaluate the ecosystem in catchment/basin from the viewpoint of meta-ecosystem by considering global–local interaction and hydrology–ecology linkage. He is also constructing an integrated assessment system toward improving biogeochemical cycles along the terrestrial–aquatic continuum for global environmental change and tipping point early warning systems for sustainable development under the constraint of global security and environmental degradation. He is also on the editorial board of two journals *Ecohydrology* published by Wiley and "Ecohydrology & Hydrobiology" published by Elsevier.

PREFACE

Japanese governments started a nature conservation project to recover the original mire ecosystem because various anthropogenic stressors have caused mire degradation in subarctic northern Japan, such as drying and the invasion of an alder-dominant shrub forest. Here, the process-based NICE model is further developed to include the feedback between hydrologic cycle, elevation change, and vegetation succession in the Kushiro Mire. NICE reproduces this nonlinear interaction showing a clear correlation between drying and alder invasion at longer time scales. This integrated system throws some light on the refinement of tipping point early warning system for sustainable development and the improvement in boundless biogeochemical cycle along terrestrial–aquatic continuum for global environmental change.

16.1 Introduction

Vegetation change and succession are closely related to nonlinear interactions between vegetation and groundwater in the mire [7,9,49,50]. Some research showed that lowering of the water table in salt marshes was due not to drainage but to water uptake by roots and suggested a feedback between plant growth, water uptake, and sediment oxidation [7]. Others [9,50] investigated whether the removal of riparian vegetation and clear cutting of forested wetlands caused a water table rise known as "watering up" and showed that this was caused mainly by reduced evapotranspiration and increased interception. As a result, the clear cutting of riparian and wetland vegetation might lead to ecosystem conversion such as encroachment [49,60]. We can also see various spatial vegetation patterns such as string and maze in the mire, which underlies self-organization and a positive feedback between hydrogeological change and plant growth [12,50,57]. Some of my group's previous studies clarified that spatial vegetation pattern and succession have affected the hydrogeological cycle at short time scales, although they didn't include feedback processes at longer time scales [29,38,39]. Other studies simulated the evolution of vegetation patterns as a function of hydrogeological redistribution and estimated that the patterns emerge from facilitation and competition feedback of limiting water resource [26,27,48,50,59]. However, it is generally difficult to simulate spatial heterogeneity and succession because these models don't include dynamic effect of sediment deposition/erosion processes and their feedback effects on hydrology and vegetation dynamics. Recent research has shown the effect of dynamic vegetation on sediment deposition/erosion and landform evolution in humid areas [18] and arid areas [51]. It is important to further develop these models through observed information about the interaction between hydrological processes and ecological dynamics [2,20] to clarify the nonlinear interaction between geomorphology and eco-hydrology and its relation to succession and spatial patterning.

The frequent occurrence of succession and positive feedback can be associated with the emergence of multiple stable states [49,52,60], where wetland ecosystems are vulnerable to disturbances and may respond to biomass losses with irreversible catastrophic shifts. Previous studies sketched a theoretical framework through observation in which these drastic switches to contrasting alternative states are related to the loss of equilibrium and resilience [52,53]. A conceptual framework explained how vegetation–water table feedback may lead to the emergence of multiple stable states and helped to investigate analytically the vegetation susceptibility to catastrophic shifts [49]. So, the occurrence of succession and alder invasion can be considered a possible sign of alternative stable states when a small change in external forcing/conditions makes the current ecosystem equilibrium unstable [49,60]. Deterministic and mechanistic models consisting of partial differential equations describing plant biomass, geomorphology, and eco-hydrology are effective not only for the reproduction of spatial vegetation patterns but also for the evaluation of catastrophic responses to alternative stable states or major change in external forcing [48,50,59]. In this way, succession and the spatial patterns are closely related to the nonlinear interaction between geomorphology and eco-hydrology in the mire. Once the alder invasion reaches a critical threshold known as "catastrophic bifurcation" [52], disturbances caused by positive feedback may invoke a cascade of stable states in reed–sedge vegetation, resulting in a sudden catastrophic shift to the alder invasion. It is necessary to clarify these relations to recover and restore the mire ecosystem.

The objective of this study was to clarify the nonlinear interaction between hydrology and ecology and evaluate the positive feedback mechanism of alder invasion for predicting the impact of river restoration on the shrinkage of alder-dominated shrub forest and the recovery of the Kushiro Mire in Japan by combining a distributed, process-based model with the current understanding of the processes involved. The process-based National Integrated Catchment-based Eco-hydrology (NICE) model series [26–45] was further expanded to evaluate the nonlinear relationship between drying and alder invasion including the competition between invasive alder-dominant shrub forest and indigenous reed–sedge vegetation.

16.2 Study Area

The Kushiro Mire (176.7 km^2, the largest wetland in Japan) and the Kushiro River catchment (2204.7 km^2, located in northern Japan; Figure 16.1) [8] have been changed by conversion to urban or agricultural uses since 1884 though the mire itself has been protected under the Ramsar Convention since 1980 and was declared a national park in 1987. The annual mean temperature is about 5°C–6°C and precipitation is

FIGURE 16.1 Land cover and observation points in study area (Kucyoro River catchment, a tributary of Kushiro River catchment) in 1976 and 1997, after channelization downstream of Kucyoro River. The simulation area is 22 km wide × 46 km long, covering the whole Kucyoro River catchment. In particular, hydrogeology and vegetation dynamics feedback process was simulated in detail within the black border in the downstream area of Kucyoro River catchment.

about 1100 mm. In summer, the mean temperature is 17°C–19°C, and fog is common. Mire is a wetland type that accumulates acidic peat, a deposit of dead plant material—usually mosses—but also lichens in Arctic climates. The dominant species in this mire include alder (*Alnus japonica*), reed (*Phragmites australis*), moss (*Polytrichum* spp., *Sphagnum* spp.), sedge (*Carex lasiocarpa, C. lyngbyei*), willow (*Salix* spp.), Japanese ash (*Fraxinus mandshurica var. japonica*), and meadow sweet (*Spiraea salicifolia*) [56]. Although the meandering former river channel in the northern part of the mire was straightened in the 1970s–1980s in order to smoothly drain runoff and to protect the farmland from flooding, runoff containing nutrients and sediment from farmland and shortcut channels has flowed directly into the mire and been deposited there as flood-borne sediment, most of which was previously deposited upstream before the channelization [25,38]. Alder, a deciduous tree forming swamp forests up to 15 m tall, has propagated widely after the channelization and resulted in drying and mire shrinkage [24] (Figure 16.1). The Japanese government has previously collaborated with organizations and foundations to cut the alder there as a form of adaptive management to preserve invaluable species. For example, this felling helps the red-crowned crane (*Grus japonensis*), designated as a precious natural treasure in Japan and "endangered" in the International Union for Conservation of Nature and Natural Resources (IUCN) Red List [17], to build nests there. In 2002, the government started a new project, Kushiro Mire Conservation Plan, to restore about 2 km of meandering former river, arrest sediment-load influx, and allow mire recovery [24]. It is important to assess the impacts of river channelization, to predict the recovery of groundwater recharge by river restoration, and to prevent sediment loading to riparian forests, which are sensitive to succession to alder, reeds, and willows.

16.3 Methods

16.3.1 Hydrologic Cycle Model

The NICE model [26–45] is a 3D process-based model that includes surface–unsaturated–saturated water processes and assimilates land-surface processes describing vegetation phenology derived from satellite data. The surface hydrological process consists of a hillslope model based on kinematic wave theory and a distributed stream network model based on kinematic/dynamic wave theories. The hillslope model includes water and thermal energy budget model and a surface runoff model. The unsaturated layer divides canopy into two layers and soil into three layers in the vertical dimension in the Simple Biosphere model 2 (SiB2) [54]. With respect to the saturated layer, NICE solves a partial differential equation describing 3D groundwater flow for both unconfined and confined aquifers. The models connect submodels from beneath the surface to the surface by considering water and heat fluxes, for example, (1) recharge calculated from the gradient of hydraulic potential between the deepest layer of unsaturated flow and groundwater level; (2) effective precipitation calculated from precipitation, infiltration, and evapotranspiration; and (3) seepage (exchange) between river and groundwater calculated as the difference in hydraulic potentials between river and groundwater head [38].

16.3.2 Elevation Change Model

The elevation change model constitutes diffusion model and riverbed continuity equation. The 2D diffusion model including sedimentation simulates mass transport on a hillslope after simulation of the constitutes hillslope runoff model including snowmelt runoff [29,39]:

$$\frac{\partial}{\partial t}\{D\cdot C\}+\frac{\partial(M\cdot C)}{\partial x}+\frac{\partial(N\cdot C)}{\partial y}$$

$$=\frac{\partial}{\partial x}\left\{K\hat{x}\cdot D\cdot\frac{\partial C}{\partial x}\right\}+\frac{\partial}{\partial y}\left\{K\hat{y}\cdot D\cdot\frac{\partial C}{\partial y}\right\}+D\cdot L(C)+q_{su}-w_f\cdot c_b \tag{16.1}$$

where

M and N are discharges per unit width along the x and y coordinate axes
H (m) is the elevation of the hillslope bottom
S (m) is the hydraulic head
D (m) is the flow depth ($= S - H$)
C (mg/L) is the mass concentration
K_x and K_y (m/s) are the kinematic eddy diffusivities along the x and y coordinate axes
$L(C)$ is concentration
q_{su} (m/s) is the rate of resuspension from the hillslope bed
w_f (m/s) is the sedimentation rate (settling velocity) of the suspended load
c_b (mg/L) is the standard concentration

The riverbed sediment continuity equation is expressed by using the finer suspended load q_{su} in diffusive mass movement and the coarser bed material load q_B in fluvial sediment transport [29]:

$$\frac{\partial z_B}{\partial t} = -\frac{1}{1-\lambda}\left(q_{su} - w_f c_b + \frac{\partial q_B}{\partial x}\right) \tag{16.2}$$

where

z_B (m) is the bed elevation
λ is the porosity of the bed material (void ratio)
w_f (m/s) is the sedimentation rate
c_b is the concentration at b (m) above the riverbed

The bed material load is expressed by Einstein's formula [10], and the suspended load is expressed by a theory based on the energy equation of solid–liquid two-phase flow, the Monin–Obukhov length, and Rubey's formula [55]. Previous studies of positive feedback between sediment accumulation and plant growth in mire ecosystems simulated only 1D hydraulic head without elevation change [50] or only 1D elevation change without the difference between suspended load and bed material load and without diffusion effect [59]. Equation 16.2 expands the empirical relation that deposition becomes maximum (minimum) at no elevation change (maximum sediment elevation) and that erosion has its maximum value at low plant density [59], by including the equations of both bed material load and suspended load and by including heterogeneous roughness coefficients of alder and reed, which has a nonlinear feedback process between hydrologic cycle and vegetation dynamics [51]. Furthermore, 1D diffusion and elevation change equations relevant to a river channel can be expressed in the same way after simulation of the stream network model (kinematic or dynamic wave theory) [38]. The simulated lateral inflow of mass was input into this 1D diffusion model on the assumption that this inflow simulated by the hillslope model enters along the side of river channel. Because the study area lies at a low elevation (Figure 16.1) and is greatly affected by groundwater level change and sediment deposition/erosion, both of which influence the spatial distribution of vegetation change in the mire, it is crucial to include groundwater level change, sediment diffusion, and deposition/erosion simultaneously in the model.

16.3.3 Vegetation Dynamics and Succession Model

The vegetation dynamics model used in this study is based on the gap-phase ZELIG model [6]. ZELIG simulates seedling establishment, mortality, and regeneration stochastically by using Monte Carlo simulations and simulates the growth stage of each tree deterministically on an array of model plots. Adjacent cells interact through light interception at low sun angles. The competitive environment of the plot is defined by the height, leaf area, and woody biomass of each individual tree determined by allometric relationships with diameter [6,26,27]. The model includes vegetation succession and competition between alder and reed in addition to stochastic simulation of seedling establishment, mortality, and regeneration, which improves the previous approaches to the dynamic interaction between hydrologic and vegetation processes [48,51,59]. The predominant vegetation type was defined as a representative one at each cell, indicating how the model determined the shift from one species to another. The simulated distribution of each species was compared with the vegetation cover of graphic information system (GIS) data (Figure 16.1) categorized against an aerial photograph [8].

Tree diameter growth assumes a logistic curve (JABOWA-derived gap model) and $L = cD^2$, where L is leaf area and c is a parameter (0.160694):

$$\frac{dD}{dt} = \frac{GD(1 - DH/D_{max}H_{max})}{(274 + 3b_2D - 4b_3D^2)}, \quad \text{where} \quad b_2 = 2\frac{H_{max} - 137}{D_{max}} \quad \text{and} \quad b_3 = \frac{H_{max} - 137}{D_{max}^2} \tag{16.3}$$

where

D (cm) is diameter at breast height (DBH)
H (cm) is height ($= 137 + b_2D - b_3D^2$)
D_{max} and H_{max} are the maximum tree dimensions
G is a growth rate parameter, which is calculated as a function of several growth reduction factors, as described below

Previous research [26,27] applied a stepwise procedure of a multiplicative approach and Liebig's law in the expansion of [3] to conform several requirements for the growth reduction factor (r) for light (Q_h), nutrients (F), soil moisture (M), temperature (T), and submerged depth of vegetation from the ground surface (S_D):

$$G = G_{max}\left[r(Q_h)r(F)r(M)r(T)r(S_D)\right]^{1/3} \tag{16.4}$$

where

G_{max} is the maximum growth rate
M is the soil moisture in the unsaturated layer simulated by the hydrologic model
S_D (m) is simulated by hydrologic and elevation change models

M and S_D are the primary determinants of the hydrologic regime, the so-called threshold for the transition, at sites that might support a vegetation community, linear response, or alternative state [60]. This means that vegetation dynamics also have a simultaneous nonlinear feedback process between the hydrologic cycle and sedimentation.

16.3.4 Feedback Process of Models

Previous studies of positive feedback among sediment accumulation, groundwater level change, and plant growth in mire ecosystems consisted of a logistic equation describing vegetation dynamics (plant biomass) and hydrogeologic equations (surface water, soil moisture, hydraulic head, nutrient availability, and sediment accumulation). They evaluated alternative stable states by two partial differential equations of a logistic curve and the relation between carrying capacity and water table [49], estimated the cascade of vegetation collapse by two partial differential equations of plant biomass and sediment accumulation [59], and reproduced string and maze patterns by three partial differential equations of plant biomass, hydraulic head, and nutrient availability [50]. Although the latter two models are spatially explicit and can predict spatial bifurcations from one stable state to another, unlike the former conceptual model and earlier models [48], it is generally difficult to simulate spatial heterogeneity and vegetation succession. Because NICE simulates water/heat budget, sediment and nutrient transport, and vegetation succession iteratively, it is possible to estimate the impact of river channelization versus

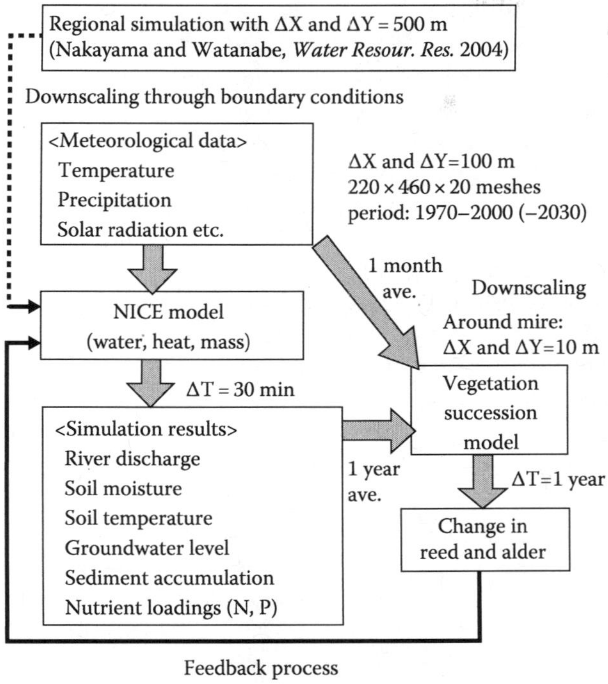

FIGURE 16.2 Flow diagram of NICE model in the expansion of previous version. The new model iteratively simulates nonlinear interactions among the hydrologic cycle, elevation change, and vegetation succession, including the competition between two vegetation types (bold arrows).

meandering and land-cover change on vegetation change downstream in the Kushiro Mire by adding the current understanding of the relationships among water, heat, nutrients, sediment, and vegetation from the previous studies (Figure 16.2). First, the model simulates the water and heat values such as evapotranspiration, river discharge (*RV*), soil moisture (*SM*), groundwater level (*GW*), and soil temperature (*ST*) at each time step by inputting meteorological forcing data [38,39]. Second, it simulates total nitrogen (*TN*) and suspended sediment (*SS*) from point and nonpoint sources such as farms, forest, and livestock by using unit flux data [29]. Third, it simulates vegetation succession [26,27]. These three processes interact nonlinearly and feed back iteratively in the model (Figure 16.2), which shows great improvement from previous studies [48–51,59].

16.4 Model Parameterization and Analysis

16.4.1 Model Input and Observed Data

Hourly observation data of radiation, precipitation, atmospheric pressure, air temperature, humidity, and wind speed at a reference level were calculated from Automated Meteorological Data Acquisition System (AMeDAS) data recorded at Tsurui observation station (Figure 16.1; 43°13′48″N, 144°19′30″E, mean elevation 42 m) [19], and these values were input in each grid because the study area is not large. Mean elevation data [13], soil texture data [14], vegetation class data [11], and land-cover data [8] were converted and input into NICE at resolutions of 100 and 500 m. The simulation area in the vertical direction was divided into 20 layers with a weighting factor of 1.1 (finer at the upper layers). Geological structure was divided into four types on the basis of hydraulic conductivity, the specific storage of porous material, and specific yield by using soil samples taken at two depths (0.1 and 1.0 m) and previous data from 150 sample data points [46]. At the upstream boundaries where there are no observed data, the reflecting condition on the hydraulic head was used for the groundwater flow submodel. In the model simulating hillslope hydrology, the flow depth and the discharge at the uppermost ridges of the mountains were set to zero. The model assumed that the various land covers have the same constant mass fluxes per unit area in farm, forest, and livestock breeding [22,29,58]. Important data about alder and reed were summarized from previous studies [4,5,16,24,47] and input into the model [26,27]. The model was validated using groundwater level meters, flow depth meters, and a water sampler/turbidity sensor during 2000–2003 in the Kucyoro River (Figure 16.1) [38,39]. The water level (KADEC, MIZU-II) was automatically recorded to data loggers (LS-3000 PtV) at hourly intervals. *H–Q* curves were used to convert water level into river discharge in rivers. The water turbidity concentration and water temperature (CTI, C105) were automatically recorded to data loggers at 10 min intervals from July 2002 to June 2003. Samples of the turbid water below the water surface were automatically collected in sample bottles (ISCO, Model-3700) at hourly intervals on several times during high-precipitation periods. Details are described in my group's previous researches [29,38,39].

16.4.2 Calibration and Validation

The water/heat budget and mass transport were firstly simulated in a regional area covering the entire Kushiro River catchment (an area 50 km wide × 80 km long) with a grid spacing of 500 m and then downscaled into a local area covering the whole Kucyoro River catchment (an area 22 km wide × 46 km long) with a grid spacing of 100 m, which is sufficiently fine to describe the shapes of meanders (Figures 16.1 and 16.2). The simulation was conducted from 1970 to 2030 in three situations with a time step of $\Delta t = 30$ min to facilitate numerical stability: (1) before straightening (1970s; meandering channel), (2) before restoration (present channelized rivers), and (3) after restoration (return to previous meanders). After the calibration and validation of hydrological values in 2001–2002 ($r_{RV}^2 = 0.78$, $r_{SM}^2 = 0.74$, $r_{GW}^2 = 0.63$, $r_{ST}^2 = 0.80$) [26,27,38,39], the simulation could backcast the water/heat budget during 1970–2000 and then forecast it during 2000–2030. Then mass transport was simulated by using the unit flux data [22,55,58] with time steps of $\Delta t = 50$ and 10 s through the calibration and validation in 2002–2003 ($r_{TN}^2 = 0.56$, $r_{SS}^2 = 0.96$) [29]. Kinematic eddy diffusivity at each layer was set to satisfy the stability condition of the diffusion term and to set the time step as large as possible.

After the earlier simulation in the Kucyoro River catchment, the hydrogeology and vegetation dynamics feedback process was simulated in detail around the Kushiro Mire through the nesting of simulated boundary data (Figure 16.1). The 2D depth-averaged simulations for shallow water theory and the advection–diffusion equation were calculated to simulate the elevation change due to sediment aggradation from 1970 to 2003 with a time step of $\Delta t = 5$ s [29]. Then simulated hydrogeological results averaged in every month were input into the vegetation succession model to simulate vegetation growth every year (Figure 16.2). Each plot size was 10 × 10 m, sufficient to represent vegetation colonies [56]. The simulated tree height and tree age were calibrated by values obtained in previous studies [15,56] ($r_{TH}^2 = 0.64$), and finally the vegetation change was simulated during 1970–2030 after a warm-up simulation of 300 years [26,27]. The simulated vegetation distributions were also reflected to the next step of the hydrogeology simulation, which means that nonlinear interaction and feedback processes are included explicitly in the model (Figure 16.2).

16.5 Results

16.5.1 Verification of Scale-Dependent Hydrogeologic Changes

The author further improved the model to include downscaling process from regional simulation in entire Kushiro River catchment at resolution of 500 m to local simulation in Kucyoro River catchment at finer resolution of 100 m (Figure 16.3). The result improved local heterogeneity about complex eco-hydrological processes depending on the resolution of input data. In particular, we can see more

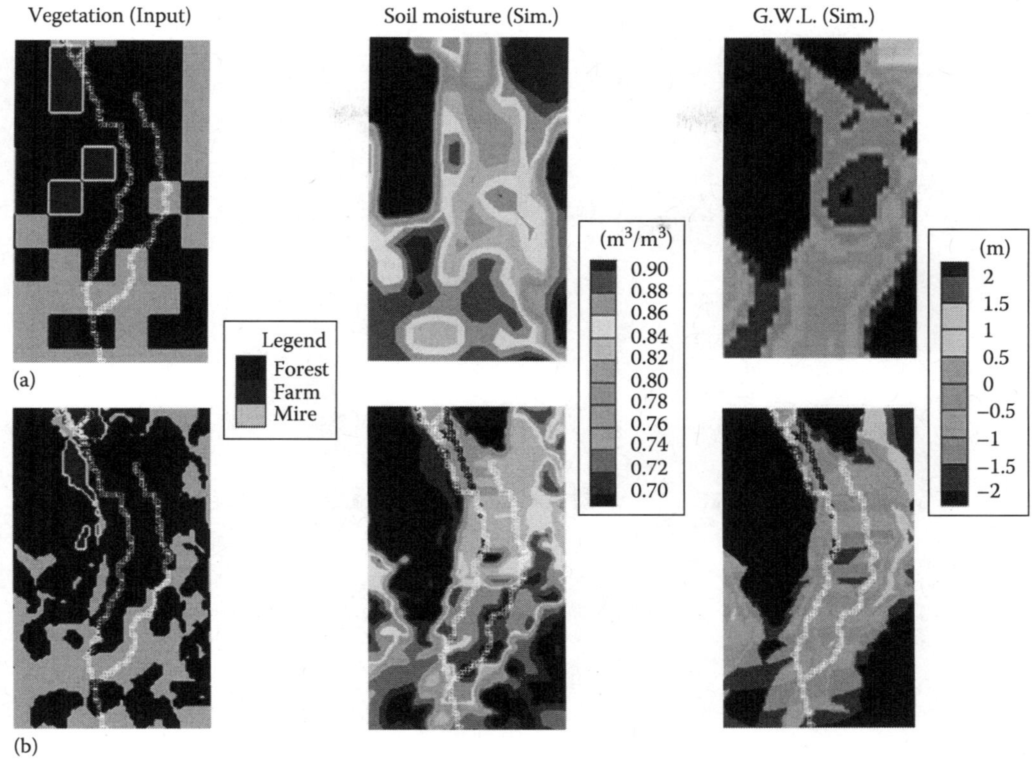

FIGURE 16.3 Improvement in hydrologic cycle by using finer resolution. Simulated results in (a) the regional area in Kushiro River catchment at resolution of 500 m and (b) downscaled local area in Kucyoro River catchment at resolution of 100 m.

clearly the local peak of soil moisture and groundwater level in the finer resolution. This improvement is important for the further evaluation of not only hydrologic change but also vegetation succession in the mire. Because boreal and subarctic peatlands store about 15%–30% of the world's soil carbon as peat [23] and affect the dynamics of greenhouse gases such as methane, this result would help to construct the improvement in boundless biogeochemical model with finer resolution [1]. This improvement was verified spatiotemporally by the observed and GIS data [26,27,29,33]. Because the elevation change by sediment deposition was a dominant factor that affects hydrologic change in the long term, simulated values of river discharge and *SS* concentration were compared with the observed values downstream of the Kucyoro River entering the mire (Figure 16.1) during 2002–2003 (Figure 16.4). The discharge simulated by the NICE agreed well with the observed discharge and the reported value [24] not only in the summer (July 2–22, 2002) but also in the spring snowmelt season (April 23–May 13, 2003) because this model includes the snowmelt process [39]. The snowmelt runoff has a greater value of base flow than the summer runoff, and this was also reproduced well by the NICE. Although the correlation obtained using the annual averaged coefficients of the empirical sediment-rating curve reproduces well the observed *SS* concentration in the summer season, it greatly underestimates the observed value for the spring snowmelt season. The model reproduces better the *SS* value for the entire year because it includes flush-out of the mass transport process at higher precipitation levels, which indicates that this flush-out process is very important in the spring snowmelt runoff [29]. The model also reproduced well the spatial distribution of elevation change during the last 30 years and showed that the aggradation is predominant downstream of the rivers flowing into the mire because most of the inflowing sediments are deposited there mainly by the rapid decrease in elevation gradients and by the increase in drag coefficients in the alder groves [29].

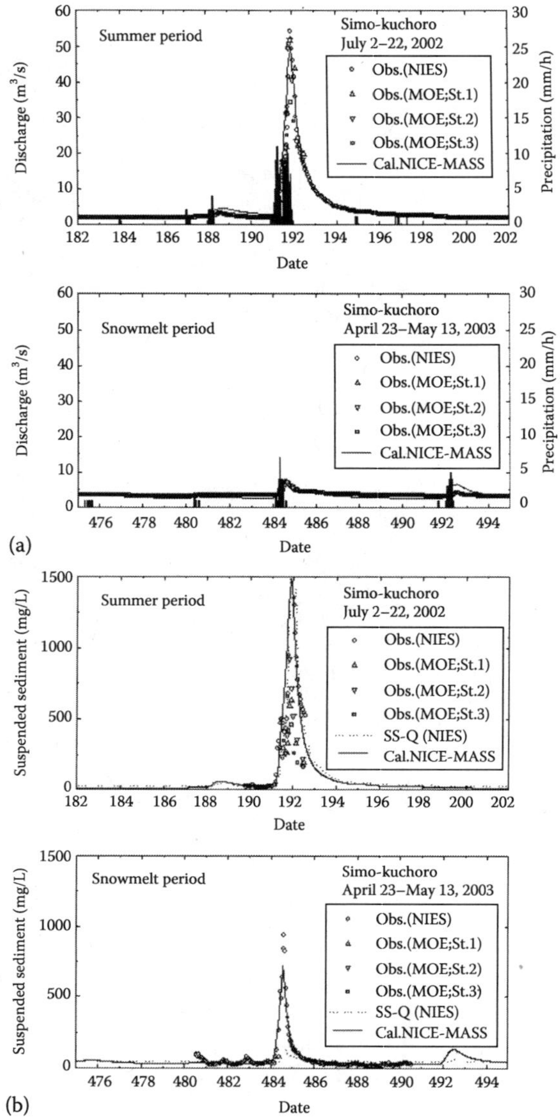

FIGURE 16.4 Verification of (a) river discharge and (b) SS concentration in the downstream of Kucyoro River shown in Figure 16.1. (From Nakayama, T., 2010, *River Res. and Appl.*, 26(3), 305, 2010.) Open circles are observed values, and open triangle and open square are the previous observation. (From Ministry of Environment, Nature restoration project in Kushiro Shitsugen Wetland, Kushiro, Japan, 2004.) Dashed lines are SS concentration estimated by empirical sediment-rating curve. Bold lines are simulated values by the NICE. Precipitation at Tsurui observation station is also plotted in Figure 16.4a.

16.5.2 Relationship between Drying Phenomenon and Alder Invasion

The long-term simulation results show that the soil moisture and groundwater level have decreased around the alder invasion area in the mire, owing mainly to the river channelization and the increased sediment load (Figure 16.5a) [25,29,38]. This is supported by previous findings that the drying, which is opposite to watering up [9], is caused primarily by hydrologic changes such as decreased recharge rate, increased water uptake, lowering of the groundwater level, increase in surface runoff, sedimentation,

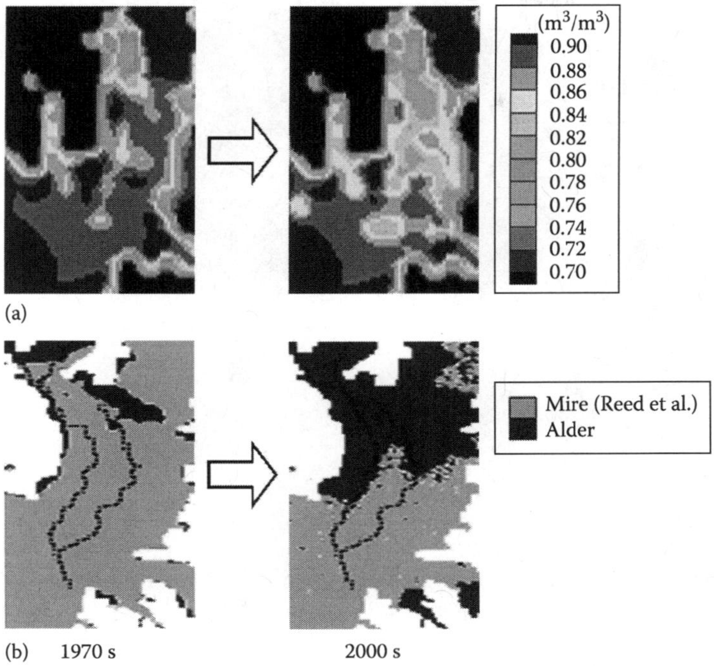

FIGURE 16.5 Simulation results of (a) soil moisture [38] and (b) alder invasion [26] in the box in Figure 16.1.

and increased nutrient input [7,24,25,38,39]. NICE also reproduces quantitatively the spatial distribution of alder invasion up to the present time (Figure 16.5b) accompanied by the hydrogeologic change [26]. The GIS data show that reed was predominant in 1976 (reed, 85.3%; alder, 14.7%) and that alder spread in 1997 (reed, 52.9%; alder, 47.1%) (Figure 16.1). This was reproduced accurately by NICE in both 1976 (84.0% and 16.0%) and 1997 (67.0% and 33.0%). Invasion of alder is very sensitive to water table depth, nutrient input, and other factors. Alder usually grows on local mounds in the mire in the shallow groundwater. In some parts, this phenomenon is included as a reduction factor for water table depth in Equation 16.4. Because the spatial scale of a mound is often smaller than the grid spacing of the simulation, it will be necessary to simulate it by using a finer grid in order to reproduce the local heterogeneity of groundwater flow [20]. This two-way interaction between vegetation and the water table is one of the key mechanisms affecting the dynamics of mire ecosystems [49].

16.5.3 Nonlinear Interaction between Geomorphology and Ecohydrology

The simulated result indicates a positive feedback between geomorphology and eco-hydrology on irregular slopes and in heterogeneous vegetation (Figure 16.6), whereas most previous studies have considered regular slopes and their relation to string or maze patterns [50,51,59]. The increased sediment deposition around the outlets of the rivers, causing drier conditions and higher nutrient content, has promoted the invasion of alder, and the expanded groves have increased friction to hydrodynamic flow and reduced water sediment transport capacity, increasing aggradation [59]. This nonlinear effect in a transient simulation is also related to the formation of concave-upward elements [51], hummock–hollow topography [12], or a string–flark pattern [57], in which runoff is higher in areas with lower biomass density (sediment erosion) and lower in areas with higher biomass (sediment deposition) on regular slopes [51]. Further, more sediment is deposited around ridges (flow divergence) and eroded around depressions (flow convergence) on irregular slopes, which makes the heterogeneity of vegetation more complex. This can be explained by the concept of derivation of the area drained over the slope, in which

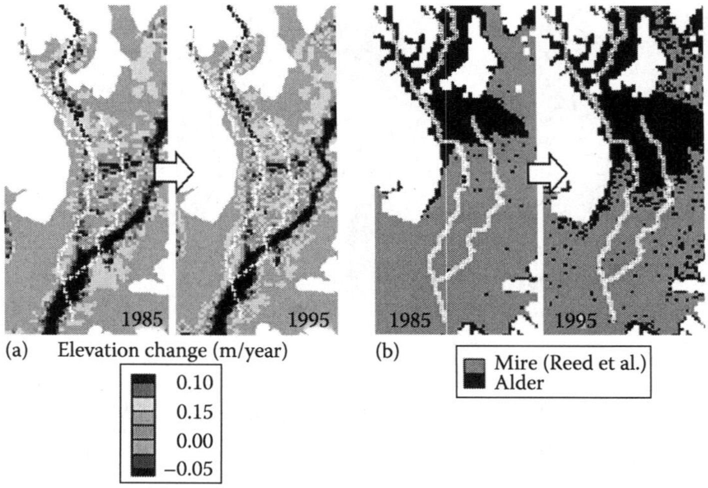

FIGURE 16.6 Simulated result of feedback process between geomorphology and eco-hydrology in the mire; (a) elevation change rate and (b) spatial distribution of alder and reed in 1985 and 1995. (From Nakayama, T., *Hydrol. Process.*, 26(16), 2455, 2012.) White grid, meandering river before channelization in 1976; black grid, straightened river after channelization in 1997 as shown in Figure 16.1.

the accumulation of water and sediment by any hillslope drainage process can be related to the ratio of area drained per unit contour length to slope gradient [21]. Furthermore, the positive feedback on short time scales may induce self-organization within the mire ecosystem, which buffers the strong physical gradient that characterizes the two vegetation types (invasive alder-dominant shrub forest and indigenous reed–sedge vegetation). This positive feedback might have developed toward a drastic change through external forcing or a catastrophic bifurcation [52] due to drying and sediment aggradation. Further research is necessary to evaluate the model's sensitivity to additional nonlinear effects: for example, the possible hysteresis effect arising from vegetation impacts on soil properties [51]. The model has to include detailed information about the particle size distribution or differences in zonation and vegetation growth in order to improve the reproduction of the local heterogeneity of elevation change in the mire. Nitrogen fixation by seedlings also plays an important role in the growth of alder [4], resulting in longer time needed to recover the mire. These are also related to the necessity to replace the linear succession model with an alternative state and transition model to explain multiple or divergent successional pathways [60], which will be of practical use as an early warning system in the estimation of ecosystem shift.

16.6 Summary and Conclusions

In this study, the process-based NICE model was further developed to include the mutual interaction and feedback of hydrogeologic and vegetation dynamics in the Kushiro Mire. The model iteratively simulated the hydrologic cycle, elevation change, and vegetation succession, including the competition between two vegetation types, to evaluate the nonlinear relationship between drying and alder invasion in the mire and to examine a method for mire restoration. Sediment erosion/deposition mechanisms change the topography, affecting water redistribution and the resultant vegetation succession. The vegetation change alters friction to hydrodynamic flow and the following sediment transport process. NICE reproduced this positive feedback, showing a clear correlation between drying and alder invasion at longer time scales. This result also implies that alder invasion is related to multiple gradients about hydrology, alkalinity, and nutrients that limit plant growth in the mire.

Acknowledgments

Some of the simulations were run on NEC SX-8 supercomputer at the Center for Global Environmental Research (CGER), NIES. The author wishes to thank the Kushiro Development and Construction Department, Hokkaido Regional Development Bureau, for providing discharge and agricultural data at the Kushiro River catchment.

References

1. Battin, T.J., Luyssaert, S., Kaplan, L.A., Aufdenkampe, A.K., Richter, A., and Tranvik, L.J. 2009. The boundless carbon cycle. *Nature Geoscience* 2: 598–600.
2. Bridgham, S.D., Pastor, J., Janssens, J.A., Chapin, C., and Malterer, T.J. 1996. Multiple limiting gradients in peatlands: A call for a new paradigm. *Wetlands* 16(1): 45–65.
3. Bugmann, H.K.M. 1996. A simplified forest model to study species composition along climate gradients. *Ecology* 77(7): 2055–2074.
4. Burgess, D. and Peterson, R.L. 1987. Effect of nutrient conditions on root nodule development in *Alnus japonica. Canadian Journal of Botany* 65: 1658–1670.
5. Chapin, F.S., Walker, L.R., Fastie, C.L., and Sharman, L.C. 1994. Mechanisms of primary succession following deglaciation at Glacier Bay, Alaska. *Ecological Monographs* 64(2): 149–175.
6. Cumming, S.G. and Burton, P.J. 1994. Zelig++ v0.9 documentation, user notes and installation guide. Univ. British Columbia, Canada. http://dino.wiz.uni-kassel.de/model_db/mdb/zelig.html
7. Dacey, J.W.H. and Howes, B.H. 1984. Water uptake by roots controls water table movement and sediment oxidation in short spartina marsh. *Science* 224: 487–489.
8. Digital National Land Information GIS Data of Japan. 1997. Database of land cover in Japan. Ministry of Land Infrastructure and Transport of Japan. http://nlftp.mlit.go.jp/ksj/
9. Dube, S., Plamondon, A.P., and Rothwell, R.L. 1995. Watering up after clear-cutting on forested wetlands of the St. Lawrence lowland. *Water Resources Research* 31(7): 1741–1750.
10. Einstein, H.A. 1950. The bed load function for sediment transportation in open channel flows. *Technical Bulletin* No. 1026, USDA Soil Conservation Service, Washington, DC, p. 71.
11. Environment Agency of Japan. 1993. Vegetation class data in Hokkaido, Japan. Japan Integrated Biodiversity Information System. http://www.biodic.go.jp/J-IBIS.html
12. Foster, D.R., King, G.A., Glaser, P.H., and Wright, Jr. H.E. 1983. Origin of string patterns in boreal peatlands. *Nature* 306: 256–258.
13. Geographical Survey Institute of Japan. 1999. *Digital map 50 m grid (elevation), Nippon-II* (CD-ROM), Ibaraki, Japan.
14. Hokkaido National Agricultural Experiment Station. 1985. *Soil map of arable land in Hokkaido*, scale 1:600,000, Hokkaido, Japan.
15. Hokkaido Regional Development Bureau. 2006. Nature restoration project in Kushiro Mire—Enforcement planning for measuring sediment inflow (Kucyoro River). http://www.ks.hkd.mlit.go.jp/kasen/15/pdf/2.pdf (in Japanese).
16. Hotes, S., Poschlod, P., Sakai, H., and Inoue, T. 2001. Vegetation, hydrology, and development of a coastal mire in Hokkaido, Japan, affected by flooding and tephra deposition. *Canadian Journal of Botany* 79: 341–361.
17. International Union for Conservation of Nature and Natural Resources (IUCN). 2001. IUCN Red List of Threatened Species (Ver. 3.1). http://www.iucnredlist.org/
18. Istanbulluoglu, E. and Bras, R.L. 2005. Vegetation-modulated landscape evolution: Effects of vegetation on landscape processes, drainage density, and topography. *Journal of Geophysical Research* 110: F02012. DOI:10.1029/2004JF000249.

19. Japan Meteorological Agency (JMA). 1970–2003. *AMeDAS (Automated Meteorological Data Acquisition System) Annual Reports at 1970–2003*. Japan Meteorological Business Support Center (CD-ROM).
20. Kikuchi, A., Nakagoshi, N., and Onda, Y. 2003. Hydrological setting of infertile species-rich wetland—A case study the warm temperate Japan. *Journal of Environmental Sciences* 15(2): 279–283.
21. Kirkby, M.J. 1978. *Hillslope Hydrology*. Wiley, Chichester, U.K., 389pp.
22. Kushiro Branch Office. 2002. *Agriculture of Kushiro in 2002*. Hokkaido Prefecture (in Japanese), Hokkaido, Japan.
23. Limpens, J., Berendse, F., Blodau, C., Canadell, J.G., Freeman, C., Holden, J., Roulet, N., Rydin, H., and Schaepman-Strub, G. 2008. Peatlands and the carbon cycle: From local processes to global implications—A synthesis. *Biogeosciences* 5: 1475–1491.
24. Ministry of Environment. 2004. Nature restoration project in Kushiro Shitsugen Wetland, Kushiro, Japan. http://www.kushiro.env.gr.jp/saisei/english/top_e.html
25. Nakamura, F., Sudo, T., Kameyama, S., and Jitsu, M. 1997. Influences of channelization on discharge of suspended sediment and wetland vegetation in Kushiro Marsh, northern Japan. *Geomorphology* 18: 279–289.
26. Nakayama, T. 2008. Factors controlling vegetation succession in Kushiro Mire. *Ecological Modelling* 215: 225–236. DOI:10.1016/j.ecolmodel.2008.02.017.
27. Nakayama, T. 2008. Shrinkage of shrub forest and recovery of mire ecosystem by river restoration in northern Japan. *Forest Ecology and Management* 256: 1927–1938. DOI:10.1016/j.foreco.2008.07.017.
28. Nakayama, T. 2009. Simulation of ecosystem degradation and its application for effective policy-making in regional scale. In *River Pollution Research Progress*, eds. Mattia N. Gallo and Marco H. Ferrari, pp. 1–89. Nova Science Pub., Inc., New York.
29. Nakayama, T. 2010. Simulation of hydrologic and geomorphic changes affecting a shrinking mire. *River Research and Applications* 26(3): 305–321. DOI:10.1002/rra.1253.
30. Nakayama, T. 2011. Simulation of complicated and diverse water system accompanied by human intervention in the North China Plain. *Hydrological Processes* 25: 2679–2693. DOI:10.1002/hyp.8009.
31. Nakayama, T. 2011. Simulation of the effect of irrigation on the hydrologic cycle in the highly cultivated Yellow River Basin. *Agricultural and Forest Meteorology* 151: 314–327. DOI:10.1016/j.agrformet.2010.11.006.
32. Nakayama, T. 2012. Impact of anthropogenic activity on eco-hydrological process in continental scales. *Procedia Environmental Sciences* 13: 87–94. DOI:10.1016/j.proenv.2012.01.008.
33. Nakayama, T. 2012. Feedback and regime shift of mire ecosystem in northern Japan. *Hydrological Processes* 26(16): 2455–2469. DOI: 10.1002/hyp.9347.
34. Nakayama, T. 2012. Visualization of missing role of hydrothermal interactions in Japanese megalopolis for win-win solution. *Water Science and Technology* 66(2): 409–414. DOI:10.2166/wst.2012.205.
35. Nakayama, T. and Fujita, T. 2010. Cooling effect of water-holding pavements made of new materials on water and heat budgets in urban areas. *Landscape and Urban Planning* 96: 57–67. DOI:10.1016/j.landurbplan.2010.02.003.
36. Nakayama, T. and Hashimoto, S. 2011. Analysis of the ability of water resources to reduce the urban heat island in the Tokyo megalopolis. *Environmental Pollution* 159: 2164–2173. DOI:10.1016/j.envpol.2010.11.016.
37. Nakayama, T. and Shankman, D. 2013. Impact of the Three-Gorges Dam and water transfer project on Changjiang floods. *Global and Planetary Change* 100: 38–50. DOI:10.1016/j.gloplacha.2012.10.004.
38. Nakayama, T. and Watanabe, M. 2004. Simulation of drying phenomena associated with vegetation change caused by invasion of alder (*Alnus japonica*) in Kushiro Mire. *Water Resources Research* 40(8): W08402. DOI:10.1029/2004WR003174.

39. Nakayama, T. and Watanabe, M. 2006. Simulation of spring snowmelt runoff by considering micro-topography and phase changes in soil layer. *Hydrology and Earth System Sciences Discussions* 3: 2101–2144.
40. Nakayama, T. and Watanabe, M. 2008. Missing role of groundwater in water and nutrient cycles in the shallow eutrophic Lake Kasumigaura, Japan. *Hydrological Processes* 22: 1150–1172. DOI:10.1002/hyp.6684.
41. Nakayama, T. and Watanabe, M. 2008. Role of flood storage ability of lakes in the Changjiang River catchment. *Global and Planetary Change* 63: 9–22. DOI:10.1016/j.gloplacha.2008.04.002.
42. Nakayama, T., Yang, Y., Watanabe, M., and Zhang, X. 2006. Simulation of groundwater dynamics in North China Plain by coupled hydrology and agricultural models. *Hydrological Processes* 20(16): 3441–3466. DOI:10.1002/hyp.6142.
43. Nakayama, T., Watanabe, M., Tanji, K., and Morioka, T. 2007. Effect of underground urban structures on eutrophic coastal environment. *Science of the Total Environment* 373(1): 270–288. DOI:10.1016/j.scitotenv.2006.11.033.
44. Nakayama, T., Sun, Y., and Geng, Y. 2010. Simulation of water resource and its relation to urban activity in Dalian City, Northern China. *Global and Planetary Change* 73: 172–185. DOI:10.1016/j.gloplacha.2010.06.001.
45. Nakayama, T., Hashimoto, S., and Hamano, H. 2012. Multi-scaled analysis of hydrothermal dynamics in Japanese megalopolis by using integrated approach. *Hydrological Processes* 26(16): 2431–2444. DOI:10.1002/hyp.9290.
46. Ohara, T., Matsushita, K., Futamase, K. et al. 1975. *Explanatory Text of the Hydrogeological Maps of Hokkado—Kushiro*. Geological Survey of Hokkaido, Hokkaido, Japan.
47. Ohmi Environment Preservation Foundation. 2001. *Research about Vegetation Condition for Reclamation of Reed in Lake Biwa*. Research Report, Kyoto, Japan. http://nippon.zaidan.info/jigyo/2001/0000024546/jigyo_info.html (in Japanese).
48. Pastor, J., Peckham, B., Bridgham, S., Weltzin, J., and Chen, J. 2002. Plant community dynamics, nutrient cycling, and alternative stable equilibria in peatlands. *American Naturalist* 160: 553–568.
49. Ridolfi, L., D'Odorico, P., and Laio, F. 2006. Effect of vegetation–water table feedbacks on the stability and resilience of plant ecosystems. *Water Resources Research* 42: W01201. DOI:10.1029/2005WR004444.
50. Rietkerk, M., Dekker, S.C., Wassen, M.J., Verkroost, A.W.M., and Bierkens, M.F.P. 2004. A putative mechanism for bog pattering. *American Naturalist* 163: 699–708.
51. Saco, P.M., Willgoose, G.R., and Hancock, G.R. 2007. Eco-geomorphology of banded vegetation patterns in arid and semi-arid regions. *Hydrology and Earth System Sciences* 11: 1717–1730.
52. Scheffer, M., Carpenter, S., Foley, J.A., Folke, C., and Walker, B.H. 2001. Catastrophic shifts in ecosystems. *Nature* 413: 591–596.
53. Schroder, A., Persson, L., and De Roos, M.D. 2005. Direct experimental evidence for alternative stable states: A review. *Oikos* 110(1): 3–19.
54. Sellers, P.J., Randall, D.A., Collatz, G.J., Berry, J.A., Field, C.B., Dazlich, D.A., Zhang, C., Collelo, G.D., and Bounoua, L. 1996. A revised land surface parameterization (SiB2) for atmospheric GCMs. Part I: Model formulation. *Journal of Climate* 9: 676–705.
55. Shimizu, Y. and Arai, N. 1988. Numerical simulation of flow and bed variations in the river mouth region. *Research Report of Civil Engineering Research Institute of Hokkaido* 419: 5–36.
56. Shinsho, H. 1982. Note on the alder thickets—*Alnus japonica* Steud—in Kushiro Moor, eastern Hokkaido II. *Research Report of Kushiro City Museum* 9: 27–36 (in Japanese, with English abstract).
57. Swanson, D.K. and Grigal, D.F. 1988. A simulation model of mire pattering. *Oikos* 53: 309–314.

58. Toda, H., Nakasa, N., Hirano, K., Uemura, Y., Okino, T., and Kawashima, H. 2002. Nitrogen cycling in the watershed of the Chikuma River. *Journal of the Japanese Agricultural Systems Society* 18(2): 90–99 (in Japanese, with English abstract).
59. Van de Koppel, J., van der Wal, D., Bakker, J.P., and Herman, P.M.J. 2005. Self-organization and vegetation collapse in salt marsh ecosystems. *American Naturalist* 165: E1–E12.
60. Wright, J.M. and Chambers, J.C. 2002. Restoring riparian meadows currently dominated by Artemisa using alternative state concepts—Above-ground vegetation response. *Applied Vegetation Science* 5: 237–246.

17

Isotope Hydrogeology

Adam Porowski
Institute of Geological Sciences PAS

AUTHOR

Adam Porowski is a licensed hydrogeologist at the Institute of Geological Sciences of the Polish Academy of Sciences (ING PAN) in Warsaw, Poland. His principal research interests concern the application of isotopic methods in hydrogeology and hydrogeochemistry. He obtained a hydrogeological license according to the Polish Geological and Mining Law from the Ministry of Environment of Poland. Dr. Porowski was granted doctoral scholarships at the Kent State University, United States, and the Helmholtz Centre for Environmental Research, Germany. He completed many courses concerning various aspects of applied hydrogeology and modern methods of groundwater sampling, groundwater dating, borehole measurements, hydrogeological and geochemical data acquisition, and processing and interpretation, for example, summer school of geothermal exploration (Izmir, Turkey), tracers in hydrogeology (University of Adam Mickiewicz, Poland), ICDP drilling course (Windischeschenbach, Germany), and dating of geological materials (GADAM Centre, Gliwice, Poland). He is the author of many scientific papers, coeditor of one book, leader or principal investigator of three research grants sponsored by the National Committee for Scientific Research and the Ministry of Science and Higher Education of Poland, and author and coauthor of several studies performed for the industrial sector. Actually, he is a secretary of the Commission on Mineral and Thermal Waters of IAH and one of the leaders of the EU FP7 ATLAB Project performed at the ING PAN, where he is responsible for the development of collaboration between science and industry.

PREFACE

Technological development in mass spectrometry and spectroscopy in the last decades makes it possible to determine on the routine basis of the isotopic composition of various types of natural waters and different organic and inorganic compounds dissolved in waters. Nowadays, the application of isotopic methods and use of isotopic data are indispensible in modern hydrogeology and hydrology in order to understand many problems concerning formation of groundwater resources, and protection of their quality and their sustainable use. In this chapter, the bases of environmental isotopes in hydrogeology are described. The isotopic composition of water and the main factors affecting its fractionation during descending parts of hydrological cycle and underground circulation are explained. The most useful applications of selected environmental isotopes (i.e., O, H, N, S, ^{3}H) are shown.

17.1 Introduction

The application of naturally occurring stable isotopes in hydrology dates back to 1935, when oxygen isotopic composition of water from Lake Michigan was determined by Dole (1935) (after Edmunds 2007). However, the beginning of modern stable isotope geochemistry is commonly connected with the landmark paper of Harold Urey published in 1947, where the basis of isotope fractionation was discovered, which subsequently opened up huge possibilities for stable isotope application in a variety of environmental studies. Several very well-written up-to-date reviews on the subject of application of environmental isotopes in hydrogeology and hydrology have been published, for example, Fritz and Fontes (1980, 1986), Clark and Fritz (1997), Criss (1999), Cook and Herczeg (2000), or Leibundgut et al. (2009).

The application of isotopic methods and isotopic data in hydrogeological studies noted a fast increase in the last decades. Scientists, hydrogeologists, and environmental engineers have understood that isotopic data give unique information and knowledge complementary to that obtained by classical geochemistry and classical hydrogeology and hydraulics.

Isotopes are atoms of a single element whose nuclei contain the same number of protons but a different number of neutrons. All isotopes can be divided into three fundamental kinds, namely, radioactive, radiogenic, and stable (e.g., Drever, 1997).

In hydrogeology and hydrology, the most widely used are stable or radioactive isotopes called environmental isotopes, that is, the naturally occurring isotopes of principal elements found in abundance in Earth's environment like H, O, N, C, and S. The relative mass differences between isotopes of these elements are large and show measurable fractionation during physical, chemical, or biological reactions encountered in our environment. That is why the isotopic composition of different compounds of these elements can be used for tracing the water in hydrological cycle, as well as tracing the cycle of nutrients, carbon, nitrogen, and sulfur.

Environmental isotopes are used not only for tracing the isotopic signature of different compounds or groundwater origin but also for tracing and explaining various biogeochemical processes encountered in our environment and their rates. Isotopic methods are significant tool for studying recharge processes, soil–water–atmosphere interactions, and contaminant transport and sink.

Watershed characterization and catchment studies concerning basin budget (i.e., assessment of groundwater recharge, storage, and discharge) are also the most important fields where isotopic methods are successfully applied.

This chapter discusses the basic issues of the isotopes in environmental studies and summarizes their most useful applications.

17.2 Basis of Isotopes in Environmental Studies: Definitions, Abundances, and Measurements

Isotopes are atoms (i.e., nuclides) of a single element whose nuclei contain the same number of protons but a different number of neutrons. The number of protons uniquely defines the element, whereas the number of neutrons defines the isotope of that element. It means that isotopes of a particular element occupy the same position in the periodic table: they have the same atomic number but different atomic weights.

Normally simplified form of isotope identification is used: ^{M}X, for example, ^{18}O, ^{2}H.

Isotopes can be divided into three fundamental kinds, namely, radioactive, radiogenic, and stable. The characteristic feature of radioactive isotopes is their decay into different forms called "daughter" atoms, at statistically predictable rates (Criss, 1999). Such decay is accompanied by the emission of radiation such as α-particles ($^{4}He^{+}$), β-particles (i.e., electrons or positrons), or γ-rays (i.e., electromagnetic radiation). On the other hand, stable isotopes do not decay. However, the term "stable" is relative, and it is possible that some nuclides have half-lives so long that their radioactivity has not been detected. The radiogenic isotopes are continuously formed by the decay of radioactive parent nuclides, and they may be stable or radioactive. Their abundance in the environment depends on the time period, the amount of parent nuclides present, as well as diffusion processes. Nonradiogenic stable isotopes (e.g., ^{18}O, ^{13}C, ^{34}S) are those whose overall abundance is not a function of time, and variations in their abundance are caused by certain physicochemical processes. That is why they are of primary importance in various hydrological, hydrogeological, and geochemical applications. For example, they are used to understand the origin of water or processes that have affected water since it was formed (i.e., entered the particular aquifer). Radioactive isotopes in water studies are used primarily to measure age or water residence time. Up to date about 270 stable nuclides and over 1700 radioactive nuclides have been identified (Clark and Fritz, 1997). Only 21 elements are pure elements, in the sense that they have only one stable isotope (Hoefs, 1997). All other elements are mixtures of at least two isotopes. It means that the atomic weight of any particular element represents the weighted average of the atomic weights of constituent isotopes (Criss, 1999):

$$A_w = \sum_i B_i^{\star} W_i \tag{17.1}$$

where

A_w is the atomic weight of the element

B_i is the relative abundance

W_i is the intrinsic weight of the ith nuclide

Such isotopically controlled variations of atomic weights of the elements have important consequences for physicochemical properties of elements and molecules.

17.2.1 Isotope Effects and Fractionations

The chemical properties of an element are generally determined by atomic number. It means that the chemical behavior of different isotopes of the same element is almost identical. However, the replacement of any atom in a molecule by one of its isotopes produces some variations in atomic mass and, in consequence, differences in thermodynamic properties of the element and the molecules of which they may be composed. Such differences in physicochemical properties arising from variations in atomic mass of the elements are usually called isotope effects, which source arises as a result of quantum mechanical effects. More detailed description of this subject can be found, for example, in Bigeleisen (1965),

TABLE 17.1 Differentiation of Physical Properties of Water of Different Isotopic Composition

Property	$^1H_2{}^{16}O$	$^2H_2{}^{16}O$	$^1H_2{}^{18}O$
Density (20°C, in g/cm³)	0.0997	1.1051	1.1106
Temperature of greatest density (°C)	3.98	11.24	4.30
Melting point (1 atm, in °C)	0.00	3.81	0.28
Boiling point (1 atm, in °C)	100.00	101.42	100.14
Viscosity (20°C centipoise)	1.002	1.247	1.056
Vapor pressure (at 100°C, in Torr)	760.00	721.60	

Source: After Hoefs, J., *Stable Isotope Geochemistry*, Springer-Verlag, Berlin, Germany, 197 pp, 1997.

Kyser (1987), O'Neil (1987a), and Criss (1999). To demonstrate such isotopic effects, a good example is the properties of water composed of different stable isotopes of oxygen and hydrogen (Table 17.1).

The relative mass differences are significant only among isotopes of the lighter elements, in which the difference in mass is a large fraction of the total mass of the atom. For example, deuterium (^2H) has 100% larger mass than its isotope protium (^1H); mass difference between ^{18}O and ^{16}O is 12.5%, whereas between ^{81}Br and ^{79}Br is only 2.5%; heavy water, $^2H_2{}^{16}O$, has a mass of 20 compared to normal water, $^1H_2{}^{16}O$, which has a mass of 18.

The most important here is the fact that molecules with differences in mass have different reaction rates, which leads to isotope fractionation—that is, unequal distribution of isotopes between two coexisting substances or two phases of the same substance. Therefore, isotope fractionation occurs in any thermodynamic reaction that causes isotopic ratios in particular phases to differ from one another. It is expressed by the fractionation factor α, which is defined as the ratio of the isotope ratios for the reactant and product (Clark and Fritz, 1997):

$$\alpha = \frac{R_{reactant}}{R_{product}} \tag{17.2}$$

for example,

$$\alpha^2 H_{water-vapor} = \frac{\left(^2H/^1H\right)_{water}}{\left(^2H/^1H\right)_{vapor}} \tag{17.3}$$

where R is the ratio of the heavy (usually rare) isotope to the light (usually common) stable isotope in reactant and product.

Isotopic fractionation effects are small, and α-values in most systems are close to unity, for example,

$$\alpha^{18} O_{water-vapor} = 1.0094 \text{ at } 25°\text{C}$$

$$\alpha^2 H_{water-vapor} = 1.079 \text{ at } 25°\text{C}$$

However, hydrogen exhibits the largest fractionations observed in nature mostly due to two times difference in mass between ^1H and ^2H. In the artificial system in which hydrogen gas is equilibrated with water using a platinum catalyst in room temperature fractionation, factor α is equal to 3.7—which is not close to unity (Criss, 1999).

Isotope fractionation is generally considered for equilibrium conditions for which reliable fractionation factors can be calculated or experimentally measured (Clark and Fritz, 1997). In case, when isotopes are randomly distributed over all the positions in substrates and products, α is related to the equilibrium constant, K, for isotope exchange reactions (Kyser, 1987):

$$\alpha = K^{\frac{1}{n}} \tag{17.4}$$

where n is the number of atoms exchanged (usually $n = 1$ and $\alpha = K$).

All thermodynamic processes producing isotope fractionation can be divided in two main groups (Kyser, 1987; O'Neil, 1987a; Hoefs, 1997; Criss, 1999):

1. Isotope exchange reactions
2. Kinetic processes

Isotope exchange reactions are equilibrium reactions in which the isotopes of a single element are exchanged between two substances (O'Neil, 1987a,b).

They can be simple physical changes of state,

$$H_2O_{water} \leftrightarrow H_2O_{vapor}, \alpha\,^{18}O_{water-vapor} = \frac{\left(^{18}O/^{16}O\right)_{water}}{\left(^{18}O/^{16}O\right)_{vapor}} \tag{17.5}$$

or chemical transformation,

$$CO_2 + H_2O \leftrightarrow H_2CO_3, \alpha\,^{13}C_{CO_2\text{-}H_2CO_3} = \frac{\left(^{13}C/^{12}C\right)_{CO_2}}{\left(^{13}C/^{12}C\right)_{H_2CO_3}} \tag{17.6}$$

or other aqueous, mineral–solution, and gas solution reactions.

The basis of such physicochemical isotopic fractionation is the difference in the strength of bonds formed by the light or the heavy isotopes of a given element, which, in consequence, provide differences in their reaction rates.

Kinetic isotope effects are also common in nature. They are usually associated with fast, incomplete, or unidirectional processes like evaporation, diffusion, dissociation, or metabolic processes (i.e., especially bacterially mediated irreversible reactions, e.g., sulfate reduction, methanogenesis). In case of diffusion or evaporation, such kinetic effects are explained by the different translation velocities of isotopic molecules moving through a phase or across a phase boundary, which arise from mass differences between isotopes. Whereas in case of dissociation reactions, the strength of bonds in molecules containing the heavy or light isotope plays the key role, molecules containing the heavy isotope are more stable and characterize higher dissociation energies (O'Neil, 1987a).

Mass differences that influence on the bond strengths between heavy and light isotopes as well as on the translation velocities and reaction rates cause the heavy isotopes to preferentially fractionate into more condense phase, that is, into the solid phase in mineral–solution reactions, or into the aqueous phase in gas (or vapor)–liquid reactions. The most important and useful fact is that the isotopic fractionation is strongly dependent on the temperature and not on any other variable of state (Criss, 1999). The fractionation factor α is very close to 1 at high temperature (i.e., if $\alpha = 1.000$, no fractionation occurs; system is well mixed); with decreasing temperature, α departs from unity and isotopic fractionation between substrates and products occurs. Equilibrium fractionation factors can be determined at different temperatures experimentally and can be also calculated from the partitioning functions.

17.2.2 Isotopic Ratios, Delta (δ) Notation, and Standards

In geochemical and environmental applications, the stable isotopes are measured as the ratio of the two most abundant isotopes of a given element. However, an absolute isotope ratio R is not used to report stable isotope measurements or to report isotopic composition of a given element. It is common practice to report the measured difference in the isotopic composition of the sample and of the standard (i.e., known reference) in terms of δ-values according to the formula

$$\delta_S = \frac{R_S - R_{ST}}{R_{ST}} * 1000 = \left(\frac{R_S}{R_{ST}} - 1\right) * 1000 \tag{17.7}$$

where R refers to the isotope ratio in the sample S and in the internationally accepted reference (standard) ST; the coefficient 1000 converts the δ-values to per mil (‰), as the fractionation processes do not cause huge variation in isotopic composition. Isotopic international standards are assigned to the value of zero per mil on the δ-scale of interest. The appropriate reference standards for the main stable environmental isotopes are listed in Table 17.2.

Fractionation factor α between two substances A and B in terms of δ-values is expressed by the equation (O'Neil, 1987b)

$$\alpha_{A-B} = \frac{1 + \frac{\delta_A}{1000}}{1 + \frac{\delta_B}{1000}} = \frac{1000 + \delta_A}{1000 + \delta_B} \tag{17.8}$$

$$\delta_A - \delta_B \approx 1000 \ln \alpha = \Delta_{A-B} \tag{17.9}$$

The value of Δ_{A-B} represents the difference in isotopic composition between substance A and B. This difference is also an approximation of the term $1000 ln\alpha$, which is the widely used and convenient

TABLE 17.2 Ratios, Abundances, and References for the Main Stable Environmental Isotopes

Isotope	Ratio	International Reference Standard (Abundance Ratio)	Material	Name of Reference
^2H	$^2H/^1H$	VSMOW ($1.5575*10^{-4}$)	Water	VSMOW2
^{18}O	$^{18}O/^{16}O$	VSMOW2 ($2.0051*10^{-3}$)	Water	VSMOW2
^{18}O	$^{18}O/^{16}O$	VPDB ($2.0672*10^{-3}$)	Calcite	VPDB
^{13}C	$^{13}C/^{12}C$	VPDB ($1.1237*10^{-2}$)	Calcite	VPDB
^{15}N	$^{15}N/^{14}N$	AIR N_2 ($3.677*10^{-4}$)	Gas	Air nitrogen
^{34}S	$^{34}S/^{32}S$	VCDT ($4.5005*10^{-2}$)	Troilite	VCDT

Source: Abundance Ratio Taken from Clark, I. and Fritz, P., *Environmental Isotopes in Hydrogeology*, Lewis Publishers, New York, 328 pp, 1997.

expression of fractionation factor α in per mil convention of δ-values. This expression arises from the correlation between α and absolute temperature:

$$\ln\alpha_{A-B} = aT^{-2} + bT^{-1} + c \tag{17.10}$$

where

A and *B* are reacting substances
a, b, c are correlation coefficients
T is the temperature in K

This relation is linear when plotting $\ln\alpha$ against the inverse of absolute temperature.

The most useful standard for oxygen and hydrogen isotopes is standard mean ocean water (SMOW), defined by Craig (1961a). For example, notation $\delta^{18}O_S = -3.5‰$ vs. SMOW means that our sample *S* is 3.5‰ depleted in oxygen-18 isotope in relation to SMOW. To be more specific, isotopic ratio $^{18}O/^{16}O$ in sample *S* reveals depletion in concentration of oxygen-18 in relation to $^{18}O/^{16}O$ ratio in SMOW. When absolute isotopic ratios of reference materials are known, it is easy to calculate absolute isotope concentration ratio of a given sample of interest. However, δ-values are used in stable isotope geochemistry to report isotopic composition and compare the results of isotopic measurements.

Over the past few decades, appropriate materials have been established as internationally accepted reference standards, the so-called certified reference materials, due to the fact that stocks of international standards have been already exhausted, that is, Vienna Canyon Diablo Troilite (VCDT) or Vienna Pee Dee Belemnite (VPDB) is no longer available. The International Atomic Energy Agency (IAEA) has taken steps to produce replacements (e.g., certified reference materials or reference materials) that are almost the same in isotopic composition (e.g., Vienna standard mean ocean water (VSMOW2)) or very close and carefully calibrated according to international standards. These references, however, are also limited in quantities and cannot be used by laboratory on an each day basis. Therefore, it is important that each laboratory has its own internal "working standards" calibrated according to internationally accepted reference materials and standards. Two organizations collaborate on the calibration, cataloguing, and distribution of international reference materials: (1) the IAEA and (2) the National Institute of Standard and Technology (NIST, formerly the National Bureau of Standards (NBS)). Reference materials and further information are available through both of these organizations. The useful websites are as follows:

IAEA: www.iaea.org; for reference materials, see page http://nucleus.iaea.org/rpst/index.htm
NIST: www.nist.gov; for reference materials, see page http://www.nist.gov/srm/

17.3 Environmental Isotopes in Hydrogeological Applications

17.3.1 Stable Isotopes of Oxygen and Hydrogen in Hydrological Cycle

Hydrological cycle is a process of continuous movement (cycling) of water masses between the Earth's lithosphere, atmosphere, hydrosphere, and biosphere. It moves from one reservoir to another by way of processes like evaporation, condensation, precipitation, deposition, runoff, infiltration, sublimation, transpiration, melting, and groundwater flow. During its movement the water changes state among liquid, vapor, and ice. Each step of this cycle and phase transformation causes specific fractionation of oxygen and hydrogen isotopes in water molecules. In consequence, the oxygen and hydrogen isotopic ratios provide conservative tracers that elucidate origin, phase transitions, and transport of water during hydrological cycle.

17.3.1.1 Precipitation and Global Meteoric Waterline

The first step of hydrological cycle, the formation of atmospheric water vapor by evaporation from oceans, is a nonequilibrium process due to the effects of low humidity and mixing of different vapor masses (Gonfiantini, 1986; Clark and Fritz, 1997).

TABLE 17.3 Typical $\delta^{18}O$ and δ^2H Values of Water in Major Reservoirs and Their Relative Volumes

Reservoir	Volume (%)	δ^2H	$\delta^{18}O$
Ocean	97.2	0 ± 5	0 ± 1
Ice caps and glaciers	2.15	−230 ± 120	
Groundwaters	0.62		
Vadose water		−40 ± 70	−5 ± 15
Dilute groundwater		−50 ± 60	−8 ± 7
Brines		−75 ± 50	0 ± 4
Surface waters	0.017		
Freshwater lakes		−50 ± 60	−8 ± 7
Saline lakes and inland seas		−40 ± 60	−2 ± 5
River and stream channels		−50 ± 60	−8 ± 7
Atmospheric water	0.001	−150 ± 80	−20 ± 10

Source: After Criss, R.E., *Principles of Stable Isotope Distribution*, Oxford University Press, New York, 254pp, 1999.

The reverse process—condensation of atmospheric water vapor and formation of precipitation and meteoric waters—takes place in equilibrium maintained between vapor and condensed water in particular air temperature. This process is the most important for hydrological studies because it produces atmospheric precipitation and meteoric waters—that is, all waters that originate from atmospheric precipitation.

To produce rain, the vapor masses evaporated over the ocean have to be cooled down. Cooling occurs by adiabatic expansion (i.e., no loss of enthalpy) as warm air rises to lower pressures or by radiative heat loss (Clark and Fritz, 1997). When the dew point is passed (i.e., the temperature at which humidity is 100%), water vapor condenses to maintain thermodynamic equilibrium, and it will rain (or snow). The evaporative flux of ocean water to the atmosphere is mainly controlled by temperature, and more than 70% of water vapor is produced from evaporation over oceans in warm, subtropical regions. That is why the air masses with water vapor normally move from its vapor source area to higher latitudes, and over continents, it cools and loses water vapor along this way as precipitation. Within the cloud, equilibrium fractionation between vapor and the condensing water preferentially fractionates ^{18}O and ^2H into the rain or snow. Such process of vapor mass evolution toward colder regions and toward isotopically depleted precipitation is called as rainout. The process of condensation and rainout causes large variation in the isotopic composition of meteoric water. The total range of $\delta^{18}O$ values in natural precipitation is about +4‰ to −62‰, while that of δ^2H values is about +40‰ to −500‰ (Criss, 1999), (See Table 17.3).

However, analyses of isotopic composition of meteoric water from different parts of the world revealed that their ^{18}O and ^2H isotopes vary in characteristic, predictable manner and undergo linear correlation defined by the equation

$$\delta^2H = 8\delta^{18}O + 10 \tag{17.11}$$

The previous equation became known as the "global meteoric waterline (GMWL)." This relationship, derived first by Craig (1961b), traces the isotopic compositions of all natural waters originating from atmospheric precipitation and not subjected to surface evaporation (Figure 17.1).

Nowadays, based on the isotopic composition of precipitation provided by the Global Network of Isotopes in Precipitation (GNIP)—a worldwide network of meteorological stations that monitor isotopic composition of precipitation coordinated by the IAEA since 1961—from 1961 to 2000, the GMWL is defined more precisely by the equation (Gourcy et al., 2007)

$$\delta^2H = 8.14\ (\pm 0.02)\ \delta^{18}O + 10.9\ (\pm 0.2),\ R^2 = 0.98 \tag{17.12}$$

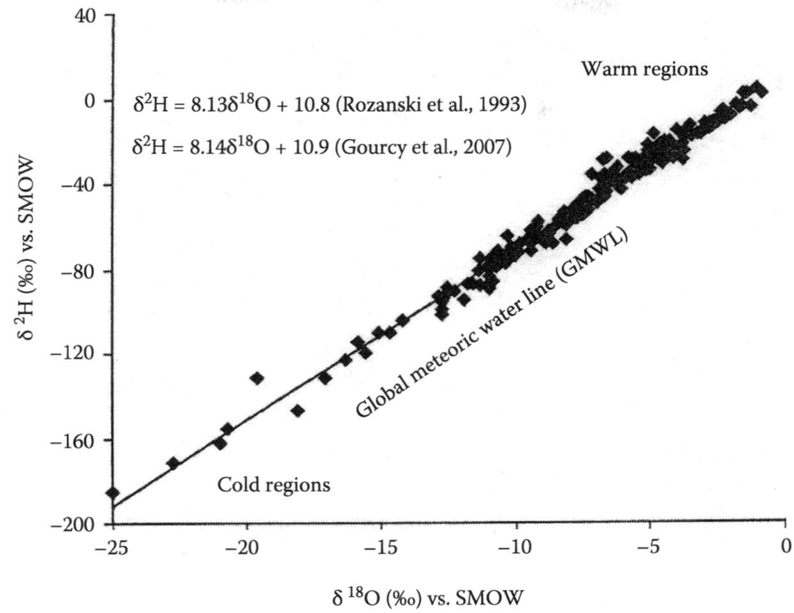

FIGURE 17.1 GMWL derived for weighted average annual values for precipitation monitored at GNIP stations (From Clark, I. and Fritz, P., *Environmental Isotopes in Hydrogeology*, Lewis Publishers, New York, 328pp, 1997.) Modified by the author.

The slope of the GMWL is close to eight due to the in-cloud ratio of the water–vapor equilibrium fractionation factors for ^2H and ^{18}O:

$$s = \frac{1000\ln\alpha^2 H_{water-vapor}}{1000\ln\alpha^{18} O_{water-vapor}} = 8.2 \text{ at } 25°\text{C} \tag{17.13}$$

Deuterium excess factor *d*, proposed by Dansgaard (1964), is defined as

$$d = \delta^2\text{H} - 8\delta^{18}\text{O} \tag{17.14}$$

and characterizes the deuterium excess in global precipitation for which the average value of *d* is about 10‰. The deuterium excess factor explains, for example, the displacement of GMWL from the position of ocean water, that is, VSMOW (Figure 17.1).

On a regional scale, minor deviations from GMWL are observed as a result of local meteorological conditions. It means that any location on the globe will have its own local meteoric waterline (LMWL) with particular slope *s* and *d* factor (Figure 17.2).

The largest deuterium excesses in precipitation are found in the Eastern Mediterranean region (Criss, 1999). For example, the eastern meteoric waterline (EMWL) introduced by Gat and Carmi (1970) is described by the following equation:

$$\delta^2\text{H} = 8\delta^{18}\text{O} + 22 \tag{17.15}$$

This fact clearly shows that for detailed stable isotope studies within a particular geographical region, the LMWL should be established to compare surface and groundwater data and to avoid interpretation errors. Generally, the strong correlation between temperature and ^{18}O and ^2H determines the position of precipitation on the MWLs. The *T*–δ (i.e., temperature–isotope) correlation

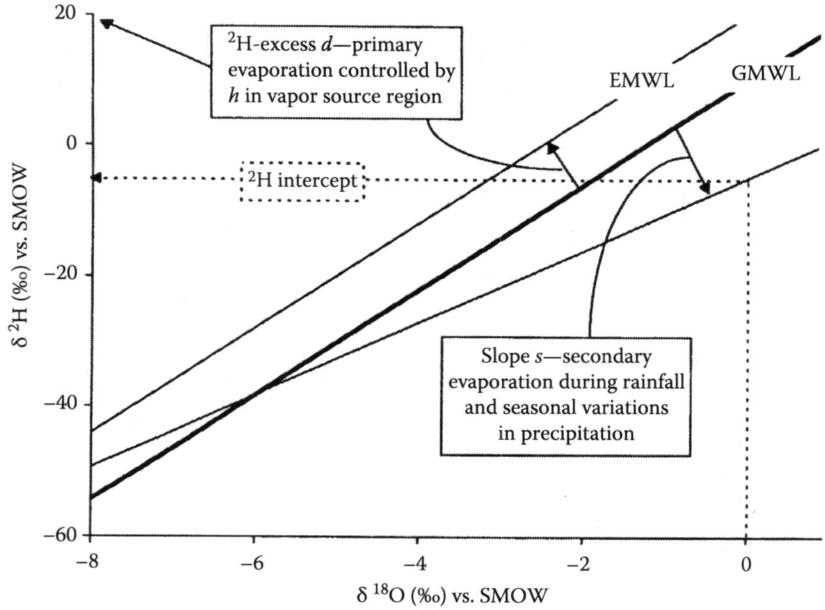

FIGURE 17.2 Main factors affecting the slope *s* and deuterium excess factor *d* of MWLs. (From Clark, I. and Fritz, P., *Environmental Isotopes in Hydrogeology*, Lewis Publishers, New York, 328pp, 1997.) There is some distinction between deuterium excess, which is calculated for slope eight, and the deuterium intercept, calculated for the actual slope and for $\delta^{18}O = 0$.

and evolution of isotopic composition during rainout can be described and modeled according to the Rayleigh distillation equation:

$$R = R_o f^{(\alpha-1)} \tag{17.16}$$

where

- R_o is the initial isotope ratio of the vapor
- R is the isotope ratio of the vapor when given fraction f of the vapor had reacted (i.e., rained out)
- f is the residual fraction of the vapor in the cloud
- α is the fractionation factor for equilibrium water–vapor isotopic exchange at ambient in-cloud temperature

For *T*–δ correlation it is possible to derive isotope effects due to seasons, altitude, latitude, continentality, and paleoclimates. The linear relationships between surface air temperatures and ^{18}O and ^{2}H of mean annual precipitation on a global scale were derived by Dansgaard (1964):

$$\delta^{18}O = 0.695T - 3.6‰$$

$$\delta^{2}H = 5.6T - 100‰$$

where T refers to annual surface air temperature in °C.

Departures from these relationships can occur on local and regional scale and when monitoring data are too short or when event scale is investigated (i.e., individual storm).

From this correlation however, it can be deduced that precipitation at higher latitudes tends to have more negative $\delta^{18}O$ and δ^2H values (the so-called latitude effect: $\delta^{18}O$ and δ^2H contents decrease with rainout of air masses); it is important to notice that polar regions are situated at the end of the Rayleigh distillation process (Clark and Fritz, 1997). Land masses and continents induce the rainout from vapor masses and more rapid evolution of isotopic composition of precipitation (continental effect), mostly due to influence of topography and temperature extremes that occur over the continents. Altitude effects (due to diversify relief of lands, ^{18}O and ^{2}H contents decrease with increasing altitude) and seasonal effects (i.e., winter precipitation is usually depleted in ^{18}O and ^{2}H relative to summer precipitation) are characteristic for particular local settings and meteorological conditions.

17.3.1.2 Identification of Groundwater Origin

All groundwater encountered in the Earth's crust can be classified into three categories depending on the origin (Gat, 1981): (1) meteoric waters, (2) paleowaters, and (3) formation waters.

17.3.1.2.1 Meteoric Waters

Meteoric waters can be defined as waters that originate from atmospheric precipitation (e.g., rain, snow, rivers, lakes, most of shallow groundwater of low-temperature environment) and actively take part in the modern hydrological cycle. These types of waters constitute the vast majority of fresh and potable water reservoirs. From the origin point of view, any type of water found in geological environment where $\delta^{18}O$ and $\delta^{2}H$ isotopic values cause its location along the GMWL or LMWL should be considered as meteoric in origin. The shallow groundwaters usually belong to meteoric waters and lie along or close to LMWL or GMWL. It means that their isotopic composition is equal to the mean weighted annual composition of precipitation. However, in some cases, the important deviations from isotopic composition of MWL are found in groundwaters, which may complicate proper interpretation of their origin. Such deviations are most likely where depth to the groundwater table is small, reflecting isotopic variability in rainstorms or recharge. There are three main reasons of such deviations: (1) propagation of seasonal variations from precipitation, (2) seasonal or local biases in recharge, and (3) evaporation processes.

In the unsaturated zone, between surface and groundwater level of saturation zone, infiltration waters reveal seasonal variation of the isotopic composition. However, these variations usually are attenuated with depth by the geological environment. The attenuation scheme is presented in Figure 17.3.

This attenuation arises from hydrodynamic dispersion within the soil, where the infiltration travels at different velocities through pores of different sizes and in some cases also through cracks, fissures, or

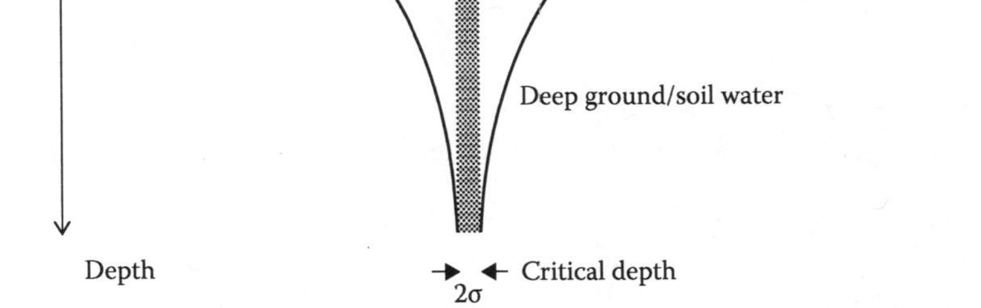

FIGURE 17.3 Scheme of the attenuation of seasonal variation of ^{18}O and ^{2}H isotopic composition of precipitation during infiltration through the unsaturated zone and filtration within saturated zone up to a critical depth below which isotopic variability is less than the analytical precision. (From Clark, I. and Fritz, P., *Environmental Isotopes in Hydrogeology*, Lewis Publishers, New York, 328pp, 1997.)

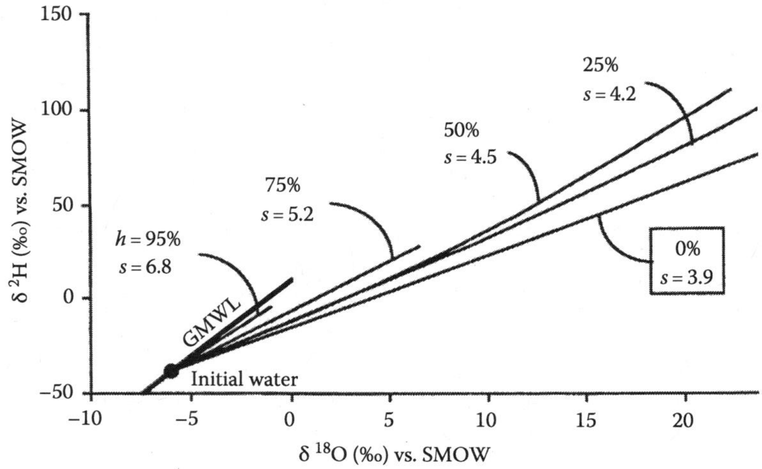

FIGURE 17.4 Isotopic enrichment in evaporating water and the effect of humidity. Heavy line refers to GMWL. (From Gonfiantini, R., Environmental isotopes in lake studies, in Fritz, P. and Fontes, J.Ch. (ed.), *Handbook of Environmental Isotope Geochemistry, Volume II. The Terrestrial Environment*, Elsevier, Amsterdam, the Netherlands, 113pp, 1986.) At low humidity *h*, kinetic evaporation maximizes and the slope of the evaporation line will be low.

macropores (Leibundgut et al., 2009). In fine-grained soils, the depth at which the seasonal variations are damped out is less than in coarse-grained soils. In fractured or karstified rocks, attenuation is expected to be slow and depth of full attenuation relatively large. If the infiltration water reaches the water table of saturated zone before the seasonal signal has been fully attenuated (i.e., in case where thickness of unsaturated zone is less than depth of attenuation), this fluctuations will also be found in the groundwaters.

Seasonal or local biases in recharge can also be responsible for deviation between weighted average annual precipitation and groundwater. It regards the situation when recharge rates at one season of the year are greater than at other times of the year. For example, the recharge from snowmelt may be responsible for observed slight depletion in ^{18}O and ^{2}H of shallow groundwater. Moreover, in arid climates only the largest and longest-lived storm events recharge the groundwater systems, and the $\delta^{18}O$ and $\delta^{2}H$ values of those storms may differ from average values of MWL in arid regions.

Subaerial evaporation from water body results in a deviation from the MWL along a line with a lower slope, which depends mostly on the relative humidity (Figure 17.4) (Gonfiantini, 1986).

The same process of evaporation can remove moisture from soil of unsaturated zone, especially in arid and semiarid regions (see Figure 17.12).

Geyh and Ploethner (1997) proposed useful correction of $\delta^{18}O$ for evaporative enrichment:

$$\delta^{18}O_{corrected} = \frac{\delta^{2}H_{measured} - e\delta^{18}O_{measured} - d}{8 - e} \left(\text{Leibundgut et al., 2009}\right) \tag{17.17}$$

where

e corresponds to the slope of the evaporation line determined
d is the deuterium excess factor for global precipitation

Corrected values indicate the isotopic composition of water without evaporative enrichment. Once the $\delta^{18}O$ has been corrected, the corresponding $\delta^{2}H$ value can be obtained by solving the equation for GMWL, as the evaporation line intersects the GMWL in the point of corrected $\delta^{18}O$.

17.3.1.2.2 Paleowaters

Paleowaters (or more adequately—paleogroundwaters) are also meteoric in origin but were recharged in the geological past under similar or different climatic conditions. In many regional flow systems,

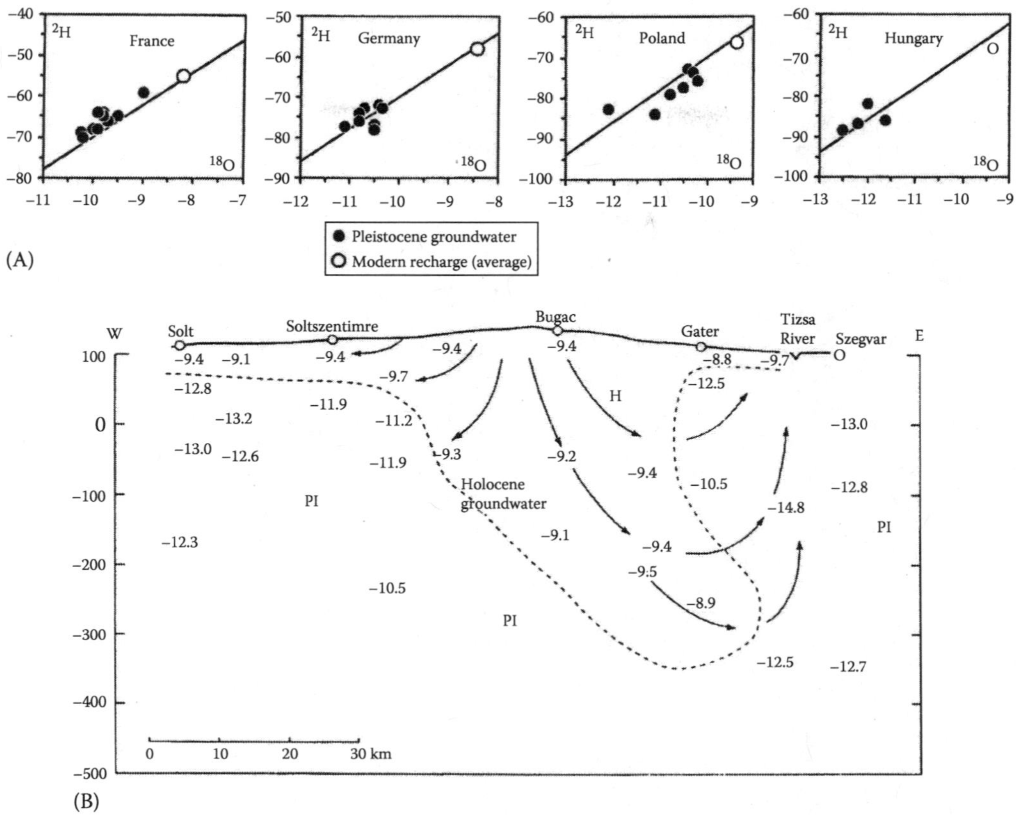

FIGURE 17.5 Exemplary isotopic composition of Pleistocene paleogroundwaters in selected countries of Europe (A) and cross section through the Great Hungarian Plain quaternary aquifer showing $\delta^{18}O$ values of groundwater recharged in the Pleistocene and Holocene (B). (Figure compiled after data from Clark, I. and Fritz, P., *Environmental Isotopes in Hydrogeology*, Lewis Publishers, New York, 328pp, 1997; Coplen, T.B. et al., Isotope engineering—Using stable isotopes of the water molecule to solve practical problems, in Cook, P. and Herczeg, A.L. (eds.), *Environmental Tracers in Subsurface Hydrology*, Kluwer Academic Publishers, Boston, MA, 79pp, 2000.)

groundwaters qualified as paleogroundwaters may be old due to their very low filtration velocities and long flow paths, but they may actually receive modern inputs in the recharge areas (Clark and Fritz, 1997) (Figure 17.5B). Paleogroundwaters may also form isolated reservoirs that are not in contact with modern hydrological cycle; their resources are then finite (e.g., Figures 17.6 and 17.7). Majority of paleogroundwaters are not part of active flow systems.

The basis of climate change is the change in air temperature and precipitation patterns. Taking into account the existence of clear correlation between isotopic composition of precipitation and groundwater of meteoric origin, we can expect that also paleogroundwaters reveal this correlation and their isotopic composition records past changes in climatic conditions under which the water has entered geological environment. Thus, the "paleoclimatic effect" is one of the most important tools in identifying paleogroundwaters (Clark and Fritz, 1997).

In temperate latitudes the dominant paleoclimatic effect is connected with Pleistocene glaciation and recharge in much colder climatic conditions. Late Pleistocene paleogroundwaters are isotopically depleted with respect to modern meteoric groundwaters and shifted along GMWL toward negative values (Figure 17.5A).

A very good illustration of Holocene- vs. Pleistocene-recharged groundwaters is the confined quaternary aquifer of the Great Hungarian Plain (Figure 17.5B). Groundwaters recharged during Pleistocene

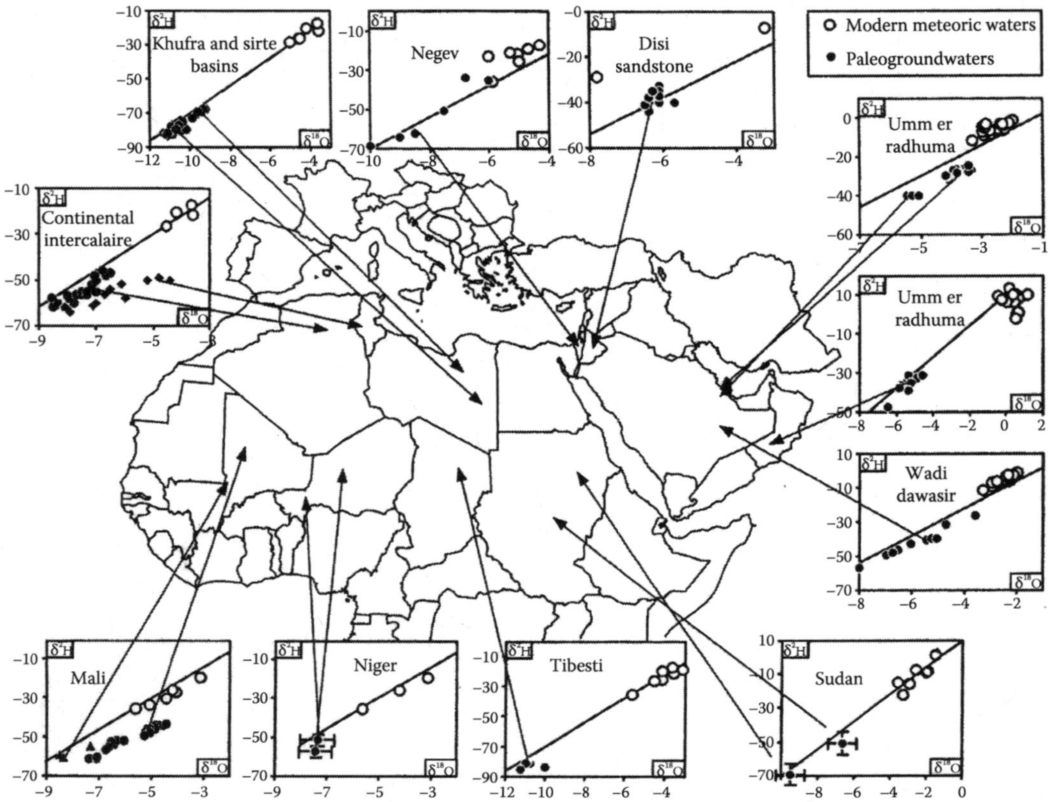

FIGURE 17.6 Examples of paleogroundwaters from the Middle East and North Africa. (From Clark, I. and Fritz, P., *Environmental Isotopes in Hydrogeology*, Lewis Publishers, New York, 328pp, 1997.) Data compiled from many authors.

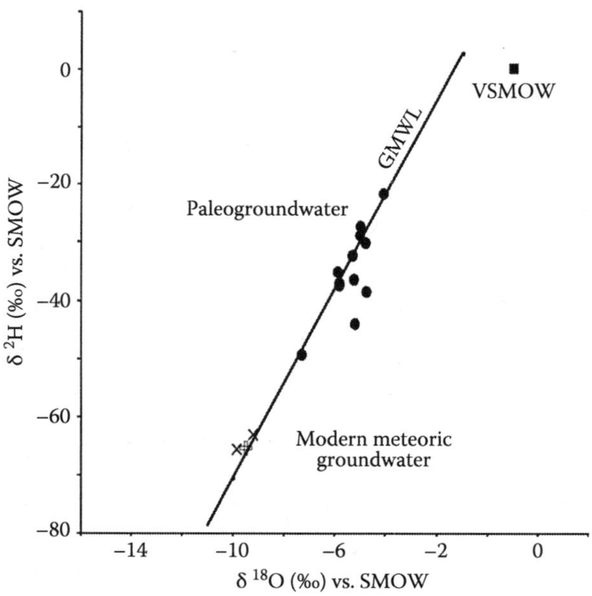

FIGURE 17.7 Example of paleogroundwaters (mine drainage waters) from Permian sediments of Foresudetic Monocline, Poland (Porowski, unpublished).

times in this region have $\delta^{18}O$ values more negative than −10‰. Moreover, content of ^{14}C of groundwater decreases along flow lines and reaches values less than 10 pmC (i.e., more than 20 ka BP) for water recharged in Pleistocene times (Deak and Coplen, 1996).

On the other hand, in arid regions, like the Eastern Mediterranean and North Africa, the paleoclimatic effect connected with the Late Pleistocene—Early Holocene is manifested by a displacement of the MWL. This is because the humid climates have characterized these regions in the past. The groundwaters recharged during pluvial climatic conditions in the Late Pleistocene or Early Holocene have isotopic composition, which locates them on or even below the modern GMWL ($d \approx 10‰$), whereas modern meteoric waters in these regions are located along MWL with deuterium excess factors of 15‰–30‰ (Figure 17.6).

In geological environment of temperate latitudes, it is also possible to find paleogroundwaters recharged in much warmer climatic conditions than nowadays in a particular geographical region (Figure 17.7).

Such waters are isotopically enriched with respect to modern waters and shifted toward positive $\delta^{18}O$ and $\delta^{2}H$ values along GMWL. They are much older than Pleistocene ones, presumably remained in pores of sediments since recharge in geological past and were disconnected completely from the modern hydrological cycle.

17.3.1.2.3 Formation Waters

Formation waters are those found in interstices and pores of crystalline and sedimentary rocks at greater depths. They are usually very slow moving or even stagnant, with heightened salinity, and their specific ^{18}O and ^{2}H isotopic composition is a result of water–rock interactions over a long time scale usually in elevated temperatures (Gat, 1981; Sheppard, 1986; Welhan, 1987). The term “formation water” is wide and has no implication on the origin or age of the water. Nevertheless, some waters formed in particular geochemical process can be identified as separate types with specific isotopic composition, for example, metamorphic waters and magmatic waters (Figure 17.8).

The majority of formation waters presented in Figure 17.8A and B come from depths between 500 and 3700 m; most of the reservoir rocks in sedimentary basins are of marine origin, within the Canadian Shield—crystalline rocks; presented waters are saline or brines (Sheppard, 1986). Waters plotted on or near MWL are the least saline and reveal meteoric origin. The trends of increasing salinity with enrichment in ^{18}O and/or ^{2}H are observed. As can be seen (Figure 17.8), unlike most modern meteoric waters and paleogroundwaters, in which stable isotopes of H and O behave conservatively, formation waters have been considerably affected by isotopic exchange with rock-forming minerals, usually in elevated temperatures, and presumably also by secondary mixing processes.

The characteristic ^{18}O-shift to the right of the MWL (Figure 17.8A) is predominantly due to isotopic exchange with ^{18}O-rich rock-forming minerals and particularly carbonates; the δD variations are less understood, and few possibilities are usually considered during interpretation, for example mixing of different waters (e.g., meteoric and metamorphic), exchange with H_2S (which is depleted in ^{2}H relative to water), fractionation during membrane filtration, or exchange with hydrous minerals. Some rare formation waters like those observed in oceanic crust and Canadian Shield can reveal enrichment in ^{2}H, which shifts them to the left of the MWL (Figure 17.8B). Fritz and Reardon (1979) attributed enrichment in ^{2}H to water participation in the process of hydration of primary silicates under closed-system conditions or low-temperature exchange with feldspars. The observed trend between some allochthonous end member (i.e., the most ^{2}H-enriched water) and meteoric water suggests mixing line rather than line of evolution of ^{18}O and ^{2}H composition (Clark and Fritz, 1997). Pore waters in oceanic sediments at depths of 100 m or more (Figure 17.8B) are often depleted in ^{18}O and ^{2}H in relation to modern ocean water. Such changes in isotopic composition of pore waters are related to the low-temperature alteration of basalts or of volcanic ash in the sediments (Muehlenbachs and Clayton, 1971).

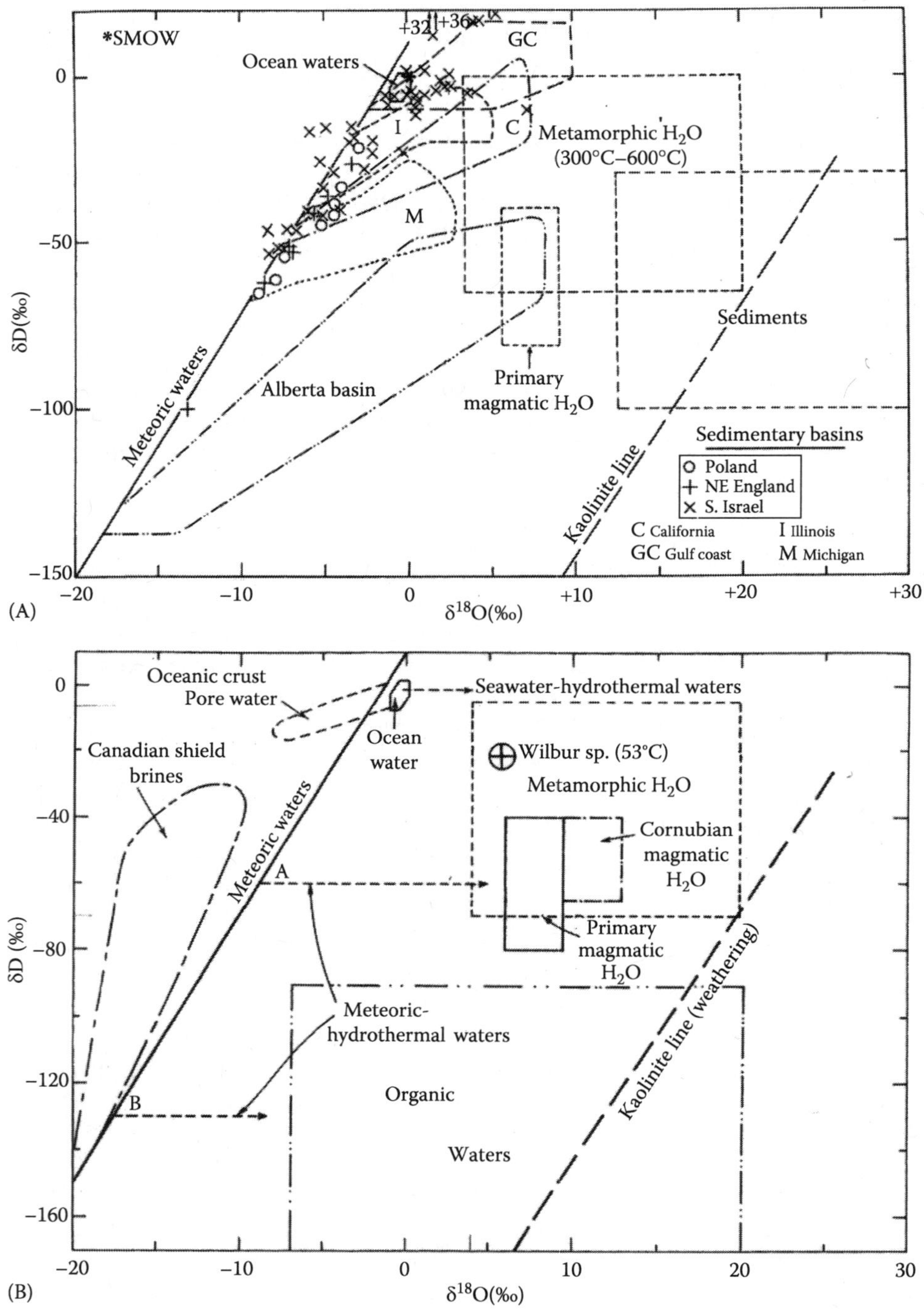

FIGURE 17.8 Isotopic composition and fields of formation waters in sedimentary basins, crystalline rocks, and other types of specific groundwaters encountered in deep geological environment. The field for common sedimentary rocks, kaolinite weathering line, and MWL is plotted for reference. Trends for the ^{18}O-shift due to water–rock interaction in hydrothermal systems for thermal waters of meteoric (A and B) and seawater origin (dashed arrows) were also shown. Plots after Sheppard (1986); data after many authors cited in Sheppard (1986). See text for additional explanation.

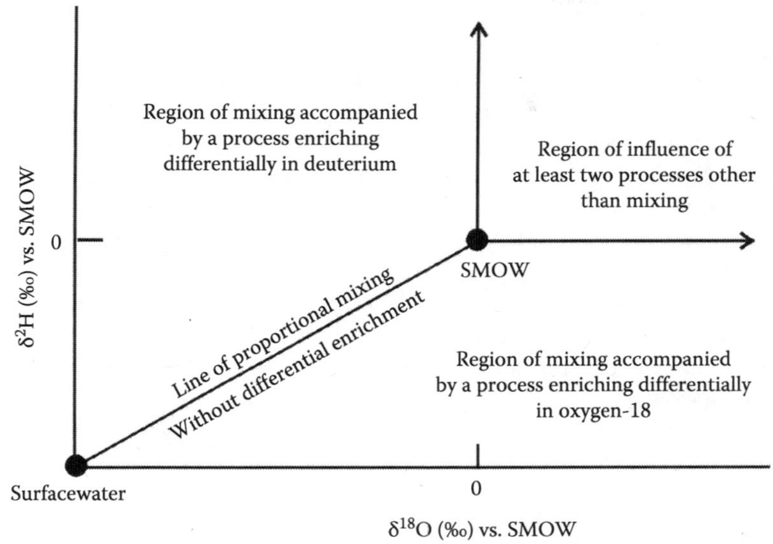

FIGURE 17.9 Schematic plot of δ^2H and $\delta^{18}O$ indicating the effects of various mixing and fractionation processes. (From Hitchon, B. and Friedman, I., *Geochim. Cosmochim. Acta*, 33, 1321, 1969.)

Other specific types of waters indicated on the plots (Figure 17.8) are usually connected with particular geochemical processes of their formation. Sheppard (1986) provides comprehensive description of these type of waters:

Metamorphic waters—are defined as waters that equilibrated with or were released from metamorphic rocks undergoing dehydration.

Magmatic waters—waters that were equilibrated with magma, regardless of their ultimate origin; it refers to separate aqueous phase and not to all types of water of any speciation that are physically dissolved in the magma.

Juvenile waters—waters that originate from the Earth's mantle or core and that have never been involved in the hydrosphere; hypothetical waters that have never been convincingly recognized.

Organic waters—those whose $^2H/^1H$ ratios are derived from the direct or indirect transformation of organic matter, bitumen, coal, kerogen, petroleum, organic gases, etc., by processes such as dehydration, dehydrogenation, oxidation, and/or exchange; the H isotope composition of organic water is related to that of their organic source material by the appropriate isotopic fractionation.

As can be seen on the plot (Figure 17.8), the fields for $\delta^{18}O$ and δ^2H of different waters may partly overlap each other, which may cause problems with proper interpretation and unambiguous water identification.

Figure 17.9 presents a general scheme that summarizes the interpretation of the observed trends in isotopic composition of waters, which deviate considerably from MWL.

The idea of differential fractionation of O and H during water reactions with rocks as well as mixing of waters of different origins is discussed in the next chapters.

17.3.1.3 Water–Rock Interaction and Hydrothermal Systems

The processes of water–rock interactions are of primary importance in modification and formation of specific isotopic composition of groundwater in deep parts of the geological environment, in high-temperature conditions, and over a geological time scale. Some evidence of water–rock interaction in low temperatures in shallow groundwater flow systems is also known. The alteration of the ^{18}O and ^{2}H composition of groundwater gives the insight into its origin, subsurface history, and geochemical reactions.

The isotopic exchange between water molecules and minerals is restricted to ^{18}O and ^{2}H. Isotope exchange is achieved by dissolution and reprecipitation (i.e., during chemical reactions) or by direct exchange of isotopes between water and the mineral crystal lattice (Clark and Fritz, 1997).

The ratios of $^{2}H/^{1}H$ and $^{18}O/^{16}O$ in composition of waters provide two independent "labels" or traces of their origin. The water-to-rock ratio, that is, the amount of exchangeable oxygen and hydrogen atoms in water and in the rock, determines the differential fractionation of O and H isotopes between reacting phases. In ordinary rocks oxygen is the most abundant element, making up about 50% by weight. In contrast, hydrogen is usually a trace element in a rock with concentration less than 2000 ppm (e.g., a rock with 1% of water by weight contains only 0.11% of hydrogen) (Sheppard, 1986). During water–rock interaction, both H and O isotope exchange may occur among the phases. In systems where the water is of external origin, this water is generally the dominant hydrogen reservoir, and it will therefore control the $^{2}H/^{1}H$ ratio of the system. The $^{18}O/^{16}O$ ratio of the water, however, may be determined by that of the exchangeable mineral oxygen and the temperature of exchange unless water-to-rock ratios are very large or the exchange reaction is ineffective for kinetic or other reasons.

The characteristic ^{18}O-shift, generally to more positive $\delta^{18}O$ values (Figure 17.9), is a result of the water trying to attain O isotope equilibrium with the ^{18}O-rich silicates and carbonates. The size of the shift is related to (1) the ratio of the quantity of oxygen in the exchangeable minerals to that in the water, (2) the temperature of exchange, and (3) the initial isotopic composition of the phases (Sheppard, 1986). Thus, the $\delta^{18}O$ of the water may not retain its "label" or memory of its source. For these reasons, the $\delta^{2}H$ of the water is often a better parameter to determine the source of the water. This is a crucial point in interpreting the origin of thermal waters (Figure 17.10).

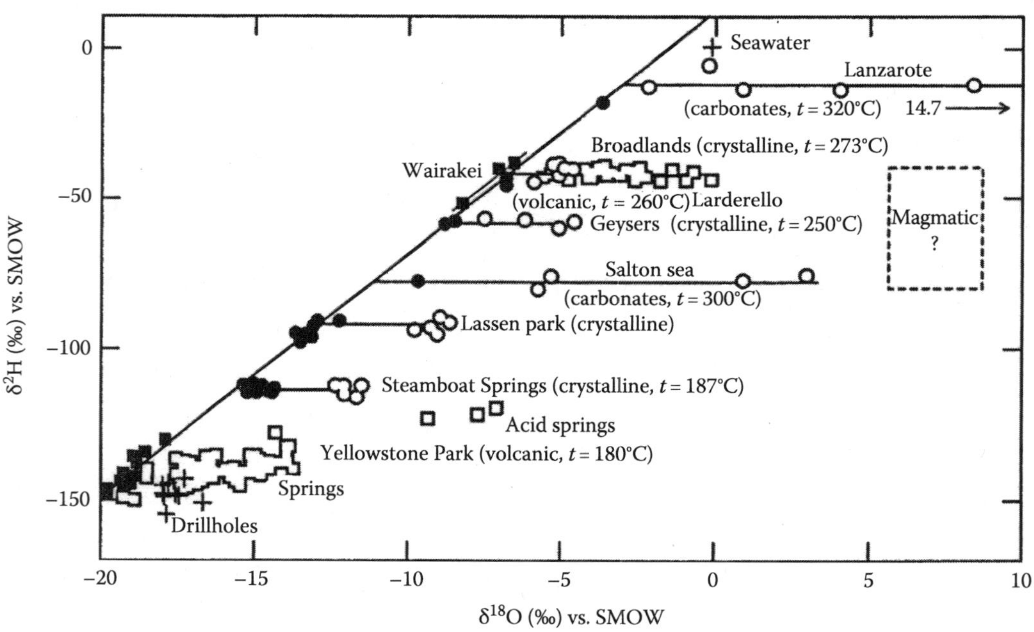

FIGURE 17.10 Isotopic composition of thermal waters and steam from various regions derived from meteoric water (open symbols) and of meteoric water local to each system (closed symbols). Thermal water demonstrates the strong ^{18}O-shift due to isotopic exchange with rocks. (From Truesdell, A.H. and Hulston, J.R., Isotopic evidence on environments of geothermal systems, Chapter 5, in Fritz, P. and Fontes, J.Ch. (eds.), *Handbook of Environmental Isotope Geochemistry, Volume 1, The Terrestrial Environment, A.* Elsevier, Amsterdam, the Netherlands, 179pp, 1980.) Modified by the author.

Craig (1963) showed that high-temperature geothermal systems have waters with the same δ^2H values as local precipitation but have variable enrichment in ^{18}O. Such ^{18}O-shift is attributed to exchange of oxygen in water with oxygen in silicate and carbonate minerals. Groundwaters of meteoric origin are generally far from isotopic equilibrium with minerals of the host rocks. Observed ^{18}O-shift in thermal waters reflects isotopic evolution toward mineral–water equilibrium (Clark and Fritz, 1997). The initial isotopic difference between the rocks and the recharge water, and the water-to-rock ratio, is also important: The ^{18}O contents of crystalline and carbonate rocks are much higher than in most meteoric waters; in case of low water–rock ratio, even a minor exchange will impart a measurable shift to the water. Thermal waters in systems with high water–rock ratios will have a diminished shift.

In low-temperature environments, another shift may be observed: fractionation between water and minerals may result in ^2H-enriched and ^{18}O-depleted waters located to the left of the MWL. Two dominant exchange reactions are usually considered in such case: (1) retrograde exchange between water and primary silicate minerals and (2) hydration of primary silicates (Clark and Fritz, 1997). To observe significant water enrichment in ^2H, very low water–rock ratios are required together with the geological time scale, because rates of exchange and hydration are very slow at low temperatures.

17.3.1.4 Groundwater Mixing

Mixing processes between different types of groundwaters occur at various scales in various hydrogeological conditions. The process of attenuation of seasonal variations in groundwater, described previously, can be an example of microscale mixing. Nevertheless, the isotopic methods can be used to quantify groundwater mixing at the local and regional scales, where mixing between groundwaters of different recharge origins, from different aquifers, and different flow systems can occur.

Mixing between two different groundwaters *A* and *B* (i.e., two-component mixing) can be quantified by a simple linear equation using δ^2H and/or $\delta^{18}O$:

$$\delta_{AB} = f^* \delta_A + (1 - f)^* \delta_B \tag{17.18}$$

where

δ_A and δ_B refer to δ^2H or $\delta^{18}O$ in water *A* and *B*, respectively
δ_{AB} refers to δ^2H or $\delta^{18}O$ in mixture of two waters
f refers to the fraction of water *A* in mixture *AB*

To solve the equation, all δ-values should be known. A graphical presentation of the two-component mixing is shown in Figure 17.11.

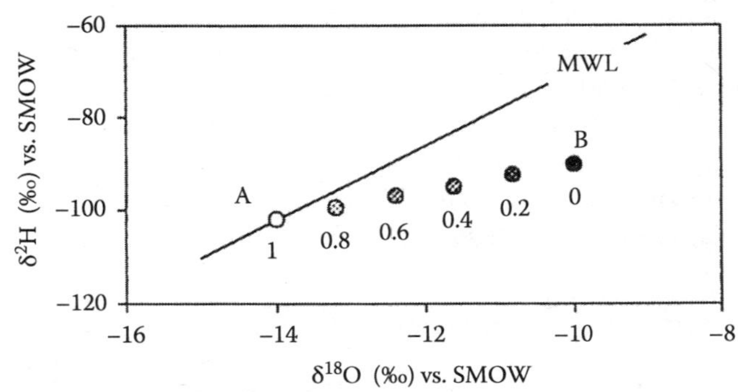

FIGURE 17.11 Two-component mixing quantified on the basis of isotopic compositions of groundwater *A* and *B*. The mixing line shows fraction of water *A* in *AB* mixture. (From Clark, I. and Fritz, P., *Environmental Isotopes in Hydrogeology*, Lewis Publishers, New York, 328pp, 1997.)

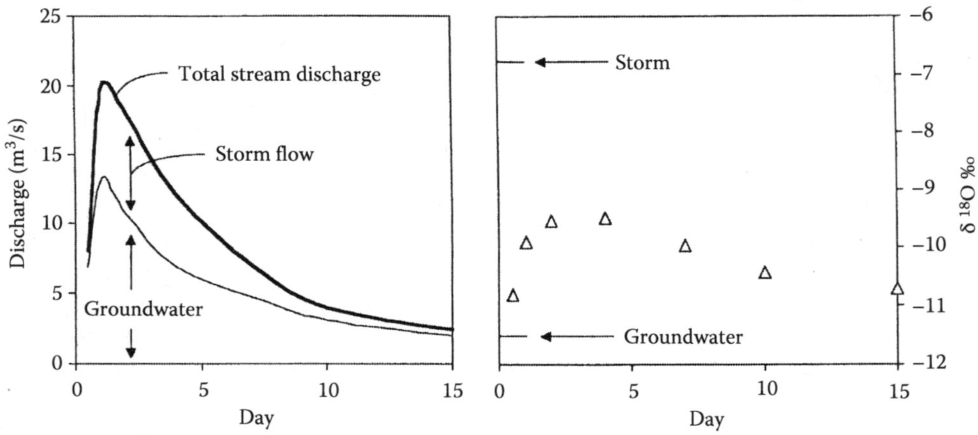

FIGURE 17.12 Example of ^{18}O isotope stream hydrograph separation for a two-component water system: storm event and groundwater. (From Clark, I. and Fritz, P., *Environmental Isotopes in Hydrogeology*, Lewis Publishers, New York, 328pp, 1997.)

Unlike most geochemical tracers, the ^{18}O and ^{2}H isotopes reveal a conservative behavior in mixing relationships and preserve the mixing ratio. That is why the mixing lines can be observed in various types of formation waters in sedimentary basins, which strongly suggest the admixture of water of meteoric origin even in greater depths (Figure 17.8).

Mixing process of two components can be observed in unsaturated zone between rain that infiltrates into the soil and evaporated soil moisture when low recharge rates occur (Allison et al., 1984). In such case, isotopic composition of infiltrating groundwater follows a line parallel to MWL but displaced to the right.

Catchment study concerning groundwater recharge, storage, and discharge is also one the most important fields where ^{18}O and ^{2}H isotopes are used to separate between mixing waters constituting, for example, baseflow and surface runoff. In the case of basing budget studies, the key issue is to assess the proportion of water resulting from storm events, which recharges groundwater and which is lost by surface runoff. Such isotopic method of the stream hydrograph separation into baseflow component and surface runoff component is based on different isotopic compositions of the groundwater in the basin and that of a given storm (Figure 17.12).

The groundwater should have an isotopic composition that reflects long-term average input value for the whole basin, whereas the isotopic composition of storm water will reflect actual conditions of falling in particular time and space.

In regional flow systems, sometimes three-component mixing can be identified. A well-known example comes from the northern Canadian Shield and was identified by Douglas et al. (2000) (Figure 17.13).

In this example, mixing between shield brines and meteoric waters was demonstrated by the decrease in salinity over time and curtained sampling sites. Additionally, a third type of groundwaters of depleted isotopic composition and intermediate salinity was identified, which deviated from the simple two-component mixing line.

Such regional scale mixing processes can be observed in deep groundwaters within basins as well as in shallow but confined systems with long subsurface flow paths (Clark and Fritz, 1997).

17.3.2 Groundwater Quality and Origin of Dissolved Compounds

The dissolved compounds in groundwater and their chemical types and amounts (i.e., salinity) are components of water quality. Application of stable isotope techniques in many cases allows tracing the origin of dissolved compounds, their formation, transformation, and sinks during different geochemical processes in the hydrosphere. This chapter is focused on the isotopic composition of dissolved sulfates

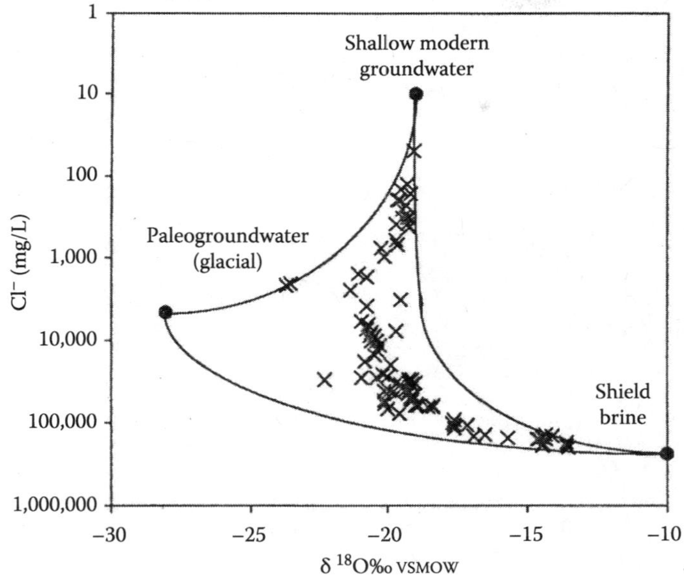

FIGURE 17.13 Three-component mixing for groundwaters from crystalline rocks of the Canadian Shield. The glacial meltwater end member was identified based on the depleted isotopic composition, its $\delta^{18}O$–Cl^- relationship, and ^3H-free. (From Douglas, M. et al., *J. Hydrol.*, 235, 88, 2000.)

(i.e., $\delta^{34}S$ and $\delta^{18}O$ of SO_4^{2-}) and nitrates (i.e., $\delta^{15}N$ and $\delta^{18}O$ of NO_3^-) as they are common and important constituents of surface water and groundwater. Identifying their sources and biogeochemical processes that govern their occurrence and transformation in aquatic environment is very important knowledge for water resources and water quality management.

17.3.2.1 Isotopic Composition and Sources of Dissolved Sulfates

Main forms of sulfur in geological environment include sulfate and sulfide minerals, sulfate (SO_4^{2-}) and sulfide (HS^-) dissolved in water, and hydrogen sulfide gas (H_2S). Under oxidizing conditions in water, sulfate is the dominant sulfur species, while under reducing conditions, sulfides prevail. Organic sulfur is a component of organic compounds such as humic substances, kerogen, and hydrocarbons. Atmospheric sulfur sources include natural and industrial SO_2, particulate sulfur, and aerosols of marine sulfate. Sulfur compounds from these various sources participate in the geochemical evolution of groundwater.

Sulfur has four stable isotopes, ^{32}S, ^{33}S, ^{34}S, and ^{36}S, occurring naturally with respective absolute abundances of about 95.02%, 0.75%, 4.21%, and 0.02% (MacNamara and Thode, 1950). Normally $^{34}S/^{32}S$ and $^{18}O/^{16}O$ are measured to obtain isotopic composition of sulfates. On a δ-scale isotopic composition of sulfate is defined as follows:

$$\delta^{34}S = \left(\frac{\dfrac{^{34}S}{^{32}S\,sample}}{\dfrac{^{34}S}{^{32}S\,VCDT}} - 1 \right) * 1000 \qquad (17.19)$$

$$\delta^{18}O = \left(\frac{^{18}O/^{16}O\,sample}{^{18}O/^{16}O\,VSMOW} - 1 \right) * 1000 \qquad (17.20)$$

Typical ranges of $\delta^{34}S$ and $\delta^{18}O$ for main sulfate sources are presented in Figure 17.14.

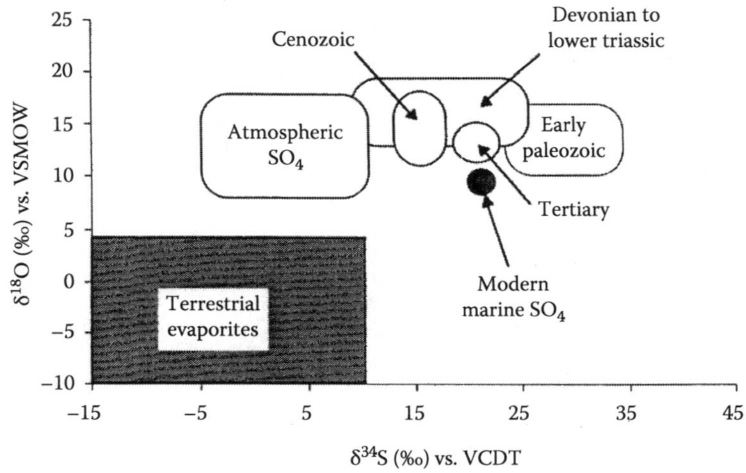

FIGURE 17.14 Isotopic composition of main sources of sulfates: evaporate, modern marine, terrestrial, and atmospheric. (From Clark, I. and Fritz, P., *Environmental Isotopes in Hydrogeology*, Lewis Publishers, New York, 328pp, 1997.) For additional explanation, see the text.

The first input of sulfates to surface water and groundwater is connected with atmospheric inputs, although this contribution is usually minor in comparison to the amount of sulfates, which can be gained in subsurface environment. The ^{34}S isotopic composition of atmospheric sulfates is controlled mainly by emissions from fossil fuel combustion and by sulfate from sea spray—at least in the coastal areas. The $\delta^{34}S$ values of sulfur from combustion of both petroleum and coal typically range from slightly negative to +10‰ vs. VCDT (usually from −3‰ to +9‰) and exceptionally can reach values as high as +16‰ from the combustion of Paleozoic crude oils in North America (Clark and Fritz, 1997). The ^{18}O isotopic composition of atmospheric sulfates is dependent on the oxidation conditions in the atmosphere and the ^{18}O composition of the moisture involved in the oxidation of atmospheric SO_2. In temperate regions $\delta^{18}O$ values of sulfate in atmospheric deposits range between +5‰ and +17‰ vs. VSMOW (Cortecci and Longinelli, 1970).

Major sources of sulfates in groundwater are connected with geological sources of sulfur compounds (minerals) in igneous, metamorphic, and sedimentary rocks. Especially, the sedimentary rocks are the most important source of groundwater sulfates as they may contain large amounts of oxidized and/or reduced sulfur minerals. Gypsum and anhydrite are principal constituents of marine evaporites in the sedimentary strata; they are very well soluble in water and are usually major contributors to groundwater sulfates. Gypsum also accumulates in the soil of arid regions and usually occurs together with limestone and dolomite sequences. The dissolution of gypsum and anhydrite occurs without measurable isotope fractionation. This allows to use directly the isotopic composition of SO_4^{2-} as tracer for sulfate origin as well as the origin of sulfate salinity. On the other hand, crystallization of sulfate minerals only slightly favors the heavier isotopes in the precipitates, for example, for gypsum, the $\delta^{34}S$ and $\delta^{18}O$ values have been measured to be 1.7‰ and 3.6‰ higher than for dissolved sulfate, from which they form (Szaran et al., 1998). The isotopic composition of SO_4^{2-} of modern ocean water has a well-defined value of $\delta^{34}S_{SO4} = 21$‰ VCDT and $\delta^{18}O_{SO4} = 9.5$‰ VSMOW. However, this composition was not constant in the past. Based on marine evaporite deposits, the measured $\delta^{34}S$ values varied over long-term trends with a maximum near +35‰ in the Cambrian and minimum of less than +10‰ in the Permian; the respective $\delta^{18}O$ varied between +20‰ and +7‰ (Claypool et al., 1980). Comparison of the isotopic composition of sulfate in groundwaters with the appropriate geological period (i.e., with determined isotopic composition of sulfate evaporates) allows to distinguish geogenic sources from other sources of sulfate salinity.

Another important source of sulfate in groundwater is oxidation of aqueous sulfide in surface or near-surface environment and oxidation of sulfide minerals as pyrite and others. Especially oxidation of sulfide minerals can contribute considerable amounts of sulfates to surface and groundwater (plus metals and acidity) and can produce gypsum crusts in arid regions. When dissolved sulfides are oxidized chemically by O_2, about 5‰ depletion is observed in $\delta^{34}S$ of accumulating sulfate phase; when oxidation process is mediated biologically involving aerobic bacteria (e.g., *Thiobacillus concretivorus*), sulfate may be depleted by up to 20‰ from the sulfide (Kaplan and Rittenberg, 1964; Fry et al., 1988). The ^{18}O content of sulfate from oxidation of dissolved sulfides is dependent on the reaction pathway and is similar to sulfate from sulfide minerals (Clark and Fritz, 1997). The $\delta^{34}S$ values of sulfate originating from biologically mediated oxidation of base-metal sulfides are depleted by 2‰–5.5‰ (Toran and Harris, 1989).

All the sulfates that form in this way in the hydrosphere contribute to the "terrestrial sulfates" field distinguished in Figure 17.14. In some cases, a plot of $\delta^{34}S$–$\delta^{18}O$ values may reveal the mixing of sulfates from two sources. However, simultaneous monitoring of spatial or temporal trends in concentrations and isotopic compositions of sulfate in groundwater and surface water is a much more effective approach for revealing addition of sulfates from various sources and for identifying sulfate transformation processes (Figure 17.15).

One of the most important processes that modify the sulfate isotopic composition and concentration in aquifers is bacterial (dissimilatory) sulfate reduction. This biologically mediated process causes the decrease of sulfate concentration and preferential enrichment in ^{34}S and ^{18}O in the residual sulfate (e.g., Mizutani and Rafter, 1969). The typical trend of $\delta^{34}S$ and $\delta^{18}O$ of the remaining dissolved sulfates resulting from bacterial sulfate reduction is shown in Figure 17.16.

17.3.2.2 Isotopic Composition and Sources of Dissolved Nitrates

In the aqueous environment, nitrogen occurs usually as NO_3^-, NO_2^-, NH_4^+, dissolved N_2 gas, and amino acids. The most stable form of nitrogen in most surface waters and groundwaters is dissolved N_2 gas and nitrates (NO_3^-). The latter is the most abundant nitrogen species in aqueous environment, and ammonia is noticeable only in reducing environments such as peat bogs and sewers (Letolle, 1980; Clark and Fritz, 1997).

Nitrogen has two stable isotopes, ^{14}N and ^{15}N, occurring naturally with respective abundances of 99.6337% and 0.3663%. Isotopic composition of nitrates on a δ-scale isotopic is expressed as follows:

$$\delta^{15}N = \left(\frac{^{15}N/^{14}N\,sample}{^{15}N/^{14}N\,AIR} - 1 \right) * 1000 \tag{17.21}$$

$$\delta^{18}O = \left(\frac{^{18}O/^{16}O\,sample}{^{18}O/^{16}O\,VSMOW} - 1 \right) * 1000 \tag{17.22}$$

Typical ranges of $\delta^{15}N$ and $\delta^{18}O$ for main nitrate sources are presented in Figure 17.17.

Nitrogen is a biologically active element and participates in a number of reactions that are important to life but also affect water quality. Decay of biomass releases organic nitrogen, which oxidizes to NO_3^-, which, in turn, is a contaminant in drinking water. The sources of nitrate in groundwater can be classified as diffusive (synthetic and organic fertilizers) and point sources (septic tanks, sewerage lagoons, manure piles) and are mainly associated with agricultural activities (Kendall and Aravena, 2000). Identification of the nitrate sources and biogeochemical processes that affect its formation, transport, evolution, and attenuation in geological environment is the main task of the application of isotopic studies.

Two factors control the $\delta^{15}N$ values of any N-bearing compound in the subsurface: (1) variations in the $\delta^{15}N$ values of inputs (i.e., sources) and outputs (i.e., sinks) of the compound in the subsurface and

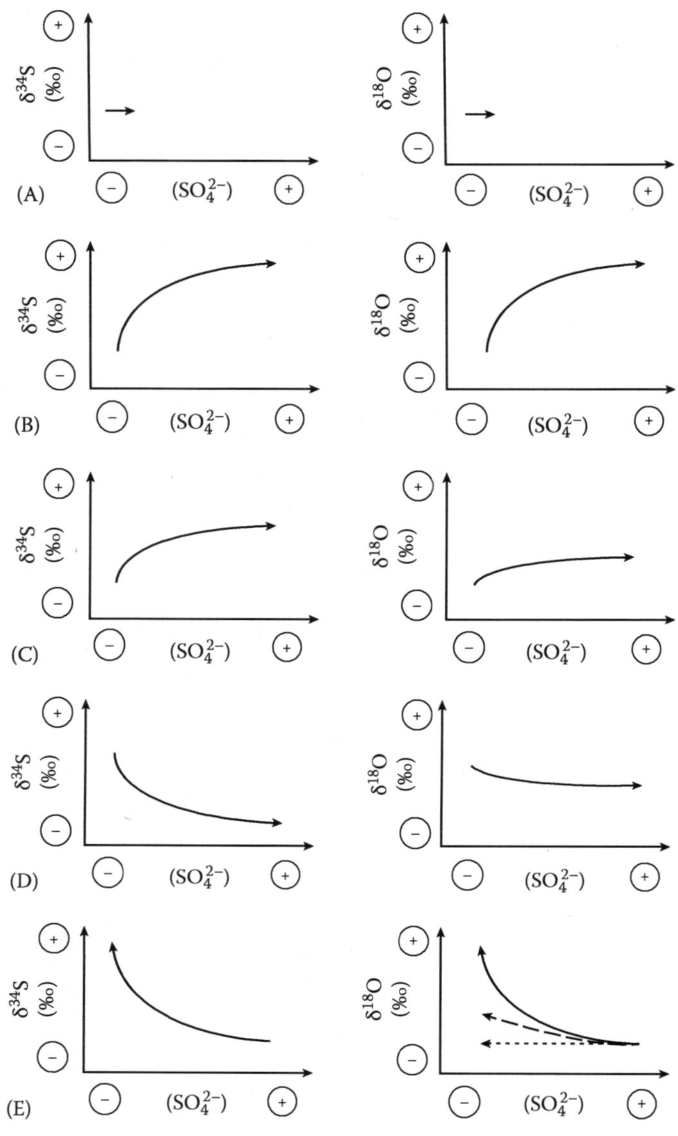

FIGURE 17.15 Trends of concentration and isotopic composition of sulfates during evaporation (A), admixture of evaporite sulfate (B), admixture of oceanic sulfate (C), admixture of sulfate from sulfide oxidation (D), and bacterial dissimilatory sulfate reduction (E). (Plots after Mayer, B., Assessing sources and transformations of sulfate and nitrate in the hydrosphere using isotope techniques, in Aggarwal, P.K., Gat, J.R., and Froehlich, K.F.O. (eds.), *Isotopes in the Water Cycle: Past, Present and Future of Developing Science*, Springer, Dordrecht, the Netherlands, 67pp, 2007.)

(2) chemical, physical, and biological transformations of materials within the soil and groundwater that produce or remove the compound (Kendall and Aravena, 2000).

Atmospheric input is a minor source of NO_3^- to groundwater systems. Moreover, $\delta^{15}N$ and $\delta^{18}O$ values of atmospheric nitrate reveal a large range of variation due to many different chemical reactions and compound involved, seasonality, meteorological conditions, ratio of NH_4 to NO_3 in precipitation, types of anthropogenic inputs, proximity to pollution sources, etc. (Hubner, 1986). Usually $\delta^{15}N$ values of atmospheric NO_3 and NH_4 are in the range of −15‰ to +15‰; $\delta^{18}O$ values reveal larger variability

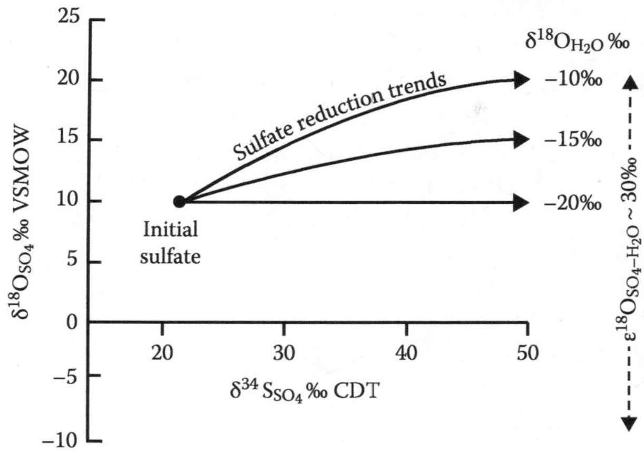

FIGURE 17.16 Effects of sulfate reduction on the isotopic composition of sulfates. The $\delta^{34}S$ values of the residual sulfate in closed system follow a Rayleigh enrichment when sulfate is consumed (A). The $\delta^{18}O$ values approach a constant value that closely corresponds to the thermodynamic equilibrium between water and sulfate, that is, about 30‰ in normal groundwater temperature (B). (From Fritz, P. et al., *Chem. Geol. (Isot. Geosci. Sect.)*, 79, 99, 1989.)

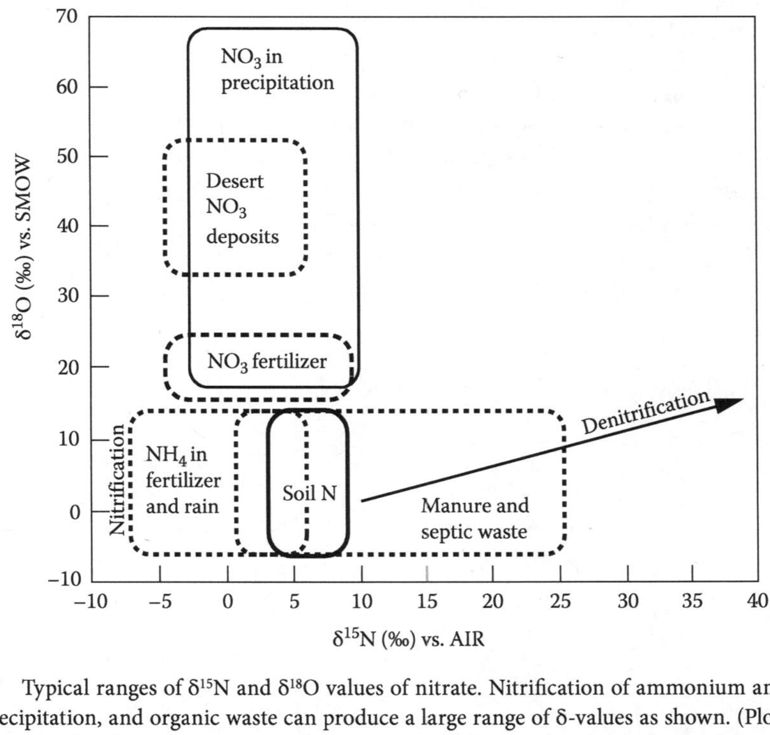

FIGURE 17.17 Typical ranges of $\delta^{15}N$ and $\delta^{18}O$ values of nitrate. Nitrification of ammonium and/or organic N in fertilizer, precipitation, and organic waste can produce a large range of δ-values as shown. (Plot after Kendall, C., Tracing sources and cycling of nitrate in catchments, in Kendall, C. and McDonnel, J.J. (eds.), *Isotope Tracers in Catchment Hydrology*, Elsevier, Amsterdam, the Netherlands, 519pp, 1998.)

from +18‰ to +70‰ as was observed in rainwater, snow, and snowmelt from forested watersheds in the United States (Kendall and Aravena, 2000).

High concentration of NO_3^- in surface waters and groundwaters in agricultural areas is connected with input from fertilizers. Synthetic fertilizers have $\delta^{15}N$ in the range of −4‰ to +4‰ and $\delta^{18}O$ values between +18‰ and +22‰ (Hubner, 1986; Amberger and Schmidt, 1987). Organic fertilizers reveal higher $\delta^{15}N$ values and a wider range of compositions, generally from −2‰ to +30‰; their $\delta^{18}O$ values range between +38‰ and +50‰ (Hubner, 1986; Bohlke et al., 1997; Kendall and Aravena, 2000).

Nitrate derived from manure or sewage is usually characterized by $\delta^{15}N$ values between +10‰ and +20‰ and $\delta^{18}O$ values usually less than +15‰ (Aravena et al., 1993).

Tracing the isotopic composition of nitrate and nitrogen-containing compounds in subsurface environment (i.e., water unsaturated and saturated zones) is not straightforward since numerous transformation processes in the nitrogen cycle are associated with significant both kinetic and thermodynamic fractionation effects. Moreover, large variations in isotopic composition of nitrates and overlapping of similar $\delta^{15}N$ and $\delta^{18}O$ values from different processes make interpretation sometimes very difficult.

The main biologically mediated reactions that control nitrogen dynamics in the soil and groundwater are (1) assimilation, (2) nitrification, and (3) denitrification (Kendall and Aravena, 2000). The most important are the two latter processes.

Nitrification is a multistep process of organic N-compound oxidation by several different autotrophic organisms for the purpose of deriving metabolic energy; the process produces acidity and NO_3^- together with other intermediate species like NH_4^+, NO_2^-, NO, and N_2O (Kendall and Aravena, 2000):

$$\text{Organic} - \text{N} \rightarrow NH_4^+ \rightarrow NO_2^- \rightarrow NO_3^- \tag{17.23}$$

Nitrification in the unsaturated zone can increase the concentration of NO_3^- significantly in groundwater. The source of N can be soil organic N-compounds, organic and inorganic fertilizer, and N-compounds from septic systems and sewerage lagoons.

On the other hand, the denitrification is one of the most important processes of nitrate attenuation in groundwater. This is a multistep process involving various nitrogen oxides (e.g., N_2O, NO) as intermediate compounds resulting from a biological-mediated reduction of NO_3^- to N_2. Denitrification occurs in anaerobic conditions and causes the $\delta^{15}N$ and $\delta^{18}O$ of the residual nitrate to increase exponentially as nitrate concentrations decrease (Hubner, 1986; Kendall and Aravena, 2000) (Figure 17.18).

The changes in $\delta^{15}N$ value of the residual nitrate during denitrification can be expressed using the Rayleigh distillation equation (similar like during sulfate reduction process). Since mixing and denitrification curves can be similar, it is often difficult to distinguish whether mixing or denitrification has

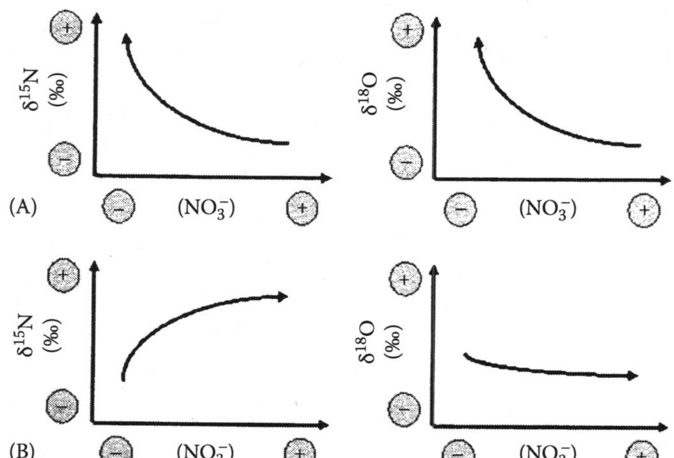

FIGURE 17.18 Trends of concentration and isotopic composition of nitrate during denitrification (A) and admixture of nitrate from manure or sewage (B). (From Mayer, B., Assessing sources and transformations of sulfate and nitrate in the hydrosphere using isotope techniques, in Aggarwal, P.K., Gat, J.R., and Froehlich, K.F.O. (eds.), *Isotopes in the Water Cycle: Past, Present and Future of Developing Science*, Springer, Dordrecht, the Netherlands, 67pp, 2007.)

occurred. In such case, data should be additionally considered by plotting $\delta^{15}N$ vs. $1/[NO_3^-]$, which will yield a straight line for mixtures of two sources. The plot $\delta^{15}N$ vs. $ln[NO_3^-]$ will yield a straight line for any process like denitrification, which can be described using the Rayleigh equation, that is, any exponential relation (Mariotti et al., 1988; Kendall and Aravena, 2000).

17.3.3 Modern Groundwater Dating with Tritium

The dating of groundwater and determination of the time of its residence in the geological environment have important implication for water resource management, contaminant transport, determination of groundwater flow parameters, hydraulic conductivity, porosity, water velocity, etc. A number of methods based on the decay of radionuclides are used to date groundwaters. Radioactive isotopes with a long half-life are used to date paleogroundwaters, for example, ^{14}C, ^{36}Cl, and ^{39}Ar. Short-lived radioisotopes are used to date modern groundwaters, for example, ^{3}H, ^{14}C, ^{32}Si, ^{37}Ar, and ^{85}Kr. Modern groundwaters are those that take active part in modern hydrological cycle and that presumably were recharged in the past few decades.

Tritium (^{3}H) is probably the most commonly employed radioisotope used to date the modern groundwater and to identify the occurrence of modern recharge. A half-life of tritium is equal only to 12.43 years. Natural tritium is produced by cosmic radiation in the upper atmosphere (Clark and Fritz, 1997):

$$^{14}_{1}N + ^{1}_{0}n \rightarrow ^{12}_{6}C + ^{3}_{1}H \quad (17.24)$$

Concentrations of ^{3}H are expressed in tritium units (TU):

1. TU = 0.118 Bq/kg water = 3.19 pCi/kg water = 1 ^{3}H per 10^{18} hydrogen atoms

Atmospheric testing of thermonuclear bombs between 1951 and 1980 released a huge amount of the so-called anthropogenic tritium and provided the tritium input signal that defines modern water. This substantial input created a tritium reservoir in the stratosphere, which contaminated global precipitation systems for over four decades. By about 1990 most of this bomb tritium had been washed from the atmosphere and tritium concentration tends to reach natural values, that is, about 3–6 TU (Kaufman and Libby, 1954; Clark and Fritz, 1997).

It should be stressed that tritium as a part of the water molecule can really date the water in the sense of determining its age. All other dating methods rely on dissolved constituents in water; therefore, their abundance is controlled by physicochemical and biological processes. However, only in well-defined and usually regional artesian aquifers, age gradient of the water will be preserved along the flow path. Thus, it is much safer to use the term "mean residence time" than "age" in relation to water dating.

As a component of water molecules, ^{3}H can be used as natural tracer of water in different stages of the hydrological cycle. In addition to conventional hydrological methods, it may be applied to determine the origin of water and mixing processes, the storage properties of the catchment, groundwater dynamics, and interactions between surface water and groundwater. The quantitative evaluation of these data is most often based on application of lumped parameter models. The advantage of these models is their relative conceptual simplicity, because they require only the knowledge of the tracer concentration in the recharge area (i.e., input function, $C_{in}(t)$) and some records of the tracer data at the observation site (i.e., output function, $C_{out}(t)$) (Figure 17.20). The lumped parameter models were described in detail by Maloszewski and Zuber (1982, 1996), Zuber and Maloszewski (2001), and Maloszewski et al. (2004). Generally, in lumped parameter approach, the groundwater system is assumed to be closed, sufficiently homogeneous, being in a steady state, and having a defined input (i.e., recharge or infiltration area) and output (pumping wells, springs, or streams draining the system) (Figure 17.19).

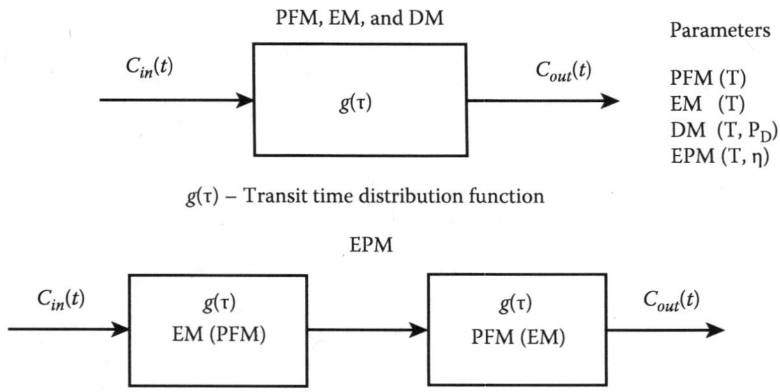

FIGURE 17.19 Schematic presentation of the lumped parameter models and their parameters. (From Zuber, A. (ed.), *Tracer Methods in Hydrogeological Applications*, Wrocław University of Technology, Wrocław, Poland, 402pp, 2007, Modified by the author.) For explanation, see the text.

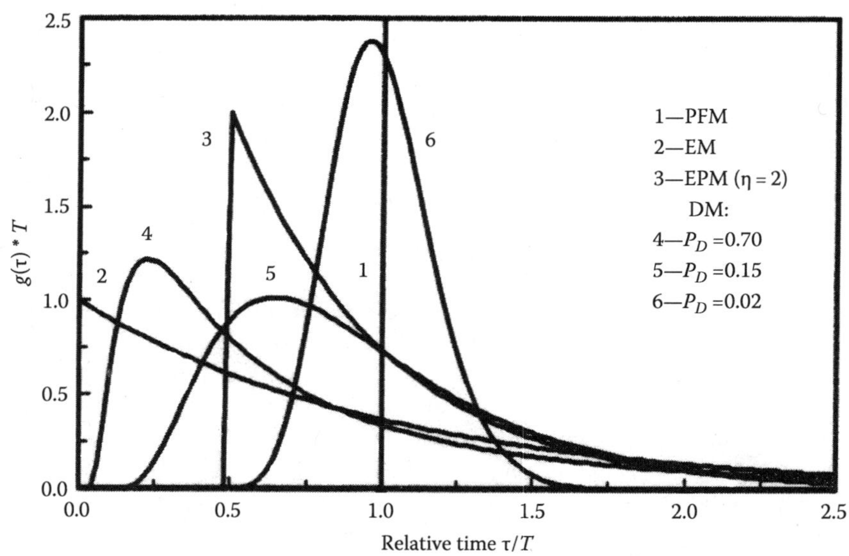

FIGURE 17.20 Examples of the RTD function $g(\tau)$ of tracer (i.e., tritium) through groundwater system calculated for different lumped parameter mathematical models. PFM, piston flow model; EM, exponential model; DM, dispersion model; EPM, combined exponential-piston flow model. (From Maloszewski, P. et al., *Isot. Environ. Health Stud.*, 40(1), 21, 2004.) For explanation, see the text.

Each lumped parameter model has its own function that describes transit time distribution in the system $g(\tau)$, sometimes called residence time distribution (RTD) function (Figure 17.20).

The function $g(\tau)$ has to be known or assumed for particular studied system based on the hydrogeological information about this system. In most cases the $g(\tau)$ function has one or two unknown (fitting) parameters, in which values should be determined by calibrating the model to the experimental data observed in the output from the system. The main parameter of all models in monoporous media is the mean transit time of water T. It refers to the free, mobile water in the system and relates to its volume V_m and flow rate Q:

$$V_m = QT \tag{17.25}$$

To determine the T value, the temporal variation of the measured tritium input concentration, $C_{in}(t)$, is used to calculate the tritium output concentration, $C_{out}(t)$, which is then compared with real concentrations measured in the output from the system (Maloszewski and Zuber, 1982; Zuber, 2007). The input concentration of the tracer (i.e., tritium) is directly measured or has to be calculated from known hydrological and isotope data as it was shown by Maloszewski and Zuber (1982, 1996) and Grabczak et al. (1984).

The relationship between $C_{in}(t)$ and $C_{out}(t)$ is described by the following convolution integral:

$$C_{out}(t) = \int_0^{\infty} C_{in}(t-\tau)\, g(\tau) \exp(-\lambda\tau)\, dt \tag{17.26}$$

where

$C_{out}(t)$ is the output ^3H concentration (i.e., measured based on sampling)
$C_{in}(t)$ is the input ^3H concentration
t is the calendar time
τ is the transit time of individual streamlines
$t-\tau$ is the input time of individual streamlines
$g(\tau)$ is the RTD function (i.e., function showing response of the system for fitted parameters)
λ is the decay constant; for ^3H, $\lambda = \dfrac{\ln 2}{T_{1/2}} = 0.0558$

The integral over the $g(\tau)$ for all possible transit times τ, that is, from zero to infinity, has to be equal to one. For most frequently applied models used for the interpretation of isotope data in the so-called monoporous media (i.e., no stagnant water in the system, no diffusion of tracer), the mean value of τ has to be equal to the mean transit time of water T.

The most common types of the transit time distribution function $g(\tau)$ used for monoporous media are as follows:

a. Piston flow model (PFM)

$$g(\tau) = \delta(\tau - T) \tag{17.27}$$

b. Exponential model (EM)

$$g(\tau) = \left(\frac{1}{T}\right)\exp\left(-\frac{\tau}{T}\right) \tag{17.28}$$

c. Dispersion model (DM)

$$g(\tau) = \left(\frac{1}{T}\right)\left[\left(\frac{4\pi P_D}{\left(\frac{\tau}{T}\right)^3}\right)\right]^{-\frac{1}{2}} \exp\left[-\frac{\left(1-\frac{\tau}{T}\right)^2}{\frac{P_D\tau}{T}}\right] \tag{17.29}$$

where

P_D is the dispersion parameter (i.e., fitted parameter; see Figure 17.20)

The selection of the proper model depends on the hydrogeological conditions. Some examples of popular hydrogeological situation are presented in Figure 17.21.

Generally, the dispersion and PFMs are applicable for confined or partly confined aquifers (Figure 17.21A through C). EM can be used only for unconfined aquifers, where the unsaturated zone is small and can be neglected (Figure 17.21B).

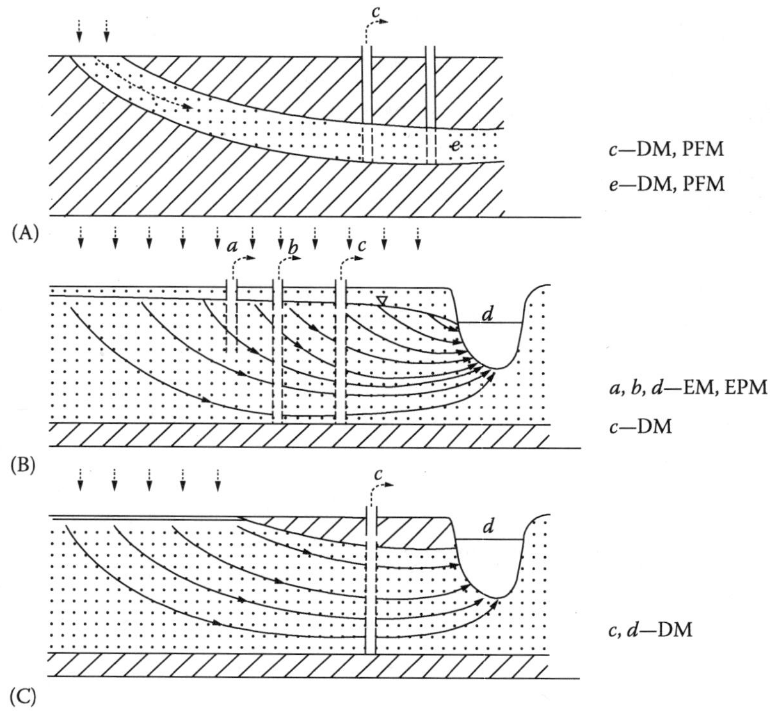

FIGURE 17.21 An example of different hydrogeological situations in which particular models can be applied. (From Maloszewski, P. et al., *Isot. Environ. Health Stud.*, 40(1), 21, 2004.) See text for explanation.

When we have a confined aquifer with a narrow recharge zone and sampling point located far away from the recharge (Figure 17.21A), the DM can be applied assuming that the confined part is impermeable with a neglectable portion of stagnant water. If the flow distance is very long (which permits to assume $P_D \approx 0$), then the PFM can be applied.

In case of unconfined porous aquifer (Figure 17.21B), where the unsaturated zone is small, then the EM can be applied for the sampling points a, b, and d. Otherwise, when the unsaturated zone has to be taken into account, the combined exponential-PFM should be applied (not mentioned in this chapter; please refer, e.g., to Maloszewski and Zuber, 1996). In case of partly confined porous aquifer (Figure 17.21C), the DM will be most applicable.

17.4 Summary and Conclusions

The most important isotopes for hydrological and hydrogeological applications are those usually called as environmental. They are naturally occurring isotopes of elements found in abundance in our environment such as C, O, H, S, and N, that is, principal elements of hydrogeological, geological, and biological systems. The principal application of stable environmental isotopes is tracing of water, carbon, nutrient, and solute cycling, as well as understanding fundamental physicochemical processes governing isotope transport and fractionation. Especially isotopes of water (i.e., H and O) and their conservative nature in low-temperature environment constitute a powerful tool for tracing the origin, formation, movement, mixing, and geochemical evolution of various types of waters encountered in the Earth's hydrosphere. Isotopic studies of dissolved biologically active compounds in aquatic environment give tremendous possibilities for tracing water quality and identification of contaminant formation, transport, and sink. On the other hand, radioactive environmental isotopes, for example, ^{3}H or ^{14}C, are used mostly for determining the age of groundwaters. This allows groundwater recharge rates and flow velocities to be determined, often more accurately than with traditional hydraulic methods, and is very helpful in the management of groundwater reservoirs.

References

Allison, G.B., Barnes, C.J., Huges, M.W., and Leaney, M.W.J. 1984. Effects of climate and vegetation on oxygen-18 and deuterium profiles in soils. In: *Isotope Hydrology 1983, Proceedings Series IAEA*, Symposium 270pp, September 1983, Vienna, Austria, pp. 105–123.

Amberger, A. and Schmidt, H.L. 1987. Naturliche isotopengehalte von nitrat als indikatoren fur dessen herkunft. *Geochimica et Cosmochimica Acta* 62: 577–584.

Aravena, R., Evans, M.L., and Cherry, J.A. 1993. Stable isotope of oxygen and nitrogen in source identification of nitrate from septic systems. *Ground Water* 31: 180–186.

Bigeleisen, J. 1965. Chemistry of isotopes. *Science* 147: 463–471.

Bohlke, J.K., Eriksen, G.E., and Revesz, K. 1997. Stable isotope evidence for an atmospheric origin of desert nitrate deposits in northern Chile and southern California, USA. *Chemical Geology.* (*Isotope Geoscience Section*) 136: 135–152.

Clark, I. and Fritz, P. 1997. *Environmental Isotopes in Hydrogeology*. New York: Lewis Publishers, 328pp.

Claypool, G.E., Holser, W.T., Kaplan, I.R., Sakai, H., and Zak, I. 1980. The age curves of sulfur and oxygen isotopes in marine sulfate and their mutual interpretation. *Chemical Geology* 28: 199–260.

Cook, P.G. and Herczeg, A.L. 2000. *Environmental Tracers in Subsurface Hydrology*. Boston, MA: Kluwer, 529pp.

Coplen, T.B., Herczeg, A.L., and Barnes, C. 2000. Isotope engineering—Using stable isotopes of the water molecule to solve practical problems. In: Cook, P. and Herczeg, A.L. (eds.), *Environmental Tracers in Subsurface Hydrology*. Boston, MA: Kluwer Academic Publishers, pp. 79–110.

Cortecci, G. and Longinelli, A. 1970. Isotopic composition of sulfate in rain water, Pisa, Italy. *Earth and Planetary Science Letters* 8: 36–40.

Craig, H. 1961a. Standard for reporting concentration of deuterium and oxygen-18 in natural waters. *Science* 133: 1833–1834.

Craig, H. 1961b. Isotopic variations in meteoric waters. *Science* 133: 1702–1703.

Craig, H. 1963. The isotopic geochemistry of water and carbon in geothermal areas. In: Tongiorgi, E. (ed.), *Nuclear Geology on Geothermal Areas*, Spaleto, Consiglio Nazionale delle Ricerche, Laboratorio di Geologia Nucleare, Pisa, Italy, pp. 17–53.

Criss, R.E. 1999. *Principles of Stable Isotope Distribution*. New York: Oxford University Press, 254pp.

Dansgaard, W. 1964. Stable isotopes in precipitation. *Tellus* 16: 436–468.

Deak, J. and Coplen, T.B. 1996. Identification of Holocene and Pleistocene groundwaters in Hungary using oxygen and hydrogen isotopic ratios. *Isotopes in Water Resources Management* 1. Vienna, Austria: IAEA, 438pp.

Dole, M. 1935. The relative atomic weight of oxygen in water and air. *Journal of American Chemical Society* 57: 2731.

Douglas, M., Clark, I.D., Raven, K., and Bottomley, D. 2000. Groundwater mixing dynamics at a Canadian Shield mine. *Journal of Hydrology* 235: 88–103.

Drever, J.I. 1997. *The Geochemistry of Natural Waters*. Upper Saddle River, NJ: Prentice-Hall Inc, 436pp.

Edmunds, W.M. 2007. Contribution of isotopic and nuclear tracers to study of groundwaters. In: Aggarwal, P.K., Gat, J.R., and Froehlich, K.F.O. (eds.), *Isotopes in the Water Cycle: Past, Present and Future of Developing Science*. Dordrecht, the Netherlands: Springer, pp. 171–192.

Fritz, P., Basharmal, G.M., Drimmie, R.J., Ibsen, J., and Qureshi, R.M. 1989. Oxygen isotope exchange between sulfate and water during bacterial reduction of sulfate. *Chemical Geology* (*Isotope Geoscience Section*) 79: 99–105.

Fritz, P. and Fontes, J.Ch. 1980. *Handbook of Environmental Isotope Geochemistry. Volume I, The Terrestrial Environment, A*. Amsterdam, the Netherlands: Elsevier, 545pp.

Fritz, P. and Fontes, J.Ch. 1986. *Handbook of Environmental Isotope Geochemistry. Volume II, The Terrestrial Environment B*. Amsterdam, the Netherlands: Elsevier, 557pp.

Fritz, P. and Reardon, E.J. 1979. Isotopic and chemical characteristics of mine water in the Sudbury area. AECL Technical Report 35, Chalk River, Ontario, Canada: Atomic Energy of Canada Limited, 37pp.

Fry, B., Ruf, W., Gest, H., and Hayes, J.M. 1988. Sulfur isotope effects associated with oxidation of sulfide by O_2 in aqueous solution. *Chemical Geology (Isotope Geoscience Section)* 73: 205–210.

Gat, J.R. 1981. Groundwater. In *Stable Isotope Hydrology: Deuterium and Oxygen-18 in the Water Cycle*. Vienna, Austria: IAEA Technical Report Series 210, pp. 223–240.

Gat, J.R. and Carmi, I. 1970. Evolution of the isotopic composition of atmospheric waters in the Mediterranean Sea area. *Journal of Geophysical Research* 75: 3039–3048.

Geyh, M.A. and Ploethner, D. 1997. *Isotope Hydrological Study in Eastern Owambo, Etosha Pan, Otavi Mountain Land and Central Omatako Catchment including Waterberg Plateau. German-Namibian Groundwater Exploration Project*. Technical cooperation project No 89.2034.0. Flow-up report, 2. Federal Institute for Geosciences and Natural Resources, Hannover, Germany, 34pp.

Gonfiantini, R. 1986. Environmental isotopes in lake studies. In: Fritz, P. and Fontes, J.Ch. (eds.), *Handbook of Environmental Isotope Geochemistry. Volume II, The Terrestrial Environment*. Amsterdam, the Netherlands: Elsevier, pp.113–168.

Gourcy, L.L., Groening, M., and Aggarwal, P.K. 2007. Stable oxygen and hydrogen isotopes in precipitation. In: Aggarwal, P.K., Gat, J.R., and Froehlich, K.F.O. (eds.), *Isotopes in the Water Cycle: Past, Present and Future of Developing Science*. Dordrecht, the Netherlands: Springer, pp. 39–51.

Grabczak, J., Maloszewski, P., Rozanski, K., and Zuber, A. 1984. Estimation of the tritium input function with the aid of stable isotopes. *Catena* 11: 105–114.

Hitchon, B. and Friedman, I. 1969. Geochemistry and origin of formation waters in the western Canada sedimentary basin—I. Stable isotopes of hydrogen and oxygen. *Geochimica et Cosmochimica Acta* 33: 1321–1349.

Hoefs, J. 1997. *Stable Isotope Geochemistry*. Berlin, Germany: Springer-Verlag, 197pp.

Hubner, H. 1986. Isotope effects of nitrogen in the soil and biosphere. In: Fritz, P. and Fontes, J.Ch. (eds.), *Handbook of Environmental Isotope Geochemistry. Volume II, The Terrestrial Environment*. Amsterdam, the Netherlands: Elsevier, pp. 361–425.

Kaplan, I.R. and Rittenberg, S.C. 1964. Microbiological fractionation of sulfur isotopes. *Journal of General Microbiology* 34: 195–212.

Kaufman, S. and Libby, W.F. 1954. The natural distribution of tritium. *Physical Review* 93: 1337–1344.

Kendall, C. 1998. Tracing sources and cycling of nitrate in catchments. In: Kendall, C. and McDonnel J.J. (eds.), *Isotope Tracers in Catchment Hydrology*. Amsterdam, the Netherlands: Elsevier, pp. 519–576.

Kendall, C. and Aravena, R. 2000. Nitrate isotopes in groundwater systems. In: Cook, P.G. and Hercheg, A.L. (eds.), *Environmental Tracers in Subsurface Hydrology*. Boston, MA: Kluwer Academic Publishers, pp. 261–297.

Kyser, T.K. 1987. Equilibrium fractionation factors for stable isotopes. In: Kyser, T.K. (ed.), *Short Course in Stable Isotope Geochemistry of Low Temperature Fluids 13*. Saskatoon, Saskatchewan, Canada: Mineralogical Association of Canada, pp. 1–85.

Leibundgut, Ch., Maloszewski, P., and Kulls, Ch. 2009. *Tracers in Hydrology*. Chichester, U.K.: Wiley-Blackwell, 415pp.

Letolle, R. 1980. Nitrogen N-15 in the natural environment. In: Fritz, P. and Fontes, J.Ch. (eds.), *Handbook of Environmental Isotope Geochemistry. Volume I, The Terrestrial Environment*. Amsterdam, the Netherlands: Elsevier, pp. 407–433.

MacNamara, J. and Thode, H.G. 1950. Comparison of the isotopic composition of terrestrial and meteoritic sulfur. *The Physical Review* 78: 307–308.

Maloszewski, P. and Zuber, A. 1982. Determining the turnover time of groundwater systems with the aid of environmental tracers. I. Models and their applicability. *Journal of Hydrology* 57: 207–231.

Maloszewski, P. and Zuber, A. 1996. Lumped parameter models for the interpretation of environmental tracers data. In: *Manual of Mathematical Models in Isotope Hydrology*, IAEA-TECDOC 910, Vienna, Austria, pp. 9–58.

Maloszewski, P., Stichler, W., and Zuber, A. 2004. Interpretation of environmental tracers in groundwater system with stagnant water zones. *Isotopes in Environmental and Health Studies* 40(1): 21–33.

Mariotti, A., Landreau, A., and Simon, B. 1988. ^{15}N isotope biogeochemistry and natural denitrification process in groundwater: Application to the chalk aquifer of northern France. *Geochimica et Cosmochimica Acta* 52: 1869–1878.

Mayer, B. 2007. Assessing sources and transformations of sulphate and nitrate in the hydrosphere using isotope techniques. In: Aggarwal, P.K., Gat, J.R., and Froehlich, K.F.O. (eds.), *Isotopes in the Water Cycle: Past, Present and Future of Developing Science*. Dordrecht, the Netherlands: Springer, pp. 67–89.

Mizutani, Y. and Rafter, T.A. 1969. Oxygen isotopic composition of sulfates. Part 4. Bacterial fractionation of oxygen isotopes in reduction of sulfate and in the oxidation of sulfur. *New Zealand Journal of Science* 12: 60–68.

Muehlenbachs, K. and Clayton, R.N. 1971. Oxygen isotope ratios of submarine diorites and their constituent minerals. *Canadian Journal of Earth Sciences* 8: 1591–1595.

O'Neil, J.R. 1987a. Theoretical and experimental aspects of isotopic fractionation. In: Valley, J.W., Taylor, H.P., and O'Neil, J.R. (eds.), *Stable Isotopes in High Temperature Geological Processes. Reviews in Mineralogy 16*. Chelsea, MI: Book Crafters Inc., Mineralogical Association of America, pp. 1–40.

O'Neil, J.R. 1987b. Appendix: Terminology and standards. In: Valley, J.W., Taylor, H.P., and O'Neil, J.R. (eds.), *Stable Isotopes in High Temperature Geological Processes. Reviews in Mineralogy 16*. Chelsea, MI: Book Crafters Inc., Mineralogical Association of America, pp. 561–570.

Rozanski, K., Araguas-Araguas, L., and Gonfiantini, R. 1993. Isotopic patterns in modern global precipitation. In: Stewart, P.K., Lohmann, K.C., McKenzie, J., and Savin, S. (eds.), *Climate Change in Continental Isotopic Records. Geophysical Monograph*. Washington, DC: American Geophysical Union, 76. pp. 1–36.

Sheppard, S.M.F. 1986. Characterization and isotopic variations in natural waters. In: Valley, J.W., Taylor, H.P., and O'Neil, J.R. (eds.), *Stable Isotopes in High Temperature Geological Processes. Reviews in Mineralogy 16*. Chelsea, MI: Book Crafters Inc., Mineralogical Association of America, pp. 165–183.

Szaran, J., Niezgoda, H., and Halas, S. 1998. New determination of oxygen and sulfur isotope fractionation between gypsum and dissolved sulfate. *RMZ—Materials and Geoenvironment* 45: 180–182.

Toran, L. and Harris, R.F. 1989. Interpretation of sulfur and oxygen isotopes in biological and abiological sulfide oxidation. *Geochimica et Cosmochimica Acta* 53: 2341–2348.

Truesdell, A.H. and Hulston, J.R. 1980. Isotopic evidence on environments of geothermal systems. Chapter 5. In: Fritz, P. and Fontes, J.Ch. (eds.), *Handbook of Environmental Isotope Geochemistry. Volume 1: The Terrestrial Environment, A*. Amsterdam, the Netherlands: Elsevier, pp. 179–226.

Urey, H.C. 1947. The thermodynamic properties of isotopic substances. *Journal of Chemical Society 69*: 562–581.

Welhan, J. 1987. Stable isotope hydrology. In: Kyser, T.K. (ed.), *Short Course in Stable Isotope Geochemistry of Low Temperature Fluids 13*. Saskatoon, Saskatchewan, Canada: Mineralogical Association of Canada, pp. 129–161.

Zuber, A. (ed.) 2007. *Tracer Methods in Hydrogeological Applications*. Wrocław, Poland: Wrocław University of Technology, 402pp. (in Polish only).

Zuber, A. and Maloszewski, P. 2001. Lumped parameter models. In: *Environmental Isotopes in Hydrological Cycle, Volume VI*. Paris, France: IAEA/UNESCO, pp. 5–36.

18

Karst Hydrogeology

Pierre-Yves Jeannin
Swiss Institute for Speleology and Karst-studies

AUTHOR

Pierre-Yves Jeannin is a geologist and hydrogeologist with 30 years of experience in karst research. He has been the CEO of the Swiss Institute for Speleology and Karst Studies (SISKA) since its foundation in 2000. SISKA is a private nonprofit institute dedicated to the improvement of karst management in Switzerland. Its activities cover fundamental research, consulting, education (courses, field trips, booklets), and karst management activities, including fieldwork such as doline and cave depollution. SISKA employs 14 specialists, mostly geologists and hydrogeologists. Dr. Jeannin has also been a lecturer and a research project leader at the Federal Institute of Technology of Lausanne (EPFL) and at the University of Neuchâtel.

PREFACE

Along the last century, the observation and understanding of karst was mainly conducted by geologists and geographers. Their aim was to explain the fascinating aspects of karst landscape (e.g., large caves) and karst hydrology (e.g., impressive swallow holes). "Karst Science" was thus mainly developed by natural scientists. In parallel, the management of karst waters was mostly carried out by engineers, who applied usual techniques or sometimes adjusted them to cope with karst-specific characteristics. However, engineers published by far less of their work than natural scientists. As a result, the gap between the official and recognized "Karst Science" and the engineering practices increased. For example, the development of computer modeling tools for karst hydrology evidenced some fundamental divergences in concepts between these two "ways of thinking": karst scientists explain how bad the models are, but do not provide a way to improve them! The present chapter is an attempt to include results and concepts of "Karst Science" into approaches and methods, which can be applied for the management of practical situations. This chapter will not fill the gap but will hopefully be a contribution to it.

18.1 Introduction

Karst regions are lands of caves and swallowing rivers suggesting flavor of wonder and magic! But karst hydrology is not magic: only nonconventional compared to other regions. Usual investigation and management techniques do not apply to karst, and specific approaches are required. However, not any unconventional approach is adequate, and a sound understanding of karst hydrogeology is necessary to develop and apply meaningful approaches.

The aim of the present chapter is to provide background on karst hydrogeology for engineers, including water uses and management, as well as investigation approaches.

To reach this goal, the chapter includes three sections:

1. The special aspects of karst hydrology compared to other terrains
2. The most usual utilizations of karst waters and what typical problems occur in this environment
3. A presentation of dedicated approaches or methods in order to address practical questions in a karst environment

18.2 Special Aspects of Karst Hydrology

18.2.1 Definition of Karst

"Karst is the term used to describe a special style of landscape containing caves and extensive underground water systems that is developed on especially soluble rocks such as limestone, marble, and gypsum" [22]. The term "karst" denotes on the one hand a landform with specific superficial and subterranean morphology, created by dissolution, and on the other hand a hydrogeological system, in which drainage is organized by a set of karst tubes (caves), which are also created by dissolution (Figure 18.1). Karst drainage systems collect waters over catchment areas with a size typically ranging between 1 and 1000 km^2.

The system of karst conduits includes not only caves penetrable by humans but also interconnected voids with diameters down to about 1 cm. Karst conduits form a connected, 3D, and hierarchically structured network that may comprise hundreds of kilometers of passages (Figure 18.2).

Karst conduits may be empty or filled with water or sediments. Clastic sediments (clay to boulders) are widespread, but chemical sediments, called speleothems (flowstone, stalagmites and the like), are very characteristic and widely known in caves.

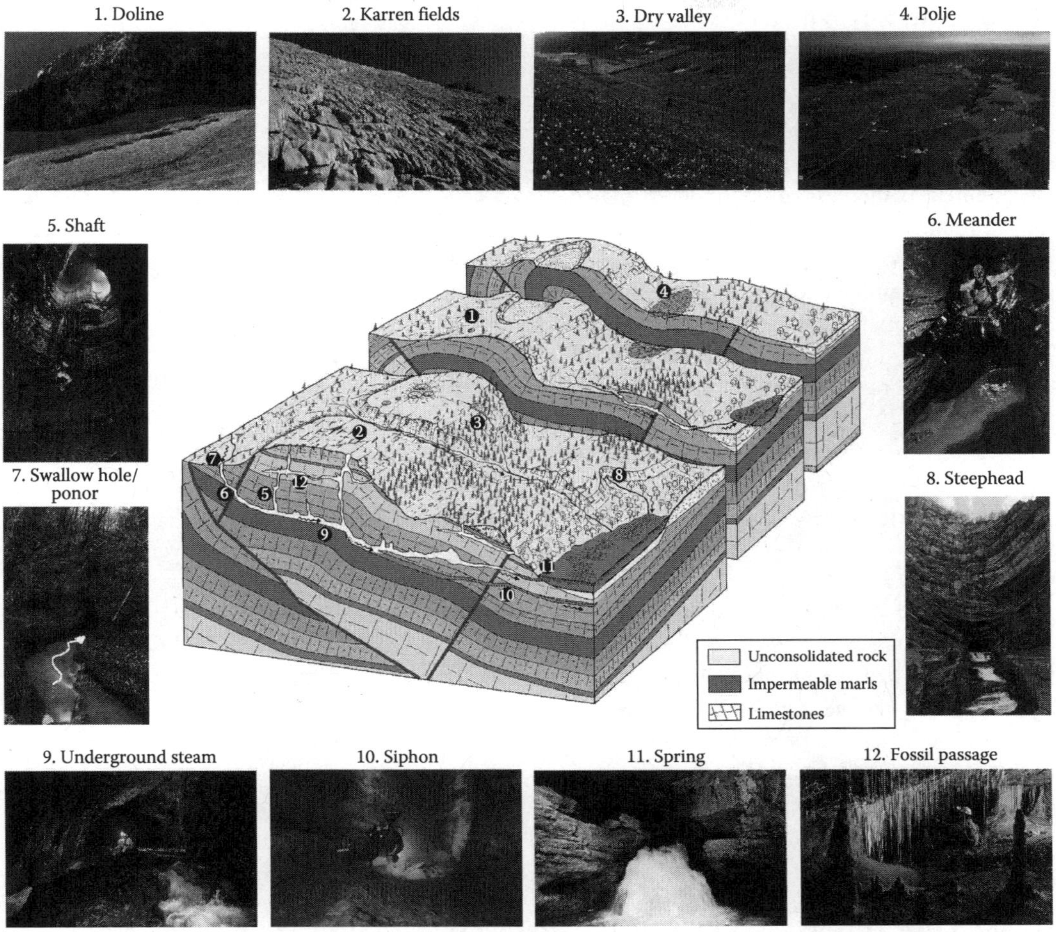

FIGURE 18.1 Typical karst phenomena (1–12), characterized by both hydrological and geomorphological features.

FIGURE 18.2 Example of a cave conduit network: Cave system of the Siebenhengste (Canton Bern, Switzerland) where cavers mapped 157 km of cave passages from 35 entrances.

18.2.2 Karst or Not Karst?

Altogether, karst is present over about 20% of the Earth's surface. Maps of karst outcrops in various countries can be found on http://web.env.auckland.ac.nz/our_research/karst/#karst1.

The exact delineation between karst and fissured systems is often difficult to set, and a region without any karst morphological evidence (doline or cave) can enclose a well-developed karst system! Karst development is therefore often underestimated, which can induce a significant risk in civil engineering. To remain on the safe side, the paradigm should be: "Any exposed karstifiable rock is assumed to be karstified (even if covered by some sediment) unless real evidences prove the opposite." Limestone, gypsum, and salt have to be considered as karstifiable in any case. Dolostone and chalk may develop some special karstification but are anyhow mostly karstified.

The presence of large springs and a reduced (or even lacking) network of surface streams are both evidences of karst hydrological systems. The reason is that the network of karst conduits is highly permeable, and, in most cases, all precipitation can be absorbed underground. Superficial watercourses (stream network) are thus generally absent (or reduced) in a mature karst area. The conduit network converges toward a few outlet points where infiltrated waters come again to land surface, feeding significant springs located at low elevation points (base level). These criteria (reduced stream network and presence of large springs) may however be difficult to recognize in semiarid or arid climates.

Telogenic (or meteoric water) karst systems are formed by the rock dissolution by infiltrated rainwater. Water traverses a more or less developed soil, becomes aggressive through CO_2 enrichment, penetrates the bedrock through cracks, and dissolves it. In some circumstances, water origin and/or aggressiveness may have another origin (e.g., deep CO_2 spring or mixing of fresh and seawater) and eogenic or hypogenic karst may develop, displaying some different characteristics. Details can be found in [12,30,37,38,48,58], or [21].

Background, problems, and methods presented in this chapter apply to telogenic karst. Adjustments should be considered for applications in other types of karst (eogenic or hypogenic).

18.2.3 Flow through a Typical (Telogenic) Karst Hydrogeological System

Rain falling on bare limestone (mostly karren fields) is quickly absorbed by karstified fissures even though soil or other quaternary cover may retain the water for some time. The first meters at the top of the limestone are often more or less fractured and cracked and can thus absorb a large part of the precipitation. This zone is called epikarst (Figure 18.3). At the bottom of the epikarst, a hanging groundwater body can develop, which feed fractures below over weeks or months. During strong precipitation events, water in the epikarst raises and overflows directly into larger cracks and conduits of the vadose zone, where cascades tumble down vertical shafts and steep passages. The epikarst is thus simultaneously a direct link between surface and karst conduits (quick flow) and an area of water storage (slow flow). If a river is swallowed into a sinkhole, it most often directly goes into the conduits of the vadose zone. The epikarst is therefore missing at this point.

In the vadose zone, flow is mainly driven by gravity and is therefore mainly vertical. Flow concentrated at land surface (swallow holes) and in the epikarst (quick overflow) feeds large vertical conduits (named shafts) connected by short meandering canyons. In most cases, water from the epikarst and the surface flows almost vertically through the vadose zone down to the karst water table or an impermeable rock layer. A vertical hierarchic structure of dissolution conduits (shafts) and a concentration of the water flows occur. The thickness of the vadose zone strongly depends on the relief and may exceed 2 km (Krubera Cave, Abkhazia). If the water reaches an impermeable rock layer before reaching the phreatic zone, it then follows dip down to the karst water table (or the spring).

Karst conduits of the permanently water-filled zone (phreatic zone) represent the lower part of the karst system. Flow is controlled by hydraulic gradients toward the karst spring, which are very

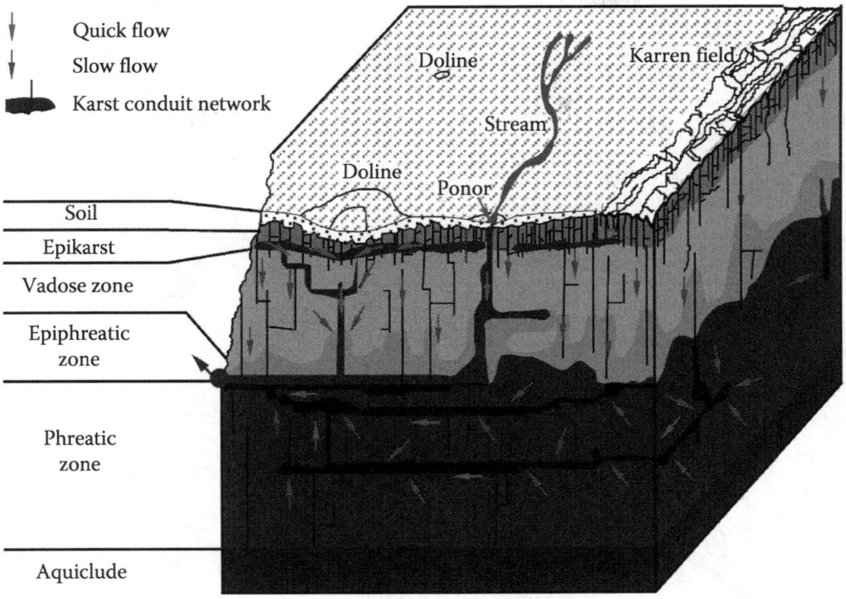

FIGURE 18.3 Water storage and flow in a karst massif. (Modified from Doerfliger, N. and Zwahlen, F., *Cartographie de la vulnérabilité en régions karstiques (EPIK). Guide ratique*, p. 58. Office fédéral de l'environnement, des forêts et du paysage, Berne, Suisse, 1998.) Soil, epikarst, and the vadose, epiphreatic, and phreatic zones are subsystems with specific characteristics, including quick- and slow-flow components.

low (nearly horizontal) in low and medium flow stages and controlled by the elevation of the spring. Conduits usually develop within the first 10–100 m below the top of the phreatic zone but may be deeper in some circumstances. Due to the convergence toward the spring, discharge in conduits may be considerable (up to 50–100 m^3/s).

During high water conditions, large amounts of water quickly enter the karst system, which may lead to significant flooding. Flooding heights of 10–100 m are usual. In some cases, they may reach 500 m (350 m in Hölloch, Switzerland; 500 m in Luire Cave, France) during a few days. The flooding zone is called epiphreatic zone. Epiphreatic conduits may be inactive for months or even years and suddenly become active (traversed by several m^3/s) during high floods.

All along the flow through a karst hydrogeological system, the medium presents an extreme heterogeneity in structure and behavior, influencing discharge patterns. The highly permeable conduit network induces quick flow and is surrounded by much less permeable bedrock, whose permeability is determined by the matrix and fracture geometry, inducing slow flow [35,36].

Below this highly karstified zone lies a zone of karst initiation, where dissolution takes place more homogeneously along all fractures. Heterogeneity of flow characteristics is not as extreme as in developed karst and this part should rather be considered as a fractured media.

In summary, karst flow systems have four specific characteristics:

1. Presence of a high permeability subterraneous conduit network providing a very efficient drainage
2. Presence of a nearly flat phreatic zone under low and medium water stages, that is, presence of a thick unsaturated zone (up to 2 km thick)
3. Presence of potentially quick and strong flooding during high water situations (rising of up to several hundred meters in elevation)
4. Presence of a "duality" of flow conditions with diffuse recharge through the epikarst versus concentrated recharge into swallow holes, storage versus quick flow in the epikarst, and quick conduit flow versus slow matrix flow in the saturated zone

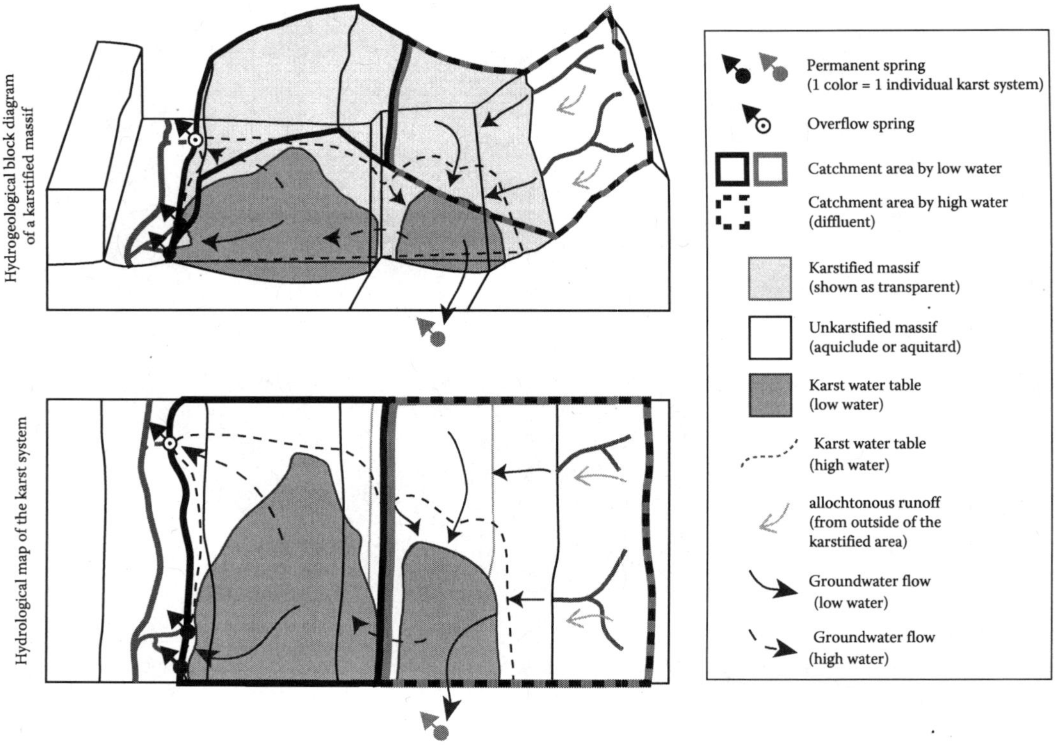

FIGURE 18.4 Schematic view of a karst hydrogeological system (details in text).

These characteristics imply that the definition of hydrogeological flow systems (i.e., of catchment areas) is not easy in karst regions. The same aquifer usually has several outlets (karst springs) with catchment parts belonging simultaneously or alternatively to two or three different springs. This leads to a different definition of catchment delineation compared to other hydrogeological systems, because catchment areas almost never correspond to topographical divides.

Figure 18.4 describes such a situation. At low water conditions, the black and gray springs are fed by the gray and black catchment areas, respectively. At high water stage, the phreatic zone raises up to the dashed line, feeding the overflow spring and linking the phreatic zone of the gray spring to that of the black ones. In these conditions, the black springs are also partially fed by the catchment area of the gray spring.

18.2.4 Typical Regime of a Karst System

Because of short to moderate storage, the regime of karst systems is more or less similar to that of a surface hydrological system with a similar catchment area.

The discharge rate of a karst system ranges between a few l/s in small systems up to several hundred m^3/s (e.g., Fontaine de Vaucluse, France, with 200 m^3/s). Peak discharge rates are 20–1000 times higher than low-flow discharge. Usually, one single karst system discharges through several springs, including overflow springs with a higher discharge rate than perennial one. The presence of overflow springs induces nonlinear relationships between system discharge and heads in karst conduits, which is quite typical for karst systems. As the name suggests, perennial springs do not dry out. This is due to the storage capacity of the epikarst and the fissured and matrix volume of the phreatic zone, draining out slowly during drought periods. If no perennial spring is visible, it may be situated below a riverbed or lake, or

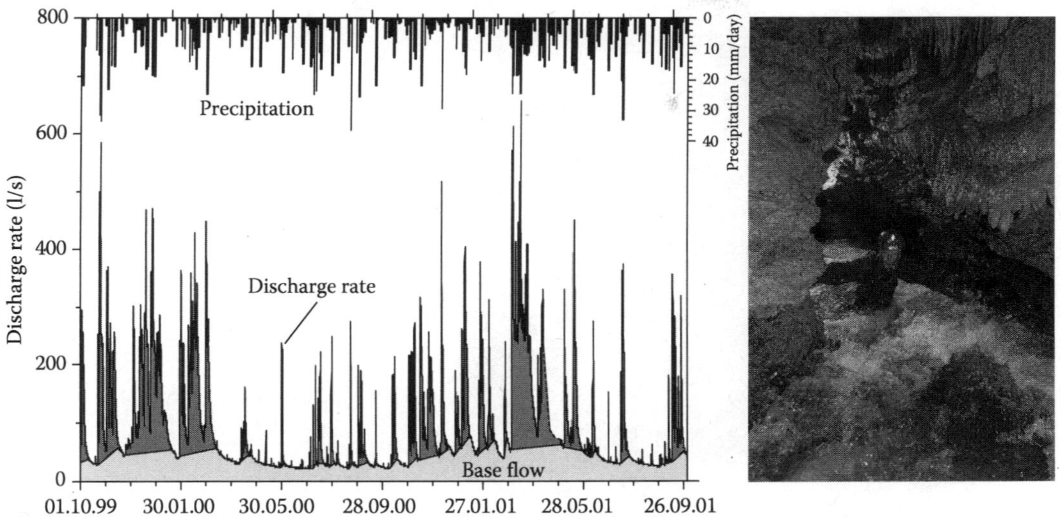

FIGURE 18.5 Hydrograph of a karst stream (Milandre, Canton of Jura, Switzerland). The hydrograph shows a quick-flow component reacting swiftly after storm events, and a baseflow component increasing in winter and decreasing in summer, due to evapotranspiration.

within alluvial pebbles. This, evidently, may not be true in very arid karst areas such as Saudi Arabia, where system discharge may fall to zero.

While springs from porous media generally only show seasonal variations, karst springs react swiftly, often within hours, to storm events (Figure 18.5).

Climate change has differentiated effects on karst spring regime depending on climate. The first studies published (e.g., [31,44]) on this problematic tend to show that effect on karst water quantity will be similar to effect on surface waters. In systems with glacial, nival, and even pluvio-nival regimes, contrasts will be smoothed: winter droughts will become shorter and less extreme, as well as summer high water conditions. Systems with a marked drought period in summer, related to the absence of any precipitation, will probably suffer a stronger drought related to an increase of evapotranspiration (temperature increase). However, each system has to be investigated and modeled according to the expected climate scenario for its particular region and according to its proper characteristics in order to make a sound prevision. Models have to explicitly include evapotranspiration and snowmelt in order to reasonably link meteorological parameters to discharge rates of the karst systems (see also Section 18.4.7) [12,22,30,38,41,42].

18.3 Engineering Hydrology in a Karst Environment

Six issues related to water management in karst regions are briefly sketched hereafter: drinking water supply, flood management, hydropower exploitation, water evacuation, geothermal utilization, and human impacts on karst waters.

18.3.1 Drinking Water

18.3.1.1 Tapping

As sketched in Figure 18.6, if the community to be supplied is located close to a karst spring, it is meaningful to tap water at the karst spring. However, if the community is located within the catchment area of a karst system, it may be more advantageous to drill a well and pump water directly from the aquifer (if this one is present!). Considering water sensitivity to pollution (especially to organic compounds), it

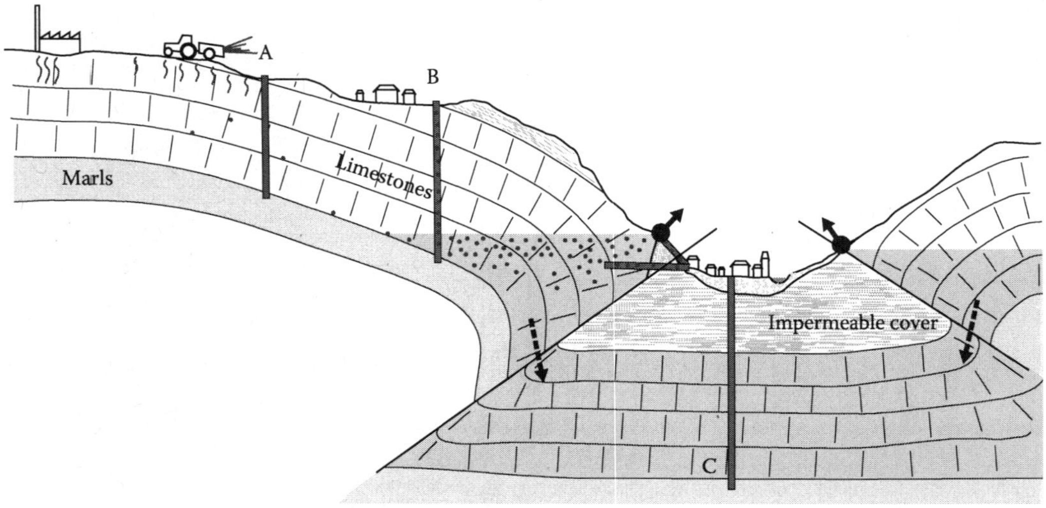

FIGURE 18.6 Schematic vertical profile across a karst massif showing various possible locations for water supply wells or tunnels. Drill hole **A** is an example of misplaced well that does not reach groundwater. Drill hole **B** and tunnel do reach groundwater. Dots represent pollution. The best tapping location is represented by well **C** that reaches below the impermeable cover.

may be interesting to drill a well in an area where the aquifer is confined below low permeability layers, even if this implies a longer transportation of the water to reach the consumer.

Before digging a well, it is absolutely necessary to ensure the presence of a phreatic zone at depth. To address this question, KARSYS (see Section 18.4.2) is a valuable approach.

Usual exploration, testing, and equipment procedures for pumping wells can be applied, although some adjustments in interpretation schemes are highly recommended. For instance, due to the extreme heterogeneity of the medium, effects of a pumping test may be visible within a few seconds at a spring or a piezometer located several kilometers away, and the same pumping test may not be visible in a piezometer located 30 m from the pumping well! It is therefore recommended to involve karst specialists in the interpretation of such tests.

If production rates of an exploration or pumping well are too low, "well development" techniques can be attempted in order to increase the permeability of the rock surrounding the well. In carbonate aquifers, injection of a controlled amount of hydrochloric acid (HCl) is usually quite effective.

18.3.1.2 Quality Management

Karst springs have strong discharge variations and related potential quality problems, especially at high water conditions and if the aquifer is recharged by many swallowing streams. Land use must also be considered, especially in the catchment areas of the swallowing streams. A significant turbidity increase is quite frequent during high water events and should be taken into account in the equipment of the spring.

The complexity and cost of water quality remediation should convince well planners not to try saving on the study of the hydrogeological system and not to necessarily aim for the highest well productivity. Indeed, wells showing the highest productivity rates are mostly directly connected to the karst conduit network, and, in most cases, this situation raises concerns on water quality similar to those mentioned for karst springs. One prevention option consists of drilling into the matrix of fissured limestone, for example, between branches of the conduit network, or in the deep part of a carbonate aquifer, below the expected karstified part of the phreatic zone. This target, however, may reveal difficult to reach.

Quality monitoring of karst waters is made difficult by the quick variations observed for most parameters. A continuous monitoring of critical potential pollutants should be set up, with a measurement time step of maximum 1 h. Fluorescence of organic compounds as well as turbidity can be advantageously monitored every few minutes at reasonable cost. Alarms and automatic switch off of supply pumps and tapping devices should be connected to the monitoring equipment. Additionally, well equipment (pump, pipes, shelter, and fence around the near protection perimeter) should be designed in order to prevent any pollutant to reach tapped water directly.

A further step in water quality management is the delineation of protection perimeters of groundwater extraction plants, which is necessary according to water regulations in many countries.

This topic was extensively investigated within the framework of several European projects. Results are available mainly in COST-65 (COST-65 1995) and COST-620 [66] final reports.

In most approaches, protection zones are based on the assessment of "vulnerability" of the respective flow routes feeding a spring (or a well). It follows the concept that a flow route with short residence time is more susceptible to be polluted than another with a longer residence time. This corresponds more or less to the regulations applied in various countries, where "high protection" zones are regions where a pollution by bacteria may reach the spring (or well) and "low protection" zones are regions where most bacteria are supposed to die before reaching the spring. *Intrinsic vulnerability* is supposed to represent a quantification of the attenuation of a conservative tracer, that is, being solely the effect of the transit through the groundwater flow system. *Specific vulnerability* is the quantification of attenuation related to the specific behavior (degradation, adsorption, retardation, etc.) of each tracer or pollutant.

European countries agreed on a "European approach" for the assessment of groundwater vulnerability, and several methods have been developed according to the various laws and regulations of different countries. These methods all use a multicriteria approach, combining maps of parameters assumed to control flow and transport in karst systems. Simple formulae are used in GIS tools for combining maps and producing a resulting vulnerability map, from which groundwater protection zones are derived (Figure 18.7).

Some of the methods include a map of the potential pollution sources (hazard map) and combine it with the vulnerability map. The resulting map thus represents the "pollution danger map" of a spring (or a well). Some methods include as well a risk assessment by combining the "danger map" to the evaluated value of the resources.

The specific behavior of many pollutants was investigated and weighted in order to be combined with intrinsic vulnerability assessment.

Methods of this kind have been developed according to national regulations and following different concepts, leading to a large palette of names: EPIK and VULK in Switzerland, LEA in England and Wales, PI and GLA methods in Germany, RISK and PAPRIKA in France, COP in Spain, Time-Input in Slovakia, VURAAS in Austria, etc.

Limitations are observed in the application of such vulnerability methods. The main one is that "intrinsic vulnerability" can hardly be translated in usual and measurable values, such as concentration, duration of a pollution, and attenuation. In fact, a spring has to be protected against the most probable pollutions, likely to occur in its catchment. This may be a point source and short duration input of pollution or a diffuse and long-lasting one. Criteria may thus not be same in all cases [9] or [8] suggested a different concept of vulnerability, in which the latter is defined by three parameters: dispersion (how long will the spring be polluted?), attenuation (what will the highest pollution concentration be?), and retardation (how long does it take for the pollution to reach the spring?) [13,14,42,54,66].

18.3.1.3 Quantity Management

Although karst resources tend to be quite large, they may prove insufficient during long droughts. Exploitation can be optimized in various manners, for instance, by storing water underground upstream of the spring during high water stages or by "overexploiting" the spring during low water periods (Figure 18.8).

Temporary overexploitation usually consists of pumping water below the spring water table, in this way extracting water stored in the aquifer below the spring. This technique may, however, fail if karst

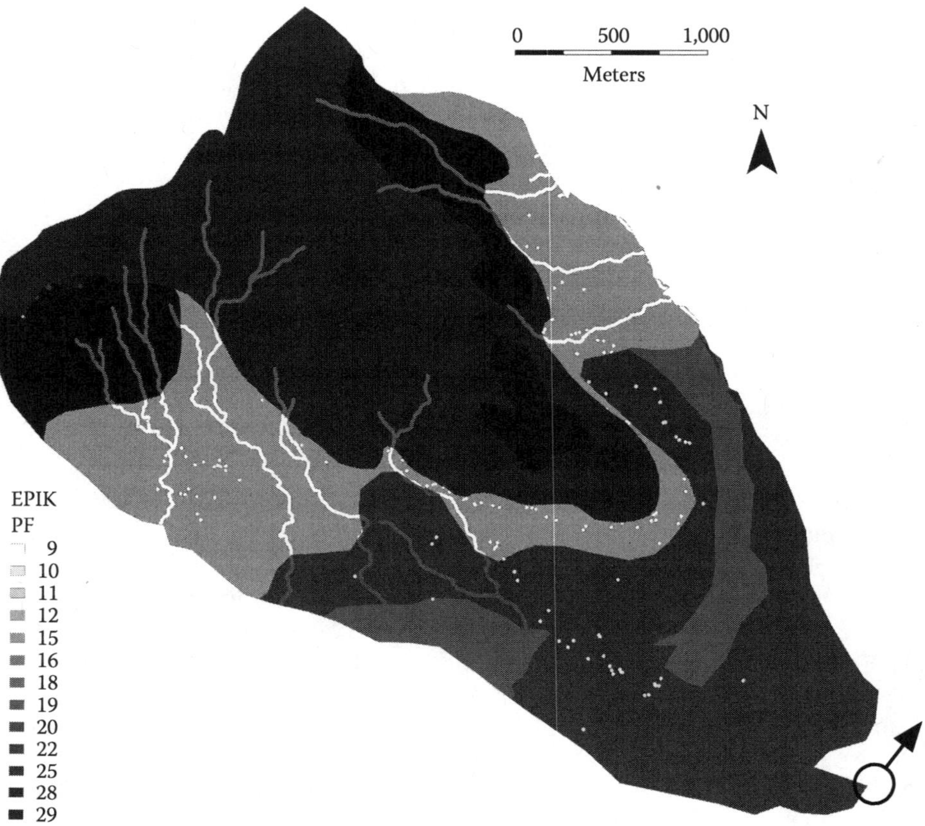

FIGURE 18.7 Example of a vulnerability map produced with the EPIK method. Numbers relate to "protection factors" (PF) qualifying the intrinsic vulnerability. Low values mean high vulnerability of the spring to pollution. Resulting maps are often referred to as "leopard skins."

conduits close to the spring loop up and down (Figure 18.8). Tests should be performed before a pumping plant (or a pumping well) is set up. In order to manage the water resource in a sustainable way, the overexploited quantity during drought periods has to be at least compensated during recharge periods. Reference [7] recommends a maximum extraction of 30% of annual recharge. Overexploitation may lead to terrain subsidence and cracks in the catchment area, especially in case of a strong water table drop. Examples can be found in the literature, especially in the proceedings of the "Multidisciplinary Conferences on Sinkholes" taking place every few years since 1984.

Water storage is obtained by damming some karst conduits in order retain water within the epiphreatic zone of the aquifer. Several dangers are however related to such damming: leakage toward an unknown and untapped outlet, land flooding, and doline collapsing, especially if the groundwater table rises close to the surface in some locations.

Reference [47] presents a series of case studies including problems and solutions concerning artificial storage in karst.

Overexploitation of coastal karst aquifers should be managed with great care in order to prevent saltwater intrusions [6,58]. Underground submarine dams have been constructed to control the ingress of saltwater and allow extraction of freshwater (e.g., Port Miou, France, http://fr.wikipedia.org/wiki/Exsurgence_de_Port-Miou).

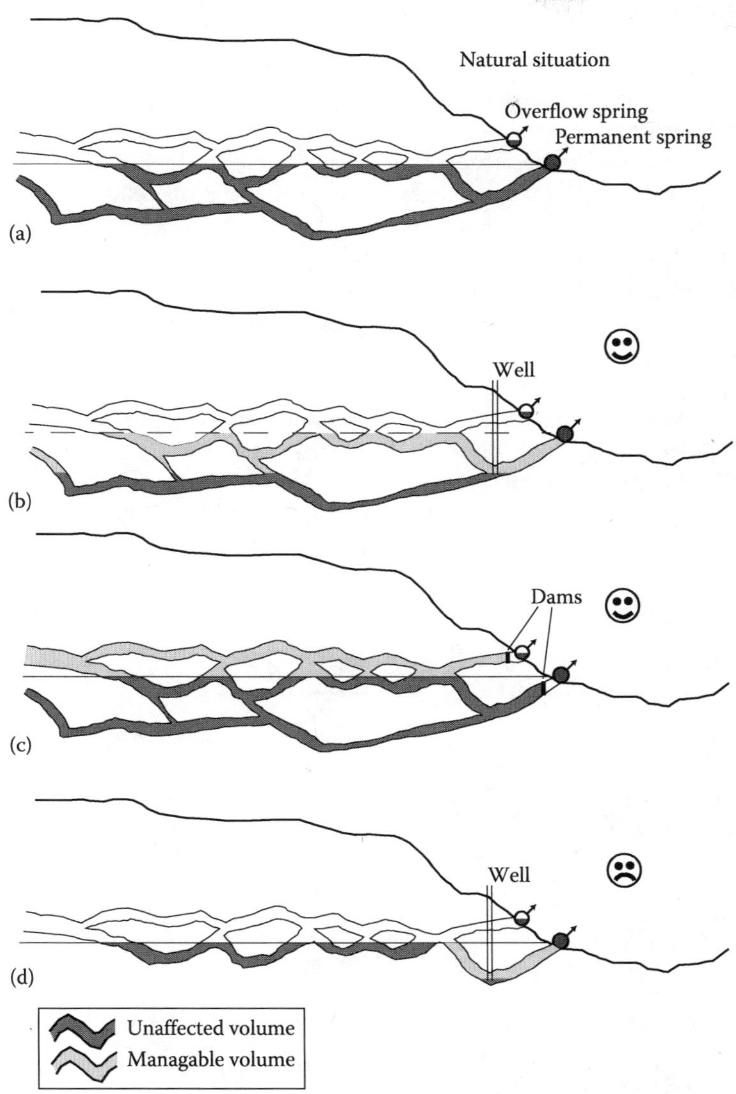

FIGURE 18.8 Example of spring discharge management: (a) natural state, (b) overexploitation by pumping below the spring level, (c) storage behind an underground dam, and (d) example of unsuccessful overexploitation due to conduit loops close to the top of the phreatic zone. In the latter case, only a small volume of water can be exploited.

18.3.1.4 Environmental Issues

Any construction designed for the management of water quantity may have severe consequences for the environment, which should be carefully evaluated!

For instance, the biodiversity and ecological value of the karst itself, as well as of the streams and rivers downstream of the tapping location, may be endangered by any water derivation. This must be taken into account in the environmental assessment of any tapping project.

These aspects are presented with some further details in Section 18.3.7 [7,41,42,47].

18.3.2 Flood Management in Karst Areas

Flood problematic is not the same for a location situated within the catchment area of a karst system or downstream of a karst spring (or group of springs).

Inside the catchment area, in normal conditions, a significant part of the precipitations can be absorbed by the epikarst and routed through the karst system to the karst spring(s). The situation becomes problematic either if the infiltration rate is exceeded, that is, the whole precipitation cannot be absorbed, or if the water table rises so high that it reaches land surface at some low points ("karst overflow," Figure 18.9). These two cases must be distinguished because the management of the two situations is completely different, as presented in the following section.

Downstream from a karst system, critical situations may be observed during extreme conditions because the reaction of a karst system is known to be nonlinear and to display thresholds. The discharge rate of a spring may, for example, suddenly increase dramatically more (or less) than what was expected from the known catchment area. If a water underground storage occurs from a certain water level inside the aquifer, the discharge rates tend to increase less than what the precipitation intensity would suggest. The opposite situation is also observed with a dramatic increase of the discharge rate related to the sudden activation of an overflow spring and/or the sudden contribution to the spring discharge of a part of the catchment area, which contributes to another spring discharge at low water stage (Figure 18.9).

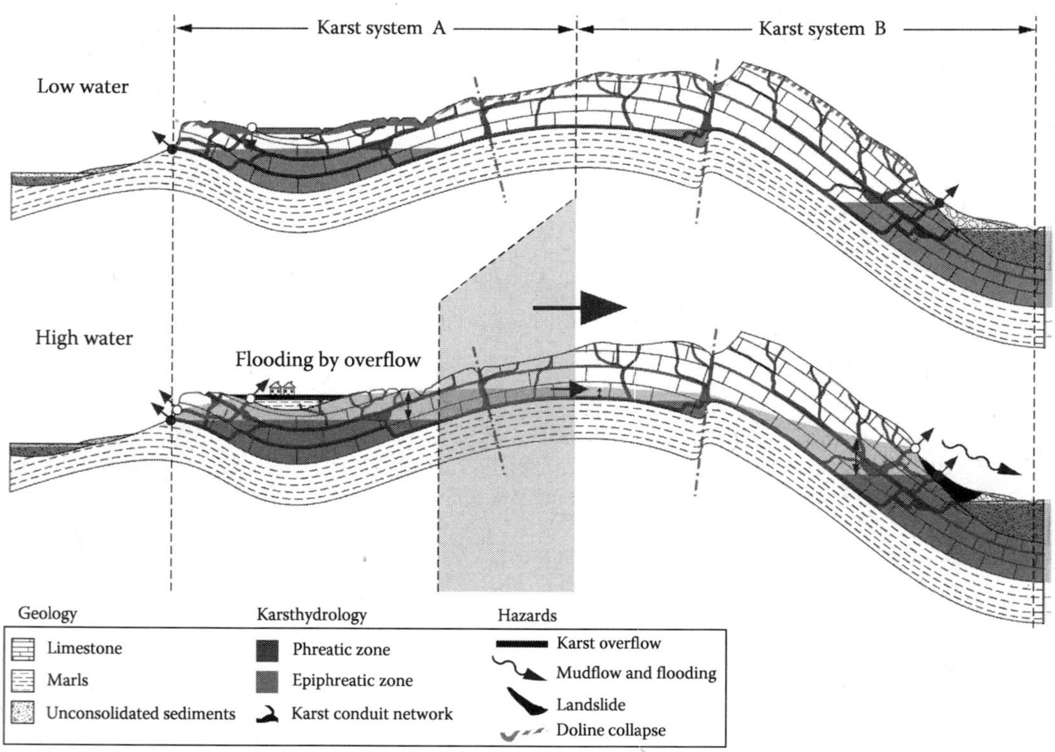

FIGURE 18.9 Flood-related hazards in a karst environment. System A (left) presents a "karst overflow" leading to the flooding of the poljé. During high water events, above a certain threshold, system B (right) receives water from system A, suddenly increasing the discharge of perennial karst spring B and inducing the activation of an overflow spring, potentially leading to flooding or landslide hazards. Doline collapses are enhanced in regions of "karst overflow."

18.3.2.1 Assessment of Potential Karst Water-Table Rise (Karst Overflow)

The "karst overflow" situation can be very critical. The strongest evidence of such an event is the presence of caves that become flooded, but which are normally not traversed by a significant stream. Head measurements in these caves inform about the water table and help to apprehend hydraulic gradients during flood events. If the water table reaches the elevation of land surface at some location, it can be postulated that flooding of land surface is related to "karst overflow." If no caves are known, measurements in boreholes may provide the same type of results. Without caves or boreholes, the assessment of a "karst overflow" may be possible but remains usually uncertain.

The mitigation of flooding due to a karst overflow is very difficult. Two strategies can be envisaged: The reduction of infiltrations into the karst system with the derivation (or storage) of water from one part of the catchment area, for example, of the stream feeding a significant ponor. The other strategy is to increase the transmissivity of the karst aquifer. This can be achieved by increasing the size of natural karst conduits or by building a drainage tunnel. In many situations, both strategies may be unrealistic!

The danger of "karst overflow" has to be taken into consideration in a region with increasing urbanization, where the rate and speed of infiltration can be drastically increased compared to natural conditions. Recharge peaks may therefore lead to a "karst overflow" in a place where it never happened under natural conditions. Consequences can be unexpectedly disastrous.

18.3.2.2 Assessment of Potential Exceeding of Infiltration Capacity

The same data as for the assessment of "karst overflow" (head measurements in caves and drill holes) help to detect if flooding is related to the exceeding of the infiltration capacity of a ponor or of the epikarst. Simultaneous measurements of heads at the ponor and at the spring can normally show whether they are hydraulically connected. If so, variations at the ponor should be visible within seconds or a few minutes at the spring. However, this observation may be difficult if the spring is large and the ponor rather small.

If the ponor can be entered at low water stage, a hydraulic assessment, as well as measurements directly inside the ponor cave, may be very useful. One or two tight passages are often sufficient to produce upstream flooding, and an enlargement of these passages may be a solution. However, one should first assess whether a quicker inflow of water will not produce a general "karst overflow!"

18.3.2.3 Assessment of Potential Flooding Areas

A good knowledge of hydraulic heads in the karst network for various high water conditions, as well as a good view of the aquifer geometry (KARSYS approach), allows the identification of areas where karst groundwater is suspected to reach land surface during extreme events. Hydraulic modeling, even strongly simplified, is usually helpful for this type of assessment.

For locations where flooding is suspected to be related to an exceeding of the infiltration capacity (typically for a ponor), discharge and head measurement in the ponor itself, and a simple hydraulic model, including the storage volume of the ponor, may help in assessing the situation for extreme events.

18.3.2.4 Assessment of Potential Related Hazards

However, groundwater level in karst systems may rise hundreds of meters during extreme high water events and can potentially induce other hazards. Two major hazards have to be mentioned here (Figure 18.9): (1) landslide and (2) terrain collapses.

If karst water is dammed behind a geological unstable impervious barrier (e.g., glacial tills), in extreme situations, the hydraulic head in the underlying karst may be high enough to "push up" the unstable mass that can be set in motion by gravity and form a landslide. The head peak may last only a few hours (or even less) and can only be identified with continuous monitoring in the underlying karst aquifer.

Abrupt terrain collapses in karst usually occur in regions where karst bedrock is covered by thin unconsolidated sediments, usually soil, but also tills or alluvial deposits. The collapsing process is strongly enhanced when the bottom of the unconsolidated cover lies within the epiphreatic zone. The up-and-down movement of water "washes out" sediments, leading to the erosion of the cover from its base, up to the breakpoint where the soil collapses. This process may be related to the movement of the karst water table but sometimes also to a perched water table of the epikarst. This kind of hazard has to be taken into account when designing an artificial infiltration in a karst region (see Section 18.3.4.) [7,11,47,65].

18.3.3 Hydropower

18.3.3.1 Exploitation of Karst Groundwater

Karst groundwater can be exploited for hydropower following three different settings (Figure 18.10):

18.3.3.1.1 Type 1: Perched Spring

Some karst springs are located several meters or dozens of meters above the bottom of the valley. This is a classical situation, which basically does not require specific knowledge about karst. However, in some cases, it can be interesting to manage the discharge or even to dam the spring in order to store water and increase the effective height for the exploitation. Such projects require a strong experience specific to karst systems in order to minimize the risk of failure and the risk for the environment. The Ombla spring project (Croatia) is a typical example of hydropower exploitation of a perched spring [57].

18.3.3.1.2 Type 2: Derivation of an Underground Stream or of a Perched Phreatic Zone

This setting, which is less usual than the first type, aims at diverting an underground stream into an artificial tunnel, which leads the water to a pressure pipe and a turbine. The stream should ideally be the output of a perched aquifer, which storage can be used for discharge regulation.

The Pierre Saint Martin/Verna power plant (France) is a typical example of this type of exploitation.

18.3.3.1.3 Type 3: Underground Turbine

At locations where a surface stream swallows into the ground (ponor) and if a vertical cave is present, it may be possible to deviate the swallowing stream into a pressure pipe and to set up a turbine

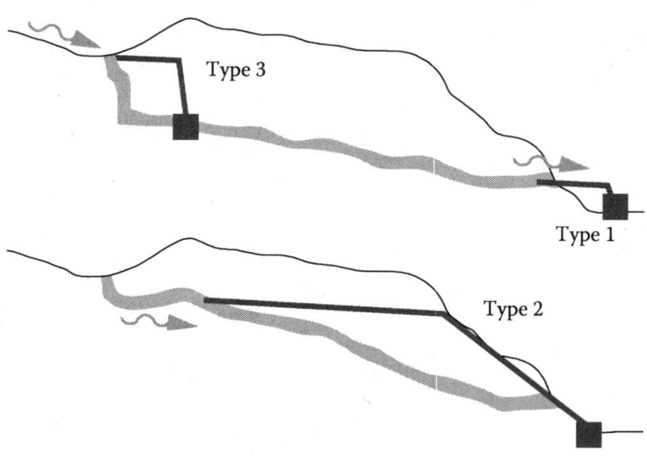

FIGURE 18.10 Various types of hydropower exploitation of karst groundwater (see text).

at the bottom of the cave. The pressure pipe is usually installed in a drill hole reaching the cave at its bottom.

Small power plants of this type are known all over the world.

18.3.3.2 Surface Storage

Because karst is a medium with a high permeability, it is always difficult to build large water storage on karst areas. However, because of the presence of significant perched depressions (dolines, poljés) in karst regions, several dams have been built in these regions. Leaks are always a problem in such dams and can sometimes reach impressive values (e.g., 26 m^3/s in Keban, Turkey, from [47]). Detailed geological and hydrogeological observations, as well as geophysical measurements, can help to localize and understand the potential outflows. Grout injections or surface sealing can be used to reduce the leaking discharge. Reference [47] provides an extensive overview of problems and solutions (which are mostly rather expensive!).

18.3.3.3 Underground Storage

In some cases, cave passages can be dammed in order to create an underground storage. This should however be attempted only in regions where the karst system is largely accessible, especially if the expected head increase is larger than a few meters. The risk of failure is significant because many cave systems have several passages and water may bypass the dam. A detailed hydraulic study with precise head and discharge measurements is required before starting such a construction. An attempt was made in Hölloch (Switzerland) but was unsuccessful. Reference [47] provides some other examples [7,47,59].

18.3.4 Water Infiltration

The absence of surface streams in the catchment area of most karst regions is often problematic for the evacuation of waters generated in urbanized zones. Outside karstified areas, waters from impervious zones (roofs, roads, parking lots, etc.) are collected and usually diverted toward rivers, lakes, or the sea. In karst regions, due to the absence of such efficient outlets, these runoffs are either, where possible, infiltrated into the ground or diverted in long pipes outside of the karst catchment area. The success rate of runoff infiltration is higher when small quantities are infiltrated in many points rather than the whole flow rate at only a few points. An assessment of the capacity of the aquifer to really absorb and route waters quickly enough to the spring(s) should be carried out in order to avoid a “karst overflow” (Section 18.3.2).

18.3.4.1 Assessment of the Infiltration Capacity

Artificial infiltration can be concentrated (ponor) or diffuse (over a surface of soil).

The infiltration capacity of a site can be assessed by some experiments and monitoring campaigns. For diffuse infiltration, the monitoring of rain intensity coupled to observations during high intensity events is of great help to assess the intensity above which runoff starts and how far this runoff flows before infiltrating into a fissure (or if water actually stays at land surface). Artificial sprinkling experiment can also provide valuable data, although they require to be very large and to use large amounts of water in order to approach extreme rain conditions. Local-sale infiltration tests (double-ring infiltrometer) are often more practicable but should be undertaken over a sufficiently long duration to approach conditions of soil (and possibly epikarst) saturation. If local-scale experiments bring out very high infiltration values compared to expected rain intensities, this test may be sufficient. Values higher than 50 or even 100 mm/h are quite common on karst soils.

The infiltration capacity of ponors must be assessed as described in Section 18.3.2.

18.3.4.2 Infiltration Plants

For diffuse infiltration, it may be recommended to infiltrate water on top of a soil cover, as potential pollutants in the infiltrated waters would be retarded and partially degraded by the high biodegradation capacity of active soils. Depending on the discharge rates that need to be infiltrated and on infiltration capacity of the soils, water should be distributed over a sufficiently large piece of land using pipes. Small dams can be constructed in order to maintain water within the defined area. If large infiltration areas are not available, infiltration pools on top of a soil cover can be used, allowing for the storage of 1–2 m of water.

18.3.4.3 Related Hazards

The probability of soil collapse is high within a distance of 30–50 m around artificial infiltration points or areas. This phenomenon is related to subsurface flow and storage in the soil and the epikarst. This problem can be addressed by infiltrating water directly at depth in a cave, but the protection effect of the soil cover against potential pollutions is then lost.

18.3.5 Geothermal Exploitation

18.3.5.1 Low-Temperature Geothermal Exploitation (Heat Pumps)

Karst springs are the outlets of a karst conduit system draining water and heat of the whole catchment area. This means that heat flux out of a karst spring is usually high or even very high, especially in temperate and humid regions. Therefore, where possible, heat pumps can advantageously be placed in the water of karst springs. Diverting a fraction of the spring water into a distribution network supplying water to heat pumps in individual houses and buildings may be a cheap and effective solution.

Down-hole heat exchangers can also be installed in wells drilled in a karstified massif, but their efficiency is strongly correlated to the location of the well within the system. The vadose zone in karst may be very thick and has a low temperature gradient (about 0.5°C/100 m), mainly because of heat drainage by the phreatic zone. In cold regions, the rock mass around the well may freeze if too much heat is taken, and the design of the installation has to be done carefully.

Similarly to the infiltration vadose zone, the upper part of the phreatic zone often shows a very low temperature gradient (<0.5°C). However, thanks to the important water flow, wells located close to active karst conduits can extract much more energy per well length unit, compared to wells located in low porosity zones. The exact location of karst conduits is usually not well known and one should consider a consequent safety margin when designing the installation.

One critical aspect of ground source heat pumps in karst environments is their environmental impact. The environmental impact may be significant if karst openings, especially water bearing ones, are drilled through. Drilling also increases the pollution hazard for aquifers exploited for freshwater supply. As a consequence, well drilling for geothermal heat exchangers is rather incompatible with an exploitation of drinking water, unless very strict conditions can be established and verified.

18.3.5.2 Medium- and High-Temperature Geothermal

Large carbonate reservoirs are located at depth and can represent interesting targets for geothermal energy exploitation, either for direct use or electric power generation. Some of these reservoirs are well karstified while others are less so. It is beyond the scope of this chapter to describe characteristics of these aquifers and strategies for their exploitation [1,28,45].

18.3.6 Human Impacts on Karst Waters

18.3.6.1 Constructions (Tunnels, Roads, Bridges, etc.)

Any construction in a karst environment has an impact on the caves, karst groundwater, and karst biotope. As karst has very specific characteristics compared to other environments, this impact has to be assessed with adequate models and approaches. This means that environmental assessment studies have to be carried out by specialists, following dedicated methods. The various kinds of exploitations of the karst presented in the previous sections have also an impact on the whole system and have to be assessed in the same way!

The environmental assessment usually includes at least the following points:

1. Relationships between the construction and the nearby karst hydrogeological systems: The karst hydrogeological system must be understood (e.g., KARSYS) and described together with the projected construction. All hydraulic connections (known or potential) have to be explicitly listed.
2. Assessment of karst hazards for the construction: From relationships evidenced in point 1, all potential karst-related hazards for the construction can be assessed, for example, the potential presence of a large void below a bridge pile. Further investigations can be undertaken according to the ratio between hazard and risk.
3. Assessment of the impact of the construction on the karst (hydrology, caves, biology, etc.): all the potential influences on the karst environment should be assessed, for example, a significant pollution hazard of a drinking water supply during the construction phase. Further investigation or mitigation measures can be undertaken considering the ratio between hazard and risk, as well as the existing regulations.

The Swiss Speleological Society edited a practical guide for the assessment of construction projects in karst regions, available in French and German [55].

18.3.6.2 Other Activities (Agriculture, Urbanization, Industry, Waste Deposits, etc.)

Any human activity influencing the water regime or inducing pollution is expected to have an impact on karst in general and groundwater in particular. The assessment of this impact is often made difficult because of the size of karst catchment areas, which increases the difficulty to localize the origin of a pollution, particularly for diffuse or multiple sources of pollution. The methods presented in Chapter 4 may help to consider several hypotheses and test their conclusions [15,52,59,60].

18.4 Investigation Methods

A series of investigation methods dedicated to karst hydrogeology have been developed in the last decades. Some of them are more effective than others. This section sketches out a standard way to approach karst hydrogeological systems, starting with methods that are the most effective for the understanding of spatial and temporal characteristics and continuing with more sophisticated, time-consuming, and targeted methods.

18.4.1 Geological and Hydrological Data Compilation

Flow in karst is controlled by the geology and by the hydrological conditions. The first step in a karst hydrogeological investigation is usually to assess the following points first:

1. Hydrostratigraphy: A compilation of hydrogeological characteristics of all rock formations, which will help distinguish which can be considered as karst aquifers (exposed karstifiable rocks), as aquicludes (e.g., mostly shales, thick marls, siliceous rocks), and as potentially intermediate rocks (e.g., marly limestone, sandstones with calcareous cement). Descriptions of the lithology and fracture characteristics are welcome.
2. Aquifer geometry: All existing geological maps and profiles have to be compiled in order to assess the degree of knowledge existing about the considered region. Drill hole data are very welcome as well. The scale of existing maps or documents has to be somehow adapted to the level of detail needed to address the question.

3. Hydrological conditions: All data on spring locations and discharge should be compiled, as well as data on the distribution of precipitation. Similarly, data on surface stream network, swallow holes, or swallowing streams/rivers, as well as the position of outcropping karst (potential recharge area), should be collected. Head measurements in drill holes are very helpful.

A first interpretation model can be derived from this compilation, which will provide an order of magnitude of the degree of karstification in the investigation region, as well as a first sketch of recharge and discharge areas. Details on this step can be found in [25].

18.4.2 KARSYS Approach: Aquifer Geometry, System Definition, and Hydrology

KARSYS is a direct approach aiming at building up a descriptive model of the hydrogeological behavior of a karst system. This synthetic approach is very effective to get a first overall picture of a karst system. The method, presented in [33], combines a series of available data into a 3D conceptual model of flow characteristics of the karst system.

Based on the data compilation described in the previous paragraph, a 3D model of the aquifer geometry (geological model) can be constructed (Figure 18.11). Ideally, the model should cover a large area in which all neighboring karst systems are also included. If this latter condition is not feasible, the area should at least cover a few kilometers around the supposed catchment of the investigated system.

The next step is the inclusion of all hydrological information in the 3D model. As the most evident information available on a karst system are the springs, their approximate discharge and their elevation, established as precisely as possible, are included in the model. Any piezometer, borehole, cave, or location where the presence of water has been identified and measured should be included in the model as well, with an indication of the height of the water table.

With these information, the water table can be modeled (Figure 18.12), at first as a flat water table extending from the springs and extrapolated throughout the aquifer volume (in 3D), if no indication about a gradient is available. This gives a first idea of the extension of the phreatic zone at low water stage. If desired, slight gradients may be assumed and introduced (usually between 0.01% and 0.5%). If water level data are available, one should first carefully assess their liability: Water level data in caves are measured directly in the conduits but are mostly characterized by a significant uncertainty on elevation. Values in drill holes can be measured more precisely but may be several meters apart from the head in the conduit network. An example is presented in Figure 18.12.

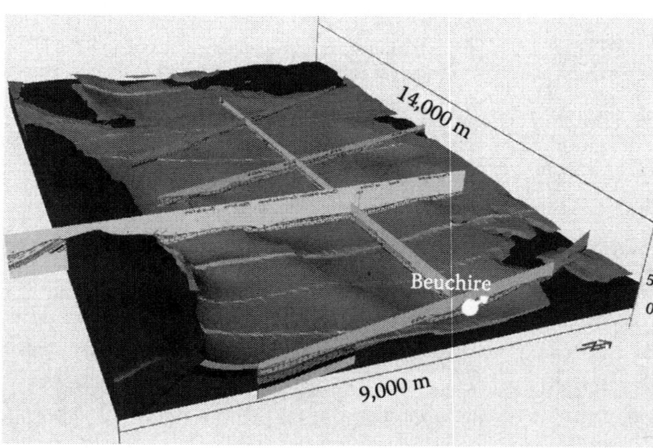

FIGURE 18.11 Three-dimensional geological model of the bottom of a karst aquifer, based on geological cross sections and geological map.

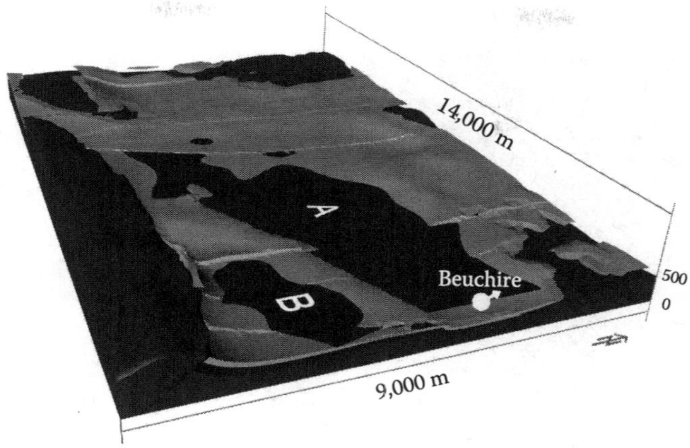

FIGURE 18.12 Spring and extension of groundwater bodies by low water conditions. Groundwater bodies A and B are not connected.

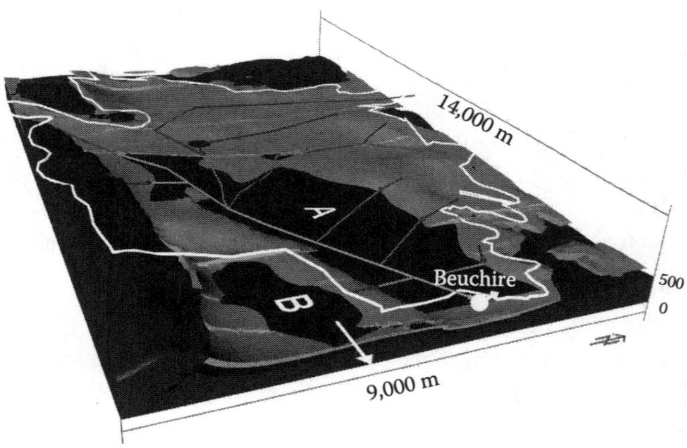

FIGURE 18.13 Hydrogeological conceptual model of karst system A by low water conditions, including the extension of the groundwater body, the estimated catchment area, and the estimated main flow paths (vadose/phreatic).

Once the representation of the water table is considered as reasonably consistent, the catchment area can be delineated by assuming a vertical flow through the unsaturated zone, down-dip flow on the top of aquicludes, and flow toward the spring(s) in the phreatic zone. Flow paths from various parts of the catchment can be constructed following the same concept (Figure 18.13). If two springs flow out of a single phreatic body, one has to decide whether both can be considered as two outlets of the same system or if they are both an outlet of their own system. In the latter case, a limit is established at the intersection line between water tables extrapolated from the two springs.

At this stage of the building of the model, the system has only been considered at low water conditions. If one considers this model as sufficient, the obtained model has to be validated. It must therefore be compared to a series of other existing data such as the following:

- Results of tracer experiments (if available), keeping in mind that only tracer tests carried out during low water conditions can be consistently compared to the model.
- Another important validation can be drawn from the minimum and average discharge flow rates out of the systems. The catchment area delineated by the KARSYS approach compared to the measured (or estimated) system discharge must be in the same range as those of nearby karst

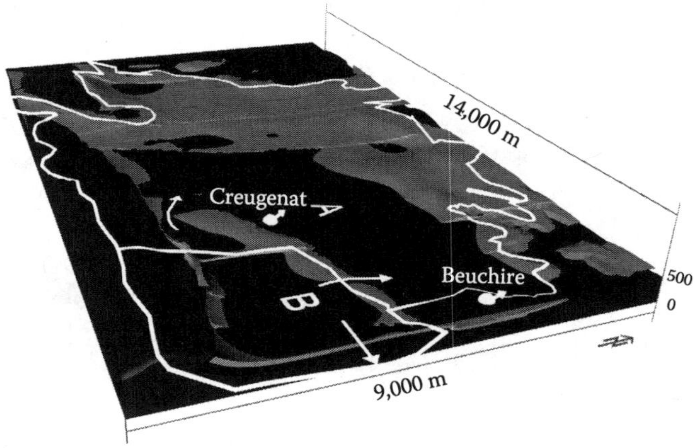

FIGURE 18.14 Hydrogeological conceptual model of karst system A at high water conditions. One overflow spring (Creugenat) become active inside the catchment area, and groundwater body B overflows toward system A. The extension of the GW body and the related catchment area significantly differ from the low-water situation presented in Figure 18.13.

systems. A karstic catchment may appear to be much larger or smaller than expected in a first approximation, which is often influenced by the topographic—or hydrological—catchment. The analysis of specific discharge and the comparison with other similar catchment (e.g., [24]) may highlight important errors in the catchment delineations (assuming that discharge rates are correctly measured).

- All other data such as hydrographs and chemical or microbiological analyses can be used as a control tool and should not be inconsistent with the proposed model.

In case of clear inconsistencies, all aspects of the model should be evaluated again (including geology), and decisions must be taken after having considered the respective data liability and significance. This procedure may bring fundamental changes to the model.

The model presented in Figure 18.13 was validated by all available tracer experiments and discharge data.

The next step of the approach is to consider data at high water conditions (Figure 18.14). The elevation of overflow springs as well as any data on heads in the karst network at high water should be included in the 3D model and used to inter/extrapolate the karst conduit water table at high water conditions. The catchment area analysis can then be carried out again and will in most cases result in different catchment boundaries than those defined for low water stage. Tracer test results at high water can be compared to the model as a validation tool.

This approach is very efficient and synthetic. It can be implemented iteratively, starting with a very approximate model, which is then refined along with the collection of new data. The time investment depends on data availability and on the complexity and size of the system. One should plan about 1–4 man-week work for one karst system.

Applications of this approach in Switzerland can be viewed on www.swisskarst.ch. Similar approaches have already been suggested by [2] or [10,33].

18.4.3 Tracing Experiments

Tracer tests are largely used in hydrogeology, and particularly in karst, because they provide reliable results, unmatched by any other method. They are very effective for the verification of specific flow routes and to test boundaries defined by the KARSYS (or similar) approach.

18.4.3.1 Definition and Aim

A tracing experiment (or tracer test) is the introduction of a substance into the system at one location, followed by the monitoring of the concentration of this substance at potential output points. In most cases, artificial substances (mainly dye) are used, because they can be detected at very low concentrations in springs or wells and are not present naturally in the water. An ideal tracer is conservative, that is, neither degraded nor absorbed in the aquifer and its behavior perfectly mimics water flow. In reality, only a few tracers are close to ideal. Tracing methods advantageously provide proofs of hydrogeological connections. Multitracing experiments are tests with several different tracers injected at the same time in various locations.

Most tracing experiments are carried out in order to identify a connection between a specific point within a catchment and a spring or a well. The aim is thus to validate a guess of a catchment delineation. In good conditions, values such as tracer recovery, tracer velocity, and tracer dispersion can be drawn from the test and can provide useful information about the tracer flow path. The shape of tracer breakthrough curves may also provide some information, but interpretation models are usually quite ambiguous.

As for any experiment, the execution of a tracing experiment can be divided in three main phases: preparation, experimentation (injection, monitoring, and analyses), and interpretation.

Tracing experiments in karst must be organized by experienced hydrogeologists in order to avoid wrong interpretations and inadequate procedures.

18.4.3.2 Preparation of a Tracing Experiment

The first step is to make a conceptual model of the flow route(s) to be tested. Injection points can then be selected according to the question to be addressed and to the conceptual model. One has to decide in what flow conditions the test will be carried out (high/low water), as well as the type and quantity of tracer that will be injected.

The next step consists of designing the monitoring network, based on the conceptual model of the studied karst system.

A strong effort must be dedicated to identify without any ambiguity the recovery of the tracer (presence/absence at a given location). Getting an idea of the tracer velocity and recovery rate can be valuable for the interpretation of the results. To answer some specific questions, detailed breakthrough curves can be obtained from the tracer monitoring.

18.4.3.3 Injection, Monitoring, and Analyses

Injection of a tracer should never be carried out by people who are involved in the monitoring process, because of a high risk of contamination that would eventually ruin any recovery results.

Tracer monitoring should start right after injection and ideally at every known spring potentially connected to the injection location. Several water samples should be collected every day at each monitored spring (or well). The sampling rate can be reduced to about one sample a day after one week and one sample every 3 days after one month. The management of the monitoring during a tracing experiment requires a significant amount of work, involving many people during several weeks (sometimes months). For large tracing experiments, the number of samples easily reaches one thousand or even more.

Tracer detection can be achieved by analyzing in the laboratory the samples collected at the springs or, for dye tracers, by field fluorometers or by using charcoal bags that are later analyzed in the laboratory. Much care should be taken to apply the right analytical procedure and to avoid by any means sample contaminations. A laboratory experienced with tracer analysis should be preferred.

Raw results must be presented as tables and maps showing all tested connections, including those where the tracer was not found (Figure 18.15).

A detailed documentation of analytical techniques and injection and sampling locations is important for the assessment of result liability.

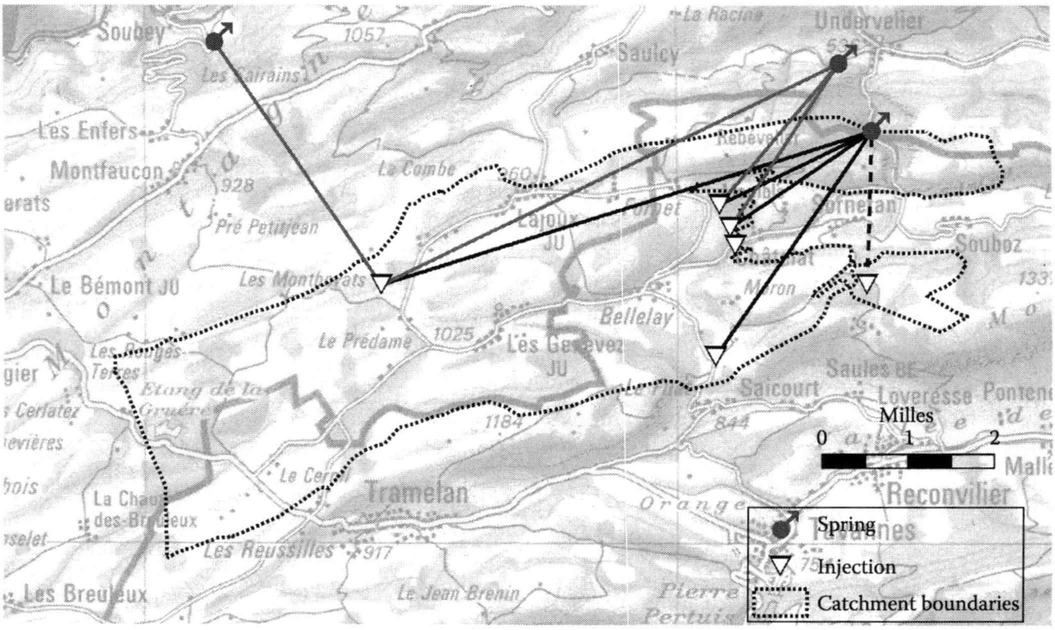

FIGURE 18.15 Example of representation of raw recovery results of a tracer test. Black and gray arrows represent proved and disproved connections, respectively. Dotted lines (black or gray) represent doubtful results. Each injection point must therefore be linked to all monitored springs (or wells) by either a black or a gray arrow.

18.4.3.4 Interpretation

In most cases, the main goal of a tracing experiment is to prove or disprove a connection between an injection point and one or several springs or wells. A strong effort must be made to obtain unambiguous results.

In karst systems, diffluences are commonly observed, that is, the tracer from one injection point is detected at several springs (or wells).

The breakthrough velocity of a tracer strongly depends on flow conditions (increase by a factor 5 to 10 in high flow conditions). Tracer velocity alone is not a good indicator of a degree of karst development. The analysis of thousands of tracing experiments made by [63] showed that tracer velocities range between 0.0001 and 1 m/s, but that about 90% range between 0.002 and 0.2 m/s (7.2 and 720 m/h) with an average velocity of 0.022 m/s (79 m/h).

Tracer recovery rate and tracer velocity are useful values to qualify the "degree of connection": a connection with a recovery of 0.01% of the injected mass cannot be directly compared to another with 60%. A recovery rate of more than 10% or 20% is usually considered as a high rate, corresponding to a very good connection. Lower recovery rates are more difficult to interpret. Low rates are often the result of an injection on top of soil and/or epikarst but may also be a consequence of the presence of large phreatic zones, diffluences, or of any kind of tracer degradation or trapping. The interpretation of the recovery rate is also highly dependent on the type of tracer used and on the context.

The shape of the breakthrough curve has been the object of many academic studies, as it can, in the best case scenario, provide information on the characteristics of the flow path (e.g., vadose or saturated flow, large or small conduits, existence of several parallel routes). Some quantitative approaches have been developed to model tracer breakthrough curves in karst. Benischke et al. [5] refers to a series of modeling tools, among which QTRACER2 [16,17] is one of the most recent and complete [4,5,27,43,56,64].

18.4.4 Prediction of Karst Occurrences within a Karst Massif (KarstALEA Method)

In many circumstances, the knowledge of the location of karst conduits within a karst massif would be very valuable. Sometimes data collected by cavers provide such information but, in most cases, existing caves have not yet been found and explored. KarstALEA is a method synthesizing all available data of a karst massif in order to predict, in a probabilistic manner, the position and main characteristics of karst conduits. Such predictions are crucial for engineering projects, such as tunneling activities, which are very sensitive to the presence of karst conduits (voids, water inflows, loose sediments).

"KarstALEA" ("alea" means hazard) provides a first practical method for the assessment of conditions in karst massifs. A complete guideline of KarstALEA is available in [20], unfortunately so far only in German. It provides probabilities of occurrence of karst voids and conduits within a karst massif and also gives an insight on the expected characteristics of the anticipated karst features (size, shape, water pressure, water discharge, presence and type of sediments, etc.).

The method is based on a series of general laws known to control the development of karst conduit networks (speleogenesis) and groundwater flow in karst massifs (karst hydrogeology).

KarstALEA is applied in seven steps (Figure 18.16). The two first steps (1 and 2 on Figure 18.16) consist in applying the KARSYS approach, that is, the construction of geological and hydrogeological 3D models of the massif. Novel elements of KarstALEA are the introduction of a speleogenetic model (3) and of an inception-horizon model (4). The main reason for including these lies in the fact that the study of cave networks surveyed by cavers shows that karst conduits preferentially develop along a restricted number of geological planes and features [19]. These planes are most commonly stratigraphic horizons but may also be joints or faults. They are very likely to favor karst development, and experience shows that 70% of the conduits may develop along typically three to five discrete horizons representing less that 10% of the limestone volume. These planes are known as inception horizons. It is furthermore well known that karst conduit networks develop with specific density and characteristics within the so-called speleogenetic domains. Notably, horizontal networks of caves mainly develop close to the regional base level, that is, close to the floors of the main surrounding valleys. Progressive valley entrenchment over time may therefore have induced a series of successive levels of subhorizontal cave passages. Observations in existing cave can thus provide useful information for the construction of models 3 and 4. The probability of occurrence of unknown cave passage is the highest at intersections between identified cave levels (model 3) and inception horizons (model 4).

The combination of speleogenetic models and inception horizons therefore provides a probabilistic assessment of the "karst conduit density" (step 5). In step 6, models 2 (hydrogeology) and 3 (speleogenesis) are combined in order to characterize the voids and obtain the KarstALEA zones that describe some characteristics of the anticipated karst conduits (notably, the size and shape of voids, the amount and type of sediments, the presence of water and its expected pressure and discharge rate, etc.). Both indications (karst conduit density and KarstALEA zones) are combined in step 7 in order to assess karst hazards, risks for the construction and for the environment. This step also provides some indications on appropriate mitigation measures.

Data needed at each step require some specific investigations but can mainly be derived from usual observations related to the project [18–20,38,49,50].

18.4.5 Global Approaches (Hydrographs, Chemographs, etc.)

Global approaches consist of analyses of time series measured at karst system outlets (mainly springs). The word "global" reminds us that variations result from the mixing of conditions and processes taking place within the whole system. Besides a simple description of the functional behavior of a karst system, the scope of such analyses is to infer some spatial characteristics of the karst hydrogeological system from a typical signature visible in a time series. Global approaches are purely inductive.

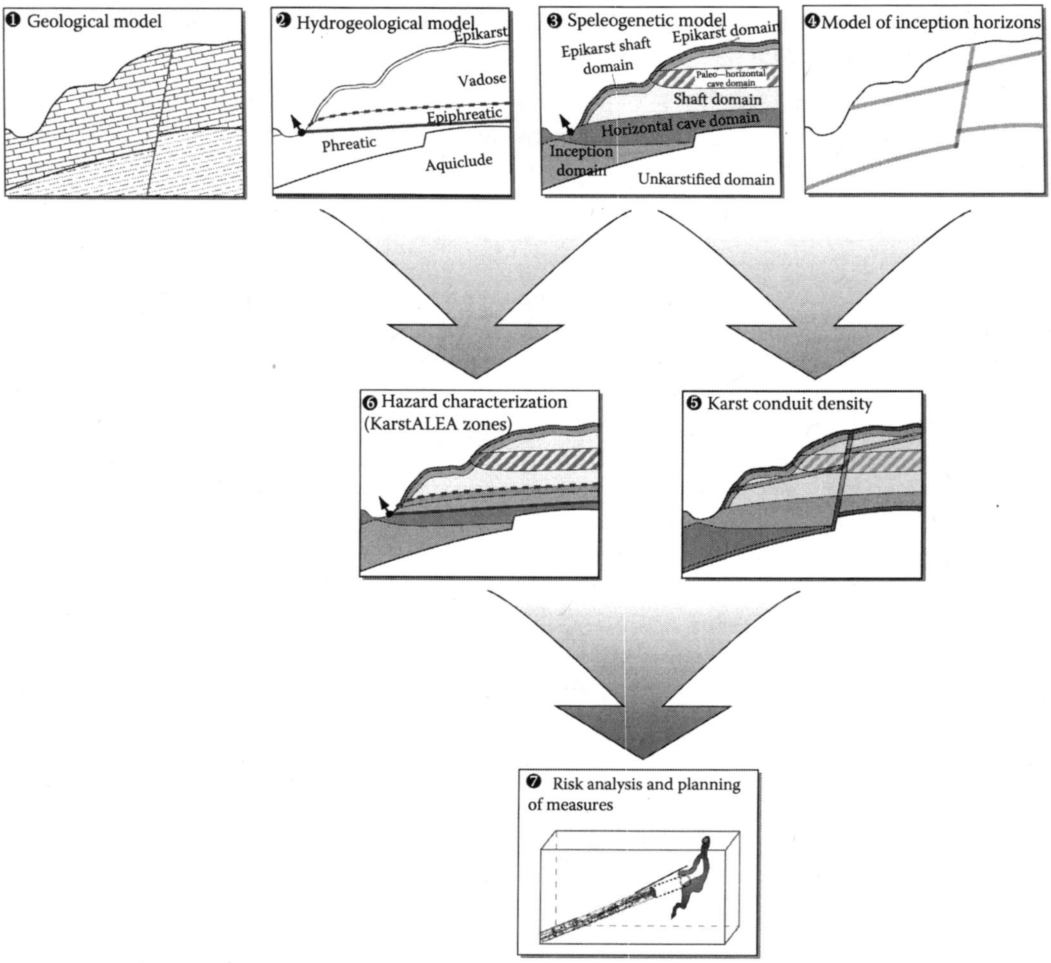

FIGURE 18.16 Principle of the KarstALEA method in seven successive steps. Four 3D representations of the massif (models) are constructed in steps 1–4. The zones with the highest expected densities of karst conduits are interpreted together with their characteristics in steps 5 and 6. In step 7, hazards, risks, and mitigation measures are identified and discussed between the project geologist and the project engineer.

Discharge rates have been recorded (hydrographs) for many decades and various analysis methods have been created along with the development of "karst hydrogeology" since the beginning of the twentieth century. Temperature and electrical conductivity probes coupled to data loggers are very common since about 1985. Turbidity can also be monitored but requires some more sophisticated equipment. Probes selectively measuring dissolved ions (e.g., Cl^- or Ca^{++}), pH, dissolved O_2, CO_2, ^{222}Rn, etc., became available during the last decades. They do require, however, more maintenance work and are not always completely liable. Sampling and laboratory analyses are still useful for chemical monitoring of major elements, isotopes, and trace elements. Monitoring of biological compounds also became possible in the last 20 years, and new techniques are steadily being developed, although most analyses remain expensive.

A common aspect to all "global studies" is to obtain a "representative" time series, that is, a record in which measurement frequency, resolution, and precision (stability) are sufficiently high in order to detect real variations of the natural system. As karst systems do react extremely fast to condition changes, time steps must be short enough to make sure that short peaks remain visible. Time steps larger than 1 h (for small karst systems) or 1 day (for large ones) are not adequate. For measurements with data loggers, usual time steps range between 5 and 60 min.

Recording time should also be reasonably long in order to allow a meaningful analysis of the time series. Depending on the objective of the project, time series of 10 years or more are very valuable, while in many cases, 1 year of good quality data can provide interesting information. For some interpretations, a few months, including at least two or three storm events, may already be useful.

Finally, one should mention that all parameters measured at a karst spring are sensitive to discharge variations, and the interpretation of time series is therefore only possible if discharge have also been recorded.

Due to the requirements of time series recording and unless data are already available, the financial cost of global approaches is relatively high, and so is the investment in work time.

More detailed information on global approaches can be found in the literature. Whether global approaches are feasible and worth, the investment in engineering projects has to be assessed from case to case, depending on the available data and on the questions to be addressed. Hydrographs, however, remain a very fundamental data set for interpreting all other parameters [2,22,26,42,61].

18.4.6 Further Field Investigation Methods

Other methods that are frequently used include geophysical methods, investigations in drill holes, and investigations in caves.

18.4.6.1 Geophysics

One of the main challenges in karst hydrogeology is the detection of the main branches of the karst conduit network. Considering the fact that conduits are often located deeper than 20 or 30 m below ground, that the size of the conduits typically ranges between 1 and 5 m in diameter, and that conduits may be partially or totally full of water and sediments, the detection from the surface is a difficult task. This information is critical in many engineering projects, mainly as a tool to assess the danger of rock collapse but also for karst hydrology.

Most geophysical methods can detect underground voids only at shallow depth, unless they are very large. Georadar, electrical tomography, seismics, or microgravimetry can provide good results. The combination of two or three methods used simultaneously increases the reliability of the results. The liability and resolution of these techniques can also be significantly improved by tomographic settings (e.g., for georadar or seismics). This can be applied for targeted investigations (e.g., tunneling in a very hazardous section).

As of today, the detection of deeper conduits (>20 or 30 m) cannot be guaranteed by any method. Literature does present some examples of deep cave detection, but only few of these could be confirmed by direct observations (drilling or cave discovery). Furthermore, the number of cases for which existing conduits were not detected by the applied geophysical methods is simply unknown!

Geophysical methods are often more useful when applied to the description of the geological settings, that is, as input in methods described in paragraphs 18.4.1 to 18.4.4.

Another application of geophysics is the positioning of conduits explored by cavers by using localization systems [3,32,34].

18.4.6.2 Investigations in Drill Holes

The objectives of drill hole investigations are the following: (1) to assess the aquifer behavior mainly through recording water levels or heads at various locations in the aquifer and (2) to determine aquifer parameters, including hydraulic conductivity, aquifer transmissivity or storativity.

Due to the very high heterogeneity of permeability distribution in karst, usual approaches, developed for intergranular porous media, may not be adapted and should at least be interpreted with care! For example, hydraulic heads may vary by several meters along a single borehole. In other words, the water level measured in a naked borehole may be quite different from the head of the regional flow system (i.e., of the nearest karst conduit). Therefore, in order to assess the head of the karst conduit network, it is suggested to first identify the most permeable zone in the drill hole, to isolate it, and finally, to measure the head in the isolated section.

In practice, karst investigations must often be focused on the identification of high permeability features (open fractures and conduits) and on the measurement of hydraulic heads. However, in some situations (e.g., a pumping well), other investigations must be conducted.

Borehole logging is often carried out (especially flowmeter, temperature, electrical conductivity, video logging, and scan logging). Lists of available techniques can be found in Bechtel et al. [3] or can be provided by any borehole logging company. Interpretations must always take into account the fact that flow in karst aquifers is mainly controlled by the karst conduit network.

18.4.6.2.1 Hydraulic Conductivity Measurements (and Aquifer Transmissivity)

Hydraulic conductivity and aquifer transmissivity are concepts derived from flow descriptions in intergranular aquifers. They do not really apply to karst where flow is concentrated in conduits. Hydraulic conductivity is however often applied to karst systems, but it should be kept in mind that this parameter summarizes characteristics of an equivalent porous medium, unless Darcy–Weisbach or Strickler equations for flow in pipes are applied. Exceptions can be found in carbonate aquifers with significant matrix porosity (chalk, dolomite, reef limestone, etc.). As a drill hole usually cuts through sections with very contrasting hydraulic conductivities, measurements performed between packers provide more specific information.

Usual methods to determine the hydraulic conductivity are slug tests, injection tests, or pumping tests.

Slug tests are quite effective, but precise measurements in sections with a hydraulic conductivity lower than 1 × 10E-06 m/s are time-consuming (hours to days) and require a rather sophisticated equipment. Slug tests in naked boreholes or long sections (>5 m) are usually difficult to interpret but can be used as a first identification of the presence of high hydraulic conductivity values in a drill hole.

Pressure tests are applied on isolated sections (packers) of 0.5–5 m of length. They consist of measuring the head while injecting or pumping water with a known discharge rate. After a certain time, equilibrium is reached between the injected/pumped rate and the measured head. The hydraulic conductivity is then directly obtained with a simple formulae: $K = Q/(Cs \bullet r \bullet H)$ with K is the hydraulic conductivity, r is the drill hole radius, H is the head, Cs is the constant depending in section length and radius. Measurements in sections with a hydraulic conductivity lower than 1 × 10E-06 m/s are time-consuming, because steady-state conditions are only attained after a long period of time.

Lugeon tests are normally used in consolidated rocks. Formulae have been developed for transforming Lugeon units in permeability values, and most of these formulae are based on steady-state conditions. However, classical Lugeon tests are conducted over a fixed duration (10 min) and steady state is therefore not approached if the K-value is lower than 1 × 10E-06 m/s. For this reason, results for lower values are strongly biased.

Injection/pumping tests in naked drill holes consist of injecting or pumping a known discharge rate of water into/from a drill hole and to measure the water level. The measurement time step must be in the order of 1–5 s. Such test can be used to identify high permeability zone in the borehole and may also give an indication on its depth, as well as an order of magnitude of its hydraulic conductivity. Injection is usually easier and more efficient than pumping because it also allows testing the vadose zone. The conduction of injection tests is much easier and cheaper than packer tests. The results are less informative but may be sufficient depending on the required precision and, thanks to the limited cost, can be and should be carried out in any well or piezometer available in a karst region. The vadose zone can only be partially investigated with this technique unless air is used as fluid for the tests.

18.4.6.3 Head Measurements

The *water level* in a naked borehole does not indicate hydraulic head in such a heterogeneous media as karst. The water level gives a broad indication about the head resulting from natural flows occurring along the borehole from sections with highest heads toward sections with lower heads. At low water stage, head differences along the hole are in the range of 1–10 m (from low permeability to high permeability sections). During high water events, they may be much larger (sometimes >100 m) and in the opposite direction (from high permeability to low permeability sections). Heads should therefore preferably be measured along short sections (0.5–5 m) isolated between packers. The measurement technique is rather simple, but after the section has been isolated, one has to wait until the head in the section is equilibrated with the surrounding rock, which can be time-consuming for low permeability sections (day or weeks for very low permeability sections) [40,41].

18.4.6.4 Investigations in Caves

In karst regions, groundwater flow paths can sometimes be directly explored and mapped by speleologists. This may be very useful for understanding, measuring, and monitoring flow characteristics as well as identifying and characterizing transport processes. Characteristics of subcatchments can be measured directly, leading to a better understanding of the processes generating an observed global response at karst spring (e.g., the mixing of waters coming from different subcatchments) [53]. Investigations in caves may also reveal helpful at finding the main source of pollution if one tributary to the cave stream is more polluted than others.

Observations in caves are particularly valuable for determining heads in the aquifer and their variations during storm events.

Besides direct observations of water in caves, useful information can be acquired concerning geology, characteristics of caves, as well as cave genesis, which can significantly support the elaboration of a representation of the karst conduit network in a region (e.g., through the KarstALEA approach). Additionally, observations of cave sediments provide an understanding of water turbidity issues.

One should assess from case to case if observation in caves can be a valuable input for addressing specific engineering issues [12,29,30,34].

18.4.7 Computer Simulation of Flow and Transport

Flow in karst is controlled by processes taking place in groundwater flow as well as in surface water flow. However, karst hydrology has been mainly dealt with by hydrogeologists and many models applied to karst thus rather belong to hydrogeological modeling than to surface hydrology.

Yet, the features controlling flow (and transport) in most karst systems is the conduit network, which is, from a heuristic point of view, the equivalent of a network of streams and rivers. Therefore, at a regional scale, flow (and transport) is mainly controlled by the characteristics of the karst conduit network.

A functional consequence of this structure is that karst has quite instantaneous and intense reactions to storm events although the average transit time of water is usually longer than in surface streams. This behavior is the result of complex exchanges between features favoring quick flow and those favoring storage (e.g., between the conduit network and the rock matrix in the phreatic zone or between subsurface storage and small conduits in the epikarst).

Those characteristics make it impossible to develop a perfect deterministic model of karst systems, able to simulate flow and transport adequately, because the level of detail on the distribution of the physical parameters could never be sufficient to build a "perfect model." This point is more critical than in porous media because of the extreme heterogeneity and of the large range of characteristics (infiltration, vadose flow, epiphreatic, and phreatic) encountered in karst.

Models representing karst systems simplify the reality to an even higher level than models of other environments. An extreme case of simplification is the simulation of karst spring response with an equivalent porous media. Unless matrix porosity is very high, this approach is almost never adequate, even under steady-state conditions, and head–discharge relationships using this type of models are usually meaningless. A pure black-box approach, using a nonphysical mathematical relationship, would even be more adequate.

Numerical models should at least include a quick- and a slow-flow component in order to provide reasonable results for the simulation of the functional response of karst systems (spring regime). Often four components have to be considered: a recharge model and a routing model having each both a quick- and a slow-flow component. In systems with contrasted characteristics in space, this four-component model must be spatially distributed. Two main categories of such models are found: (1) hydrological models where routing is assumed to be somehow similar to streamflow and where recharge is assumed to be similar to that of runoff generation and (2) hydrogeological models in which routing is assumed to be a groundwater flow simulated with more or less sophisticated schemes and where recharge is often strongly simplified. The latter can also be simulated with more or less sophisticated models of vertical flow through partially saturated soils. The first category of models (hydrological ones) tends to consider less parameter and can thus be calibrated more easily than those of the second category.

Unfortunately, no "karst modeling tool" for engineering hydrology is available as a product on the software market, most tools being developed for academic purposes. It is therefore difficult to recommend specific software or even a particular approach for the simulation of hydrological processes in karst systems. Simulation tools and approaches must be selected according to the problem that has to be addressed, the field that will be investigated, and the know-how available in the involved team [23,36,39,41,46,51,62].

18.5 Summary and Conclusions

Besides an overview of karst-specific characteristics, the aforementioned chapter provides a description of typical problems encountered in a karst environment, as well as some methods that can help engineers and land managers to manage water resources in an adequate and sustainable way.

Karst hydrology includes aspects of both surface and groundwater hydrology: On the one hand, peaky hydrographs and the presence of a drainage network are similar to concepts of surface hydrology, and on the other hand, storage characteristics and the presence of a slow-flow component are more related to groundwater science. Fluctuations in catchment delineation and considerable water level rise during storm events are karst specific and must be assessed accordingly. These characteristics induce specific engineering issues and investigation methods.

Water-related engineering aspects discussed in this chapter include drinking water, flood management, hydropower, water infiltration, geothermal exploitation, and human impact on karst waters. Each of these domains requires a karst-specific approach and cannot be managed with techniques developed for other environments. A good understanding of the flow system geometry and its variations in time is a key to manage water resources in a karst environment.

As flow and transport processes taking place in a karst system are mostly invisible, one has to imagine them, which today is mostly done by computer modeling. Investigation methods must therefore focus on gathering geometrical information of the flow system feeding each spring, with the aim to construct a 3D model of the flow paths. This part of the work generally starts with the delineation of the catchment of a spring, which is in most cases completely different from a surface (topographic) catchment and which can therefore not be sketched with an automated procedure in a GIS. The next step then consists of refining the model according to the questions that need to be addressed.

References

1. Badino, G. 2005. Underground drainage systems and geothermal flux. *Acta Carsologica* 34(2):277–316.
2. Bakalowicz, M. 2005. Karst groundwater: A challenge for new resources. *Journal of Hydrology* 13:148–160.

3. Bechtel, T.D., Bosch, F.P., and Gurk, M. 2007. Geophysical methods. In *Methods in Karst Hydrogeology*, (eds.) Goldscheider, N. and Drew, D., pp. 171–199. Taylor & Francis, London, U.K.
4. Behrens, H., Beims, U., Dieter, H. et al. 2001. Toxicological and ecotoxicological assessment of water tracers. *Hydrogeology Journal* 9(3):321–325.
5. Benischke, R., Goldscheider, N., and Smart, C. 2007. Tracer techniques. In *Methods in Karst Hydrogeology*, (eds.) Goldscheider, N. and Drew, D., pp. 147–170. Taylor & Francis, London, U.K.
6. Biondic, B., Dukaric, F., and Biondic, R. 1999. Impact of the Sea on the Perilo abstraction site in Bakar Bay–Croatia. In *Karst Hydrogeology and Human Activities: Impacts, Consequences and Implications*, (eds.) Drew, D. and Hötzl, H., pp. 244–251. A.A Balkema, Rotterdam, the Netherlands.
7. Bonacci, O. 1987. *Karst Hydrology, with Special Reference to the Dinaric Karst*. Springer-Verlag, Heidelberg, Germany.
8. Brouyères, S. 2004. A quantitative point of view of the concept of vulnerability. In *COST Action 620. Vulnerability and Risk Mapping for the Protection of Carbonate (Karst) Aquifers–Final Report*, (ed.) Zwahlen, F., pp. 10–15. EU Publications Office (OPOCE), Luxembourg, Germany.
9. Brouyères, S., Jeannin, P.Y., Dassargues, A. et al. 2001. Evaluation and validation of vulnerability concepts using a physically based approach. In *Proceedings of the 7th Conference on Limestone Hydrology and Fissured Media*, Besançon, France, September 2001 pp. 67–72.
10. Butscher, C. and Huggenberger, P. 2007. Implications for karst hydrology from 3D geological modeling using the aquifer base gradient approach. *Journal of Hydrology* 342:184–198.
11. Cooper, A., Farrant, A., Adlam, K. et al. 2001. The development of a national geographic information system (GIS) for British karst geohazards and risk Assessment. In *Geotechnical and Environmental Applications of Karst Geology and Hydrogeology. Proceedings of the 8th Multidisciplinary Conference on Sinkholes and the Engineering and Environmental Impacts of Karst*, April 1–4, Louisville, Kentucky. American Society of Civil Engineers.
12. Culver, D. and White, W. 2005. *Encyclopedia of Caves*. Elsevier Academic Press, Amsterdam, the Netherlands.
13. Daly, D., Dassargues, A., Drew, D. et al. 2002. Main concepts of the "European approach" to karst-groundwater-vulnerability assessment and mapping. *Hydrogeology Journal* 10(2):340–345.
14. Doerfliger, N. and Zwahlen, F. 1998. *Cartographie de la vulnérabilité en régions karstiques (EPIK). Guide ratique*, p. 58. Office fédéral de l'environnement, des forêts et du paysage, Berne, Suisse.
15. Drew, D. and Hoetzl, H. 1999. *Karst Hydrogeology and Human Activities: Impacts, Consequences and Implications*. A.A Balkema, Rotterdam, the Netherlands.
16. Field, M. 2002. *The QTRACER2 Program for Tracer-Breakthrough Curve Analysis for Tracer Tests in Karstic Aquifers and Other Hydrologic Systems*, p. 179. US Environmental Protection Agency, Washington, DC. 600/R-02/001.
17. Field, M. 2003. A review of some tracer-test design equations for tracer-mass estimation and sample collection frequency. *Environmental Geology* 43:867–881.
18. Filipponi, M. and Jeannin, P.Y. 2010. Karst-ALEA: A scientific based karst risk assessment for underground engineering. In *Advances in Research in Karst Media*, (eds.) Andreo, B., Carrasco, F., Durán, J.J., and LaMoreaux, J.W., pp. 435–439. Springer Environmental Earth Sciences, Berlin, Germany.
19. Filipponi, M., Jeannin, P.Y., and Tacher, L. 2009. Evidence of inception horizons in karst conduit networks. *Geomorphology* 106:86–99.
20. Filipponi, M., Schmassmann, S., Jeannin, P.Y. et al. 2012. *KarstALEA: Wegleitung zur Prognose von karstspezifischen Gefahren im Untertagbau–Forschungsprojekt FGU 2009/003 des Bundesamt für Strassen ASTRA*. Schweizerischer Verband der Strassen- und Verkehrsfachleute VSS, Zürich, Schweiz. Available at http://www.isska.ch/pdf/kars+ALEA_d.pdf.
21. Florea, L.J. and Vacher, H.L. 2006. Springflow hydrographs: Eogenetic vs. telogenetic karst. *Ground Water* 44(3):352–361.
22. Ford, D. and Williams, P. 2007. *Karst Hydrogeology and Geomorphology*. John Wiley & Sons, Chichester, England.

23. Ghasemizadeh, R., Hellweger, F., Butscher, C. et al. 2012. Review: Groundwater flow and transport modeling of karst aquifers, with particular reference to the North Coast Limestone aquifer system of Puerto Rico. *Hydrogeology Journal* 20:1441–1461.
24. Ghazavi, G. and Eslamian, S. 2006. Runoff in an Iranian karstic watershed as compared with a neighbor non-karstic watershed. In *Actes du 8e colloque d'hydrogéologie en pays calcaire, 2006*. Neuchâtel, Suisse, Presses universitaires de Franche-Comté, Université de Franche-Comté, Besançon, France.
25. Goldscheider, N. and Andreo, B. 2007. The geological and geomorphological framework. In *Methods in Karst Hydrogeology*, (eds.) Goldscheider, N. and Drew, D., pp. 9–23. Taylor & Francis, London, U.K.
26. Goldscheider, N. and Drew, D. 2007. *Methods in Karst Hydrogeology*. Taylor & Francis, London, U.K.
27. Goldscheider, N., Meiman, J., Pronk, M. et al. 2008. Tracer tests in karst hydrogeology and speleology. *International Journal of Speleology* 37(1):27–40.
28. Goldscheider, N., Mádl-Szönyi, J., Eröss, A. et al. 2010. Review: Thermal water resources in carbonate rock aquifers. *Hydrogeology Journal* 18:1303–1318.
29. Groves, C. 2007. Hydrogeological methods. In *Methods in Karst Hydrogeology*, (eds.) Goldscheider, N. and Drew, D., pp. 45–64. Taylor & Francis, London, U.K.
30. Gunn, J. 2004. *Encyclopedia of Caves and Karst Science*. Routledge, London, U.K.
31. Hao, Y., Wang, Y., Zhu, Y. et al. 2009. Response of karst springs to climate change and anthropogenic activities: The Niangziguan Springs, China. *Progress in Physical Geography* 33:634–649.
32. Hoover, R.A. 2003. Geophysical choices for karst investigations. In *Proceedings of the 9th Multidisciplinary Conference on Sinkholes & the Engineering and Environmental Impact of Karst*, (ed.) Beck, B.F., pp. 529–538. American Society of Civil Engineers, Reston, VA.
33. Jeannin, P.Y., Eichenberger, U., Sinreich, M. et al. 2012. KARSYS: A pragmatic approach to karst hydrogeological system conceptualisation. Assessment of groundwater reserves and resources in Switzerland. *Environmental Earth Sciences* 69(3):999–1013. DOI 10.1007/s12665-012-1983-6.
34. Jeannin, P.Y., Groves, C., and Haeuselmann, P. 2007. Speleological investigations. In *Methods in Karst Hydrogeology*, (eds.) Goldscheider, N. and Drew, D., pp. 25–44. Taylor & Francis, London, U.K.
35. Kiraly, L. 1998. Modelling karst aquifers by the combined discrete channel and continuum approach. *Bulletin d'Hydrogéologie* 16:77–98.
36. Kiraly, L. 2003. Karstification and groundwater flow. *Speleogenesis and Evolution of Karst Aquifers* 1(3):1–26.
37. Klimchouk, A.B. 2007. *Hypogene Speleogenesis: Hydrogeological and Morphogenetic Perspective*. National Cave and Karst Research Institute, Carlsbad, NM.
38. Klimchouk, A.B., Ford, D.C., Palmer, A.N. et al. 2000. *Speleogenesis. Evolution of Karst Aquifers*. National Speleological Society Inc., Huntsville, AL.
39. Kovacs, A. and Sauter, M. 2007. Modelling karst hydrodynamics. In *Methods in Karst Hydrogeology*, (eds.) Goldscheider, N. and Drew, D., pp. 201–222. Taylor & Francis, London, U.K.
40. Kresic, N. 2007. Hydraulic methods. In *Methods in Karst Hydrogeology*, (eds.) Goldscheider, N. and Drew, D., pp. 65–92. Taylor & Francis, London, U.K.
41. Kresic, N. 2013. *Water in Karst*. McGraw-Hill Education, New York.
42. Kresic, N. and Stevanovic, Z. 2010. *Groundwater Hydrology of Springs, Engineering, Theory, Management, and Sustainability*. BH (Elsevier), Amsterdam, the Netherlands.
43. Käss, W. 1998. *Tracing Technique in Geohydrology*. A.A. Balkema, Rotterdam, the Netherlands.
44. Loaiciga, H., Maidment, D., and Valdes, J. 2000. Climate-change impacts in a regional karst aquifer, Texas, USA. *Journal of Hydrology* 227:173–194.
45. Luetscher, M. and Jeannin, P.Y. 2004. Temperature distribution in karst systems: The role of air and water fluxes. *Terra Nova* 16:344–350.
46. Martin, J.B. and White, W.B. 2008. *Frontiers of Karst Research: Proceeding and Recommendations of the Workshop Held May 3-5, 2007 in San Antonio, Texas, USA*, p. 118. Karst Waters Institute, Leesburg, VA.

47. Milanovic, P.T. 2004. *Water Resources Engineering in Karst*. CRC Press LLC, Baco Raton, FL.
48. Mylroie, J.E., Carew, J.L., and Vacher, H.L. 1995. Karst development in the Bahamas and Bermuda. In *Terrestrial and Shallow Marine Geology of the Bahamas and Bermuda, GSA Special Paper 300*, (eds.) Curran, H.A. and White, B., pp. 251–267. Geological Society of America, Boulder, CO.
49. Palmer, A. 1991. Origin and morphology of limestone caves. *Geological Society of America Bulletin* 103:1–21.
50. Palmer, A.N. 1989. Stratigraphic and structural control of cave development and groundwater flow in the Mammoth Cave region. In *Karst Hydrology, Concepts from the Mammoth Cave Area*, (eds.) White, W.B. and White, E.L., pp. 293–316. Von Nostrand Reinhold, New York.
51. Palmer, A., Palmer, M., and Sasowsky, I. 1999. *Karst Modeling*. Karst Water Institute Special Publication 5.
52. Parise, M. and Gunn, J. 2007. *Natural and Anthropogenic Hazards in Karst Areas*. Geological Society Special Publication 279.
53. Perrin, J., Jeannin, P.Y., and Cornaton, F. 2007. The role of tributary mixing in chemical variations at a karst spring, Milandre, Switzerland. *Journal of Hydrology* 332:158–173.
54. Ravbar, N. 2007. The protection of karst waters. A comprehensive Slovene approach to vulnerability and contamination risk mapping. *Carsologica* 6:256.
55. Schmassmann, S. and Hitz, O. 2010. Wegleitung zur Beurteilung von Projekten in Karstgebieten. Schweizerische Gesellschaft für Höhlenforschung SSS|SGH: 38p. Published by the Swiss speleological Society and available online at http://www.speleo.ch/~site/images/stories/documents/instrpratiqkarsteie_v1.0_f.pdf
56. Schudel, B., Biaggi, D., Dervey, T. et al. 2002. *Utilisation des traceurs artificiels en hydrogéologie. Guide pratique*, p. 75. Office fédéral de l'environnement, Berne, Suisse.
57. Sever, Z. and Misetic, S. 1999. Ombla hydropower plant, environmental impact study, Summary, unpublished report for Hrvatska elektroprivreda d.d., available from http://en.wikipedia.org/wiki/Ombla.
58. Tulipano, L. 2005. *Action Number: 621, Groundwater Management of Coastal Karstic Aquifers–Final Report*, p. 336. EU Publications Office (OPOCE), Luxembourg, Germany.
59. Van Beynen, P.E. 2011. *Karst Management*. Springer, Dordrecht, the Netherlands; Heidelberg, Germany; London, U.K.; New York.
60. Veni, G. 2005. Environmental impact assessment. In *Encyclopedia of Caves and Karst Science*, (ed.) Gunn, J., pp. 319–321. Fitzroy Dearborn (Taylor & Francis), London, U.K.
61. White, W.B. 1988. *Geomorphology and Hydrology of Karst Terrains*. Oxford University Press, New York.
62. White, W.B. 2003. Conceptual models for karstic aquifers. *Speleogenesis and Evolution of Karst Aquifers* 1:1–6.
63. Worthington, S. 2007. Groundwater residence times in unconfined carbonate aquifers. *Journal of Cave and Karst Studies* 69:94–102.
64. Worthington, S.R.H. and Smart, C.C. 2003. Empirical equations for determining tracer mass for sink to spring tracer testing in karst. In *Sinkholes and the Engineering and Environmental Impacts of Karst*, (ed.) Beck, B.F., pp. 287–295. American Society of Civil Engineers, Reston, VA.
65. Zhou, W. 2007. Drainage and flooding in karst terranes. *Environmental Geology* 51:963–973.
66. Zwahlen, F. 2003. *COST Action 620. Vulnerability and Risk Mapping for the Protection of Carbonate (Karst) Aquifers–Final Report*. EU Publications Office (OPOCE), Luxembourg, Germany.

19

Long-Term Generation Scheduling of Hydro Plants

Mônica de Souza Zambelli
State University of Campinas

Secundino Soares Filho
State University of Campinas

Leonardo Silveira de Alburqueque Martins
State University of Campinas

Anibal Tavares de Azevedo
State University of Campinas

AUTHORS

Mônica de Souza Zambelli received her BSc in computer engineering in 2003 from the Federal University of Espirito Santo (UFES), Brazil, and her MSc and PhD in electrical engineering in 2006 and 2009 from the State University of Campinas (UNICAMP), Campinas, Brazil. Her research interests are electric energy generation, acting on the following subjects: optimization, dynamic programming, simulation, and operation of hydrothermal systems. Two of her recent publications include "Hydropower scheduling in large scale power systems," *In: InTech. (Org.). Hydropower–Practice and Application*, and "Deterministic versus stochastic dynamic programming for long term hydropower scheduling," *In: PowerTech, Trondheim.*

Secundino Soares Filho received his BSc in mechanical engineering in 1972 from the Aeronautical Institute of Technology (ITA). He received his MSc and his PhD in electrical engineering in 1974 and 1978 from UNICAMP, Campinas, Brazil. He joined the staff of UNICAMP in 1976. From 1989 to 1990, he was a visiting associate professor with the Department of Electrical Engineering at McGill University in Canada and is currently a professor in the Electrical and Computer Engineering School at UNICAMP. His research interests concern planning and operation of electrical energy systems. Two of his most cited publications are the following: "Short term hydroelectric scheduling combining network flow and interior point approaches," *International Journal of Electrical Power & Energy Systems*, and

"Comparison between closed-loop and partial open loop feedback control polices in long term hydro-thermal scheduling," *IEEE Transactions on Power Systems.*

Leonardo Silveira de Albuquerque Martins received his BSc and MSc in computer science and engineering in 2002 and 2005, respectively, from the Federal University of Góias (UFG), Brazil. He received his PhD in electrical engineering in 2009 from UNICAMP, Campinas, Brazil. His research interests are hydro- and thermal power generation in large-scale systems, nonlinear optimization, interior-point methods, energy-efficient hydropower unit commitment, and pump scheduling in water supply systems. Two of his recent publications include "Short-term hydropower scheduling via optimization simulation decomposition approach," *in PowerTech, Bucharest*, and "A nonlinear model for the long-term hydro-thermal generation scheduling problem over multiple areas with transmission constraints," *In: Proc. PSCE, Seattle.*

Anibal Tavares de Azevedo received his BSc in applied computational mathematics in 1999, as well as an MSc and a PhD in electrical engineering in 2002 and 2006 from UNICAMP, Campinas, Brazil. He was a professor at the State University of São Paulo (UNESP) from 2007 to 2011. Currently, he is a professor at UNICAMP. His research interests are linear, integer, and nonlinear optimization, heuristics, electric power system operation, and hydrothermal operation planning. Two of his recent publications include "Automatic guided vehicle simulation in MATLAB by using genetic algorithm," *MATLAB for Engineers—Applications in Control, Electrical Engineering, IT and Robotics*, and "Interior point method for long-term generation scheduling of large-scale hydrothermal systems," *Annals of Operations Research.*

PREFACE

How could the power transmission constraints influence long-term reservoir operation in power systems connected over multiple areas?

Finding the answer to this question is not an easy task, especially dealing with a system that comprises 147 hydro plants, distributed in several cascades, and 144 thermal plants with real operative constraints that are distributed into four areas representing the north (N), northeast (NE), south (S), and southeast/center-western (SE/CO) regions of the country.

The techniques proposed by the literature, until now, do not consider the possibility of exchanging hydropower, in cascade detail, with a multiple-area system where transmission constraints are enforced. The authors faced this problem by turning it into an optimization problem with nonlinear constraints solved by a primal–dual interior-point method with linear search filter.

To better understand the model behavior and decisions, three real physical data from the Brazilian power system are analyzed highlighting the operation of four reservoirs under different electrical connectivity conditions—zero transmission capacity and both constrained and unconstrained transmission limits. In them, it is possible to assess the benefits of increasing transmission capacity in terms of water savings in the form of increased efficiency.

19.1 Introduction

The operational planning of hydrothermal power systems is designed to provide a reliable and economic operational policy that embraces time frames ranging from the long-term management of hydropower resources to the electric and hydraulic aspects of daily generation dispatch. Due to the large scale and complexity of this problem, it has usually been approached by decomposition into two hierarchically connected problems, long-term generation scheduling (LTGS) and short-term generation scheduling (STGS).

LTGS covers a planning period of one or more years, divided into monthly intervals, with the first month being divided into weeks. The objective of this planning is to control the water storage in the reservoirs in order to maximize hydroelectric production and, as a result, minimize the cost of nonhydraulic generation sources. LTGS planning provides a weekly generation target for each hydro plant.

STGS covers the next week ahead, usually divided into hourly intervals. Its goal is to obtain a generation scheduling that matches the long-term generation targets for each hydro plant and optimizes the efficiency of hydropower conversion. This planning should comprise a detailed representation of hydro plant operational characteristics and the security operation of the electrical transmission system [21].

To cope with the stochastic nature of the LTGS problem, one common approach is to consider the randomness of inflows by their probability distribution functions and apply classical optimization techniques based on stochastic dynamic programming (SDP). This approach requires some sort of simplification, however, due to the well-known "curse of dimensionality" associated with SDP problems [4]. This simplification can be in the representation of the multireservoir system using aggregated composite models [2] or in the solution procedure, such as the use of Bender's decomposition [17], which requires problem linearization [19]. Another approach, based on deterministic optimization models, is the open-loop feedback control, as presented in [14], which is capable of handling the stochastic nature of streamflows at lower computational costs.

Thus, in the framework of a scenario technique for LTGS problems, a robust and efficient tool is crucial for solving the deterministic version of the problem since it must be applied several times for different inflow scenarios. Moreover, for the purpose of simulation, the decision process has to be repeated for each time interval in order to update the storage in the reservoirs on an open-loop feedback control framework [14]. Several techniques have been suggested for the solution of this problem, including nonlinear programming [10] and network flow approaches [6,20]. More recently, the option of using interior-point methods (IPMs), which are especially efficient for solving large-scale optimization problems, has been suggested also for the LTGS problem [16].

The aforementioned techniques, however, do not consider the possibility of exchanging hydropower in a multiple-area system where transmission constraints are enforced. One first attempt is introduced in [11] by employing linear production functions, though neither operational nor economic analysis of the results is presented by the authors. Several other works have also proposed hydrothermal scheduling models with network constraints over the years. However, such approaches have dealt with short-term problems only, that is, [22], and are not capable of rendering conclusions on the effects of network constraints over reservoir operation in the long run.

The present chapter is concerned with the development of an open-loop feedback control framework that employs an efficient IPM for solving the deterministic version of the LTGS problem with transmission constraints for large-scale hydrothermal systems. The problem is formulated precisely by considering nonlinear functions for hydroelectric generation, nonhydraulic operational costs, and transmission limits. To overcome difficulties related to the precise representation of nonlinear objective function and related constraints, a specialized IPM capable of exploiting the sparse structure of the problem and handling its Hessian indefiniteness efficiently is necessary.

The robustness of the proposed approach has been tested on a case study simulating the monthly operation schedule from June 2011 to December 2015 for the whole Brazilian power system. It comprises 147 hydro plants (96 GW of power capacity) distributed in several cascades and 144 thermal plants (33 GW of power capacity) with real operative constraints that are distributed into the Brazilian interconnected power system composed of four areas representing the north (N), northeast (NE), south (S), and southeast/center-western (SE/CO) regions of the country.

19.2 Single Reservoir Systems: Closed-Loop versus Open-Loop Optimal Control Approach

19.2.1 Problem Formulation

Without loss of generality, the hydropower scheduling problem considered here attempts to minimize thermal fuel costs associated to the dispatch of thermal plants in order to complement the load demand on a hydro-dominant power system. In this case, the LTHS problem for a single hydro plant system can be formulated as the following stochastic nonlinear programming problem:

Objective:

$$\min_{q} \mathrm{E}_y \left\{ \sum_{t=1}^{T} \psi_t(D_t - p_t) \right\} \tag{19.1}$$

subject to

$$p_t = k \cdot h_t \cdot q_t, \quad \forall t \tag{19.2}$$

$$h_t = \phi(a_t) - \theta(u_t) - \xi(q_t), \quad \forall t \tag{19.3}$$

$$a_t = a_{t-1} + \left(b_t - u_t\right) \cdot \gamma_t, \quad \forall t \tag{19.4}$$

$$u_t = q_t + v_t, \quad \forall i, t \tag{19.5}$$

$$A_t^{\min} \le a_t \le A_t^{\max}, \quad \forall i, t \tag{19.6}$$

$$u_t \ge U_t^{\min}, \quad \forall i, t \tag{19.7}$$

$$q_t \le Q_t^{\max}, \quad \forall i, t \tag{19.8}$$

$$v_t \ge 0, \quad \forall i, t \tag{19.9}$$

$$x_0 \quad \text{given} \tag{19.10}$$

where

E_y is the expected value with respect to inflows
t is the time stage index (months)
T is the number of time stages in planning period
ψ is the thermal fuel cost ($)
D is the load demand (MW)
p is the hydropower generation (MW)
a is the reservoir storage at end of stage (hm^3)
$A^{\min}$, $A^{\max}$ are the limits for reservoir storage (hm^3)
u is the release from reservoir (m^3/s)
$U^{\min}$ is the minimum release (m^3/s)
q is the discharge through turbines (m^3/s)
$Q^{\max}$ is the maximum discharge function (m^3/s)

v is the spillage from reservoir (m^3/s)
k is the efficiency constant factor ($MW/(m^3/s)\cdot m$)
φ is the forebay elevation function (m)
θ is the tailrace elevation function (m)
ξ is the penstock head loss function (m)
b is the inflow into reservoir (in m^3/s)
γ is the stage duration in seconds divided by 10^6

The objective function to be minimized in (19.1) is the expected thermal fuel costs, represented by a decreasing function of hydropower generation p, for a given load demand D. Hydro generation at stage t, represented by (19.2), is a nonlinear function of water storage in the reservoir, water discharge through the turbines, and water spillage from the reservoir. The constant k represents the product of the water density, the gravity acceleration, and average turbine/generator efficiency. The equality constraint in (19.5) represents the water balance in the reservoir. Lower and upper bounds on variables, expressed by constraints (19.6) through (19.10) are imposed by physical operational constraints of the hydro plant, as well as by constraints associated with multiple uses of water.

19.2.2 Closed-Loop Control Approach

Dynamic programming (DP) [4] is an optimization method concerned with the efficient solution of closed-loop discrete-time minimization problems. For solving problem (19.1) through (19.10) by DP, the optimization problem is embedded into a family of similar problems with the objective of minimizing the expected costs from stage t until the end of the planning horizon T subject to constraints (19.2) through (19.10), for a known previous state a_{t-1}:

$$\alpha_{t-1}(a_{t-1}) = \min_{q} \mathrm{E}_y \left\{ \sum_{k=t}^{T} \psi_k(D_k - p_k) \right\} \tag{19.11}$$

To simplify the problem, since spillage does not produce energy and therefore it does not reduce thermal fuel costs, it will not be handled as a decision variable. This means that the minimization in (19.11) will be performed only with respect to the discharge variable q. Indeed, spillage will only be different from zero if, during the simulation, the discharge decisions would take the reservoir storage above the maximum.

In closed-loop approaches, the decisions are placed at the last possible moment to take into account information about the costs incurred by those decisions until that moment. In mathematical terms, these amounts determine a policy or control rule $\pi = \{\beta_m\}$, described by functions $\beta_m = \beta_m(a)$, $1 \le m \le 12$, where the function value is the water to be discharged at month m if the system is at state a. For each control decision β, there is a cost function α that evaluates its consequences on the system operation.

According to Bellman's optimality principle [4], the solution for problems (19.1) through (19.11) can be obtained by backward resolution of the recursive equation (19.12) at each stage j, where C_p, the present cost, given by Equation 19.13, represents the cost associated with thermal generation at stage j and C_f, the future cost, represents the minimum expected future cost associated with future thermal generation:

$$\alpha_{j-1}(a_{j-1}) = \min_{q_t} \{C_p + C_f\} \tag{19.12}$$

$$C_p = \psi_j(D_j - p_j) \tag{19.13}$$

19.2.2.1 Deterministic Dynamic Programming

In the deterministic dynamic programming (DDP) approach, the inflow is considered to be known at the beginning of each stage j and is given by the average inflow $\bar{b}_j$ at that stage. In this case, the future cost function C_f is given by (19.14) where the state variable a_j is the reservoir's level at the end of stage j and is calculated by the water balance equation (19.15), considering the expected value of inflow $\bar{b}_j$:

$$C_f = \alpha_j(a^j) \tag{19.14}$$

$$a_j = a_{j-1} + (\bar{b}_j + q_j)\gamma_j \tag{19.15}$$

19.2.2.2 Stochastic Dynamic Programming

In the SDP approach, the inflow at each stage is represented by a probability distribution function, generally with some correlation assumed between inflows of consecutive months [5]. For this particular analysis, a Lag-1 model will be considered.

In this case, the future cost function is given by (19.16) where the state variable a^k_j is calculated by the water balance equation (19.17) and is associated with an inflow b^k_j that can assume Nb discrete values in its range of variation, each of them with a probability p_k, previously calculated from the conditional probability density function $f(b_m/b_{m-1})$ of the random variable b_m for each month of the year, given the occurrence of inflow b_{m-1} on the previous month:

$$C_f = \sum_{k=1}^{Nb} \alpha_j(a_j^k, b_j^k) \cdot p_k \tag{19.16}$$

$$a_j^k = a_{j-1} + (b_j^k + q_j)\gamma_j \tag{19.17}$$

Note that the state variable changes to include the inflow of the previous stage. The solution process starts at the final stage T, where the cost $\alpha_T(a_T)$ is assumed to be known, and works back to the initial stage $j = 0$, solving the recursive equation iteratively until the convergence of the discharge decision rules for each month of the year within a certain range of tolerance.

19.2.3 Open-Loop Control Approach

In open-loop approaches, strategic decisions are not calculated prior to the simulation process but rather determined at the beginning of the time stages.

The open-loop control approach applied to the LTHS problem, also known as model predictive control (MPC) [7], is a feedback control scheme accomplished by a two-step procedure: In the first step, the future inflows are forecasted for a limited time horizon, then in the second step, the system is optimized for the forecasted inflows by a nonlinear deterministic optimization model.

The simulation of the system's operation with such an open-loop approach results on a "forecast–optimize–update" scheme that runs for all stages until the end of the planning horizon. The state resulting from a decision by the end of each time stage is sent back to the control system to update its information and improve decision making on further stages.

Previous tests with this approach, focusing specifically on the uncertainty of inflows, evaluated the results for single reservoir systems where dimension is not an issue [14,24]. The approach has shown results equivalent to those of standard methods based on SDP. Figure 19.1 presents a diagram for the predictive control operating policy at a certain stage t.

During the simulation, for each stage t of the planning horizon, the reservoir storage level at the end of the previous stage is observed and used as an initial condition for solving a deterministic optimization problem for a specific optimization horizon $[t, T^*]$.

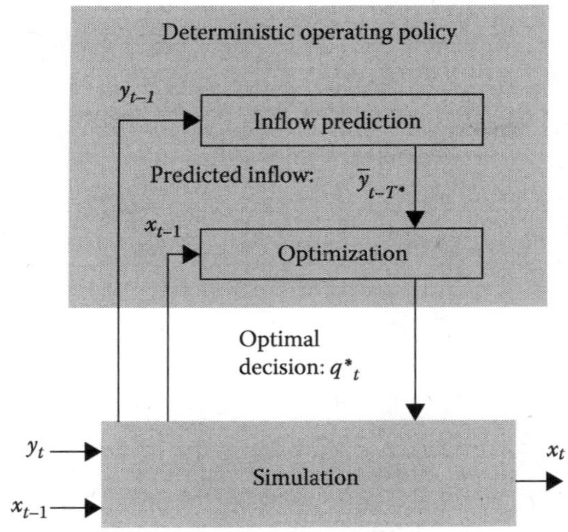

FIGURE 19.1 MPC scheme.

This optimization considers a series of predicted values for the unknown parameters to be considered, in this case water inflows, determined by the inflow prediction module, based on past observed values.

The optimization module then provides optimal releases for each hydro plant for the optimization horizon but only the discharge solution for the first stage q^*_t is selected and submitted to a simulation model. The latter tests the feasibility of results and makes the necessary corrections if needed, according to formulation (19.2) through (19.8). Corrections are frequently needed due to differences between the predicted inflow series $\bar{b}$ and the simulated inflow series b and eventually due to subsystems' interchange constraints.

In the next stage $t+1$, the storage level of the reservoirs resulting from simulation is observed, and a new forecasting of inflows can be given based on the latest available information. This procedure of "forecast–optimize–update" is repeated until the end of the planning horizon T.

In this chapter, the monthly average of historical inflow records (MLT) was adopted as the forecasting values for the inflow series during optimization, which neglects the benefits of the updating scheme. This means that the performance of the approach implemented can probably be improved significantly with the use of more efficient forecasting techniques.

To assure the MPC approach's efficiency, the deterministic nonlinear optimization model (19.1) through (19.10) must be solved using specialized optimization techniques such as network flow algorithms with capacitated arcs [6] or IPMs [3]. The latter was used in this case study since it allows a detailed representation of the hydropower system, including specific aspects of individual plants and nonlinear production functions.

The optimization horizon T^* adopted starts at the current stage t and consists on a rolling window of at least 13 months and at most 24 months, depending on the current monthly stage. Full reservoir storage is imposed as a boundary condition at the end of this horizon, adjusted to match the month of April, since reservoirs are expected to be close to full by the end of the wet seasons. These two parameters were estimated based on successive simulation tests with various considerations and have shown to maximize the approach performance.

19.2.4 Comparison Results

Case studies with single reservoir systems are reported in this section. The results are given in detail for Furnas hydro plant, located on the Grande River, with 1312 MW of installed capacity. For the other plants mentioned, the results are presented as a summary.

TABLE 19.1 Furnas Hydro Plant Operation Bounds

k	U^{min}	Q^{max}	A^{min}	A^{max}
0.008633	204	1692	5733	22,950

TABLE 19.2 Furnas Hydro Plant Polynomial Functions

		$1E{-}3{\cdot}x$	$1E{-}7{\cdot}x^2$	$1E{-}12{\cdot}x^3$	$1E{-}17{\cdot}x^4$
φ	735.25	3.4966	−1.9744	6.9170	−9.7736
θ	671.63	1.0174	−1.7997	0.2513	
ξ	0	0	366.8		

Tables 19.1 and 19.2 show the data for the Furnas hydro plant. The former consists of operation bounds for the variables in Equations 19.6 through 19.9 and the latter consists of the functions used to determine the waterhead in (19.3), where the forebay φ and tailrace θ elevations are calculated by fourth-degree polynomial functions and the penstock head loss ξ by a quadratic function.

A thermal plant with an installed capacity equal to that of Furnas was considered to compose the hypothetical hydrothermal system. The operation costs are calculated using the following quadratic function:

$$\psi_t = 0.0002(D_t - p_t)^2 \tag{19.18}$$

Load demand was assumed constant and equal to half the installed capacity of the system. The closed-loop and open-loop methods were evaluated by simulation on a monthly basis using the historical inflow records from May 1931 to April 2009. For the SDP method, the inflow probability p_k was determined by log-normal probability distribution functions, adjusted over the historical records for each month. For the DDP method, average inflows were considered, with the monthly values presented in Figure 19.2 along with the standard deviation.

Maximum initial water storage of the reservoir is assumed. To establish a basis for comparison, an optimal solution (OS) with perfect foresight of inflows was obtained by deterministic optimization over the historical inflow records. Simulation results for the problem formulation and procedure described so far are summarized on Table 19.3 using average values of operational cost, hydropower generation, spillage, and efficiency.

From the average values presented, it is clear that OS represents the lower bound with a much better performance. This however can only be achieved with perfect knowledge of inflows, which, of course, is not possible in real operation. This will thus be used to compute a certain "cost of uncertainty," which

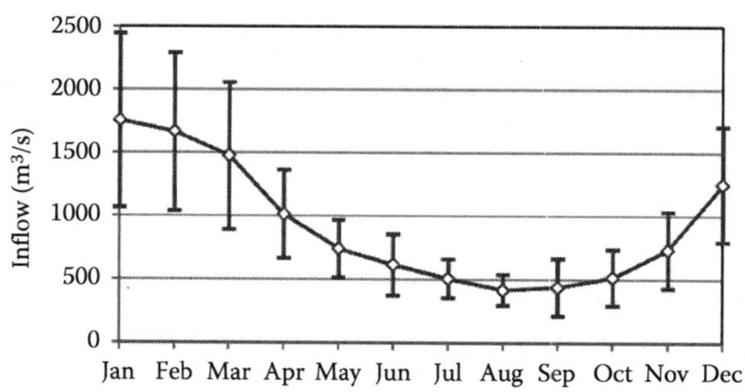

FIGURE 19.2 Average and standard deviations of monthly inflows for Furnas.

TABLE 19.3 Simulation Results

	Cost ($)	Generation (MW)	Spillage (m^3/s)	Efficiency (MW/m^3/s)
OS	4,084,031	712.0	5.0	0.794
DDP	4,487,582	687.5	36.7	0.789
SDP	4,474,563	686.9	33.7	0.785
MPC	4,488,032	686.6	31.0	0.782

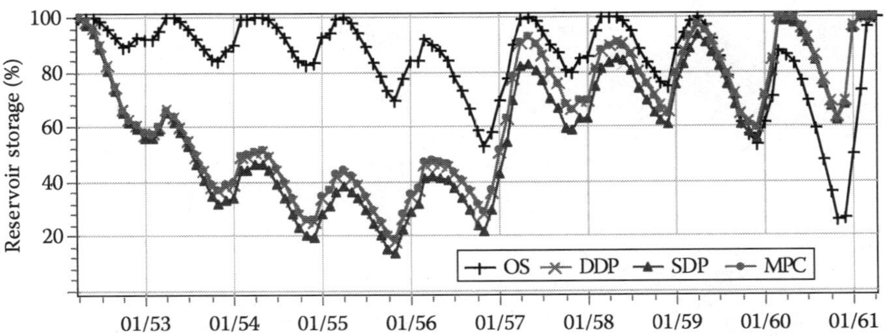

FIGURE 19.3 Reservoir evolution in critical inflow period.

is the relative increase in costs resulting from simulation with real operating policies in relation to that resulting from optimization assuming perfect inflow foresight.

In this case study, SDP performed slightly better than DDP, presenting a 0.3% cost reduction. This is in accordance to the expectation that considering stochastic inflow information in the decision making leads to better operation. MPC performance was equivalent to DDP with differences in cost around 0.01%. This can be explained for both methods that take into consideration monthly long-term averages (LTA) as expected inflows during optimization. In the open-loop approach, the use of specialized inflow forecasting methods can increase the method's performance to a level equal or better than SDPs [24]. There was no significant difference between average hydro generations resulting from the three methods tested, which was 3.5% lower than that of OS. One main concern in practical operation is to guarantee efficient system operation during critical inflows. To illustrate strategic differences in reservoir operation according to the simulated approaches, focusing on this situation, the resulting storage evolution of the reservoir during a critical inflow period is presented in Figure 19.3.

Similar tests with other hydro plants [25] reinforce the conclusion that open-loop approaches can perform equivalent to closed-loop ones with the advantage that they can be applied to large-scale multireservoir systems with no additional simplifications.

19.3 Multiple Reservoir Systems: Nonlinear Optimization Model Considering Electric Connections

19.3.1 Problem Formulation

The long-term hydrothermal scheduling problem over multiple areas can be formulated as the continuous nonlinear optimization problem [15] of finding

$$\min_{a,q,v,y,z} f(z) = \sum_{t=1}^{T} \Psi_t(z_{t,j}) \tag{19.19}$$

subject to water conservation equations

$$a_{i,j} = a_{i,j-1} + \frac{\Delta_j}{\gamma}\left(b_{i,j} + \sum_{\forall i' \in \Omega_i}(q_{i',j} + v_{i',j}) - q_{i,j} - v_{i,j}\right), \quad \forall i,j, \tag{19.20}$$

DC power flow and Kirchhoff's second law

$$d_{k,j} = \sum_{\forall i \in k} p_{i,j}(x^{i,j}) + \sum_{\forall t \in k} z_{t,j} + \sum_{\forall n \in N_k} y_{(k,n),j} - y_{(n,k),j}, \quad \forall k,j, \tag{19.21}$$

$$Cy = 0,$$

and operational constraints

$$\begin{aligned} \underline{x}_{i,j} &\leq x \leq \overline{x}_{i,j} \\ \underline{y}_{i,j} &\leq y \leq \overline{y}_{i,j}, \\ \underline{z}_{i,j} &\leq z \leq \overline{z}_{i,j} \end{aligned} \tag{19.22}$$

where the set of hydro variables a, q, v represents water storage, discharge, and spillage, respectively. Power flow is represented by y variables, and z is the set of complementary thermal power generation. The objective of the program is to minimize the costs associated with thermal power generation at each plant t over all stages j of the planning horizon. It is assumed here that Ψ is convex, separable, and twice differentiable. Water conservation states that the storage at reservoir i at stage j equals its storage at stage $j-1$ plus controllable upstream and incremental inflows $b_{i,}j$ discounted by composite water release. Both hydro- and thermal power generations are subject to a DC power flow model composed of a set of nonlinear equations representing Kirchhoff's first law, where $d_{k,}j$ is the load demanded by the power system area k at stage j, $p_{i,}j(x^{i,}j)$ is the hydropower generation as a nonlinear function of the hydraulic operational variables at plant i and stage j, and $y_(k,n),j$ and $y_(n,k),j$ represent outbound and inbound power flow between k and neighboring areas n, respectively. Finally, operational constraints define lower and upper bounds on problem variables.

19.3.2 Solution by a Primal–Dual Interior-Point Method

In order to solve the optimization problem with nonlinear constraints given by (19.19) through (19.22), a primal–dual IPM with linear search filter is used. Some notation simplification of the problem formulation is thus necessary for easier algorithmic deduction. If matrix notation is to be adopted, the model can be rewritten as follows:

$$\begin{aligned} &\min_{x,y,z} f(z) \\ &\text{s.t. } Ax = b \\ &\qquad By = P(x) + Q(z) - d \\ &\qquad Cy = 0 \\ &x, y, z \in X, Y, Z \end{aligned} \tag{19.23}$$

where $x = (a,q,v)^T$ is the set of hydro variables. The optimization model (19.23) can be further simplified such that it is rewritten as follows:

$$\begin{aligned} &\min_w f(w) \\ &\text{s.t. } g(w) = 0 \\ &l \le w \le u \end{aligned} \tag{19.24}$$

where $w = (x^T, y^T, z^T)^T$ is the set of variables and

$$g(w) \overset{\Delta}{=} \begin{pmatrix} Ax - b \\ By - P(x) - Q(z) + d \\ Cy \end{pmatrix} = 0 \tag{19.25}$$

represents the system of nonlinear equations that describe problem constraints. If slack variables are added to the operational constraints such that inequalities are removed from the formulation and a barrier parameter is introduced into the objective function in order to eliminate nonnegativity constraints on the newly added slack variables, Problems (19.24) and (19.25) can finally be rewritten in an equivalent form as follows:

$$\begin{aligned} &\min_{\tilde{w}} \varphi_\mu(\tilde{w}) \\ &\text{s.t. } \tilde{g}(\tilde{w}) = 0 \end{aligned} \tag{19.26}$$

where $\tilde{w} = (w^T, s^T, t^T)^T$ and

$$\tilde{g}(\tilde{w}) \overset{\Delta}{=} \begin{pmatrix} g(w) \\ w + s - u \\ w - t - l \end{pmatrix} \tag{19.27}$$

where $s, t > 0$ and

$$\varphi_\mu(\tilde{w}) = f(w) - \mu \left(\sum_{i=1}^{n} \ln s_i + \ln t_i \right) \tag{19.28}$$

where μ is the scalar barrier parameter.

This way, one can write the Lagrangian function using (19.26) through (19.28):

$$\ell_\mu(\tilde{w}, \lambda, \pi, \zeta) \overset{\Delta}{=} \varphi_\mu(\tilde{w}) + g(w)^T \lambda + (w + s - u)^T \pi + (t - w + l)^T \zeta, \tag{19.29}$$

and thus, the first-order optimality conditions identifying a stationary point of the Lagrangian (19.29) are met when the following primal–dual system is solved:

$$L_\mu(\tilde{w}, \lambda, \pi, \zeta) \overset{\Delta}{=} \begin{pmatrix} \nabla f(w) + \nabla g(w)^T \lambda + \pi - \zeta \\ g(w) \\ w + s - u \\ t - w + l \\ \pi - \mu S^{-1} e \\ \zeta - \mu T^{-1} e \end{pmatrix} = 0 \tag{19.30}$$

The last two matrix equations of the system (19.30) define the complementarity conditions. They can be equivalently rewritten as two nonlinear systems of equations:

$$\begin{aligned} S\Pi e &= \mu e \\ TZe &= \mu e \end{aligned} \tag{19.31}$$

where S, T, Π, Z are strictly positive diagonal matrices corresponding to vectors s, t, π, z, respectively. Complementarity conditions are perturbed by barrier parameter μ and require nonnegativity of π and ζ, which is consistent with the fact that these Lagrange multipliers were originally associated with inequalities. Finally, $e \in \Re^n$ is a vector of 1s.

For the solution of the system of equations to be equal to the OS of the original problem in the last iteration k*, there must be a sequence $\{\mu^{(k)}\} \to 0$ for at least one $\mu^{(0)} > 0$ when $k \to \infty$. At every iteration k, a Newton direction is computed so that the system of nonlinear equations is solved by a Taylor series of linear approximations around $(\tilde{w}^{(k)},\lambda^{(k)},\pi^{(k)},\zeta^{(k)})$:

$$\Delta^{(k)} = -\left[\nabla L_{\mu^{(k)}}\left(\tilde{w}^{(k)},\lambda^{(k)},\pi^{(k)},\zeta^{(k)}\right)\right]^{-1} L_{\mu^{(k)}}\left(\tilde{w}^{(k)},\lambda^{(k)},\pi^{(k)},\zeta^{(k)}\right) \tag{19.32}$$

In order for the search direction to be computed, it is required that $\nabla L_{\mu^{(k)}}$ be nonsingular. For computational efficiency reasons, though, LDLT factorization is used to solve the Newton system. Moreover, for the filter line search to function properly, a matrix is added to$\nabla L_{\mu^{(k)}}$, such that its projection over the null space of $\nabla g(w^{(k)})^T$ is positive definite [23].

19.3.3 Numerical Results

To better understand how power transmission constraints influence reservoir operation in power systems connected over multiple areas, a few numerical examples are presented in the succeeding text. In them, it is possible to assess the benefits of increasing transmission capacity in terms of water savings in the form of increased efficiency. Although the cases are hypothetical, they present real physical data from the Brazilian power system. A total of three cases are analyzed highlighting the operation of four reservoirs under different electrical connectivity conditions—zero transmission capacity and both constrained and unconstrained transmission limits. At each reservoir site, there are exactly one corresponding hydro station and one thermal station capable of meeting the local load, which, in all cases, is constant over time and equals 85% of the local installed hydro capacity. The underlying power network topology mimics that of the Brazilian power system for what long-term scheduling is concerned, consisting of four load centers, one transport bus, and five transmission lines, as illustrated in Figure 19.4.

Planning horizon is monthly discretized and is 12 months long, starting in May and ending in April of next year. The cost of thermal complementarity is $1/2 z_{t,j}^2 s$ for every t, j. Incremental water inflow is equal to the monthly long-term average calculated between 1931 and 2006. Water reservoirs are set to start and end at full capacity. Table 19.4 provides additional common information about system configuration for all case studies considered.

If no power transmission capacity is present, each system seeks to minimize the complementary thermal power generation given its local hydropower availability, such that the optimization problem is equivalent to four independent subproblems. In this case, hydro- and thermal power generation are complementary to each other. Moreover, for the systems with larger water storage capacity, such complementarity is almost constant over time. This can be explained by the use of quadratic cost functions that impose minimized variance of the marginal costs in the OS. In the case of Machadinho plant, because of its small capacity of reservoir regulation and its particular hydrological characteristics, a higher thermal variance is observed. The operation results for the case with no power transmission capacity are illustrated in Figure 19.5.

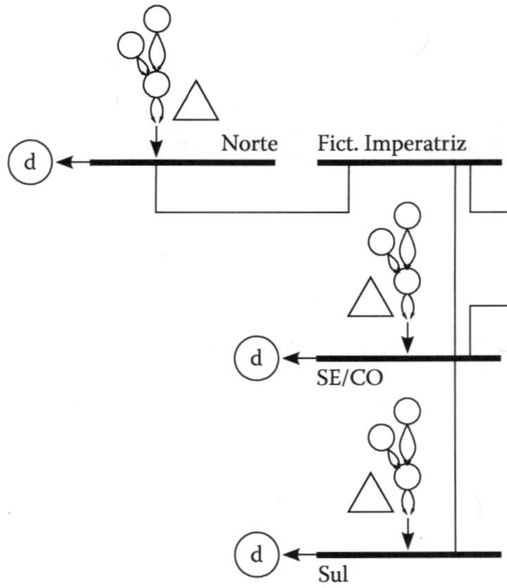

FIGURE 19.4 Power network topology.

TABLE 19.4 System Configuration

Plant	System	Vol. (hm³)	Capacity (MW)	Load (MW/m)
Furnas	SE/CO	17,217	1312	1115
Machadinho	Sul	1,056	1140	969
Sobradinho	Nordeste	28,669	1050	892
Serra da Mesa	Norte	43,250	1275	1084

The difference in reservoir capacity between the systems is highlighted by the thermal power generation. Serra da Mesa, with its large reservoir, enables the Norte system to request only 22% of load demand to be met with complementary thermal power in average, while the Sul system, given its small reservoir, has its demand met with thermal power at a 30% share base in average. Thus, the operation of large reservoirs is based on the regulation of inflows, such that at the OS, average water discharge values are kept at approximately average inflow.

On the other hand, when no power transmission constraints are imposed over the connectivity of the systems, the resulting operation is completely different in what thermal power generation is concerned. In such cases, because the quadratic cost functions are equal among the systems, thermal power generation is equally set and its variance is minimized. Moreover, hydropower complementarity is filled by energy imports in all systems, as illustrated in Figure 19.6, where negative import values mean that energy is exported to other systems. Such scenario is observed in traditional models that do not account for power transmission network topology and constraints and, thus, overestimate hydro generation capacity.

Figure 19.7 illustrates how the introduction of power transmission constraints affects the operation of reservoirs. Apart from the case with no transmission capacity, which solely serves didactic purposes, it can be observed that simply accounting for an underlying power transmission network with bounds suffices to cause reasonable changes in reservoir operation in all systems. Average hydropower output and reservoir storage results for all cases are provided in Table 19.5 to further emphasize the effects of these constraints on reservoir operation. Another side benefit of this model representation scheme is the possibility that one has to compute the benefits of increased connectivity in terms of both higher hydropower generation and reservoir storage levels.

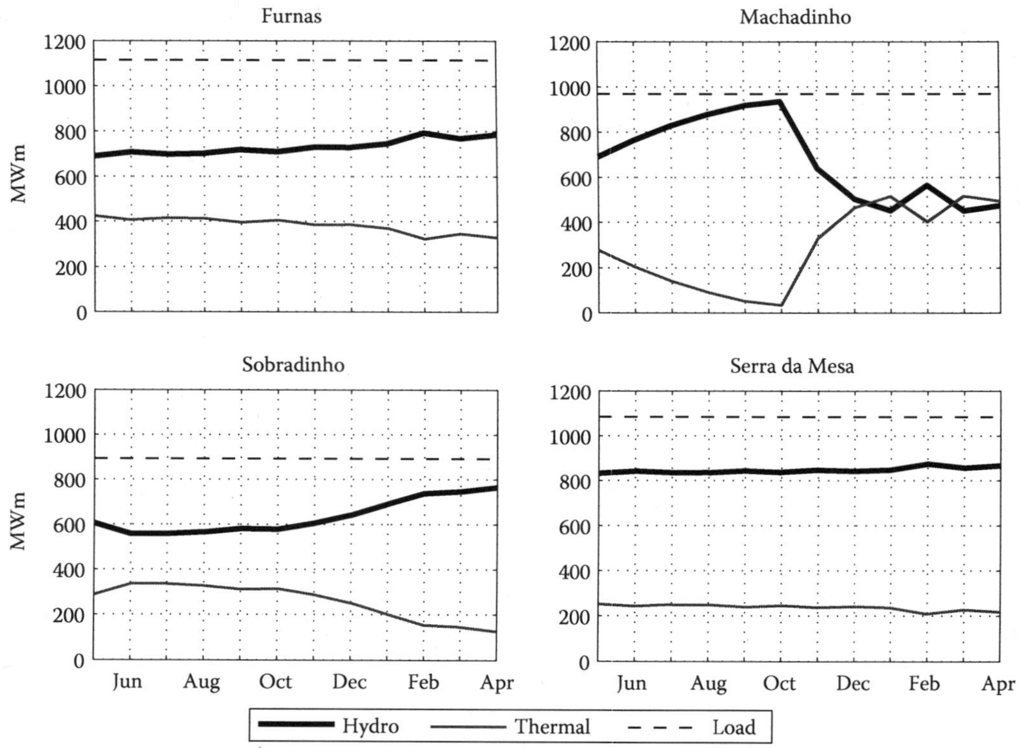

FIGURE 19.5 System operation results with no transmission capacity.

19.4 Large-Scale Hydrothermal Systems: A Case Study on the Brazilian Interconnected Power System

19.4.1 Brazilian Interconnected Power System

In Brazil as abroad, the world energy market has been through a lot of change toward a more competitive and deregulated model. The operation planning of the National Interconnected System (SIN), however, remains regulated and is coordinated by the National System Operator (ONS), with the support of a chain of models whose main objective is to minimize system operation costs while attending specified energy shortage risks [13]. Concerned with the stochastic nature of inflows, stochastic models have been adopted, and specifically for the LTHS, a stochastic dual dynamic programming (SDDP) algorithm was developed that tries to overcome the dimensionality issue by using Bender's cuts in a stochastic, multistage decomposition framework, approximating the future expected costs as piecewise linear functions [17].

Due to the continental dimension of the system and the DP computational limitations, an aggregated composite model [2] is used to represent SIN as four interconnected composite energy reservoirs. Although proving only a rough simplification of the hydropower system operation, this approach has been applied in many hydropower systems around the world since it allows the straightforward application of SDP-based methods. A sketch map of the SIN divided into four subsystems is presented in Figure 19.8, where SE/CO comprises the main load centers and is composed by plants located in the SE/CO regions, and S, NE, and N are composed by power plants located in the S, NE, and N regions of Brazil, respectively.

One widely used alternative for the LTHS problem is based on a deterministic optimization model within the framework of scenario techniques. In this approach, the operational policy is based on the

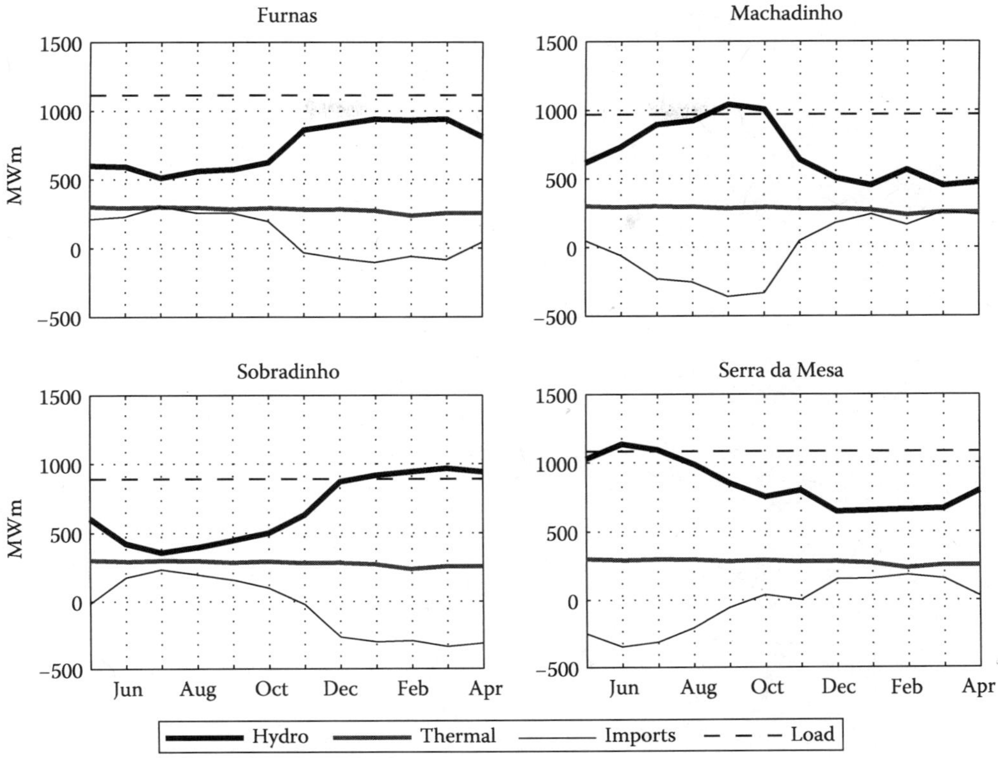

FIGURE 19.6 System operation under unconstrained transmission conditions.

solutions of a deterministic optimization model for a set of different scenarios used to represent the stochastic nature of inflows [8,9].

Under this framework, the MPC is a powerful approach for the LTHS since its performance in benchmark systems is equivalent to that of SDP-based models, and the power system can be modeled more precisely, with individual plants' nonlinear characteristics preserved even for very-large-scale power systems.

19.4.2 Long-Term Power Generation Scheduling Results

A case study simulating the monthly operation schedule on March 2011 for the whole Brazilian power system was considered, comprising 147 hydro plants (96 GW of power capacity) and 144 thermal plants (33 GW of power capacity) with real operative constraints and taking into account the expansion plan for the next 2 years and the interchange capacity between the four subsystems.

MPC approach was implemented and simulated on a monthly basis in a framework of scenarios. Seventy-five inflow series were taken from historical database, extending from 1932 to 2009, each of them composed with 22 months. System data used for the case study were taken from official data source (www.ccee.org.br) and reflect the system configuration and expansion schedule from March 2011 up to December 2012.

All constraints presented in the formulation of the optimization problems (19.1) through (19.10), have been considered in the simulation. The forebay φ and tailrace θ elevations were represented by fourth-degree polynomial functions and the penstock head loss ξ by a linear function.

The expected values of inflows used by the optimization model within the MPC framework were given by average of monthly historical records. Although naive, average expectation is easy to implement and provides an acceptable reference to the optimization of the LTHS problem.

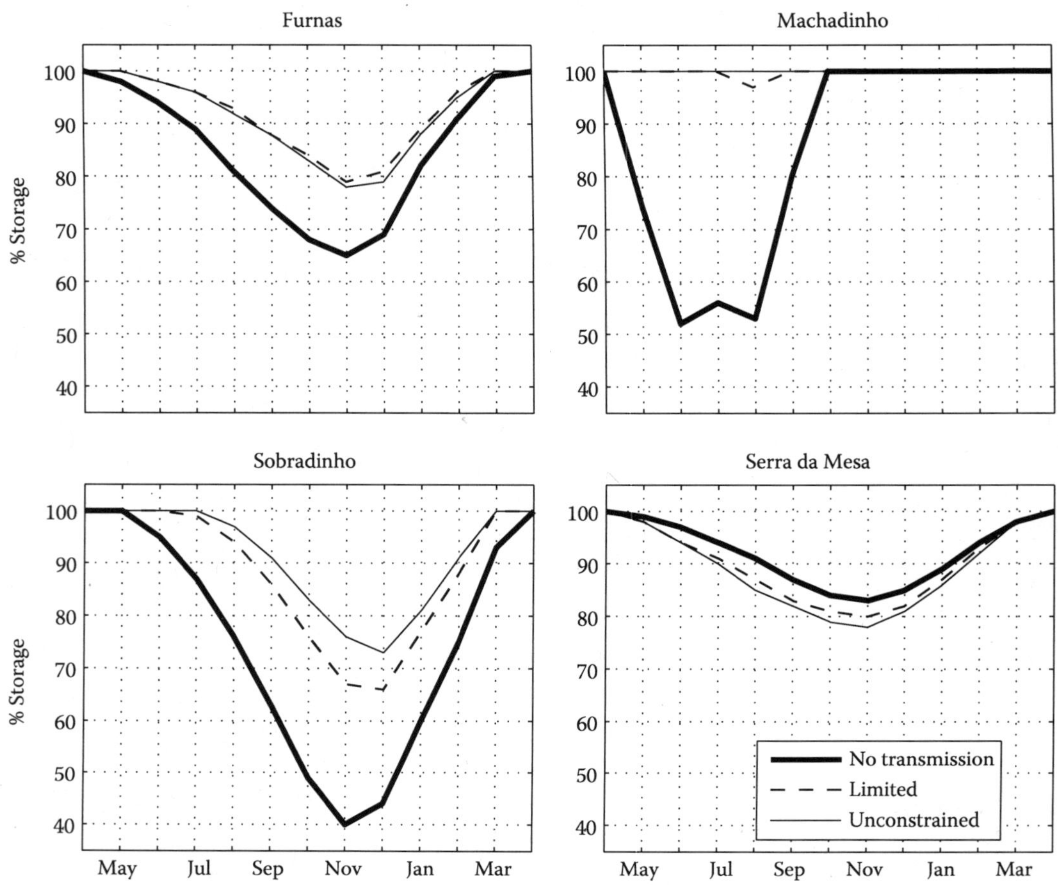

FIGURE 19.7 Reservoir operation under different network connectivity conditions.

TABLE 19.5 Average Hydropower Output and Reservoir Storage Results

	No Transmission		Limited Transmission		Unconstrained	
Plant	Output (MW)	Storage (%)	Output (MW)	Storage (%)	Output (MW)	Storage (%)
Furnas	731	84.2	740	92.0	739	91.4
Machadinho	675	84.6	693	99.7	693	100.0
Sobradinho	636	73.5	663	87.7	668	91.0
Serra da Mesa	847	91.7	843	89.5	841	88.6
Total	2890		2938		2941	

In order to validate the proposed approach, the results are compared to those provided by the stochastic approach actually in use by the Brazilian electric sector, where an SDDP model obtains the expected future cost functions for the aggregated subsystems and then a linear optimization model calculates the generation decisions for individualized plants.

19.4.2.1 Case Study Results for the Global System

Table 19.6 summarizes the simulation results in terms of average values of operational cost, hydro and thermal generation, and final stored energy in the system, for the 75 scenario simulations.

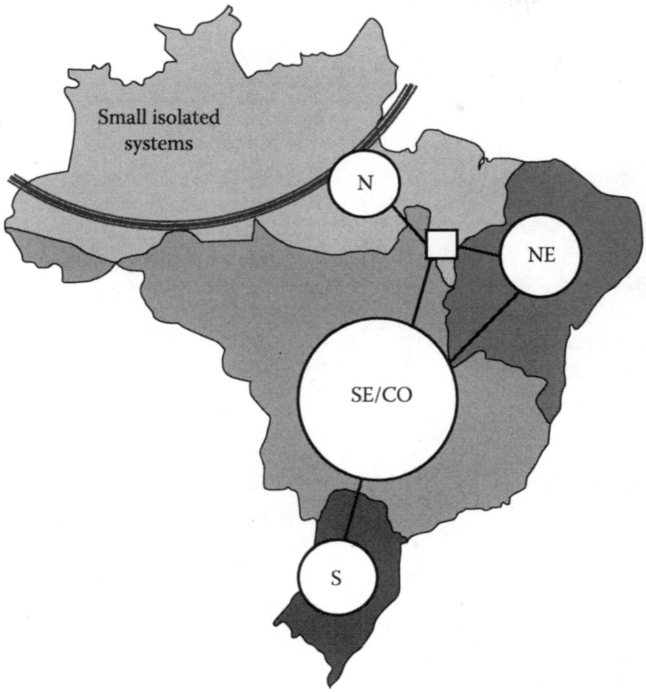

FIGURE 19.8 Brazilian subsystems.

TABLE 19.6 Average Simulation Results

	Cost (Million US$)	Hydropower Generation (MW)	Thermal Power Generation (MW)	Final Stored Energy (MW-Months)
MPC	176.60	49,003.7	9983.8	146,300.0
SDDP	182.51	49,284.4	9750.0	133,526.0

The MPC approach presented a 3.2% cost reduction with respect to the stochastic approach and saved more water in the reservoirs as indicated by the 9.6% greater stored energy by the end of planning horizon (December 2012).

In average, the MPC approach presented lower hydropower generation (0.6%) and greater thermal generation (2.4%), indicating that this operation strategy gives higher priority to the water reserves in order to better guarantee the load attendance in the future.

The stored energy is a sum of the reservoirs' useful volume pounded by the plants' average efficiency factor accumulated along the downstream cascade. The system's average stored energy evolution is presented in Figure 19.9 where it is possible to notice that, especially in the last year of this planning horizon, the SDDP approach leaves the system with reduced reserves as it makes more intense use of water resources. This can also be inferred by the observation, in Figure 19.10, of the average hydro generation trajectory resulting from both MPC and SDDP approaches, where the load demand for the planning horizon also figures.

19.4.2.2 Case Study Results for the Subsystems

It is important to mention that the subsystems are very different from each other, from its generating resources to its load demands. The SE/CO subsystem is the biggest one, composed by more than

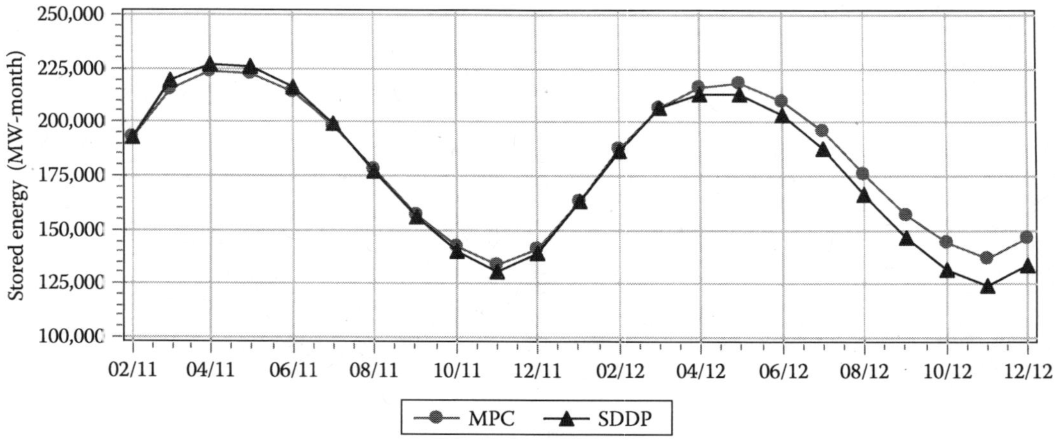

FIGURE 19.9 Average stored energy.

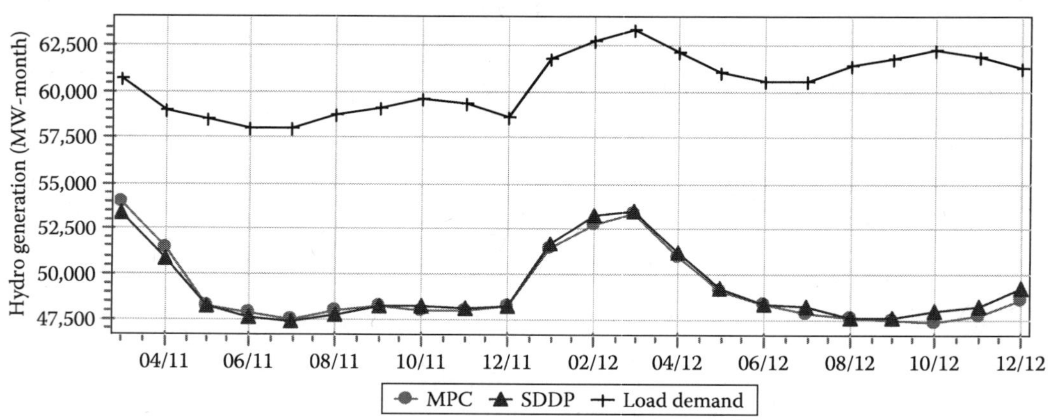

FIGURE 19.10 Average hydro generation.

100 hydro plants,* half of them with regulation reservoirs, located in more than six river basins located on the SE/CO region of Brazil. Its total installed capacity is near 60,000 MW. The S subsystem groups around 30 hydro plants are located in the S region with a total installed capacity of 15,000 MW. The NE subsystem is composed of seven hydro plants located in the NE region and sums over 10,000 MW of installed capacity. The N subsystem has four operating hydro plants with near 9000 MW of installed capacity, but as this subsystem comprises the river basins of the Amazon River in the N region of Brazil, there are several projects under study, and at least four other hydro plants are to be delivered by the end of year 2015 [1].

The load demands are about the same proportion, and SE/CO accounts for around 62% of the total SIN demand, whereas S, NE, and N take 16%, 15%, and 7%, respectively.

Regarding thermal generation, SE/CO subsystem has 50% of the total installed capacity on 52 thermal plants. NE subsystem has 63 plants, corresponding to 34% of the total thermal installed capacity. Subsystems S and N correspond to 11% and 5%, respectively.

* Hydro plants with less than 30 MW of installed capacity are considered small and do not participate in the integrated power scheduling. There are over 400 small hydro plants in operation and their estimated power generations are subtracted from the subsystems' demands.

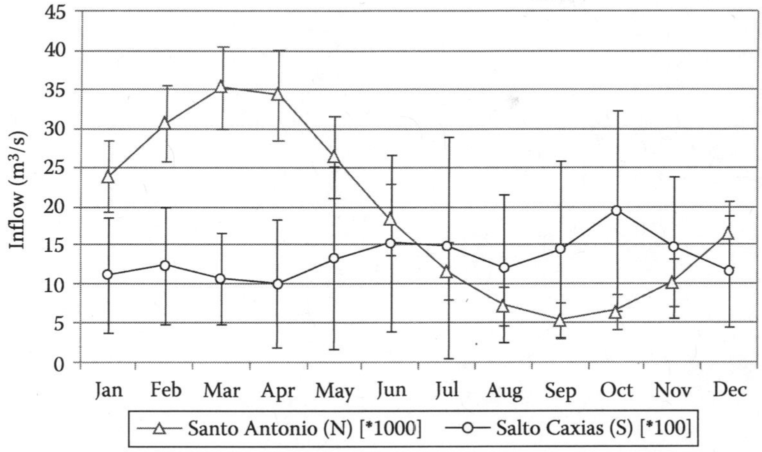

FIGURE 19.11 Monthly inflow profile for two hydro plants located in N and S subsystems.

Despite these structural characteristics, hydrology aspects are distinct among regions. Figure 19.11 illustrates that in subsystem N, hydro plant Santo Antonio located in the Amazon River Basin receives 10 times more water flow in average than does Salto Caxias, in the S subsystem. Moreover, river basins in the SE/CO, NE, and N subsystems follow a seasonal pattern with well-defined wet and dry seasons, while the river basins in the S subsystem do not.

Coordinated hydropower scheduling should take advantage from different seasons so that power generation among regions can be complementary, if the transmission network supports the resulting power flow.

Observing the average stored energy of the four subsystems in Figure 19.12, it is possible to see how they make use of their reservoirs along the planning period.

Meanwhile in subsystem S, stored energy goes down in the first semester and keeps stable near 60% of its maximum, and the others react to natural seasons.

Although big upstream reservoirs generally work to regulate their river flows with a pattern complementary to that of natural inflows, since the majority of plants have relatively small reservoirs used for short-term control only, this effect is not observed globally.

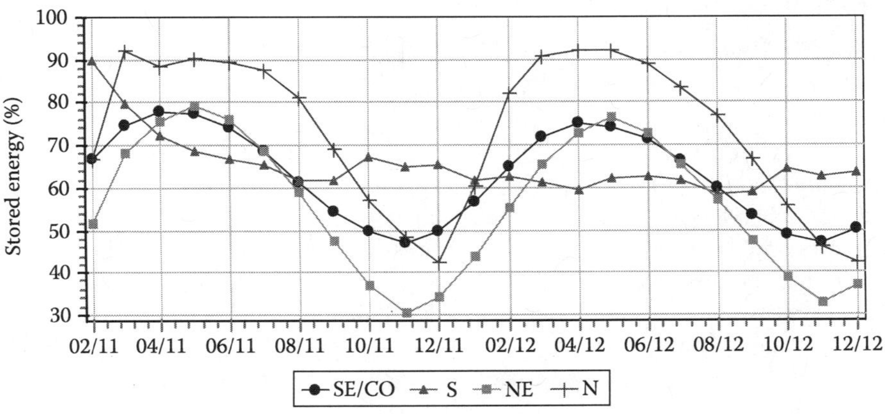

FIGURE 19.12 Subsystem's average stored energy.

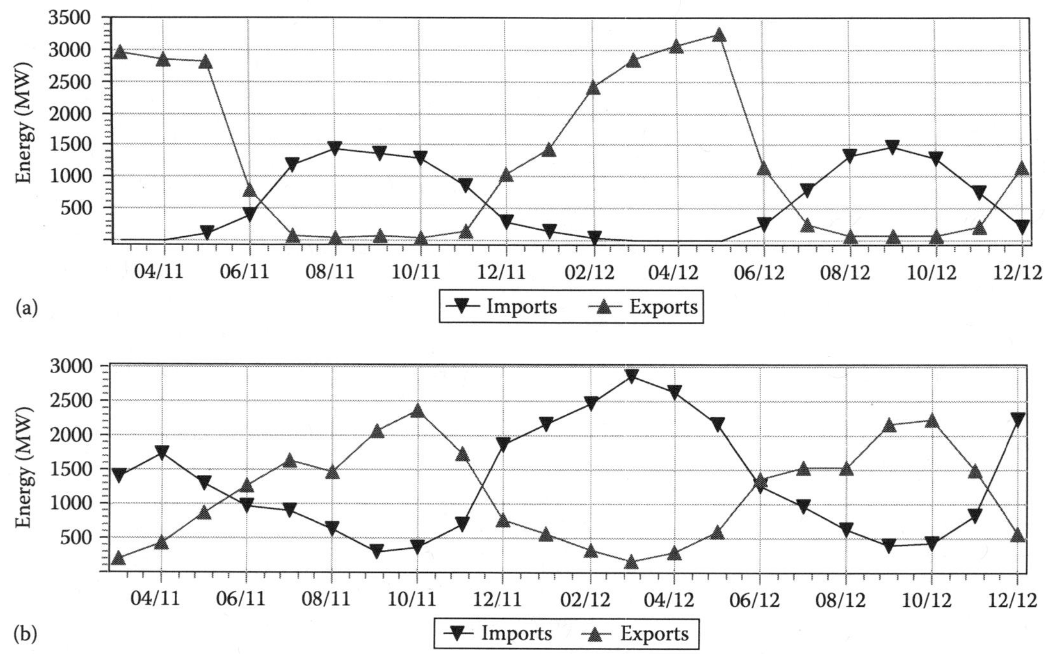

FIGURE 19.13 Average imports and exports of energy for subsystems N (a) and S (b).

Another important aspect that has motivated the development of a model that can handle network connections explicitly is that the network transmission capacity, if constrained, can severely affect optimal scheduling.

In Figure 19.13, the evolution of average imports and exports for subsystems N and S is presented. It is clear that from December to May, with high hydropower generation due to its favorable inflows, subsystem N exports energy to supply demands on other subsystems (Figure 19.13a); a big part of it crosses the whole country to help supply S subsystem's demand (Figure 19.13b).

The MPC approach with the IPM optimization model presented in Section 19.3 can help identify bottlenecks in the transmission network and evaluate the benefits of investments in new lines. This can be very useful for the system expansion planning studies, especially with several new plants already planned to be constructed in the northern region of Brazil.

19.5 Summary and Conclusions

The present chapter developed and solved an open-loop feedback control framework that employs an efficient IPM for solving the deterministic version of the LTGS for hydro plant problem with transmission constraints for large-scale hydrothermal systems. The problem is formulated precisely by considering nonlinear functions for hydroelectric generation, nonhydraulic operational costs, and transmission limits.

The proposed open-loop approach has been evaluated and compared with classical closed-loop approach, represented by the SDP technique, for the case of single reservoir systems. The results show similar performances, justifying the use of the open-loop approach in large-scale multireservoir hydrothermal systems since the closed-loop approach will require some sort of simplifications in order to overcome the "curse of dimensionality."

The performance of the proposed open-loop approach has been evaluated in a case study simulating the monthly operation schedule from June 2011 to December 2015 for the whole Brazilian interconnected power system. The system comprises 147 hydro plants (96 GW of power capacity) and

144 thermal plants (33 GW of power capacity) distributed into four areas representing the N, NE, S, and SE/CO regions of the country.

The results show that the system performance can greatly benefit from the interconnection between regions as it explores the differences on hydraulic behavior of the corresponding river basins. Comparison with the methodology in use in Brazil, which is based on SDDP, shows that the proposed approach reduces the operational cost while saving considerable amount of water at the end of the planning period.

Further work will include the implementation of inflow forecasting models in order to improve the open-loop approach here presented that has used the average inflows as prediction and the representation of the whole power grid instead of the simple interconnections between regions.

Acknowledgments

This research was supported by the Foundation for the Support of Research of the State of São Paulo (FAPESP), Brazilian Council for the Development of Science and Technology (CNPq), and Brazilian Projects and Studies Funding (FINEP).

References

1. ANEEL-Brazilian Electricity Regulatory Agency. http://www.aneel.gov.br
2. Aravanitidis, N.V. and Rosing, J. 1970. Composite representation of a multireservoir hydroelectric power system. *IEEE Transactions on Power Apparatus and Systems* PAS-89(2), 327–335.
3. Azevedo, A.T., Oliveira, A.R.L., and Soares, S. 2009. Interior point method for long-term generation scheduling of large-scale hydrothermal systems. *Annals of Operations Research*, 169(1): 55–80.
4. Bellman, R. 1962. *Dynamic Programming*. Princeton, NJ: Princeton University Press.
5. Bertsekas, D.P. 1987. *Dynamic Programming: Deterministic and Stochastic Models*. New York: Academic Press.
6. Carvalho, M. and Soares, S. 1987. An efficient hydrothermal scheduling algorithm. *IEEE Transactions on Power Apparatus and Systems* PWRS-2(3), 537–542.
7. Camacho, E.F. and Bordons, C. 2004. *Model Predictive Control*. Berlin, Germany: Springer.
8. Dembo, R.S. 1991. Scenario optimization. *Annals of Operations Research* 30(1), 63–80.
9. Escudero, L.F., Garcia, C., De La Fuente, J.L., and Pietro, F.J. 1996. Hydropower generation management under uncertainty via scenario analysis and parallel computation. *IEEE Transactions on Power Apparatus and Systems* 11(2), 683–689.
10. Gagnon, C.R., Hicks, R.H., Jacoby, S.L.S., and Kowalik, J.S. 1974. A nonlinear programming approach to a very large hydroelectric system optimization. *IEEE Transactions on Power Apparatus and Systems* 6(1), 28–41.
11. Gorenstein, B., Campodonico, N., Costa, J., and Pereira, M. 1992. Stochastic optimization of a hydrothermal system including network constraints. *IEEE Transactions on Power Systems* 7(2), 791–797.
12. Labadie, J.W. 2004. Optimal operation of multireservoir systems: State-of-the-art review. *Journal of Water Resources Planning and Management* 130(2), 93–111.
13. Maceira, M.E.P., Terry, L.A., Costa, F.S., Damázio, J.M., and Melo, A.C.G. 2002. Chain of optimization models for setting the energy dispatch and spot price in the Brazilian system, *14th PSCC Proceedings*, Sevilla, Spain, June 24–29, pp. 1–7.
14. Martinez, L. and Soares, S. 2002. Comparison between closed-loop and partial open-loop feedback control policies in long term hydrothermal scheduling. *IEEE Transactions on Power Apparatus and Systems* 17(2), 330–336.
15. Martins, L.S.A., Soares, S., and Azevedo, A.T. 2009. A nonlinear model for the long-term hydrothermal scheduling problem over multiple areas with transmission constraints, *PSCE 2009 Proceedings*, Seattle, WA, pp. 1–7.

16. Medina, J., Quintana, V., and Conejo, A. 1999. A clipping-off interior-point technique for medium-term hydro-thermal coordination. *IEEE Transactions on Power Apparatus and Systems* 14(1), 266–273.
17. Pereira, M.V.F. and Pinto, L.M.V.G. 1991. Multi-stage stochastic optimization applied to energy planning. *Mathematical Programming* 52(2), 359–375.
18. Oliveira, G.G. and Soares, S. 1995. A second-order network flow algorithm for hydrothermal scheduling. *IEEE Transactions on Power Systems* 10(3), 1635–1641.
19. Ponnambalam, K. 2002. Optimization in water reservoir systems. In P.M. Pardalos and G.C.M. Mauricio (eds.), *Handbook of Applied Optimization*, pp. 933–943. Oxford, U.K.: Oxford University Press.
20. Rosenthal, R.E. 1981. A nonlinear network flow algorithm for maximization of benefits in a hydro-electric power system. *Operations Research* 29(4), 763–785.
21. Soares, S. and Salmazo, C.T. 1997. Minimum loss predispatch model for hydroelectric power systems. *IEEE Transactions on Power Apparatus and Systems* 12(3), 1220–1228.
22. Wang, C. and Shahidehpour, S.M. 1993. Power generation scheduling for multi-area hydro-thermal systems with tie line constraints, cascaded reservoirs and uncertain data. *IEEE Transactions on Power Systems* 8(3), 1333–1340.
23. Wächter, A. and Biegler, L.T. 2005. Line search filter methods for nonlinear programming: Motivation and global convergence. *SIAM Journal on Optimization* 16(1), 1–31.
24. Zambelli, M., Luna, I., and Soares, S. 2009. Long-term hydropower scheduling based on deterministic nonlinear optimization and annual inflow forecasting models. *IEEE 2009 PowerTech Conference*, Bucharest, RO, pp. 1–8.
25. Zambelli, M., Siqueira, T.G., Cicogna, M., and Soares, S. 2006. Deterministic versus stochastic models for long term hydrothermal scheduling. *IEEE Power Engineering Society General Meeting*, Montreal, Quebec, Canada, pp. 1–7.

20

Low-Flow Hydrology

Mehdi Vafakhah
Tarbiat Modares University

Saeid Eslamian
Isfahan University of Technology

Saeid Khosrobeigi Bozchaloei
Tarbiat Modares University

AUTHORS

Mehdi Vafakhah received his BSc from the Department of Rangeland and Watershed Management at Agricultural Sciences and Natural Resources University of Gorgan, Gorgan, Iran, in 1996, and his MSc and PhD from the Department of Watershed Management at Tarbiat Modares University and from the Department of Rehabilitation of Arid and Mountainous Regions at Tehran University in 1999 and 2008, respectively. He has been lecturing in the Department of Watershed Management at Tarbiat Modares University since 1998. His research interests include surface hydrology, snow hydrology, geostatistical and parameter estimation with artificial neural networks, adaptive neurofuzzy inference system, and several data-driven techniques. He has published many national and international scientific papers and is also involved in many national watershed management projects.

Saeid Eslamian received his PhD from University of New South Wales, Australia, with Professor David Pilgrim. He was a visiting professor in Princeton University, United States, and ETH Zurich, Switzerland. He is currently an associate professor of hydrology in Isfahan University of Technology. He is the founder and chief editor of the *Journal of Flood Engineering* and the *International Journal of Hydrology Science and Technology*. He has published more than 200 publications, mainly in statistical and environmental hydrology and hydrometeorology.

Saeid Khosrobeigi Bozchaloei received his BSc from the Department of Rehabilitation of Arid and Mountainous Regions at Tehran University, Iran, in 2011. He is currently an MSc student of watershed management in Tarbiat Modares University, Iran.

PREFACE

During periods of low flow, most stream habitats are reduced in extent and water quality, and biota can be affected. During summer months, the effects of increased water demand for agricultural, household, recreational, and energy uses can exacerbate natural low-flow conditions. Information on low-flow characteristics provides threshold values for different water-based activities and is required for such water resource management issues as water supply, irrigation, and water quality and quantity estimates. Low-flow characteristics of rivers have been increasingly utilized in recent years as most nations are experiencing population growth, resource depletion, and the overextraction of water. Therefore, periods of low flow are critical for managing.

Many factors have an impact on the low-flow regimes of rivers. Whenever, land-use and climate change happen, and these can affect the water available in rivers. More recently, there have been growing concerns over relationship between water and the environment. This chapter, namely, low-flow hydrology, is thus a paramount importance for the development and implementation of sustainable water resource management practices. The chapter comprises eight sections including introduction, hydrograph recession curve and low flow, effective factors on low flow, flow duration curve (FDC), low-flow measures and indices, low-flow estimation methods in ungauged catchments, river low flow and environmental flow in hydroecology, and case studies.

20.1 Introduction

Low-flow statistics is important for water quality management, water supply planning, hydropower design, cooling-plant facility design, sanitary landfill allocation, aquatic conservation, reservoir storage design, and interbasin transfers of water and allowable basin withdrawal decision making [31]. Low flows are normally derived from groundwater discharge or surface discharge from lakes, marshes, or melting glaciers. Lowest annual flow usually occurs in the same season each year.

Low flow as defined by International Glossary of Hydrology is the "flow of water in a stream during prolonged dry weather" [52,57]. This definition does not make a clear distinction between low flows and droughts. Low flow is a seasonal phenomenon and an integral component of a flow regime of any river. Some researchers define low flow as "minimum flow in a river during the dry periods of the year" [33,45].

20.2 Hydrograph Recession Curve and Low Flow

During the dry weather periods, the gradual depletion of discharge constitutes "recession." The recession curve is the specific part of the hydrograph after the crest where the streamflow diminishes. The recession curve contains valuable information about the natural storages feeding the stream.

The shape of recession curve is the integrated response of all runoff processes in the watershed and depends on topography, drainage pattern, soils, and the geology of the watershed. Different runoff components, overland flow, interflow, and groundwater flow, have their own characteristic recession rates. In a low-flow context, baseflow is the most important component, which is defined by Hall [24] as "the portion of flow that comes from groundwater or other delayed sources." The numerical estimation of recession indices involves the selection of an analytical expression to fit the curve, the determination of characteristic recession, and the optimization of the recession parameters. Recession analysis is carried out by fitting an analytical expression to outflow function:

$$Q = Q(t) \tag{20.1}$$

where

Q is the rate of flow
t is the time

Many authors [6,7,11,12,20,27,30,32,36,39] have proposed baseflow recession using various mathematical equations. The most commonly used model is the exponential recession model:

$$Q_t = Q_0 k^t \tag{20.2}$$

where

Q_t is the flow at time t
Q_0 is the flow when $t = 0$
k is constant

The k constant can be estimated from the slope of a master recession by constructing a mean curve. The most commonly used methods for the construction of a master recession curve are the matching strip, the correlation, and the graphical and tabulating methods.

In correlation method, the discharge at one time (Q_t) is plotted against discharge one time interval later (Q_{t+dt}) for the selected recession periods as a straight line with slope k, where

$$\frac{Q_{t+dt}}{Q_t} \tag{20.3}$$

Recession equations can easily be applied to forecast low flows in gauged rivers [2,45]. The length of the achievable forecast period depends on the size of the drainage basin and its degree of storage; it is generally possible to derive forecasts between 1 and 20 days. The maximum length of the forecast period is evident from slope of the derived recession curve and the degree of pessimism concerning the likely length of time before rain is expected [58]. The simple nature of a recession-based forecast is shown in Figure 20.1. This figure illustrates a master recession curve for a drainage basin with a continuous reduction in flow over a 20-day period starting for around 90 mL/d. Let us assume that the current

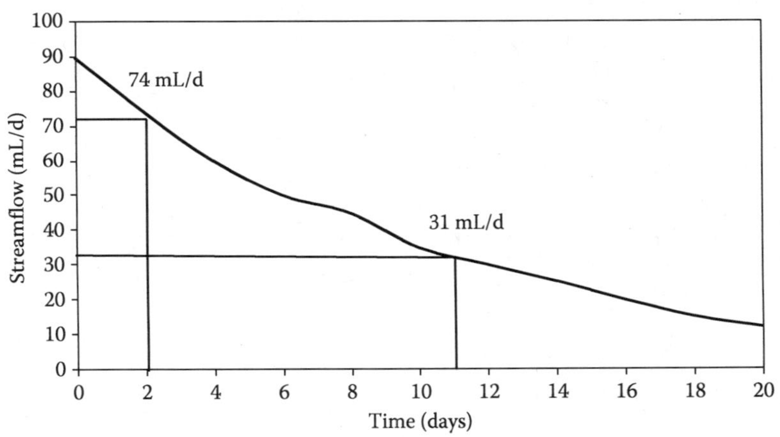

FIGURE 20.1 Forecast based on a master recession curve.

flow in the river is 74 mL/d and that an estimate is needed of how long it might take before streamflows are reduced to a threshold of 31 mL/d. The forecast might be undertaken graphically as shown in Figure 20.1, where the number of days before streamflows is reduced to the threshold of interest and can be estimated directly as a difference between t and t_0, which in this case is 9 (11.2) days. Of course, if the master recession curve was fitted to a mathematical function, then the forecast could be carried out analytically. The forecast is conservatively low, in that it is based on the assumption that no rain occurs over the intervening period. Today, low-flow forecasting is commonly performed as an integrated part of real-time simulation of runoff using calibrated rainfall–runoff models. Recession behavior and low flows are determined by the storage volumes and drainage functions selected in the model, and the approaches are numerous [17]. A comprehensive review of recession analysis is provided by Hall [24] and Tallaksen [50].

20.3 Effective Factors on Low Flow

The flow of the river is as a result of complex processes that occur in scale basin. Low flows happen therefore after periods of no rain or when precipitation falls as snow. Besides, this decreases the water stored in the soils and also the outflow to the river. The time of depletion depends on hydrological processes that work near the channel as the storage properties within the catchment. Catchment geology influences the capability of the catchment's precipitation absorption and releases as low flow. The various factors have an effect on low flow including climate, physiographic characteristics, geology, land use, human activity, and storage characteristics.

20.3.1 Climate

Catchment input is precipitation or snowmelt, so low flow can be decreased due to different causes, as

- Extended dry periods, when potential evaporation is higher than precipitation
- Extended periods of low temperature, when precipitation falls as snow

Warm and dry weather periods are associated with high-pressure systems, when high temperatures, high radiation input, low humidity, and wind increase evaporation and transpiration rates. Conversely, snowfall and snow storage happen during periods of low temperatures, associated with cold, polar air masses, and decreasing temperatures. In this situation, the absence of snowmelt creates an accumulation of solid water and then low flows [26]. Winter low flows or summer low flows in mid- and high-latitude climates occur both annually. In low-latitude climates, different dry seasons and then distinct low-flow periods can occur during the year. In arid and semiarid climates, the combination of low precipitation and higher evaporation causes minimal river networks and ephemeral rivers. These kinds of basins are characterized by prolonged periods of zero flows and episodic high flows, often in the form of flash floods. Climate processes have then a big influence on low flows, due to the influence on the magnitude and the variation of temperature, the potential evaporation, and the precipitation that modifies the input water in the catchment. Climate maps furnish good information to understand low-flow conditions of an area. However, climate varies spatially, particularly in mountain regions where there are strong altitude-dependent temperature and precipitation gradients, and temporally, at interannual, decadal, and centennial time scales.

20.3.2 Physiographic Characteristics

20.3.2.1 Topography

Topography determines the rate that the runoff will reach a river. Obviously, precipitation that falls in sheer mountainous areas will arrive at the river earlier than flat or gently sloping areas.

20.3.2.2 Area

The area of watershed is also identified as the drainage region, and it is the most important watershed attribute for hydrological analysis. It reflects the volume of water that can be generated from a rainfall. Thus, the drainage area is required as input to models ranging from simple linear prediction equations to complex computer models. Once the watershed has been delineated, its area can be determined, either by approximate map methods or by geographic information system (GIS).

20.3.2.3 Orientation

The orientation of a watershed influences the melting speed of snow and influences low flow. Watersheds developed especially in the north–south direction have an alternative exposure to sunrays, the melting speed of snow thus being smaller than in cases of watersheds developed toward east–west. For a precise determination of the influence of watershed orientation, it is necessary to know the direction and frequency of the dominant wind.

20.3.2.4 Altitudes

The extreme altitudes of the watershed, such as minimum and maximum, are obtained as a starting step for topographic maps. The maximum altitude is the elevation of the highest point of the watershed, while the minimum altitude is the elevation of the lowest point, this being generally the outlet section of the watershed. These values determine the altimetry amplitude of a watershed and help to calculate the slope. The average altitude of a watershed can be deduced directly from the hypsographical curve or from reading the topographic map.

20.3.2.5 Characteristic of Slopes and Length for Hydrographic Network

The steep slope of a watercourse favoritism accelerates the runoff, while a small slope gives the water the necessary time to infiltrate totally or partially into the soil. The calculation of the average slope is obtained from the longitudinal profile of the main stream and its tributaries. The most frequent method used to calculate the longitudinal slope of a watercourse consists of correlating the difference of altitude of the extreme points of the stream with its length.

20.3.2.6 Length of Watershed

Practically, this is the space traveled by the surface drainage and sometimes more suitably labeled as hydrological length. This length is usually used in computing a time parameter, which is a measure of the travel time of water through a watershed. The watershed length is therefore measured along the principal flow path from the watershed outlet to the basin boundary. Since the channel does not extend up to the basin boundary, it is necessary to extend a line from the end of the channel to the basin boundary. The measurement follows a path where the greatest volume of water would travel.

20.3.2.7 Shape of Watershed

Basin shape is not usually used straight in hydrological design methods; however, the parameters that reflect basin shape are used occasionally and have a conceptual basis. Watersheds have an infinite variety of shapes, and the shape supposedly reflects the way that runoff will "bunch up" at the outlet. A circular watershed would result in runoff from various parts of the watershed reaching the outlet at the same time. An elliptical watershed having the outlet at one end of the major axis and having the same area as the circular watershed would cause the runoff to be spread out over time, thus producing a smaller flood peak than that of the circular watershed. A number of watershed parameters have been developed to reflect basin shape.

20.3.2.8 Slope

Land slope is connected to several factors that also influence low flow. Generally, the steeper the stream slope (slope of the basin along the stream), the less regolith or overburden available to store precipitation.

Runoff of precipitation is more rapid. These factors allow less time for infiltration. As a stream flows out of the mountains into valleys, the stream slope decreases. When the stream slope decreases rapidly, a stream at baseflow can disappear into the coarse-grained alluvium. The stream may disappear and reappear as the streambed rises above and falls below the groundwater table. Side slope (slope of the basin perpendicular to the stream) also affects low flow. Shallow side slopes may allow more evapotranspiration because the groundwater table is closer to land surface, while steeper side slopes may reduce evapotranspiration and allow more seepage into a stream or creek.

20.3.2.9 Drainage Density

The drainage density (D) is the ratio of the total length of streams within a watershed to the total area of the watershed; thus, D has units of the reciprocal of length. A high value of D would indicate a relatively high density of streams and thus a rapid storm response. D affects the response of the watershed to rainfall. High densities usually allow fast runoff removal. Therefore, greater peaks and hydrographs with shorter durations are expected for watersheds with higher drainage densities. The effect of D on runoff volume is associated with the time during which the runoff remains in the watershed. Low densities allow for long residence times; therefore, abstraction mechanisms have more time to remove water.

20.3.2.10 Channel Length

The efficient length of a channel depends on flow greatness. Large flows overtop the banks and fill the floodplain whose length is usually shorter than that of the meandering streambed. A long drainage channel usually shows a long runoff removal time. Therefore, longer channels cause a response to rainfall slower than for shorter channels.

20.3.3 Geology

Rock type is a major influence on streamflow during low-flow situation. The rock type and the depth and characteristics of the overload establish the ability of the basin to accept, store, and transmit water. These properties influence the quantity and rate at which precipitation enters the shallow groundwater structure and how it is released to maintain streamflow. Geology obviously can at the reverse cause losses of water from the basin. These processes can be synthesized in

a. Groundwater recharge from streamflow where the prelatic surface lies below the channel
b. Bed losses, where unconsolidated alluvial material underlies the river channel
c. Losses to relatively dry soils forming the banks of streams [45]

20.3.4 Land Use

Land use can influence low-flow characteristics more than any other factor. Humans can change natural groundwater and surface water hydrology to the extent that low-flow characteristics are drastically altered. The change can occur almost immediately or very gradually. Evapotranspiration, permeation, and runoff rates are all affected by changes in land use. Examples of such changes are the decreases in forest areas due to cutting for agricultural use or urban development, decreases in wetland areas due to draining or filling, and newly constructed lakes or ponds. Much of the precipitation that previously may have infiltrated into the groundwater system in undeveloped areas now becomes surface runoff. As a result, the groundwater contribution to streamflow during low-flow conditions may be reduced. Many localities that are underlaying by highly permeable limestone or sand and gravel deposits use retention ponds to dispose of runoff from developed areas [43]. Retention ponds not only help recharge the groundwater system but also keep the increased runoff from overloading small streams during heavy rainfall periods [40].

20.3.5 Human Activity

The anthropoid influence is really important for low flows and can cause water increase or decrease in the river. The more important human impact on low flows is abstraction within subsurface drainage area. This decreases the level of prelatic surfaces and therefore potential reemergence for groundwater in stream channels. Localized reductions in the level of the water table may affect either hydraulic gradients or the length of channel that intersects the phreatic surface. The effects of groundwater pumping near the head of a perennial river may result in groundwater table depletion through the interception of recharge water and induced recharge of the aquifer from the river itself [45]. Various studies concern groundwater abstractions on low flow [3,10,15,22,38]. Another important human impact of the flow is the change in forestation. This affects mainly evapotranspiration losses from the soil, affecting gain and losses to alluvial storage. In addition, the presence of vegetation increases interception losses and disturbance of the soil structure. Land-use changes affect strongly low flows since it changes infiltration or evaporation characteristics and involve the groundwater recharge [16]. All these actions indirectly affect low flows, but there are processes that directly act on low flows, removing or adding water in the stream. A synthesis of these processes is given here. Water abstractions for industrial, agricultural and domestic use. These processes decrease the amount of water in the river and affect mainly the dry season and the frequency of this period [14,29,56]. It is important to remember that irrigation returns water to river channels and irrigation return flows can be a big part of a stream's water balance [5].

The construction of dams and river flow regulation regime can increase or decrease low-flow discharge levels depending on the operational management of the reservoir. This can be seen as the most important impact on a river's low-flow regime [45]. Different human impacts can affect low-flow regimes, and many rivers of the world with perennial regime have become intermittent, while many rivers have been created artificially.

20.3.6 Storage Characteristics

While climatic variables influence the rising or declining of water in the catchment, in terms of deficit in one season or surplus in another, the storage characteristics affect how these surplus and deficit propagate to the streamflow. The most significant properties are soil moisture and groundwater storage, aquifer properties, and hydraulic resistance between aquifer and river. As the soil is saturated, water may flow vertically towards the aquifer to recharge groundwater storage or movement laterally through permeable soil layers toward the stream. Available soil moisture capacity is really important for low flows because it furnishes a support to high annual transpiration. By means of macropores, cracks on the rocks, and pipes, water can recharge aquifer or move toward the stream without recharging the soil layer. In semiarid or arid aquifers, recharge occurs through river beds of ephemeral rivers and originates from high precipitation on mountain regions. Water arriving from the soil to the aquifers increases the groundwater level. The groundwater discharge to the stream occurs where stream channels intersect the main phreatic surface in a draining aquifer. For low flows to be sustainable,

a. The draining aquifer must be recharged seasonally with adequate amounts of moisture.
b. The water table must be shallow enough to be intersected by the stream.
c. The aquifer's size and hydraulic properties must be sufficient to maintain flows throughout the dry season [45].

Lakes and reservoirs hydraulically connected with rivers usually have a strong impact on low flows. The adequate water level in a lake should be maintained during the dry season to allow lateral outflow into a stream. Exhaustive studies that deepen the effects of lakes on low flows are not wide enough, but the importance of lakes for low flows in some regions may be derived from the inclusion of lake-related parameters in prediction models for low flows [48].

20.4 Flow Duration Curve

The flow duration curve (FDC) is defined by cumulative frequency curve that shows the percentage of moment in time that a given discharge is equaled or exceeded during a fixed period. It is one of the most informative methods of displaying the flow characteristics of a stream throughout the range of discharge, without regard to the sequence of occurrence. It is possible to observe this curve as the relationship between magnitude and frequency [45]. FDC may be constructed using different time resolutions of streamflow data: annual, monthly, or daily. The selection of the time interval depends on the purpose of the study. As the time interval increases, the range of the curve decreases (Figure 20.2). The production of the FDC can also be done using other time resolutions, more suitable to the studied problem by using the moving average method and then calculating these curves using day or month average time series. The details of the variations in flows are obscured if the time unit is long [44]. Using monthly or annual data could be unsatisfactory for many streams because it does not permit to evaluate the variability of the flow. While daily flow rates of small storms are useful for the pondage studies in a runoff river power development plant, monthly flow rates for a number of years are useful in power development plants from a large storage reservoir. The FDC is actually a river discharge frequency curve, and the longer the period of record, the more accurate is the indication of the long-term yield of a stream. A flat curve indicates a river with a few floods with large groundwater contribution, while a steep curve indicates frequent floods and dry periods with little groundwater contribution. Since the area under the curve represents the volume of flow, the storage will affect the FDC, that is, reducing the extreme flows and increasing the very low flows. This graph in the usual construction does not have a probability meaning because discharge is correlated to successive time intervals and discharge characteristics are dependent on the season; hence, the probability that discharge on a particular day exceeds a specified value depends on the discharge on preceding days and of the time of the year [18,37].

20.4.1 Construction of FDCs

Two main methods exist to construct FDCs, and they are the following:

1. The total period method (FDC)
2. The calendar year method (annual flow duration curve (AFDC))

An FDC is constructed by ranking the data in a decreasing order of magnitude, assigning flow values to class intervals, and calculating the number of occurrence in each class. Afterward, it is feasible to

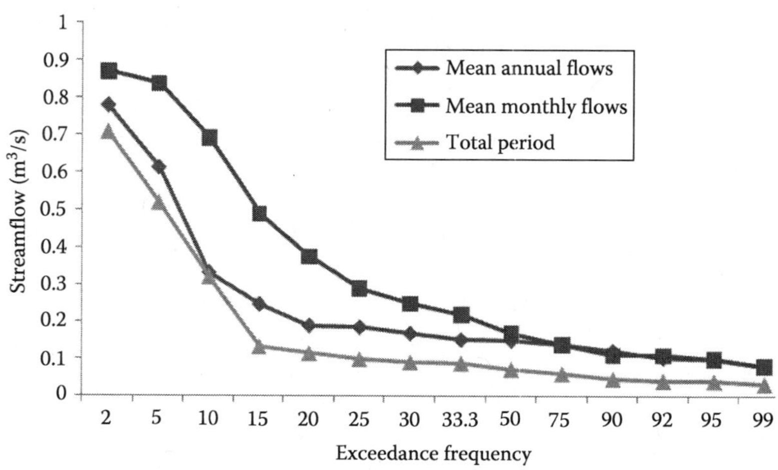

FIGURE 20.2 FDCs in Sarbhedeh station, Iran, at different time scales.

calculate cumulated class frequency and express them as a percentage of the total number of the time steps in the record period. At last, lower boundaries of every discharge class are plotted against the percentage point [44]. Another method to calculate these curves is through the definition of FDC as the complement of the cumulative distribution function based on the complete period of record. Rank the observed discharge data in ascending order FDC streamflow rates, using daily data to calculate the percentage of the discharge Weibull equation:

$$P = \frac{m}{n+1} \tag{20.4}$$

where

P is the percent of time lower value of streamflow equaled or exceeded
m specifies the number of rows
n is the number of data (Table 20.1) [8].

20.5 Low-Flow Measures and Indices

A variety of indices can be estimated by FDC. Of particular importance is the low-flow section of the FDC, which is the part of the curve below the median flow or Q_{50}. This part of the curve gives in order about the storage capacity of the river and the contribution of the baseflow to the streamflow. The slope of the curve, steep or flat, indicates the baseflow contribution and the permeability of the basin. Other indices can be extracted from FDC. In particular, Q_{95}, the 95 percentile flow, or rather the flow that is exceeded 95% of the period of record is a key index of low flow. The percentile used as low-flow index depends on the type of the river being studied. However, the flows within the range of 70%–99% time exceedance are usually widely used as design low flow. Some used example indices are as follows: one- or

TABLE 20.1 FDC of Mean Daily Flow Data of a River in 1-Year Period

Data Ranking	Streamflow (m^3/s)	Ranking	Exceedance Frequency
1	0.1	0.9	0.3
2	0.11	0.87	0.5
3	0.11	0.87	0.8
4	0.12	0.87	1.1
5	0.09	0.87	1.4
6	0.1	0.87	1.6
7	0.12	0.87	1.9
8	0.12	0.87	2.2
9	0.12	0.87	2.5
10	0.12	0.87	2.7
—	—	—	—
—	—	—	—
—	—	—	—
—	—	—	—
360	0.1	0.09	98.4
361	0.1	0.09	98.6
362	0.1	0.08	98.9
363	0.09	0.08	99.2
364	0.1	0.07	99.5
365	0.1	0.06	99.7

n-day discharges exceeded 75%, 90%, and 95% of the time called $Q_{75}(7)$, $Q_{75}(10)$, $Q_{90}(1)$, $Q_{95}(1)$, and $Q_{95}(10)$. Other indices can be estimated to evaluate streamflow variability such as the ratio Q_{20}/Q_{50} and the ratio Q_{50}/Q_{90}, which in particular represent the low-flow variability. The percentage of time that a stream is at zero flow conditions indicates the intermittency characteristics of a river [19,47]. Most statistical methods are concerned with extreme values, in meticulous with minima values in low-flow analysis. The low-flow frequency analysis obtains the proportion of years when a flow is exceeded (return period or average recurrence interval). The low-flow frequency curve (LFFC) is constructed based on the historical account of annual flow minima (daily or monthly minimum discharges of flow volume). Usually different theoretical distributions are used to forecast beyond the limits of observed probabilities and to improve low-flow estimates because the available observed flow records are not sufficient. In this case, it is not possible to know the true parent distribution for low flows, but it is important to associate reasonable distribution and estimate its parameters. The fitting procedure consists of different graphical and statistical methods that are used to decide the best theoretical distribution function. The distributions usually applied are different forms of Weibull, Gumbel, Pearson type III and log-normal distributions. Literature largely studies what are the best distributions to fitting minima data in some regions and using different averaging intervals.

Many of these studies are concerned with evaluating the lower limit of the distribution. A universally defined distribution to fit low-flow data is not yet identified although many studies consider this problem [41,55]. Various studies suggest the use of nonparametric methods for low-flow frequency analysis in order to avoid the specification of a parent distribution [34,35,51].

One of the main issues of low-flow frequency analysis is the attendance of observed streamflow time series of zero flow data. Zero data are typical of arid climates and very cold regions where the streams can be frozen in winter. Zero values can be present in a record also because the gauging station has a streamflow limit and the river level is below this threshold (censored data). The presence of zero flows has an important effect, and it is not possible to ignore it in the statistical analysis of low-flow series. The distribution fitted to series with zero flows will result in a positive probability of negative streamflows unless the distribution is explicitly constrained to have a lower bound of zero. Such results are physically meaningless [45]. In addition, the flexibility of distribution is reduced constraining it to zero lower bound. Haan [23] suggested a procedure based on conditional probability adjustment. This procedure adds a new parameter that represents the probability that an observation is zero, while continuous distribution is fitted on nonzero data. The result is adjusted with the new parameter. Durrans et al. [13] studied zero data caused by zero flow or censored data and suggested to treat them as censored data in uncertainty case. In arid and semiarid regions, rivers are often intermittent or ephemeral, and the annual minimum value is often zero so in these cases, it is not possible to apply a low-flow frequency analysis. It is possible to obtain many information and indices by LFFC. The more important indices estimated by LFFC are

- Slope of LFFC
- Break in the curve near the modal value
- 7-day 10-year low flow ($7Q_{10}$) and 7-day 2-year low flow ($7Q_2$)
- Dry weather flow

The slope of LFFC can be estimated as the difference between two flow values, normalized by catchment area, from low- and high-probability domain. The steepness of the slope indicates the variability of low-flow regime. The LFFC can present a break in the curve near the modal value. Velz and Gannon [54] identified this as the point where a change in drought characteristics occurs. Flows characterized by higher frequencies are those that have a tendency to normal conditions and cannot be considered as drought flows. Though not a general feature of the LFFC, this may be interpreted as a condition at which a river starts getting water exclusively from a deep subsurface storage [45]. The 7-day 10-year low flow ($7Q_{10}$) and 7-day 2-year low flow ($7Q_2$) are the most used indices in United States. These values represent the lowest average flows that occur for a consecutive 7-day period at the recurrence interval of 10 and 2 years. The dry weather flow is the annual series of minimum 7-day average flow [21,25,42]. It is used mainly in the United Kingdom for abstraction licensing. The 7-day average permits to eliminate the problem of day-by-day variability of the river flow and also measurement errors.

Low-flow frequency analysis belongs to extreme event frequency analysis, but its study is limited in comparison with flood frequency analysis. The low-flow studies report [28] recommended using the proportion of baseflow in the river defined by a baseflow index (BFI), for indexing the effect of low flows. Unlike the low-flow studies report [28], BFI is not used directly as a key variable linking low-flow statistics and catchment characteristics. BFI has been used primarily as a general index of catchment response.

20.5.1 Calculation of Annual Baseflow Index

There are two alternative methods for calculating annual BFIs. The first is to compute the separation for the entire record and then estimate BFI for each year. The second is to run the separation program on year 1 and then on year 2, starting each year as an entirely new record.

20.5.1.1 Calculation of Period of BFI Record

A mean value of BFI can be calculated from a series of annual BFIs, but this will be different from a single value calculated from the period of record. BFI is defined as the ratio V_B/V_A where V_B represents the volume beneath the baseflow separation line and V_A represents the mean flow beneath the hydrograph. Ten years of record will provide 10 annual values of BFI. However, the mean of these 10 values will not in general equate to the single value obtained from the entire 10 years of record.

20.5.1.2 Hydrometric Errors and Missing Data

A study was carried out to assess the influence of hydrometric errors and missing data on BFI. A bias was applied to the gauged flow data by increasing or decreasing daily flow data to simulate hydrometric error. The sensitivity of the derived BFI to this simulated error was then examined. A similar sensitivity analysis was carried out to evaluate the impact of missing data on calculated values of BFI.

20.5.2 Low-Flow Statistics

20.5.2.1 Mean Flows

The mean flow is one of the most commonly used statistics in hydrology and water resource planning. It can be estimated from a time series of gauged data by summing all daily discharges and dividing by the number of days in the record. It is normally calculate for complete calendar or hydrological years of data. It can be also calculated for specific months or seasons. The mean flow provides an estimate of the total water resources available and is key variable in all water resource assessment investigations. When a percentage of mean flow and the available head are used to give an estimate of the potential generating capacity of the scheme.

20.5.2.2 Ninety-Five Percentile Flow

This is one of the most common low-flow indices used operationally and is defined as the flow exceeded for 95% of all values. Q_{95} can also be derived from individual months, for groups of month, or any specified periods from the FDC. Other percentiles can similarly be derived. Q_{95} is often used to assess the amount of water available at low flow. For public water supply, a constant abstraction is often required, perhaps with seasonal variability. A second common application of both Q_{95} and the mean of the 7- or 10-day annual minima is that of irrigation scheme design. This involves estimating the potential area that can be irrigated by the supply river for schemes without storage. Q_{95} is also used to assess the availability of water for dilution of industrial or domestic effluents.

20.5.2.3 Mean Annual Minima (MAM) (n-Day)

Annual minima can be derived from a daily flow every year and the mean of the minima calculated. Minima of different durations can be determined, 1, 7, 10, 30, and 90 days being commonly used. The annual minima can be used to determine a destruction function for estimating the frequency or return period of low flow. In temperate climates, the mean annual 7-day minima are numerically similar to Q_{95} for most flow records.

20.6 Low-Flow Estimation Methods in Ungauged Catchments

Many management and engineering decisions are requisite in catchments where measured data are not available. Hydrological networks have declined in many regions of the world, and the system for collecting water information is not always existent or inadequate. This situation surely needs to be solved, and different possible approaches to classify these methods exist and surely all arbitrary [45].

20.6.1 Regional Analysis Approach

This is perhaps the most extensively used method in low-flow estimation at ungauged sites. It normally includes more than a few steps that are described in Chapter 23 of this handbook, *Regionalization of Hydrological Variables.*

20.6.2 Regional Mapping of Low-Flow Indices

The interpolation of flow character is based on a theory of existence of flow and its relationship with physiographic, climate, geology, and land use. A low-flow characteristic estimated at any gauged location in a region is assumed to be representative for the whole catchment above the gauge. Therefore, calculated flow values are usually assigned to the cancroids of gauged catchments. Flow contour lines are then constructed either physically or by available computer packages. Automated contouring has compensation of efficiency and reproducibility, whereas manual contouring allows the exercise of potentially more accurate expert local knowledge, where it exists. The alternative way of mapping is to delineate spatial units (subareas) with constant values or narrow ranges of low-flow characteristics [45]. The dependability of flow estimates obtained from maps depends upon a number of factors: the already mentioned density of gauging network and quality of flow data used, variability of flow characteristic being mapped in time and space, the scale of the map and the contour interval, etc. At the same time, maps of flow characteristics provide an easy way of estimating required flows at ungauged sites, indicate the quantity of water resources available in a region, and may be a valuable water resource planning tool (e.g., [1]). Alternative ways of presenting spatially changing low-flow characteristics normally concentrate on the river itself, rather than on a catchment or region.

20.6.3 Statistical Approach

Low flow is often measured as deterministic objects that can replicate the regime feature of a river in a given section. Conversely, the statistical models have a probabilistic approach to FDC and are characterized by the view of these curves as the complement of the cumulative frequency distribution. Although Ganora et al. [18] showed that theoretically FDC could not be considered as a probability curve since discharge is correlated between successive time intervals and discharge characteristics are dependent on the season, these methods have a known big distribution.

The common steps for the application of this approach are as follows:

- A suitable frequency distribution is chosen as the parent distribution for the data.
- The parameters of the region are estimated through the observed data at the gauging stations.

- The evaluation of FDC obtained in ungauged basins is done through different methods. In particular, the multiple regression model is widely used to connect region characteristics to curve parameters in order to estimate their values in ungauged basins.

Usually, statistical procedures to regionalize FDC use stochastic models to represent them, and besides many frequency distributions are fitted to the FDC. The most used parent distribution adopted to fit the data is the log-normal frequency distribution [8]. The readers can find more information about this method in References 59, 60.

20.6.4 Graphical Approaches

The graphical approach is based on these steps:

- First of all, it is necessary to standardize the FDC for each basin for all gauge driver basins, through division for a mean or median discharge.
- After that through average of the standardized FDC in different sites, the graphical regional FDC is estimated.
- Finally, the FDC for a particular basin can be estimated multiplying the mean or median discharge (flow index) estimated in gauged stations to the dimensionless FDC.
- The estimation on ungauged basins can be done relating the flow index to geomorphoclimatic characteristic of the area through regression models.

The approach in this form was defined by Smakhtin et al. [46], but different papers are involved in the definition of graphical methods to estimate regional FDCs. They model the FDC as nonparametric object and to regionalize it, and they consider the dissimilarity between curves. A linear norm is in fact used to obtain a dissimilarity matrix between the curves. This matrix is related to another one composed by the differences between all possible values of each descriptor within the set of basins. A cluster analysis is then used to identify homogeneous regions. Every region is characterized by a single dimensionless FDC [8].

20.7 River Low Flow and Environmental Flow in Hydroecology

River flow regime affects the water quality, power, biotic communications, and environment of rivers. Rivers, streams, and swamplands need certain amounts of water at certain times to maintain healthy river ecosystems. In rivers that have been dammed or are being used for irrigation, this normal flow is changed. In other situations, where water is added to a river, such as outflow from a sewage treatment plant, this also alters the natural flow of the river. To recompense for changes in flow, water may be released from dams or protected from abstraction (this is where water is removed from a river for irrigation or some other purposes) at certain times to allow rivers to function normally. For rivers where more water is being added, it may be essential to let for additional concepts to control the flow. There are two wide-ranging classes of ecological flows: releases of water below dams and safety of flows in tolerant rivers. Environmental flows are designed to imitate the natural condition of rivers. It is about not only the quantity of water but also the timing and quality. Environmental flows are naturally low during summer and autumn and are much higher during winter and spring. Rivers also naturally experience periods of very low or no flow, and at other times there are floods. It is important that environmental flows mimic this variability of flows.

The quality of water released below dams can sometimes be compromised by lower than normal water temperature, low dissolved oxygen levels, or other water quality parameters. Releasing substandard water quality can severely impair the functioning of aquatic ecosystems. Hydrologically based design flows are determined by performing extreme value statistical analysis of the single lowest flow event in each of the X years of record. Biologically based design flows are determined by analyzing the absolute X lowest flow events in the combined X years of record. The biologically based flow event calculation may therefore include multiple low-flow events in a single year and no events from other years. The rationale for the two methods is also different. The hydrologically based plan flow method was at first developed to answer questions relating to water supply, such as "On average, in how many years out of ten will the flow be below a certain level?" The biologically based method was developed to make easy the use of two averaging periods specified in the two concentrations used to state water living criterion in calculating devise flows. Biologically based propose flows are intended to compute the real amount of low-flow events by means of value to both the duration and frequency (i.e., the number of days that water life is subjected to flows below a certain level within a period of several years). Although the extreme value methodical techniques used to determine hydrologically based design flows have been used widely in the field of hydrology and in state water quality standards, these methods do not capture the cumulative nature of effects of low-flow events because they only consider the most extreme low flow in any given year. By considering all low-flow events with a year, the biologically based design flow method accounts for the cumulative nature of the biological effects related to low-flow events. The range from the 25th to 75th percentiles could be considered the adaptive range for the river ecosystem. If the annual or seasonal FDC of a given year is located below the 25th percentile FDC, the area between the 25th percentile FDC and the annual or seasonal FDC was defined as ecodeficit. This value represents the amount of water deficiency to the river ecosystem requirement. Conversely, if the annual or seasonal FDC of a given year is located above the 75th percentile FDC, the area between the 75th percentile FDC and the annual or seasonal FDC was called the ecosurplus, which represents the amount of water affluence to the river ecosystem requirement (Figure 20.3). The values of the and ecodeficit were divided by the annual mean or seasonal mean flow amount to quantify the fractions of ecosurplus and ecodeficit, respectively.

The indicators of hydrological alteration (IHA) metrics comprise 33 parameters, as listed in Table 20.2. When using a shortened water year, IHA metrics are calculated using only the data in the shortened period. Monthly average flows that overlap the boundary of the shortened period will be calculated only for the portion of the month that is in the period. The 3-, 7-, 30-, and 90-day minimums and maximums

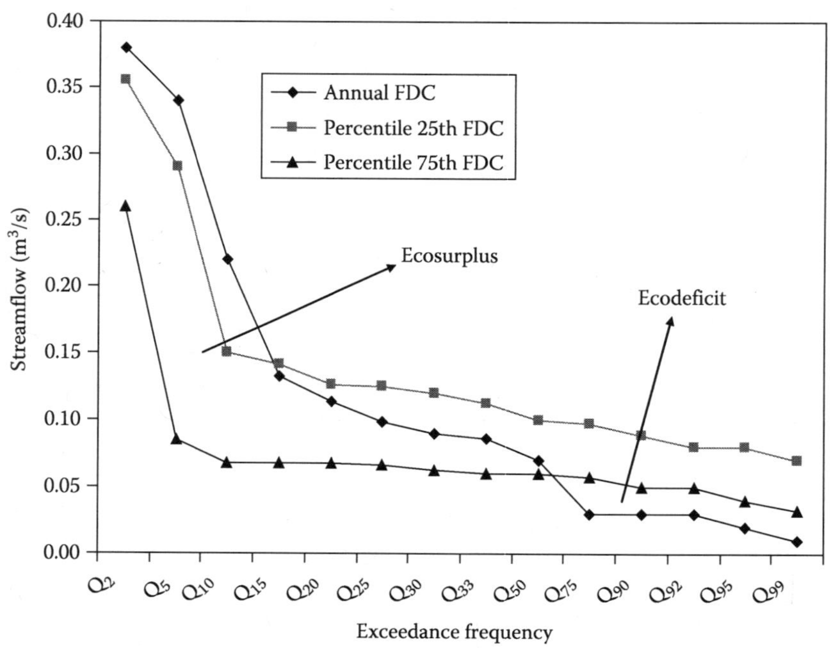

FIGURE 20.3 Determination of ecosurplus and ecodeficit in the flow duration curve in Mortezaabad station, Iran.

TABLE 20.2 Thirty-Three IHA and Their Ecosystem Influences

Indicator of Hydrological Alteration	Ecosystem Influences
Mean or median flow in January	Habitat availability for aquatic organisms
Mean or median flow in February	Soil moisture availability for plants
Mean or median flow in March	Availability of water for terrestrial animals
Mean or median flow in April	Availability of food/cover for fur-bearing mammals
Mean or median flow in May	Reliability of water supplies for terrestrial animals
Mean or median flow in June	Access by predators to nesting sites
Mean or median flow in July	Influences water temperature, oxygen levels, photosynthesis in water column
Mean or median flow in August	
Mean or median flow in September	
Mean or median flow in October	
Mean or median flow in November	
Mean or median flow in December	
BFI	Balance of competitive, ruderal, and stress-tolerant organisms
1-day minimum	
3-day minimum	Creation of sites for plant colonization
7-day minimum	Structuring of aquatic ecosystems by abiotic vs. biotic factors
30-day minimum	
90-day minimum	Structuring of river channel morphology and physical habitat conditions
1-day maximum	
3-day maximum	Soil moisture stress in plants
7-day maximum	Dehydration in animals
30-day maximum	Anaerobic stress in plants
90-day maximum	Volume of nutrient exchanges between rivers and floodplains
Number of zero flow days	
	Duration of stressful conditions such as low oxygen and concentrated chemicals in aquatic environments
	Distribution of plant communities in lakes, ponds, and floodplains
	Duration of high flows for waste disposal, aeration of spawning beds in channel sediments
Number of low pulses within each water year	Frequency and magnitude of soil moisture stress for plants
Mean or median duration of low pulses (days)	Frequency and duration of anaerobic stress for plants
Number of high pulses within each water year	Availability of floodplain habitats for aquatic organisms
Mean or median duration of high pulses (days)	Nutrient and organic matter exchanges between river and floodplain
	Soil mineral availability
	Access for waterbirds to feeding, resting, reproduction sites
	Influences bed load transport, channel sediment textures, and duration of substrate disturbance (high pulses)
Rise rates: mean or median of all positive differences between consecutive daily values	Drought stress on plants (falling levels)
	Entrapment of organisms on islands, floodplains (rising levels)
Fall rates: mean or median of all negative differences between consecutive daily values	Desiccation stress on low-mobility stream edge (varial zone) organisms

(*continued*)

TABLE 20.2 (continued) Thirty-Three IHA and Their Ecosystem Influences

Indicator of Hydrological Alteration	Ecosystem Influences
Number of hydrological reversals	
Julian date of each annual 1-day maximum	Compatibility with life cycles of organisms
Julian date of each annual 1-day minimum	Predictability/avoidability of stress for organisms
	Access to special habitats during reproduction or to avoid predation
	Spawning cues for migratory fish
	Evolution of life history strategies, behavioral mechanisms

Source: The Nature Conservancy, Indicators of hydrologic alteration version 7 user's manual, http://www.nature.org/initiatives/freshwater/files/ihav7.pdf, 2006.

are taken from moving averages of the appropriate length calculated for every possible period that is completely within the water year. High or low pulses are defined as those periods within a year in which the daily mean flow either rises above the 75th percentile (high pulse) or drops below the 25th percentile (low pulse) of all daily flows. BFI is given by the 7-day minimum flow divided by the annual mean flow. Reversals are calculated by dividing the hydrological record into "rising" and "falling" periods, which correspond to periods in which daily changes in flows are either positive or negative, respectively. For date of maximum and minimum parameters, if there are multiple days in the water year with the same flow value, the earliest date is reported. The detailed definition of all these IHA metrics could be found at the user manual of IHA [53].

20.8 Case Studies on Low Flow

20.8.1 Low-Flow Estimation in the United Kingdom

This report contains procedures for the calculation of two low-flow measures (the flow duration and flow frequency curve) from observed data and for their estimation at ungauged locations throughout the United Kingdom [21]. There are four principal improvements in the hydrological database available to this report, which benefit the updated estimation procedures: (1) more gauging stations; (2) most gauging stations possess extended data series; (3) the extended data series include recent extreme drought events, in particular 1975/1976, 1984, and 1989; and (4) data from Northern Ireland are analyzed. A comprehensive classification of the quality of low-flow data at 1643 gauging stations in the United Kingdom has been completed in conjunction with the hydrometric authorities. A total of 1643 gauging stations were assessed, of which 1366 were classified. Methods for the calculation of low-flow measures from gauged flow data are described, particularly the mean flow, BFI, FDC, and low-flow frequency. The methods described have been applied to all gauging stations with usable periods of data and calculated values of low-flow statistics and catchment characteristics. Relationships are developed between catchment characteristics and flow statistics, specifically the mean flow and two key low-flow statistics, $Q_{95}(1)$, the 95th percentile from the 1 day FDC, and MAM(7), the mean annual 7-day minimum, representing the flow duration and low-flow frequency measures, respectively. The development of internal relationships between $Q_{95}(1)$ and other flow duration percentiles and between MAM(7) and other frequency statistics enables the construction of flow duration and LFFCs at ungauged sites. Other internal relationships allow the construction of frequency curves of annual minima of different durations. The procedures for the incorporation of local data are presented, and the sequence of steps in the estimation procedure is summarized, with guidance according to data availability at, or adjacent to, the ungauged site.

20.8.2 Impacts of Climate Change on Low Flows in the Rhine Basin

This research is about the amount of low flows, which is expected to increase as a consequence of climate change [4]. For the determination of a low flow, two indicators are used: the total summer deficit and the threshold for navigation purposes. The threshold for navigation is set at a 1000 m^3/s at Lobith, the Netherlands. These indicators are based on the low critical values. The four different climate scenarios have been run to assess how the two indicators will develop. In Aladin Arpege, the average deficit decreases, but this can be explained because an extreme event has occurred in the period 1962–1992. This influences this indicator quite a lot. The changes of the number of days with a discharge below 1000 m^3/s are within the variability of the climate. This can be explained because this data series is only to 2050; by then, the effects are not quite visible; the effects are the most visible in the period 2050–2100.

For the three climate scenarios with a runtime to 2100, the annual deficit increases with 460%–2810% if the period 1962–1992 is compared with the period 2070–2100. Hereby, it is questionable whether such an increase is realistic; a part of the increase can be explained by low starting values. In the number of days with a discharge below 1000 m^3/s a year, the increase fluctuates from 112% to 249% if the period 1962–1992 is compared with the period 2070–2100; this is more realistic than the increase of the deficit. The indicator of the number of days below 1000 m^3/s is more robust than the indicator of the deficit.

The probability on a deficit during a year increases from 80% to 87% in the different scenarios for the period 2070–2100. The probability that a flow below 1000 m^3/s will occur in a specific year is between 80% and 96% in the period 2070–2100. This is quite high and it looks like a yearly event.

A comparison between the Aladin Arpege and the Hirham Arpege is made; another comparison is made between the Racmo Echam5 and the Remo Echam5 scenario. By comparing those scenarios, it is possible to determine the effects of a certain regional climate model (RCM). It can be concluded that the temperature and precipitation development in the Rhine basin is dependent on the chosen general circulation models (GCMs) for the same emission scenario. The Aladin Arpege scenario has a small increase in precipitation, and the Hirham Arpege scenario has a decrease in precipitation. A reason for this difference is that two different GCM runs for the Arpege scenarios are used. For the GCM Echam5, the precipitation is developing in the same direction, in both RCM scenarios.

The developments in the precipitation amount can be the reason why there is a difference in the development of the different indicators for the GCM of Arpege. In the Hirham Arpege scenario, the severity of a low flow increases; in the Aladin Arpege scenario, the severity decreases. In the Echam5 scenario, the development of the severity of a low flow is about the same for both RCMs. The differences in the low-flow indicators are therefore not only due to different RCMs but also influenced by the differences between these GCM runs, resembling the natural variability of climate.

20.8.3 Geostatistical Regionalization of Low-Flow Indices: Physiographic Space-Based Interpolation and Top-Kriging

Castiglioni et al. [9] applied two innovative geostatistical approaches for the prediction of low flows in ungauged basins. The first one, named physiographic space-based interpolation (PSBI), performs the spatial interpolation of the desired streamflow index (e.g., annual streamflow, low-flow index, flood quantile) in the space of catchment descriptors. The second technique, named topological kriging or top-kriging, predicts the variable of interest along river networks taking both the area and nested nature of catchments into account. PSBI 10 and top-kriging are applied for the regionalization of Q_{355} (i.e., the streamflow that is equaled or exceeded 355 days in a year, on average) over a broad geographic region in central Italy, which contains 51 gauged catchments.

For each river basin, a number of physiographic and climatic descriptors are available, namely, drainage area, A (km^2); main channel length, L (km); percentage of permeable area, P (%); maximum, mean, and minimum elevations, H_{max}, H_{mean}, and H_{min} (m a.s.l.); average elevation relative to

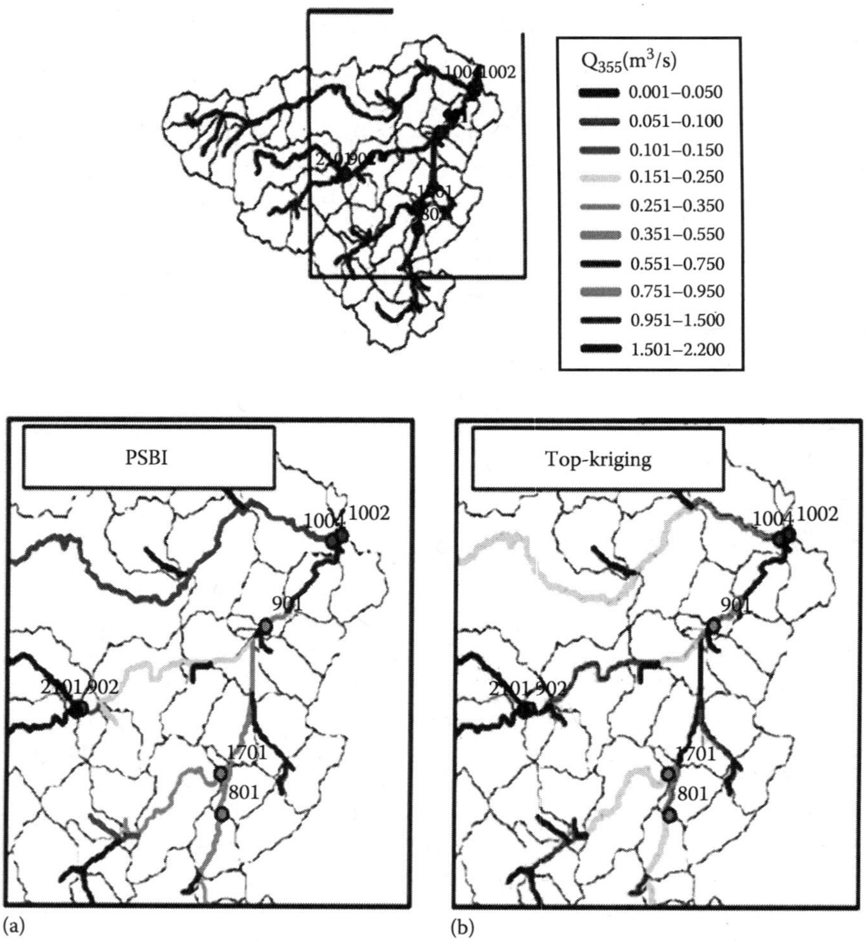

FIGURE 20.4 Cross-validation estimates of Q_{355} along the river network for PSBI (a) and top-kriging (b): observed values are shown for available stream gauges as black and gray dots; estimates are represented as black and gray lines.

H_{min}, $\Delta H = H_{mean} - H_{min}$ (m); concentration time, τc (hours), estimated according to Giandotti's (1934) empirical formula; and mean annual precipitation, MAP (mm/year).

They applied PSBI and top-kriging for predicting Q_{355} along Metauro's river network. The Metauro's basin morphology is analyzed on the basis of the SRTM 90 m digital elevation model (DEM). The soil properties of Metauro's basin are retrieved from a pedolithological map of Marche and Abruzzo. Figure 20.4 shows the cross-validation predictions of Q_{355} for PSBI (on the left) and top-kriging (on the right). The estimated values are shown on stream network, and the empirical values are shown at the 7 stream gauges where observed series of discharge 25 are available.

20.9 Summary and Conclusions

This chapter provided the methods for studying low-flow hydrology. The low flow was defined, and relationships between hydrograph recession curve and low-flow characteristics were discussed. A separate section illustrates the effective factors on low flow. The chapter provided the methods for estimating low-flow indices and their applications. Different methods were provided for low-flow estimation in a stream gauging station or ungauged river where no data are available. The methods were described for determining FDC for stream gauging stations. Four methods were presented for estimating low-flow statistics for ungauged river catchment: regional analysis, regional mapping of low-flow indices, and statistical and graphical approaches. In both developed and developing countries, both groundwater and surface water are under pressure during low flows, and these pressures will increase with increasing population growth and climate and land-use changes. Finally, the chapter presented specific applications of low-flow data in river ecology studies and environmental flow management.

References

1. Arnell, N.W. 1995. Grid mapping of river discharge. *Journal of Hydrology* 167(1):39–56.
2. Bako, M.D. and A. Owoade. 1988. Field application of a numerical method for the derivation of baseflow recession constant. *Hydrological Processes* 2(4):331–336.
3. Bickerton, M., G. Petts, P. Armitage, and E. Castella. 2006. Assessing the ecological effects of groundwater abstraction on chalk streams: Three examples from eastern England. *Regulated Rivers: Research and Management* 8(1–2):121–134.
4. Bisterbosch, J. 2010. Impacts of climate change on low flows in the Rhine basin, MSc thesis, University of Twente, Enschede, the Netherlands, 63pp.
5. Blodgett, J.C., J.R. Walters, and J.W. Borchers. 1992. *Streamflow Gains and Losses and Selected Flow Characteristics of Cottonwood Creek, North-Central California, 1982–85*. US Department of the Interior, U.S. Geological Survey, Sacramento, CA.
6. Boussinesq, J. 1877. *Essai sur la théorie des eaux courantes*. Imprimerie nationale, Paris, France.
7. Boussinesq, V.J. 1904. Recherches théoriques sur l'écoulement des nappes d'eau infiltrées dans le sol et sur le débit des sources Journal de Mathématiques Pures et Appliquées 5th Ser., 10: 5–78, 363–394.
8. Castellarin, A., G. Galeati, L. Brandimarte, A. Montanari, and A. Brath. 2004. Regional flow-duration curves: Reliability for ungauged basins. *Advances in Water Resources* 27(10):953–965.
9. Castiglioni, S., A. Castellarin, A. Montanari, J.O Skøien, G. Laaha, and G. Blöschl. 2011. Smooth regional estimation of low-flow indices: Physiographical space based interpolation and top-kriging. *Hydrology and Earth System Sciences* 15(3):715–727.
10. Clausen, B. and C.P. Pearson. 1995. Regional frequency analysis of annual maximum streamflow drought. *Journal of Hydrology* 173(1):111–130.
11. Cooper, H.H. and M.I. Rorabaugh. 1963. *Ground-Water Movements and Bank Storage due to Flood Stages in Surface Streams*. U.S. Government Printing Office, Washington, DC.
12. Coutagne, A. 1968. Les variations de débit en période non influencée par les précipitations. *La Houille Blanche* 3:416–436.
13. Durrans, S.R., T.B.M.J. Ouarda, P.F. Rasmussen, and B. Bobée. 1999. Treatment of zeroes in tail modeling of low flows. *Journal of Hydrologic Engineering* 4(1):19–27.
14. Eheart, J.W., and D.W. Tornil. 1999. Low-flow frequency exacerbation by irrigation withdrawals in the agricultural Midwest under various climate change scenarios. *Water Resources Research* 35(7):2237–2246.
15. Fendekova, M. and P. Nemethy. 1994. Low flows of Starohorsky potok brook and their affecting factors. FRIEND: Flow regimes from International Experimental and Network Data. *IAHS Publication* (221):163–169.
16. Ferguson, B.K. and P.W. Suckling. 1990. Changing rainfall-runoff relationships in the urbanizing peachtree creek watershed, Atlanta, Georgia. *JAWRA Journal of the American Water Resources Association* 26(2):313–322.
17. Franchini, M. and M. Pacciani. 1991. Comparative analysis of several conceptual rainfall-runoff models. *Journal of Hydrology* 122(1):161–219.
18. Ganora, D., P. Claps, F. Laio, and A. Viglione. 2009. An approach to estimate nonparametric flow duration curves in ungauged basins. *Water Resources Research* 45(10):W10418.
19. Gorgens, A.H.M. and D.A. Hughes. 1982. Synthesis of streamflow information relating to the semi-arid Karoo biome of South Africa. *South African Journal of Science* 78(2):58–68.

20. Griffiths, G.A. and B. Clausen. 1997. Streamflow recession in basins with multiple water storages. *Journal of Hydrology* 190(1):60–74.
21. Gustard, A., A. Bullock, and J.M. Dixon. 1992. *Low Flow Estimation in the United Kingdom*. Report No. 108, Institute of Hydrology, Wallingford, U.K.
22. Gustard, A. and A.J. Wesselink. 1993. Impact of land-use change on water resources: Balquhidder catchments. *Journal of Hydrology* 145(3):389–401.
23. Haan, C.T. 1977. *Statistical Methods in Hydrology*. Iowa State University Press, Ames, IA, pp. 1–20.
24. Hall, F.R. 1968. Base-flow recessions—A review. *Water Resources Research* 4(5):973–983.
25. Hindley, D.R. 1973. The definition of dry weather flow in river flow measurement. *Journal of the Institution of Water Engineers and Scientist* 27:438–440.
26. Hopkinson, C. and G.J. Young. 1998. The effect of glacier wastage on the flow of the Bow River at Banff, Alberta, 1951–1993. *Hydrological Processes* 12(10–11):1745–1762.
27. Horton, R.E. 1933. The role of infiltration in the hydrologic cycle. *Transaction American Geophysical Union* 14:446–460.
28. Institute of Hydrology. 1980. *Low Flow Studies*. Institute of Hydrology, Wallingford, U.K., pp. 1–4.
29. Kottegoda, N.T. and L. Natale. 1994. Two-component log-normal distribution of irrigation-affected low flows. *Journal of Hydrology* 158(1):187–199.
30. Kovács, A. 2003. Geometry and hydraulic parameters of karst aquifers: A hydrodynamic modeling approach. PhD thesis, University of Neuchatel, Neuchâtel, Switzerland.
31. Kroll, C.N. and J.R. Stedinger. 1996. Estimation of moments and quantiles using censored data. *Water Resources Research* 32(4):1005–1012.
32. Kullman, E. and I. Imro. 1990. *Krasovo-puklinové vody*. Geologický ústav Dionýza Štúra, Bratislava, Slovakia.
33. Laaha, G. and G. Blöschl. 2007. A national low flow estimation procedure for Austria. *Hydrological Sciences Journal* 52(4):625–644.
34. Lall, U. 1995. Recent advances in non-parametric function estimation: Hydrological applications. *Reviews of Geophysics* 33:1093–1102.
35. Loaiciga, H.A. and M.A. Marino. 1988. Fitting minima of flows via maximum likelihood. *Journal of Water Resources Planning and Management* 114(1):78–90.
36. Maillet, E.T. 1905. Essais d'hydraulique souterraine & fluviale. A. Hermann, Paris, France.
37. Mosley, M.P. and A.I. McKerchar. 1993. Streamflow. In: *Handbook of Hydrology*, Maidment, D.R. (ed.). McGraw-Hill, New York.
38. Owen, M. 1991. Groundwater abstraction and river flows. *Journal of the Institution of Water and Environmental Management* 5(6):697–702.
39. Padilla, A., A. Pulido Bosch, and A. Mangin. 2005. Relative importance of baseflow and quickflow from hydrographs of karst spring. *Ground Water* 32(2):267–277.
40. Parker, G.G., G.E. Ferguson, and S.K. Love. 1955. *Water Resources of Southeastern Florida*. U.S. Geological Survey Water-Supply Paper 1255, Washington, DC, p. 965.
41. Pearson, C.P. 1995. Regional frequency analysis of low flows in New Zealand rivers. *Journal of Hydrology* (New Zealand) 33(2):94–122.
42. Pirt, J., M. Simpson, and S.T.W. Authority. 1983. *The Estimation of River Flows*. Severn Trent Water Authority, Coventry, U.K.
43. Seaburn, G.E. and D.A. Aronson. 1974. *Influence of Recharge Basins on the Hydrology of Nassau and Suffolk Counties, Long Island, New York*. U.S. Government Printing Office, Washington, DC.
44. Searcy, J.K. 1959. *Flow-Duration Curves*. U.S. Government Printing Office, Washington, DC.
45. Smakhtin, V.U. 2001. Low flow hydrology: A review. *Journal of Hydrology* 240(3):147–186.
46. Smakhtin, V.U., D.A. Hughes, and E. Creuse-Naudin. 1997. Regionalization of daily flow characteristics in part of the Eastern Cape, South Africa. *Hydrological Sciences Journal* 42(6):919–936.
47. Smakhtin, V.U., D.A. Watkins, and D.A. Hughes. 1995. Preliminary analysis of low-flow characteristics of South African rivers. *Water S. A.* 21(3):201–210.

48. Sokolov, B.L. 1974. Method for reducing single discharge measurements to the average discharge of the low-flow period. *Soviet Hydrology* (Selected Papers) 4, 235–241.
49. Spruill, C.A., S.R. Workman, and J.L. Taraba. 2000. Simulation of daily and monthly stream discharge from small watersheds using the SWAT model. *Transactions of the ASAE* 43(6):1431–1439.
50. Tallaksen, L.M. 1995. A review of baseflow recession analysis. *Journal of Hydrology* 165(1):349–370.
51. Tasker, G.D. 1987. A comparison of methods for estimating low flow characteristics of streams. *JAWRA Journal of the American Water Resources Association* 23(6):1077–1083.
52. WMO/Unesco Panel on Terminology. 1970. *International Glossary of Hydrology: 2d Draft of Definitions in English (1969)*. Secretariat of the World Meteorological Organization, Geneva, Switzerland.
53. The Nature Conservancy. 2006. Indicators of hydrologic alteration version 7 user's manual. http://www.nature.org/initiatives/freshwater/files/ihav7.pdf
54. Velz, C.J. and J.J. Gannon. 1953. Low flow characteristics of streams. *Ohio State University Studies Engineering Survey* 22(4):138–157.
55. Vogel, R.M. and C.N. Kroll. 1990. Generalized low flow frequency relationships for ungaged sites in Massachusetts. *JAWRA Journal of the American Water Resources Association* 26(2):241–253.
56. Wilber, D.H., R.E. Tighe, and L.J. O'Neil. 1996. Associations between changes in agriculture and hydrology in the Cache River basin, Arkansas, USA. *Wetlands* 16(3):366–378.
57. World Meteorological Organization (WMO). 1974. *International Glossary of Hydrology*. WMO, Geneva, Switzerland.
58. World Meteorological Organization (WMO). 2008. *Manual on Low Flow Estimation and Prediction*. WMO, Geneva, Switzerland.
59. Eslamian, S. S. and M. Biabanaki, 2009. Low flow regionalizationmodels. *International Journal of Ecological Economic & Statistics*. Special Issue on Stream Ecology and Low Flows (SELF) 12(F08): 82–97.
60. Eslamian, S. S. 2009. Editorial: An ecologically based low flowreview. *International Journal of Ecological Economic & Statistics*. Special Issue on Stream Ecology and Low Flows (SELF). 12(F08): 1–6.

21

Modern Flood Prediction and Warning Systems

Zheng Fang
Rice University

Antonia Sebastian
Rice University

Philip B. Bedient
Rice University

AUTHORS

Zheng Fang obtained his PhD in civil and environmental engineering from Rice University. He has been working on surface water and groundwater problems for over 10 years, including floodplain studies, hydrologic/hydraulic modeling, water treatment, hydrodynamic simulation, storm water management modeling, and water quality assessment for a number of watersheds and areas in Texas, Florida, Connecticut, California, and Louisiana. Not only has he accomplished many projects in drainage modeling and design, but he has also actively worked in the area of hydrologic/hydraulic analysis for flood prediction and warning in real-time mode. Dr. Fang has enhanced a radar-based flood warning system to achieve more accurate and timely flood forecasts. He recently developed advanced features for a real-time flood alert system (FAS) for the Texas Medical Center (TMC) based on the use of NEXRAD radar data. He currently works as a project manager and assistant to the director in the Severe Storm Prediction, Education, and Evacuation from Disasters (SSPEED) Center at Rice University.

Antonia Sebastian obtained her BS in civil and environmental engineering in 2011 and is currently a graduate student in the Environmental Engineering Department at Rice University. She has completed various international projects in ground- and surface water during her five-year tenure as a research assistant to Dr. Philip Bedient, Herman Brown Professor of Engineering, including surface and groundwater hydrology, geographical information, flood warning and alert systems, flood control and water quality strategies, hydrologic and hydraulic modeling/design, storm water management, open channel flow analysis, and storm surge analysis.

Philip B. Bedient is the Herman Brown Professor of Engineering in the Department of Civil and Environmental Engineering at Rice University. He teaches and performs research in surface and groundwater hydrology, disaster management, and flood prediction systems. He served as chair of environmental engineering from 1992 to 1999. He has worked on surface water problems for over 33 years, including major floodplain studies, water quality assessments, and hydrologic modeling for a number of watersheds in Texas, Florida, and Louisiana. He has been actively involved in the area of hydrologic analysis for flood prediction and warning and recently teamed with others to develop a real-time FAS for the TMC. This is based on converting NEXRAD radar directly to rainfall via a geographic information systems (GIS) system, which is then used in hydrologic models for predicting peak floods. The system is currently delivered real time on a website that has been tested on major flood events in Houston, including TS Allison. He has directed 55 research projects over the past 33 years and has written over 180 articles in journals and conference proceedings. He is the lead author on a text on "Hydrology and Floodplain Analysis" (Prentice Hall, 5th edn., 2012) used in over 75 universities across the United States. He also has a second text on "Groundwater Contamination" (Prentice Hall, 2nd edn., 1999). Dr. Bedient received the Herman Brown endowed Chair of Engineering in 2002 at Rice University. He was elected to Fellow ASCE in 2006 and received the prestigious C.V. Theis Award from the American Institute of Hydrology in 2007.

PREFACE

Of any weather-related event, flooding presents the greatest danger to infrastructure, private property, and public safety. Flood warning systems, as one type of nonstructural flood protection, have been widely adopted in some flood-prone areas because they initiate preventive measures to reduce the risk of damage and loss of life in the case of an impending flood. Over the past decades, modern flood alert systems (FASs) have evolved to meet the specific needs of different clients, regions, or even nations, differing greatly in complexity, size, and scope. FASs range from basic river gauge monitoring networks to complex forecasters, employing radar rainfall data and computer modeling techniques to produce lead times ranging from hours to days. Along with gauge networks, this chapter will discuss in detail the flash flood forecasting provided by the National Weather Service (NWS), Rice University and the Texas Medical Center's FAS, and the European Flood Awareness System. In addition, an ongoing update to many of these systems—real-time inundation mapping—will also be discussed as well as its role in improving the public's reaction and comprehension of flood warnings. With the introduction of all different flood warning systems, this chapter also established a background understanding of flooding and flood control, radar rainfall measurement, and hydrologic computer modeling.

21.1 Introduction

Of any weather-related event, flooding presents the greatest danger to infrastructure, private property, and public safety. Within the United States, inundation alone accounts annually for 136 deaths and over $4 billion in damages [12], mirroring European totals of 88 deaths and $3.6 billion [1,7]. Hurricanes, as one kind of natural disasters, bring oceanic moisture into inland to generate severe floods, causing billions of dollars of loss in the United States every year. Table 21.1 shows selected major hurricanes occurring between 1992 and 2008 in the United States. Globally, these statistics can be more severe and will continue to worsen as global warming increases the frequency and magnitude of these events and urbanization forces new populations into floodplains. However, despite presenting these serious problems, flood protection methodologies consistently failed to match the development of hurricane and tornado forecasting technologies for much of the twentieth century. This sentiment changed when, in recent decades, a wrath of major flooding events in the United States and Europe prompted national and local investment in improving flood forecasting and alert systems. It was realized that accurate and timely warning as necessary information can enable the public to be prepared for the next flood event.

TABLE 21.1 Selected Major Hurricanes Affecting the United States, 1992–2008

Name	Date	Category	Landfall	Damages	Deaths	Records
Andrew	August 1992	4 3	Dade County, FL; Central Louisiana	$26 billion	23	Second costliest; fourth most intense at landfall
Alberto	July 1994	TS	Florida Panhandle	$.75 billion	30	
Opal	September 1995	3	Pensacola, FL	$3 billion	9	
Floyd	September 1999	2	Cape Fear, NC	$6 billion	56	
Allison	June 2001	TS	Freeport, TX	$5 billion	41	Most costly and deadly TS
Isabel	September 2003	2	Outer Banks, NC	$3 billion	17	
Charley	August 2004	4 1	Fort Myers, FL; McClellanville, SC	$15 billion	15	
Frances	September 2004	2	Stuart, FL	$9 billion	23	
Ivan	September 2004	3	Gulf Shores, AL	$14 billion	43	
Jeanne	September 2004	3	Stuart, FL	$7 billion	10	
Dennis	July 2005	3	Pensacola, FL	$2 billion	9	
Katrina	August 2005	1 3	North Miami Beach, FL; Buras, LA	$75 billion	1336	Costliest U.S. hurricane; third deadliest; third most intense at landfall
Rita	September 2005	3	Texas/Louisiana border	$10 billion	7	Fourth lowest minimum pressure
Wilma	October 2005	3	Naples, FL	$1 billion	6	Lowest minimum pressure
Ike	September 2008	2	Houston, TX	$24.9 billion	20	Third costliest

In essence, the function of a flood alert system (FAS) is to "collect, handle, analyze, and disseminate hydrologic information, ideally in real time, for the purpose of providing accurate advance warnings of an impending flood condition" [2]. However, despite sharing this basic function, modern FASs have evolved to meet the specific needs of different clients, regions, or even nations. Consequentially, these systems, and the organizations that oversee them, differ greatly in complexity, size, and scope. FASs range from basic river gauge monitoring networks to complex forecasters, employing radio detection and ranging (radar) rainfall data and computer modeling techniques to produce lead times fluctuating from hours to days. Along with gauge networks, this chapter will discuss in detail the flash flood forecasting provided by the National Weather Service (NWS), Rice University and the Texas Medical Center's (TMC) FAS, and the European Flood Awareness System (EFAS). In addition, an ongoing update to many of these systems—real-time inundation mapping—will also be discussed as well as its role in improving the public's reaction and comprehension of flood warnings. However, before this information can be covered, the following sections will establish a background understanding of flooding and flood control, radar rainfall measurement, and hydrologic computer modeling.

21.2 Flooding and the Rainfall–Runoff Process

In a natural watershed, precipitation undergoes several steps in its conversion into runoff. At the beginning of a storm event, the first drops of rainfall are captured by vegetation in a process known as interception storage. After this stage, excess rainfall interfaces with the soil where it is subject to infiltration losses. The magnitude of these losses normally decays over the course of a storm event, eventually achieving a constant value or final infiltration rate [3]. As long as precipitation intensities exceed rates of interception and infiltration, excess rainfall volume—the runoff—fills depression storage or is received by a stream or other bodies of water. If runoff volumes during a storm exceed the capacity of natural channels or reservoirs, flooding results, and water spills into low-lying floodplains. Figure 21.1 demonstrates the process from rainfall to loss and overland runoff.

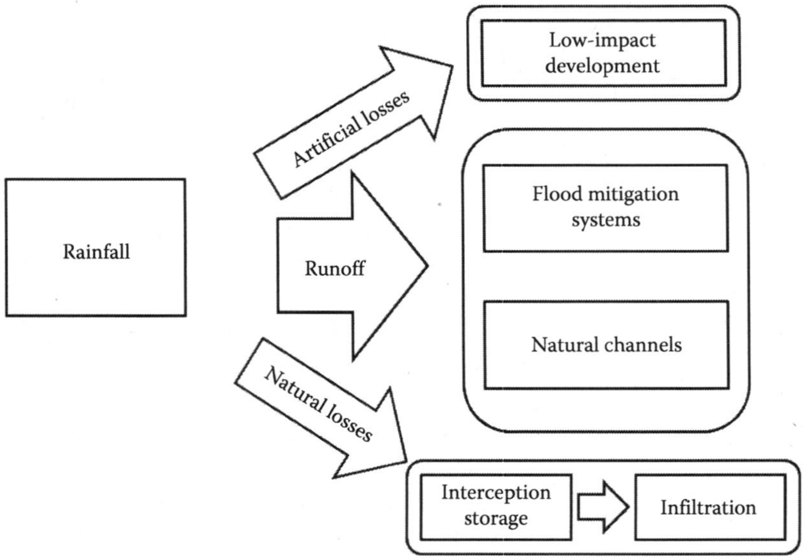

FIGURE 21.1 Hydrologic process from rainfall to loss and over-land runoff.

The specific volumes associated with the rainfall–runoff process are unique to each watershed. Often characterized by the runoff coefficient—the ratio between the volume of runoff observed and a given amount of precipitation—the response of a watershed largely depends on the size, shape, slope, soil type, storage, and land use of the area [2]. Changes in land use are especially correlated with variations in the runoff coefficient. As land use within a watershed trends toward commercial and residential development, natural depression storage and surfaces once present for infiltration are replaced with impermeable pavement and rooftops, effectively eliminating absorption into the soil. As a result, urban runoff coefficients are normally much higher than natural watersheds, resulting in floods with higher peak flow volumes and shorter response times.

To accommodate increased runoff volumes, urban authorities have two strategies: installing or upgrading conventional systems and attenuating runoff volumes through low-impact development (LID) practices. Conventional runoff systems include street gutters, storm sewers, retention ponds, detention ponds, and urban channels. These features are designed to either divert storm water from overland flow toward larger bodies of water or attenuate peak flows with temporary or semipermanent storage. While effective in dealing with smaller precipitation volumes, many of these systems are only designed to two-year storm specifications that can easily be exceeded. For example, in coastal Texas, a mesoscale weather system moving off of the Gulf of Mexico can produce localized accumulations above this value [9]. When overtaxed, the rapid response time of these storm sewer systems causes urban flash flooding or isolated rapid rises in water level. These floods, and not high winds or lightning, are often the most dangerous aspects of a storm.

For new construction, LID practices present a contemporary solution to the urban runoff problem. At its core, LID focuses on attenuating runoff volumes and reducing runoff coefficients with increased urban infiltration rates. This is accomplished through the installation of a variety of design features, including green roofs, bio-swales, and permeable pavements. Because of its localized approach to managing urban runoff, LID often presents significant cost advantages to conventional systems that can require extensive installation and modification of infrastructure [19].

However, regardless of control strategy, when runoff volumes exceed the capacity of channel systems, flooding will occur. At this point, the most effective way to mitigate damage is through timely and accurate warning of impending flood conditions, often via a dedicated FAS. Flood events can vary from flash flooding in small streams and streets to the inundation of major river systems, impacting individual watersheds or multiple nations. Consequentially, a large variety of food alert systems have been created to meet the demands of these diverse flood situations. The subsequent sections will describe, in detail, these organizations, these systems, and the tools and methods they employ.

21.3 River Gauging Flood Alert Systems

Despite appearing relatively antiquated with the emergence of forecast systems, river gauging networks are still commonly employed at many areas for flood warning purpose. Typically, these systems consist of a network of interconnected river and rain gauges and automatically disseminate alerts when weather parameters, flow volumes, or river stages exceed prespecified threshold values. While these systems may not be able to estimate a final flood stage or forecast its arrival, their warnings still provide valuable advisory of the location of occurring or imminent flood conditions.

In the United States, one of the principle gauge-based FASs is the U.S. Geological Survey's (USGS) Water Watch. The system displays real-time, recent, and past streamflow conditions at more than 3000 locations nationwide. From a graphical interface, users can see which locations are approaching or surpassing flood stage and can click on individual gauges to view over 30 years of data [20]. In 2010, the USGS introduced Water Alert, a real-time, personalized FAS. During registration, users are able to select a specific gauge as well as notification parameters such as rainfall accumulation and channel flow volume. After this, the system will e-mail or text message daily or hourly updates of the gauge as well as send emergency notification when parameter values exceed a user-defined threshold [21].

In particularly flood-prone urban areas, existing USGS infrastructure is often augmented with additional gauges in a localized flood warning system. For example, a monitoring network known as the Harris County Flood Warning System in Houston, Texas, has 133 local gauges that are operated and maintained by the Harris County Flood Control District (HCFCD). In addition to this role, gauge data are made public through a graphical user interface on the HCFCD website [6]. The benefit of additional gauge infrastructure in this and similar systems is not only the increased data resolution but also the ability to tailor gauge placement to vulnerabilities in each urban area. Proper station location ensures critical infrastructure is monitored, allowing emergency officials to respond more intelligently and efficiently to an emerging crisis.

On a smaller scale, Fort Hood—one of the United States's largest armored posts—is home to another gauge-based FAS. The network was installed by the Backland Research Center's (BRC) Water Science Laboratory in 2002 to address chronic flooding problems. Under the control of Fort Hood's Integrated Training Area Management (ITAM) director, the system continuously measures water levels at six low water crossings of three major streams. At 10 min intervals, data are uploaded to a BRC internet server for up-to-date access. While the system is currently exclusively stream gauge-based, future plans have been created to upgrade the system to support prediction modeling, additional monitoring locations, and automated warning lights at high-risk areas. Despite its simplicity, Fort Hood's gauge warning network maybe is one of the very few programs—including Rice University and TMC's FAS—that are customized for specific clients and locations.

While not on the cutting edge of flood warning technology, river gauge-based FASs are still being utilized extensively. The networks provide adequate warning times in most situations and serve important roles in many flood management organizations. By discussing radar rainfall techniques and the hydrologic computer modeling, the subsequent sections will serve as a foundation for understanding more advanced flood warning systems.

21.4 Application of Radar Rainfall Data in Flood Alert Systems

Traditionally, precipitation values have been determined by networks of rain gauges. However, recent advancements in the accuracy and calibration of radar rainfall data have permitted their use in hydrologic applications with better precision, enabling modern FASs to provide more reliable warning information. This section will serve as a general introduction to radar rainfall measurement.

Developed in the 1930s, radar was originally designed to monitor enemy ships and aircraft in the impending global conflict. However, during World War II, radar operators noticed that precipitation often produced signal interference, obscuring targets. While measures were taken to remove the unwanted signal, this phenomenon and its associated potential for weather prediction were not forgotten. As a result, military radars were adapted for weather purposes in the years following World War II. With these early systems, empirical relations between reflectivity and rainfall accumulation were derived in conjunction with rain gauge calibration techniques. In the subsequent decades, the United States constructed several networks of weather radars, the most current being the WSR-88D NEXRAD system [2]. WSR-88D radar technology was developed in the late 1980s and consists of an S-band Doppler radar with a 250 nm range [13]. Figure 21.2 shows that the NEXRAD radar network is composed of 160 locations distributed across the United States, and despite some lapses in the mountainous West, it provides excellent coverage for the entire nation [14]. Figure 21.3 shows a radar dome located at the WSR-88D Radar Operations Center in Norman, Oklahoma. In Europe, the equivalent system is the European Weather Network. The network is based on the Operational Program on the Exchange of Weather Radar Information (OPERA), which takes the radar products of the variety of weather radar systems utilized by 29 European countries and aggregates them into a single, useful product [17].

When compared to traditional rain gauge information, the radar rainfall data normally offer improved spatial and temporal resolution as well as rainfall totals [2]. From a flood prediction perspective, radar rainfall data more realistically represent a storm event than gauge data, because its increased

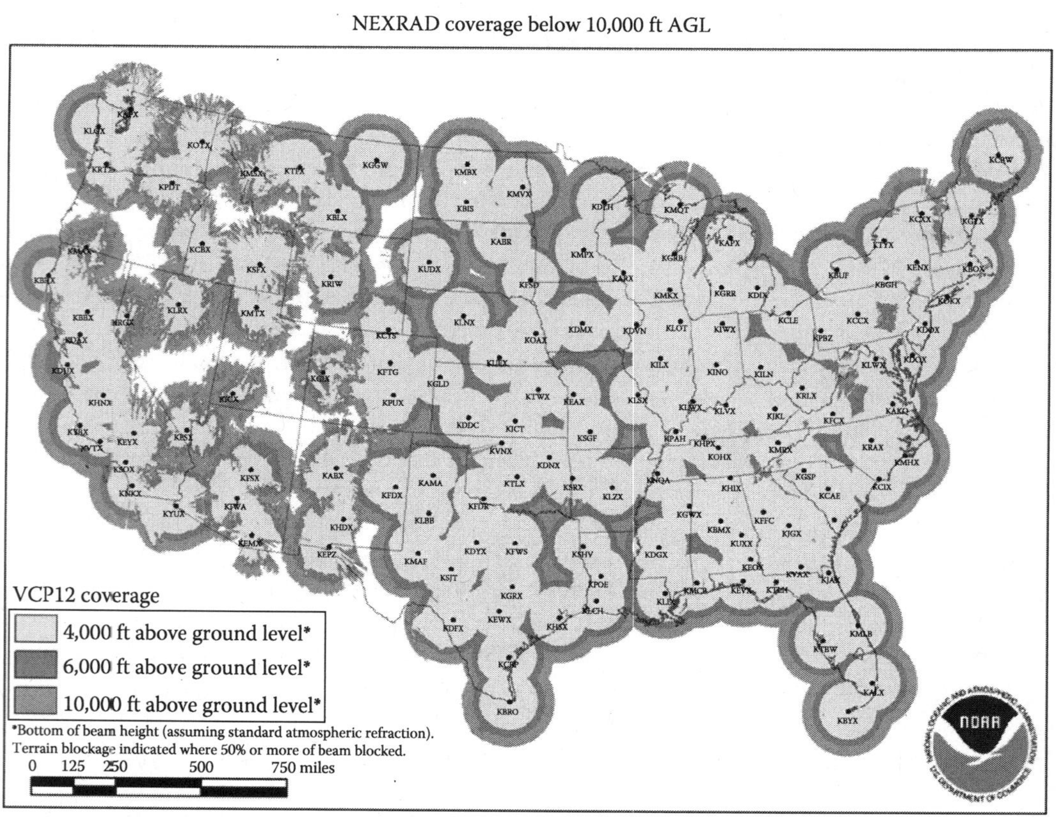

FIGURE 21.2 NEXRAD radar network across the United States. (From National Weather Service [NWS], National weather service advanced hydrologic prediction service, http://www.nws.noaa.gov/oh/ahps/presentations/ahps_FY2005/AHPS%20FY05%20Fact%20Sheet.pdf, 2012, accessed July 11, 2012.)

FIGURE 21.3 A WSR-88D radar dome at the NWS Radar Operations Center in Norman, Oklahoma.

resolution is able to capture the intricacies within a rainfall pattern. This is because in a rain gauge network, precipitation values are spatially averaged, or "lumped," to represent large areas. Often, this traditional method fails to account for localized variations in rainfall intensity, leading to inaccurate calculations of rainfall volumes by computer programs. As a result, most modern FASs tend to incorporate radar rainfall data into their models for the greatest level of accuracy.

21.5 Hydrologic and Hydraulic Modeling

Essential to forecasting flooding, hydrologic and hydraulic modeling software was originally developed in the 1970s. The computer has vastly improved our capacity to understand and predict the behavior of complex watersheds in the past decades. This section will describe the differences between two types of models, lumped-parameter and distributed, which are commonly used in current flood alert applications. In addition, 2-D hydraulic models, often used for inundation mapping, will also be introduced in this section.

One of the lumped-parameter models is HEC-1, developed at the Hydrologic Engineering Center (HEC) by the U.S. Army Corps of Engineers in 1973. Since the models are characterized by the fact that data inputs are "lumped," or a large area's worth of data are spatially averaged and represented as a single point, or subbasin, they offer effectively no accommodation for the spatial variation of input parameters. Rainfall–runoff conversion is generally accomplished with synthetic unit hydrograph techniques, and the runoff response of an entire watershed is generated through aggregating the individual responses of a network of subbasins. Over the past three decades, the original HEC-1 model has evolved itself into a graphic user interface (GUI) version, the Hydrologic Modeling System (HMS) model, which is an example of a commonly used contemporary lumped-parameter model. Due to their scriptability, both HEC-1 and HEC-HMS models can be incorporated into flood warning systems as modeling engines to perform hydrologic computations for real-time flood predication.

Spurred by the increased availability of GIS data and digital elevation mapping after the mid-1990s, distributed-parameter models accommodate spatial variations in hydrologic parameters and use physics-based equations—often kinematic wave methods—to determine overland and channel flow. These models can easily handle spatially varied slope, hydraulic roughness, infiltration, and rainfall input. While much more computationally intensive than lumped-parameter models, the physically distributed model parameters combining with spatially varied radar rainfall data often produce relatively accurate results. And because of their distributed nature, the models can produce output anywhere on the modeling domain, not just channel outlets, which allows for improved inundation forecasting and mapping [2]. Vflo™, developed by Vieux, Inc., is a contemporary example of a commonly used distributed hydrologic model [22].

2-D hydraulic models have abilities to depict the spatial distribution of a flood and serve integral roles in delineating floodplain areas, which makes 2-D models indispensable for floodplain delineation as current trend in floodplain mapping. Many 2-D modeling programs can intake modeled results from a 1-D model because the 1-D model such as the HEC-River Analysis System model (RAS) can still do a very good job on simulating the channel and pipe networks for typical urban settings. Combining two kinds of models can provide sound simulated results for flood-prone zones [23]. Current examples of 2-D models include XP-SWMM as well as MIKE FLOOD and MIKE URBAN.

Besides these examples, there are many other types of hydrologic and hydraulic models. However, these three software categories—lumped, distributed, and 2-D—include most modeling programs and therefore serve as a solid introduction for flood warning. The subsequent sections will consist of descriptions of the NWS, Rice University and TMC's FAS, the EFAS, and inundation mapping programs.

21.6 National Weather Service

The primary flash flood forecasting agency within the United States is the NWS. The NWS delegates the responsibility of generating and disseminating flash flood guidance to a local Weather Forecast Office (WFO). Raw weather data from the NEXRAD network are transmitted to WFOs in the form of the Digital Hybrid Reflectivity (DHR) product, a digital file of reflectivity values oriented on 1 degree by 1 km grid [11]. These data are input into one of two main software programs at the NWS's disposal: Flash Flood Monitoring and Prediction (FFMP) and the Site-Specific Hydrologic Prediction Model (SSHPM).

FFMP is the first, GIS-based program shown in Figure 21.4 [16]. Using the DHR and gridded flash flood guidance from River Forecast Centers (RFCs), the software is able to conduct precipitation analysis for customizable drainage basins as small as a few square miles. Figure 21.5 depicts the data flow process for the FFMP program. Flow direction between basins can also be determined during the customization process. Based on what areas are receiving significant rainfall, FFMP is able to generate short-term warning guidance for the affected basins. WFOs analyze these data and determine what to advise the public. After this decision is issued, communication is maintained with the affected areas to verify predictions and receive updates. FFMP data as well as local stream gauge readings (if available) are reassessed throughout the course of the rainfall, and the public is relayed updated information.

Along with FFMP, WFOs also employ the Site-Specific Hydrologic Predictor Model (SSHPM) to forecast flash flooding. The model focuses on small streams and creeks that exhibit the rapid response time characteristic of flash flood-prone environments. At its core, the software incorporates a calibrated hydrologic model for a particular drainage basin. The program accepts inputs of rainfall totals, gauge data, as well as meteorologist estimates of future precipitation. The program outputs a hydrograph and the data are analyzed by the WFO, and the appropriate information is then issued to the public.

WFO statements fall into one of three categories: flash flood watches, flash flood warnings, and urban and small stream flood advisories. Figure 21.6 shows the deterministic procedure of the statements. Hours before a storm, the WFO issues a flash flood watch, indicating that developing conditions are favorable for producing flooding. The watch is upgraded to a warning once models indicate flash flooding is imminent, or it is reported. During the event, the watches and warnings are supplemented with flash flood statements that give specific details about flood locations, flood magnitudes, radar, and predicted rainfall totals. When flooding presents an inconvenience or is not dangerous to life or property, the WFO issues an urban and small stream flood advisory.

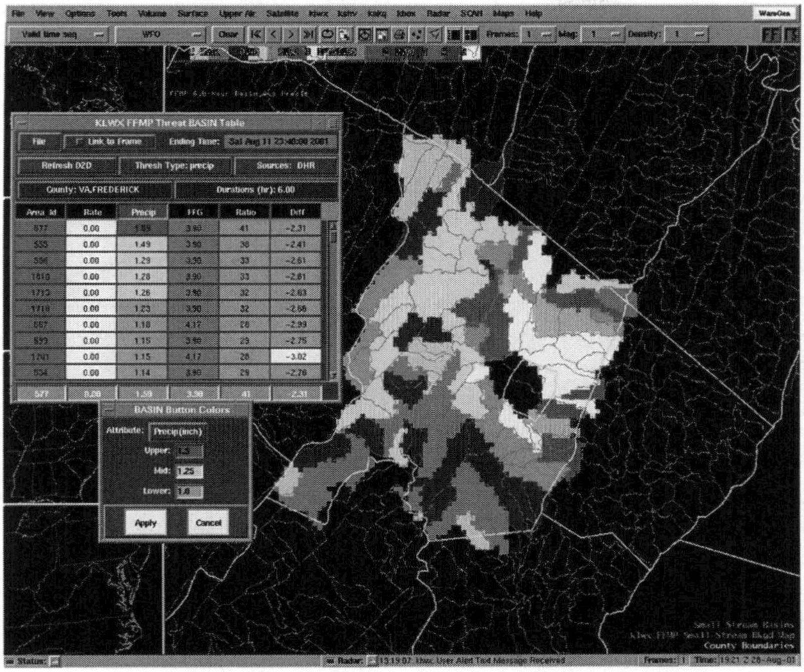

FIGURE 21.4 FFMP interface. (From National Weather Service [NWS], National weather service advanced hydrologic prediction service, http://www.nws.noaa.gov/oh/ahps/presentations/ahps_FY2005/AHPS%20FY05%20Fact%20Sheet.pdf, 2012, accessed July 11, 2012.)

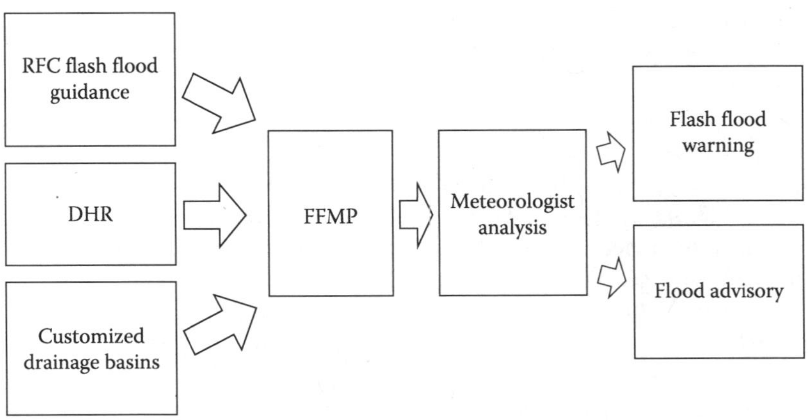

FIGURE 21.5 Data flowchart for the FFMP program.

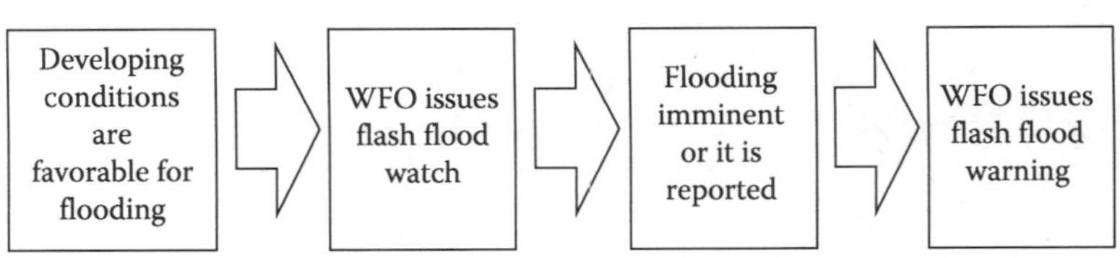

FIGURE 21.6 Deterministic procedure of the WFOs statements: flash flood watches, flash flood warnings, and urban/small stream flood advisories.

21.7 Customized Flood Alert Systems

Although large agencies, such as the NWS, are able to provide accurate and timely flood warnings at county levels for most areas in the nation, they have difficulty in providing precise warning information for small urban catchments because the rapid runoff response time is often too small of an interval to effectively run complex computer simulations. For especially time-sensitive, flood-prone targets within these urban watersheds, a customized approach to FASs is more effective and accurate. A radar-based FAS created by Rice University and the TMC exemplifies this growing trend.

The FAS was originally created in 1997 to gather radar rain data, to determine rainfall volume over a highly developed watershed—Brays Bayou in southeast of Houston, Texas—and to predict peak flows based on hydrologic nomograph. Between 1998 and 2004, the FAS successfully proved itself with providing accurate predictions to the 22 institutions in the TMC during many storm events including Tropical storm (TS) Frances (1998) and TS Allison (2001). Information generated by the original FAS allowed for the TMC to not only deal with imminent flood threats but also develop plans for future emergency responses and implement these procedures.

The second generation of the FAS (FAS2) was developed in 2004 with many improvements including 5 min NEXRAD radar rainfall data (Level II) as well as two real-time hydrologic models: a lumped model (RTHEC-1) and a distributed model (Vflo). After inputting radar rainfall data in to the models, FAS2 was able to provide TMC emergency personnel with more accurate flow estimates and flood forecasts within 2–3 h of lead time. The real-time hydrologic models used in FAS2 were able to take advantage of radar rainfall data to increase the lead time and accuracy of the flow predictions. Figure 21.7 shows the dashboard of the current FAS (FAS3) consisting of real-time rainfall/floodplain mapping module, hydrologic prediction module, NEXRAD radar viewer, and bayou camera viewer. The improvement of accuracy was found over the original FAS through increased radar resolution, rain gauge calibration, and the inclusion of the real-time models. In addition, notification capacity was increased from the internet to include cellular phones, pagers, and e-mail.

It is found over the last decade that the FAS2 system can generally forecast floods with average differences of 0.87 h in peak timing and 2.8% in peak flows. Also, the FAS2 system inclines as desired to better predict shapes, peak values, and peak timing for severe storms than for small and/or moderate events. The strength of the system was exemplified during Hurricane Ike (2008), when FAS2 predicted a peak flow value of 26,811 ft^3/s, only 5% larger than USGS observed value of 25,500 ft^3/s. Hurricane Ike made landfall as Category 2 hurricane at the island city of Galveston, Texas, at about 2:10 a.m. on September 13, 2008, generating an extremely moist atmosphere with more than 8 in. of rainfall in 24 h. Hurricane Ike caused more than \$30 billion damage and about 100 deaths as the third costliest Atlantic hurricane of all time. With its proven performance in the past, the FAS was upgraded to the third generation (FAS3) in 2010. Table 21.2 shows the overall performance of the FAS during the last 14 major events with the differences of 35 min and 4.21% in peak timing and peak flow, respectively. The FAS3 system has been found to better predict shapes, peak values, and the peak timings for bigger storms events than those for small and/or moderate events. The major improvements include updates to hydrologic models as well as the introduction of a new hydraulic prediction tool, the Floodplain Map Library (FPML), which will be described in detail in a later section. Overall, accurate and reliable performance has proven FAS essential to TMC and Rice University officials, and the similar technology and methodology has been applied to expand FAS to other flood-prone areas in Texas.

21.8 European Flood Awareness System

After the devastating, multinational floods of the Rhine–Meuse in 1993 and 1995 as well as the 2002 floods of the Elbe and Danube, the European Commission recognized the need for increased cooperation between member states in flood forecasting and warning. In addition, the Commission also realized the necessity of standardizing the quality and accuracy of flood warnings across the many national

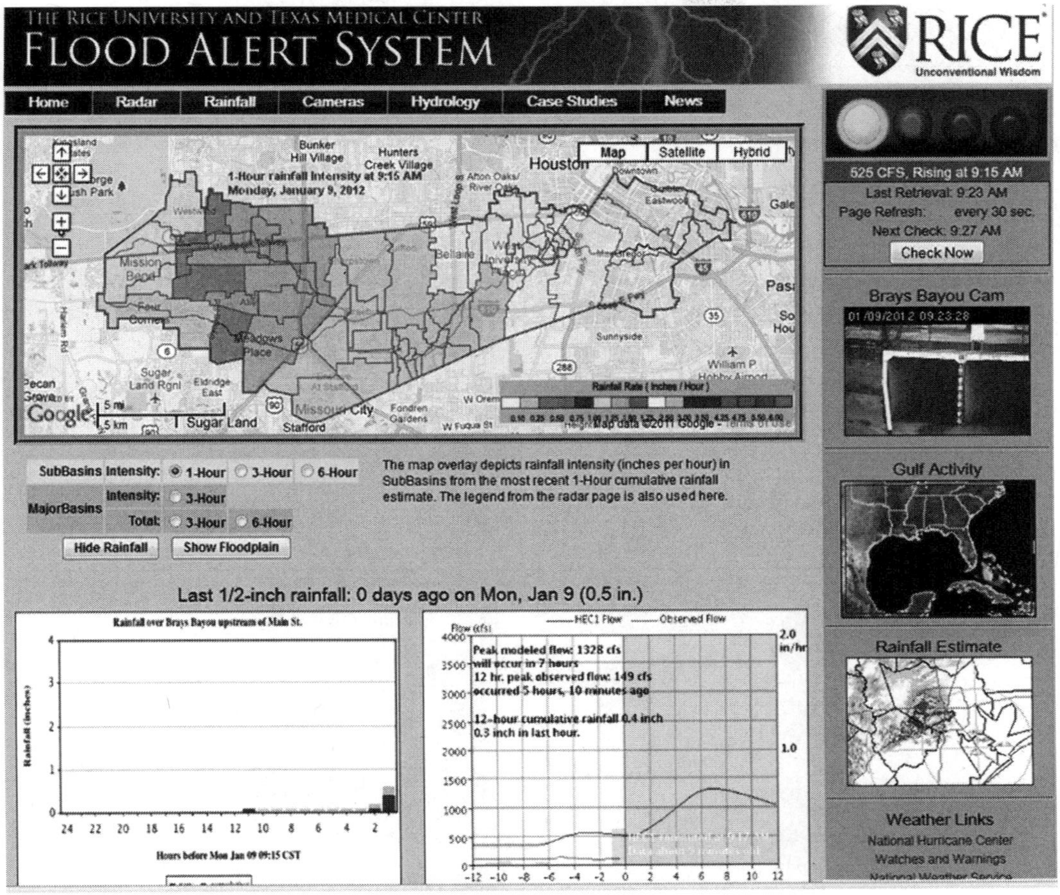

FIGURE 21.7 The dashboard of the Rice University/TMC's FAS (FAS3).

TABLE 21.2 Summary of the Overall Performance for the Rice University and TMC's FAS

Storm Event	Observed Peak Flow (cfs)	Predicted Peak Flow (cfs)	Time Difference (min)	Peak Flow Difference (%)
8/15/2002	17,900	18,097	+50	1.10
9/12/2003	10,445	11,949	+120	14.40
10/9/2003	16,000	13,504	+85	−15.60
11/17/2003	25,500	24,990	+30	−2
4/9/2006	7,990	8,123	+68	1.67
7/26/2006	10,400	11,524	−45	10.81
10/16/2006	17,700	16,088	+65	−9.11
8/21/2008	10,138	12,064	+50	19.00
9/13/2008	25,500	26,775	+45	5.00
4/28/2009	19,442	22,553	+35	16.00
7/2/2010	18,800	17,691	+60	−5.90
1/9/2012	25,300	27,317	−5	7.40
1/25/2012	7,794	9,930	+45	21.51
2/4/2012	12,000	11,397	−85	−5.29
Average			+37	4.21

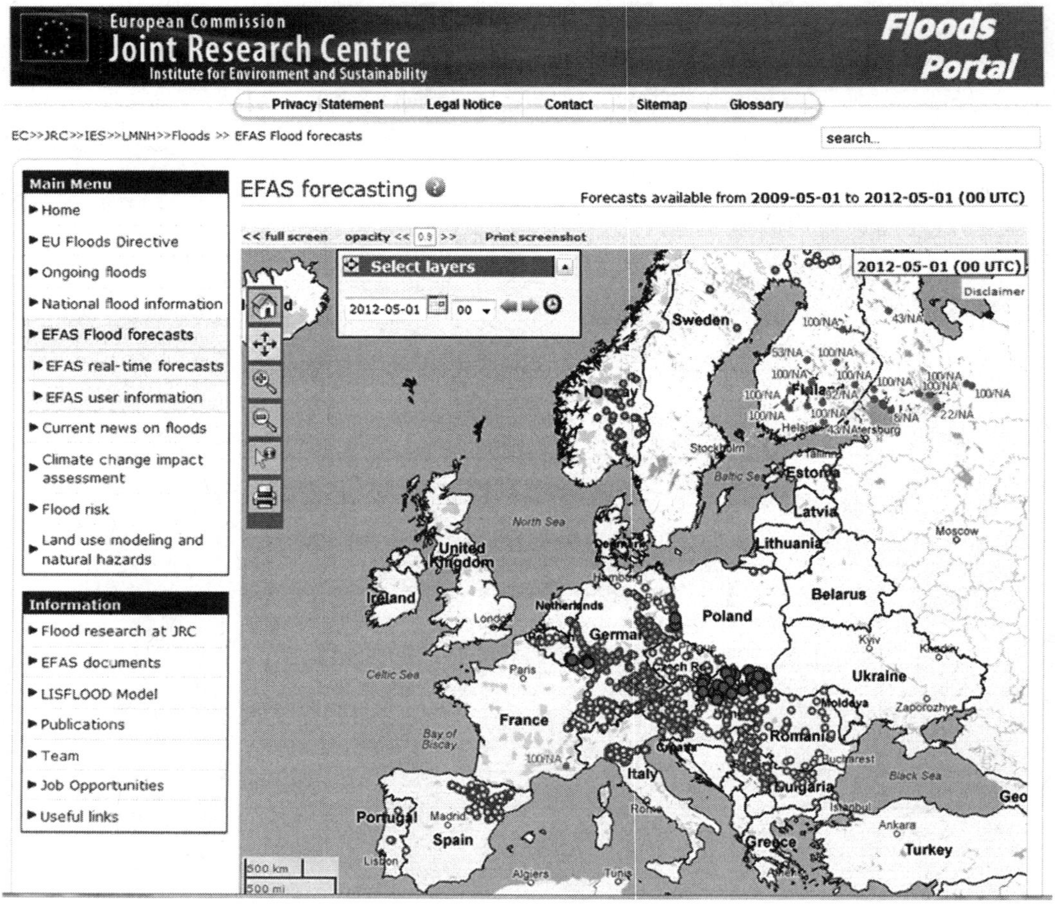

FIGURE 21.8 Portal interface of the EFAS.

flood agencies within the European Union. To achieve these goals, in 2003, the European Commission's Joint Research Center's (JRC) Institute for Environment and Sustainability (IES) began the development of the EFAS, designing it to provide accurate, medium-range forecasts of flood conditions. The forecast range was achieved via augmenting existing deterministic forecasts data with ensemble probabilistic predictions. Ensemble forecasting was included; the process is much more accurate over long periods than deterministic methods, functioning on a 3–10 day timescale versus only 2–3 days [18]. Figure 21.8 shows the portal interface of the EFAS.

Figure 21.9 demonstrates the data flow process behind the EFAS. Generating flood warnings with the EFAS begins with receiving deterministic and ensemble forecasts generated by the European Center for Medium-Range Weather Forecasts and Deutsche Wetterdienst. The forecast data are then input into the EFAS's hydrologic model or LISFLOOD. LISFLOOD is a GIS-based, distributed hydrologic runoff and routing model designed specifically for large, international river basins. After running the model with the forecast data, LISFLOOD-simulated discharges are compared to EFAS created critical thresholds. If simulations using ensemble and deterministic forecast values persistently generate flow values above these thresholds, the EFAS notifies its network of 29 national and regional flood agencies, which determine how to prepare and advise the public.

Since its creation, EFAS alerts have proven to be roughly 70%–80% correct in their magnitudes, timings, and locations [8]. While this value does not approach the accuracy afforded by customized

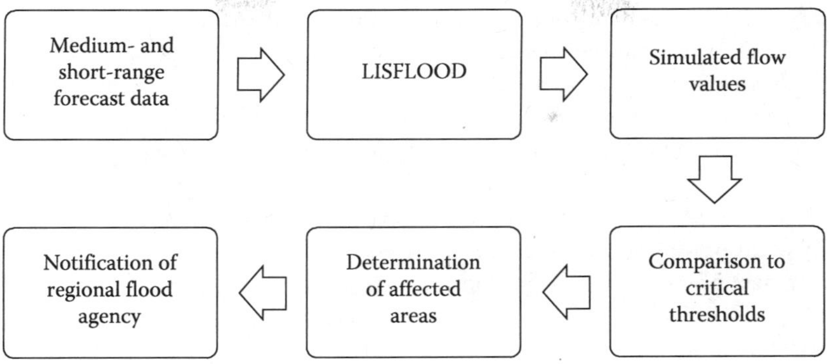

FIGURE 21.9 Data flowchart and operation process of the EFAS.

systems like Rice University and TMC's FAS, for the size of the coverage area as well as its excellent lead time, the EFAS's performance has proven to be quite impressive. The EFAS is designed as an early warning system, and its primary function is to prompt regional and national flood agencies to monitor their own forecast systems and develop their own predictions. The 3–10 days of warning offered by the system are extremely helpful before large-scale flood events, where coordinating and communicating with regional and local authorities and taking major preventative actions can be time consuming. While currently the EFAS represents the only fully operational system capable of issuing accurate medium-range flood forecasts, future years will probably see increased application of ensemble prediction methods in other FASs.

21.9 Improving Flood Warning: Inundation Mapping

With increasingly popular mapping visualizers like Google Maps available online, the future of flood prediction and warning lies in disseminating flood information in a visual user interface that can be easily interpreted and understood by the public. Improving public comprehension of flood warning is seen as especially important to flood forecasting agencies because even with accurate and timely warning, public response is often the limiting factor in reaction to a flood emergency.

The Advanced Hydrologic Prediction Service (AHPS) constitutes the NWS's action toward improving its publically accessible data. After the devastating flooding in 1993, the project originated in the Des Moines basin, growing to be currently available at 2500 sites across the United Sates [15]. The project aims to provide users with a geographical user interface (GUI) depicting real-time observations of river stage and flow volumes as well as predictions of how high the water will rise, when the river will reach its peak, where property will be flooded, and how long flooding with continue. From the NWS website, all of this information can be accessed by clicking on gauge locations on a Google Maps-based interface. In terms of describing where inundation will occur, for most locations, the AHPS provides written descriptions of what areas will experience flooding at a given river stage. However, in some especially flood-prone parts of Texas, the Appalachians, Alabama, Pennsylvania, and Ohio, the AHPS has been upgraded to include inundation mapping. The maps present users with a geographic delineation of where flooding will occur at the near, minor, moderate, and major flood stages as well as real-time indication of where flood conditions are located for observed and forecasted stages. In the future, the NWS hopes to reduce the costs associated with developing inundation map libraries, so the practice will be adopted at more locations [10].

On a smaller scale, a similar hydraulic prediction tool has been developed as part of the improvements of the Rice University and TMC's FASs. Known as the FPML, the hydraulic prediction module has now been implemented into FAS3 in 2010; it is a system that visualizes flood inundations in real time to aid emergency personnel to better understand watershed response under severe weather conditions [4,5]. The FPML system consists of the inundation maps that were predelineated based on various rainfall totals incorporating frequencies, durations, and spatial variations. By associating the observed rainfall pattern with a corresponding floodplain map, the system is able to automatically display inundation maps for events between the no flood and 100-year categories. The information is analyzed by emergency officials and made publically accessible from the front page of the FAS website. This presentation format has assisted the TMC and the Texas Department of Transportation in visualizing a flood's impact on infrastructure, assisting emergency officials in adapting their emergency responses on an event-by-event basis. Regardless of size, visual delineation of inundation levels represents a significant improvement over existing notification systems. This mapping feature is useful to many critical facilities such as the TMC because it enables them to understand conditions and initiate appropriate evacuation strategies at many levels to deal with emerging issues. In the future, increased prevalence of readily accessible flood warning information will hopefully lead to improved public response during future floods, reducing loss of life and property.

21.10 Summary and Conclusions

Of any weather-related event, flooding presents the greatest danger to infrastructure, private property, and public safety. Flood warning systems, as one type of nonstructural flood protection, have been widely adopted in some flood-prone areas because they initiate preventive measures to reduce the risk of damage and loss of life in the case of an impending flood. Over the past decades, modern FASs have evolved to meet the specific needs of different clients, regions, or even nations, differing greatly in complexity, size, and scope. FASs range from basic river gauge monitoring networks to complex forecasters, employing radar rainfall data and computer modeling techniques to produce lead times ranging from hours to days. Along with gauge networks, this chapter discussed in detail the flash flood forecasting provided by the NWS, Rice University and the TMC's FAS, and the EFAS. In addition, an ongoing update to many of these systems—real-time inundation mapping—was also discussed as well as its role in improving the public's reaction and comprehension of flood warnings. With introducing all different flood warning systems, this chapter also established a background understanding of flooding and flood control, radar rainfall measurement, and hydrologic computer modeling.

Infrastructure alone cannot solve urban flooding problems because it is impossible for design to accommodate any event. When flooding inevitably occurs, the best option to mitigate damages to life and property is to obtain accurate and timely warning from emergency officials. Although many existing FASs vary in size, complexity, and forecasting ability, they are all designed to provide flood warnings as great lifesaving tools. With recent advances in technology, forecast accuracy, and lead time, flood agencies are still striving to improve how they communicate threats to the public who often form the weak link in flood mitigation. The authors believe that the concerning state of current flood damages could become a thing of the past with continued investment in the FASs with better sustainable urban development practices and global adoption.

References

1. Barredo, J.I. 2009. Normalized flood losses in Europe: 1970–2006. *Natural Hazards and Earth System Sciences*, 9: 97–104.
2. Bedient, P.B., W.C. Wayne, and B.E. Vieux. 2013. *Hydrology and Floodplain Analysis*. Upper Saddle River, NJ: Pearson Education, Inc.
3. Critchley, W., K. Siegert, and C. Chapman. 1991. *A Manual for the Design and Construction of Water Harvesting Schemes for Plant Production*. New York: Natural Resources Management and Environment Department, The United Nations.
4. Fang, Z., P.B. Bedient, J.A. Benavidas, and A.L. Zimmer. 2008. Enhanced radar-based flood alert system and floodplain map library. *Journal of Hydrologic Engineering*, 13(10): 926–938.

5. Fang, Z., P.B. Bedient, and B. Buzcu-Guven. 2011. Long-term performance of a flood alert system and upgrade to FAS3: A Houston Texas case study. *Journal of Hydrologic Engineering*, 16(10): 818–828.
6. Harris County Flood Control District (HCFCD). 2012. About flood warning system. http://www.harriscountyfws.org/About (accessed July 1, 2012).
7. Hoyois, P. and D. Guha-Sapir. 2003. Three decades of floods in Europe: A preliminary analysis of EMDAT data. http://www.emdat.be/publication/three-decades-floods-europe-preliminary-analysis-emdat-data (accessed July 18, 2012).
8. Joint Research Centre (JRC). 2012. European flood awareness system. http://ies.jrc.ec.europa.eu/european-flood-alert-system (accessed July 12, 2012).
9. Lindner, J., D.C. Schwertz, P.B. Bedient, and N. Fang. 2012. Flood prediction and flood warning systems. In *Lessons from Hurricane Ike*, ed. P.B. Bedient, pp. 38–49. College Station, TX: Texas A&M University Press.
10. National Weather Service (NWS). 2008. NWS flood inundation mapping services. Presented B. Weiger. http://www.srh.noaa.gov/media/lch/outreach/052808/6BenWeiger.pdf (accessed July 11, 2012).
11. National Weather Service (NWS). 2012. Digital hybrid scan reflectivity (DHR). http://www.nws.noaa.gov/oh/hrl/wsr88d_prods/docs_bld05/DHR.pdf (July 2, 2012).
12. National Weather Service (NWS). 2012. Advanced Hydrologic Prediction Services (AHPS). http://water.weather.gov/ahps/about/about.php (accessed June 5, 2012).
13. National Weather Service (NWS). 2012. Radar Operations Center (ROC). http://www.roc.noaa.gov/WSR88D/About.aspx (accessed July 12, 2012).
14. National Weather Service (NWS). 2012. Radar Operations Center (ROC). http://www.roc.noaa.gov/WSR88D/Images/WSR-88DCONUSCoverage2011.jpg (accessed July 1, 2012).
15. National Weather Service (NWS). 2012. National weather service advanced hydrologic prediction service. http://www.nws.noaa.gov/oh/ahps/presentations/ahps_FY2005/AHPS%20FY05%20Fact%20Sheet.pdf (accessed July 11, 2012).
16. National Weather Service (NWS). 2012. Flash flood monitoring and prediction (FFMP). http://www.nws.noaa.gov/mdl/ffmp/images/ffmpUG20fig1.gif (accessed July 11, 2012).
17. European Meteorological Network (EUMETNET). 2013. Operational Programme for the Exchange of weather Radar information (OPERA). 2013. http://www.eumetnet.eu/opera (accessed July 16, 2013).
18. Thielen, J., J. Bartholmes, M.H. Ramos, and A. de Roo. 2009. The European flood awareness system. *Hydrology and Earth System Sciences*, 13: 125–140.
19. United States Environmental Protection Agency (USEPA). 2012. Low impact development. http://water.epa.gov/polwaste/green/index.cfm (accessed July 20, 2012).
20. U.S. Geological Survey (USGS). 2012. Water watch. http://waterwatch.usgs.gov/index.php?id=ww_about (accessed July 15, 2012).
21. U.S. Geological Survey (USGS). 2012. Water alert. http://water.usgs.gov/wateralert/ (accessed July 15, 2012).
22. Vieux, B.E. 2004. *Distributed Hydrologic Modeling Using GIS*. Dordrecht, the Netherlands: Kluwer Academic Publisher.
23. XP Software. 2012. XPSWMM technical description. http://www.xpsoftware.com/wp-content/uploads/xpswmm-techdesc.pdf (accessed July 8, 2012).

22

Optimum Hydrometric Site Selection

Jonathan Peter Cox
Caribbean Institute for Meteorology and Hydrology

Sara Shaeri Karimi
Dezab Consulting Engineers Company

Saeid Eslamian
Isfahan University of Technology

AUTHORS

Jonathan Peter Cox received his PhD from the Department of Civil Engineering and Construction, University of Salford, Manchester, United Kingdom. He later took on a postdoctorate research position under the Mediterranean Desertification and Land Use (MEDALUS) Program of the European Community at the University of Murcia, Spain. He was appointed hydrologist for the Automatic Hydrological Information System (SAIH) at the Confederacion Hidrografica del Segura, Murcia, in 1995, a post that he occupied for 15 years before transferring to the Comisaria of Aguas, Murcia, as head of hydrometry and statistics. In parallel, Jonathan has developed a university lecturing career and since 2001 has been a member of the Department of Civil Engineering at the Catholic University of Murcia, Spain. In 2011, he initiated collaboration with the German hydrometry masters OTT, as a consultant for their Iberian office. Jonathan has recently taken up a position as senior lecturer in hydrology at the Caribbean Institute for Meteorology and Hydrology.

Sara Shaeri Karimi graduated in water engineering from Shahid Chamran University of Ahvaz, Iran. She received her master's degree in hydraulic structures from Urmia University, Urmia, Iran. Currently, she is working as a hydraulic engineer for the Irrigation and Drainage Division at Dezab Consulting Engineers Company, Ahvaz, Iran. Sara has collaborated in numerous national and international publications and has keen research interests in environmental flow, sustainable water resources management, and climatic change. In the future, she plans to complement her professional career as an engineer with a doctorate in water resources.

Saeid Eslamian received his PhD from the University of New South Wales, Australia, with Professor David Pilgrim. He was a visiting professor in Princeton University, United States, and ETH Zurich, Switzerland. He is currently an associate professor of hydrology in Isfahan University of Technology.

He is founder and chief editor of the *Journal of Flood Engineering* and the *International Journal of Hydrology Science and Technology*. He has published more than 200 publications, mainly in statistical and environmental hydrology and hydrometeorology.

PREFACE

Data without exactitude are of little use in modern hydrometry. Improvements in sensor precision and reliability have brought consequential real-time hydrometric fiscality within the reach of all water resources managers. Nevertheless, optimum measuring structures, sensor commissioning, and maintenance are all essential to consolidate this goal. This chapter details the development of hydrometric perfection from a practical perspective based on real experience of the authors.

22.1 Introduction

Hydrometry is an important branch of hydrological engineering dating back to more than 4000 years to installations in the river Nile, Egypt; nonetheless, in many countries this science has often taken second place in hydraulic works. All too frequently one finds that the budget for hydrometric installations is left for the remnant of a project, and if any preinstallation for sensors and scales has been foreseen, it is usually unsatisfactory. In addition, all systems and structures require postwork maintenance. This is valid for hydrometric elements too. A hydrometric maintenance contract is the only way to ensure that the objective of the civil work is maintained and valid discharge data are measured. In the case of water quality sensors, a more exhaustive maintenance schedule must be undertaken to ensure reliable data for the life of the station. First-class water management requires accurate water measurement [14]; second-class water management is not acceptable.

Hydrometry is directly concerned with the monitoring of the components of the hydrological cycle, including precipitation, both solid and liquid, groundwater level and its respective chemical constitution, surface water discharge, storage, and quality characteristics.

Hydrometry is all around us, from the smallest of creeks or brooks to the largest reservoir, from the Dasht-e Lut salt desert in southeastern Iran with negligible precipitation to Mawsynram, northeastern India, reportedly the wettest place on Earth, with an annual rainfall of 11,872 mm.

Hydrometry initially was implemented as a method to tax the population in function of discharge, thus in a dissuasive manner, facilitated management of available basin resources [1]. All the same, in spite of its early beginnings, growth of the science was comparatively sluggish when scrutinized against other areas of hydrology. Significant developments were scarce until the 1830s, when a landmark was passed with the first automatic hydrometric measuring system installed on the river Thames at Sheerness in the United Kingdom. From this moment forward, developments have been substantial in the field. Above all over the last 30 years, new and emerging developments in flow instrumentation have significantly improved our capabilities to measure surface water discharge and flow dynamics of rivers [8]. Coupled with this, data storage and transmission has also improved dramatically moving from paper tape-punching records through magnetic media and more recently electronic memory storage coupled with direct tele-transmission of data in real time or near real time via radio, GSM, GPRS, WiMAX, TETRA, and satellite networks. With improved measuring techniques and transmission protocol, managers are demanding data that are up to date and can be trusted.

22.2 Necessity to Measure

The data recorded in the bulk of the gaging stations consist in one or two measurements of water level, be it surface or subterranean, a register for precipitation, and, in the case of reservoirs, water column on the upper face of the dam that together with gate opening measurements permits the precise calculation of discharge data from the outflow structures. Additional data such as evaporation, temperature, and wind speed are commonplace, enabling a complete theoretical calculation of reservoir water balance and basin management.

Hydrometry has become more important in recent years for a number of clear motives, principally due to the greater demand on the static volume of the resource by an ever-escalating world population. Combining this with the possible outcome of climate change, the mounting occurrence, and detection of extreme events, such as drought and floods, the consequence is a much higher interest in the science. Furthermore, as a result of the implementation of the Water Framework Directive, international legislation in countless regions necessitates a strict control and quantification of water resources, to the extent that even minor abstractions must be considered and recorded at the points of abstraction and return to the public domain. To this end, in Spain the Ministerial Order 1312/2009, of May 20, 2009, was published to regulate abstractions from the public domain and subsequent returns [7]. The deed holder for the abstraction is now obliged to install and maintain a hydrometric system with capacity to store data and calculate volumes abstracted over variable time intervals. The expenses of installation and operation of this system are borne by the deed holder.

22.2.1 Economical Aspects

Although discharge is continuous, when water is transported from point *A* to point *B* via infrastructure, such as a pipeline or a canal, there are expenses incurred. For this reason, and as a method of wastage dissuasion, water users are levied a charge or tax. In many regions, governments may subsidize this charge to enable competitive prices for the end users' final product. This illustrates the necessity and importance of precise discharge and volume measurement, thereby accounting for individual water use combined with pricing policies that penalize excessive use [14].

22.2.2 Ecological and Environmental Flows

To ensure that a particular reach of a watercourse maintains the natural or ecological discharge conditions, where the discharge reproduces the hydrograph for the reach prior to anthropogenic activities, reaches must be monitored by hydrometric systems at the initiation and termination of each reach. Environmental flow additionally describes the quantity, the timing, and the quality required to sustain freshwater and estuarine ecosystems. Through the implementation of ecological and environmental flows, water managers strive to achieve a discharge that provides for anthropogenic activities, such as water supply, energy, recreation, or flood control, and maintains the essential equilibrium to support a healthy river ecosystem. Once more, this requires data in real, or near real, time.

22.2.3 Political View

International, national, and regional borders are extremely important sites to measure discharge particularly when water scarcity is a problem. This can be unquestionably illustrated through an analysis of interbasin transfers. There are abundant large interbasin transfers around the world, most of them concentrated in Australia, Canada, China, India, and the United States. The oldest interbasin transfers date back to the late nineteenth century, with an exceptionally old example being the Roman gold mine at *Las Médulas* [4] in Spain. Their primary purpose usually is either to alleviate water scarcity or to generate hydropower. In Spain, one of the most important and politically sensitive operational interbasin transfers connects the Tagus basin in the center of the Iberian Peninsula with the Segura basin in the southeast. This structure was constructed in the 1970s to irrigate the arid Segura basin and complement potable water supply. The first transfer was received in the Segura basin on March 31, 1979 [9]. The volume to be transferred is calculated quarterly and is variable depending on the resources available in the donor Tagus basin. For the period from July to September in 2012, a volume of 114 million cubic metres (MCM) has been agreed by the ministry concerned. In case of programmed civil works in the canal, the volume corresponding to the days of disruption must be previously transferred. This visibly illustrates the need to be able to accurately measure discharge at strategic points on the infrastructure, ensuring the rapid detection of leaks and illegal abstractions [14] and more importantly being able to certify the volume received in the receptor basin [3]. To this end, sensored hydraulic structures were incorporated into the design of the canal. At the entrance to the Segura basin, there is a Parshall flume with a width of 7.62 m. The flume has a capacity to measure up to 26 m^3/s. This is complimented with two thin plate weirs coupled downstream of a butterfly valve to precisely gage low flows and regulate the discharge from the main canal. The Parshall flume is sensored with noninvasive radar and a pneumatic sensor situated in a stilling well. In addition the canal has a vertical staff gage mounted at one-third the length of the converging section of the Parshall flume. The weirs are sensored with radars calibrated with staff gages referenced with the crest of the weir plate. Measurements of the water column are taken every 5 min and via hydraulic formula the associated discharge calculated. To ensure reliability of the discharge calculations, hydrometrists carry out field verification at least once a month. Any discrepancies are investigated and when deemed necessary notified to the corresponding governing body. The error associated with the aforementioned structures is estimated at $\pm 3\%$ of the discharge for the full range. Parshall flumes were frequently employed in the Segura basin; nevertheless, the advantage of the theoretical predictability in hydraulic performance of long-throat flumes, ease of and cost of construction, and posterior verification favors the use of the latter over Parshall flumes for new stations in canals [2]. Under hydrologically favorable conditions in the donor basin, the receptor Segura basin can legally be sent up to 600 MCM per hydrological year [6]; nevertheless, this has only come to pass over one hydrological year in the 30 of operation. In drought years, insignificant volumes or only canal runoff is transferred, illustrating the precarious state of the receptor basin.

22.2.4 Network Density

Network density may vary in function of political, physical, and ecological parameters. There is a reference from the World Meteorological Organization (WMO) [15] based on topographical factors, yet political and ecological factors can significantly alter these guidelines. From a topographical perspective, it is convenient to have a gaging station downstream of any significant confluence. One could qualify the significance as more than 5% of the discharge upstream of the confluence. In addition, it is recommended to install a gaging station upstream of any significant urban area, with a view to the mitigation of floods, and downstream to register drainage discharge from the metropolitan area. Supplementary anthropogenic factors such as irrigation zone divisions, reservoirs, are also prime points to install gaging stations. There are also several statistical techniques proposed by the author regarding network density, including Husain [5]. Evidently these methods play a role in network density optimization; however, professional knowledge and a clear understanding of the stakeholders' needs can often be more important.

22.3 Specific Characteristics of the Station

Each station requires individual analysis to determine the optimum configuration. Primarily, it is necessary to appraise if the data are to be used for real-time monitoring and control or as an offline station to complete basin statistics and assist management over longer time intervals. Either way, sensors are required with satisfactory installation and maintenance. As previously stated, advances in telecommunications cover 99% of potential stations; nevertheless, financial or operational limitations could present barriers. Satellite transmission is expensive, and in some cases essential maintenance can cause downtime for the network. Nonetheless, it is probable that satellite communications will see an increase in hydrometric networks over the next decade as more complex stations are installed and the demand for quality real-time data by water managers and stakeholders becomes more standard. In this way, it is certain that web-enabled stations will also be a major growth area in the field as will be wider use of digital multichannel dataloggers.

Accessibility is another important issue. Clearly environmental factors come into play here. Firstly, if the site is in an environmentally sensitive zone, this could be a severe restriction, especially if civil works are indispensable to be able to access the selected site and ensure a stable control section. Secondly, it is likely that access could be limited to specific periods of the year due to breeding seasons of protected species, for instance, or physical limitations imposed by seasonal flooding or land instability.

Other important physical factors include the following [15]:

- A straight stretch of river of at least 150 m must be sought for a gaging station.
- At the precise point of measurement, it is convenient to have just one channel.
- Water should not be filtering through the bedrock.
- Vegetation should be avoided to reduce deviation from the discharge curve due to the growth of flora in the control section.

The reader is also recommended to consult the following guidelines: USGS Stage Measurement at Gaging Stations: Techniques and Methods [10].

Regarding optimum sensorization, there are two divisions: quality and quantity. Firstly, we will consider quantity or discharge measurements. Traditionally, river gaging stations have measured the hydrostatic pressure or water column. The level recorded is then applied to a stage discharge curve to obtain a discharge measurement. Verification of measured levels can be carried out at the station or remotely via cameras contrasting the measurement with the staff gage reading. Redundancy of sensors is also a positive strategy to ensure quality data. In the case of classic hydraulic measuring structures in rivers, such as weirs or flumes, the level recorded is used to calculate the discharge via the appropriate equation of the structure concerned. Nevertheless, in both cases, it is imperative to have contrasted measurements on a regular basis to ensure the quality of the calculated discharge data is adequate. These measurements are carried out by a team of hydrometrists having specialized knowledge of discharge measurement procedures and applying the methodology of the World Meteorological Organisation. It is not uncommon for sensors to lose precision with time or for textbook hydraulic structures to decalibrate due to changes in the roughness parameter of the construction material or settlement of the structure over time. Deficiencies in workmanship are also commonplace; therefore, usually it's a false economy to purchase mediocre sensors. In recent years as a result of more efficient manufacturing processes and consequent reduction in price, open-channel flowmeters are starting to be installed alongside more traditional measurement systems. These sensors, when installed correctly and adequately calibrated, can reduce uncertainties associated with more traditional systems and also, in numerous cases, reduce the necessity for costly hydraulic control structures in the watercourse. Nonetheless, the new technology is by no means faultless. Exceptional care must be taken when evaluating the quality of the data, especially under extreme conditions. It is not uncommon that flowmeters stop measuring when sedimentation load increases as is the case under flood conditions or the water level falls below 30 cm as could be the case in drought or low-flow regimes. As a recommendation, it is fundamental that a qualified hydrologist is consulted before selecting hydraulic structures or sensors to ensure that the quality of the data recorded will be high enough to justify the economic inversion in network construction, installation, management, and maintenance.

Turning our focus to quality parameters, one must consider article 8 of the Water Framework Directive, related to the quality of surface waters. This article foresees the monitoring of the ecological and chemical state of water masses and of ecological potential in modified masses of water, providing a coherent and complete ecological and chemical vision of each reach under analysis in a basin, in so doing permitting international classification according to the state of the masses of water. It is important to reinforce that good water measurement and management practice prevents excess runoff and deep percolation, which can damage crops and pollute groundwater with chemicals and pesticides [14].

The control levels established by the directive for surface continental waters are the following [12]:

- To increase the knowledge of the impacts of anthropogenic origin on the masses of water in the basin
- To facilitate the future elaboration of new control programs
- To evaluate the long-term changes of the natural conditions of surface waters

The control of surveillance should measure the following:

- The representative parameters of all the biological quality indicators
- The representative parameters of all the hydromorphological quality indicators
- The representative parameters of all the physical–chemical quality general indicators

The operative control should measure the following:

- The parameters corresponding to more sensitive biological quality indicators to which the masses of water are subjected to
- All the spilled high-priority substances and the other pollutants spilled in significant amounts
- The parameters corresponding to the indicator of the hydromorphological quality more sensitive to the detected pressure

The traditional configuration of a quality monitoring station in continental waters includes the sensors of a standard hydrometric station, along with sensors to analyze chemical and physical quality indicators such as turbidity, pH, conductivity, dissolved oxygen, temperature, carbon dioxide, ammonium, phosphates, nitrates, and certain heavy metals. These data are usually required in real time to give support to the mitigation of pollution incidents [11].

22.4 GIS in Hydrometric Site Selection

There are several applications that can assist with optimal hydrometric site selection. This section will outline the Hydrologic Engineering Center's Geospatial Hydrologic Modeling System (HEC-GeoHMS) by means of a practical case.

22.4.1 HEC-GeoHMS in HSS

The HEC-GeoHMS has been developed as a geospatial hydrology toolkit for engineers and hydrologists. HEC-GeoHMS uses ArcView and the Spatial Analyst extension to develop a number of hydrologic modeling inputs for the Hydrologic Engineering Center's Hydrologic Modeling System (HEC-HMS). Analyzing digital terrain data, HEC-GeoHMS transforms the drainage paths and watershed boundaries into a hydrologic data structure that represents the drainage network. The extension allows users to visualize spatial information, document watershed characteristics, perform spatial analysis, and delineate subbasins and streams. Working with HEC-GeoHMS allows the user to create hydrologic inputs for HEC-HMS [13].

To illustrate the use of the software, an area 4 km^2, located on the Mediterranean coast in the province of Murcia, southeastern Spain, has been selected. In this case, the client required the optimum location of gaging stations to position stilling basins with a view to torrential runoff recycling. The objective of this example is to demonstrate how HEC-GeoHMS can be used as a decision support tool to calculate the morphological parameters of basins and provide information for the hydrologist or project engineer in selecting the optimum installation location for the future river gaging station.

22.4.1.1 Geographical Situation

All analysis related to the installation of a new hydrometric station should be initiated with cartography in the office. In this case the client provided a political limit for the study area. This can be appreciated in Figure 22.1.

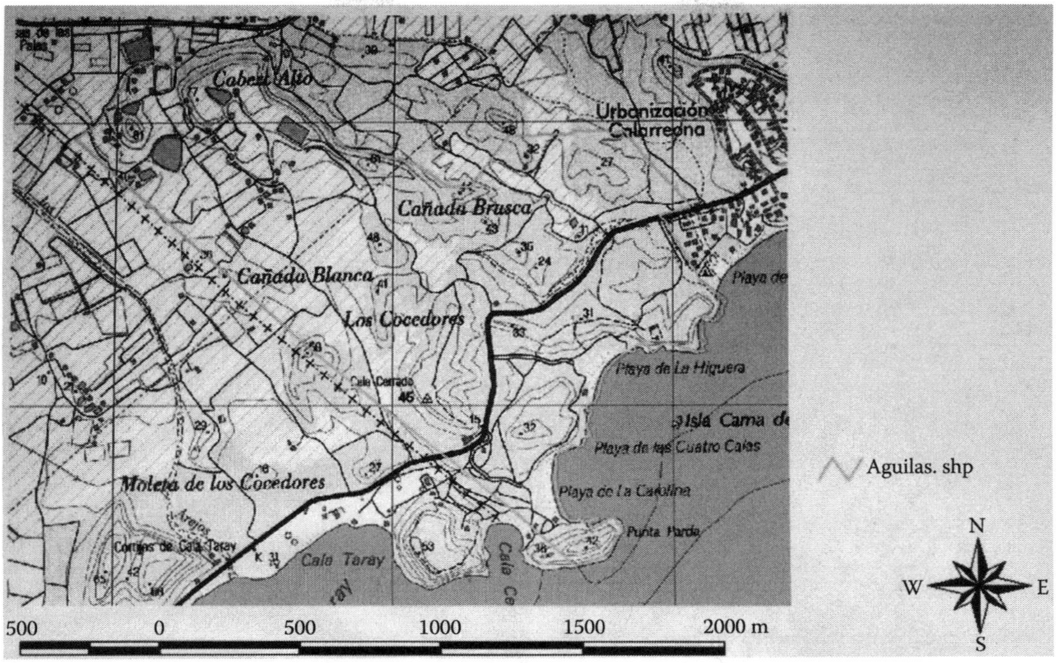

FIGURE 22.1 Location on the study zone.

Other information such as an aerial photo can also be very useful to clarify issues identified from the cartography. Today with the use of widely available programs such as Google Earth, it is possible to see, with varying degrees of resolution, the specific details of potential sites for hydrometric stations (Figure 22.2).

22.4.1.2 Drainage Network and Basin Delimitation

The physical characteristics of the basin, for example, form, relief, drainage network, soil type, vegetation, geology, and land use, are readily available from numerous sources. These parameters can influence decisively upon the location of the gaging station. HEC-GeoHMS is compatible with numerous extensions written by users and professional developers to aid the analysis of the terrain from a hydrological perspective. It is fundamental to have a good-quality digital terrain model (DTM) as a starting point. It may be necessary to correct the DTM, eliminating spurious data and ensuring that the representation of the land surface is true. The terrain can be classified by gradient, orientation, flow direction, etc., thereby enabling the calculation of the flow length in each cell of the DTM. The spatial resolution of the DTM is also important. The user must look for equilibrium between velocity and precision. Optimum spatial resolution could be around 20 m. Nevertheless, in certain areas, above all, areas with little difference in altitude, such as the outlet of the basin, higher spatial resolution data are a distinct advantage. As a guide for a basin of 20,000 km^2 and a DTM resolution of 100 m, the calculation time could be as little as 20 min on a professional PC. If however, the resolution is increased to 20 m, the calculations could easily take 4 or 5 h. Care must be taken not to generate unnecessary data files. The drainage network calculated from the DTM may not be perfect, and the result should be treated with caution and if possible verified with other sources (Figure 22.3).

22.4.1.3 Flow Direction

This is the direction where the maximum gradient is encountered from the eight possible adjoining cells. The altitude in each cell is compared to that of the adjacent cells using the DTM as the data source.

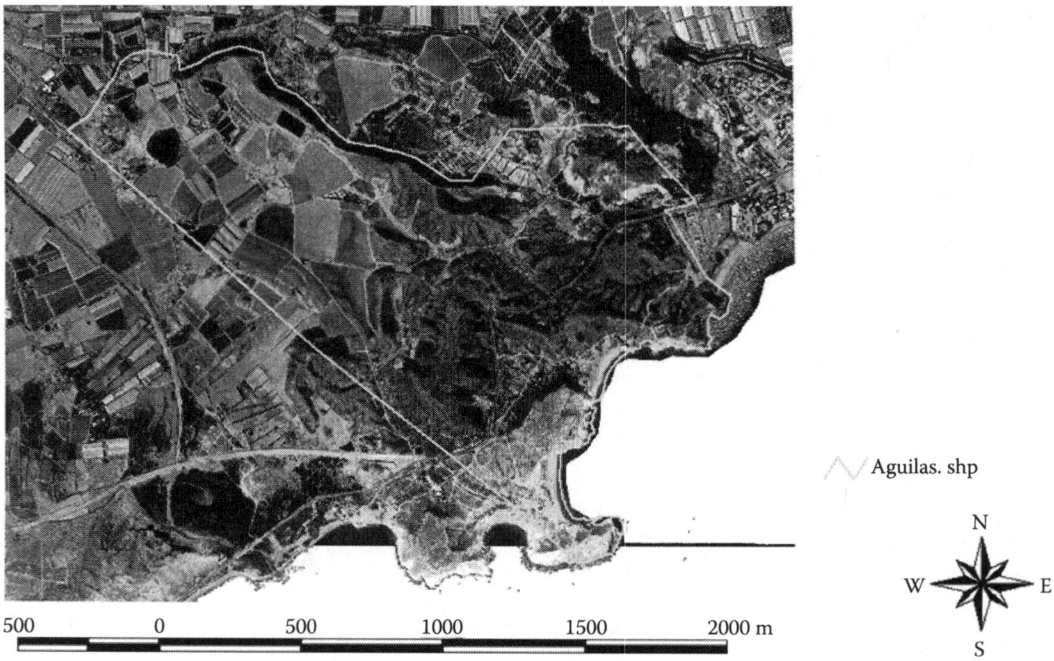

FIGURE 22.2 Aerial photo of the study zone.

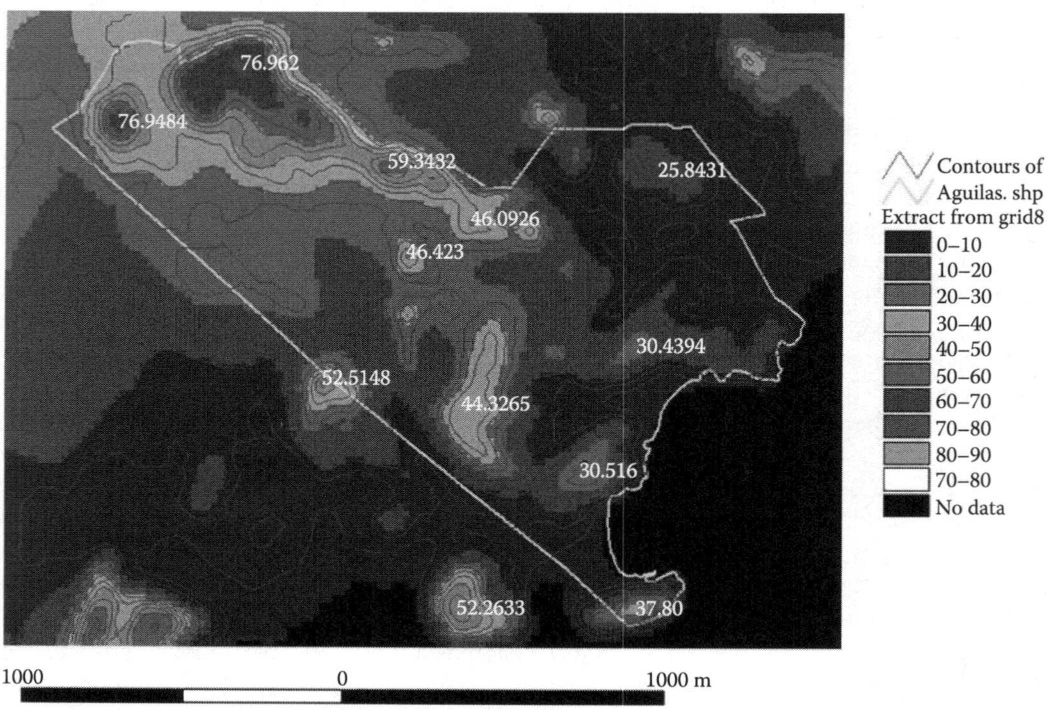

FIGURE 22.3 Altitude data for the study zone.

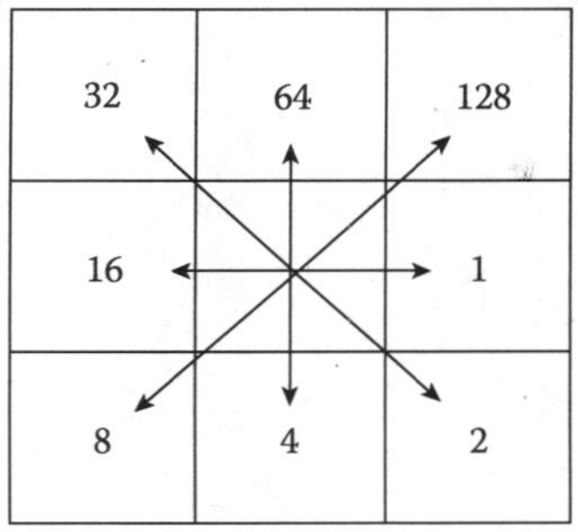

FIGURE 22.4 Flow direction coding.

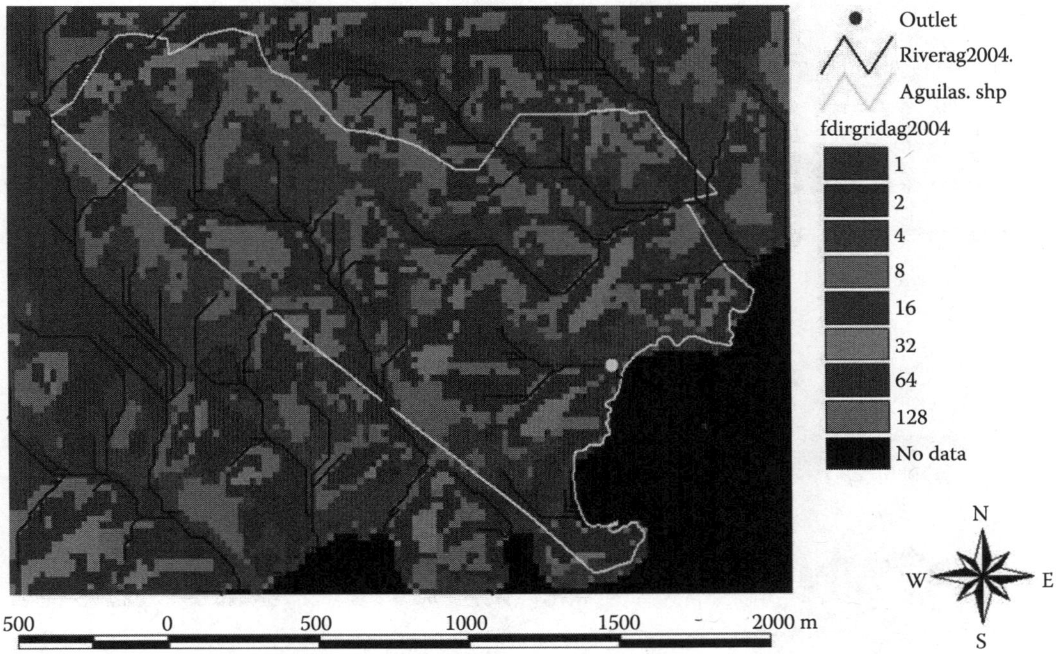

FIGURE 22.5 Flow direction in the study area.

The output is a layer with coded information depending on the flow direction. This is detailed in Figure 22.4. If the flow direction is to the east, a value of 1 will be returned; however, if the flow direction is to the north, a value of 64 will be assigned.

Once the aforementioned process has been executed on the terrain data in question, a map can be generated illustrating the orientation and flow direction of each DTM cell. See Figure 22.5.

22.4.1.4 Gradient Determination

The gradient of each cell is calculated, from the maximum of the ratio between the distinct altitude and the distance that separates the center of the cell and the 100 adjacent cells. The mean gradient of the watercourse is simply calculated by dividing difference in altitude of the source and outlet of the watercourse by the calculated flow length (Figure 22.6).

22.4.1.5 Drainage Network

Via the calculations on the flow direction layer, the drainage network is calculated always considering those cells with a predefined user-specified area threshold, for example, 40 ha.

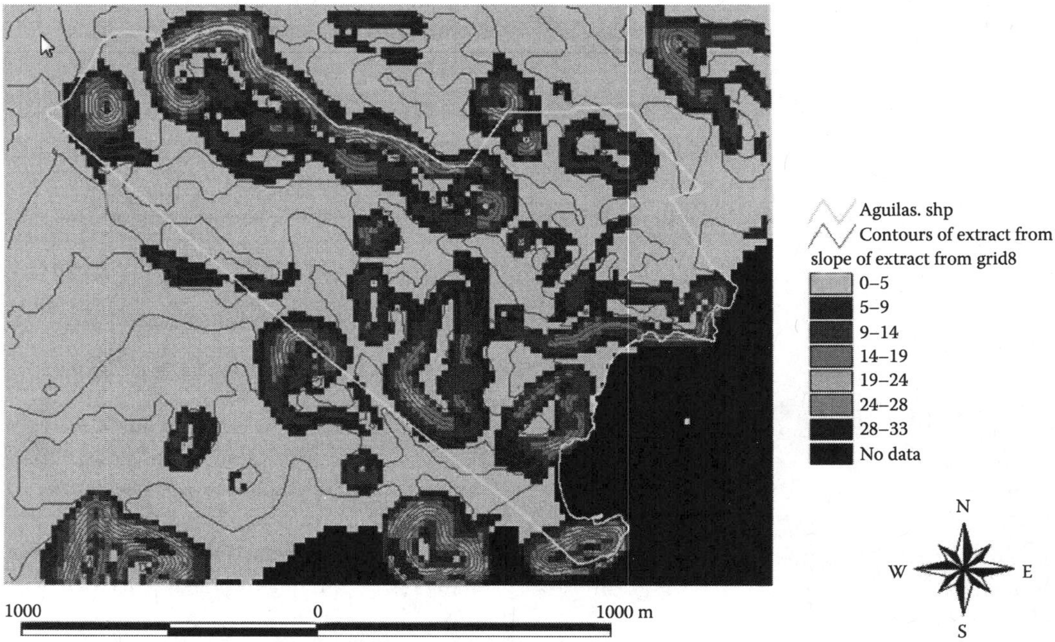

FIGURE 22.6 Calculated gradient in the study zone.

FIGURE 22.7 Identification of the drainage network.

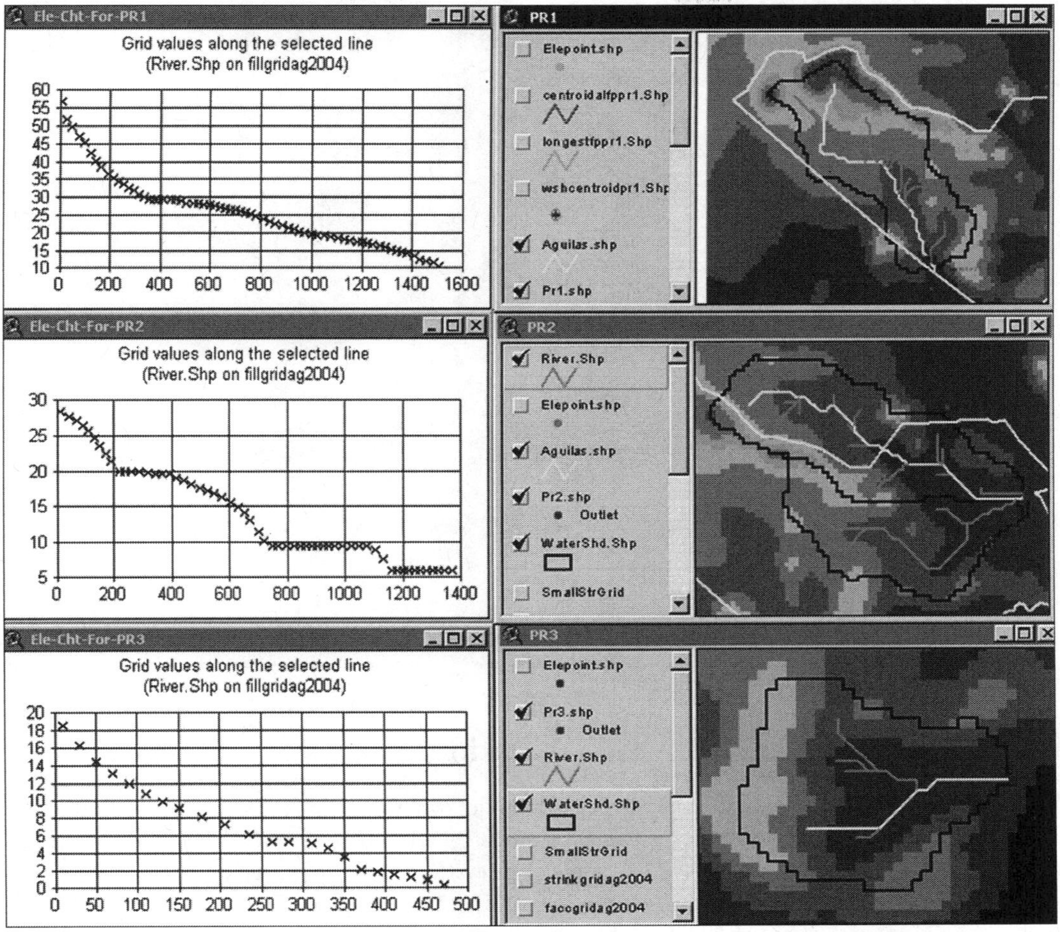

FIGURE 22.8 Principal characteristics of the identified basins.

TABLE 22.1 Parameters Associated to the Subbasins

Basin Units	Area (m^2)	Z_{Max} (m)	Z_{Min} (m)	Delta h (M)	Length of Watercourse (m)	Gradient (%)
PR1	586,800	56	10	46	1500	0.0306
PR2	712,400	28	6	22	1390	0.0158
PR3	170,000	19	1	18	475	0.0379

From the layer generated, the drainage network is presented (Figure 22.7).

Within the study area in question, three distinct basins can be identified. The hydrologist or engineer would need to decide if it is necessary to install one station on the coast, thereby measuring the total discharge of the three basins or alternatively install a central station at the outlet of the largest basin, with remote sensors linked via radio or cable to provide data from the adjacent basins. The details of each basin are shown in Figure 22.8 and presented in Table 22.1.

In some instances it may be productive to present the geographical information in 3D, thereby providing the engineer with an additional perspective (Figure 22.9).

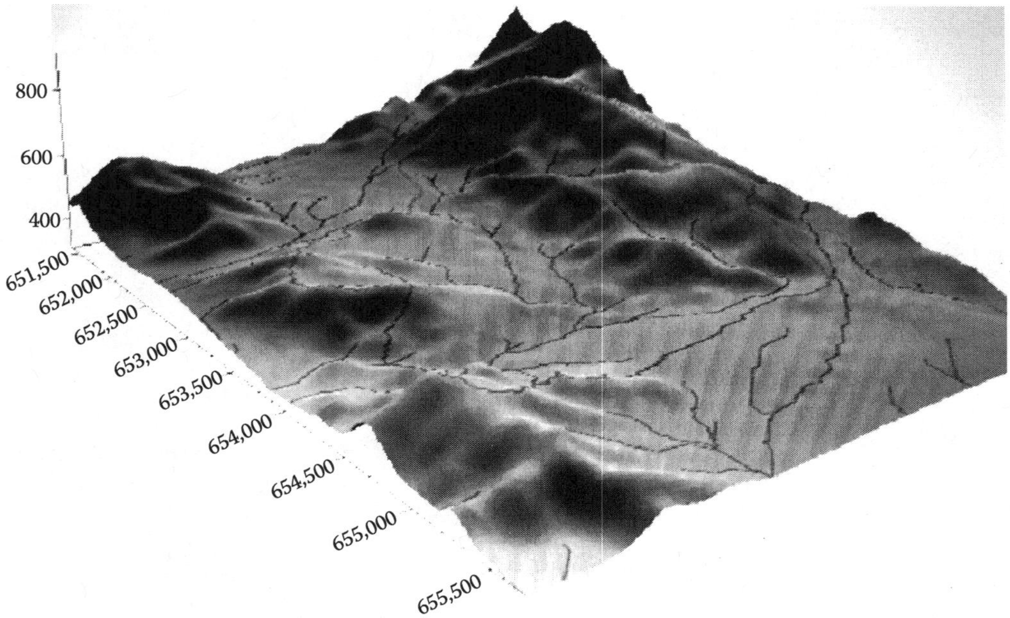

FIGURE 22.9 A 3D representation of the basin.

22.5 Station Calibration and Operation

Once a station has been constructed, even if the station follows a standard design, direct measurement using hydrographers and specialized equipment is extremely recommendable. One should try to get a full range of measurement and thereby evaluate the theoretical discharge curve for each station based on real measurements. This process may take time. As a recommendation, a minimum of 10 measurements should be carried out under distinct discharge conditions. Once the discharge curve has been calculated, it should remain static for a fixed documented period of time. Normally this would coincide with the hydrological year applicable to the basin in question. By adopting this protocol and recording each change, it is possible to reconstruct data sets using distinct criteria. This could be especially important under water scarcity conditions.

22.6 Summary and Conclusions

Hydrometry has evolved considerably over the last 30 years; nevertheless, changes should be incorporated with care and attention to detail, always permitting the restitution of historic data sets. Good water measurement practices facilitate accurate and equitable distribution of water [14]. Field hydrometric contrasts performed on a regular basis are imperative for data quality. Maintenance contracts must be foreseen when designing networks to ensure maximum return on civil work investments and guarantee the quality of data recorded.

By employing up-to-date spatial analysis tools, optimum hydrometric site location can be a relatively straightforward task. This together with the development of higher-quality, more robust sensors of greater precision, and drastically improved telecommunication methods, means that accurate real-time measurements are easily attainable in most offices. Nevertheless, failures can always happen. For this reason, redundant or duplicate sensors are recommendable, as is the use of traditional mechanical backup systems.

References

1. Caesperlein, A. 1974. Historical development of hydrometry, *Proceedings of the Three Centuries of Scientific Hydrology Symposium*, Paris, France, pp. 54–63.
2. Clemmens, A.J., Wahl, T.L., Bos, M.G., Replogle, J.A. 2001. *Water Measurement with Flumes and Weirs*. International Institute for Land Reclamation and Improvement (ILRI), Wageningen, the Netherlands, ISBN 90-70754-55-X.
3. Cox, J.P., Cabañas, A., Toledano, F., Iborra, A. 2004. La Calidad en la Hidrometría y en la Telehidrometría en la Cuenca del Segura. *II Jornadas Sobre Sistemas de Ayuda a la Decisión Ante Problemas Hidráulicos e Hidrológicos en Tiempo Real 2004*. Centro de Estudios Hidrográficos del CEDEX, 28005 Madrid, Spain, NIPO:1630400309.
4. Gobierno de España, Ministrio of Economía Y Competitividad. 2003. Las Médulas Patrimonio de la Humanidad Guía didáctica, Real Jardín Botánico, Madrid, Spain. http://museovirtual.csic.es/salas/paisajes/guia_medulas/gm9.htm
5. Husain, T. 1987. Hydrologic network design formulation. *Canadian Water Resources Journal* 12(1), 44–63.
6. Melgarejo Moreno, J. 1998. La política tarifaria del trasvase tajo-segura. *I congreso ibérico sobre gestión y planificación aguas, Zaragoza 1998*. El agua a debate desde la Universidad, Por una Nueva Cultura del Agua.
7. Ministerio de Medio Ambiente, y Medio Rural y Marino. 2009. Boletín Oficial del Estado, núm. 128, Order ARM/1312/2009, por la que se regulan los sistemas para realizar el control efectivo de los volúmenes de agua utilizados por los aprovechamientos de agua del dominio público hidráulico, de los retornos al citado dominio público hidráulico y de los vertidos al mismo, pp. 43940–43966.
8. Muste, M., Kim, W., Fulford, J.M. 2008. Developments in hydrometric technology: New and emerging instruments for mapping river hydrodynamics. *WMO Bulletin* 57(3), 163–169.
9. Region of Murcia, Historia del trasvase Tajo-Segura. http://www.regmurcia.com/servlet/s.Sl?sit=c,373,m,1915&r=ReP-27526-DETALLE_REPORTAJES
10. Sauer, V.B., Turnipseed, D.P. 2010. *Stage Measurement at Gaging Stations: U.S. Geological Survey Techniques and Methods*, Book 3, Chapter A7, 45pp, ISBN 978-1-4113-2989-8. http://pubs.usgs.gov/tm/tm3-a7/pdf/tm3-a7.pdf
11. System of Alert and Information of Water Quality, SAICA. http://www.magrama.gob.es/eu/ceneam/grupos-de-trabajo-y-seminarios/seguimiento-largo-plazored-parques-nacionales/2012_-_E.Barrios._MAGRAMA_tcm9-247965.pdf
12. The European Parliament and the Council of the European Union. 2000. Establishing a framework for community action in the field of water policy. Directive 2000/60/EC of the European Parliament and of the Council of 23 October 2000.
13. U.S. Army Corps of Engineers. Hydrologic engineering center HEC hydrologic modeling system. http://www.hec.usace.army.mil/software/hec-geohms/documentation/HEC-GeoHMS11.pdf
14. U.S. Department of the Interior, Bureau of Reclamation. 2001. *Water Measurement Manual*, Water Resources Technical Publication, 317pp. http://www.usbr.gov/pmts/hydraulics_lab/pubs/wmm/wmm.html
15. World Meteorological Organization (WMO). 2008. *Guide to Hydrological Practices: Volume I, Hydrology—From Measurement to Hydrological Information*, 6th edn. WMO Publication, 168, 296pp.

23

Procedures for Selection of Check Dam Site in Rainwater Harvesting

Saumitra Mukherjee
Jawaharlal Nehru University

AUTHOR

Saumitra Mukherjee is a professor at the Jawaharlal Nehru University, New Delhi, India. He received his MSc and PhD in geology from Banaras Hindu University. He started his scientific career from the Central Groundwater Board, Government of India, in 1985 as a hydrogeologist. In 1989, he joined as a scientist in the Remote Sensing Applications Center, UP, in the Department of Space, Government of India. While working in these two organizations, he explored the difficult terrain of India including the Himalayas, islands, Sundarban delta, and ravines of Bundelkhand for groundwater exploration and rainwater harvesting. In 1992, he joined in Jawaharlal Nehru University to establish the teaching of remote sensing applications in geology. He was a pioneer in Delhi and instrumental in the construction of check dams in the university to solve the water crisis in the water-starved terrain. He has guided 16 doctorate students in different aspects of hydrogeology and remote sensing who are well placed in various organizations in India. He has authored 7 books and more than 75 international papers including journals. He was the guest editor of *Geocarto International Journal* on the theme "Remote Sensing Applications in Geosciences" in 2012. He is Commonwealth Fellow (United Kingdom) and Executive Fellow of Earth Sciences (India). Based on his academic excellence, he has been selected as an academic excellence committee member of the American Geophysical Union (AGU) (United States). He is in active collaboration with National Aeronautics and Space Administration (NASA) (United States), Indian Space Research Organisation (ISRO), Geological Survey of India (GSI), Central Groundwater Board (CGWB), Department of Science and Technology (DST), Ministry of Environment and Forest (MOEF), National Informatics Center (NIC), Defence Terrain Research Laboratory (DTRL), etc. He has worked on various themes on terrestrial and extraterrestrial remote sensing to infer natural resource exploration and natural hazard investigations, including earthquakes, tsunamis, and forest fires in different parts of India, the United Kingdom, the United States, and Africa. Besides these, he has been invited to identify the hidden water resources within south pole of the Moon using Chandrayan-1 data during the 2012–2014 period.

PREFACE

Selection of check dams is based on high spectral reflectance with low resistivity and magnetic value are suggestive of suitability of the terrain to infiltrate the rain water. Delhi quartzite of Precambrian age has restricted groundwater availability. Due to improper land use, the area faced acute scarcity of drinking water resources. It was possible to improve the water scenario by the selection of three sites of rainwater harvesting structures in the uneven quartzite.

23.1 Introduction

The management of a resource not only includes insuring a sustainable input of the resources to a place where it is in fewer amounts from a place of excess. But to maintain its supply, there should be exhaustive planning to channelize the resources and in the mean time considering the safety of the environment. So the management includes bringing a balance to the resource availability at a place ensuring the benefit of the local population while keeping in view the well-being of the environment. Besides the selection of rainwater, harvesting sites should have sufficient information about its seismotectonic potentiality [6].

The Jawaharlal Nehru University (JNU) and Research and Referral Hospital (RRH) case studies invariably speak in favor of the earlier approach. The study area is situated in a part of the Delhi Ridge, which is geologically a quartzite mass. This makes the possibilities of finding the groundwater much scarce. Interestingly in such a terrain, the movement and availability of groundwater is depending on the geological as well as structural inhomogeneity. Implementation of the project and popularizing the technique in those areas where depth of the water level is more than 25 m below ground level have a declining trend. Information on existing land use pattern, the spatial distribution, and its changes is required for planning utilization and formulation of policies and programs for sustainable development of depleted water resources. Groundwater exploration in hard rock terrain is a challenging task. The occurrence and amount of groundwater in hard rocks are related to the complex interactions of lithological, structural, geomorphological, and pedological and climatological factors [9]. Crystalline rocks, with little intergranular porosity, are generally impervious. Movement and storage of water can occur in the network of joints, fractures, and faults and in weathered zones. The thickness and nature of the weathered layer are also related to lithological (Table 23.1) and structural characters and climatological conditions.

Remote sensing techniques using satellite images have become a powerful tool in groundwater exploration to supplement conventional methods [5,11]. Importance of remote sensing in groundwater studies is based on the fact that images help in identifying morphological and structural features that influence groundwater movements and occurrences [3]. Delineation of lineaments and fractures,

TABLE 23.1 Stratigraphy of Delhi

Period	Formation	Description
Quaternary	Newer alluvium	Unconsolidated interbedded line of sand, silt, gravel, and clay confined to floodplains of the Yamuna River
	Older alluvium	Unconsolidated interbedded, interfingering deposit of sand, clay and kankar, moderately sorted thickness variables, at places more than 300 m
Precambrian	Alwar quartzite	Well-stratified thick-bedded brown to buff color, hard to compact, intruded locally by pegmatite and quartz vein interbedded with mica sheets

drainage pattern, weathered zones, vegetation anomalies, and their mutual relationships have been carried out utilizing the resistivity and magnetic data. This would in turn help in locating groundwater discharge and recharge areas and exploratory sites.

23.2 Data Used and Methodology

Multispectral and multitemporal data merged with land use, geological, geomorphologic, hydrogeological, and magnetic data have potential for identification of suitable areas for the construction of check dams across drainage at appropriate locations is one of the successful method of artificial recharge and selection of drilling sites for groundwater exploration.

The result derived through the satellite image analysis and other collateral data were integrated together to synthesize and mark potential areas for artificial recharge.

Satellite data products inferred land use, geological, hydromorphological, and ecological information. The satellite image covering the area was analyzed digitally to prepare geomorphologic map on a 1:12,500 scale. The following satellite data products were used in analysis:

1. Indian Remote Sensing (IRS) 1D Panchromatic (PAN) + Linear Image Self Scanning (LISS) III merged acquired on PAN on February 15, 2002, and LISS February 15, 2002
2. IRS 1D PAN + LISS III merged acquired on PAN on March 6, 2004, and LISS on February 23, 2004
3. SPOT MSS and PAN of March 1989

Part of the RRH encompasses the western and southwestern parts of the national capital territory of Delhi, while JNU is situated in the southern part of Delhi. However, the rock types of RRH and JNU are the same and are mostly Aravali quartzite. The terrain is basically having undulating topography with regional slope toward north–northwest. The low-lying area is filled up with buried pediment plain, which is shallow to medium. Due to this heterogeneity in aquifer material, the water resources management in the RRH and JNU area needed a holistic approach of geological information supplemented with remotely sensed data, detailed resistivity, and magnetic surveys.

Sustainable fresh groundwater availability in the area can only be expected in the shallow freshwater aquifers, which are regularly recharged over suitable geological formations. In order to find the remedial measures, it has become essential to identify suitable areas where optimum rainwater harvesting can be done.

23.3 Location of the Study Area

The JNU campus is situated on a low-relief hill, northwest of Mehrauli and southwest of Hauz Khas in South Delhi, falling in topographical sheet No. 53 H/2 (SE quadrant). Latitude 28° 32′ 30″ E and longitude 77° 10′ 00″ N pass through the middle of the campus.

Physiographically, Delhi is at the northeastern culmination of the Mewat branch of the Aravali mountain system. Here, this branch of the Aravalis is called the Delhi Ridge. JNU is situated on an easterly projecting spur of this ridge and forms an undulating and dissected plateau of quartzite. The general slope is toward north to northeast.

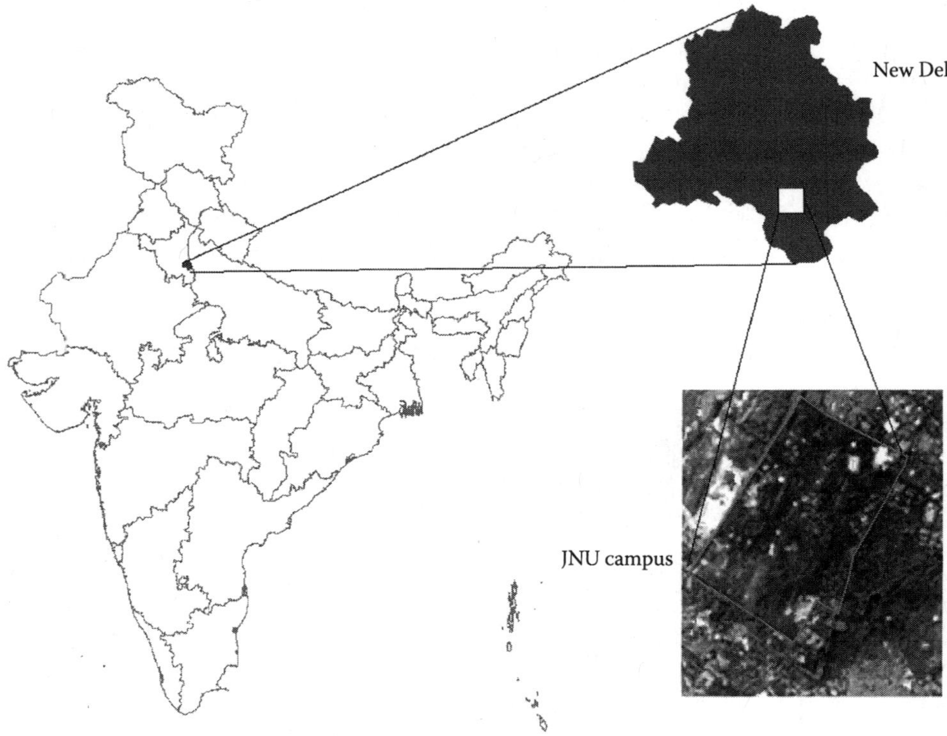

FIGURE 23.1 Check dam site selection in Aravali quartzite rock JNU campus.

The campus area is drained by three drainage systems, forming structurally controlled subdendritic to subparallel pattern. The main streams are controlled by formational strike (N–S to NE–SW), while the minor drainages by fractures and joints. All the drainages are dry except in rainy season, and the entire rainwater flows out of the campus in the absence of any structure to control the runoff. There are a few shallow ponds along the drainage, but they are mostly silted and remain dry.

RRH complex is located SW from Dhaula Kuan and is adjacent to Indira Gandhi International Airport in the west. JNU is situated SSE to the hospital. To its west is NH-8 going to join Dhaula Kuan. The place is interestingly located on the line joining the measure portion of ridge forest area of Delhi. This linear fracture of ridge is reaching Yamuna at one end that is NE from the hospital.

The area is covered by latitude 28° 34′30″ to 28°35′30″ and longitude 77° 8′30″ to 77°10′10″ (Figure 23.1). The RRH complex boundary consists of RRH itself as well as a residential area, also consisting parade ground and a lot of xerophytic vegetation characteristic of arid to semiarid conditions.

The hospital premises can be visited from two gates: the main gate at the Dhaula Kuan and a second gate near Basant Gaon. The complex has an area of 146 acres (0.590862 km^2) situated in the southwest district of the national capital territory of Delhi.

23.4 Geology of the Area

Geologically, the area is occupied by alluvium/buried pediment plain, comprising of sand, silt, clay, and kankar (agglomeration of clay and sand) underlain by bedded, highly weathered Alwar quartzite constituting a north–south-trending low ridge in the area [1]. A clear lineament has been observed in the north–south direction and few lineaments are having diagonal relationship with the main one. The sequence of the rock formation in the area is

Quaternary alluvium/buried pediment plain
Post-Delhi intrusive—pegmatite and basic intrusive
Delhi Supergroup—Alwar quartzite (ferruginous/siliceous)

Delhi region is a part of the Indo-Gangetic alluvial/buried pediment plain at an elevation ranging from 198 to 270 m above mean sea level. A quartzite ridge extending roughly from north–northeast to the south–southwestern part of the area transects this area. The thickness of alluvium on the eastern and western side of the ridge is variable. Ridge it is generally thicken (>300 m) toward west of the covering part of southwest districts (Table 23.1).

23.5 Hydrogeomorphology

From the hydrogeomorphological investigation of JNU and RRH of the Delhi area, it is clear that it comprises the following features:

1. Low residual/structural hills: These are parts of the Delhi Ridge forming N–S to N–S–SW-trending structural ridges, tors, and mounds and composed of folded and jointed quartzite. Joints allow only limited groundwater infiltration. This unit constitutes the surface runoff and therefore has very poor prospects of groundwater. They are mostly barren, with scanty vegetation along joints and slopes.
2. Pediment: The undulating, eroded and dissected, shallow, buried planer surface along the fringe and slopes of ridges and tors forms this unit. The main drainage systems are developed in this unit. Weathering is shallow and soil thickness varies, the maximum being in the valleys near the streams. The soil is generally clayey and fine silty and partly gritty and gravelly. Drainage dissection is quite intense at places, often developing gullies. Weathering is more intense in coarse gritty or arkosic quartzite. Groundwater potential is generally low due to poor infiltration and high runoff resulting from varying slopes and clay mantle.
3. Buried pediment: This unit forms the almost flat terrain in the northeastern part of the RRH premises. In JNU this type of feature exists in the northeastern part mainly, while in between the ridges the depth of the buried pediment is shallow (Figure 23.1). It has a shallow to moderately thick soil cover, which is mainly silty and clayey, and at places gritty and gravelly. The surface slopes gently toward the northeast and merges with deeply buried pediment beyond the campus and with the Yamuna alluvial plain further east. This unit forms a moderate to good groundwater potential especially along fracture and drainage.

23.6 Groundwater: How Does It Occur

It is difficult to visualize water underground. Some people believe that groundwater collects in underground lakes or flows in underground rivers. In fact, groundwater is simply the subsurface water that fully saturates pores or cracks in soils and rocks. Groundwater is replenished by precipitation and, depending on the local climate and geology, is unevenly distributed in both quality and quantity. When rain falls, some of the water evaporates, some is transpired by plants, some flows over land and collects in streams, and some infiltrates into the pores or cracks of soil and rocks. After the water requirements for plant and soil are satisfied, any excess water will infiltrate to the water table—the top of the zone below which the opening in rocks is saturated.

23.7 Role of Remote Sensing in Groundwater Studies

Space technology in the form of remote sensing can play a useful role in hydrological studies. Remote sensing is defined as the science of deriving information from measurements made at a distance from the object without the sensor actually coming in contact with it. Remote sensing though is a fledging phenomenon; seem to a shot in arm by either substituting or complementing or supplementing the conventional technology with reasonably faster, efficient, and accurate methods of survey in the domain of water resource planning, conservation, development, management, and utilization.

Remote sensing by virtue of its synoptic coverage, spectral behavior, repeatability, and availability offers an effective first-hand tool in mapping and monitoring resources in a reasonably short time frame. Synoptic view facilitates the study of objects and their relationships. Spectral signatures permit identification of various features, while the temporal aspect allows change detection in the environment. The real advantage is the real-time measurement that facilitates constant and effective monitoring. The main advantage of remote sensing is that the data are in the digital form and can be analyzed easily with the help of computers. The wavelengths widely used in remote sensing are visible and near infrared in the wavelength of 0.4–3 μm.

The application areas for remote sensing data are both wide and varied. Radiometric data potentially represent a very useful source of information in pedological research, and in the study of water quality though, remote sensing cannot be used for groundwater studies. But the remote sensing data allow us to make indirect references regarding subsurface through sacrificial expression of the aquifer. The subsurface hydrological conditions are inferred based on identification and correlation of surface phenomenon involving geological features and structures, geomorphology, surface hydrology, soils and soil moisture anomalies, vegetation types and distribution, land use, and many others as indicators. The benefits that accrue in the use of remotely sensed data are usually greatest when they are applied for large-scale preliminary investigation of groundwater reserves.

Although remote sensing technique can never replace conventional hydrological observation network [4], remote sensing data have two distinct advantages. Remote sensing platforms provide data with high resolution in space and data can be obtained for areas that have no record of measurements (e.g., remote areas); therefore, remote sensing data, particularly satellite data, can be most helpful for design and operation purposes if they are used in combination with ground truth. The disadvantages of satellite data are unfavorable combination of resolution in time, and space airborne geophysical exploration is highly used in groundwater prospects. Conventional prospecting tools, namely, hydrogeological and geophysical instruments, generally do not yield the relevant details and occasionally exhibit lack of resolution. Integration of satellite data and vertical electrical sounding (VES) data as well as magnetic intensity data, indirectly giving the potential fracture zones using other collateral data generated from the imagery as well as collected from various institutions, is used to access the groundwater potentiality of various geomorphic units.

In the present work, more thrust has been given on the detection of an area that has high groundwater potential and area that can serve as good point to recharge to the groundwater; also undertaken in this exercise, an endeavor is made to generate a model to find the trend of flow of the surface water during rainy months so that this water can be diverted toward the areas selected for recharging the groundwater. All this has been achieved using geophysical technique, namely, resistivity survey, magnetic survey, soil analysis, and drawdown test of existing pumps, and using remotely sensed data to correlate the data as well as to generate some new data like trend of the lineament as well as to detect the reflectance values and to generate a model to find the trend of flow and help in locating a suitable site for check dams (Figure 23.1).

23.8 Electrical Resistivity Method

The electric resistivity of a rock formation limits the amount of current passing through the formation when an electric potential is applied. It may be defined as the resistance in ohms between opposite faces of a unit cube of the material. If a material of resistance R has a cross-sectional area A and a length L, then its resistivity can be expressed as

$$\rho=\frac{RA}{L} \tag{23.1}$$

Units of resistivity (ρ) are Ω-m.

Resistivity of rock formations varies over a wide range, depending on the material, density, porosity, pore size and shape, water content and quality, and temperature. There are no fixed limits for resistivity of various rocks. In relatively porous formations, the resistivity is controlled more by water content and quality within the formation than by the rock resistivity. For aquifers composed of unconsolidated materials, the resistivity decreases with the degree of saturation and the salinity of groundwater. Clay mineral conducts electric current through their matrix; therefore, clayey formation tends to display lower resistivity than do permeable alluvial aquifers.

Actual resistivity is determined from apparent resistivity, which is computed from measurements of current and potential differences between pairs of electrodes placed in the ground surface. The procedure involves measuring a potential difference between two electrodes (potential electrode) resulting from an applied current through two other electrodes (current electrode) outside but in line with the potential electrode. If the resistivity is everywhere uniform in the subsurface zone beneath the electrodes, the current and equipotential lines will form an orthogonal network of circular arcs. The measured potential difference is a weighted value over a subsurface region controlled by the shape of the network. Thus, the measured current and potential differences yield an apparent resistivity over an unspecified depth. If the spacing between the electrodes is increased, a deeper penetration of the electric field occurs and a different apparent resistivity is obtained. In general, actual subsurface resistivity varies with depth; therefore, apparent resistivity will change as electrode spacing are increased, but not in a like manner. Because changes of resistivity at great depths have only a slight effect on the apparent resistivity compared to those at shallow depths, the method is seldom effective for determining actual resistivity below a few hundred meters.

Electrodes consist of metal stakes driven into the ground. In practice various standard electrode spacing arrangements have been adopted; the most common are the Wenner and Schlumberger arrangements.

The Wenner arrangement has the potential electrode located at the third points between the current electrodes. The apparent resistivity is given by the ratio of voltage to current times a spacing factor. For the Wenner arrangement, the apparent resistivity

$$\rho_a = 2\pi a \frac{V}{I} \tag{23.2}$$

where

a is the distance between the adjacent electrodes
V is voltage difference between the potential electrodes
I is apparent current

The Schlumberger arrangement used for this study has the potential electrodes close together. The apparent resistivity is given by

$$\rho_a = \pi\left(\left(\frac{L}{2}\right)^2 - \left(\frac{b}{2}\right)^2 \frac{V}{b} * 1\right) \tag{23.3}$$

where L and b are current and potential electrode spacing, respectively. Theoretically, $L >> b$, but for practical application good results can be obtained if $L >= 5b$.

23.8.1 Typical Resistivity of the Geological Material

Resistivity near surface materials is heavily affected by groundwater, and water is a low-resistivity material. In general finer-grained sediments have low resistivity, and bedrock has high resistivity. Resistivity values for different rock types are shown in Table 23.2.

TABLE 23.2 Typical Resistivity of Various Geological Materials of Aravali Quartzite of Delhi

S.No.	Lithology	Resistivity (Ω)
1	Silt	10–100
2	Sand–gravel	300–8000
3	Freshwater sand	50–100
4	Argillaceous sand	25–50
5	Saltwater sand	0.4–1.3
6	Pebble aquifer	100–several hundred
7	Limestone	80–several hundred
8	Clay marls	Several to 50

Resistivities are reduced by

- Increasing porosity
- Increasing ion content of groundwater
- Increasing content of clay
- Decreasing grain size

23.8.2 Magnetic Anomaly Studies

Magnetic anomalies can be a useful geophysical component in groundwater management. Magnetic survey was carried out in and around the RRH complex, using the instrument called proton precession magnetometer modal PM-600 manufactured by Integrated Geo Instruments and Services (P) Ltd.

Proton precession magnetometer utilizes the spinning of protons or nuclei of the hydrogen atoms in a sample of hydrocarbon fluid to measure the total magnetic field intensity. Water, kerosene, alcohol, etc., are taken as samples. The protons in these fluids behave as small spinning magnetic dipoles. These magnetic dipoles are temporarily aligned (polarized) by application of strong uniform magnetic field by sending a current through a coil wound on the bottle containing the hydrocarbon fluid/water. When the current is removed, that is, when the applied field is removed, the spin of protons causes frequency that is proportional to the ambient field intensity. The total magnetic intensity as measured by a proton precession magnetometer is a scalar measurement, that is, it gives simply magnitude of the total Earth's magnetic field independent of its direction. The area, that is, the RRH complex, comes in a region with normal magnetic intensity around 47,000 gamma.

In place of trespassing it can be passing through the hospital complex inferred by IRS-1D PAN+LISS III merged, acquired on March 06, 2004, LISS March 23, 2004, respectively, and the second data by IRS-1D PAN+LISS merged February 15, 2002, and February 15, 2002, respectively. It shows a strong trend in NNE–SSW direction, which passes directly through the hospital premises. On its N–NE side from the hospital, the lineament is showing a water body. This place is Subroto Park. And also on the SSW of the hospital, the lineament is having a water body inside the hospital premises. Measurements were taken along this lineament, which is inferred as sudden decrease in magnetic values (average magnetic value of Delhi region 47,000 gamma). Based on the spot magnetic values, a contour map was made along the profile. A contour line has been drawn at every 3000 gamma interval. Low magnetic values were noticed along lineament and places with fractured ferruginous quartzite. Selection of check dams was based on the points inferred by a magnetometer showing low magnetic values and interconnected lineaments.

Keeping all these points in consideration, it is suggested that the recommendation given can be implemented for the integrated water resource management in the terrain. Rainwater harvesting and groundwater exploration in selected areas may lead to sustainability to this area.

23.9 Rainwater Harvesting in JNU Campus, New Delhi, India

The study area, JNU, is situated in the southwest of Delhi metropolis (Figure 23.1). Information on the existing land use pattern, the spatial distribution, and its changes is required for planning, utilization, and formulation of policies and programs for sustainable development. Humans transform the land for different activities based on their needs, and quantifiable information on these dominating activities is necessary to develop future plans. Soil properties and land use patterns are major contributing factors to the hydromorphogeology of a particular area. Historical monuments and other urban features have been analyzed by satellite imagery. Remote sensing data are helpful in the studies of brick kiln areas that are located in the recharge areas of JNU new campus for groundwater replenishment. Multispectral and multitemporal data from SPOT, IRS-1A, IRS-1B, and IRS-1C (Figures 23.2 and 23.3), when integrated with land use, geological, geomorphological, hydrogeological, and magnetic data, have potentiality for identification of suitable areas for construction of check dams (Figure 23.4).

Interception of surface runoff by check dams across drainage at appropriate locations is one method for artificial recharge.

In general, JNU area is lacking sufficient surface water bodies and paleochannels. Very thin soil cover in this area is also not supporting the dug well structures. Groundwater occurrence is restricted to the deep-seated fracture zones only. For selection of artificial recharge areas, the radiance values of pixels in near-infrared region of IRS satellite data were studied. Due to high recharge, the soil moisture is less as well as the vegetation. Lineaments are passing through the check dam numbers 1, 2, and 3, which were selected based on their low spectral reflectance and low magnetic values over the weathered ferruginous quartzite, which are prevailing geologic suite of the study area. Groundwater level went down in this area, which increased after the artificial recharge. Plantations of suitable area-specific species are restoring eco-conservation in this campus.

The area under study is a 2 km × 2.5 km region situated on a low-relief hill, northwest of Mehrauli and southwest of Hauz Khas in South Delhi, falling in topographical sheet No. 53 H/2 (SE quadrant). Latitude 28 32′ 30″ E and longitude 77 10′ 00″ N pass through the middle of the campus.

Satellite data products inferred land use, geological, hydromorphogeological, and ecological information. The satellite imageries covering the area were analyzed digitally to prepare a geomorphological map on a 1:50,000 scale.

FIGURE 23.2 Improvement of Jawaharlal Nehru vegetation health after recharge from check dam.

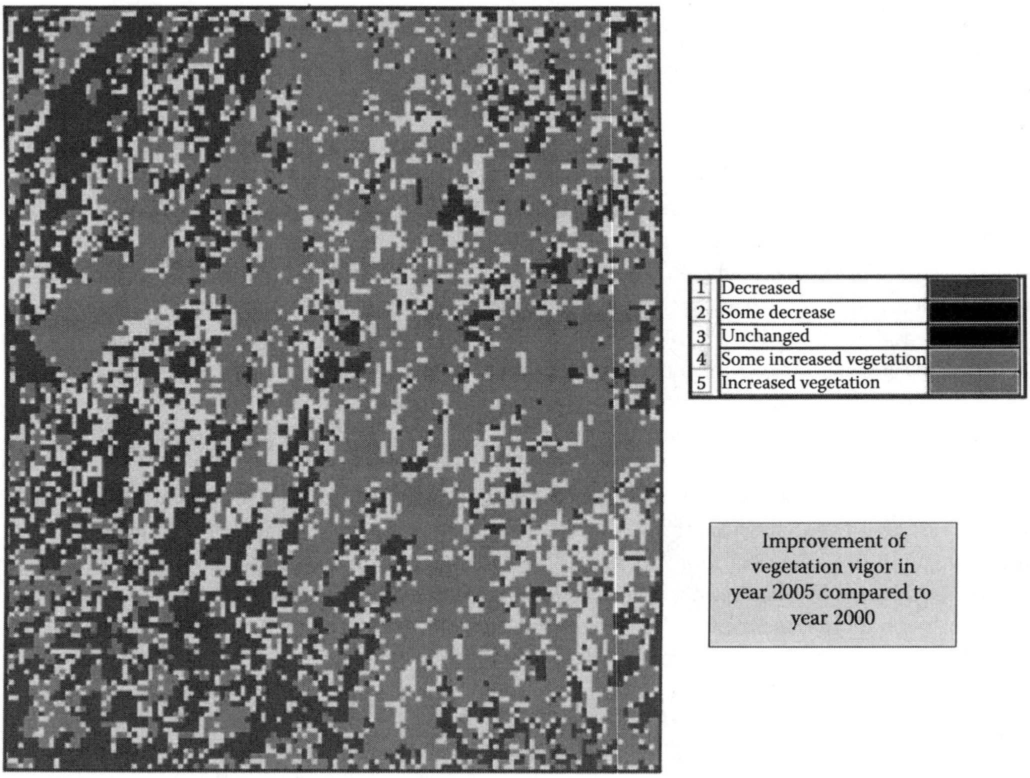

FIGURE 23.3 Vegetation vigor improved after check dam construction.

23.10 Land Use Studies

Environmental impact of anthropogenic activities on its surrounding area is dramatic [2,10]. Land use change and overexploitation of groundwater has reduced the water table in JNU and its surrounding area. These changed land use patterns were inferred from the high albedo structure in satellite images. It is clearly mentioned in a map showing fluctuation in water level in a part of South Delhi that fall in water level is found during 1994–1995 in Indian Institute of Mass Communication (IIMC), Munirka Vihar, and JNU old campus, Ber Sarai, National Institute of Health and Family Welfare (NIHFW), and Admn. Block of JNU and Vasant Kunj. The recharge area for these points is Masoodpur and its surroundings that are inferred by lineament and geological attitude of quartzite rocks [8]. Further changes in these features may lead to lowering down of the water level that may even lead to land subsidence due to increased effective geostatic pressure.

23.11 Geological Studies

Rocks with ferruginous stains or siliceous character containing bands of mica schist, grits and argillaceous material. Intrusions of pegmatites are present in quartzite along the shear zones, and when altered, they give rise to morrum and china clay.

Geological formation exposed in the JNU area belongs to the Alwar group of the Delhi Supergroup of the Precambrian age. They are predominantly thick-bedded, massive, compact, gray-colored quartzite with subordinate argillites. They are fine to medium grained in general but form gritty, arkosic, and ferruginous quartzite, often giving brownish to reddish colors and streaks.

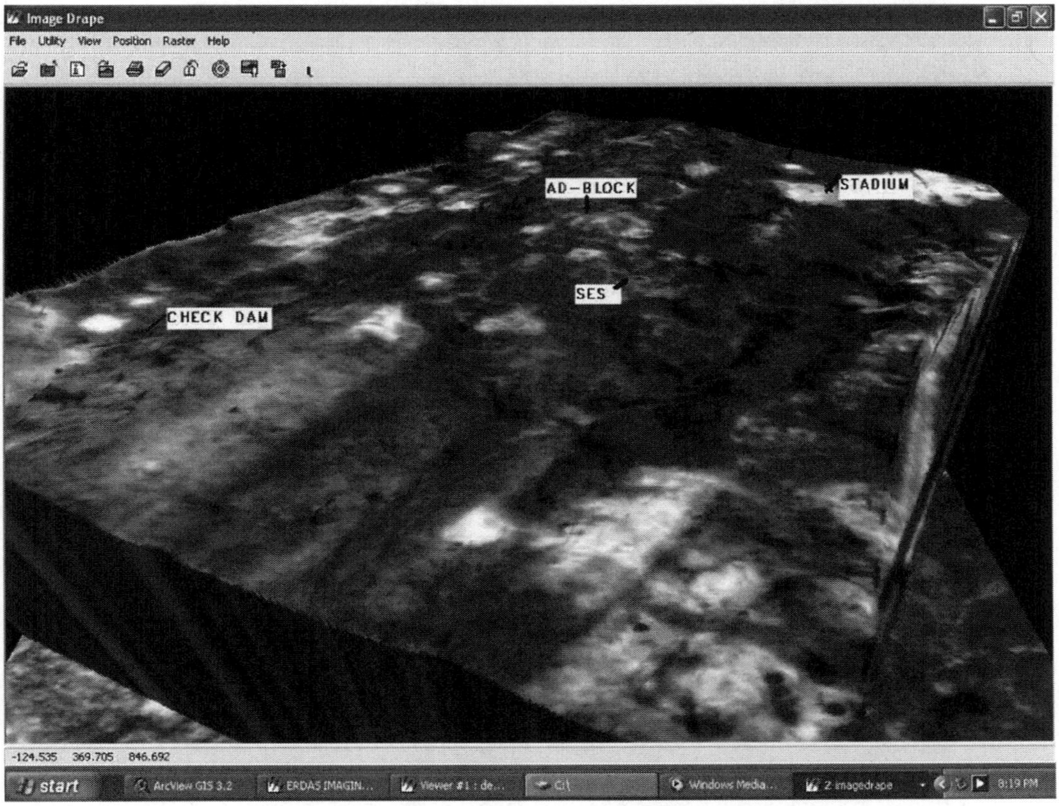

FIGURE 23.4 Selection of check dam in JNU based on remote sensing and GIS study.

23.12 Structural Geology by Remote Sensing Techniques

The formation of the study area strikes along N–S to NE–SW and dips steeply on either side indicating isoclinal folds. In JNU satellite, images show a series of folds in the western parts with N–S axes.

The quartzite is characterized by several sets of well-developed fractures and joints. Of these, the following sets are more prominent in the study area:

- NNE–SSW (parallel to bedding) with steep easterly dips
- E–W to ENE–WSW, vertical
- NE–SE, vertical, perpendicular to strike

A few lineaments have also been identified in the area from satellite images. Of these, the two nearly E–W-trending lineaments traced in the northern part of the campus are prominent. The less conspicuous lineaments/fractures identified in the images are along NW–SE. These lineaments have a control over the drainage in the area.

Vein quartz and pegmatite have been emplaced along fractures and joints in quartzite. Pegmatite contains feldspars, muscovite, quartz, and tourmaline. Clay (kaolin) derived from weathering of pegmatite is extensively mined in the area west of the campus.

The compact, gray quartzite is highly resistant to weathering, and they stand out as jointed, blocky, or surrounded boulders as a result of intersecting joints. But weathering is more prominent along sheared gritty, arkosic, and ferruginous quartzite, which results in sandy or gritty, red soil.

Over the eroded surfaces of Alwar quartzite and in the elongated depressions between the quartzite ridges, the weathered colluvial and alluvial materials and soil are deposited, which include clay, loam, silt, grit, gravel, and kankar of recent to subrecent origin.

JNU lineaments inferred by IRS-1A, SPOT, and IRS-1C data show a strong trend in the NE–SW direction, which passes though the JNU sports complex (stadium). Measurements were taken along this lineament that is inferred as a sudden decrease in magnetic values (average magnetic value of Delhi region is 47,000 gamma). Based on the spot magnetic values in and around JNU, 11 profiles were done. Contour maps were made along the profile. Contour lines were drawn at every 10 gamma interval; low magnetic values were noticed in lineaments on ferruginous quartzite. Selection of check dams was based on the points inferred by the magnetometer showing low magnetic values and interconnected lineaments.

A geographic information system (GIS) map was prepared for the JNU area (Figure 23.5) showing geology, lineament, and magnetic anomaly.

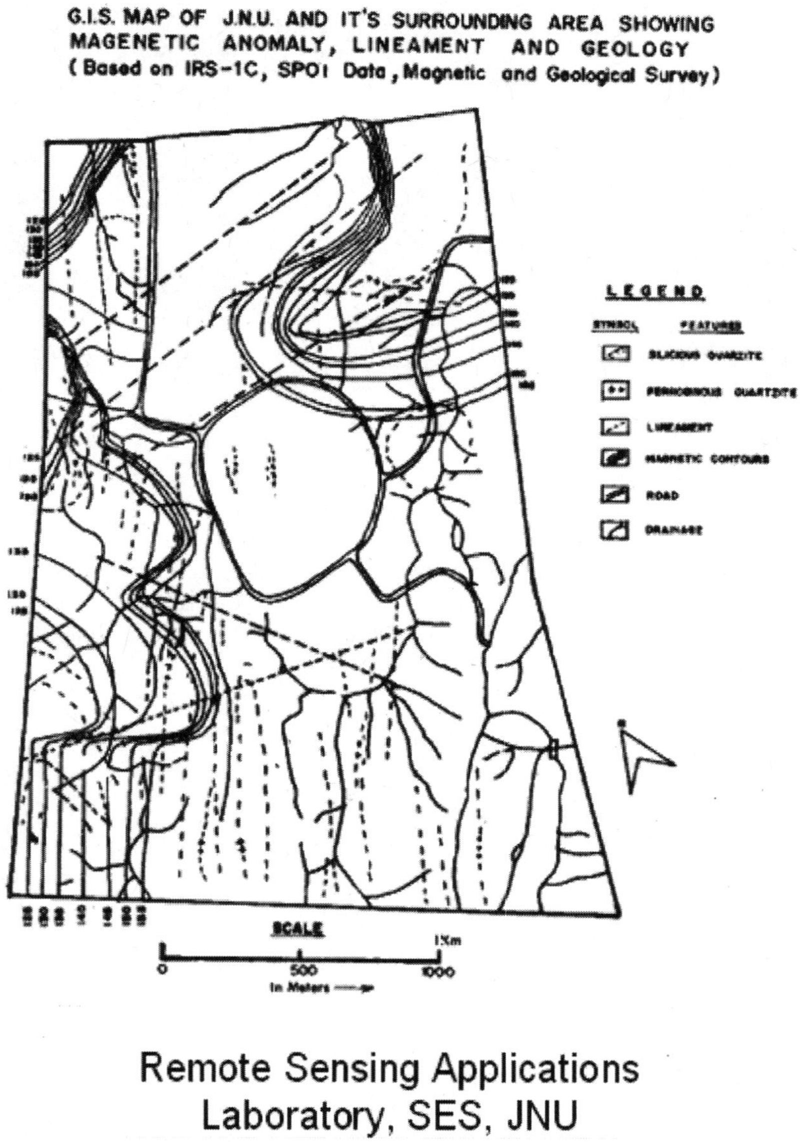

FIGURE 23.5 Showing geology, lineament, and magnetic anomaly.

23.13 Surface Water Management: Rainwater Harvesting

Interception of surface runoff by check dams across drainage at appropriate locations is one method of surface water management. Three check dams have been constructed in JNU.

In the JNU campus, a series of check bunds with pipe outlets have been proposed to construct for rainwater harvesting. Locations for such structures have been selected where the valley section is narrow and maximum spreading up of steam with least bearing on the height of the structure is possible. Thus, 14 sites have been selected for check dams, extending outside the campus.

Out of these, three check dams are already constructed in the western stream and another in the northern stream. The rise in water level in 16 piezometers is being monitored every month. The results are encouraging. It is observed that there is substantial increase in water level in the JNU campus after the recharge through check dams.

23.14 Ecoconservation of JNU

The Delhi master plan of August 1990 clearly states that the ridge "should be conserved with utmost care and should be afforested with indigenous species with minimum of artificial landscape." Within JNU, while some areas of the ridge have successfully retained and regenerated their biodiversity and also sustained a considerable variety of fauna, including some rare species, other parts have suffered substantial denudation, soil erosion, etc.

The water level in some parts of JNU has already risen by over 10 m in less than 2 years, and the water discharge in bore wells in neighboring areas of JNU has shown remarkable improvement. Also the check dams are creating water bodies that the master plan says "should be developed to act as major lung spaces and to attract migratory birds and for improving the microclimate."

Multidate satellite images infer changes in vegetation canopy cover from 1987 to 1997. This appears to be due to increase in soil moisture by artificial recharge in the JNU campus. There has been about 1.104 km^2 (46.96%) increase in dense vegetation canopy cover due to increase in groundwater level. This increase is mainly reported in the JNU valley and Sanjay Van and its eastern parts.

It is now planned to recreate the forest cover of the large denuded parts of the JNU ridge through planting of new trees of indigenous varieties, which are native to the ridge. Based on the soil type, thickness, moisture-retaining capacity, and slope of the area, plantation scheme has already started. The areas where it was decided to do planting on a priority basis were

1. On sides of roads where no trees have been planted so far or where trees have died
2. In selected forest areas especially around the check dams so as to increase the percolation of water and to prevent silting through soil erosion
3. Any areas where excessive denudation has occurred

Being situated on a structural hill made up of hard, massive quartzite rocks and buried pediment, the flora of this campus is unique. An attempt was made to classify the suitable hydromorphogeological niche for new plantation. IRS-1C and SPOT data were used for this purpose. Ground truth was done by resistivity meter and field spec ground truth radiometer. In areas where the resistivity value is ranging between 20 and 300 Ω-m, the geomorphological unit is classified as buried pediment. Pediments and ridges show relatively higher resistivity values. A new terminology is being introduced here, "eco-hydromorphogeology" that infers suitable hydrology, geology, and morphology for the ecology of the area. Some species selected for roadside planting are *Prosopis cineraria* (khejri) for shallow buried pediment, *Acacia leucophloea* (safed kikar), *Acacia senegal* (kumta), *Cordia rothii* (gondi), etc. Suitable plants for sandy soil near check dam sites are *Salvadora persica* (pilu), *Boswellia serrata* (salai), etc. Suitable draught-resistant plant would be *Acacia modesta* (phulanhi). Fast-growing plants in buried pediment plain (deep) are suitable for *Zizyphus nummularia* (kokanber). The eco-hydromorphogeology of the JNU campus (Table 23.3) gives a clear guideline of the suitability of plant species in this varied terrain [7].

TABLE 23.3 Ecohydromorphogeology of the JNU Campus for Suitability of Check Dam Construction

Geomorphic Unit	Landform	Hydrogeology/Suitability for Check Dam Construction	Suitable Flora
Residual/ structural hills	Rocky ridges, tors and mounds	Massive compact jointed quartzite. Poor groundwater	*Prosopis juliflora* (kabuli kikar), *Azadirachta indica* (neem), *Mitragyna parviflora* (kadamb)
Pediment	Undulating, eroded and dissected, shallow, buried pediment with rock exposures. Thickly vegetated with scrub	Weathered, coarse, gritty or arkosic quartzite with cover of clayey and silty soil along stream course. Moderate to good groundwater prospects along fracture and shear zones	*A. senegal* (kumta), *Wrightia tinctoria* (dudhi), *Balanites aegyptiaca* (hingot), *Sterculia urens* (kullu), *B. serrata* (salai)
Buried pediment	Plain to gently sloping ground with occasional rock outcrops	Silty, clayey and at places gravelly soil derived from weathering of arkosic and gritty quartzite. Good groundwater prospects	*Ficus benghalensis* (bargad), *Cassia fistula* (amaltas), *Albizia lebbeck* (siris), *Ficus religiosa* (peepal), *Ficus infectoria* (pilkhan), *Terminalia arjuna* (arjun), *Bauhinia variegata* (kachnar)

23.15 Qualitative Improvement in Groundwater by Rainwater Harvesting

The quality of water for domestic consumption is of paramount significance as the chemical and microbiological contamination of potable water can lead to serious health hazards/body disorders through a waterborne disease of toxic chemicals. Qualitative improvement in groundwater quality in the JNU campus can be quantified by pre- and post-rainwater harvesting water quality studies. Table 23.4 shows the qualitative improvement of groundwater after artificial recharge [5]. This work has a long-term economic importance as well. For improvement of groundwater quality otherwise, the cost of water treatment would have levied measurable recurring financial burden on the university.

From 1996 to 1998, four rainfall cycles were identified each year, and their effects of recharging to groundwater regime were observed in 16 piezometers, tapping shallow and deep aquifers. These piezometers were installed downstream along its course and lateral to it. The depth to water level in the area prior to check dam construction was ranging between 17 and 22 m below land surface. From 1996, the rise of water level was noticed between 5.26 and 12.50 m in the piezometers about 166 and 44 m from the dam, respectively. It was computed that 90, 000 m^3 of water was recharged per year to the shallow and deep aquifers and 2000 m^3 were lost due to evaporation. Out of the total quantity of water resource available, about 98% had been recharged to the aquifers. Due to increase in soil moisture, the total area under dense and sparse vegetation canopy cover has increased. Dense vegetation has increased 46.96%; sparse vegetation has increased 24.30%. Area without vegetation has shown 2.14% vegetation. Eco-hydromorphogeological plantations in this area will be useful in reducing the air pollution. Other parts of Delhi also require a similar type of rainwater harvesting practices to improve

TABLE 23.4 Improvement in Groundwater Quality in the JNU Campus after Artificial Recharge (between 1996 and 2001)

pH		EC (μmhos/cm)		Hardness $CaCO_3$ (mg/L)		Ca (mg/L)		Mg (mg/L)		NO_3 (mg/L)		F (mg/L)	
Prior	After	Prior	After	Prior	After	Prior	After	Prior	After	Prior	After	Prior	After
8.5	7.0	694	137	460	394.5	92	148	73.3	22.06	296	148.2	0.90	0.90

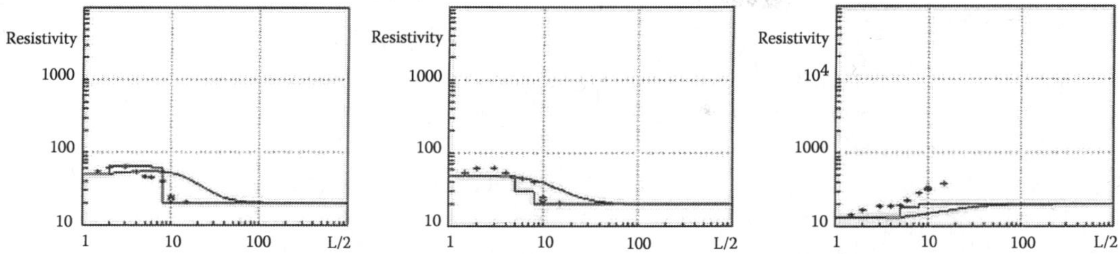

FIGURE 23.6 Resistivity survey suggesting construction of recharge pits for rooftop rainwater harvesting sites in JNU.

groundwater environment and for eco-conservation [7]. In the next phase of rainwater harvesting, the rooftop rainwater is being allowed to recharge the aquifer through the pits selected by resistivity surveys. The work was initiated in 2006 and its feasibility mapping and subsurface information confirmed the selection of sites. In 2012, the rooftop rainwater harvesting has been implemented in this area based on the findings of resistivity surveys (Figure 23.6). It is expected that this technique will be useful for the rise in the water level in this water-starved area.

23.16 Discussions

Land use/land cover and water resources management are directly linked with each other. It is clear that selection of the site for RRH and JNU was not done keeping this point in mind. However, during that period, latest knowledge of remote sensing as well as its integration with geophysical data was not in consideration. The study area is on a series of parallel lineaments. These lineaments are connecting conduit of occasional surface water. In JNU, the surface water is going out of the campus through seismically generated lineaments, while in RRH, the water logging takes place in the basement of the hospital. JNU area has shown remarkable development (Figure 23.2) in its environment after the water resource management program, and similar development is reported in RRH areas also.

23.17 Summary and Conclusions

Information on existing land use pattern, the spatial distribution, and its changes are required for planning, utilization, and implementation of groundwater exploration and rainwater harvesting. Multispectral and multitemporal satellite data have the potential to delineate the sites for water resource management. Magnetic and resistivity survey can further confirm the subsurface layers of aquifers. Due to heterogeneity of aquifer material, it is essential to infer a holistic approach of geological information supplemented with remotely sensed data supported by resistivity and magnetic anomaly detection in locations of higher spectral reflectance followed by the intense lineament density to sustain the recharge of all layers of the aquifer system. This technique was used in the Aravali quartzite terrain of Delhi in RRH and JNU areas. Multispectral and multitemporal data from SPOT, IRS IC, and IRS-1D that were interpreted to infer land use, geological, geomorphologic, hydrogeological, resistivity, and magnetic data were used. These interpretations have potential for identification of suitable areas for construction of check dams, rooftop rainwater harvesting pits, and drilling sites in the difficult terrain of Aravali quartzite. In this water-starved terrain within the premises of JNU, three check dams were constructed during the 1992–1997 period. With the great success of rise in water level, this work was repeated in RRH, Indira Gandhi National Open University (IGNOU), and other parts of Delhi. During 2008–2012, further studies on rooftop rainwater pit location were inferred by using resistivity survey. New school buildings of the university have been equipped with the rooftop rainwater harvesting structures, and in August 2012, rain is being utilized to further recharge the aquifer system.

References

1. Bhanumurty, Y.R., Dimri, D.B., and Hassan, Y. (1978). Geophysical investigations for the geotechnical project of Delhi metropolitan area. Geological Survey of India, Miscellaneous Publication No. 43, pp. 35–43.
2. Jaiswal, R.K. and Mukherjee, S. (1999). Application of remote sensing technology for land use land cover change analysis, *International Journal of Remote Sensing*, 27(2), 123–128.
3. Kale, P. (1992). Sustainable development: Critical issues, *Journal of Indian Society of Remote Sensing*, 20(4), 183–186.
4. Mukherjee, S. (2009). Sensible measures to guard India's groundwater supply. *Nature*, 462, 276 (November 19, 2009), doi: 10/462276d. http://www.nature.com/nature/journal/v462/n7271/full/462276d.html
5. Mukherjee, S. and Mukherjee, A. (2001). Quantitative and qualitative improvement in groundwater by artificial recharge; a case study in JNU, New Delhi, Rain water catchment systems, Margrav-Verlag, Mannheim, Germany, pp. 35–39.
6. Mukherjee, S. (1997). Change in groundwater environment with land use pattern in a part of South Delhi: A remote sensing approach, *Asian-Pacific Remote Sensing and GIS Journal*, 9(2), 9–14.
7. Mukherjee, S. (1998). Eco-conservation of a part of JNU campus, by GIS analysis, *Proceedings of the National Symposium on Artificial Recharge of Groundwater*, New Delhi, India.
8. Mukherjee, S. (1997). Re-evaluation of seismogenic potentiality of Delhi-Rohtak area using remote sensing and seismological data. Unpublished DST Project Report.
9. Mukherjee, S. (2004). *Textbook of Environmental Remote Sensing*, Macmillan India Ltd, Bangalore, India.
10. Rao, L.K.M. (1995). Remote sensing for land use planning, *International Journal of Remote Sensing*, 16(1), 53–60.
11. Saraf, A.K. and Chaudhary, P.R. (1998). Integrated remote sensing and GIS for groundwater exploration and identification of artificial recharges sites, *International Journal of Remote Sensing*, 19(10), 1825–1841.

24

Quality Control and Homogenization of Climatological Series

José A. Guijarro
State Meteorological Agency (AEMET)

AUTHOR

José A. Guijarro is a meteorologist in charge of the Mediterranean Meteorological Studies unit at the State Meteorological Agency (AEMET) office in the Balearic Islands (Spain). He worked on various posts (radiosounding operator, aeronautical forecaster), but most of his work has been devoted to different fields of climatology, which was also the subject of his PhD dissertation. He has published over 75 papers and some book chapters and has lectured in several international courses on applied and statistical climatology. He is currently a member of an Open Panel of Experts of the Commission for Climatology (World Meteorological Organization [WMO]) and leader of the Data Rescue Task Team of the Working Group on Climatology and Hydrology (WMO Regional Association VI).

PREFACE

This chapter addresses the importance of using homogenized and quality-controlled series in order to obtain reliable estimations of the trends and variability of any climatological element, including those relevant for hydrological studies. Currently used methodologies are discussed, focusing on available computer packages, and a case application example is presented to better illustrate the benefits of applying these procedures.

24.1 Introduction

Meteorological observatories, through their observational programs, collect the data that build the climatological series that can be seen as strands lengthening year after year along the operation life of the observatories. In any geographical area we have normally several meteorological stations, and when the climatologist weaves their different strands, he or she obtains a carpet depicting the spatial and temporal variability of the involved climatic elements in that area. However, observations may be affected by many perturbations that introduce point errors or changes in the statistical properties of the series, hence compromising the reliability of any conclusion derived from their analysis. Apart from human errors in the observation or mechanization of manual instruments, observational practices may change along with time, and miscalibrations, instrumentation replacement, station relocations, and changes in the surrounding land use are common events in the history of any observatory, introducing *inhomogeneities* in their series. From a statistical point of view, a series would be inhomogeneous whenever their statistical properties change along with time, but this would include actual changes in the climate. Therefore, climatological series inhomogeneity is referred only to changes introduced by perturbations alien to the climate variability.

Concern about inhomogeneities existing in the climatological series is as old as the series themselves. Many methods have been proposed to detect and remove these unwanted perturbations from the series, and thorough reviews and advice about their use can be seen in Peterson et al. [10], Aguilar et al. [1], and Beaulieu et al. [3]. On the other hand, the Hungarian Meteoro-Hydrological Service has been celebrating a series of International Seminars on Homogenization and Quality Control of Climatological Databases that are also a good source of information about past and current efforts on this subject (see the more recent proceedings in WMO [15–17]), as well as the website of a COST Action [4] that emerged from one of these seminars.

The next sections will address the quality control approaches, and then homogenization algorithms and their implementation in publicly available computer packages will be discussed, to end with an example application of one of them.

24.2 Quality Control Procedures

Quality control of climatological data is normally first carried out at the institutions that collect, mechanize, and archive the observations and depends greatly on the observed element and the instrumentation used. For example, cooperative observers may be manually recording daily extreme temperatures with a Six–Bellani thermometer, and errors can originate by reading the wrong side of the index or when confusing the nearest 5°C thick tick mark of the scale. In contrast, common manual precipitation data errors are assigning yesterday's 07–07 h total to today or accumulating the precipitation of several missing observations into one. Therefore, it is the institution managing the observing network the best suited to know all the peculiarities of the sites and their observing practices and to control the quality of their data accordingly.

Nevertheless, even when this first-stage quality control is properly applied, some errors will inevitably pass the filters or can be originated afterword, during the data transfer between different devices or archive systems, and therefore data users should perform additional quality controls whenever possible. These quality controls will be more effective when applied to original data, because derived data may smooth the errors making them more difficult to detect. For example, a 10°C error in a manual reading of a daily maximum temperature becomes a mere 0.33°C error in the monthly maximum average and a still less noticeable 0.17°C error in the monthly mean temperature.

Quality controls can be grouped in four types:

1. Physical plausibility, that is, checking whether any value is outside its expected range (e.g., negative precipitation). It is often adjusted to the local climate: 0°C can be a usual temperature in temperate zone winters but unexpected in a tropical coastal site.
2. Temporal increments: Limits may be imposed to the maximum and minimum change between two or more consecutive measures.
3. Internal consistency between two or more variables: Minimum temperature greater than the maximum, precipitation amounts when no precipitating meteor is reported, etc.

4. Spatial consistency: The series of nearby stations are expected to present a similar behavior, and hence their comparison is a powerful method for detecting errors. This is not applicable to sub-daily data unless sensors are close enough as to guarantee the series synchronization (time shifts could be applied, but this would complicate the method). A previous normalization of the series is advised in areas with complex orography where climatic variables may display different ranges of values in varied altitudes and topographic locations.

The two first controls are applied over single series, and the others require two at least (from the same observatory in the third case and from several in the fourth type of controls). More information about their implementation can be found in the literature, as in Shearman [13]; Peterson et al. [11]; Guijarro [7]; and Zahumenský [18].

Homogenization packages discussed in the following section only include very basic quality controls, such as the possibility to remove outliers greater than a prescribed threshold, although visual inspection of their graphic outputs may help in detecting other kind of problems in the series.

24.3 Series Homogenization

Series homogenization is the process by which any influence unrelated to climate and weather is removed from climatological data sets. In this section we will address the variety of these unwanted perturbations and the different strategies to eliminate them from the observational series.

24.3.1 Causes and Consequences of Inhomogeneities

As advanced in the introduction, any change in the conditions of observation or in the environment of the observatory may have a significant impact on the measured data. The first ranges from changes of instrumentation (including accessories such as wind or radiation shields) or observer, observing schedules and formulas used to compute daily values from observations at fixed hours, to relocation of the instruments (both horizontal or vertical displacements) or the whole observatory. And the environment may change in a variety of scales, from growing trees or new buildings or other obstacles in the vicinity to general urbanization or other land use changes in the area (e.g., introducing irrigation practices or new crops). Some changes will produce abrupt variations in the series, while others will introduce gradual shifts. For example, a forest fire affecting the area will suddenly change the albedo and hydric balance to slowly return to the initial conditions as vegetation recovers.

The impact of these changes depends on the climatic element being considered. Trees and other obstacles will affect wind speed and rainfall, this last much modified by whirls in the rain gauge, but pressure will not be influenced, while any relocation implying changes in altitude will be clearly detected by the latter. Temperature is one of the more important elements and at the same time very sensitive to all kind of changes. Since a perfectly calibrated thermometer yields just its own temperature, we must grant that this is also the temperature of the air by allowing a perfect thermal equilibrium between both. This is accomplished by letting the air flow as freely as possible around the thermometer bulb, while it is concealed from any kind of radiation exchange (from the sun, direct or reflected by the ground, but also from infrared radiation imbalances with heated surfaces or the coldness of a cloudless sky). It is clear that both requirements are incompatible, but different thermometer screens have been developed in the past that try to approximate to these ideal conditions. In the first observatories, thermometers were merely mounted on a north wall, but then they were exposed to (mild but noticeable) sunshine in the early and late hours of the summer days. Therefore, they were shielded from the sun with simple shading panels. Afterward it was realized that the thermal inertia of the walls and their north orientation had a microclimatic influence, and thermometers begun to be installed either on terraces or open ground. But as the widest diurnal temperature range takes place on the ground (due to its strong radiation exchanges), air temperature bears important vertical gradients, mainly on calm sunny days, and therefore the height at which the thermometer is placed has an important influence on the measured temperatures. That ended in thermometers being installed on the ground, at a fixed height (although not the same in all countries: 1.5, 1.8, 2 m, etc.). Even repainting a thermometer shield can change its albedo and have an (presumably small, but still noticeable) influence on its temperature.

Therefore, observational series may bear anomalous shifts or trends that will mislead any conclusions about their variability, unless provisions are made to correct them, as discussed in the following subsections.

24.3.2 Homogeneity Test Algorithms

The first approaches to investigating the homogeneity of meteoro-hydrological series consisted in expert judgment on visual inspection of the plotted series. It soon became evident that absolute homogeneity was incompatible with long-term climatic variability, and then cumulative values of the problem series were compared with those of an assumed homogeneous reference (double-mass analysis). Another common approach is to compute differences or quotients between the problem series and the reference, allowing objective detection methods to be applied on them. The history of the observatories provides invaluable information about possible changes, hence helping in deciding which series have suffered them and which can serve as references because of their continuity. But these histories (*metadata*) are often incomplete, if not absent, and therefore no series can be fully trusted of being homogeneous, making difficult to ascribe an inhomogeneity to the problem or to the reference station. Two main strategies have been adopted to overcome this difficulty: to compare the problem series with a *composite reference* series made of several neighbor or well-correlated stations and to make *multiple comparisons* with all the potential references, looking for repeating inhomogeneities at the same time step. Composite references, on their side, can be built in different ways, but generally a number of stations in the area of the problem station are chosen based on proximity and/or correlation. Data of the reference series can be normalized to avoid too different ranges, and the composite series can be a plain mean of all the candidates or an average weighted by distance or correlation. Other methodologies compute the reference series with a multiple regression model that estimates the problem data as a function of other available data in the study area.

When a composite reference series has been obtained, differences with the problem series are computed when the variable has a near-normal frequency distribution as, for example, temperature. But quotients are sometimes preferred with elements with a zero lower limit (the case of precipitation or wind speed), frequently presenting an L-shape biased distribution. Nevertheless, quotients become problematic in arid areas, where months with very low or null precipitation are not uncommon. In these cases differences can be better than quotients, and a previous transformation of the data (e.g., through a cubic root) can be used to approximate the frequency distribution to the normal. When both the problem and the reference series are homogeneous, the difference (or quotient) series should be normally distributed, and therefore statistical methods to test this normality can be applied, and this hypothesis can be rejected when the results are unexpected at a chosen significant level.

These tests can be parametric, as the classical t-test, the standard normal homogeneity test (SNHT) [2], the maximum likelihood ratio, or the serial correlation Durbin–Watson test, or nonparametric, as the Wilcoxon–Mann–Whitney [6], and can be applied on the whole series or on running windows (to improve the detection of multiple break points). Other algorithms compute trends before and after a moving time step, as is the case of two-phase regression [5], looking for significant differences between both models. Most of these algorithms provide indication of the location of the inhomogeneous shifts, and it is advisable to compare them with the metadata that can help in confirming or better adjusting the date of the shift.

The following step is to correct the detected inhomogeneities. Several statistical techniques can be used, from computation of the mean differences before and after the break to analysis of variance (ANOVA) or direct computation of the missing values after splitting the series into their homogeneous segments.

The sensitivity of these detection algorithms relies on a good signal-to-noise ratio: with series with a high variability, only big shifts can be detected reliably. Some climatic elements are more variable than others, and daily data have a much higher variability than monthly, seasonal, or annual means or totals. As a result, less inhomogeneities are normally detected on precipitation than in temperature series, and homogenization techniques are usually not applied to daily data but rather to monthly series, whose correction terms are afterward transferred (with some smoothing interpolation scheme) to the daily series.

24.3.3 Homogenization Computer Packages

We have seen that there exist a variety of algorithms among which climatologists can choose the most appropriate for their research, depending on the climatic element, the density of the measuring stations, and the geographical characteristics of the study area. When dealing with a reduced amount of series, these algorithms may be applied manually with the help of spreadsheets or other elementary computing aids. Otherwise they must be laboriously implemented in more complex computer programs. Fortunately several climatologists that have already made this task offer their developments to other users by preparing and documenting ready to use computer packages and distributing them in the Internet or by other means.

Tables 24.1 through 24.3 summarize the characteristics of the currently available homogenization packages, most of them in active stage of development to refine their procedures or to include new capabilities (updates to this information can be followed at http://www.climatol.eu/DARE/index.html). Although developed for the same purpose, they differ in a variety of ways that allow users to choose them according to their preferences, beginning by the operating system on which they run, or whether they are closed or open source. Interactivity or automatism is also an important feature: some users may love the possibility to fully interact with every step of the process through a sophisticated graphical user interface (GUI) and many algorithms to choose (as is the case in AnClim), while others will prefer the ability to run the software in a fully automatic way, either from a terminal or from a script. Most packages are able to compute or select the reference series by themselves, but that is not the case of RHTestV3 that, working on single series, leaves the provision of the references entirely to the user. HOMER is even more interactive in its design, since it is addressed to trained users that can make use of their expertise.

Another important aspect is the tolerance to missing data in the series to be homogenized. When homogenization was applied by manual means, investigators restrained themselves to use only the longest and/or more complete series available for their study. But now that automatic methods allow processing hundreds or even thousands of series, a high tolerance to missing data permits a denser network of stations to be homogenized. The benefits are double: on one side, each problem station will have nearer and better correlated reference data, improving the quality of the results. On the other, stations with short functioning periods will have their series homogenized and their missing data estimated, making them useful for many applications.

Most packages that are primarily devised to operate on monthly data can be applied to daily data as well. This is advisable for the detection of isolated errors in the form of outliers, but not for diagnosing nor correcting inhomogeneities. Current homogenization strategies for daily data relay on different methods of adjusting the monthly correction terms or factors to them. Some packages are programmed in R, a popular statistical computer environment [12], hence allowing its use on virtually any platform or operating system. USHCNv2 is also provided in source code, but in Fortran, and requires a nontrivial

TABLE 24.1 Homogenization Packages: Software Types and Availability

Package	Version	License	Open Source	Operating System	Program Type	Primary Operation	Availability
ACMANT	1.2	Freeware	No	DOS/ Windows	Binary	Automatic	http://www.c3.urv.cat/ members/softpeter.html
AnClim/ ProClimDB	?	Freeware	No	Windows	Binary	Interactive (and automatic)	http://www.climahom.eu/
Climatol	2.2	GPL	Yes	(Most)	R source	Automatic	http://www.climatol.eu/
GAHMDI/ HOMAD	?	GPL	Yes	(Most) Linux	R source R/Fortran	Automatic Interactive	Mail to andrea.toreti@ giub.unibe.ch
HOMER	?	?	Yes	(Most)	R source	Interactive	Mail to olivier.mestre@ meteo.fr
MASH	3.03	Freeware	No	DOS/ Windows	Binary	Automatic (and interactive)	Mail to szentimrey.t@ met.hu
ReDistribution test	?	Freeware	Yes	(Most)	R source	Interactive	Mail to predrag. petrovic@hidmet.gov.rs
RHTestV3	3	Freeware	Yes	(Most)	R source	Interactive	cccma.seos.uvic.ca/ ETCCDMI/software. shtml
USHCNv2	2	Freeware	Yes	Some Linux versions	Fortran source	Automatic	ftp://ftp.ncdc.noaa.gov/ pub/data/ ushcn/v2/monthly/ software

Note: An interrogation sign marks unknown or doubtful items.

TABLE 24.2 Homogenization Packages: Data Types and Detection Methods

Package	GUI	Time Resolution	Input Format	Metadata Use	Detection Method	Ref. Series Selection	Detection Statistic	Climatic Variables
ACMANT	No	Monthly	Text	No	Reference	Correlation	Caussinus–Lyazrhi	Temperature
AnClim/ ProClimDB	Yes	Any	Text DBF	Yes	Ref. and pairwise	Correlation and distance	Several	Any
Climatol	No	Monthly	Text	No	Reference	Distance	SNHT	Any
GAHMDI/ HOMAD	No	Monthly Daily	Text	Yes	Pairwise	Correlation	New method	Any Temperature
HOMER	No	Monthly	Text	Yes	Pairwise	Correlation	Penalized likelihood	Temperature
MASH	No	Monthly and daily	Text	Yes	Multiple references	Correlation	Muliple linear regression (MLR) and hypothesis test	Any
ReDistribution Test	No	Subdaily	Text	No	Distribution	None	SNHT-like	Wind speed and direction
RHTestV3	Yes	Monthly and daily	Text	Yes	Reference	Correlation	Penalized max. t and F tests	Any
USHCNv2	No	Monthly	Text	Yes	Pairwise	Correlation	MLR	Temperature

TABLE 24.3 Homogenization Packages: Correction Methods and Outputs

Package	Correction Method	Missing Data Tolerance	Outputs				Documentation
			Homogenized Series	Corrected Outliers	Corrected Breaks	Graphics	
ACMANT	ANOVA	75%–80% (?)	Yes	Yes	Yes	No	User's guide
AnClim/ ProClimDB	Several	User defined	Yes	Yes	Yes	Yes	Manuals
Climatol	Missing data filling	Very high	Yes	Yes	Yes	Yes	User's guide
GAHMDI/ HOMAD	New method	?	Yes	No	Yes	Yes	None
HOMER	ANOVA	?	Yes	?	Yes	Yes	None
MASH	Multiple comparisons	30%	Yes	Yes	Yes	Yes	User's guide
ReDistribution test	None	10%–20%	No	No	Detected breaks	No	None
RHTestV3	Multiphase regression	?	Yes	No	Yes	Yes	User's guide
USHCNv2	Multiple comparisons	Very high	Yes	?	Yes	No	Plain text notes

Note: An interrogation sign marks unknown or doubtful items.

compilation step before its use (unless specific Linux versions are used: RedHat EL4/EL5 and Ubuntu Hardy Heron, according to their developers).

Apart from preferences about the operating system, interactivity, or whether a GUI is available, potential users will be interested in the performance of these packages in homogenizing data sets. The most thorough comparison exercise was made during the COST Action ES0601, its results being summarized in Venema et al. [14], but as this field is in active development, many packages have improved since those Action results and more intercomparisons are needed with the new versions. It is very difficult to repeat that exercise in the same way, which included time-consuming manual methods, but automatic benchmarking experiments are currently under way, although restrained to fully automatic runs of the packages. The first results of a comparison on ten simulated temperature series with one or two shifts in the mean of ±2°C in five of them show a general good and similar performance of ACMANT, Climatol, RHTestV3 relative with constant correction, and HOMER. All of them corrected quite well the anomalous trends of the unhomogenized series. RHTestV3 with quantile matching gave worse results, although not as bad as absolute homogenization, which should be avoided whenever possible [9]. The referred packages were those easier to automatize, but others will be included in the future and more complex homogenization problems will be tried as well (see http://www.climatol.eu/DARE/testhomog.html).

24.3.4 Application Example

For a better illustration of the use of these packages, we can proceed with a case study in which, from a set of raw monthly precipitation data, we want to obtain average values and trends after a homogenization process. The data come from 37 real stations located in a 115 × 65 km rectangular area centered at 41.3°N, 1.7°E (a coastal area including Barcelona, Spain), and they cover a period of 60 years (1951–2010).

As is usually the case, these series are far from being complete, as can be seen in Figure 24.1. Therefore, and due to the strong variability of the precipitation, missing data must be estimated before computing consistent averages of those stations. This task will be accomplished with the *Climatol*

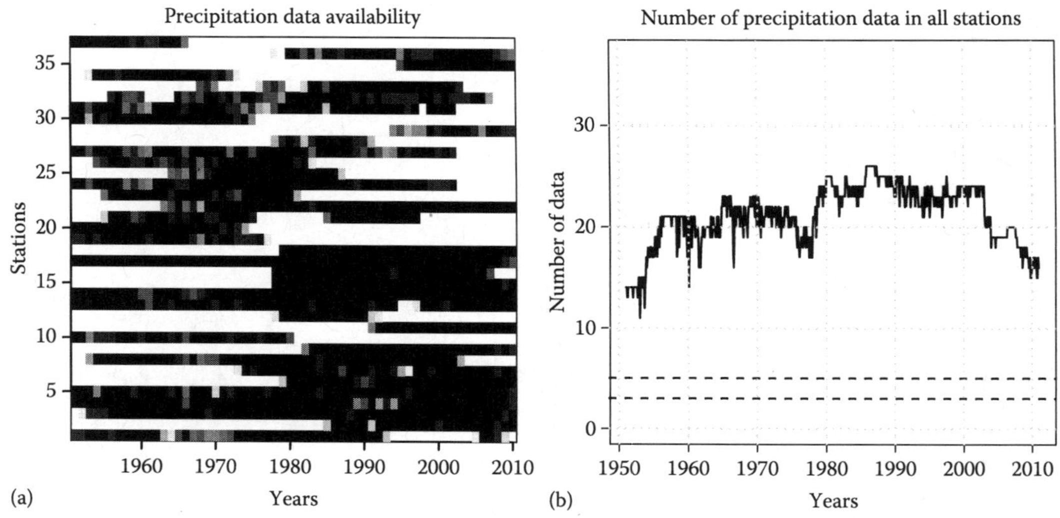

FIGURE 24.1 Data availability in the example area. (a) Data coverage in each station (lighter gray indicate less data in that year). (b) Number of data in every time step for the whole area.

package that has a high tolerance to missing data because reference stations are chosen by proximity and not by correlation (which cannot be reliably computed without a minimum number of simultaneous observations).

On the other hand, the area has a typical Mediterranean climate, with a marked minimum in July, although null precipitations may happen eventually in any month. Therefore, monthly precipitations exhibit a strong L-shape frequency distribution (Figure 24.2a), with values ranging from 0 to 730 mm. *Climatol* offers two possibilities in these situations, either to normalize the data using rate to normals (instead of the default standardization with the mean and the standard deviation) or to apply a root transformation to the data in order to approximate their distribution to the normal. A cubic root transformation has been chosen in our example, producing a near-normal distribution (Figure 24.2b) if we

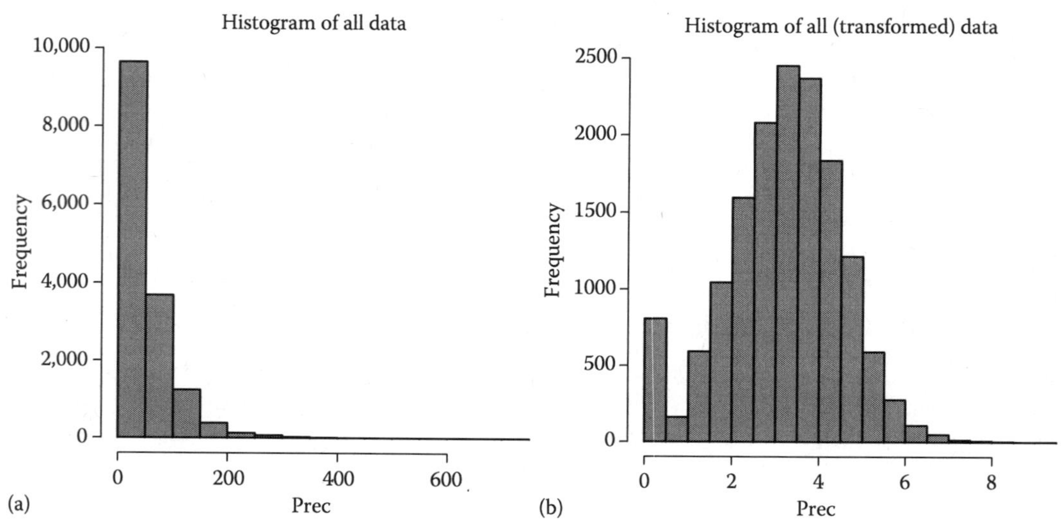

FIGURE 24.2 Frequency distributions of the precipitation data used. (a) Original. (b) After a cubic root transformation.

leave apart the zero-precipitation months. To apply this package to our case, we need to have installed the free R statistical environment and the *Climatol* contributed package (see the R help for installation instructions).

The second step is either to read the data directly from our archives or databases with our own R procedures or to save them in the *Climatol* plain text input formats, which is the option followed here. Only two files are needed in this case: one for the station coordinates and the other for the data. If we use "Prec" as short for precipitation, then the name of the first file must be "Prec_1951–2010.est," and it lists the X(°), Y(°), Z(m) coordinates, the codes, and the names of the stations, as shown in these first lines:

```
1.054 41.074 19 "0013" "CAMBRILS"
1.052 41.141 123 "0015I" "RIUDOMS (CAMARA AGRARIA)"
1.179 41.15 68 "0016A" "REUS (AEROPORT)"
1.145 41.112 53 "0017" "VILASECA DE SOLCINA"
...
```

The data are stored in a file named "Prec_1951–2010.dat," also as plain text, containing the data of all the stations in the same order as in the previous file and in chronological order within each station, from January 1951 to December 2010 (including special codes or values for any missing data). The first lines of this file, corresponding to the years 1951–1953 of Cambrils, are (NA stands for "not available," the standard missing data code in R) as follows:

```
42 16.6 NA 59.9 56.2 27.8 11.9 45.7 146.2 122.5 59 36
21 21 47 42 34 49 31 3 NA 104.7 7.5 9.1
0 2.1 40.1 13 20 140.2 13 37 58 62.5 28.7 64.4
...
```

After having prepared these two files, we start R in our computer and issue these two commands to load the *Climatol* library in memory and homogenize our example data (for detailed explanations on the command parameters, see the user's guide [8]):

```
library(climatol)
homogen("Prec", 1951, 2010, rtrans=3, vmin=0, gp=4)
```

After several minutes (depending on the processor speed) we get four output files, with the same base name as the input files and with extensions:

esh: Output coordinates and names of the stations, with supplementary information
dah: Output homogenized data
txt: Log of all console output during the process
pdf: Collection of diagnostic graphics (123 pages in our example)

All output files are plain text files except the PDF. These graphics show us the course of the homogenization process, in which, after normalizing all series, anomalies computed as differences between each one and an average of their closest neighbors are subject to the reliable and well-known standard normal homogeneity test, and the more outstanding inhomogeneities result in splitting the series at that point in an iterative process, until no inhomogeneity remains greater than a prescribed threshold. Figures 24.3 and 24.4 show one of those anomaly graphics and the reconstruction of both series resulting from splitting the original. Eight splits were made in our example exercise, two of them in the same station.

The homogenized output data in "Prec_1951–2010.dah" has the same text format as the input data, which is not much user friendly because of the lack of any time or station reference apart from the position of each data item in the file. Fortunately *Climatol* has a post-processing function to help in

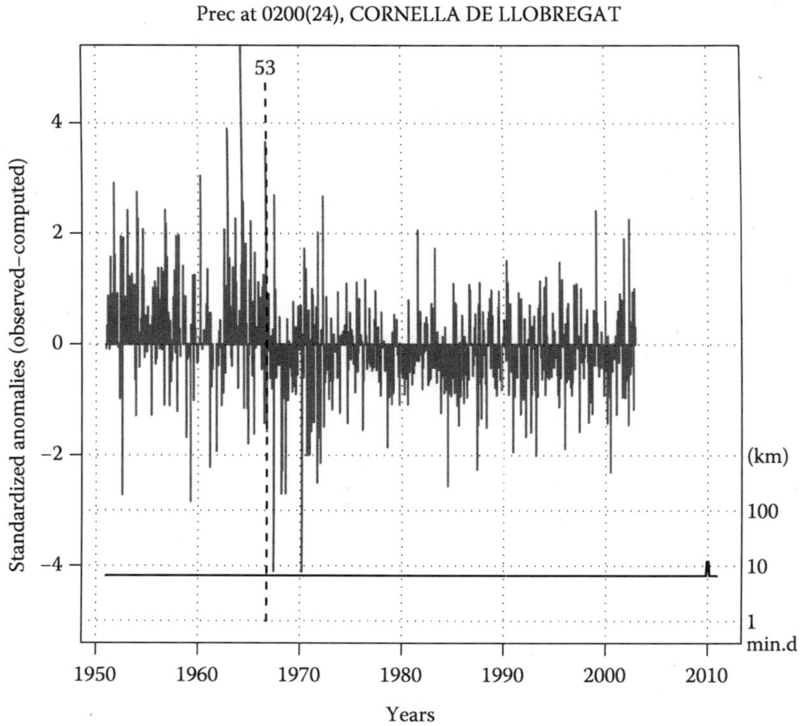

FIGURE 24.3 Shift in the mean detected with SNHT = 53 on the standardized anomaly series. (The closest reference data are at about 9 km).

obtaining various indexes from the homogenized data. If we are interested in the trends of the last 50 years (1961–2010), we issue the following command:

```
dahstat('Prec', 1951, 2010, 1961, 2010, out='tnd', vala=1, long=TRUE)
```

which yields monthly and annual trends in mm/century, computed as regression coefficients with time, of the 37 stations (choosing the longest period with original data when they have been split). In order to assess the effect of the homogenization, we can ask *Climatol* to compute the missing data without splitting any series and then get a new set of trends without homogenization. Figure 24.5 displays the frequency histograms of both trend sets, and it is evident that trends are more consistent and less spread after homogenization. Not only the trend range is reduced as a result of correcting the inhomogeneous series, but all trends are affected through the missing data interpolation process: the mean trend changes from −40.7 to −21.1 mm/century, and the more robust median, from −36.9 to −22.1 mm/century. The same *homogen()* function can be used to compute 1961–1990 normals or any empirical percentile for, for example, categorize the forthcoming new monthly values for climate monitoring purposes, and these categories (dry, normal, wet, etc.) can vary significantly if the percentiles are computed with inhomogeneous data.

24.4 Summary and Conclusions

This chapter has presented the variety of unwanted perturbations that can affect climatological series in such a way as to induce to incorrect evaluations about their trend and variability, hence compromising the quality of the decisions made by the user.

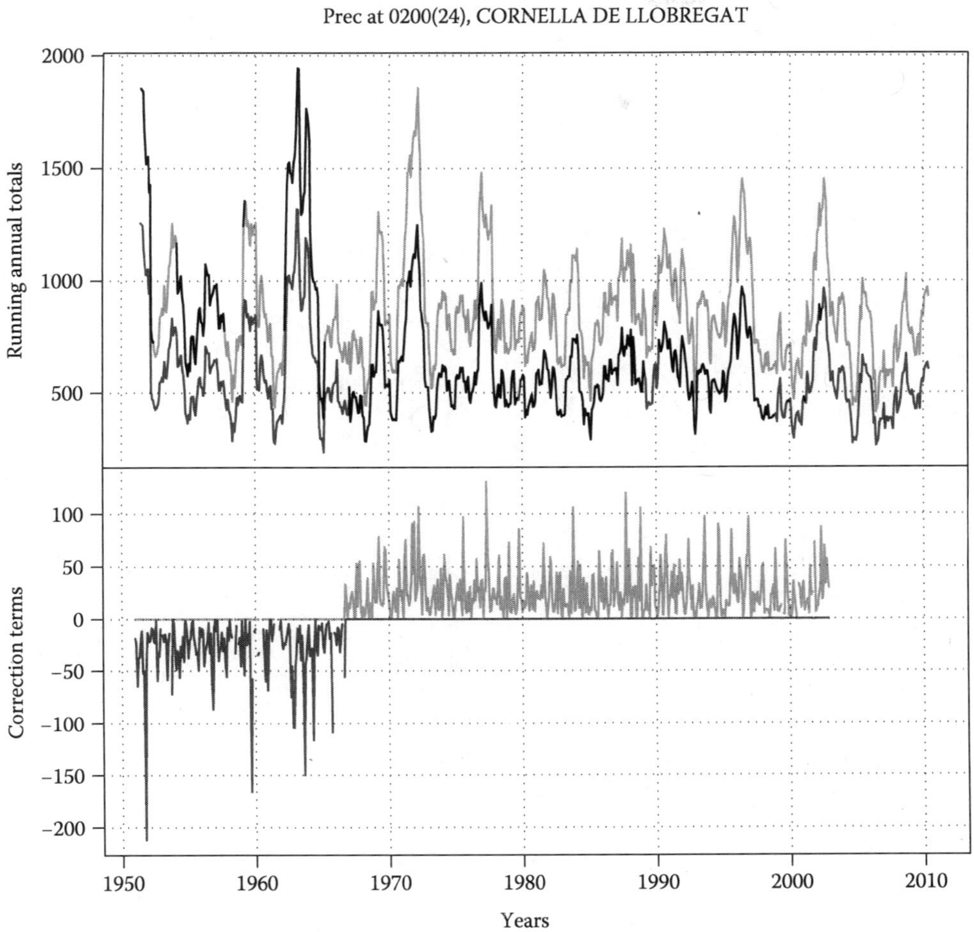

FIGURE 24.4 Reconstruction of the series from the two homogeneous sections (original data in black).

Different techniques to correct these inhomogeneities have been listed, but as their application to a big number of series can be very time consuming, attention has been focused on different implementations in the form of computer packages that are freely available. Their characteristics have been summarized in synoptic tables, and an example application to a real case has been illustrated, showing the higher consistency of the trends by comparison to the values obtained from the raw series.

The same may happen with other statistical characteristics of the series, and although the homogenized series obtained by different methods may differ, any good homogenization methodology will be worth being applied instead of using the original inhomogeneous data.

Ideally, the best situation would be to have homogeneous observational series, making unnecessary the use of these techniques. Unfortunately this is often not the case, but network managers can help in minimizing the problem by trying to keep observation conditions unchanged as much as possible, logging detailed and easily available records of unavoidable changes to provide good metadata to the users and avoiding by all means any change affecting a whole network at the same time, since these would invalidate the use of even the best relative homogenization methods.

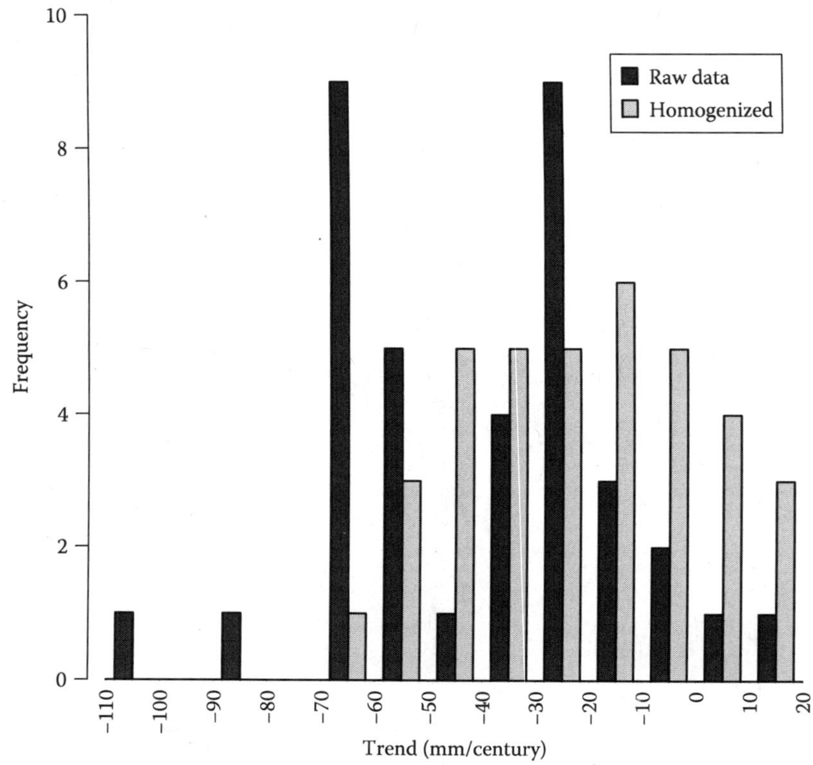

FIGURE 24.5 Example of precipitation trends computed before and after homogenization.

References

1. Aguilar, E., Auer, I., Brunet, M., Peterson, T.C., and Wieringa, J. 2003. *Guidelines on Climate Metadata and Homogenization*. WCDMP-No. 53, WMO-TD No. 1186, 53pp. World Meteorological Organization, Geneva, Switzerland.
2. Alexandersson, H. 1986. A homogeneity test applied to precipitation data. *J. Climatol.*, 6(6): 661–675.
3. Beaulieu, C., Seidou, O., Ouarda, T.B.M.J., Zhang, X., Boulet, G., and Yagouti, A. 2008. Intercomparison of homogenization techniques for precipitation data. *Water Resour. Res.*, 44: 20.
4. COST Action ES0601. Advances in homogenisation methods of climate series: An integrated approach HOME. http://www.homogenisation.org/
5. Easterling, D.R. and Peterson, T.C. 1995. A new method for detecting undocumented discontinuities in climatological time series. *Int. J. Climatol.*, 15(4): 369–377.
6. Gérard-Marchant, P.G.F. and Stooksbury, D.E. 2008. Methods for starting the detection of undocumented multiple changepoints. *J. Clim.*, 21(18): 4887–4899.
7. Guijarro, J.A. 1998. Data quality control to the Spanish surface automatic meteorological stations in PYREX. *Proceedings of the 1st Workshop on Quality Control of Meteorological data in MAP*, Vienna, Austria, January 26–27, pp. 64–71.
8. Guijarro, J.A. 2011. User's guide to climatol, 40pp. http://www.climatol.eu/climatol-guide.pdf (accessed on July 15, 2013).
9. Guijarro, J.A. 2013. Influence of network density on homogenization performance. *Proceedings of the Seventh Seminar for Homogenization and Quality Control in Climatological Databases*, Budapest, Hungary, October 24–28, 2011, WCDMP/WMO-TD series, in press.

10. Peterson, T.C., Easterling, D.R., Karl, T.R., Groisman, P., Nicholls, N., Plummer, N., Torok, S. et al. 1998. Homogeneity adjustments of in situ atmospheric climate data: A review. *Int. J. Climatol.*, 18(13): 1493–1517.
11. Peterson, T.C., Vose, R., Schmoyer, R., and Razuvaёv, V. 1998. Global historical climatology network (GHCN) quality control of monthly temperature data. *Int. J. Climatol.*, 18(11): 1169–1180.
12. R Development Core Team. 2010. *R: A Language and Environment for Statistical Computing.* R Foundation for Statistical Computing, Vienna, Austria. ISBN 3-900051-07-0, http://www.R-project.org/
13. Shearman, R.J. 1975. Computer quality control of daily and monthly rainfall data. *Meteorol. Mag.*, 104: 102–108.
14. Venema, V., Mestre, O., Aguilar, E., Auer, I., Guijarro, J.A., Domonkos, P., Vertacnik, G. et al. 2012. Benchmarking homogenization algorithms for monthly data. *Clim. Past*, 8: 89–115.
15. W.M.O. 2004. *Fourth Seminar for Homogenization and Quality Control in Climatological Databases*, Budapest, Hungary, October 6–10, 2003, WCDMP-No. 56, WMO-TD No. 1236, 248pp.
16. W.M.O. 2008. *Proceedings of the Fifth Seminar for Homogenization and Quality Control in Climatological Databases*, Budapest, Hungary, May 29–June 2, 2006, WCDMP-No. 71, WMO-TD No. 1493, 217pp.
17. W.M.O. 2010. *Proceedings of the Sixth Seminar for Homogenization and Quality Control in Climatological Databases*, Budapest, Hungary, May 26–30, 2008, WCDMP-No. 76, WMO-TD No. 1576, 121pp.
18. Zahumenský, I. 2005. Guidelines on quality control procedures for data from automatic weather stations. *WMO Technical Conference on Meteorological and Environmental Instruments and Methods of Observation*, TECO-2005, Bucharest, Romania, May 4–7, 2005.

25

Satellite-Based Systems for Flood Monitoring and Warning

Reza Khanbilvardi
The City University of New York

Marouane Temimi
The City University of New York

Jonathan Gourley
National Weather Center (NWC)

Ali Zahraee
The City University of New York

AUTHORS

Reza Khanbilvardi is a National Oceanic and Atmospheric Administration (NOAA) chair professor of civil engineering, director of NOAA-Cooperative Remote Sensing Science and Technology (CREST) Center by the City University of New York (CUNY), and executive director of CREST Institute at City College/CUNY. He is the editor in chief of the *International Journal of Water*. He is the author of 10 books in water resources and over 100 papers.

Marouane Temimi is a research associate professor at CREST Institute at City College/CUNY. Dr. Temimi's research interests cover the determination of land surface parameters from space using satellite imagery. Dr. Temimi has published several peer-reviewed articles in scholarly journals of international circulation and contributed to three books.

Jonathan Gourley is a research hydrologist with the NOAA National Severe Storms Laboratory (NSSL) in Norman, OK, and holds and affiliate associated professor position with the University of Oklahoma's School of Meteorology. Dr. Gourley's expertise is in remote sensing of precipitation and hydrological modeling. He has published over 50 peer-reviewed articles.

Ali Zahraee is a research associate at NOAA-CREST and City College/CUNY. He received his PhD from the University of California–Irvine regarding precipitation estimation and forecasting using remote-sensing information. Dr. Zahraei was also a visiting scholar at NOAA NSSL.

PREFACE

The conventional tools to estimate hydrological parameters of a flood may often fail to record an extreme event given the fact that climate changes affect the magnitude and the frequency of these events. The extent and frequency of catastrophic events suggest using satellite technology as the key tool for flood monitoring and warning. The main core in this field is the application of satellite technology to measure required data for hydrological modeling. This chapter therefore summarizes the existing satellite technologies, relevant hydrological information, and their applications. In addition, satellite capabilities in quantitative precipitation estimation (QPE) and forecasting (QPF) have been discussed. Given the importance of uncertainty quantification in hydrological models (HyMODs), the review of the ensemble approach to assimilate satellite observations into HyMODs closes this chapter.

25.1 Introduction

Major flood events with substantial damages are observed across the United States every year. Climate change and variability affect the magnitude and the frequency of these events. Nowadays extreme meteorological events are more destructive and more frequent. In 2008, for instance, much of central and eastern Iowa was affected by a 500-year flood, which was classified as the worst in the history of the region [45]. Many other similar events have occurred in the United States and around the world. The recent major flood in Pakistan in 2010 is another example. It is therefore important to develop an accurate and clear understanding of the dynamics of extreme hydrological events.

Three main elements are necessary for a reliable forecast and monitoring of hydrological processes, namely, good-quality precipitation data, accurate characterization of surface conditions, and robust, accurate, and detailed HyMODs. This chapter addresses these three elements and analyzes their role in delivering reliable forecast of hydrological events. A particular focus is placed on satellite-based land surface and precipitation products and their integration into operational HyMODs through data assimilation systems. The following questions will be addressed:

How could satellite-based precipitation products improve the performance of flood monitoring and forecasting systems?

Is it possible to accurately determine surface conditions and hydraulic parameters from space?

What is the impact of integrating satellite-based products into flood monitoring and warning systems?

In the United States, NOAA National Weather Service (NWS) issues on a daily basis maps of Flash Flood Guidance (FFG) that include the amount of rainfall that is necessary to produce a flash flood at a given location and for a specific duration of rainfall. NWS via its River Forecast Centers (RFCs) develops FFG maps for each county within the RFC domain for three durations: 1, 3, and 6 h. US Geological Survey (USGS), on the other hand, has established a vast network to observe and monitor hydraulic conditions in several rivers across the country. Discharge and water level in rivers and lakes are routinely observed and made available for end users. NOAA National Climate Data Record provides the community with satellite, radar, and rain gauge precipitations. In addition, the NOAA NSSL is developing techniques for high-resolution application over the United States and is involved in the development of global hydrological forecast systems.

Despite the investment of tremendous efforts to monitor and forecast hydrological processes and mitigate the impact of extreme events on the society, these events continue to impact larger areas and occur on regional and local scales. The extent and the dynamic of extreme events suggest using remote-sensing-based techniques to monitor them and understand their dynamic.

Remote-sensing data and particularly passive microwave (MW) images have been largely used to monitor inundation and major hydrological events [39]. Passive MW observations have appeared as an attractive source of information on surface and near-surface water because of their high temporal resolution, their large spatial coverage, and the capability of their signal to penetrate through clouds. Satellite imagery has been also used to determine water level in lakes and reservoirs, discharge in rivers, and even surface conditions like roughness and relief. So, it is possible to determine from space key hydrological parameters and force HyMODs with appropriate, up-to-date information on surface conditions. Hence, the integration of satellite-based products has been tested and implemented in rural and large watersheds as well as small and urban basins. The overall performance of these models still depends, however, on the quality of the precipitation product, which is the main driver of flood monitoring and forecasting systems.

The observation of the space/time variability of precipitation globally is vital for understanding global water and energy cycle. In terms of precipitation measurements, it is critical to have locally collected data over a global domain to capture spatiotemporal variability of precipitation in micro, meso, and synoptic scales. Precipitation is the most important driving force in HyMODs as it impacts the soil moisture and other important soil or land surface parameters. QPE and QPF are the application of science and technology to either estimate or predict the rainfall rate for a specified time and a given location. QPE and QPF algorithms incorporate satellite-based atmospheric observations (e.g., visible [VIS], infrared [IR], and MW channels) to retrieve rainfall rate for the current or future time(s) [49].

Hence, remote sensing of hydrometeorological variables, such as precipitation, soil moisture, water levels in large bodies of water, areal extent of inundation, and river discharge, has provided great potential to monitor and forecast flooding in many ungauged basins in the world. Such satellite-based products are derived from different observations and a number of sensors. They are generated at different spatial resolutions and derived from observations taken at different overpass times. The indirectness of the observations to the desired measured variable, limited resolution in space and time, and retrieval errors all must be considered when applying remote-sensing data for flood applications. Assessing the exact impact of the integrating remote-sensing-based products into flood forecasting systems is crucial.

This chapter comprises three sections: The first section of this chapter includes further details about the use of satellite imagery to characterize land surface conditions and determine relevant surface and subsurface parameters that can be integrated in HyMODs. The second section presents in more details satellite-based precipitation products and their use to forecast as QPF or monitor as QPE rainfall events. The third section presents results on error modeling of satellite-based rainfall estimates, HyMOD calibration that considers the pixel resolution of the rainfall product, and the use of the ensemble square root filter (EnSRF) to assimilate a proxy for river discharge based on passive MW brightness temperature signals with the overall aim of improving flood forecasts in ungauged basins.

25.2 Satellite Data for Hydrological Modeling

The detection of flooded regions typically comes from trained spotter reports collected by the NWS or from USGS stream gauges. Both sources are woefully inadequate to capture the spatial extent of flooding, particularly in sparsely populated regions. Satellite imagery greatly enhances the observational aspects of flooding and thus enables refined initialization and evaluations of distributed HyMOD outputs. Several sensors have been used to delineate flooded areas. They fall into one of the two categories. First, VIS and IR sensors, often onboard polar-orbiting satellites, provide images on a daily basis like Moderate Resolution Imaging Spectroradiometer (MODIS), VIIRS, and AVHRR at moderate spatial resolution ranging from 250 to 1000 m. Under this category, we may also find high-resolution sensors like Landsat and ASTER. Despite their long revisit cycle (16 days in the case of Landsat) that is not appropriate to monitor rapid processes like flooding events, these sensors provide images with a spatial resolution that can reach 30 m. The major disadvantage of this category of sensors is their sensitivity to clouds that hampers their use to monitor flooding conditions that often occur under cloudy sky. So, the second category of sensors, namely, MW sensors, which are capable of penetrating clouds, has been utilized as an alternative. Under this category of sensors, we may find two types of sensors, namely, passive and active MW sensors. They both penetrate through clouds. They are only sensitive to atmosphere at higher frequencies, above 19 GHz. Several passive MW sensors like special sensor microwave/imager (SSM/I), Advanced Microwave Scanning Radiometer-Earth Observing System (AMSR-E), and Advanced Microwave Sounding Unit (AMSU) have been utilized to monitor flooding events and delineate inundated areas particularly on a regional scale because of the coarse spatial resolution of passive MW observations. On the other hand, active MW sensors commonly known as radar like Radarsat, ERS, and QuickSAT can deliver images at higher spatial resolution. For instance, in the case of Radarsat-1, images captured in the fine mode have a spatial resolution of 12 m. However, the extent of scenes captured at such high spatial resolution is limited and does not provide a comprehensive coverage of flooding events that usually occurs on a watershed scale. Because of the limitation of each individual sensor, the use of a multi-sensor approach has been investigated [37]. Recent advances in sensor development and data manipulation and sharing have fostered the implementation of such approaches.

In addition to delineating flooded regions, satellite imagery has been used to determine soil moisture that is a critical parameter in HyMODs as it controls the partitioning of rainfall into infiltration and runoff. Passive MW-based approaches to determine soil wetness generally fall into one of the following categories: In the first category, radiative transfer models are generally used in an inversion process that minimizes the differences between observed and simulated brightness temperatures [35]. In the second category, simplistic approaches have been developed to determine a number of wetness indices as proxies for soil moisture [3].

Recent sensors developed for soil moisture retrieval operate in the L-band frequency that minimizes the effect of vegetation and atmosphere and penetrates to deeper soil layers. Recent missions, like SMOS and AQUARIUS, have L-band radiometer onboard. The NASA SMAP mission planned for 2014 will have both radar and radiometer operating at the 1.4 GHz frequency. Despite the great interest in low frequencies such as L- and C-bands that are more appropriate for soil moisture retrieval, other studies have explored the potential of the 37 GHz channel to monitor flood conditions. Observations from AMSR-E at this frequency were used by Temimi et al. to monitor streamflow in the Mackenzie River Basin in Canada [44]. Recently, Temimi et al. have introduced a Polarization Ratio Variation Index (PRVI) that is a MW-based wetness index that was successfully tested during the major flood in Iowa in 2008 [45]. The index was expanded on global scale and implemented as flood warning system. The index showed sensitivity to flooding conditions on a global scale and was successfully verified during the major flood in Pakistan in 2010.

The use of remote sensing was not limited to sensing surface conditions, but it was expanded to determining key hydraulic variables. Brakenridge et al. have used AMSR-E 37 GHz brightness temperature to infer river discharge [6]. Their findings were in line with those by [5] that investigated the feasibility of measuring river discharge from space. A relationship was established between discharge values and the extent of flooded regions or other river and watershed parameters like the hydraulic width [10]. Temimi et al. have estimated the time of concentration of a large watershed in Upper Mississippi by determining the phase lag between the peak of discharge downstream and the timing of the maximum of water extent detected from space [45]. Smith and Pavelsky estimated the flood wave propagation speed using remote-sensing data [41].

In northern watersheds, remote-sensing-based products have been critical in determining ice coverage on lakes and rivers that significantly affect the hydraulic and hydrological processes. During the spring melting period, breaking ice raises the water in rivers and lakes to dangerous levels. Then, the breakup releases destructive flood waves. Only in the United States, the estimated cost of damages related to these processes is in the order of $100 million [52]. Moreover, the dynamic of river ice significantly affects hydropower plants management, navigation, and even transportation since ice on rivers is usually used during winter as a bridge to make the link between northern communities. VIS, IR, and MW images have been used for ice mapping over rivers and ice jam monitoring [36]. The MODIS is providing VIS/IR images at an interesting spatial resolution of 250 and 500 m. This spatial resolution is suitable for large rivers such as the Mackenzie River in northwestern Canada, the Colville River in

Alaska, and the Lena River in Russia. However, over narrower rivers the coarse spatial resolution of these images does now allow for possible monitoring on ice. Active MW, on the other hand, that is, radar images, was used in numerous studies as an interesting tool since they can penetrate cloud and provide images at higher spatial resolution. Several studies were conducted to investigate interaction between ice and backscattering radar signal [7]. These studies showed that it is possible to retrieve ice characteristics from the spectral content of backscattered signal [32]. However, these images have a limited field of view at fine mode and do not include comprehensive information on ice dynamic and flooded area particularly over large rivers.

Finally, coastal areas and wetlands play a major role in both hydrological processes and climate trends; therefore, it is crucial to understand the hydrodynamics of these systems and the interplay between the ocean, estuaries, rivers, and surrounding wetlands. Satellite images (optical and MW) have also shown an interesting potential to monitor changes that occur over long and short time scales. Active microware sensors measure backscatters and return signal amplitude, phase, and polarization that are independent of sun angles and weather conditions (Figure 25.1).

Synthetic Aperture Radar (SAR) interferometry (InSAR) measures the corresponding phase difference resulting from the difference in distances to the same target in two SAR images. It is possible based on this concept to determine changes in water levels in coastal region and lakes and rivers as well. Combining interferometric radar observations with a digital elevation model of the river region will allow for rapid determination of water levels, especially if such assessments are done routinely. In addition, different frequency (C-, X-, L-bands) SARs provide different sensitivity and radar penetration

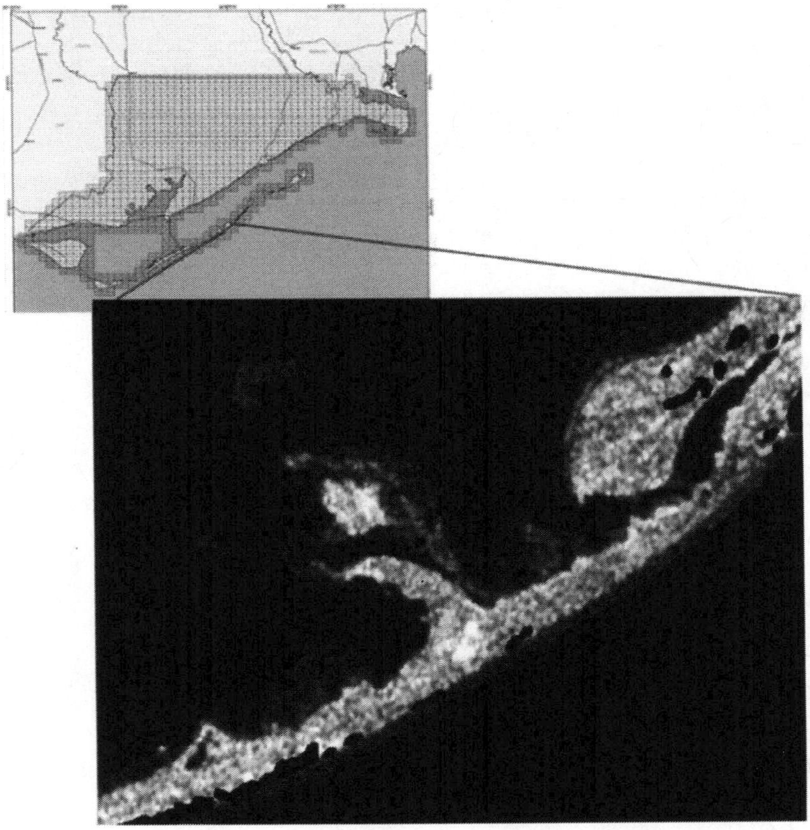

FIGURE 25.1 Coastal inundation (in dark gray) as detected by Radarsat-1 in Franklin County, FL, on January 5, 2010.

depth depending on the properties of land cover, thereby providing additional information for robust classification and morphology determination. A time series of SAR images taken over the same region at the same tide phase allows for an assessment of the morphodynamics of the study area.

25.3 Quantitative Precipitation Estimation/Nowcasting

The observation of the space/time variability of precipitation globally is vital for understanding global water and energy cycle. In terms of precipitation measurements, it is critical to have locally collected data over a global domain to capture spatiotemporal variability of precipitation in micro, meso, and synoptic scales. Precipitation, the most important driving force in HyMODs, impacts soil moisture and other important soil or land surface characteristics. QPE and QPF are the application of science and technology used to either estimate or predict the rainfall rate for a specified time and a given location. QPE and QPF algorithms incorporate satellite-based atmospheric observations (e.g., VIS, IR, and MW channels) to retrieve rainfall rate for the current or future time(s) [18,49].

25.3.1 Satellite-Based Precipitation Estimation

The satellite remote sensing started using VIS and IR sensors first to estimate rainfall [2]. The VIS-/IR-based techniques do not measure precipitation directly. Using VIS/IR sensors, the rainfall rate would be estimated from cloud characteristics such as cloud brightness temperature [28,46]. Previous studies have shown that VIS-/IR-based algorithm works well for convective-dominated areas, and not necessarily in stratiform systems [43]. As opposed, MW sensors could penetrate into the cloud. MW-based satellite precipitation measurement has been a major milestone in QPE. Considering MW sensors' relative insensitivity to the cloud cover, they provide direct rainfall measurements. The first major step in using MW-based instrument used the SSM/I that was launched in 1987 with Defense Meteorological Satellite Program (DMSP) with sun-synchronous, near orbital, and 98.8° inclination orbit, equipped with high-frequency sensor of 19.35, 22.23, 37, and 85.5 GHz [9,43].

The Tropical Rainfall Measuring Mission (TRMM) satellite and AMSR-E launched with Earth Observing System (EOS) Aqua satellite are more recent examples [23,30,42]. The TRMM satellite includes the passive TRMM Microwave Imager (TMI) sensor and the first space-based active MW radar. The precipitation radar (PR) provides a vertical distribution of precipitation that would be used for other passive MW sensors [15,23,30]. Even though the TRMM mission was a significant step ahead, the MW-based satellites have both spatially and temporally sparse coverage. Considering all MW-based satellites, they have six to eight observations per day for a given region. Following on the TRMM mission, an international measure to launch new generation of MW-based satellite sensors called Global Precipitation Measurement (GPM) mission has been initiated. The GPM will include a constellation of satellites with passive MW and next-generation active PR. Regardless of higher accuracy and improved temporal and spatial resolution of GPM precipitation products, MW-based sensors might not have the temporal and spatial resolution required for some hydrological activities [18].

Another category of high-resolution precipitation retrieval algorithms consists of a combination of MW observations along with VIS/IR instruments [1]. There are several algorithms using MW rain rate retrievals to calibrate the VIS-/IR-based algorithms [4,16,19,24,47]. In addition, there has been also a trend to use a combination of multi-sensor or multi-satellite observations to provide the intended spatial and temporal scales (e.g., TRMM Multi-Satellite Precipitation Analysis [TMPA] [21,22,34]).

25.3.2 Review of Some Operational Precipitation Retrieval Algorithms

25.3.2.1 TRMM Multisatellite Precipitation Analysis

TMPA uses a wide variety of satellite-based precipitation sensors to have a best-possible estimate of global precipitation. The estimated precipitation has a relatively fine scale (0.25° × 0.25°) every 3 h.

The precipitation has been provided in both real and post-real time. Studies have shown that the finer scale has been associated with more errors and uncertainties. It is recommended to use the TMPA fine resolution data to create a time/space average that is required for each specific application [21].

25.3.2.2 Global Precipitation Measurement Mission

The GPM mission as an international network of satellites plans to estimate the precipitation every 2–4 h globally. The GPM has a "Core" observatory including advanced active and passive MW sensors in a non-sun-synchronous orbit. The GPM will be used as a calibration reference to unify rain and snow rates from a constellation of satellites [40]. The GPM Core Spacecraft will have two precipitation sensors including the GPM Microwave Imager (GMI) and the dual-frequency PR (DPR). The aforementioned sensors will be able to measure both light and heavy rain rate along with snowfall. In comparison with previous MW-based sensors (e.g., TRMM), the GPM has the new capabilities of the GMI and DPR using high-frequency channels (165.6 and 183.3 GHz) on the GMI and a new Ka-band (35.5 GHz) radar on the DPR [40].

25.3.2.3 CMORPH: High-Resolution Precipitation Product Morphing Passive Microwave and IR Satellite Sensors

The CPC MORPHing technique (CMORPH) algorithm incorporates the most desired aspect of VIS/IR sensors that is higher temporal and spatial resolution, along with passive MW precipitation retrieval sensors that has higher quality. The CMORPH algorithm uses the advection field derived from geostationary satellite IR channels to estimate the passive MW-based precipitation propagations [24]. CMORPH provides global precipitation (60 N–60 S) with (~0.10 latitude/longitude, ½ hourly) spatial and temporal resolution, respectively. This algorithm takes advantages of direct precipitation measurement from passive MW as opposed to other merging algorithms of IR and MW observations. CMORPH also has superiority to simply averaging available MW-based precipitation retrieval techniques [24].

25.3.2.4 Self-Calibrating Multivariate Precipitation Retrieval

Self-Calibrating Multivariate Precipitation Retrieval (SCaMPR) algorithm retrieves rain rate using VIS/IR and MW bands from geostationary and earth-orbiting satellites. The MW-based rain rate is used to calibrate an algorithm framework that applies optimal VIS/IR predictors along with optimal calibration coefficients in real time [27,28]. The real-time version of SCaMPR started running from November 2004 applying IR-based data (bands 3–6) from Geostationary Operational Environmental Satellite (GOES)—West and East. The newer SCaMPR version also uses VIS data from GOES (0.6 μm).

SSM/I-based rain rate along with AMSU has been implemented to produce MW-based rain rates in real-time version of SCaMPR [9,50]. In 2009, the operational SCaMPR also began using TMI and PR for calibration [27,28].

25.3.2.5 Precipitation Estimation from Remotely Sensed Information Using Artificial Neural Network Cloud Classification System

Precipitation Estimation from Remotely Sensed Information Using Artificial Neural Network Cloud Classification System (PERSIANN-CCS) is another application of geostationary IR channels that provides high-resolution global precipitation product based on extracted cloud features. This algorithm has four consecutive steps to (1) segment images using cloud classification, (2) extract feature from IR cloud patches, (3) feature classification, and (4) rainfall estimation. The MW-based rainfall rate is used for model calibration [16]. PERSIANN-CCS provides high-resolution precipitation products (HRPPs) (0.04° × 0.0.4°) with the latency between 40 min and ~2 h over the region 180° W to 180° E and 60° S to 60° N, through the UNESCO, G-WADI data server [20].

25.3.2.6 NRL-Blend High-Resolution Precipitation Product

The Naval Research Loboratory (NRL)-blend algorithm uses a real-time collection of time/space matching of pixels from all operational geostationary VIS/IR imagers and passive MW observations. From

January of 2009, the operational model uses five geostationary satellites and twelve available passive MWs and an active radar system. This algorithm produces three hourly accumulated rain rate at a gridded spatial scale of 0.25° between ±60° latitude, updated every 3 h [47,48].

It has been documented that the satellite-based real-time precipitation estimation can significantly improve the hydrological forecasting capabilities. In addition, during the last few years, there have been some activities to use the current atmospheric situation for precipitation short-term forecasting or nowcasting. The precipitation nowcasting might extend the hydrological forecasting lead time.

25.3.3 Satellite-Based Short-Term Quantitative Precipitation Forecasting (Nowcasting)

The nowcasting algorithms improve the predictability of hydrological extreme events [13]. Extrapolation-based or "data-driven" algorithms are capable of extracting information from the ever-increasing volume of remotely sensed data and are documented to be able to produce reliable forecasts, especially within a few hours of the analysis time [11,38,53].

Typically, a nowcasting algorithm generally includes three steps: (1) storm identification, (2) storm tracking, and (3) storm projection. The identification and tracking/matching of stormy areas at the current time (t) with the corresponding storm in the previous time step(s) (e.g., $t-1$, $t-2$,...) is a major challenge for nowcasting. Due to the dynamic nature of storms, they change in terms of intensity, texture, and geometrical characteristics. They may also split or merge with other storms, which is a very complicated situation. To deal with this complexity, several storm identification and tracking techniques have been proposed [8,12,54]. The Storm Cell Identification and Tracking (SCIT) algorithm [25] and the Thunderstorm Identification, Tracking, Analysis, and Nowcasting (TITAN) algorithm [8] are two examples with hydrological applications. The integration of some of these algorithms into the NWS Warning Decision Support System (WDSS) has been reported [31].

Nowcasting algorithms usually have some significant limitations. Most of the nowcasting algorithms are designed either to work exclusively with ground-based radar observations or to nowcast greater and long-lived atmospheric events such as Mesoscale Convective Systems (MCSs) [8,50].

There have been multiple algorithms using high-resolution GOES observation to forecast cloud advection and evolution. As an example of newly developed object-based nowcasting algorithms, the Forecast and Tracking the Evolution of Cloud Cluster (ForTraCC) has been proposed to identify, track, and forecast MCSs over southern America [50].

25.3.4 Review of Operational Precipitation Nowcasting Algorithms

There are currently very few satellite-based precipitation nowcasting algorithms. A major example, the Hydro-Nowcaster (HN), provides high-resolution precipitation forecast in the next 0–3 h running on the National Environmental Satellite, Data, and Information Service (NESDIS) [29]. The HN focuses on motion of rain cells and storm cells growth and decay (GD). The HN identifies each storm, and storm will be matched together between two consecutive satellite imageries (typically 15 min). The storm position and advection will be extrapolated based on the storm tracking and matching process. The GD trend also will be used to extrapolate storm size change. Scofield et al. and Kuligowski et al. have shown the promising performance of the HN in different case studies, although the algorithm still requires more improvement [29,38].

Another recent algorithm called PERsiann-foreCAST (PERCAST) also predicts the rate of rainfall in the next 4 h using the most recent storm images to extract storm features, including advection field and changes in storm intensity and size. PERCAST algorithm is a storm nowcasting module coupled with a previously developed precipitation retrieval algorithm PERSIANN-CCS to forecast rainfall rates. The results show that, by considering storm GD trends for the prediction, the PERCAST-GD further improves the predictability of convective storms [54].

25.4 Satellite Data Assimilation in Hydrological Models

The major components of HyMODs are generally divided into forcings (i.e., rainfall, evapotranspiration), model states (e.g., soil moisture), model parameters, and outputs (i.e., streamflow). The quantification of uncertainty is required in all components because the observations are not perfect nor are the physical processes represented in the HyMOD. An ensemble approach is often preferred in modeling because it readily accommodates observed variables that have been perturbed, equally probable model parameter settings, EnSRF for state estimation, and it yields an array of forecast outcomes that can be used to derive maximum and minimum bounds, measures of central tendency (e.g., ensemble mean), and probabilistic outputs (e.g., percent chance of exceeding a threshold). The multidimensional rainfall error model called SREM2D described in Hossain and Anagnostou uses stochastic space/time formulations to characterize the spatial error structure of satellite retrievals [17]. SREM2D models the joint probability of delineating rain and non-raining areas; thus, it is capable of stochastically producing rainfall in areas where the deterministic satellite-based estimate did not detect it and removing satellite-detected rainfall in other regions. In this sense, it statistically represents missed rain detection and false alarms. A reference rainfall product, preferably gridded rainfall from ground-based radar and rain gauges, is provided to SREM2D as well as the satellite rainfall product. SREM2D then estimates parameters to arrive at an ensemble of equally probable "reference-like" rainfall products. Figure 25.2 shows an annual time series of cumulative basin-averaged rainfall estimated by three different satellite-based algorithms (TRMM 3B42RT, CMORPH, and PERSIANN) in a sub-catchment in the Tar

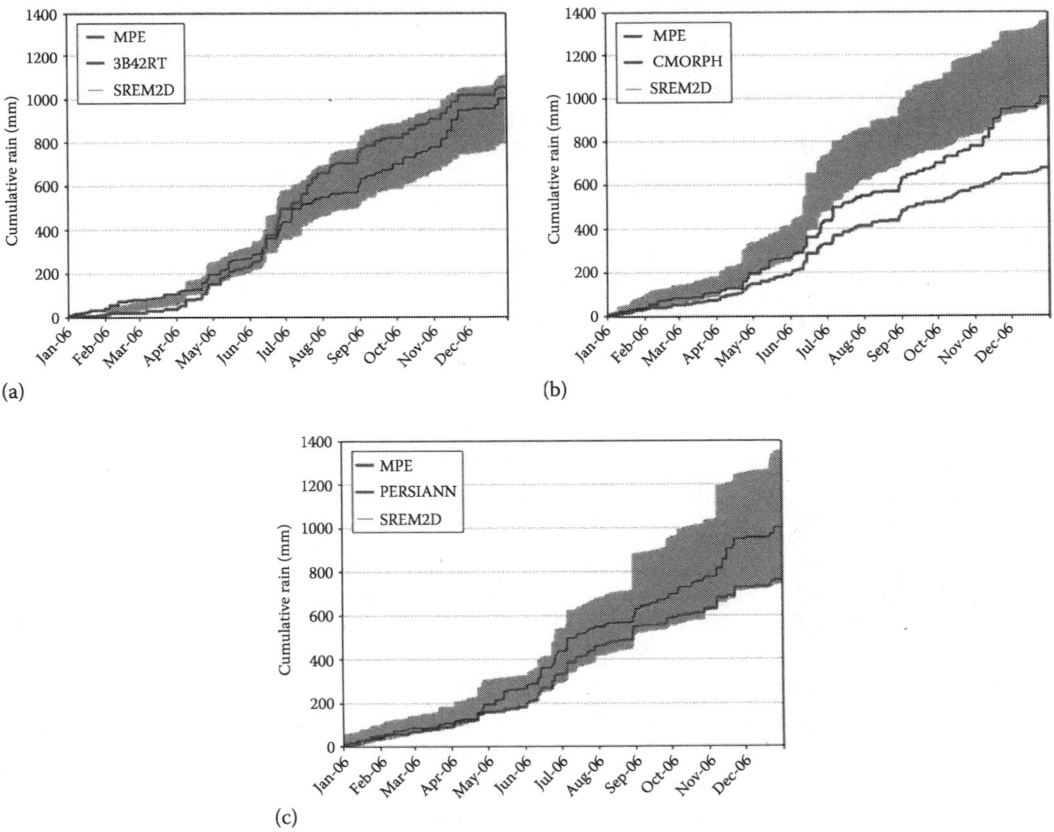

FIGURE 25.2 Cumulative rainfall from (a) TRMM 3B42RT, (b) CMORPH, and (c) PERSIANN on a subcatchment of the Tar River Basin, NC, during 2006. (Adapted from Maggioni, V. et al., *J. Hydrometeorol.*, 2012, in review.)

River Basin, NC. The MPE curve corresponds to the reference rainfall that was estimated by NEXRAD radars and adjusted using rain gauges. These results illustrate how SREM2D is capable of generating a 50-member ensemble that generally encompasses the reference rainfall and yields ensemble spread that is based on the residuals between reference and satellite rainfall. Although the underestimation from CMORPH has been largely corrected, SREM2D had a tendency to overestimate rainfall during the warm season months.

HyMODs that are used for flood prediction purposes have imperfect physical representations, which generally become more evident at small basin scales. Simulated basin responses to rainfall will generally improve following the estimation of model parameters and states. Parameter estimation, also referred to as model calibration, is the process of utilizing observations of the system behavior (i.e., streamflow response to rainfall) in order to iteratively adjust parameters to yield streamflow simulations that better match observations. This process can be quite complex as it involves interactions among model parameters, different or multiple modeling objectives in terms of objective function(s) and variables, and uncertainties in the observations. When using satellite-based algorithms, another oft-ignored feature to consider is the pixel resolution of the satellite precipitation estimates. Gourley et al. evaluated a variety of rainfall estimates from ground-based radar, satellite, rain gauges, and combinations using a HyMOD on the densely instrumented Ft. Cobb basin in OK [14]. In addition to streamflow observations from a stream gauge, they made use of a reference simulation that was forced by gauged rainfall observed by a dense Micronet in the basin. This reference simulation served as a benchmark representing the most accurate simulation to be expected for the HyMOD given the high accuracy and density of the rainfall observations.

Model parameters were automatically estimated using the reference Micronet rainfall forcing at 4 km/h and then aggregated up to the 0.25°/3 h resolution corresponding to TRMM level 3 rainfall algorithms. These two models were thus calibrated with the "best" reference rainfall at two different resolutions. It turns out the HyMOD parameters were sensitive to the rainfall pixel and corresponding model grid cell resolution. It is useful to define a metric that is similar to the Nash–Sutcliffe Coefficient of Efficiency (NSCE) but provides information about the simulation skill in reference to the Micronet-forced simulation rather than the mean of the observed streamflow. The Micronet-relative efficiency (MRE) is thus defined as follows:

$$\mathrm{MRE} = 1 - \frac{\sum_{i=0}^{N}\left(Q_i^{\mathrm{R}} - Q_i^{\mathrm{obs}}\right)^2}{\sum_{i=0}^{N}\left(Q_i^{\mathrm{ARS\_Micronet}} - Q_i^{\mathrm{obs}}\right)^2} \tag{25.1}$$

where

Q corresponds to streamflow

the superscript R is for the different rainfall algorithms

obs is for observed

Figure 25.3 shows the model performance relative to the simulations that used the reference Micronet rainfall forcing at the 4 km/h resolution. As expected, there is degradation in MRE when the HyMOD was forced with rainfall values that were different than the reference values used in calibration. The curves noted with "RESAMPLE" correspond to the simulations that were forced by TRMM 3B42RT and 3B42V6 into the model calibrated at the coarser satellite resolution. By considering this resolution difference, we now see the RESAMPLE simulations have improved markedly. In fact, the RT-RESAMPLE simulation has equivalent skill to the V6 forcing into the model calibrated at the finer resolution. This result indicates that the consideration of the satellite product resolution has a considerable impact on hydrological simulation, perhaps as important as the errors in the rainfall retrievals themselves.

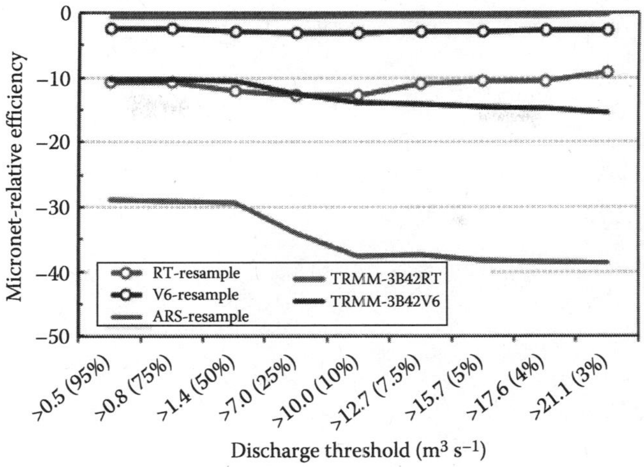

FIGURE 25.3 MRE (see Equation 25.1) of streamflow simulations forced by the rainfall algorithms indicated in the legend. Scores are plotted as a function of flow exceedance threshold. Results show hydrological impacts from rescaling reference rainfall to 0.25° resolution of TRMM. (Courtesy of Gourley, J.J. et al., *J. Hydrometeorol.*, 12, 973, 2011.)

Data assimilation methods such as variational approaches and the extended Kalman filter (EKF) are well suited for updating and improving model state estimation, such as soil moisture, and have shown great potential for improving hydrological forecasts. The technique discussed here is the EnSRF [51], which is a deterministic Monte Carlo simplification of the EKF. The EnSRF can be designed to account for uncertainties in the precipitation forcing (e.g., using a perturbations from SREM2D), in model states, and in the observed system behavior (i.e., streamflow) when generating hydrological forecasts. The variables commonly assimilated in HyMODs include soil moisture and streamflow, observed by gauges or from spaceborne instruments. The following section will focus on the assimilation of remotely sensed discharge using passive MW brightness temperature signals from AMSR-E onboard the Aqua satellite.

A time series of streamflow observed by the Rundu stream gauge and the AMSR-E signal at a collocated pixel at the basin outlet of the Cubango River located in southwest Africa was found to be very well correlated with a Pearson linear correlation coefficient of 0.94. It is noted, however, that the AMSR-E signal is in dimensionless units and shows poor sensitivity during low-flow periods. Additional procedures, such as comparing the signal to observed discharge, must be implemented in order to calibrate the signal to represent fluxes (in units of m^3/s). Khan et al. introduced an alternative, called the exceedance probability approach, in hydrological modeling to normalize observations in the frequency domain by utilizing the period of recorded observations [26]. The technique can be applied to any time series data and has been recently adopted in order to assimilate AMSR-E signal data as a proxy for observed streamflow. This approach essentially converts the AMSR-E signal ratios into probability of exceedance (%), thus improving the signal's applicability without the requirement of in situ streamflow measurements.

The dimensionless AMSR-E signal data were first converted into probability of exceedance using the available data archive. This proxy for river discharge measurements (now in the frequency domain) was subsequently used to automatically estimate parameters in the lumped, parsimonious HyMOD. This procedure required that the simulated streamflow data were converted into the frequency domain as well so that the simulated and proxy-observed data were both expressed as probability of exceedance. The conversion also enabled the assimilation of AMSR-E signals in order to update internal model states. Figure 25.4a shows simulated streamflow from the calibrated HyMOD with and without assimilation of the AMSR-E data. Simulated streamflow with data assimilation matches peak flows better

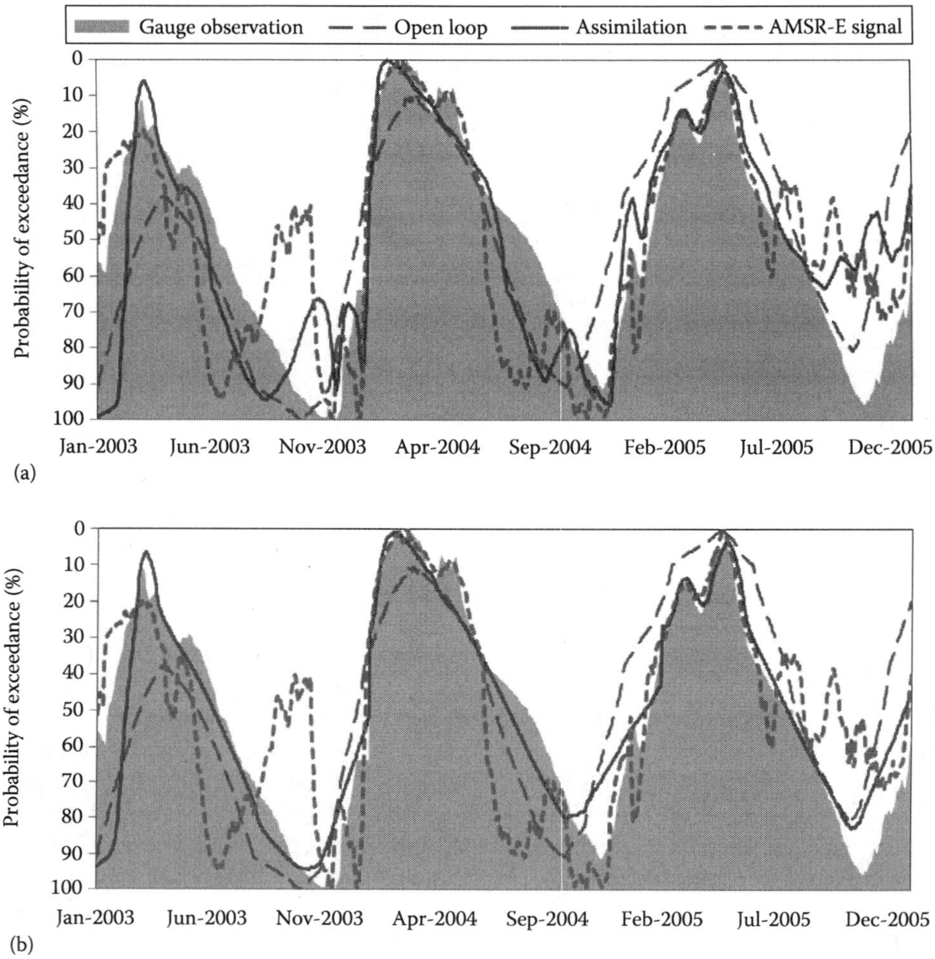

FIGURE 25.4 Time series of gauge streamflow (shaded in gray), AMSR-E signal (short-dashed curve), simulated streamflow without data assimilation (long-dashed curve), and simulated streamflow with data assimilation (solid curve) when using the AMSR-E signal (a) all the time and (b) only when the probability of exceedance was less than 30%.

than without data assimilation. However, there are apparently spurious AMSR-E signals during low-flow periods (i.e., November 2003) that result in degraded performance following data assimilation. The statistical performance of these experiments was compared to simulations from the HyMOD that was calibrated with streamflow observations from the Rundu gauge and also assimilated these streamflow observations; the benchmark was from a traditional gauge-calibrated HyMOD that incorporated data assimilation.

The NSCE and root-mean-square error (RMSE) with the benchmark assimilation were 0.88 and 42.0%. When the AMSR-E data were used as the streamflow proxy for calibration and assimilation, the performance dropped to 0.60 and 52.3% for NSCE and RMSE, respectively. In order to address the low-flow sensitivity problem with AMSR-E signal data, a "high-flow threshold" was applied to assimilate AMSR-E data only if the probability of exceedance was less than 30%. Figure 25.4b shows that the "assimilation" curve now performs much better during low-flow periods and is still capable of simulating the peak flows. This latter experiment had an NSCE of 0.85 and RMSE of 47.4%. These values are quite comparable to the benchmark simulation that used stream gauge observations for model calibration and data assimilation. This study promotes the concept of precipitation estimation, model calibration, data assimilation, and flood forecasting based entirely on remote-sensing data. Provided near-global coverage of remote-sensing data, possibilities for prediction in ungauged basins are becoming increasingly viable.

25.5 Summary and Conclusions

This chapter includes examples of studies that demonstrate the feasibility of assessing inundation extent from space. This can be done at regional and global scales as well as local scale like coastal regions, for instance. Information on inundation extent is valuable for HyMODs. In addition, the retrieval of information on precipitation was discussed, and methodologies to QPE and QPF were addressed. Several approaches that are based on radar and satellite measurements were compared. Finally, the last section of this chapter demonstrates the important role that information on land surface parameters and precipitation have in the development of accurate understanding of hydrological processes, particularly when integrated through data assimilation processes in gridded land surface models.

References

1. Anagnostou, E.N., 2004. Overview of overland satellite rainfall estimation for hydro meteorological applications. *Surveys in Geophysics*, 25: 511–537.
2. Barrett, E.C. and D.W. Martin, 1981. *The Use of Satellite Data in Rainfall Monitoring*. Academic Press, New York, 340pp.
3. Basist, A., N.C. Grody, T.C. Peterson, and C.N. Williams, 1998. Using the special sensor microwave/imager to monitor land surface temperatures, wetness, and snow cover. *Journal of Applied Meteorology*, 37 (9): 888–911.
4. Bellerby, T., M. Todd, D. Kniveton, and C. Kidd, 2000. Rainfall estimation from a combination of TRMM precipitation radar and GOES multispectral satellite imagery through the use of an artificial neural network. *Journal of Applied Meteorology*, 39: 2115–2128.
5. Bjerklie, D.M., S.L. Dingman, C.J. Vorosmarty, C.H. Bolster, and R.G. Congalton, 2003. Evaluating the potential for measuring river discharge from space. *Journal of Hydrology*, 278 (1–4): 17–38.
6. Brakenridge, G.R., S.V. Nghiem, E. Anderson, and R. Mic, 2007. Orbital microwave measurement of river discharge and ice status. *Water Resources Research*, 43 (4), doi: 10.1029/2006wr005238.
7. Chokmani, K., T.B.M.J. Ouarda, S. Hamilton, M.H. Ghedira, and H. Gingras, 2008. Comparison of ice-affected streamflow estimates computed using artificial neural networks and multiple regression techniques. *Journal of Hydrology*, 349 (3–4): 383–396, doi: 10.1016/j.jhydrol.2007.11.024.
8. Dixon, M. and G. Wiener, 1993. TITAN: Thunderstorm identification, tracking, analysis and nowcasting—A radar-based methodology. *Journal of Atmospheric and Oceanic Technology*, 10: 785–797.
9. Ferraro, R.R., 1997. Special sensor microwave imager derived global rainfall estimates for climatological applications. *Journal of Geophysical Research*, 102: 16715–16735.
10. Frazier, P., K. Page, J. Louis, S. Briggs, and A.I. Robertson, 2003. Relating wetland inundation to river flow using Landsat TM data. *International Journal of Remote Sensing*, 24 (19): 3755–3770.
11. Ganguly, A.R. and R.L. Bras, 2003. Distributed quantitative precipitation forecasting using information from radar and numerical weather prediction models. *Journal of Hydrometeorology*, 4(6): 1168–1180.
12. Germann, U. and I. Zawadzki, 2002. Scale-dependence of the predictability of precipitation from continental radar images. Part I: Description of the methodology. *Monthly Weather Review*, 130(12): 2859–2873.

13. Golding, B.W., 1998. Nimrod: A system for generating automated very short-range forecasts. *Meteorological Applications*, 5(1): 1–16.
14. Gourley, J.J., Y. Hong, Z.L. Flamig, J. Wang, H. Vergara, and E.N. Anagnostou, 2011. Hydrologic evaluation of rainfall estimates from radar, satellite, gauge, and combinations on Ft. Cobb basin, Oklahoma. *Journal of Hydrometeorology*, 12: 973–988, doi: 10.1175/2011JHM1287.1.
15. Grecu, M., W.S. Olson, and E.N. Anagnostou, 2004. Retrieval of precipitation profiles from multi-resolution, multifrequency, active and passive micro-wave observations. *Journal of Applied Meteorology*, 43: 562–575.
16. Hong, Y., K. Hsu, S. Sorooshian, and X. Gao, 2004. Precipitation estimation from remotely sensed imagery using an artificial neural network cloud classification system. *Journal of Applied Meteorology*, 43(12): 1834–1853.
17. Hossain, F. and E.N. Anagnostou, 2006. A two-dimensional satellite rainfall error model. *IEEE Transactions in Geosciences and Remote Sensing*, 44(6): 1511–1522.
18. Hossain, F. and D. Lettenmaier, 2006. Flood prediction in the future: Recognizing hydrologic issues in anticipation of the Global Precipitation Measurement mission. *Journal of Water Resources Research*, 42: W11301, doi:10. 1029/2006WR005202.
19. Hsu, K.L., H. Gupta, X. Gao, and S. Sorooshian, 1999. Estimation of physical variables from multichannel remotely sensed imagery using a neural network: Application to rainfall estimation. *Water Resources Research*, 35: 1605–1618.
20. Hsu, K.L., A. Behrangi, B. Imam, and S. Sorooshian, 2009. Extreme precipitation estimation using satellite-based PERSIANN-CCS algorithm. In *Satellite Rainfall Applications for Surface Hydrology*, eds. M. Gebremichael and F. Hossain, pp. 49–68, Springer, New York.
21. Huffman, G.J., R.F. Adler, D.T. Bolvin et al., 2007. The TRMM multi-satellite precipitation analysis: Quasi-global, multi-year, combined-sensor precipitation estimates at fine scale. *Journal of Hydrometeorology*, 8: 38–55.
22. Huffman, G.J., R.F. Adler, M. Morrissey et al., 2001. Global precipitation at one-degree daily resolution from multi-satellite observations. *Journal of Hydrometeorology*, 2: 36–50.
23. Iguchi, T., T. Kozu, R. Meneghini, J. Awaka, and K. Okamoto, 2000. Rain-profiling algorithm for TRMM precipitation radar. *Journal of Applied Meteorology*, 39: 2038–2052.
24. Joyce, R.J., J.E. Janowiak, P.A. Arkin, and P. Xie, 2004. CMORPH: A method that produces global precipitation estimates from passive microwave and infrared data at high spatial and temporal resolution. *Journal of Hydrometeorology*, 5(3): 487–503.
25. Johnson, J.T., P.L. MacKeen, A. Witt, E.D. Mitchell, G.J. Stumpf, M.D. Eilts, and K.W. Thomas, 1998. The storm cell identification and tracking algorithm: An enhanced WSR-88D algorithm. *Weather and Forecasting*, 13 (2): 263–276.
26. Khan, S.I., Y. Hong, H.J. Vergara, J.J. Gourley, G.R. Brakenridge, T. De Groeve, Z.L. Flamig, F.S. Policelli, and B. Yong, 2012. Microwave satellite data for hydrologic modeling in ungauged basins. *IEEE Geoscience and Remote Sensing Letters*, 9: 663–667, doi: 10.1109/LGRS.2011.2177807.
27. Kuligowski, R.J., 2009. The self-calibrating multivariate precipitation retrieval (SCaMPR) for high-resolution, low-latency satellite-based rainfall estimation. In *Satellite Rainfall Applications for Surface Hydrology*, eds. M. Gebremichael and F. Hossain, pp. 39–48, Springer, New York.
28. Kuligowski, R.J., 2002. A self-calibrating real-time GOES rainfall algorithm for short-term rainfall estimation. *Journal of Hydrometeorology*, 3: 112–130.
29. Kuligowski, R., R.A. Scofield, and J.C. Davenport, 2004. The hydro-nowcaster: Recent improvements and future plans in: *World Weather Research Programme International Symposium on Nowcasting and Very Short Range Forecasting*, Toulouse, France. Available at: www.meteo.fr/cic/win05/resumes-longs/5.15–203.pdf (accessed on September 5–9, 2005).
30. Kummerow, C., Y. Hong, W.S. Olson et al., 2001. The evolution of the Goddard profiling algorithm (GPROF) for rainfall estimation from passive microwave sensors. *Journal of Applied Meteorology*, 40: 1801–1820.

31. Lakshmanan, V., K. Hondl, and R. Rabin, 2009. An efficient, general-purpose technique for identifying storm cells in geospatial images. *Journal of Atmospheric and Oceanic Technology*, 26: 523–537.
32. Leconte, R., S. Daly, Y. Gauthier, N. Yankielun, F. Berube, and M. Bernier, 2009. A controlled experiment to retrieve freshwater ice characteristics from an FM-CW radar system. *Cold Regions Science and Technology*, 55 (2): 212–220, doi: 10.1016/j.coldregions.2008.04.003.
33. Maggioni, V., H.J. Vergara, E.N. Anagnostou, J.J. Gourley, Y. Hong, and D. Stampoulis, 2013. Investigating the applicability of error correction ensembles of satellite rainfall products in river flow simulations. *Journal of Hydrometeorology* (in press), doi: http://dx.doi.org/10.1175/JHM–D–12–074.1
34. Marzano, F.S., M. Palmacci, D. Cimini, G. Giuliani, and F.J. Turk, 2004. Multivariate statistical integration of satellite infrared and microwave radio-metric measurements for rainfall retrieval at the geostationary scale. *IEEE Transactions on Geoscience and Remote Sensing*, 42: 1018–1032.
35. Njoku, E.G., T.J. Jackson, V. Lakshmi, T.K. Chan, and S.V. Nghiem, 2003. Soil moisture retrieval from AMSR-E. *IEEE Transactions on Geoscience and Remote Sensing*, 41 (2): 215–229.
36. Pavelsky, T.M. and L.C. Smith. 2004. Spatial and temporal patterns in arctic river ice breakup observed with MODIS and AVHRR time series. *Remote Sensing of Environment*, 93 (3): 328–338, doi: 10.1016/j.rse.2004.07.018.
37. Schroeder, R., M. A. Rawlins, K. C. McDonald, E. Podest, R. Zimmermann, and M. Kueppers, 2010. Satellite microwave remote sensing of north eurasian inundation dynamics: Development of coarse-resolution products and comparison with high-resolution synthetic aperture radar Data. *Environmental Research Letters* 5, 1.
38. Scofield, R.A., R.J. Kuligowski, and C. Davenport, 2004. The use of the Hydro-Nowcaster for mesoscale convective systems and the tropical rainfall nowcaster (TRaN) for landfalling tropical systems, Preprints, *Symposium on Planning, Nowcasting, and Forecasting in the Urban Zone*, Seattle, WA, American Meteorological Society, CD–ROM, 1.4.
39. Sippel, S.J., S.K. Hamilton, J.M. Melack, and B.J. Choudhury, 1994. Determination of inundation area in the Amazon river floodplain using the SMMR 37 GHz polarization difference. *Remote Sensing of Environment*, 48 (1): 70–76.
40. Smith, E.A., G. Asrar, Y. Furuhama et al., 2007. The International Global Precipitation Measurement (GPM) program and mission: An overview. In *Measuring Precipitation From Space*, 611–653, Springer, Netherlands.
41. Smith, L.C. and T.M. Pavelsky. 2008. Estimation of river discharge, propagation speed, and hydraulic geometry from space: Lena River, Siberia. *Water Resources Research*, 44 (3), doi: 10.1029/2007wr006133.
42. Stephens, G.L. and C.D. Kummerow, 2007. The remote sensing of clouds and precipitation from space: A review. *Journal of the Atmospheric Sciences*, 64: 3742–3765.
43. Tang, Q., H. Gao, H. Lu, and D. Lettenmaier, 2009. Remote sensing: Hydrology. *Journal of Progress in Physical Geography*, 33 (4): 490–509.
44. Temimi, M., R. Leconte, F. Brissette, and N. Chaouch, 2005. Flood monitoring over the Mackenzie River Basin using passive microwave data. *Remote Sensing of Environment*, 98 (2–3): 344–355.
45. Temimi, M., T. Lacava, T. Lakhankar, V. Tramutoli, H. Ghedira, R. Ata, and R. Khanbilvardi, 2011. A multi-temporal analysis of AMSR-E data for flood and discharge monitoring during the 2008 flood in Iowa. *Hydrological Processes*, 25 (16): 2623–2634, doi: 10.1002/hyp.8020.
46. Thies, B., T. Nauß, and J. Bendix, 2008. Precipitation process and rainfall intensity differentiation using Meteosat Second Generation Spinning Enhanced Visible and Infrared imager data. *Journal of Geophysical Research*, 113: D23206, doi: 10.1029/2008JD010464.
47. Turk, F.J. and S. Miller, 2005. Toward improving estimates of remotely-sensed precipitation with MODIS/AMSR-E blended data techniques. *IEEE Transactions in Geoscience Remote Sensing*, 43: 1059–1069.
48. Turk, J.T., G.V. Mostovoy, and V. Anantharaj, 2009. The NRL-blend high resolution precipitation product and its application to land surface hydrology. In *Satellite Rainfall Applications for Surface Hydrology*, eds. M. Gebremichael and F. Hossain, pp. 85–104, Springer, New York.

49. Vasiloff, S.V., D.J. Seo, K.W. Howard et al., 2007. Improving QPE and very short term QPF: An initiative for a community-wide integrated approach. *Bulletin of the American Meteorological Society*, 88: 1899–1911.
50. Vila, D.A., L.A.T. Machado, H. Laurent, and I. Velasco, 2008. Forecast and Tracking the Evolution of Cloud Clusters (ForTraCC) using satellite infrared imagery: Methodology and validation. *Weather and Forecasting*, 23: 233–245.
51. Whitaker, J.S. and T.M. Hamill, 2002. Ensemble data assimilation without perturbed observations. *Monthly Weather Review*, 130 (7): 1913–1925.
52. White, K.D. and H.J. Earnes, 1999. CRREL ice jam database. In Report 2., edited by CRREL. http://icejams.crrel.usace.army.mil/
53. Zahraei, A., K. Hsu, S. Sorooshian, J.J. Gourley, Y. Hong, V. Lakshmanan, and T. Bellerby, 2012. Quantitative precipitation nowcasting, a Lagrangian pixel-based approach. *Journal of Atmospheric Research*, 118: 418–434.
54. Zahraei, A., K.L. Hsu, S. Sorooshian, J.J. Gourley, Y. Hong, and A. Behrang, 2013. Short-term quantitative precipitation forecasting using an object-based approach. *Journal of Hydrology*, 483: 1–15.

26

Stochastic Reservoir Analysis

Khaled H. Hamed
Cairo University

AUTHOR

Khaled H. Hamed holds a BSc and MSc in civil engineering from Cairo University, Egypt, and a PhD in civil engineering from Purdue University, United States. He has 25 years of practical experience in the fields of surface hydrology, stochastic hydrology, time series analysis, and flood frequency analysis. He also has vast experience in watershed modeling, stochastic reservoir analysis, time series modeling and forecasting, detection of trends and variations due to climatic changes, and stochastic as well as physical rainfall–runoff modeling. He is also the coauthor of two textbooks on flood frequency analysis and stationarity analysis of hydrologic and environmental time series published in the United States. Professor Hamed is an associate editor of the IAHS *Hydrological Sciences Journal* (HSJ) and a reviewer for a number of prominent international scientific journals, which include *Water Resources Research*, *Journal of Hydrology*, *Hydrological Sciences Journal*, and *Climate Research*, among others.

PREFACE

Water resource systems in many parts of the world have storage reservoirs as an integral part. Storage reservoirs are essential to overcome the problem of annual and over-year fluctuations in water supply systems. Although sizing reservoirs and estimating their reliability based on the analysis of flow characteristics has been documented only since the late 1800s, it is probable that using storage reservoirs for flow regulation may be as old as civilization itself, albeit at much smaller scales.

Over the past years, different methods have been proposed in the literature to perform different types of reservoir analysis. These range from sizing reservoirs for given reliability and yield, estimating reservoir reliability, to calculating the distribution of different quantities (storage, spillage, and deficit), etc. The stochastic nature of the inflow to storage reservoirs adds to the complexity of the problem. This chapter summarizes a number of methods used in stochastic reservoir analysis, ranging from simple reservoir simulation to the exact solution of the stochastic reservoir equation.

26.1 Introduction

Reservoirs are essential elements of water resource systems. Many water supply systems depend on river flow as their main water source. Due to the inherent natural variability in river flow, direct withdrawal does not guarantee a stable supply over time, and fluctuations in river flow will reflect directly into fluctuations in yield. Storage reservoirs have therefore been used to absorb such fluctuations and guarantee a certain level of yield. The main function of storage reservoirs is to store excess water at some point in time to be used at other times when the supply is low. Due to the stochastic nature of river flows, however, a reservoir of infinite storage volume would be needed to achieve perfect regulation of flow and guarantee a yield equal to the mean flow. A finite reservoir, on the other hand, can only guarantee a yield that is less than the mean flow. It thus becomes a matter of economy to choose a reservoir of suitable size that would guarantee a certain yield with a given reliability. However, due to the limited capacity of a finite reservoir, both spillage of excess water and emptying of the reservoir are likely to occur. Of particular importance is the emptying case for which the reservoir fails to deliver the specified yield.

The elements involved in the hydrologic design of a water supply reservoir consist of the reservoir size V, the demand or draft D, and the probability of failure f. Probability of failure is usually expressed in terms of its complement, the probability of no failure, that is, $1 - f$, commonly known as the reliability of the reservoir. Design relationships known as reservoir storage–yield–reliability (S–R–Y) relationships have therefore been developed to describe the relationship between the size of a reservoir, its yield, and reliability.

Evaluation of storage, spillage, and yield for a reservoir with a given size and input sequence was investigated by Rippl [32] using a number of graphical and arithmetical techniques. Hazen [11], Sudler [34], and Barnes [2] contributed to the introduction of the use of synthetic data in reservoir studies. Hurst [14] used rescaled range analysis for sizing reservoirs with long-term storage. According to Klemeš [15], modern stochastic reservoir theory was introduced by Kritskiy and Menkel [18,19] and Savarenskiy [33], only to be independently reinvented in the mid-1950s by Moran [25]. Further significant contributions to stochastic reservoir analysis were made by Gani [4], Langbein [20], Lloyd [21], and Klemeš [16].

Several methods have been suggested in the literature to solve the stochastic reservoir equation to obtain the distribution of storage as well as S–R–Y relationships. Monte Carlo simulation (behavior analysis) is one of the widely used methods, which involves the generation of synthetic inflow data of the desired probability distribution and applying the mass balance equation at a suitable time interval for a large number of time steps. Alternatively, a class of stochastic methods involves finding the equilibrium probability distribution of the storage function given a certain reservoir size and yield for an inflow with a given probability distribution, which facilitate the calculation of different types of probabilities, including those of reservoir failure and spilling. This problem is commonly referred to as the "solution of the stochastic reservoir equation." Due to the complexity of the equilibrium probability distribution of the storage function, exact solutions are only available for a limited, yet very flexible, class of inflow distributions, which include the exponential distribution [26] and a subset of the gamma distribution with integral shape parameters [30]. Numerical solution methods based on the discretization of the stochastic reservoir equation have been suggested in the literature [16,29]. A number of approximate methods have also been proposed, such as the diffusion approximation for normally distributed inflows [3]. A comprehensive review of different methods for solving the stochastic reservoir equation can also be found in Lloyd [22].

26.2 Reservoir Simulation

For a given reservoir of size V receiving a certain inflow I with an outflow Q, the distribution of reservoir storage S can be obtained through the application of the mass conservation principle, which states that the rate of change of reservoir storage is equal to the difference between the rate of inflow (i) and the rate of outflow (q) from the reservoir. The resulting equation is the well-known continuity equation that takes the form

$$\frac{dS}{dt} = i - q \tag{26.1}$$

Selecting a time step that is suitable for the type of analysis, the evolution of reservoir storage can thus be calculated as

$$S_{t+1} = S_t + I_t - Q_t \tag{26.2}$$

where

I_t is the total inflow to the reservoir between time steps t and $t+1$

Q_t is the total outflow from the reservoir in the same period

The total outflow Q_t usually comprises the draft D (which may be constant or time dependent) and any losses from the reservoir such as evaporation and seepage losses.

Because a reservoir has a finite size V, storage in Equation 26.2 earlier is constrained by two conditions. First, if the calculated storage S^*_{t+1} exceeds the reservoir size V, water is spilled and the stored volume cannot exceed the reservoir size V. The volume of spilled water in this case is equal to the difference between the calculated storage S^*_{t+1} and the reservoir size V. Similarly, if the calculated storage S^*_{t+1} is negative, the reservoir becomes empty and the outflow is limited by the amount of available water, and a deficit equal to $-S^*_{t+1}$ occurs. These two conditions can be put in equation form as follows:

$$S_{t+1} = \min\,(S^*_{t+1}, V), \quad \text{spill} = S^*_{t+1} - S_{t+1} \tag{26.3}$$

$$S_{t+1} = \max\,(S^*_{t+1}, 0), \quad \text{deficit} = S_{t+1} - S^*_{t+1} \tag{26.4}$$

Equations 26.3 and 26.4 can actually be combined into one equation in the form

$$S_{t+1} = \max\left[0, \min\,(S^*_{t+1}, V)\right], \quad \Delta Q = S_{t+1} - S^*_{t+1} \tag{26.5}$$

where

a positive ΔQ indicates spillage

a negative ΔQ indicates deficit

Using the aforementioned equations offers a very flexible way of simulating the evolution of reservoir storage, as it allows different types of inflows and outflows to be accounted for, including outflows that depend on the state of the reservoir storage, such as evaporation and seepage losses as well as a variable draft from the reservoir.

The inflow in Equation 26.2 earlier may be a historical time series, but more often, it will be a synthetically generated inflow sequence (or a large number of sequences) with a length that is suitable to get reliable estimates of the distribution of storage and various reservoir performance indicators, such as the probability of failure or that of overtopping. The procedure in this case is commonly known as Monte Carlo simulation or as behavior analysis. Koutsoyiannis [17] shows that the number of simulation steps required to obtain an estimate of a given probability β with an acceptable error of ±ε β and confidence α with an annual time step is given by

$$n = \left(\frac{z_{(1+\alpha)/2}}{\varepsilon}\right)^2 \left(\frac{1-\beta}{\beta}\right) \tag{26.6}$$

where $z_{(1+\alpha)/2}$ is the standard normal variate corresponding to a cumulative probability of (1+α)/2, so if a 90% confidence is required, then $z_{(1+\alpha)/2} = z_{0.95} = 1.64$. For example, if the probability of failure is $\beta = 0.1$, and the error tolerance $\varepsilon = 0.01$, the required number of steps for 90% confidence in β would be equal to $n = 242{,}064$ simulation steps. It should be noted, however, that due to the stochastic nature of the problem, there would still be a 10% chance that the error in the obtained value of β exceeds the designated 0.01 tolerance. Smaller error tolerance and/or higher confidence will require much longer sequences, as is clear from Equation 26.6. Pretto et al. [31] discuss the convergence properties of reservoir storage analysis using Monte Carlo simulation.

As a simplified example, consider a reservoir of size $V = 4$ units with an exponential flow of mean of 2 units and a draft $m = 1.8$ units. Simulation results using 1×10^6 simulation steps indicate that the probability of reservoir emptying is equal to 0.212896, while the probability of spilling is equal to 0.191747. These quantities will be compared later with those obtained using other solution methods. The simulated distribution of the storage is shown in Figure 26.1.

26.3 Standard Reservoir

Gomide [6] introduced the standard reservoir convention, as he noted that reservoirs with different characteristics give the same results if the ratios of their parameters to the standard deviation of the inflow are the same. A standardized reservoir size *C* is equal to the actual reservoir size *V* divided by the standard deviation of the inflow σ as in Equation 26.7:

$$C = \frac{V}{\sigma} \tag{26.7}$$

The corresponding standardized net mean inflow to the reservoir is known as the drift ε, which was first introduced by Pegram [29] and is defined as in Equation 26.8:

$$\varepsilon = \frac{\mu - D}{\sigma} \tag{26.8}$$

where

μ is the mean of inflow

D is the draft

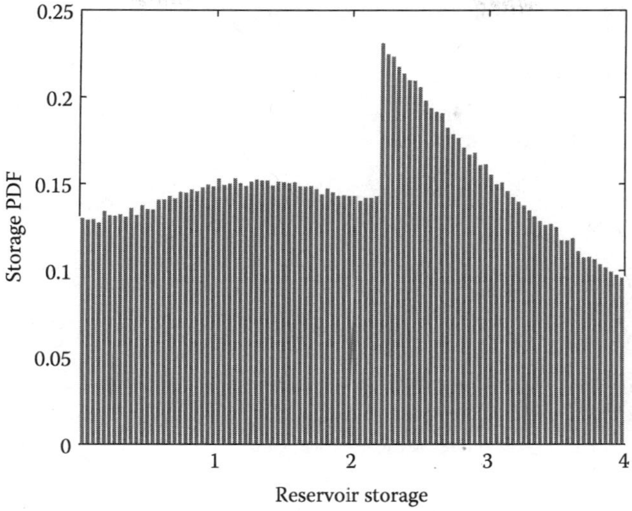

FIGURE 26.1 The PDF of storage in a reservoir of size $V=4$ units with an exponential flow of mean = 2 units and a draft $m=1.8$ units using 1×10^6 Monte Carlo simulation steps.

The net inflow to the standard reservoir is the standardized net inflow given by

$$N_t = \frac{I_t - D}{\sigma} \tag{26.9}$$

We note that adopting the standardized reservoir convention is economic, since results for reservoirs with proportional values of the quantities V, D, μ, and σ can all be expressed in terms of C and ε only with standardized inflow. As an example, the results for a standard reservoir of size $C=2$ and drift $\varepsilon=0.2$ having a standard normal inflow would be the same for a reservoir of size $V=20\times10^9$ m^3, having a draft $D=23\times10^9$ m^3, with a normal inflow with mean $\mu=25\times10^9$ m^3, and standard deviation $\sigma=10\times10^9$ m^3. The same results apply for a reservoir of size $V=10\times10^6$ m^3, having a draft $D=14\times10^6$ m^3, with a normal inflow with mean $\mu=15\times10^6$ m^3, standard deviation $\sigma=5\times10^6$ m^3, and so forth.

26.4 Stochastic Reservoir Equation

As seen in Section 26.2, the conservation of mass principle was used to derive Equations 26.2 through 26.5 that describe the evolution of storage in time. As a result of the application of these equations, the stochastic distribution of the inflow is converted into a stochastic distribution of the storage. However, the distribution of storage is obtained in this case through a sample of a limited size, and the number of simulation steps needed to attain a reasonable accuracy is usually large as detailed in Section 26.2 earlier. From a statistical point of view, Equation 26.2 represents storage S_t as a function of the random variable I_t. The distribution of a function of a random variable is a well-known problem in probability theory, although the conditions imposed by Equations 26.3 through 26.5 often prevent an explicit closed-form solution of the distribution of storage, except in a few cases as will be shown later.

Basic principles of the probability theory can be used to directly derive the distribution of storage given that of the inflow for the case of a fixed draft and independent inflow. Consider a stochastic reservoir of size C and net inflow N that is independent in time. At any time step i, the reservoir can be in any of three distinct states: empty ($S_i = 0$), full ($S_i = C$), and partially full ($0^+ < S_i < C^-$), where 0^+ and C^- are used to indicate the continuous part of the distribution between 0 and C, excluding the $S_i = 0$ and $S_i = C$ states, each of which has its own discrete probability mass.

The probability distribution of storage S_{i+1} in the following time step $i+1$ can simply be expressed [10] using the well-known total probability rule as

$$P(S_{i+1} \le s) = P(S_{i+1} < s \cap S_i = 0) + P(S_{i+1} < s \cap 0 < S_i < C) + P(S_{i+1} < s \cap S_i = C)$$

$$= P(N \le s)P(S_i = 0) + \int_{u=0^+}^{u=C^-} P(N \le s - u)P(S_i = u)\,du + P(N \le s - C)P(S_i = C) \tag{26.10}$$

The middle term in the RHS of Equation 26.10 is a convolution of the net inflow and the storage distribution functions. In essence, Equation 26.10 can be viewed as an application of the "conservation of probability" principle, in analogy to the "conservation of mass" principle.

In terms of the distribution functions $G(s)$ of the storage and $F(n)$ of the net inflow, Equation 26.10 becomes [10,29]

$$G_{i+1}(s) = F(s) \cdot G_i(0) + \int_{0^+}^{C^-} F(s - u) \cdot g_i(u)\,du + F(s - C)\left[1 - G_i(C^-)\right] \tag{26.11}$$

The sequence $G_i(s)$, $G_{i+1}(s)$, … converges to a limiting "equilibrium" distribution $G(s)$ of the storage in the reservoir [22,29] and the subscripts can be dropped. Because of the special form of Equation 26.11, where the distribution function of storage $G(s)$, as well as its derivative $g(s)$, appears on both sides of the equation, a general analytical solution is not available. It is important to note that, unlike the simulation approach, this approach does not require the generation of lengthy synthetic data to obtain the distribution of storage, since all that is needed is the distribution function of the net inflow.

As stated earlier, Equation 26.11 is valid only for reservoirs with fixed draft and independent net inflow, but extensions to accommodate variable draft and inflows having first-order Markov dependence have also been tackled in the literature [29]. Although this approach has some limitations compared to the very flexible reservoir simulation approach, it can offer tools for rapid sizing and assessment of reservoirs in the planning phase, prior to detailed simulation studies. At the same time, results obtained using this approach can provide benchmark tests for validating detailed simulation models.

26.5 Solution of the Stochastic Reservoir Equation

26.5.1 Prabhu's Exact Solution

Prabhu [30] gives the only available complete exact solution of the stochastic reservoir equation in the case of independent gamma-distributed inflow with integer shape parameters, of which exponentially distributed inflow is a special case, which had been tackled earlier by Moran [26,27]. The class of gamma inflows with integral shape parameter, also known as the "Erlangian" family, have a probability density function given by

$$dG(x) = \frac{\mu^p}{(p-1)!} e^{-\mu x} x^{p-1} dx, \quad 0 < x < \infty \tag{26.12}$$

where $\mu > 0$ and $p = 1, 2, 3, \ldots$ are the distribution parameters. The value of $p = 1$ corresponds to the exponential distribution, while integer values of $p > 1$ correspond to gamma distributions with integral shape parameter. The mean of the distribution μ_x is equal to p/μ, while the variance σ_x^2 is given by p/μ^2. The coefficient of variation and the skewness coefficient are given by $C_v = 1/\sqrt{p}$ and $\gamma = 2/\sqrt{p}$, respectively.

For this family of distributions, the stationary distribution of the reservoir storage for a reservoir of size K with a draft m was studied by Prabhu [30]. The reservoir size K in Prabhu's work is the gross reservoir size that is equal to $V+m$, that is, the reservoir size plus the draft. This convention assumes that a reservoir first receives the full inflow before the draft is initiated, which is the Moran model convention [22,29]. To keep with the "simultaneous reservoir" convention used throughout this chapter, Prabhu's equations are written here in terms of the net reservoir size V, that is, replacing K with $V+m$:

$$F(z)=\begin{cases} 1-e^{\mu(V-z)}\sum_{r=0}^{p-1}\alpha_r\sum_{q=0}^{N+1}(-\lambda)^q\dfrac{\{V-qm-z\}^{qp+r}}{(qp+r)!}, & -m\le z<V-(N+1)m \\ 1-e^{\mu(V-z)}\sum_{r=0}^{p-1}\alpha_r\sum_{q=0}^{s}(-\lambda)^q\dfrac{\{V-qm-z\}^{qp+r}}{(qp+r)!}, & V-(s+1)m\le z<V-sm, \\ & s=N,N-1,\ldots,1 \\ 1-e^{\mu(V-z)}\sum_{r=0}^{p-1}\dfrac{\alpha_r}{r!}(V-z)^r, & V-m\le z<\infty \end{cases} \tag{26.13}$$

where

$$V = N\,m+U,\ 0<U<m \tag{26.14}$$

that is, N is the integral part of (V/m) and U is the remaining part, which is a fraction of m. If V is a proper multiple of m, then U would be equal to zero.

The parameter λ in Equation 26.13 is given by

$$\lambda=(-1)^{p-1}\mu^p e^{-\mu m} \tag{26.15}$$

Also, the values α_r in Equation 26.13, where $r = 0, 1, \ldots, p-1$, can be obtained by solving the p linear equations given by

$$\alpha_r-\lambda\sum_{s=0}^{p-1}d_{rs}\alpha_s=(-\mu)^r e^{-\mu(V+m)}\sum_{s=0}^{p-r-1}\frac{[\mu(V+m)]^s}{s!},\quad r=0,1,\ldots,p-1 \tag{26.16}$$

where

$$d_{rs}=(-1)^{p+r-1}\sum_{q=0}^{N}(-\lambda)^q\int_{qm}^{V}\frac{(t-qm)^{qp+s}(t+m)^{p-r-1}}{(qp+s)!(p-r-1)!}dt,\quad r,s=0,1,\ldots,p-1 \tag{26.17}$$

The probabilities of emptying and filling, respectively, based on Equation 26.13, are thus given by

$$P_0=F(0)=1-e^{\mu V}\sum_{r=0}^{p-1}\alpha_r\sum_{q=0}^{N}(-\lambda)^q\frac{\{V-qm\}^{qp+r}}{(qp+r)!} \tag{26.18}$$

$$P_c = 1 - F(V) = \alpha_0 \tag{26.19}$$

Hamed [9] shows that Equation 26.13 gives the full distribution of storage in the reservoir, including the distribution of both deficit and spillage. The reader is referred to Hamed [9] for further discussion and examples.

For Prabhu's solution, a standardized reservoir size *C* is equal to the actual reservoir size *V* divided by the standard deviation of the inflow as in Equation 26.20:

$$c = \frac{V}{\sigma_x} = \frac{V}{\left(\sqrt{p}/\mu\right)} = \frac{\mu V}{\sqrt{p}} \tag{26.20}$$

The corresponding standardized net mean inflow to the reservoir, or the drift, is defined as in Equation 26.21:

$$\varepsilon = \frac{\mu_x - m}{\sigma_x} = \frac{p/\mu - m}{\sqrt{p}/\mu} = \frac{p - m\mu}{\sqrt{p}} \tag{26.21}$$

Now the solution can be obtained in terms of *C*, drift ε, and shape parameter *p*.

The computer program P1 in Appendix 26.A is a MATLAB® implementation of Prabhu's solution in the case of $p = 1$, that is, exponentially distributed inflow. The case of general *p* requires special calculations and the Symbolic Math toolbox in MATLAB is required to correctly perform the calculations. The reader is referred to Hamed [9] for important details of implementation. Consider the example given in Section 26.2. The application of Prabhu's exact solution gives an emptying probability of 0.211634 and a spilling probability of 0.191868, which are very close to those obtained by simulation. Furthermore, Figure 26.2 shows the PDF of the storage distribution as given by Prabhu's solution, which is similar to that obtained through simulation. Prabhu's exact solution clearly identifies the jump in the PDF near storage equal to 2.2.

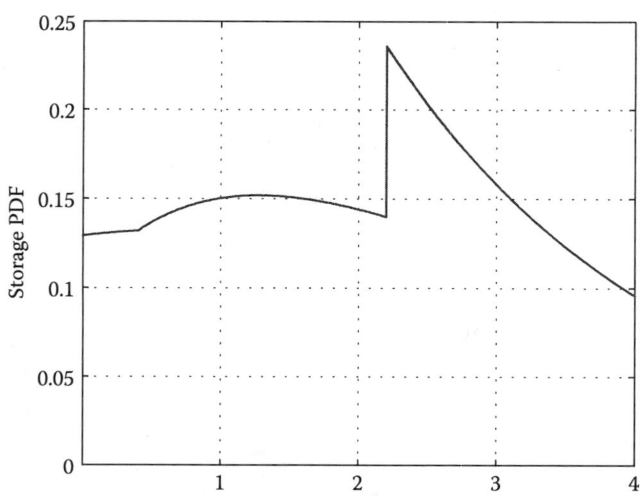

FIGURE 26.2 The PDF of storage in a reservoir of size $V = 4$ units with an exponential flow of mean = 2 units and a draft $m = 1.8$ units using Prabhu's exact solution.

26.5.2 Integral Equation Solution

The integral equation of the stochastic reservoir as discussed earlier is given as

$$G(s)=G(0)\cdot F(s)+\int_{0^+}^{C^-} g(u)F(s-u)\cdot du+\left[1-G(C^-)\right]F(s-C),\quad 0\le s\le C \tag{26.22}$$

where

- $G(.)$ and $g(.)$ are, respectively, the cumulative distribution function and the probability density function of storage
- $F(.)$ is the cumulative distribution function of the net inflow to the reservoir
- C is the reservoir size

The following simpler equation can be obtained through integration by parts [29]:

$$G(s)=F(s-C)+\int_{0^+}^{C^-} f(s-u)\cdot G(u)\cdot du \tag{26.23}$$

If a reservoir of size C is divided into k equal slices, each of width $h=C/k$, $G(s)$ can be approximated by $H(s)$ at $s=ih$, $i=0, 1, \ldots, k$. The integral on the right-hand side of Equation 26.23 can be evaluated by using a suitable quadrature formula, such as the trapezoidal or Simpson's rule, in terms of the unknown $G(u)$. If the weights of the quadrature formula are given by w_j, $j=0, 1, \ldots, k$, then Equation 26.23 can be discretized as

$$H(ih)=F(ih-c)+\sum_{j=0}^{k} w_j\cdot f[(i-j)h]\cdot H(jh) \tag{26.24}$$

Equation 26.24 gives a system of $k+1$ simultaneous equations, which can be solved to obtain the values of $H(ih)$ using a suitable method. The emptying probability is thus given by $H(0)$, while the filling probability is given by $1 - H(c)$. Once the values of $H(ih)$ are obtained, the expression in Equation 26.24 works as an interpolation formula that employs all the information in the kernel to obtain the storage distribution function at any intermediate point [29] as

$$H(s)=F(s-c)+\sum_{j=0}^{k} w_j\cdot f[s-jh]\cdot H(jh) \tag{26.25}$$

Other solutions of different forms of Equation 26.22 have also been suggested in the literature [22] using different schemes of reservoir discretization.

The computer program P2 is a MATLAB implementation of the integral equations solution of the stochastic reservoir equation for a reservoir with exponentially distributed inflow as an example.

Consider the example given in Section 26.5.1 earlier. The application of the integral equation solution with $k=1000$ stages of the reservoir gives an emptying probability of 0.212986 and a spilling probability of 0.191611, which are very close to those obtained by Prabhu's exact solution and by simulation.

26.5.3 Difference Equations Solution

The first step in this solution method is the discretization of the reservoir into k states of equal size h. With the addition of the two states of full and empty, a total of $k+2$ states are obtained for the reservoir.

Pegram [29] found that $k \leq 30$ gave a reasonable accuracy in most cases, with standardized reservoir sizes up to 8 and standardized net mean inflow up to 1. Using this discretization, the elements q_{ij} of the $k+2$ square transition matrix Q between states can be easily calculated based on the cumulative distribution function of the input $F(u)$ as

$$q_{i,0} = \begin{cases} F[0], & i=0 \\ F[ih] - F[(i-1)h], & i=1,2,\ldots,k \\ 1-F[c], & i=k+1 \end{cases}$$

$$q_{i,k+1} = \begin{cases} F[-c], & i=0 \\ F[ih-c] - F[(i-1)h-c], & i=1,2,\ldots,k \\ 1-F[0], & i=k+1 \end{cases} \tag{26.26}$$

$$q_{i,j} = \begin{cases} F[(-j+\tfrac{1}{2})h], & i=0 \\ F[(i-j+\tfrac{1}{2})h] - F[(i-j-\tfrac{1}{2})h], & i=1,2,\ldots,k \\ 1-F[c-(j-\tfrac{1}{2})h], & i=k+1 \end{cases}$$

It is to be noted that each column of Q should sum to unity, which is the total probability for each state. Based on this transition probability matrix, the $(k+2)$-element vector $v(t+1)$ that gives the probabilities of the storage states at time $t+1$ is related to $v(t)$ as [27]

$$v(t+1) = Qv(t) \tag{26.27}$$

The equilibrium storage distribution vector π can thus be obtained as

$$\pi = \lim_{t\to\infty} v(t) \tag{26.28}$$

It is to be noted that the first and last elements of $v(t)$ represent the probabilities of emptying and filling, respectively. Several methods have been proposed to solve Equation 26.28 such as the power method that starts with $v(0)$ to be any initial vector with elements that sum to unity and then successively apply Equation 26.27 for 20–30 iterations to reach the stable equilibrium [29].

The same result can be obtained by raising Q to a large power using linear algebra methods. Other methods involve the solution of the homogeneous system of equations $(I - Q)\ \pi = 0$, the calculation of the eigenvectors of $(I - Q^T)$, and the use of generalized inverses [12,13].

The computer program P3 is a MATLAB implementation of the difference equation solution of the stochastic reservoir equation for a reservoir with exponentially distributed inflow as an example. Consider the example given in Section 26.5.1 earlier. The application of the difference equation solution with $k=200$ states of the reservoir gives an emptying probability of 0.212986 and a spilling probability of 0.191611, which are very close to those obtained by Prabhu's exact solution and by simulation.

26.6 Probabilistic Approach

The solutions of the stochastic reservoir equations covered in the previous section apply only for independent inflow. Pegram [29] outlines a method for Markovian inflow, that is, inflow with first-order autocorrelation between successive values. Hamed [10] suggested a probabilistic method to calculate the probabilities of filling and emptying of reservoirs with dependent Gaussian inflow. First, let us consider the distribution of the critical period. The critical period (T^*_{C0}) of a reservoir is defined as the period of time it takes a full reservoir to completely empty without reverting to the full state [28]. According to Oğuz and Bayazit [28], the critical period concept was first introduced in reservoir analysis by Hall et al. [8] and further extended to various models such as normal and lognormal zero-, first-, and second-order Markov models as well as normal and lognormal fractional Gaussian noise [1]. Further study of the properties of the critical period has been performed by [7] as well as Loucks et al. [23], Oğuz and Bayazit [28], and Montaseri and Adeloye [24]. Because of the condition in the definition of the critical period that the reservoir goes from the full state to the empty state without reverting to the full state, it is straightforward to derive the distribution of the critical period for Gaussian inflow in terms of a number of multivariate normal integrals. Recent advances in computing power and numerical integration methods make this task less demanding than it used to be earlier, thus offering an alternative way of calculating reservoir reliability as will be shown.

Consider an over-year storage reservoir of size C with normally distributed net inflow $\mathbf{I} = [I_1, I_2, I_3, \ldots]$ and an annual time step. Let $f_T^{(n)}$ be the probability that a full reservoir becomes empty in exactly n time steps without reverting to the full state (as per the definition of the critical drawdown period). For $n = 1$ (i.e., the reservoir empties in exactly one step), the required probability is equal to the probability that I_1, the net inflow in the same time step, is less than $-C$:

$$f_T^{(1)} = P\left(I_1 < -C\right) \tag{26.29}$$

The aforementioned probability involves only a univariate normal distribution integral.

For $n = 2$ (i.e., the reservoir empties in exactly two steps), we have two conditions. The first condition is that the net inflow in the first time step must be greater than $-C$ (so as not to empty the reservoir in one time step), but less than zero (so as not to revert to the full state, as per the definition of the critical drawdown period). The second condition is that the sum of the net inflow to the reservoir in the two time steps $I_1 + I_2$ is less than $-C$:

$$f_T^{(2)} = P\left(-C < I_1 < 0, I_1 + I_2 < -C\right) \tag{26.30}$$

In this case, a bivariate normal integral is involved.

In general, for the reservoir to empty in n steps, the cumulative sums of the first $n - 1$ time steps must each lie between zero and $-C$, while the sum of net inflow for all n time steps must be less than $-C$:

$$f_T^{(n)} = P\left[\bigcap_{i=1}^{n-1}\left(-C < \sum_{j=1}^{i} I_j < 0\right), \sum_{j=1}^{n} I_j < -C\right] \tag{26.31}$$

For the Gaussian net inflow considered here, the aforementioned probabilities can be calculated by the numerical integration of the standard multivariate normal density with n variates, which is given by

$$f_I(\mathbf{I}) = \frac{1}{(2\pi)^{n/2}\left|\Sigma\right|^{1/2}} e^{-1/2\left(\mathbf{I}-\boldsymbol{\beta}\right)^T \Sigma^{-1}\left(\mathbf{I}-\boldsymbol{\beta}\right)} \tag{26.32}$$

where

$\mathbf{I}$ is the net inflow vector $[I_1, I_2, \ldots, I_n]$
$\boldsymbol{\beta}$ is the mean of the net inflow
$\boldsymbol{\Sigma}$ is the $n \times n$ covariance matrix of the net inflow

The integration is performed in the region specified by the conditions stated earlier, which can be put in a linear inequality matrix form as follows:

$$\mathbf{a} < \mathbf{A}\,\mathbf{I} < \mathbf{b} \tag{26.33}$$

where

$\mathbf{a} = [-C, -C, \ldots, -C, -\infty]^T$

$\mathbf{b} = [0, 0, \ldots, 0, -C]^T$

A is a lower-triangle matrix of ones. Numerical methods for the calculation of the multivariate normal probability with the integration region specified as a set of linear inequalities as in Equation 26.33 earlier are given by Genz [5].

Oğuz and Bayazit [28] noted that the probabilities $f_T^{(n)}$ obtained earlier need to be normalized, by dividing by their sum, to obtain proper probabilities that sum to unity, but they did not comment on the meaning of their sum. However, careful examination indicates that critical period events with all possible lengths $n = 1, 2, \ldots, \infty$ constitute a group of mutually exclusive and collectively exhaustive events. The probability $f_T^{(n)}$ represents the intersection of each of these events with the event of the reservoir being empty. Therefore, it can be conjectured that the probability of the reservoir being empty should be equal to the sum of $f_T^{(n)}$, that is,

$$G(0) = \sum_{n=1}^{\infty} f_T^{(n)} \tag{26.34}$$

It is also worth noting that storage evolution is a Markov process [22], since the storage at any time step is fully determined through the storage at the previous time step only. Although properties of the critical period (which is a special type of first-passage times as seen earlier) do not appear to have been considered in the literature related to Markov processes, the relationship in Equation 26.34 earlier has been verified by Hamed [10]. The relationship in Equation 26.34 is a very useful link between the distribution of the critical period and the classical probability of the reservoir being empty, thus offering an alternative way of computing reservoir reliability.

The sum in Equation 26.34 is an infinite sum, but the distribution decays rapidly, and thus, one can set a reasonable limit for the inclusion of terms in the sum. For example, the calculation could be stopped when a term $f_T^{(n)}$ becomes smaller than, say, 1/1000 of the largest value attained (the mode of the distribution). The main advantage of this method is that it can accommodate any arbitrary correlation structure represented by the matrix $\mathbf{\Sigma}$ in Equation 26.32, as long as $\mathbf{\Sigma}$ is positive definite to qualify as a valid covariance matrix. Calculations can be made based on the autocorrelation function only without the need for actually generating synthetic data. Another advantage of this approach is that estimates of the error in calculating the multivariate normal probabilities can also be obtained from the computer code.

The probability of spilling (or the reservoir being full), $G(C)$, can be obtained in a similar manner by considering the distribution of the period of transition from an empty to a full reservoir without reverting to the empty state. This period could be termed the "direct filling period" as mentioned earlier. It is easy to show that the integration region in this case is defined as in Equation 26.7 earlier, but with $\mathbf{a} = [0, 0, \ldots, 0, C]^T$, $\mathbf{b} = [C, C, \ldots, C, \infty]^T$, and **A** is a lower-triangle matrix of ones, as before.

The computer program P4 is a MATLAB implementation of this probabilistic method to calculate the probabilities of emptying and filling of a reservoir having a Gaussian inflow using the proposed method. The code uses the multivariate normal numerical integration function "qscmvnv.m," which is available at (http://www.math.wsu.edu/faculty/genz/homepage). As an example, consider a reservoir with standard size $C=4$, drift $m=0.1$, and standard Gaussian inflow with first-order autocorrelation coefficient $r=0.6$. The emptying probability is calculated as 0.186321 and the spilling probability as 0.282422. Monte Carlo simulation results give these probabilities as 0.185447 and 0.282248, respectively.

26.7 Summary and Conclusions

Reservoirs are important elements of many water resource systems. The reliability of storage reservoirs as well as the distribution of storage, spillage, and deficit can be obtained through various methods. The first method covered in this chapter is Monte Carlo simulation, which is the most flexible method, allowing different components of inflow and outflow with possibly variable values. Convergence of the method, however, should be carefully assessed and long sequences of inflow are usually needed to have accurate results. Direct calculation of the distribution of storage in a reservoir from the distribution of inflow through the stochastic reservoir equation is another option. An exact solution for the special case of the gamma inflow with integer shape parameter, which is known as Prabhu's solution, is presented. Two numerical solution methods of the stochastic reservoir equation, namely, the integral equations solution and the difference equations solution, have also been outlined. Finally, a probabilistic approach to calculating the probabilities filling and emptying of a reservoir fed with dependent inflow is covered.

Appendix 26.A

The following programs are written in MATLAB language. The text should be saved as a plain text file with the same function name and extension "m." For example, Program P1 should be saved as "Prabhuexp.m."

Program P1

This program calculates the probabilities of filling and emptying of a reservoir of size *c* with draft *m* and exponential inflow with parameter *mu*. The program also calculates and plots the PDF and CDF of the storage distribution.

```
function [pe pf] =Prabhuexp(c,m,mu)
n=floor(c/m);
u=c-n*m;
lambda=mu*exp(-mu*m);
d=0;
for q=0:n
y=(c-q*m)^(q+1)/(factorial(q+1));
d=d+(-lambda)^q*y;
end;
alpha=exp(-mu*(c+m))/(1-lambda*d);
pe=1;
for q=0:n
pe=pe-exp(mu*c)*alpha*(-lambda)^q*(c-q*m)^q/factorial(q);
end;
pf=alpha;
fprintf('\nPempty =%10.9f, Pfull =%10.9f \n\n',pe, pf);
%CDF and PDF
nx=100;
x=[];
```

```
f=[];
F=[];
if abs(u)>.00000001;
z=linspace(-m,-m+u,ceil((nx+1)*u/m));
[G g]=getd1(z,c,alpha,mu,lambda,m,n+1);
x=[x z];
f=[f g];
F=[F G];
end;
for s=n:-1:1
z=linspace(c-(s+1)*m,c-s*m,nx+1);
if z(end)>0 && z(1)<0; idd=find(z<0,1,'last');z=[z(1:idd) 0
z(idd+1:end)];end;
[G g]=getd1(z,c,alpha,mu,lambda,m,s);
x=[x z];
f=[f g];
F=[F G];
end;
for s=0:-1:-1
z=linspace(c-(s+1)*m,c-s*m,nx+1);
if z(end)>0 && z(1)<0; idd=find(z<0,1,'last');z=[z(1:idd) 0
z(idd+1:end)];end;
[G g]=getd2(z,c,alpha,mu);
x=[x z];
f=[f g];
F=[F G];
end;
subplot(1,2,1);
plot(x,f,'-','linewidth',1.6);
grid on;
hold on;
plot([0 0],[0 f(find(x>=0,1,'first'))],'-r','linewidth',2);
plot([c c],[0 f(find(x>=c,1,'first'))],'-r','linewidth',2);
set(gca,'xlim',[-m c+m]);
hold off;
subplot(1,2,2);
idx=find(x>=0&x<=c);
plot([0 x(idx) c],[0 F(idx) 1],'-','linewidth',1.6);
hold on;
plot(x,F,':','linewidth',2);
plot([0 0],[0 F(find(x>=0,1,'first'))],'-r','linewidth',2);
plot([c c],[0 F(find(x>=c,1,'first'))],'-r','linewidth',2);
axis([-m c+m 0 1]);
grid on;
hold off;
function [G g]=getd1(z,c,alpha,mu,lambda,m,s)
sm1=1;
sm2=mu;
for q=1:s
sm1=sm1+(-lambda)^q*(c-q*m-z).^q/factorial(q);
sm2=sm2+(-lambda)^q*(c-q*m-z).^(q-1)/factorial(q-1).*(mu*(c-q*m-z)/q+1);
end;
G=1-alpha*exp(mu*(c-z)).*sm1;
g=alpha*exp(mu*(c-z)).*sm2;
function [G g]=getd2(z,c,alpha,mu)
```

```
G=1-alpha*exp(mu*(c-z));
g=mu*alpha*exp(mu*(c-z));
```

Program P2

This program calculates the probabilities of filling and emptying of a reservoir of standard size *c*, drift *m*, and exponential inflow with parameter *mu* using a number of reservoir stages *k* using the integral equations approach.

```
function [p1, p2]=PegramIntExp(c,m,k,mu)
dk=c/k;
y=0:dk:c;
C=gencdf(y'-m-c,mu);
rp1=repmat(y,k+1,1);
rp2=repmat(y',1,k+1);
A=-genpdf(rp2-m-rp1,mu);
A(:,1)=A(:,1)/2;
A(:,end)=A(:,end)/2;
A=A*dk;
A=A+eye(res+1);
ff1=inv(A)*C;
ff=ff1(2:end-1);
ff=ff';
p1=ff1(1);
p2=1-ff1(end);
fprintf('\nPempty =%10.9f, Pfull =%10.9f \n\n',ff1(1), 1-ff1(end));
function y=gencdf(x,mu);
x=x+1/mu;
y=1-exp(-x*mu);
y(x<0)=0;
function y=genpdf(x,mu);
x=x+1/mu;
y=mu*exp(-x*mu);
y(x<0)=0;
```

Program P3

This program calculates the probabilities of filling and emptying of a reservoir of standard size *c*, drift *m*, and exponential inflow with parameter *mu* using a number of reservoir stages *k* using the difference equations approach.

```
function [p1, p2]=PegramDiff(c,m,k,mu)
h=c/k;
Q(1,1)=gencdf(-m,mu);
Q(k+2,1)=1-gencdf(c-m,mu);
Q(1,k+2)=gencdf(-m-c,mu);
Q(k+2,k+2)=1-gencdf(-m,mu);
i=1:k;
Q(2:k+1,1)=gencdf(i*h-m,mu)-gencdf((i-1)*h-m,mu);
Q(2:k+1,k+2)=gencdf(i*h-m-c,mu)-gencdf((i-1)*h-m-c,mu);
j=1:k;
Q(1,j+1)=gencdf((-j+.5)*h-m,mu);
Q(k+2,j+1)=1-gencdf(c-(j-.5)*h-m,mu);
i=repmat((1:k)',1,k);
```

```
j=i';
Q(2:k+1,2:k+1)=gencdf((i-j+.5)*h-m,mu)-gencdf((i-j-.5)*h-m,mu);
P=Q^1000;
p1=P(1);
p2=P(end);
fprintf('Pempty =%10.9f, Pfull =%10.9f \n',p1, p2);
function y=gencdf(x,mu)
x=x+1/mu;
y=1-exp(-x*mu);
y(x<0)=0;
```

Program P4

```
function [fe ff]=criticalmvnar(c,m,r)
%criticalmvnar calculates the probabilities of being empty and full of a
% reservoir with AR(1) Gaussian inflow through the
% distribution of the critical drawdown and filling periods
%
% For a reservoir with standard size c, drift m, and Gaussian inflow with
% first-order autocorrelation coefficient r, this function uses the
% multivariate normal integral to solve for the probabilities fe and ff
% of being empty and full, respectively, via the calculation of the
% distribution of the critical drawdown and filling periods.
nmx=500;
mp=50000;
cr=r.^(0:nmx);% This is the ACF for AR(1). Change for other models
% 1)Full to empty and 2)Empty to full
f11=zeros(1,nmx);
f22=zeros(1,nmx);
ee1=zeros(1,nmx);
ee2=zeros(1,nmx);
f11(1)=normcdf(-c,m,1);
f22(1)=1-normcdf(c,m,1);
hi=0;
for i=2:nmx
mu1=(1:i)*m;
ll1=[-c*ones(1,i-1) -inf]-mu1;
ul1=[zeros(1,i-1) -c]-mu1;
A=tril(ones(i,i));
[f11(i) ee1(i)]=qscmvnv(mp,toeplitz(cr(1:i)),ll1,A,ul1);
if f11(i) > hi; hi=f11(i);end;
if f11(i)<0.001*hi;break;end;
end;
hi=0;
for i=2:nmx
mu2=(1:i)*m;
ll2=[zeros(1,i-1) c]-mu2;
ul2=[c*ones(1,i-1) inf]-mu2;
A=tril(ones(i,i));
[f22(i) ee2(i)]=qscmvnv(mp,toeplitz(cr(1:i)),ll2,A,ul2);
if f22(i) > hi; hi=f22(i);end;
if f22(i)<0.001*hi;break;end;
end;
fe=sum(f11);% Probability of being empty
ff=sum(f22);% Probability of being full
```

References

1. Askew, A.J., Yeh, W.W.-G., and Hall, W.A., 1971. A comparative study of critical drought simulation. *Water Resour. Res.*, 7(1), 52–62.
2. Barnes, F.B., 1957. Storage required for a city water supply. *J. Inst. Eng. Aust.*, 26, 198–203.
3. Buchberer, S.G. and Maidment, D.R., 1989. Diffusion approximation for equilibrium distribution of reservoir storage. *Water Resour. Res.*, 25(7), 1643–1652.
4. Gani, J., 1955. Some problems in the theory of provisioning and of dams. *Biometrica*, 42, 179–200.
5. Genz, A., 1992. Numerical computation of multivariate normal probabilities. *J. Comput. Graph. Stat.*, 1, 141–149.
6. Gomide, F.L.S., 1975. Range and deficit analysis using Markov chains. Colorado State University, Fort Collins, CO Hydrology Papers, No. 79, p. 76.
7. Hall, W.A. and Dracup, J.A., 1970. *Water Resources Systems Engineering*. McGraw-Hill, New York, 372pp.
8. Hall, W.A., Askew, A.J., and Yeh, W.W.-G., 1969. Use of the critical period in reservoir analysis. *Water Resour. Res.*, 5(6), 1205–1215.
9. Hamed, K.H., 2009. On the implementation of Prabhu's exact solution of the stochastic reservoir equation. *Adv. Water Resour.*, 32, 594–606.
10. Hamed, K.H., 2012. A probabilistic approach to the calculation of the reliability of over-year storage reservoirs with persistent Gaussian inflow. *J. Hydrol.*, 448–449, 93–99.
11. Hazen, A., 1914. Storage to be provided in impounding reservoirs for municipal water supply. *Trans. ASCE*, 77, 1539–1640.
12. Heyman, D.P. and O'Leary, D.P., 1995. What is fundamental for Markov chains: First passage times, fundamental matrices, and group generalized inverses. In: *Computations With Markov Chains*, Ed. W.J. Stewart. Kluwer Academic Publishers, Boston, MA, pp. 151–161.
13. Hunter, J.J., 2005. Simple procedures for finding mean first passage times in Markov chains. *Proceedings of IWMS 2005,* Auckland, NewZealand: 29 March–1 April 2005, *Res. Lett. Inf. Math. Sci.*, Vol. 8, pp. 209–226.
14. Hurst, H.E., 1951. Long term capacity of storage reservoirs. *Trans. ASCE*, 116, 770–799.
15. Klemeš, V., 1977. Discrete representation of storage for stochastic reservoir optimization. *Water Resour. Res.*, 13, 149–158.
16. Klemeš, V., 1981. Applied stochastic theory of storage in evolution. In: *Advances in Hydroscience*, Ed. V.T. Chow. Academic, San Diego, CA, Vol. 12, pp. 79–141.
17. Koutsoyiannis, D., 2005. Reliability concepts in reservoir design. *Water Encyclopedia, Vol. 4, Surface and Agricultural Water*, Eds. J.H. Lehr and J. Keeley. Wiley, New York, pp. 259–265.
18. Kritskiy, S.N. and Menkel, M.F., 1935. Long-term streamflow regulation (In Russian). *Gidrotekh. Stroit.*, 2, 3–10.
19. Kritskiy, S.N. and Menkel, M.F., 1940. Generalized methods for runoff control computations based on mathematical statistics (In Russian). *Gidrotekh. Stroit.*, 2, 3–10.
20. Langbein, W.B., 1958. Queuing theory and water storage. *J. Hydraul. Div. ASCE*, 84(HY5), 1–24.
21. Lloyd, E.H., 1963. The theory of storage with serially correlated inputs. *J. Hydrol.*, 1, 99–128.
22. Lloyd, E.H., 1993. The stochastic reservoir: Exact and approximate evaluations of the storage distribution (Invited Review). *J. Hydrol.*, 151, 65–107.
23. Loucks, D.P., Stedinger, J.R., and Haith, D.A., 1981. *Water Resources Planning and Analysis*. Prentice Hall, Englewood Cliffs, NJ, 559pp.
24. Montaseri, M. and Adeloye, A.J., 1999. Critical period of reservoir systems for planning purposes. *J. Hydrol.*, 224, 115–136.
25. Moran, P.A.P., 1954. A probability theory of dams and storage systems. *Austr. J. Appl. Sci.*, 5, 116–124.
26. Moran, P.A.P., 1955. A probability theory of dams and storage systems: Modifications of the release rules. *Austr. J. Appl. Sci.*, 6, 117–130.

27. Moran, P.A.P., 1959. *The Theory of Storage*. Methuen, London, U.K.
28. Oğuz, B. and Bayazit, M., 1991. Statistical properties of the critical period. *J. Hydrol.*, 126, 183–194.
29. Pegram, G.G.S., 1980. On reservoir reliability. *J. Hydrol.*, 47, 269–296.
30. Prabhu, N.U., 1958. On the integral equation of the finite dam. *Q. J. Math.*, 9(35), 183–188.
31. Pretto, P.B., Chiew, F.H.S., McMahon, T.A., Vogel, R.M., and Stedinger, J.R., 1997. The (mis)behavior of behavior analysis storage estimates. *Water Resour. Res.*, 33(4), 703–709.
32. Rippl, W., 1883. Capacity of storage reservoirs for water supply. *Minutes Proc. Inst. Civil Eng.*, 71, 270–278.
33. Savarenskiy, A.D., 1940. A method for streamflow control computations (In Russian). *Gidrotekh. Stroit.*, 2, 24–28.
34. Sudler, C.E., 1927. Storage required for the regulation of stream flow. *Trans. ASCE*, 91, 622–660.

27

Sustainability in Urban Water Systems

Saeid Eslamian
Isfahan University of Technology

Seyed Sajed Motevallian
University of Tehran

AUTHORS

Saeid Eslamian received his PhD from the University of New South Wales, Australia, with Professor David Pilgrim. He was a visiting professor in Princeton University, United States, and ETH Zurich, Switzerland. He is currently an associate professor of hydrology in Isfahan University of Technology. He is the founder and chief editor of the *Journal of Flood Engineering* and *International Journal of Hydrology Science and Technology.* He has published more than 200 publications, mainly in statistical and environmental hydrology and hydrometeorology.

Seyed Sajed Motevallian is a research assistant at the University of Tehran. He received a BSc in civil engineering from the University of Mazandaran in 2008 and an MSc in civil engineering with emphasis on water engineering from the University of Tehran in 2011. He has collaborated with governmental and private institutions to undertake research projects in different fields such as integrated water resources management (IWRM) and water reuse and reclamation. His research interests include sustainability assessment of water resources systems, sustainable urban water management (SUWM), and application of systems analysis tools (e.g., multicriteria decision-making methods [MCDMs], system dynamics, driving force–pressure–state–impact–response [DPSIR]) in water resources planning and management.

PREFACE

The concept of sustainability has opened up a new discussion on where development policies and strategies lead us. Of particular interest for hydrologists and water experts may be that on how sustainability requirements can be met in urban water systems, and how they can assess the performance of urban water systems in achieving sustainability goals. This chapter concisely discusses the main topics in the context of sustainable urban water systems and provides a brief review of the state-of-the-art situation in putting sustainable urban water management (SUWM) into the practice. It also addresses the main challenges that exist in the path toward sustainable urban water systems.

27.1 Introduction

Involving a large number of stakeholders and a high degree of complexity has made the management of urban water systems a challenging task. Significant growth of urban population and the dramatic increase in water demand in recent years have exacerbated the management problem in recent years. The introduction of sustainable development and sustainability concepts in the late 1980s has inspired water scientists and experts to initiate a new scientific branch in the context of urban water management called "sustainable urban water management (SUWM)."

SUWM aims to promote the use of technologies and practices able to preserve scarce resources for the future generations while providing water and sanitation for the current users at affordable costs and without significant environmental damages to vital ecosystems. Thus, those urban water systems that fulfill sustainability requirements at expected levels may be called "sustainable urban water systems."

This chapter aims to provide readers with practical information on the basic concepts commonly discussed in the context of SUWM and sustainable urban water systems. To achieve this objective, first, different components of urban water systems and their functions in the life of urban residents are reviewed. Then, different perspectives on the meaning of sustainability and sustainable development concepts and the objectives of sustainable urban water systems are evaluated. Next, a brief review of conventional methods applied to appraise the sustainability of urban water systems including sustainability assessment methods and sustainability indicators (SIs) is presented. Finally, some experiences in implementing SUWM around the world are described, and the remaining issues in the context of SUWM are addressed.

27.1.1 Urban Water Systems

Sustaining an acceptable standard of living in urban areas requires a variety of public services such as water, energy, and transportation. Urban water systems are among those infrastructures, which serve a key role in ensuring the human health and the well-being of urban dwellers. Urban water systems include water supply system, wastewater collection and treatment system, and urban drainage. Water supply systems extract water from its source and convey it to a treatment plant in which raw water is purified and made suitable for use. Treated water is distributed via pipe networks to consumers (e.g., households, small industries) in order to satisfy their need to water for drinking, washing, etc. Wastewater collection systems convey wastewater generated by domestic and nondomestic consumers to the disposal sites. In some cases, a treatment plant exists in which wastewater is treated in order to reclaim water and nutrients for further uses such as landscape irrigation and fertilizer production. Urban drainage has the function of collecting stormwater to prevent flooding and protect the public from unhygienic conditions.

The systems described earlier are known as "conventional urban water systems" and [41] primarily refers to the centralized and large-scale nature of these systems. Conventional urban water systems have been utilized for years in more developed parts of the world; however, a majority of urban residents in less-developed countries has not yet enough access to water and sanitation, mainly due to explosive population growth, financial limitations, and poor institutional involvement [50]. These systems have been widely criticized for employing low-efficient technologies and having adverse impacts on the [44,61]. Thus, there is an increasing tendency toward employing alternative urban water systems wherever conventional systems are not affordable in terms of financial or natural resources [49].

Alternative urban water systems predominantly rely on distributed technologies and small-scale and decentralized infrastructures and aim to minimize the distance between the location of end use and water source [49]. Furthermore, these systems are primarily built in a manner to use energy and raw materials more efficiently in comparison with conventional ones. Rainwater harvesting, on-site wastewater treatment systems, ecological sanitation, and water-sensitive urban design (WSUD) are some well-known types of alternative urban water systems. To obtain more information on these systems, one can see References 20, 40.

27.1.2 Sustainability: Debate on Definition

Rapid population growth and the overuse of natural resources that substantially occurred in the second half of twentieth century have degraded the global environment significantly. They have also caused widespread concerns about having sufficient access to scarce resources (e.g., water, energy) in the future. In the last decades of the twentieth century, it was globally acknowledged that the adopted development policies that aim to maximize the economic benefits from exploited resources posed a serious threat to nature as well as to human health [44].

Primary efforts to introduce a more eco-friendly definition of development in order to progress toward a more sustainable future were made by IUCN et al. [30] in the report called "1980 World Conservation Strategy." In this report, sustainable development was defined based on three following objectives: "a) Maintenance of essential ecological processes and life support systems, b) Preservation of genetic diversity, and c) Sustainable utilization of species and the ecosystem" [30].

In 1987, the report entitled "Our Common Future" delivered by the World Commission on Environment and Development (WCED) was published in which sustainable development is defined as "a development that fulfills the needs of present generation without compromising the ability of the future generations to fulfill their basic needs" [60].

The concept of sustainable development incorporates the ecological values into the decision-making processes while recognizing the importance of human well-being and aiming to create a balance between the social, economic, and environmental goals of development. In this regard, sustainability might refer to a condition in which economic benefits and social welfare can be achieved within the carrying capacity of the environment. This "triple bottom line" (TBL) approach to the problem of sustainability that was coined by Elkington [18] refers to the "three Ps" (people, profit, and planet) or the principles of sustainability. Similarly to Elkington [18], Adams [1] believes that sustainable development is built on the "three pillars" that are economic growth, environmental protection, and social progress.

Having different perceptions about sustainable development in different regions of the world has made it difficult to reach a global consensus about the objectives of this concept. Jeppsson et al. [31] maintained that people in developed countries may consider population growth as the major threat to sustainable development, while in developing countries overconsumption may be recognized as the prime issue in the context of sustainable development.

Since the introduction of sustainable development in WCED [60], different experts and organizations have proposed various definitions of sustainability. Most of existing definitions are contextual and originated from the scientific background (e.g., urban planning, architecture, water resources, forestry) of the individual or organization, which provides the definition. As water is in the center of development plans, water scientists have made ongoing efforts to define the concept of sustainability and incorporate it into the different levels of water resources planning and management.

A prevalent approach to define water resources systems sustainability is to pertain it to the performance of the system, while the sustainable development of the system is assumed as an ongoing process. In this regard, sustainability can be assessed in terms of different social, economic, environmental, and technical objectives. Loucks [33] defined water resources sustainability based on the three conventional performance indicators (i.e., reliability, resiliency, vulnerability) mostly applied for reservoir systems operation. Howard [28] suggested that the sustainability of water resources systems should be defined in terms of risk, and sustainable water resources systems are those that maintain acceptable risks over an unlimited time horizon. Mays [42] defined water resources sustainability as "the ability to use water in sufficient quantities and quality from the local to the global scale to meet the needs of humans and ecosystems for the present and the future to sustain life, and to protect humans from the damages brought about by natural and human-caused disasters that affect sustaining life." Defining sustainability in terms of system's performance has been criticized in a few studies. Voß et al. [59] indicated that "sustainability cannot be translated into a blueprint or a defined end state from which criteria could be derived and unambiguous decisions be taken to get there." Bagheri and Hjorth [5] maintained that "sustainability is neither a state of the system to be increased or decreased, nor a static goal or target to be achieved." In their opinion, one should monitor the systems' progress using process indicators to ensure that sustainability requirements are achieved rather than measure the sustainability of the system by means of performance indicators [5].

27.1.3 Sustainable Urban Water Systems

Sustainable urban water systems are those that are designed and function along with the objectives of sustainable development. In addition to providing urban residents with basic needs such as water and sanitation, it is expected that a sustainable urban water system utilizes valuable resources efficiently and has minimal impact upon the environment. Different authors have attempted to describe "sustainable urban water system" or more generally "sustainable water resources systems" in their studies. An American Society of Civil Engineers (ASCE) Task Committee in their joint publication with UNESCO-IHP working group maintained that sustainable water resource systems are those "designed and managed to fully contribute to the objectives of society, now and in the future, while maintaining their ecological, environmental, and hydrological integrity" [2]. Hellström et al. [27] stated that a sustainable urban water system as well as any of urban infrastructures should "1) move toward a nontoxic environment, 2) improve health and hygiene, 3) save human resources, 4) conserve natural resources, 5) save financial resources."

Marsalek et al. [40] suggested the following basic goals for sustainable urban water systems:

- The supply of safe and good-tasting potable water to the urban residents in an unlimited time horizon
- Collection and treatment of wastewater in order to protect the urban dwellers from diseases and the environment from adverse impacts
- Control, collection, conveyance, and quality improvement of urban runoff in order to protect the urban environment from flooding and pollution
- Reclamation, reuse, and recycling of water and nutrients for domestic or agricultural uses in case of water shortage

27.2 Tools and Methods

Sustainability assessment refers to a systematic process in which a given system is evaluated to gain knowledge about the performance of the system meeting the sustainability requirements. It can be applied to rank the fields of activity, to provide suggestions to select appropriate technologies, and to identify solutions for implementing designed plans [49]. In this section, a brief review of tools and methods applied in the context of sustainability assessment of urban water systems is provided.

27.2.1 Sustainability Indicators

A prevalent approach to achieve a quantitative understanding of progress toward the objectives of sustainable urban water systems is to present the result of sustainability assessment studies by means of SIs. Indicators can be defined as "pieces of information, which summarize important properties, visualize phenomena of interest, quantify trends and communicate them to relevant target groups" [38]. The prime functions of indicators are "a) Assessing conditions and trends, b) Providing information for spatial comparison, c) Providing early warning information, d) Anticipating future conditions and trends" [24].

SIs can be utilized to provide information on to what extent the sustainability objectives have been accomplished; however, as the importance of different aspects of sustainability is determined by stakeholders, SIs cannot be measured in absolute terms [58]. A set of typical SIs applied to sustainability assessment of urban water systems can be found in References 6, 35. A variety of frameworks have been applied to develop and select SIs. Here, some SIs selection frameworks are introduced.

27.2.1.1 Balanced Lists (Theme Frameworks)

Balanced lists (theme frameworks) are frameworks that categorize the SIs into separate groups by means of a TBL approach. They usually include the main aspects of sustainability, social, economic, and environmental, although in some cases other aspects such as technical and institutional are also considered [37]. Balanced lists were first adopted by the United Nations Commission on Sustainable Development (UNCSD) in their report "Indicators of Sustainable Development: Framework and Methodologies" published in 1996 to suggest a first list of 134 SIs. The indicators were considered to be employed by different countries in the decision-making process of development and to create an opportunity for international comparisons [54]. Balanced lists have been widely used to evaluate the sustainability of urban water systems [27,35].

27.2.1.2 Pressure–State–Response Framework

Pressure–state–response (PSR) is an environmental assessment framework developed by the Organization of Economic Cooperation for Development (OECD) in the early 1990s, and a first set of PSR indicators were introduced in the report entitled "Towards sustainable development: Environmental indicators" [45]. In PSR, SIs are developed about a specific problem and categorized into pressure, state, and response groups. Pressure indicators refer to environmental pressures caused by human activities and degrade the environment through emissions and the depletion of natural resources. State indicators describe the current state of the environment, which is threatened by human activities in terms of quality and quantity of natural resources. Response indicators provide information about the response of human society to environmental damages. For instance, if low water quality of a given river due to the discharge of untreated industrial wastewater is considered as the problem, the volume of discharged wastewater, the concentration of heavy metals in the river, and the number of industrial factories that treat their wastewater before discharging it into the river can be considered as pressure, state, and response indicators, respectively.

27.2.1.3 Driving Force–Pressure–State–Impact–Response Framework

Driving force–pressure–state–impact–response (DPSIR) is a modified version of PSR framework developed by adding two new indicator groups (i.e., driving force, impact) to PSR indicators. DPSIR has been adopted by the European Environmental Agency (EEA) and the European Statistical Office (Eurostat) with the aim of monitoring the state of the environment in the European Union (EU) countries [17].

It adopts a cause and effect approach to provide an understanding of the relations between emissions and impacts [37]. Driving force indicators refer to the hydrologic and socioeconomic factors, which indirectly exert stress on the environment such as population growth and climatic condition. Impact indicators describe the socioeconomic impact of a degraded environment such as poverty and unemployment. For instance, for a given river, in addition to discharge of untreated industrial wastewater (pressure), low precipitation and high level of evaporation (driving forces) can diminish the quality of water. Furthermore, waterborne diseases, which spread in residential areas in vicinity of the river due to the drinking of polluted water, are an example of impact indicators.

27.2.2 Sustainability Assessment Methods

A variety of system analysis methods has been applied to assess the sustainability of urban water systems. Beck [7] defined system analysis as "The procedure and corpus of methods for providing support and guidance in the systematic analysis of decision-making problems. It may involve the development and use of mathematical models and is often associated with the notion of optimal solutions to problems." Some well-known and conventional system analysis methods applied in the context of sustainability assessment of urban water systems are cost–benefit analysis (CBA), life cycle costing (LCC), life cycle assessment (LCA), material flow analysis (MFA), exergy analysis, microbial risk assessment (MRA), and multi-criteria analysis (MCA). Here, LCA, MRA, and MCA are introduced, and the application of these methods in sustainability assessment studies is reviewed.

27.2.2.1 Life Cycle Assessment

LCA is a process in which "the environmental effects associated with any given activity from initial gathering of raw materials from the earth (petroleum, crops, ores, etc.) to the point at which all materials are returned to the earth are evaluated" [16]. In recent decades, LCA methodologies have been developed, and the global use of LCA has been promoted through the scientific work of organizations such as the Society of Environmental Toxicology and Chemistry (SETAC), International Standardization Organization (ISO), and the United Nations Environmental Program (UNEP) [26]. Currently, the LCA methodology is being standardized by ISO within the framework of ISO14000 series [31]. LCA includes the four following steps [26]:

- *Goal and scope definition*: Goal and scope definition refers to formulizing the objectives of the study in terms of a question, a specific application, etc., and determining the spatial and temporal coverage of the study, respectively.
- *Inventory analysis*: It is a process in which the input and output flows of the product system through its life cycle including the use of raw materials, energy, water, and emissions into the air, soil, and water are quantified.
- *Impact assessment*: In this phase, inventory data are associated with specific environmental impacts in order to achieve an understanding of these impacts.
- *Interpretation*: This is the final phase of LCA in which the soundness and robustness of results are evaluated.

LCA has been widely employed to assess the sustainability of urban water systems [36,38,39].

27.2.2.2 Microbial Risk Assessment

Urban water systems can potentially pose threats on humans' health due to a variety of reasons such as discharge of untreated wastewater to water bodies or accidental spills of stormwater [19]. Therefore, estimating the level of microbial risk associated with urban water systems in terms of quantitative measures is a matter of vital importance. A systematic framework to assess the risks of human disease posed by pathogens was first developed by Craun et al. [15]. MRA is an analytical method that aims to measure the disease risk associated with exposure to pathogenic microorganisms [20]. It involves the application of dose response functions—which define the relationship between the number of pathogens consumed and the likelihood of infection—in order to predict the degree of risk from an exposure in terms of a probability of infection [20]. MRA has been used to assess the sustainability of urban water systems in various real case studies such as the comparison of two wastewater treatment system alternatives in Sydney, Australia, and Hammarby Sjöstad, Sweden, by Fane [19] and Ashbolt et al. [3], respectively.

27.2.2.3 Multicriteria Analysis

MCA is a type of integrated assessment (IA) tools primarily applied in situations when there are competing criteria [43]. The main advantage of MCA is that sustainability criteria and indicators can be quantified in quite different units without the need for converting all criteria into monetary units. Furthermore, by means of MCA, different social, economic, and environmental indicators can be weighted and combined in order to provide an integrated perspective of sustainability. Multi-criteria decision-making methods (MCDMs) are a group of MCA methodologies designed to facilitate the process of decision-making for human and to provide optimum solutions for multi-criteria problems using rational principles and mathematical formulations. MCDMs are usually categorized into multi-objective decision-making methods (MODMs) and multi-attribute decision-making methods (MADMs). The latter is used to select the best alternative among a variety of alternatives with regard to different criteria, while the former is basically designed to find the optimum solution of a mathematical function in a given decision space. Goal programming (GP) and compromise programming (CP) are some conventional MODMs, and analytic hierarchical process (AHP), Technique for Order of Preference by Similarity to Ideal Solution (TOPSIS), Elimination Le Choix Traduisant la REalité (ELECTRE), and Preference Ranking Organization METHod for Enrichment of Evaluations (PROMETHEE) are some typical MADMs. In the context of sustainability assessment of urban water systems, MADMs have been widely used to integrate the results of different sustainability assessment methods and produce a sustainability index [31,35].

27.3 Implementing Sustainable Urban Water Management

One of essential prerequisites to accomplish the objectives of sustainable urban water systems is a sustainable practice for managing these systems. SUWM can be described as "integrated and biophysical systems, which consider social, economic, environmental and political contexts, provision of water for ecological and human uses, and a long-term perspective" [57]. In this section, some experiences of implementing SUWM are described.

27.3.1 SWITCH: Toward the City of the Future

Sustainable Water Management Improves Tomorrow's Cities Health (SWITCH) was an EU-funded research program implemented and co-funded by a consortium of 33 partners from 15 countries over the period 2006 to 2011. The main objective of SWITCH program was to accommodate the change toward more SUWM in the "city of the future" [58]. To achieve this objective, integrated urban water management (IUWM) was selected as the management practice.

IUWM is a popular version of SUWM applied to urban water systems around the world especially in Europe and Australia. The word "integrated" implies the integration of different aspects of urban water management such as supply and demand, sources and sinks, scale, and considering urban water systems to be a whole system rather than separate entities [41]. van der Steen and Howe [58] put forward a hypothesis that "design and management of the urban water system based on analysis and optimization of the entire urban water system (infrastructure and human organizations, water supply, sanitation, stormwater, etc.) will lead to more sustainable solutions than optimization of separate elements of the system." Mays [41] outlined two reasons for applying IUWM to urban water systems. First, different components of urban water systems (water supply, wastewater, stormwater system) are naturally connected through the hydrological cycle, and, second, the real advantages can be realized through the integrated management of these systems rather than independent action.

The SWITCH program consisted of three main parts [58]:

- *Learning alliances*: They were based on this proposition that in case of dealing with complex situations, a group of people working interactively are more likely to reach better solutions than talented individuals working separately [12]. Learning alliances were established in "SWITCH cities" including Accra, Alexandria, Beijing, Belo Horizonte, Birmingham, Cali, Hamburg, Lima, Lodz, Tel Aviv, and Zaragoza.
- *Action research*: It involved implementing IUWM in terms of real case studies in "SWITCH cities."
- *Multiple-way learning*: This refers to this premise that through the implementation of case studies, urban water experts and mangers from European countries and developing countries can share their knowledge and learn from each other.

Details about the results of implemented case studies can be found in Reference 12.

27.3.2 Integrated Resources Planning: An Australian Approach

Integrated Resources Planning (IRP) or the so-called least-cost planning [29] is "a process of planning services in a way that ensures the efficient and sustainable management of water, energy, or other resources" [22]. IRP involves making detailed demand forecasts, producing a large number of alternatives, assessing demand-side and supply-side alternatives equally, and deciding how to fill the demand–supply gap at least cost while recognizing the sustainability requirements [22]. Within a project undertaken by the Institute of Sustainable Futures (ISF), University of Technology Sydney (UTS) for the Water Services Association of Australia (WSAA), an IRP framework for urban water management was developed [52]. The key principles of the IRP framework are as follows [52]:

- *Water service provision*: It recognizes that what is important to costumers is the services required (e.g., healthy and adequate water for drinking, bathing) and not the water itself.
- *Detailed demand forecasting*: This principle involves disaggregating the water demand through an end-use analysis, which enables demand forecasting as well as determining the water conservation potential.
- *Consideration of a broad spectrum of viable options that satisfy service needs*: It recognizes that the objectives of IRP can be achieved through a wide range of options such as water reuse and supply augmentation.
- *Comparison of options using a common metric, boundary, and assumptions*: To ensure that water services are supplied to society at the lowest cost, IRP adopts a common monetary unit (net present value) to assess all options.
- *A participatory process*: As urban water systems interact with different parts of the environment and society, and involve a multitude of individuals with various preferences, IRP recognizes the importance of stakeholder participation at particular levels of the planning process in order to incorporate diverse needs and values into final decisions.
- *Adaptive management*: This principle emphasizes that IRP is a continuous learning process in which plans are developed, put into action, and evaluated iteratively.

27.3.3 Soft Path for Water: Canadian Experiences

The term "soft path" was first introduced by Lovins [34] to describe "soft path for energy." Lovins [34] maintained that people do not actually need energy resources such as electricity and oil, but they require the services provided by these resources. Wolff and Gleick [61] adopted the soft path approach to develop the concept of "soft path for water." They believed that in addition to the conventional "hard" path that primarily relies on centralized infrastructures and decision-making and adopts supply-side management practices, there is a soft path to meet water-related needs by conserving water more efficiently and matching the water services to consumers' diverse needs [61]. This approach is called "soft path" because it primarily relies on human creativeness to find solutions for water-related problems—"working with nature rather than attempting to overcome it" [8]. The soft path for water or the so-called water soft path primarily relies on following principles [9]:

- *To resolve the water supply–demand gaps from the demand side*: This principle emphasizes that before adding any new water supply sources to the existing system, it should be ensured that all feasible water conservation efforts have been made.
- *To match the quality of the water supplied to the required quality by end users*: This principle that is also known as "water quality cascading" [20] recognizes that most-important needs to water such as drinking should be satisfied by high-quality water and less-important needs such as washing should be met by lower qualities.
- *Backcasting rather than forecasting*: Backcasting starts from the desired future and works backward in order to identify the "soft paths" that connect that future to the existing situation. In this regard, the main aim of planning is not to find where current routes will take the system but how the objectives of the system can be attained.

A framework for water soft path analysis was developed by Friends of the Earth, Canada. The following steps should be taken to carry out the framework [10]:

- *Identify water services*: The first step is to make a list including all services provided by water for consumers (drinking, bathing, washing, etc.) and calculate how much water is used by each service. In this step, feasible alternatives for water conservation are also evaluated.
- *Create a "business-as-usual scenario"*: This step involves creating a scenario in which rates of water withdrawals and uses grow through the time scope of the study assuming that the size of the population and of the economy has normal growth.
- *Review water supply options*: In this step, all current water supply sources (surface water, groundwater, etc.) are identified and evaluated to determine if any are being excessively used or diminished. All new water supply sources, which threaten ecological, social, and cultural values, should be rejected.
- *Establish a desired future scenario*: In a desired future scenario, both sources and uses of water should be sustainable in the target year. In this step, effective communication with stakeholders may be beneficial in order to identify more desirable options for the future. At least two scenarios should be created for the future. In one scenario, possible cost-effective modifications should be investigated. It encompasses a multitude of policy options to make water services cost-effective for both water users and water service providers. In the second scenario, cost-effective changes should be integrated with major changes in consumers' behavior, growth rates, and the economic structure of the society. Furthermore, solutions to implement water quality cascading should be sought in this scenario.
- *Ensure desired future scenarios are sustainable*: This step involves examining the desired future scenarios to ensure the sustainability of the supply–demand can be met. Making adjustments to the scenarios is likely if any of them suffers from weaknesses that prevent them to fulfill the expected goals.
- *Adjust for expected effects of climate change*: In this step, the impacts of climate change on water resources as well as human activities should be incorporated into the future scenarios.
- *Backcast from the desired future to the actual present*: This step involves developing multiple soft paths by designing policies and programs that connect the desired future to the current situation. In other words, it should be determined what actions are required to be done to achieve the desired future.

Friends of the Earth, Canada, in cooperation with research teams from three Canadian universities (University of Victoria, Acadia University, and University of Waterloo) implemented the water soft path analysis framework in three case studies at different spatial scales including a generalized urban area, a composite watershed, and a province [10]. The case studies were conducted mainly to try out the framework, not to obtain instantly practical results [10]. More details about the results of the case studies and guidelines on conducting a water soft path analysis can be found in Reference 11.

27.4 Sustainable Urban Water Management: Remaining Issues

A large number of research studies have been implemented in the context of SUWM; however, the following issues required to be addressed more in future studies:

1. *Impact of climate change on urban water systems*: With regard to ongoing trends of greenhouse gases (GHGs) emission, climate change becomes inevitable. It creates several impacts on urban water cycle that result in urban flash floods and significant changes in the amount of urban water supply and demand [32]. In the Fourth World Water Forum held in March 2006, the Cooperative Programme on Water and Climate (CPWC) remarked that risks posed by climate change are not adequately addressed in the development and management plans of the water sector [41]. Mays [41] contended that global climate change has received inadequate attention in urban water management even in highly developed parts of the world such as the United States. A useful approach to deal with this challenge is to adjust the results of urban water planning models with regard to perceived impacts of climate change. Some important information on the implication of climate change for urban water supplies and water demand can be found in Fane et al. [21].
2. *Need for a paradigm shift in dealing with the problem of sustainability*: By a review on methods and tools applied for sustainability assessment of urban water systems, one can realize that "hard systems thinking" approach has been primarily adopted in previous studies. The word "hard systems thinking" first used in Checkland [13] describes "an approach to real world problems in which an objective or end-to-be-achieved can be taken as given." In this way, there is a desired state and a present state, and the final objective is selecting the best way of getting from the present sate to the desired state [13]. Decision-making based on the hard system thinking approach basically involves selecting a choice among a clearly defined set of options for action [46]. MCA is a typical example of analytical methods that strongly relies on the hard systems thinking approach and primarily has been used to find an optimum alternative for a given urban water system from a sustainability perspective.

   Adopting the hard systems thinking approach to assess the sustainability of socioeconomic systems such as urban water systems has drawn criticism in a few studies [5,59]. Socioeconomic systems basically consist of a large number of interactions and feedback loops makes them show complex and nonlinear behavior. Regarding this fact, adopting conventional linear approaches such as the hard systems thinking approach is logically not the most effective way to deal with the problems lie in such complicated nonlinear systems. In contrast with the hard system thinking approach, there is a "soft system thinking" approach. Soft systems thinking or soft systems methodology is "an action-oriented process of inquiry into problematic situations in which users learn their way from finding out about the situation, to taking action to improve it" [14]. Sushil [51] asserted that adopting the paradigm of "learning"—which is the basis of "soft systems thinking" approach—is a more appropriate tool to systematically study and evaluate complex socioeconomic systems in comparison to the paradigm of "optimization." This premise implies a paradigm shift in dealing with management problems. One of soft system thinking-based concepts that has gained renowned attention in the context of natural resources management is "social learning." Pahl-Wostl and Hare [47] described the social learning as "an ongoing learning and negotiation process where a high priority is given to questions of communication, perspective sharing and development of adaptive group strategies for problem solving." Few studies have been conducted using the social learning concept to deal with the problem of sustainability in urban water systems [4,48]. Bagheri and Hjorth [4] employed system dynamics methodology and the social learning concept to propose a framework for monitoring the progress of complex systems toward sustainable development and examined it in a case study for the urban water system of Tehran. Adopting the social learning concept to make the sustainable urban water systems workable seems a wise choice; however, a good deal of work is required to be done in order to fortify the position of the this concepts in urban water management.

3. *Capacity development in academic institutions, governmental and private organizations, and society*: Capacity development or capacity building is "the process through which individuals, organizations and societies obtain, strengthen and maintain the capabilities to set and achieve their own development objectives over time" [55]. It requires to be addressed at three interrelated levels as the following [53]:
   a. *Individual level*: In this level, capacity development involves providing the conditions under which government officials are able to initiate an ongoing process of learning, acquire new knowledge and skills, and use them in new ways.
   b. *Institutional level*: In this level, capacity development primarily focuses on modernizing systems and processes.
   c. *Societal level*: In this level, capacity development aims to create a more interactive practice of public administration that learns from actions as well as feedbacks received from the public.

Considerable pilot studies and research projects have been conducted to promote the application of sustainable practices of urban water management; however, sustainability and its objectives in urban water sector are not still fully acknowledged by water users and service providers in different parts of the world particularly in developing countries. A reason for this shortcoming may be poor capacity development in academic institutions, governmental and private organizations, and society. In a study conducted at the Regional Center on Urban Water Management (RCUWM) based in Tehran, Iran, in order to assess the problems of urban water management in the Middle East and Central Asia countries, it was remarked that there is an urgent need for capacity building of staff in these countries, and thus high priority should be given to training and capacity building for sustainability [23]. In a discussion paper provided by the UNEP, the following strategies are proposed to increase the effectiveness of capacity building for sustainable development: "a) Identifying needs and building on existing capacities, b) Being clear about the objectives, c) Using a wide range of capacity building approaches, d) Target the right people to build a critical mass, e) Making the training-of-trainers approach work, f) Institutionalizing capacity building programme at regional and national level" [56].

27.5 Summary and Conclusions

In this chapter, a concise review on main topics in the context of sustainability in urban water systems was presented. After a brief introduction of urban water systems and their functions, existing visions on sustainability definition and the objectives of sustainable urban water systems were reviewed. Then, conventional tools and methods applied for sustainability assessment of urban water systems were discussed, and good practices of SUWM around the world were described. Finally, this chapter addressed some important issues that need to be discussed in future studies.

Sustainability principles in urban water systems have been established well in terms of definitions, conceptual frameworks, and analytical methodologies during the past decades; however, considerable efforts are required to make these principles work in the real world. Regarding this reality that roughly 1.1 billion people lack safe drinking water, about 2.6 billion do not have access to adequate sanitation, and between 2 and 5 million people die annually from water-related diseases [25], sustainability objectives in urban water systems are still far from being reached in a global scale. To facilitate the progress toward the sustainable development of urban water systems, considerable emphasis should be placed on developing mechanisms to involve stakeholders in the decision-making processes of urban water systems that enable them to communicate their visions and share their knowledge and more importantly to learn interactively from their decisions.

References

1. Adams, W.M. 2006. *The Future of Sustainability: Re-Thinking Environment and Development in the Twenty-First Century.* Report of the IUCN Renowned Thinkers Meeting, The World Conservation Union. Retrieved July 12, 2013, from http://cmsdata.iucn.org/downloads/iucn_future_of_sustanability.pdf
2. ASCE and UNESCO-IHP. 1998. *Sustainable Criteria for Water Resources Systems.* Reston, VA: American Society of Civil Engineers (ASCE).
3. Ashbolt, N.J., S.R. Petterson, D.J. Roser, T. Westrell, J. Ottoson, C. Schönning, and T.A. Stenström. 2006. Microbial risk assessment tool to aid in the selection of sustainable urban water systems. *2nd IWA Leading-Edge on Sustainability in Water-Limited Environments Conference.* 8–10 November 2004, Sydney, Australia. Eds. B.M. Beck and A. Speers, pp. 187–195. London, U.K.: IWA Publishing.
4. Bagheri, A. and P. Hjorth. 2007. A framework for process indicators to monitor for sustainable development: Practice to an urban water system. *Environment, Development and Sustainability,* 9 (2): 143–161.
5. Bagheri, A. and P. Hjorth. 2007. Planning for sustainable development: A paradigm shift towards process-based approach. *Sustainable Development,* 15 (2): 83–96.
6. Balkema, A.J., H.A. Preisig, R. Otterpohl, and F.J.D. Lambert. 2002. Indicators for the sustainability assessment of wastewater treatment systems. *Urban Water,* 4 (2): 153–161.
7. Beck, B.M. 1997. Applying systems analysis in managing the water environment: Towards a new agenda. *Water Science and Technology,* 36 (5): 1–7.
8. Brandes, O.M. and D.B. Brooks. 2006. *The Soft Path for Water in a Nutshell.* Victoria, British Columbia, Canada: Friends of the Earth Canada and The POLIS Project on Ecological Governance at the University of Victoria.
9. Brooks, D.B. 2005. Beyond greater efficiency: The concept of water soft paths. *Canadian Water Resources Journal,* 30 (1): 83–92.
10. Brooks, D.B. and S. Holtz. 2009. Water soft path analysis: From principles to practice. *Water International,* 34 (2): 158–169.
11. Brooks, D.B., O.M. Brandes, and S. Gurman (Eds.). 2009. *Making the Most of the Water We Have: The Soft Path Approach to Water Management.* Sterling, VA: Earthscan.
12. Butterworth, J., P. McIntyre, and C. da Silva Wells (Eds.). 2011. *SWITCH in the City: Putting Urban Water Management to the Test.* The Hague, the Netherlands: IRC International Water and Sanitation Centre.
13. Checkland, P. 1981. *Systems Thinking, Systems Practice.* Chichester, U.K.: Wiley.
14. Checkland, P. and J. Poulter. 2010. Soft systems methodology. *Systems Approaches to Managing Change: A Practical Guide.* Eds. M. Reynolds and S. Holwell, pp. 191–242. Milton Keynes, U.K.: The Open University and London.
15. Craun, G., A. Dufour, J. Eisenberg et al. 1996. A conceptual framework to assess the risks of human disease following exposure to pathogens. *Risk Analysis,* 16 (6): 841–848.
16. Curran, M.A. 1996. *Environmental Life-Cycle Assessment.* New York: McGraw-Hill.
17. European Environment Agency (EEA). 2000. Environmental signals 2000. Environmental Assessment Report No. 6, EEA, København K, Denmark.
18. Elkington, J.B. 1997. *Cannibals with Forks: The Triple Bottom Line of 21st Century Business.* Oxford, U.K.: Capstone Publishing.
19. Fane, S. 2004. Life cycle microbial risk analysis of sustainable sanitation alternatives. *2nd International Ecological Sanitation Symposium.* 7–11 April 2003, Lübeck, Germany. Ed. C. Werner. pp. 389–396. Lübeck, Germany: Gesellschaft für Technische Zusammenarbeit (GTZ).

20. Fane, S. 2006. Planning for sustainable urban water: Systems—Approaches and distributed strategies. PhD thesis, Institute for Sustainable Futures, University of Technology, Sydney, New South Wales, Australia.
21. Fane, S., J. Patterson, S. Maheepala, and D. Kirono. 2010. Incorporating climate change into urban water IRP. Integrated Resource Planning for Urban Water—Resource Papers, Waterlines Report, Institute of Sustainable Futures, University of Technology Sydney, Canberra, Australia.
22. Fane, S., J. Patterson, S. Maheepala et al. 2011. Integrated resource planning for urban water—Resource papers. Waterlines Report, Institute of Sustainable Futures, University of Technology Sydney (UTS), Canberra, Australia.
23. Figueres, C.M. 2005. Urban water management in the Middle East and central Asia: Problem assessment. *IWRA 12th World Water Congress*. New Delhi: International Water Resources Association (IWRA).
24. Gallopín, G.C. 1997. *Indicators and Their Use: Information for Decision-Making. Sustainability Indicators: A Report on the Project on Indicators of Sustainable Development*. Chichester, U.K.: John Wiley & Sons.
25. Gleick, P., M. Palaniappan, C. Hunt, V. Srinivasan, M. Moench, D. Haasz, G. Wolff, C. Henges-Jeck, M. Kiparsky, and N.L. Cain. 2004. *The World's Water 2004–2005: The Biennial Report on Freshwater Resources*. Washington, DC: Island Press.
26. Guinée, J.B., M. Gorrée, R. Heijungs et al. (Eds.). 2004. *Handbook on Life Cycle Assessment: Operational Guide to the ISO Standards*. Dordrecht, the Netherlands: Kluwer Academic Publishers.
27. Hellström, D., U. Jeppsson, and E. Kärrman. 2000. A framework for system analysis of sustainable urban water management. *Environmental Impact Assessment Review*, 20 (3): 311–321.
28. Howard, C.D.D. 2002. Sustainable development-risk and uncertainty. *Journal of Water Resources Planning and Management*, 27 (5): 309–311.
29. Howe, C. and S. White. 1999. Integrated resources planning for water and wastewater: Sydney case studies. *Water International*, 24 (4): 356–362.
30. IUCN, UNEP, and WWF. 1991. *Caring for the Earth - Strategy for Sustainable Living*. London, U.K.: Earthscan.
31. Jeppsson, U., D. Hellström, and E. Kärrman. 1999. Systems analysis of sustainable urban water management. Technical Report TEIE-7135, Lund Institute of Technology, Lund, Sweden.
32. Karamouz, M., A. Moridi, and S. Nazif. 2010. *Urban Water Engineering and Management*. Boca Raton, FL: CRC Press.
33. Loucks, D.P. 1997. Quantifying trends in system sustainability. *Hydrological Sciences*, 42 (4): 513–530.
34. Lovins, A.B. 1977. *Soft Energy Paths: Toward a Durable Peace*. Cambridge, MA: Ballinger/Friends of the Earth.
35. Lundie, S., N.J. Ashbolt, D. Livingston, E. Lai, E. Kärrman, J. Blaikie, and J. Anderson. 2008. Sustainability framework—methodology for evaluating the overall sustainability of urban water systems, Occasional Paper. *Water Services Association of Australia (WSAA)*, (14): 6–99.
36. Lundie, S., G.M. Peters, and P.C. Beavis. 2004. Life cycle assessment for sustainable metropolitan water systems planning. *Environmental Science and Technology*, 38 (13): 3465–3473.
37. Lundin, M. 2002. Indicators for measuring the sustainability of urban water systems: A life cycle approach. PhD thesis, Chalmers University of Technology, Gothenburg, Sweden.
38. Lundin, M. and G.M. Morrison. 2002. A life cycle assessment based procedure for development of environmental sustainability indicators for urban water systems. *Urban Water*, 4 (2): 145–152.
39. Mahgoub, M.E.M., N.P. van der Steen, K. Abu-Zeid, and K. Vairavamoorthy. 2010. Towards sustainability in urban water: A life cycle analysis of the urban water system of Alexandria City, Egypt. *Journal of Cleaner Production*, 18 (10–11): 1100–1106.
40. Marsalek, J., B. Jiménez-Cisneros, M. Karamouz, P.A. Malmquist, J. Goldenfum, and B. Chocat. 2008. *Urban Water Cycle Processes and Interactions*. Paris, France: UNESCO Publishing and Leiden.

41. Mays, L.W. 2009. *Integrated Urban Water Management: Arid and Semi-Arid Regions*. Paris, France: UNESCO Publishing.
42. Mays, L.W. 2007. *Water Resources Sustainability*. New York: McGraw-Hill.
43. Ness, B., E. Urbel-Piirsalu, S. Anderberg, and L. Olsson. 2007. Categorising tools for sustainability assessment. *Ecological Economics*, 60 (3): 498–508.
44. Novotny, V., J. Ahern, and P. Brown. 2010. *Water Centric Sustainable Communities: Planning, Retrofitting, and Building the Next Urban Environment*. Hoboken, NJ: John Wiley & Sons.
45. OECD. 1998. *Towards Sustainable Development—Environmental Indicators*. Paris, France: OECD.
46. Pahl-Wostl, C. 2002. Towards sustainability in the water sector—The importance of human actors and processes of social learning. *Aquatic Sciences-Research across Boundaries*, 64 (4): 394–411.
47. Pahl-Wostl, C. and M. Hare. 2004. Processes of social learning in integrated resources management. *Journal of Community and Applied Social Psychology*, 14 (3): 193–206.
48. Pearson, L.J., A. Coggan, W. Proctor, and T.F. Smith. 2010. A sustainable decision support framework for urban water management. *Water Resources Management*, 24 (2): 363–376.
49. Shuping, L., L. Siuqing, B. Chocat, and S. Barraud. 2006. Requirements for sustainable management of urban water systems. *Environmental Informatics Archives*. Ed. A. Ernest, Vol. 4, pp. 116–128. Regina, saskatchewan, Canada: International Society for Environmental Information Science (ISEIS).
50. Stephenson, D.B. 2001. Problems of developing countries. *Frontiers in Urban Water Management: Deadlock or Hope*. Eds. Ĉ. Maksimoviĉ and A. Tejada-Guibert, pp. 264–315. London, U.K.: IWA Publishing.
51. Sushil. 1993. *System Dynamics: A Practical Approach for Managerial Problems*. New Delhi, India: Wiley Eastern Ltd.
52. Turner, A., J. Willetts, S. Fane, D. Giurco, A. Kazaglis, and S. White. 2008. *Guide to Demand Management*. Ultimo, New South Wales, Australia: Institute for Sustainable Futures.
53. UN. 2006. Definition of basic concepts and terminologies in governance and public administration. Note by the Secretariat, Economic and Social Council-Committee of Experts on Public Administration. New York: United Nations (UN).
54. UNCSD. 1996. *Indicators of Sustainable Development: Framework and Methodologies*. New York: United Nations Committee on Sustainable Development (UNCSD).
55. UNDP. 2009. Frequently asked questions: The UNDP approach to supporting capacity development. Briefing Paper, Capacity Development Group, Bureau for Development Policy. New York: United Nations Development Programme (UNDP).
56. UNEP. 2006. Ways to increase the effectiveness of capacity building for sustainable development. Discussion Paper, Economics and Trade Branch (ETB), Division of Technology, Industry and Economics (DTIE), *IAIA Annual Conference*, Stavanger. Norway, U.K.: United Nations Environment Programme (UNEP).
57. van de Meene, S.J., R.R. Brown, and M.A. Farrelly. 2011. Towards understanding governance for sustainable urban water management. *Global Environmental Change*, 21 (3): 1117–1127.
58. van der Steen, P. and C. Howe. 2009. Managing water in the city of the future; strategic planning and science. *Reviews in Environmental Science and Biotechnology*, 8 (2): 115–120.
59. Voß, J.P., R. Kemp, and D. Bauknecht. 2006. *Reflexive Governance for Sustainable Development*. Cheltenham, U.K.: Edward Elgar.
60. WCED. 1987. *Our Common Future*. Oxford, U.K.: Oxford University Press.
61. Wolff, G. and P.H. Gleick. 2002. The soft path for water. *The World's Water: The Biennial Report on Freshwater Resources: 2002–2003*. Ed. P.H. Gleick. pp. 1–32.Washington, DC: Island Press.

28

Urban Hydrology

Miguel A. Medina, Jr.
Duke University

AUTHOR

Miguel A. Medina, Jr. is professor of civil and environmental engineering, Pratt School of Engineering, Duke University, United States. He is a past president of the American Institute of Hydrology and a fellow of the American Society of Civil Engineers. He has been a registered professional hydrologist since 1983. He was named external evaluator of the UNESCO International Hydrological Programme from 2002 to 2004 and the lead evaluator (2007) of the UNESCO World Water Assessment Programme. Professor Medina has conducted funded research in hydrologic and water quality mathematical modeling for the U.S. Environmental Protection Agency (EPA), the National Science Foundation, the Office of Water Research and Technology, the U.S. Air Force, the U.S. Army Waterways Experiment Station, the Naval Oceanographic Office, DuPont Engineering, the U.S. Geological Survey, the North Carolina Water Resources Research Institute, and the State of North Carolina. His current research focuses on flow and solute transport surface/groundwater interactions. The results of his research are reported in such journals as *Advances in Water Resources, Water Resources Research, Journal of Hydrology, Journal of Contaminant Hydrology, Journal of Hydrologic Engineering, Journal of Hydraulic Engineering, Ground Water, Desalination, Environmental Modeling and Software, Applied Mathematics and Computation, Journal of the American Water Resources Association, Computing in Civil Engineering, Journal of Water Resources Planning and Management, Journal of Environmental Engineering, Hydrological Science and Technology,* and *Environmental Health Perspectives.*

PREFACE

The rapidly changing land use of the urban environment has led to a new subfield of hydrology. Storm water runoff generated from rainfall over this environment is concentrated into small natural channels, man-made open channels and storm sewers, natural and engineered swales, and detention structures. Analyzing long-term records of rainfall statistically requires the definition of independent storm events. The complexity of the many interactions from such runoff flowing over both pervious and impervious surfaces requires a comprehensive mathematical modeling approach, albeit limited by the availability of rainfall/runoff quantity and quality data within the study area. Quantitative methods of describing the urban hydrologic cycle are reviewed, and applications are presented.

28.1 Introduction

The transformation of precipitation into surface runoff is depicted in Figures 28.1 and 28.2. Figure 28.1 illustrates the relationship between the rainfall process, infiltration to the subsurface, generation of overland flow, and flow into the man-made drainage system. Among the many variables that describe these processes mathematically are the width, area, percent imperviousness, ground slope, roughness parameters of the land cover for both impervious and pervious fractions, and several infiltration rate parameters that depend upon methods chosen. Eventually, the runoff makes its way into inlets of the man-made drainage system. Figure 28.2 illustrates the transport of runoff through both natural and man-made conduits, as well as routing through a detention structure. Whether natural or man-made, the length, geometry, slope, and roughness characteristics must be specified for each conduit. Natural channels can be approximated with triangular, trapezoidal, and other cross-sectional shapes. For storage elements such as detention basins (there is also storage along a channel), stage–discharge curves and other geometric data are required. Any mathematical model is an abstraction of the actual physical system it attempts to simulate, and uncertainty in the values of many model parameters requires field measurements and calibration.

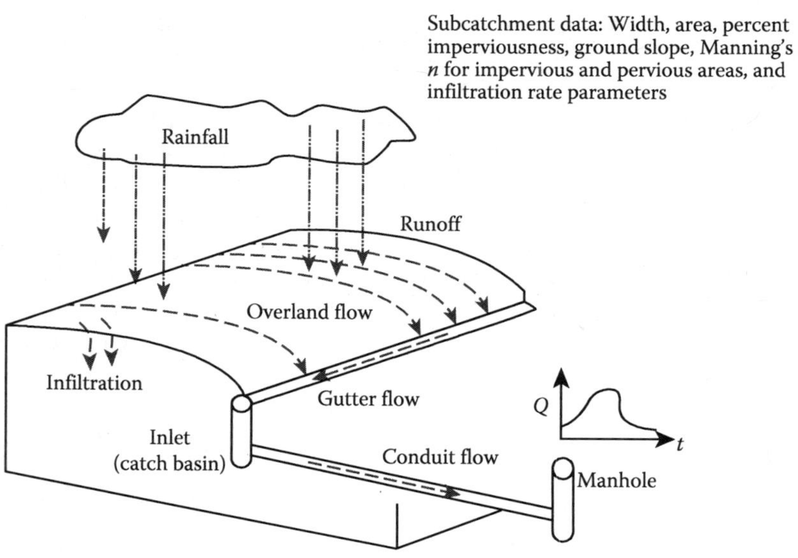

FIGURE 28.1 Rainfall, infiltration, overland flow, and flow into drainage system.

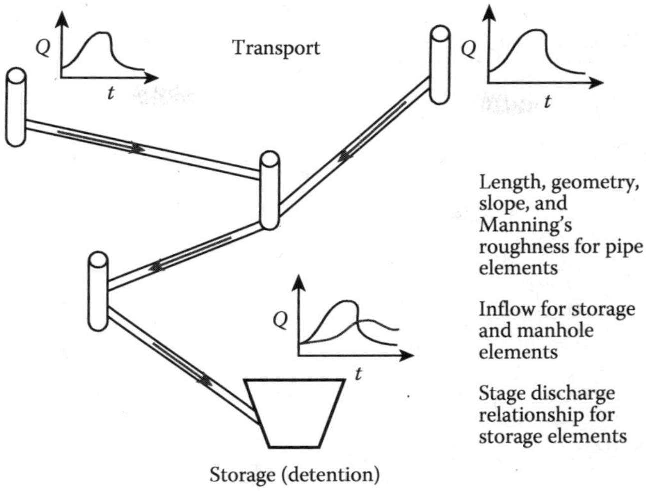

FIGURE 28.2 Transport and storage components of the drainage system.

28.2 Statistical Analysis of a Rainfall Time Series

It has been recognized for over three decades that the large number of factors that affect surface runoff quantity (rainfall duration, intensity, time between events, volume, infiltration, antecedent soil moisture, etc.) and surface runoff quality (pollutant loads and buildup between storms, surface washoff, transport, kinetic interactions, etc.) prevents the exclusive use of any single event for proper analysis and design. Regardless of whether water quantity or quality is the primary objective, long-term historical rainfall data are required. An hourly or shorter-interval precipitation time series at least 30–40 years long is desirable. The purpose in quantitative analysis of the rainfall time series is to summarize the variables of interest (depth, duration, intensity, and time between storm events) and to statistically characterize the rainfall record to assess the probability of occurrence of storm events of various magnitudes.

To properly analyze the rainfall time series, storm events must first be defined in terms of their statistical independence (see Figure 28.3). A common approach is to derive a minimum interevent time (MIT), such that the intervals between storm event midpoints are nearly *exponentially distributed* [12].

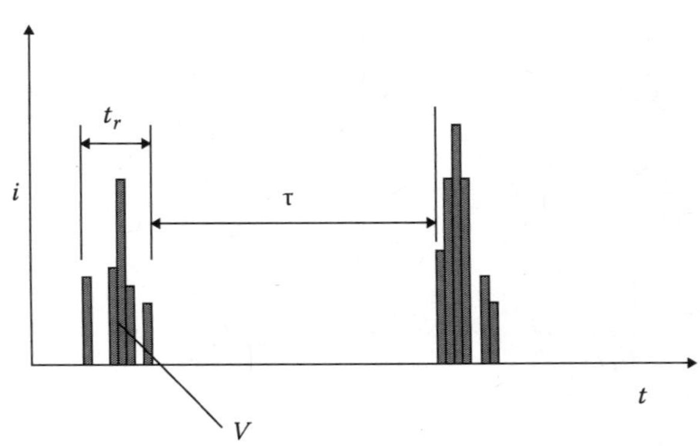

FIGURE 28.3 Statistical definition and characterization of storm events. i, Intensity, in./h; V, Volume (total depth), in.; t_r, Duration, h; T, Time between independent events, h.

TABLE 28.1 NOAA Rainfall Station Description and Statistics of Storm Events, RDU, NC (1948–2011)

Station No. 317069	Raleigh–Durham WSFO AP, Wake County Hourly rainfall	Lat: N 35:52:00 Long: W 78:47:00 Elevation: 127 m	Record Length: 64 Years, 1948 (6 Months)
	Storm Event Variable	Mean	COV
	No. of storms[a] per year	99.523	0.11
	Depth (mm)	10.668 0.425	1.394
	Intensity (mm/h)	1.778 0.071	1.340 1.336
	Duration (h)	6.873 6.915	1.097 1.093
	Time between events (h)	87.372 88.53	1.002 0.992
	Total no. of storms[a]	6309	

[a] *Note*: Number of storms is based on a MIT of *6 h*, which yielded a *COV very near 1.0 for time between events* for the rainfall time series. Interevent times have an exponential distribution (for which the mean equals the standard deviation, ∴ COV = 1.0).

Trial values of the MIT are chosen until a coefficient of variation (COV) near 1.0 is finally obtained for the time interval between event midpoints.

A reliable source of data is the National Oceanic and Atmospheric Administration (NOAA) National Climatic Data Center (NCDC): access to a geographically comprehensive number of rainfall stations and their data is available online. The historical record of hourly precipitation from 1948 to 2011 at Raleigh-Durham (RDU) Airport (located only 8 miles from Durham, NC) was analyzed to obtain storm event statistics, summarized in Table 28.1. This represents over 6300 storms using a MIT of 6 h of dry weather to separate independent storm events. The average storm duration was 6.9 h.

The annual variability in the rainfall record is presented in Figure 28.4, with a mean annual rainfall of 1057 mm. Figure 28.5 shows that the mean *monthly* rainfall is distributed rather uniformly throughout the year, with hurricane-related extremes occurring from June to October. An analysis of the frequency distribution of storm depths reveals a large number of storms between total rainfall depths 0.254 and 4.06 mm, with almost 90% of all storms in the record less than about 25.4 mm. Capturing the "first flush"

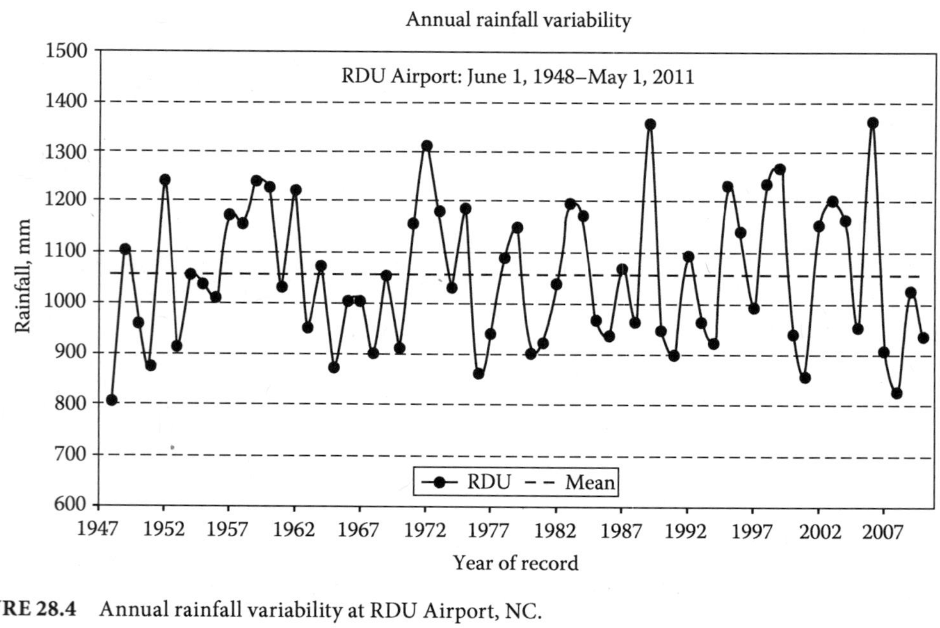

FIGURE 28.4 Annual rainfall variability at RDU Airport, NC.

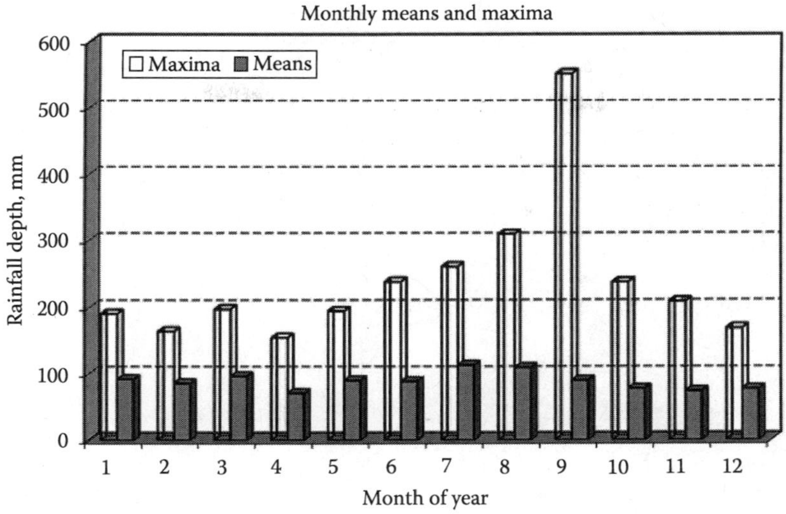

FIGURE 28.5 Monthly rainfall means and maxima.

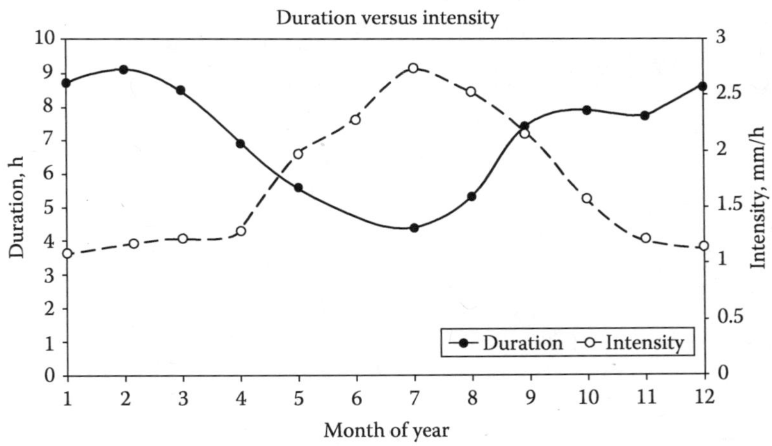

FIGURE 28.6 Storm event duration versus intensity per month of year.

would control 90% of all storm events. Figure 28.6 shows clearly that short-duration, high-intensity storms occur during the summer months. Figure 28.7 shows that storms are more frequent during the summer months, less frequent from September to December: there are also water quality implications during large intervals between storms, in terms of pollutant load accumulations on the ground. The rainfall time series statistics in Table 28.1 are an integral part of a code used to predict the expected pollutant removal performance of detention facilities based on a probabilistic model, discussed later.

28.3 Overland Flow

Overland flow is very thin sheet flow that develops after precipitation falls over a sloped land surface and infiltration occurs into the subsurface. The depth and flow rate of this thin sheet of water depends, among others, on the rainfall intensity, ground surface characteristics (e.g., roughness and slope), antecedent moisture, and immediate subsurface conditions (e.g., hydraulic conductivity). By neglecting the acceleration and pressure terms in the momentum equation for dynamic waves, the kinematic-wave

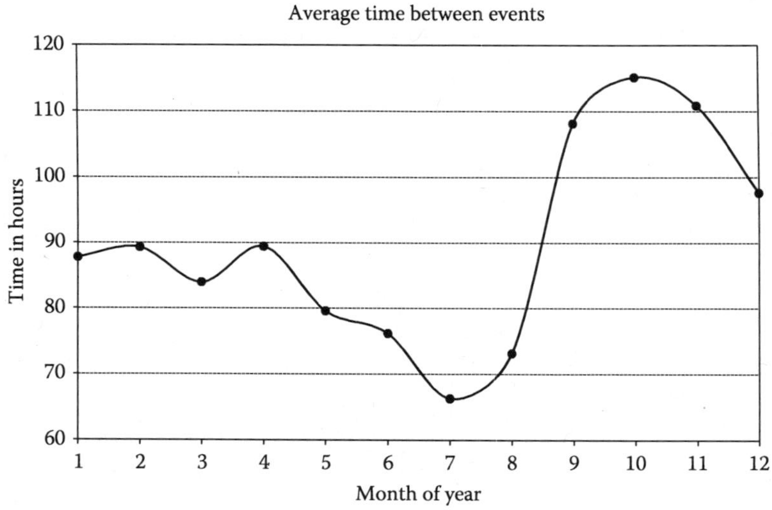

FIGURE 28.7 Time between independent storm events.

model [7] substitutes a steady uniform flow (stage–discharge) relationship for the momentum equation. However, unsteady flow is preserved through the continuity equation.

The dynamic wave equations for overland flow consist of continuity and momentum equations, as follows, respectively [4]:

$$\frac{\partial y}{\partial t}+V\frac{\partial y}{\partial x}+y\frac{\partial V}{\partial x}=i-f \tag{28.1}$$

$$\frac{\partial V}{\partial t}+V\frac{\partial V}{\partial x}+g\frac{\partial y}{\partial x}=g\left(S_0-S_f\right)+\left[(i-f)\frac{V}{y}\right] \tag{28.2}$$

where

y is depth of water (L)
S_f is friction slope (L/L)
S_0 is bed slope (L/L)
V is water velocity (L/T)
i is rainfall intensity (L/T)
f is infiltration rate (L/T)
g is acceleration of gravity (L/T^2)
t is time (T)
x is distance (L)

These equations are for moderately wide overland flow and small bottom slope. An order-of-magnitude analysis of Equation 28.2 shows that $(i-f)$ has a negligible effect on the flow dynamics [4]. The flow per unit width q (L^2/T) is equal to Vy. Thus, Equation 28.1 may be written as

$$\frac{\partial q}{\partial x}+\frac{\partial y}{\partial t}=i-f \tag{28.3}$$

The simplest distributed routing method is the *kinematic-wave* model, which neglects the local acceleration, convective acceleration, and pressure terms in the momentum equation (thus, friction and gravity forces essentially balance each other, $S_f = S_0$). The most common formulas relating water velocity, slope, and catchment roughness are [10]:

Manning formula

$$V = \frac{k}{n} R^{2/3} \sqrt{S_0} \tag{28.4}$$

Chezy formula

$$V = C\sqrt{R\,S_0} \tag{28.5}$$

where

- V is water velocity (L/T)
- n is the Manning friction coefficient (dimensionless)
- R is the hydraulic radius (L)
- C is the Chezy coefficient ($L^{1/2}$/T)
- $k = 1.486$ for English system units and 1.0 for the SI system
- the bed slope S_0 replaces the friction slope S_f
- For wide rectangular sections (e.g., overland flow), the hydraulic radius reduces to water depth y, and the momentum equation for kinematic waves reduces to

$$q = \alpha y^m \tag{28.6}$$

where, for *fully turbulent* flow,

using Manning

$$\alpha = \frac{k}{n}\sqrt{S_0} \quad m = \frac{5}{3} \tag{28.7}$$

using Chezy

$$\alpha = C\sqrt{S_0} \quad m = \frac{3}{2} \tag{28.8}$$

Values of exponent m for laminar and mixed laminar-turbulent conditions are tabulated by Ponce [10]. For overland flow problems using the kinematic wave, the continuity and momentum equations can be combined into one differential equation, as follows:

$$\left.\begin{aligned} &\frac{\partial y}{\partial t} + \frac{\partial q}{\partial x} = i - f \\ &q = \alpha y^m \end{aligned}\right\} \Rightarrow \frac{\partial y}{\partial t} + \alpha \frac{\partial \left(y^m\right)}{\partial x} = i - f \tag{28.9}$$

28.4 Channel Flow

The distributed routing models allow us to compute flow rate and water level variation through space and time: a major advantage over the lumped model in terms of the design criteria of any storage structure, such as a detention pond or reservoir. Similar to the overland flow problems, by eliminating some terms in the momentum equation of the Saint Venant equations, alternative distributed flow routing models are obtained. The dynamic wave equations for channel flow consist of continuity and momentum equations, as follows, respectively [1]:

$$\frac{\partial Q}{\partial x} + \frac{\partial A}{\partial t} = q(x,t) \tag{28.10}$$

$$\frac{1}{A}\frac{\partial Q}{\partial t}+\frac{1}{A}\frac{\partial}{\partial x}\left(\frac{Q^2}{A}\right)+g\frac{\partial y}{\partial x}-g\left(S_0-S_f\right)=0 \tag{28.11}$$

where

- y is depth of water (L)
- S_f is friction slope (L/L)
- S_0 is bed slope (L/L)
- Q is flow rate (L^3/T)
- $q\ (x, t)$ is the net lateral inflow per unit length of channel (L^2/T)
- g is acceleration of gravity (L/T^2)
- A is the channel cross-sectional area (L^2), t is time (T)
- x is distance (L)

In Equation 28.11, the first, second, and third terms represent local acceleration, convective acceleration, and pressure force, respectively. The bed slope and friction slope in the last term represent the gravity force and friction force.

Again, the simplest distributed routing method is also the *kinematic-wave* model, which neglects the local acceleration, convective acceleration, and pressure terms in the momentum equation for dynamic waves. For open channel flows, the continuity and momentum equation and their combined form for kinematic waves are given as follows:

$$\left.\begin{array}{l}\dfrac{\partial A}{\partial t}+\dfrac{\partial Q}{\partial x}=q(x,t)\\[2ex] Q=\alpha A^m\end{array}\right\}\Rightarrow\frac{\partial A}{\partial t}+\alpha\frac{\partial\left(A^m\right)}{\partial x}=q(x,t) \tag{28.12}$$

where, for *fully turbulent* flow,

Manning

$$\alpha=\frac{k}{n}\frac{\sqrt{S_0}}{P^{2/3}}\quad m=\frac{5}{3}$$

Chezy

$$\alpha=C\frac{\sqrt{S_0}}{\sqrt{P}}\quad m=\frac{3}{2} \tag{28.13}$$

Here

- A is the channel cross-sectional area (L^2)
- Q is the flow rate (L^3/T)
- n is the Manning friction coefficient (dimensionless)
- C is the Chezy coefficient ($L^{1/2}/T$)
- P is the wetted perimeter (L)
- S_0 is the bed slope (L/L)
- t is time (T)
- $q\ (x, t)$ is the net lateral inflow per unit length of channel
- x is the distance along the flow axis (L).

As before, $k = 1.486$ for English system units and 1.0 for the SI system.

28.5 Infiltration

Most urban simulation models have used the Horton equation for prediction of infiltration capacity into the soil as a function of time,

$$f_{cap} = f_\infty + (f_0 - f_\infty)e^{-\alpha t} \tag{28.14}$$

where

f_{cap} is the infiltration capacity into soil, say in./h
f_∞ is the minimum or ultimate value of f (at $t = \infty$), in./h
f_0 is the maximum or initial value of f (at $t = 0$), in./h
t is the time from beginning of storm, sec
α is the decay coefficient, 1/time

The actual infiltration is given by

$$f(t) = \min\left[f_{cap}(t), i(t)\right] \tag{28.15}$$

where

f is the actual infiltration into the soil (say in./h)
i is the rainfall intensity (say in./h)

Equation 28.15 simply states that the actual infiltration will be *the lesser* of actual rainfall and available infiltration capacity.

Typical values for parameters f_0 and f_∞ are often greater than typical rainfall intensities. Thus, when Equation 28.14 is used such that f_{cap} is a *function of time only*, then f_{cap} will *decrease even if rainfall intensities are very light*. This results in a reduction in infiltration capacity regardless of the actual amount of entry of water into the soil. Thus, to correct this problem, the integrated form of the Horton equation may be used

$$F(t_p) = \int_0^{t_p} f_{cap}\, dt = f_\infty t_p + \frac{f_0 - f_\infty}{\alpha}\left(1 - e^{-\alpha t_p}\right) \tag{28.16}$$

where F is the cumulative infiltration at time t_p (inches or mm). The true cumulative infiltration will be

$$F(t) = \int_0^{t} f(\tau)\, d\tau \tag{28.17}$$

where f is given by Equation 28.15. Equations 28.16 and 28.17 may be used to define the equivalent time, t_p. That is, the actual cumulative infiltration given by Equation 28.17 is equated to the area under the Horton curve (given by Equation 28.16), and the resulting equation is solved for t_p and serves as its definition. Unfortunately, the following equation

$$F = f_\infty t_p + \frac{f_0 - f_\infty}{\alpha}\left(1 - e^{-\alpha t_p}\right) \tag{28.18}$$

cannot be solved explicitly for t_p: It must be done iteratively. Note that $t_p \le t$ that states that the time t_p on the cumulative Horton curve will be less than or equal to the actual elapsed time. This also implies

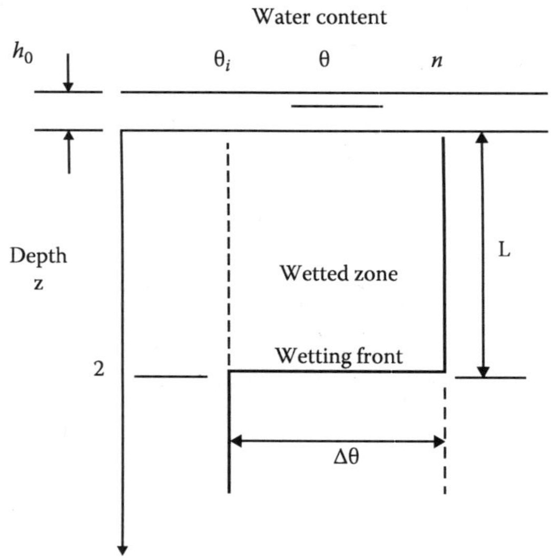

FIGURE 28.8 Schematic for Green–Ampt equation. (Modified after Chow, V.T. et al., *Applied Hydrology*, McGraw-Hill Book Company, New York, 1988.)

that available infiltration capacity $f_{cap}(t_p)$ will be greater than or equal to that given by Equation 28.14. Thus, f_{cap} will be a function of the actual water infiltrated and not a function of time only, with no other effects. The original Horton model does poorly when light rainfall falls early, since it decays in time independently of accumulated infiltration. The integrated form of the Horton equation is applicable to urban environments, particularly for relatively short storm events. It should be noted that the Horton equation is an approximate solution to the *Richards equation*: 1D, unsteady, unsaturated flow—under several simplifying assumptions.

An alternative model is the Green–Ampt formulation, better suited for nonurban environments. Green and Ampt [5] derived the first physically based equation describing the infiltration of water into a soil. It has been refined and adapted for use in computer codes. A schematic representation of the Green–Ampt model is presented in Figure 28.8. In Figure 28.8, the wetting front has penetrated to a depth L in time t since infiltration began. Water is ponded to a small depth h_0 on the soil surface. The Green–Ampt equations for *cumulative infiltration* F and infiltration rate f are given by

$$F(t)-\psi\,\Delta\theta\,\ln\left[1+\frac{F(t)}{\psi\,\Delta\theta}\right]=Kt \tag{28.19}$$

$$f(t)=K\left[\frac{\psi\,\Delta\theta}{F(t)}+1\right] \tag{28.20}$$

The parameters are defined as follows:

Ψ is the wetting front soil suction head (cm)
θ is the moisture content
K is the hydraulic conductivity (cm/h)

Equation 28.19 is commonly solved by Newton–Raphson iteration. Once F is found, then the infiltration rate is obtained. The parameters are well documented for a variety of soil classes (sand, sandy loam, loam, silt loam, sandy clay loam, clay loam, silty clay loam, sandy clay, and clay).

28.6 Detention Storage

The basic governing differential equation, the continuity equation, is given by

$$\frac{d\forall}{dt} = Q_{in}(t) - Q_{out}(h) \tag{28.21}$$

The change in volume $d\forall$ due to a change in depth dh may be expressed as

$$d\forall = A(h)dh \tag{28.22}$$

such that the continuity equation can now be expressed as

$$\frac{dh}{dt} = \frac{Q_{in}(t) - Q_{out}(h)}{A(h)} \tag{28.23}$$

A common method to numerically integrate ordinary differential equations is the *fourth-order* Runge–Kutta method, such that the head is approximated as follows:

$$h_{n+1} = h_n + \frac{1}{6}\Delta t\left[k_1 + 2k_2 + 2k_3 + k_4\right] \tag{28.24}$$

where

$$k_1 = f(h_n, t_n) \qquad k_2 = f\left(h_n + \frac{1}{2}k_1\Delta t, t_n + \frac{\Delta t}{2}\right)$$

$$k_3 = f\left(h_n + \frac{1}{2}k_2\Delta t, t_n + \frac{\Delta t}{2}\right) \qquad k_4 = f\left(h_n + k_3\Delta t, t_n + \Delta t\right)$$

More sophisticated urban hydrologic simulation programs, such as the well-known U.S. EPA Storm Water Management Model (SWMM5), use a *fifth-order* accurate Runge–Kutta method with adaptive step size control. This means that such a code adapts the step size to achieve some predetermined accuracy with minimum computational effort [11].

28.7 Surface Water Quality

In surface water quality, pollutant accumulation and washoff processes are mathematically approximated. The assumptions of these processes are as follows: (1) the amount of pollutant, which can be removed during a storm event, depends on rainfall duration and initial quantity of pollutant available; (2) chemical changes or biological degradation does not affect the pollutant decay during the runoff process; and (3) the amount of pollutants percolating into the soil by infiltration is insignificant [8]. The exponential decay of solid buildup is generally represented by

$$p_{n,j,t} = p_{\max_{n,j}}\left[1 - e^{-\lambda_{n,j}t}\right] \tag{28.25}$$

where

$p_{n,j,t}$ is the loading of pollutant j on subcatchment n at time t [lb/acre (kg/ha)]
$p_{\max_{n,j}}$ is the maximum (asymptotic) loading of pollutant j on subcatchment n [lb/acre (kg/ha)]
$\lambda_{n,j}$ is power exponent coefficient for pollutant j on subcatchment n [day^{-1}]
t is accumulation time [day]

The washoff is calculated by making it proportional to runoff rate at each time step of simulation as follows:

$$P_t = \frac{dP}{dt} = -ar^b P_o \tag{28.26}$$

where

P_t is the pollutant load washed off at time t [lb/s (kg/s)]
P_o is the load available for washoff at time t [lb (kg)]
r is the runoff rate [in/h (mm/h)]
a is the washoff coefficient, including conversion units
b is a power constant

Pollutant transport in overland flow and through urban storage/treatment systems such as pipes, channels, and detention basins may be represented by the 1D version of the classical advective–dispersive equation (28.9):

$$\frac{\partial C}{\partial t} = \frac{\partial}{\partial x}\left[E\frac{\partial C}{\partial x} - UC\right] \pm \sum_{i=1}^{n} S_i \tag{28.27}$$

where

C is the concentration of pollutant [M/L³]
t is time [T],—$E\partial C/\partial x$ is the mass flux due to longitudinal dispersion along the flow axis (x-direction) [M/L²]
UC is mass flux due to advection by the fluid containing the mass of pollutant [M/L²T]
S_i is the sources or sinks of the substance C [M/L³T].
n is the number of sources or sinks
U is the flow velocity [L/T]
E is the longitudinal dispersion coefficient [L²/T]

The mass balance equation for a well-mixed, variable volume unit is derived from Equation 28.27 and is given by Medina et al. [9]:

$$\frac{d(VC)}{dt} = I(t)C^I(t) - O(t)C(t) - KC(t)V(t) \tag{28.28}$$

where

V is the reservoir volume [L³]
C^I is the influent pollutant concentration [M/L³]
C is the effluent and reservoir pollutant concentration [M/L³]
I is the inflow rate [L³/T]
O is the outflow rate [L³/T]
t is time [T]
K is the first-order decay coefficient [1/T]

This equation is solved numerically over time for concentration, using average values for quantities that might change over a time step, such as flow rate and volume.

28.8 Detention Pond Long-Term Average Pollutant Removal Efficiency

The basic probabilistic methodology was developed by Di Toro and refined by Di Toro and Small [2] and adapted to wet ponds (sedimentation devices) and percolation (infiltration and recharge) devices by

Driscoll et al. [3]. The long-term average performance is computed from the statistical properties of the detention basin inflows and a few design characteristics. Removal due to sedimentation under *dynamic* (flow-through) conditions is given by

$$R = 1 - \left(1 + \frac{1}{n}\frac{V_s}{[Q/A]}\right)^{-n} \tag{28.29}$$

where

R is the fraction of initial solids removed ($R \times 100 =$% removal)
V_s is the settling velocity of particles
Q/A is the rate of applied flow divided by the surface area of the basin (an "overflow velocity," often designated *overflow rate*)
n is a measure of the degree of turbulence or *short circuiting*

which tends to reduce removal efficiency; typically, $1 \le n \le 5$, with $n = 3$ for good performance and $n > 5$ for very good performance. For variable runoff flows that are *gamma distributed* entering a treatment device and characterized by a mean flow and COV (CV_q), the long-term average fraction of total mass removed is

$$R_L = Z\left[\frac{r}{r - \ln(R_M / Z)}\right]^{r+1} \tag{28.30}$$

where

R_L is the long-term average fraction removed
R_M is the fraction removed at the mean runoff rate
V_q is the COV of runoff flow rates
r is the $1/CV^2$ (reciprocal of square of CV_q)
Z is the maximum fraction removed at very low rates

Removal under quiescent conditions is very significant since average storm duration in the United States is 6 h and the average interval between storms is from 3 to 4 days. The volume of a detention basin relative to the volumes of runoff events routed through becomes the principal factor in pollutant removal effectiveness. For storm volumes that are gamma distributed, the *fraction not captured* (over all storms) may be represented by

$$f_{\forall} = \frac{(r_1)^{r_1}(r_2)^{r_2}}{G(r_1)G(r_2)} \int_{q=0}^{\infty} q^{r_1} \exp[-r_1 q] \int_{\Delta=0}^{\infty} \Delta\left[\Delta + \frac{\forall}{q}\right]^{r_2-1} \exp[-r_2 \Delta]\, d\Delta\, dq$$

$$r_1 = \frac{1}{CV_q^2} \qquad r_2 = \frac{1}{CV_d^2} \tag{28.31}$$

where

CV_q is the COV of runoff flow rates
CV_d is the COV of runoff durations
q is the storm runoff flow rate
Δ is the average time interval between storm midpoints
$\forall$ is the basin effective volume, divided by mean storm runoff volume
$G(r_1)$ is the gamma function of r_1
$G(r_2)$ is the gamma function of r_2
$f_{\forall}$ is the fraction of all volumes NOT captured by the detention basin

Detention Pond Long Term Pollutant Removal Metric/SI Units

Rainfall Stats:	Mean	CV
Volume (in or mm)	10.668	1.394
Intens (in/h or mm/h)	1.778	1.340
Duration (h)	6.873	1.097
Delta (h)	87.372	1.002

Runoff COEF (Rv)	= 0.30
Drainage area (acres or hectares)	= 80.94
Basin surf area (sq ft or sq m)	= 5774.18
Basin depth (ft or m)	= 2.74
Short circ param	= 2.50
Mean runoff rate (QR) =	431.72 cu m/h
Overflow rate (QR/A) =	0.07 m/h
Vol capac factor (VB/VR) =	6.12
Basin Surface Area / Catchment Area, in %	= 0.71

Size Frac	% Remove Dynamic	% Remove Quiescent	% Remove Combined	Sett veloc M / HR	Vol factor VE / VR
1	0.7	76.8	77.5	0.01	5.36
2	3.1	79.0	82.1	0.09	6.06
3	6.9	79.2	86.1	0.46	6.12
4	7.8	79.2	87.0	2.13	6.12
5	7.9	79.2	87.0	19.81	6.12
Avg	5.3	78.7	84.0		

Avg total suspended solids removal	= 84.0
Avg total phosphorus removal	= 58.8
Avg lead removal	= 75.6
Avg tkn, bod, cu, zn removal	= 42.0

FIGURE 28.9 WetDet model output.

The double integral cannot be evaluated analytically. A numerical integration technique using Laguerre quadrature is applied to approximate the integral with a weighted polynomial. An application follows using the rainfall statistics presented earlier in Table 28.1, for a hypothetical West Campus detention basin. Model output is presented later in Figure 28.9. As expected, detention removes a significant percentage of total suspended solids and lead (found to a large extent in suspended solids) and is much less efficient in removing dissolved pollutants.

The results previously presented are specific to a ratio of detention pond surface area to drainage area contributing of 0.71%. This is a critical factor in long-term pollutant removal performance, as shown in Figure 28.10, when the detention pond surface area is varied to obtain a range of ratios.

28.9 U.S. Environmental Protection Agency Storm Water Management Model

EPA SWMM5 is a dynamic rainfall/runoff simulation model that can be used for single storm event simulation or continuous simulation of multiple storms of a specific watershed. The model calculates the flow and water quality constituents of the watershed based on hydraulic parameters set by the user and provides graphs of simulation results such as flow hydrographs and water quality concentrations. The model is widely used to plan, analyze, and control storm water runoff; to design drainage system components; and to evaluate watershed management of both urban and nonurban areas [6,13]. Images of urban areas obtained from GIS mapping software can be displayed as backdrops to visualize the physical system being simulated.

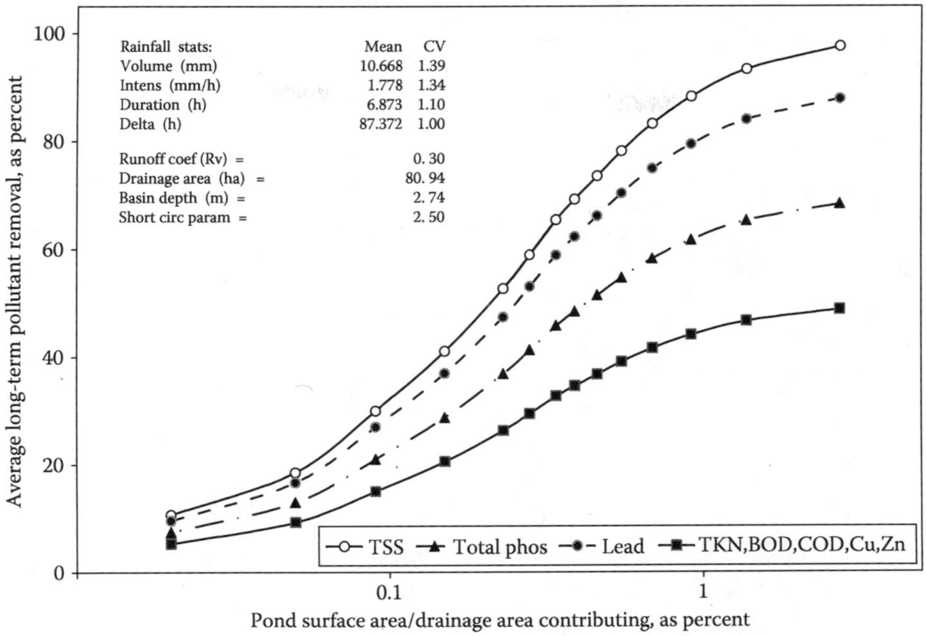

FIGURE 28.10 Detention pond performance as a function of pond surface area.

The runoff component of SWMM simulates surface runoff over subcatchment areas, as generated from rainfall input. Then, SWMM routes this runoff with the transport component through a conveyance system of pipes, channels, storage/treatment devices, pumps, and regulators. SWMM calculates the quantity and quality of runoff generated within each subcatchment and the flow rate, flow depth, and quality of water in each pipe and channel during a simulation period comprised of multiple time steps. SWMM inputs include precipitation data, subcatchment delineation, pipe system characteristics, and soil properties. The watershed profile is formed by defining input data such as slope, area, and imperviousness of subcatchments; length, cross section, height, and slope of conduits; and porosity, hydraulic conductivity, and field capacity of aquifers, into the program. Outputs of the model include change of flow rate (hydrograph) and change of concentration (pollutograph) through time and daily, monthly, annual, and total simulation summaries (for continuous simulation) and statistical frequency analysis. SWMM is also capable of viewing results in tabular and graphical form.

While calculating surface runoff, precipitation, and flow from upstream subcatchments that contribute to the inflow, infiltration and evaporation contribute to the outflow. For flow routing in conduits, steady flow, and kinematic, diffusion and dynamic wave routing options may be used. In addition, for infiltration calculation, three options exist, which are the integrated Horton method, the Green–Ampt method, and the SCS curve number method. EPA SWMM simulates surface runoff quality by using exponential, power, and saturation functions for buildup and exponential, rating curve, and event mean concentration functions for washoff.

28.10 Application to Duke University Campus Watershed in Durham, North Carolina

The Duke University West Campus watershed is located in Durham, North Carolina. Runoff from the West Campus of Duke University is simulated by using the EPA SWMM to predict surface runoff and quality and transport through the drainage system. The first formal application of EPA SWMM to the Duke Campus was completed by Mahi [8]. In order to be able to use EPA SWMM for surface runoff and

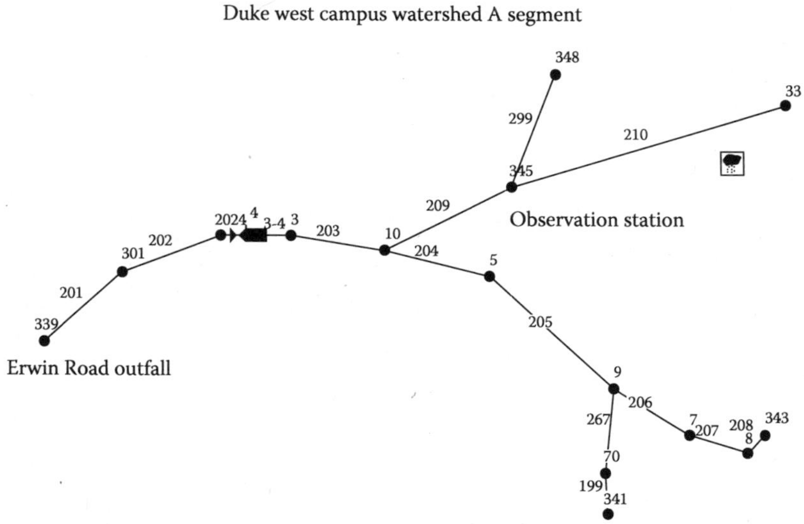

FIGURE 28.11 Watershed a discretization showing node 345 observation station.

channel flow predictions, the model needed to be calibrated against measured rainfall and runoff data. For this purpose, a rain gage was placed on the roof of Hudson Hall (Pratt School of Engineering), and several storm events were measured in 15 min intervals. About 30 storms were recorded from 1994 to 1995: these were grouped into summer and winter storms for calibration purposes. In order to measure the flow rate, an aluminum compound weir was placed at the outlet of conduit 210, near node 345, a 1.83 m (6 ft) diameter pipe. Typically, pressure transducers were located a few feet upstream of the weirs, transmitting date, time of day, and depth of flow data to electronic data loggers. Several model parameters (e.g., infiltration, depression storage, and Manning's roughness) were adjusted to match predicted versus observed flows.

Figure 28.11 illustrates the location of the weirs used for calibration (near node 345). The rapid urbanization (Medical Center) upstream of this measurement station over the past 5 years resulted in the decision to consider the design and construction of a large detention pond to control storm water flowing off-campus (at the Erwin Road outfall) into the City of Durham. Detention ponds are considered one of the most effective runoff control devices; in particular, to reduce peak flows, settle particulate matter, while also reducing some pollutant concentrations in the outflow. They are not as effective in reducing dissolved fractions of contaminants, as shown earlier. The effect of a large detention pond in capturing a relatively small storm (storm of February 15–17, 1995, for which runoff measurements were available) is presented in Figure 28.12. Conduit 203 is upstream of the simulated detention pond. Conduit 202 is downstream from the pond and shows the pond storage effects on the hydrograph. The response of the catchment at the Erwin Road outfall (node 339) to the most intense rainfall event (May 11, 2011) ever recorded in Durham, NC, is presented in Figure 28.13.

28.11 Summary and Conclusions

Although the basic fluid mechanics principles of urban hydrologic processes presented in this chapter have been understood for decades by professional hydrologists, dramatic improvements in visualization techniques and ease-of-model-execution features have accelerated the learning process for model application by more inexperienced users. This trend has its drawbacks, as unfamiliarity with the governing equations (including their limitations) and *black box* application of these numerical models will likely result in completely unrealistic predictions. Preparation of the extensive input data required for these comprehensive models can be a tedious and time-consuming exercise. Calibration of these models with

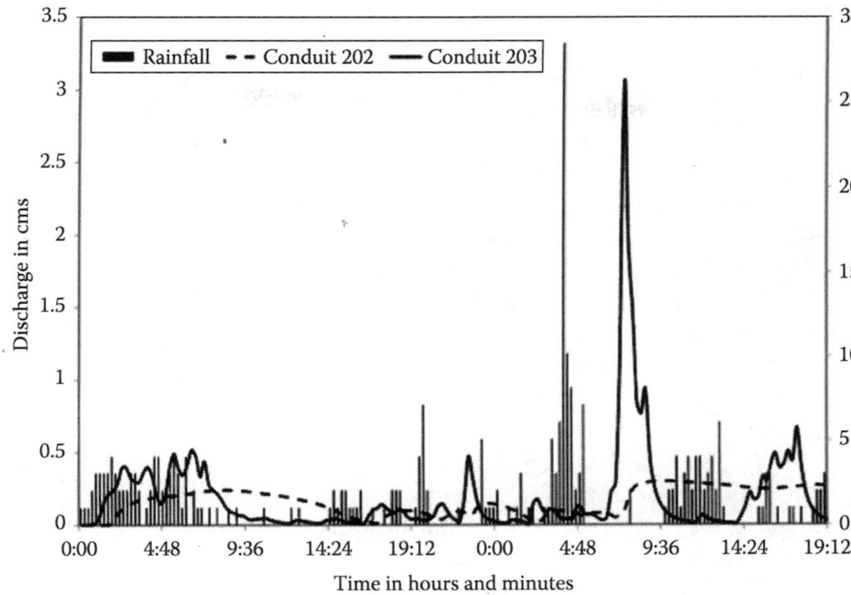

FIGURE 28.12 Hydrographs upstream and downstream of detention pond for February 15–17, 1995, storm.

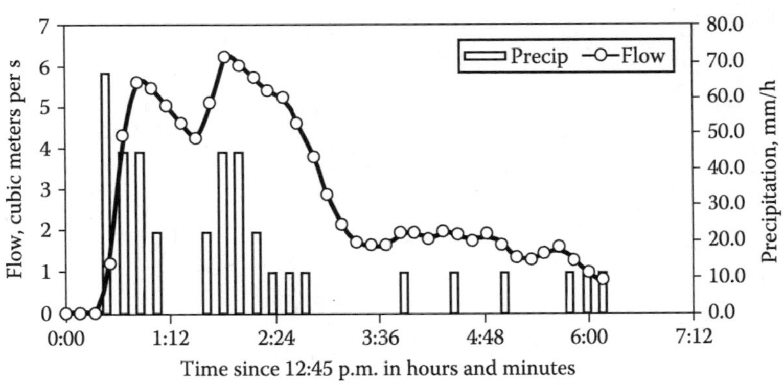

FIGURE 28.13 Hydrograph at node 339 for May 27, 2011, storm.

measured rainfall, runoff, and water quality data is an essential component of a successful simulation, as described in the previous section. Validation with data sets independent of those used for calibration is desirable.

The models are best suited for development of urban planning scenarios and comparison of runoff control alternatives for a multitude of rainfall events or continuous rainfall time series.

References

1. Chow, V.T., Maidment, D.R., and Mays, L.W., 1988. *Applied Hydrology*, McGraw-Hill Book Company, New York, ISBN 0-07-010810-2.
2. Di Toro, D.M. and Small, M.J., 1979. Stormwater interception and storage, *Journal of Environmental Engineering Division, ASCE*, 105 (EE1), 43–54.
3. Driscoll, E.D., Di Toro, D.M., Gaboury, D., and Shelley, P., 1986. *Methodology for Analysis of Detention Basins for Control of Urban Runoff Quality*, Office of Water, U.S. EPA, Washington, DC.

4. Eagleson, P.S. 1970. *Dynamic Hydrology*, McGraw-Hill, Inc., New York.
5. Green, W.H. and Ampt, G.A., 1911. Studies on soil physics, 1, The flow of air and water through soils, *Journal of Agricultural Science*, 4(1), 1–24.
6. Huber, W.C. and Dickinson, R.E., 1988. *Storm Water Management Model*, Version 4, User's Manual, Environmental Research Laboratory, Athens, GA, Office of Research and Development, U.S. Environmental Protection Agency (EPA).
7. Lighthill, M.J. and Whitham, G.B., 1955. *On Kinematic Waves. I. Flood Movement in Long Rivers, Proceedings*, Royal Society of London, London, U.K., Series A, Vol. 229, No. 1178, pp. 281–316.
8. Mahi, H., 1994. Stormwater management and modeling at Duke University, M.S. thesis, Durham, NC.
9. Medina, M.A., Huber, W.C., and Heaney, J.P., 1981. Modeling stormwater storage/treatment transients: Theory, *Journal of Environmental Engineering Division, ASCE*, 107 (EE4), 781–797.
10. Ponce, V.M., 1989. *Engineering Hydrology: Principles and Practices*, Prentice Hall, Englewood Cliffs, NJ 07632.
11. Press, W.H., Teukolsky, S.A, Vetterling, W.T., and Flannery, B.P., 1989. *Numerical Recipes in Fortran: The Art of Scientific Computing*, 2nd edn., Cambridge University Press, Cambridge, U.K.
12. Restrepo-Posada, P. and Eagleson, P.S., 1982. Identification of independent rainstorms, *Journal of Hydrology*, 55, 303–319.
13. Rossman, L.A., 2010. Storm Water Management Model, User's Manual, Version 5. Water Supply and Water Resources Division National Risk Management Research Laboratory, Cincinnati, OH, U.S. Environmental Protection Agency, EPA/600/R-05/040.

29

Wetland Hydrology

Shafi Noor Islam
Brandenburg University of Technology

Rezaul Karim
DANIDA-LGED Project

Ali Noor Islam
Bengal Golden Aviations Ltd

Saeid Eslamian
Isfahan University of Technology

Authors

Shafi Noor Islam is a lecturer in euro hydro-informatics and water management, at the Brandenburg University of Technology at Cottbus, Germany. He holds his PhD in environmental sciences. He was also working as a guest lecturer in the Master Program, Urban Management, at the Technische Universität Berlin, Germany. He has postdoctoral research experience on food security issues, and presently he is the candidate of habilitation (Dr.rer.nat.habil.) research degree on the topic "Ecosystem Services and Food Security under Threat in the Mega Deltas of Asian Coastal Regions: An Analysis on the Ganges–Brahmaputra–Meghna Delta," in the Chair of Ecosystems and Environmental Informatics, Brandenburg University of Technology at Cottbus, Germany. His main activities are in the interdisciplinary water resource research on water salinity and mangrove ecosystem analysis. Wetlands and coastal resource management and climate change impacts on food security and ecosystem services are the focus research fields. Transboundary water and char-land management strategies, encompassing groundwater quality and urban water supply and ecology management issues, are more subjective interests. He has published 42 articles in international journals and books. He has research experiences in China, Malaysia, Indonesia, Germany, Poland, Denmark, Italy, Spain, France, Netherlands, Sweden, and Mexico.

Rezaul Karim is an MBA and chartered professional accountant and consultant of rural communication and development activities. He is also a biologist. He is the chief accountant of the Rural Communication and Development Project of DANIDA, Ministry of Local Government and Engineering Department (LGED), Government of the People's Republic of Bangladesh. He has longtime working experiences on rural development projects, fisheries, and wetland development and management projects in Bangladesh. He has a number of publications on these development and management issues.

Ali Noor Islam is a geographer and freelance consultant of the development and geographical research. He is also the managing director of the Means of Transportation and Communication Firm in Dhaka. He has longtime working experiences on transportation, communication, development and natural site conservation, and development and management issues in Bangladesh.

Saeid Eslamian received his PhD from the University of New South Wales, Australia, with Professor David Pilgrim. He was a visiting professor in Princeton University, United States, and ETH Zurich, Switzerland. He is currently an associate professor of hydrology in Isfahan University of Technology. He is the founder and chief editor of *Journal of Flood Engineering* and *International Journal of Hydrology Science and Technology.* He has published more than 200 publications, mainly in statistical and environmental hydrology and hydrometeorology.

PREFACE

Wetlands and water resources are essential to human development processes and in achieving the Millennium development goals, which seek to eradicate extreme poverty and hunger and to ensure environmental sustainability. The regional hydrology and wetlands are affected by natural calamities, anthropogenic influences, and climate change impacts. Most part of the tropical and subtropical regions are facing hydrologic problems. Recurring water-related natural hazards such as floods, cyclones, drought, and decline in the water availability, combined with deteriorating water quality, have undermined the local health and livelihoods of millions worldwide especially in the developing countries. Therefore, wetlands are also suffering from different types of disturbances. The management of wetlands and the regional wetland hydrology is not in a good shape worldwide as well as in Bangladesh.

29.1 Introduction

The world wetlands are playing a potential role in economic and ecological point of view. Particular interest has been shown to a number of wetlands around the world with regard to their extent uniqueness and attributes. It includes coastal and inland deltas, riverine wetlands, salt marshes and mangroves, freshwater marshes, and peatlands. They are found on every continent except Antarctica and in very rough climates from the tropics to the tundra. A total of 750,000 km^2 of wetlands have been registered with the convention on wetlands of international importance as of 2000 (Mitsch and Gosselink, 1986).

Wetlands in different parts of the world have a great importance for the country's economic, industrial, ecological, socioeconomic, and cultural context for certain reasons (Islam and Gnauck, 2007). They contain very rich components of biodiversity of all valuable ecosystems (Kundzewicz, 2003). This ecosystem serves as habitat for a variety of resident and migratory waterfowl and endangered and commercially important species of national and international interest (Nishat, 1993; Islam and Gnauck, 2009a). Moreover, it supports a rich biodiversity of flora and fauna, substantially contributing to socioeconomic improvement for millions of people living especially in rural areas. This created livelihood sustainability is providing opportunities for employment, food and nutrition, fuel, fodder, transportation, and irrigation (Nishat, 1993). The wetlands of developing countries such as in Asia, Latin America, and African have suffered drastically from the impacts of burgeoning human population and anthropogenic activities on the wetlands. Furthermore, wetlands are recognized as a driving force for biodiversity conservation and rural socioeconomic improvement (Nishat, 1993; Ahmed and Falk, 2008). The smart use of wetlands can solve the ecosystem problems in the floodplain areas. In the South Asian Ganges–Brahmaputra–Meghna floodplains alone, approximately 2.1 million ha of wetlands has been lost due to flood control, drainage, and irrigation development (Khan et al., 1994); therefore, wetlands are facing serious challenges from environmental changes and anthropogenic impacts (Sarkar, 1993; Nair, 2004; Ahmed and Falk, 2008).

The modern approach of wetland construction is planning a potential role in socioeconomic improvement. In general, millions of industrial constructed wetlands have been used in treating wastewater or conventional water of different characteristics throughout the world in every year, including domestic wastewater, various types of industrial wastewater, agricultural wastewater, and storm water. The constructed wetlands in Bangladesh are seen in a very small scale.

Bangladesh is a land of wetlands; it lies in the eastern part of the Bengal Basin, one of the largest river floodplains in the world. Bangladesh has long depended on the basin's three formidable rivers, the Padma (Ganges), the Jamuna (Brahmaputra), and the Meghna, as well as their numerous smaller tributaries and distributaries for freshwater, transportation, and fish and for the floods that deposit fertile silt on their farmland each year. In Bangladesh, where inland water bodies constitute nearly 50% of total land area, wetlands are critical to economic development and environmental improvement. The total area of wetlands in Bangladesh is estimated to be 70,000–80,000 km^2 of national land (Akonda, 1989; Khan et al., 1994). These include rivers, estuaries, mangrove swamps, marsh (*haor*), oxbow lake (*baor*) and *beels*, water storage reservoirs, fishponds, and some other lands (Khan, 1993; Hughes et al., 1994; Gopal and Wetzel, 1995; Islam and Gnauck, 2008). The major roles of wetland are nutrient retention/removal, support for food chains, fishery production, habitat for wildlife, recreation, natural heritage values, biomass production, water transport, biodiversity presentation, and microclimate stabilization.

There are five types of wetlands available in Bangladesh, such as saltwater wetlands, freshwater wetlands, palustrine wetlands, lacustrine wetlands, and man-made wetlands. Presently, most of the wetlands in country are under threat due to unsystematic utilization, encroachments and reclamation, urbanization and drawbacks from agricultural development, and flood control actions (Nishat, 1993; Khan et al., 1994; Gopal, 1999). The protection and management status of wetlands in developing countries are more or less complex and severe due to financial, technical, social, and political decision, as well as lack of integration of those sectors (Gopal and Wetzel, 1995; Islam and Gnauck, 2008; Sultana and Thompson, 2008). A comprehensive analysis of various issues leading to wetland degradation in the case areas in Bera, Santhia, Sujanagar upazilas in the middle location of Bangladesh, and Sundarbans mangrove wetlands in the southwestern region of the country has been chosen. This comparative wetland hydrologic analysis has been highlighted through applied research findings on the freshwater and saline water wetlands in Bangladesh.

The objective of this chapter is to understand the characteristics of wetlands hydrology and to investigate the wetland biodiversity conservation to promote the sustainable management of wetland biodiversity in Bera, Santhia, and Sujanagar upazilas in Pabna district in Bangladesh. One of the key components to foster for effective implementation of Ramsar Conventional Policies of wetlands in the context of sustainable wetland and natural resource management is supporting this endeavor. The following specific objectives of the wetlands biodiversity conservation are as follows:

- A special investigation will be limited on the two case areas in Bera, Santhia, Sujanagar upazilas for freshwater wetlands, and the Sundarbans mangrove saline coastal wetlands in Bangladesh.
- To protect the wetland ecosystems and conserve its biodiversity.
- To use wisely and sustain natural resources.
- Improve the livelihoods of the poor wetland communities in the Bera, Santhia, and Sujanagar upazila of the eastern part of the Pabna district where the Ganges (Padma) River meets the Brahmaputra (Jamuna) River and southwestern region the Sundarbans mangrove area.
- Prepare some recommendations for future development of wetlands in Bangladesh and ensure local food security to the dwellers.

29.2 Hydrology of the Wetlands

Hydrologic processes occurring in wetlands are the same processes that occur outside of wetlands and collectively are referred to as the hydrologic cycle. Major components of the hydrologic cycle are precipitation, surface water flow, groundwater flow, and evapotranspiration (ET). Wetlands and uplands continually receive or lose water through exchange with the atmosphere streams and groundwater. Water supply is necessary for the existence of wetlands.

The wetland water budget is the total inflows and outflows of water from a wetland. The functions, persistence, and size of wetlands are controlled by hydrologic processes.

Wetlands constitute a part of human heritage. It has played a significant role in the development of human culture and society. Moreover, it contains very rich components of biodiversity of local, national, and regional significance. They also provide a habitat for a variety of resident and migratory waterfowl, a significant number of endangered species, and a large number of commercially important species.

However, high population density has resulted in intense pressure on both land and water resources. Rice cultivation became the overriding priority, so that flood protection embankments, irrigation canals, sluice gates, and flood drainage structures were built in the 1980s to protect agriculture from flooding, cutting off many traditional fishing grounds from rivers and blocking fish migration routes. The impact on the wetland ecosystem, including the freshwater fisheries, was in many cases devastating—fish yields decreased in some areas by as much as 75%. The global climate change is further expected to impact the north Bengal wetlands by increasing drought, monsoonal rains, floods and bank erosion, accentuate food insecurity problems, and consequently rural poverty among others. Anthropogenic influences on the surface water and groundwater pollution through arsenic contamination are a new threat for community health risk in the northwestern region of Bangladesh.

In addition, over half of all poor households catch fish for food or income; thus, the decline in wetland biodiversity and productivity has resulted in the loss of a vital safety net for poor people in the wet season. In the wetland biodiversity conservation project in three upazilas (Bera, Santhia, and Sujanagar) in the Pabna district in Bangladesh supported by the Department of Fisheries of the Ministry of Fisheries and Livestock, the Bangladesh Water Development Board of the Ministry of Water Resources is working to increase wetland biodiversity, restore vital freshwater fishery habitats, and improve the livelihoods of wetland dwellers, initially in Bera, Santhia, and Sujanagar upazilas of the eastern part of the Pabna district where the Ganges (Padma) meets the Brahmaputra River (Jamuna).

29.2.1 Special Distribution of World Wetlands

Wetlands are distributed unevenly throughout different climatic zones of the world because of the difference in geology, climate, and source of water. They occur in widely diverse settings ranging from coastal margins, where tides and river discharge are primary source of water, to high mountain valleys where rain and snowmelt are the primary source of water. Marine and saline wetlands are found at all the coastal states of the world. Estuary wetlands (where tidal saltwater and inland freshwater meet and mix) are most plentiful in the coastal region of the world.

The special distribution and differences in wetland type, vegetative composition, and soil type are caused primarily by geology, topography, and climate. In turn, the wetland soils and vegetation alter water velocities, flow paths, and chemistry. The hydrologic and water quality functions of wetlands, that is, the roles wetlands play in changing the quality or quantity of water moving through them, are related to the wetland physical setting (Carter, 2010).

There are thousands of wetlands distributed in different parts of the world (Figure 29.1). The 10 major great wetlands are distributed in the different continents of the world. The 10 large wetlands are as follows:

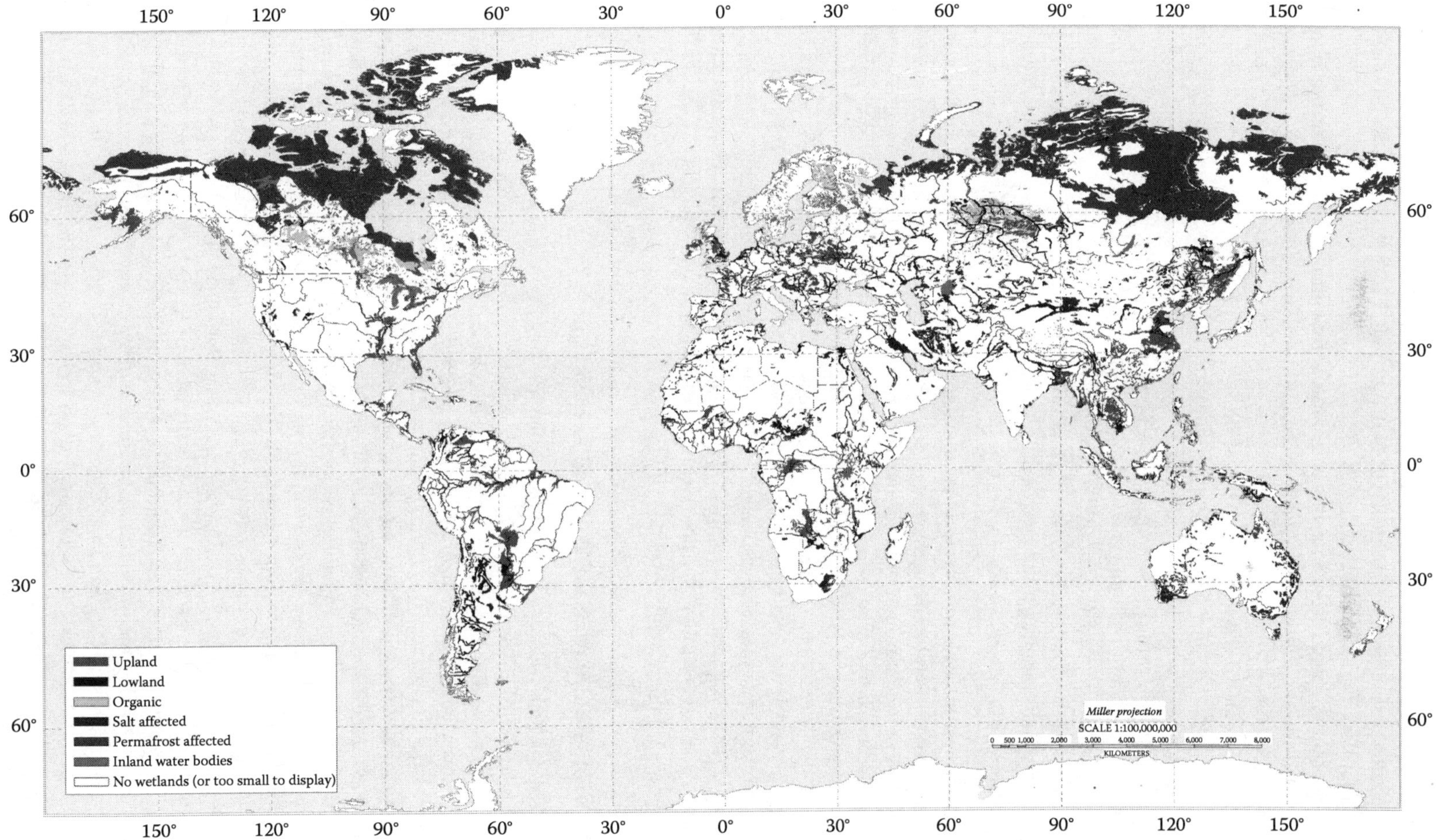

FIGURE 29.1 World wetland distribution. (From U.S. Department of Agriculture—(USDA), Natural Resources Conservation Service (NRCS)-Global distribution of wetlands map, USDA-NRCS, Soil Science division, World Soil Resources, Washington, DC, 2003.)

1. Pantanal (Brazil) is the world's largest wetlands, lying mostly in western Brazil but extending in Bolivia and Paraguay as well. It is covering a region of 140,000 km^2.
2. The Sundarbans mangrove wetlands (Bangladesh and India). It is the largest littoral mangrove wetland and covering an area of 10,000 km^2.
3. Okavango delta wetland (Botswana). It is one of the world's great inland wetlands. It is formed where the Okavango River empties onto a basin in the Kalahari Desert. The area of Okavango wetland is 15,000 km^2.
4. Everglades (Florida). It covers an area of 51,240 ha.
5. Kerala Backwaters (India) is 590 km along the coast and the largest lake area is covering an area of 200 km^2.
6. Kakadu wetlands (Australia) cover an area of 19,804 km^2.
7. Mekong delta wetlands (Vietnam) cover an area of 40,000 km^2.
8. iSimangaliso Wetland Park (South Africa). It covers an area of 3280 km^2.
9. Wasur National Park (Indonesia). It covers an area of 4138 km^2.
10. Camargo wetlands (France). It encompasses the Rhone River delta in the southeast of France and it covers an area of 150,300 ha (Figure 29.1).

These 10 world largest wetlands are playing a potential role in making a balance of wetland ecosystems and providing ecosystem services in surroundings.

In Bangladesh, the wetlands have suffered from human population, and approximately 2.1 million ha of wetlands has been lost to flood control, drainage, irrigation development, and other human developmental activities (Nishat, 1993, 2003). There are only 43 designated wetlands, and most of them are under threat from indiscriminate utilization, encroachments and reclamation, urbanization and drawbacks from agricultural development, and flood control. Almost 50% of the country's people are directly dependent on wetland resources. The vast majority of the poor people in the wetland areas are dependent on wetland resources for their nourishment (Islam, 2010). Therefore, the hydrologic modeling and natural resource management plan and policy are needed toward the conservation of the invaluable unique wetland sites in Bangladesh and other countries as well.

29.2.2 Hydrologic Balance for Wetland Ecosystems

The term *borsha* has been widely used to denote regular flooding in order to contrast it with extreme flooding, which is referred as *bonna*, as monsoonal rainfall and regular flooding do not necessarily lead to disasters. Seasonal flooding is needed to maintain the ecological and hydrologic balance, and rural people perceive this to be essential to their livelihoods. Floodplain people have developed wide-ranging and indigenous strategies for managing normal flood conditions. Extreme flooding can devastate crops, infrastructure, and livelihoods and bring misery for prolonged periods of time, creating disasters that challenge the resilience of the population (Sultana, 2010). Social process interacts with natural processes to produce the differentiated vulnerabilities and suffering that ensue, which have gendered implications. While poverty and vulnerability are not always correlated in much of rural Pabna district in Bangladesh, they interact to reinforce gendered powerlessness and suffering (Cannon, 2002). While women are often the victims, they are also resourceful agents who cope with disasters and play important roles in rebuilding, rehabilitation, care giving, and mitigation (Enarson and Morrow, 1998).

Gender power relations and socioecological transformations in each case interact with the root causes of poverty, marginalization, and inequities, but there are differences that can be highlighted. Flood control measures alter the hydrology and geomorphology of the Ganges–Jamuna River floodplain region, changing its ecology and socioecological systems. Coping strategies include marking portable stoves, saving footstock and fuel in dry places, trying processions to trees and huts, lifting belongings to platforms created just under the roof, and learning to live on the thatched roofs when the floods are too high. Huts are generally built on raised earth mounds or platforms, and this enables

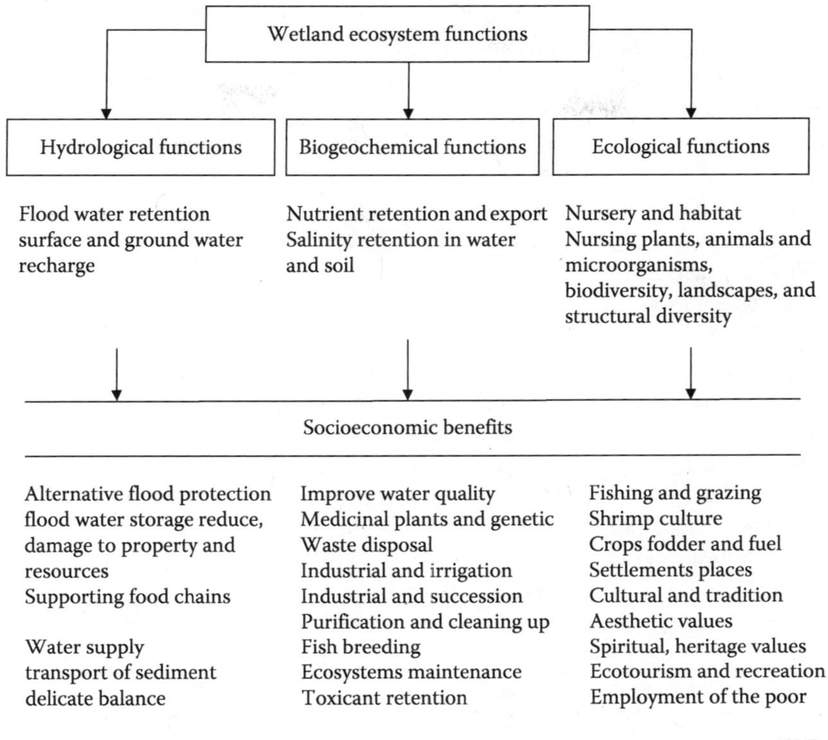

FIGURE 29.2 Different types of wetland ecosystem functions. (From Islam, S.N., *Front. Earth Sci. China*, 4(4), 438, 2010.)

most households to deal with annual flooding (Sultana, 2010). Drinking water collection is particularly difficult during times of extreme floods, as tube wells are often submerged, polluting the water. The freshwater and saline water wetlands are playing three types of functions (Figure 29.2); these are very much essential for the balance of ecosystem and its services as well as to ensure the local livelihood sustainability.

The Bera, Santhia, and Sujanagar upazilas are facing such natural calamities; furthermore, social and cultural conflicts and problems are arising. The Bera upazila has severe flooding and the people are facing the same problem almost every year (Sultana, 2010). Natural calamities and anthropogenic influences, which are damaging the wetland functions, are hampering provision of the natural ecoservices in the fresh- and saline-water wetlands in Bangladesh. As a whole, the natural resources as well as wetlands are reducing its area, losing its ecosystem and biodiversity. Considering the present vulnerability situation, the wetland biodiversity conservation project in Bera, Santhia, and Sujanagar upazilas and the coastal mangrove region in the southwestern region in Bangladesh could be a viable development and applied research project in Bangladesh, which will carry out tremendous findings and results for future development guideline framework for wetland biodiversity conservation in Bangladesh as well as similar areas for other countries.

29.2.3 Wetland Plant Species

Wetlands have often been viewed as wastelands to most of the community people. Wetlands are used for different purposes and controlled by the landowners and landlords for their own purposes. At one time, landowners were encouraged to drain "water-logged" lands, but today, the wetlands are using multifarious purposes and playing a crucial role in society and is beginning to be understood and valued. Wetlands are one of the potential sources of community livelihoods, particularly for the cultivation of food crops, vegetables, fishing, and pasturelands (Khan et al., 1994). There are some special common vegetation plants that are relatively uniform in the haor, beel, and baor areas: the names of such trees are hizal (*Barringtonia acutangula*), tamal (*Diospyros cordifolia*), barun (*Crataeva nurvala*), madar (*Erythrina variegata*), gab (*Disopyros peregrine*), dumur (*Ficus hispida*), chalta (*Dillenia indica*), dehua (*Artocarpus lacucha*), and paniphal (*Trapa bispinosa*) (Akonda, 1989; Talukdar et al., 2008). Aside from those plants, a number of medicinal plants and species are available in the wetland areas, being widely used by the community and especially by the indigenous people. Some of these special plants are polygonum, locally known as *Akanda, Kalmi, Helencha, Aarnalata, Bashok, Bishkatali, and Kukra*. The flowers and seeds of *paddo* (Indian lotus) are prescribed for piles, as a cardiac tonic and for elimination of ringworm. The flowers of water lilies are reputed as a remedy for heart ailments (Khan et al., 1994; Rahman, 1995). The wetland plants and species are traditionally used by health practitioners, harvesting these medicinal resources for their livelihoods.

The wetland plants have been contributing substantially to the livelihood of the millions of rural people as a source of food, fuel, fodder, manure, thatching materials, medicine, and so forth. Varieties of macrophysics are eaten as vegetables and some are eaten raw. During monsoon, water hyacinth extensively used as fodder in many areas and nonfood aquatic resources retain the ecological balance for the local residents. These plants are harvested mostly by the rural poors and their access for this purpose is almost free for all. Wetlands provide a wide range of economic, social, and ecological benefits. Wetlands are one of the most productive and resourceful areas, which provide food. From the perspective of a developing country, wetlands are an important source of commercial fishing, agriculture, seasonal livestock grazing, wood collection, and ecotourism.

Beyond the obvious role of providing habitat for waterfowl and shore birds, the deeper role of wetlands is also becoming more and more apparent and understood. During high water and runoff seasons, wetlands act as a natural flood control. Wetlands also act as a natural filtration system. They remove sediments and even toxic chemicals from our water supplies. Some areas of the United States are even using wetlands as a natural sewage treatment system.

29.2.4 Hydrologic Modeling

Hydrologic models are the simplified and conceptual representation of a part of the hydrologic cycle or water cycle. Primarily it is used for hydrologic prediction and monitoring for understanding the hydrologic process on a certain area. In general, the wetland areas are dependent on the local water cycle process, and the local water cycle process is dependent on the surface and groundwater supply and situation, temperature, rainfall, soil moisture, ET, etc. Toward protecting and making the balance of the hydrologic cycle or water cycle of a certain region of the wetland areas, modeling is very much essential. The model can forecast the present to future situation.

In general, there are two types of hydrologic models that can be distinguished such as the stochastic model and process-based model. The stochastic model is a model of black box systems, and this is based on data and using mathematical and statistical concepts to link a certain input to the model output. In such model development, the rainfall, runoff, neural networks, and system identification are potential considering issues. The output of this type of model is called stochastic hydrologic model. The process-based model is attempted to represent the physical process that is observed in the real world. This model represents surface runoff, subsurface flow, ET, and channel flow, but it is more complicated and it is called deterministic hydrologic model.

Figure 29.3 shows the real-world hydrologic model of the case areas in the Pabna district and the coastal mangrove wetland region in Bangladesh. It is important to note that the water cycle or hydraulic cycle or hydraulic model has been designed to use the same input and similar model components as the operational hydrologic forecast model of the case areas. The two case areas have been chosen based on the ecological characteristics as well as freshwater and saline water wetlands. In both

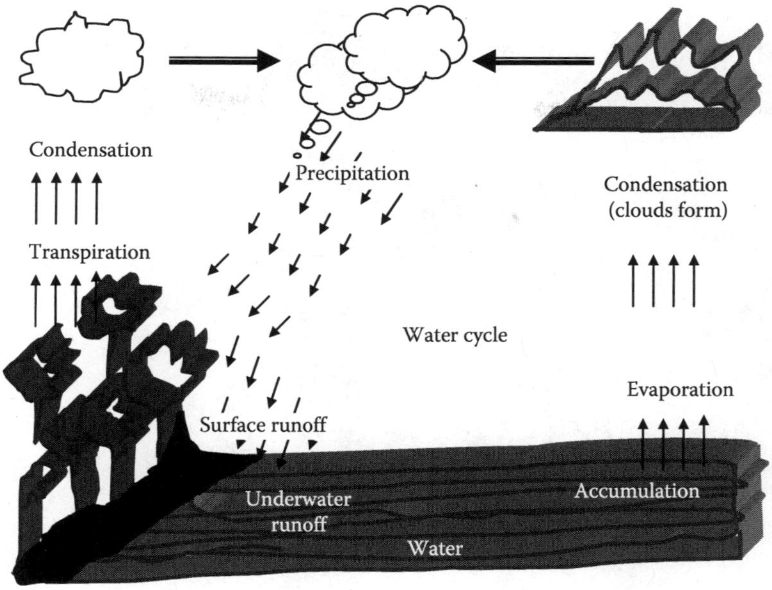

FIGURE 29.3 The hydrologic cycle (water cycle).

cases, the elements of the hydrologic cycle are hampering the balance of water cycle of that case area. Therefore, Figure 29.3 illustrates the proper functions of all elements; some elements are not working properly; this is why for understanding the real situation and forecasting, the condition of the hydrologic model is necessary for the wetland resource management and conservation, its biodiversity, and community livelihood sustainability.

29.2.5 Phytoremediation Effect

Environmental pollution with metals and xenobiotics is a global problem, and the development of phytoremediation technologies for the plant-based cleanup of contaminated soil is therefore of significant interest. Phytoremediation technologies are currently available for only a small subject of pollution problems such as arsenic and salinity intrusion. Salinity and arsenic removal employ naturally selected hyperaccumulator ferns, which accumulate very high concentrations of arsenic specifically in aboveground tissues. Elegant two-gene transgenic approaches have been designed for the development of mercury or arsenic phytoremediation technologies.

Arsenic concentration in the groundwater of the Ganges floodplain wetland areas of the Bengal delta region was first discovered in the 1980s in the west Bengal of India. Arsenic presence in well water in three districts in the northwest region of Bangladesh was discovered in 1993. A number of theories and hypothesis have been developed to explain the origin and basic cause of Arsenic calamity in the wetlands in Bangladesh (BGS and DPHE, 2001). Arsenic (As) in groundwater used for drinking water is slowly poisoning tens of millions of people all over the world. The extent of this disaster was not internationally recognized until the mid-1900s (Moore et al., 2000). A few years later, arsenic (As) was found in the Ganges deltaic floodplains to contain arsenic (As) above the 10 μg/L, which is the WHO guideline value. Figure 29.4 illustrates the arsenic-contaminated groundwater in the wetland areas in Gopalganj and Madaripur districts in Bangladesh (Chowdhury, 2009).

The chronic levels of 50 μg/L can cause health problems after 10–15 years of exposure (Moore et al., 2000). The wetlands in the Bengal delta region are affected as an estimated 35 million people have been drinking arsenic-contaminated water for the last 25–35 years (Smedley and Kinniburgh, 2002). The examination for arsenic dermatologic symptoms in 29,000 people showed that 15% had skin lesions

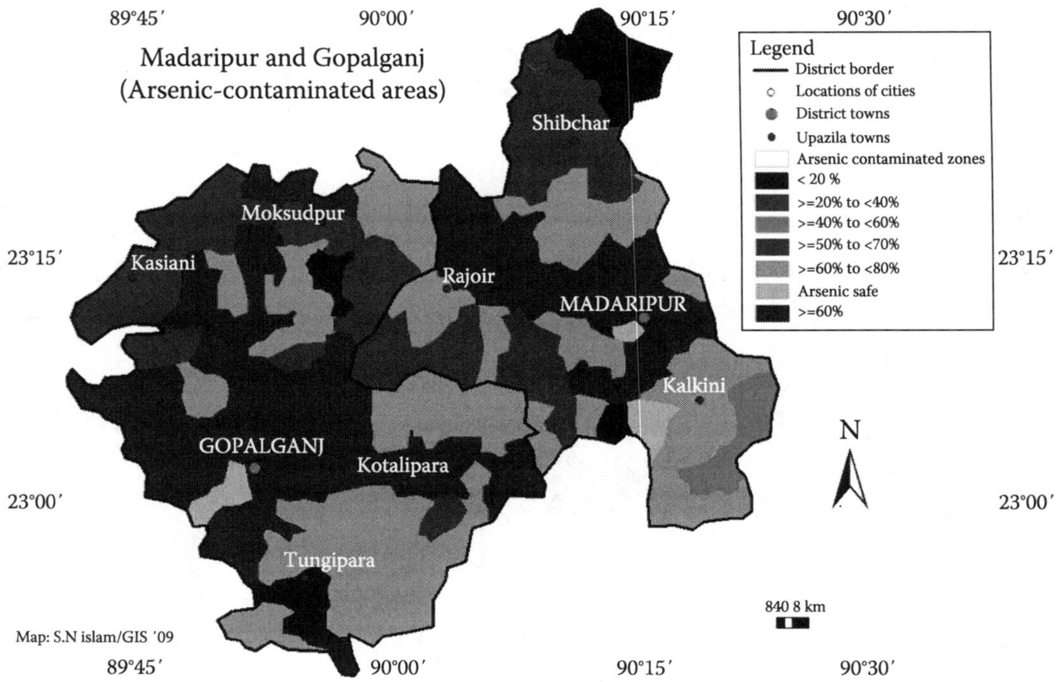

FIGURE 29.4 Phytoremediation effect by arsenic contamination in the wetlands and floodplain region in Bangladesh.

(Chowdhury, 2009). The coastal region (47,000 km^2) of Bangladesh is affected by high-saline water. Therefore, the coastal wetlands in the southern region of Bangladesh are highly contaminated and are vulnerable regions for drinking water, cultivation, and livelihood sustainability. In the development of the science, scientists come up with two ecologically friendly and economically reliable methods to tract wastewater by combining knowledge of botany and chemistry. These are phytoremediation, which uses living plants, and biosorption, which uses dead plant materials. Therefore, phytoremediation technique is used in the initial process of wastewater treatment, and constructed wetlands are used in phytoremediation technique where plants are growing in the artificial wetland.

The constructed wetland is also highly productive and can provide a useful tool to prevent eutrophication in surface waters while generating a potentially valuable crop for farmers. In Bangladesh, only the large city areas have developed the wastewater treatment plants to purify drinking water for the urban citizens. Free water surface constructed wetlands with emergent macrophyte functions as land-intensive biological treatment systems. Suspended solid removal is usually a fairly rapid physical process. Therefore, constructed wetland or conventional wetland development initiatives are not enough in Bangladesh, and it should be considered as an alternative source of quality water for drinking and irrigation. The constructed wetlands in Bangladesh could carry out a tremendous positive influence in society for fundamental development of water quality.

29.2.6 Effect of Vegetation on Wetland Flow

The water flows in the wetland areas are playing a potential role in protecting the functional activities. The water flows and quantities and quality of surface water are the main sources and balance wetland flows. The plants and vegetation of wetland areas are dependent on the surface water flows. Asian, South American, and African countries are suffering from surface water flows and wetland flow; as a result, the vegetation of such regions is damaging its diversity and succession. All over the world, the freshwater wetlands and saline water wetlands are facing similar types of natural and anthropogenic actions. The quality of river flow enriches the special types of vegetation; moreover, it contains very rich compounds of biodiversity of local, national, and regional significance.

The quality of freshwater flows also provides habitats for a variety of resident and migratory water flows, a significant number of endangered species, and a large number of commercially important species. In general, the coastal, saline wetlands are affected by sea level rise and saline water penetration. The coastal mangrove wetlands are connected with upstream freshwater flow in the world. In the case of Bangladesh, the wetlands of Bera, Santhia, and Sujanagar upazilas are the freshwater wetlands, and these are dependent on the Ganges and the Brahmaputra (Jamuna) River water flow condition. These two rivers supply freshwater to the wetlands of these three upazila wetlands in the wet season. It is in the records that almost all of the wetlands are annually inundated. Some wetlands are losing its area and water depthness because of high temperature and high ET, irrigation and agricultural land extension, and settlement development in the wetland areas in the Pabna district in Bangladesh. On the other hand, the Sundarbans mangrove wetlands and coastal wetlands are affected due to high-salinity intrusion and shrimp cultivation, settlement development, and urbanization process, because the Sundarbans mangrove wetlands are located at the Bengal coast and are crisscrossed by a vast network of rivers and channels. These occupy around 10% of the total coastal area. The combined flow of these water courses is about 140,000–200,000 m^3/s (Islam and Gnauck, 2007, 2009, 2009a). A major portion of this flow enters into the Sundarbans river estuaries and mangrove wetlands in the coastal region. The annual pattern of salinity changes inside the Sundarbans is also related with the changes of freshwater flow from upstream rivers. The peak salinity was found to be about 56,186 dS/m in 2001 and 2002, and the minimum salinity during postmonsoon was found to be about 10,805 dS/m (IWM, 2003).

29.3 Case Study in Bangladesh

According to the Ramsar 1971 Convention *("The Convention on Wetlands of International Importance especially as Waterfowl Habitat")* in Article 1.1, the term wetland links together a wide range of inland, coastal, and marine habitats, which share a number of common features (Rashid 1991; Mather et al., 1993; Kim et al. 2006). Usually, wetlands occupy transitional zones between permanently wet and generally dry environment (Davies and Claridge 1993). In Bangladesh, wetlands are permanent, and seasonal freshwater lakes and marshes of floodplains are known in different names such as (1) haor (it is a bowl-shaped depression between the natural levees of a river mostly found in the eastern region of the greater Mymensingh and Sylhet districts), (2) baor (it is called oxbow lakes and is the dead arm of the river situated in the moribund delta of the Ganges), and (3) beel (it is the lowest part of the floodplain landscape, usually saucer-shaped wetlands) (Table 29.2) (Akonda, 1989; Nokashima, and Khan 1993).

The total area of wetlands in Bangladesh is estimated to be 70,000–80,000 km^2; it is about 50% of the total national land (Akonda, 1989; Khan et al., 1994). These include rivers, estuaries, mangrove swamps, marsh (haor), oxbow lake (baor) and beels, water storage reservoirs, fishponds, and some other lands, which suffer from seasonal inundation (Khan, 1993; Hughes et al., 1994; Gopal and Wetzel, 1995; Islam and Gnauck, 2008). In the Ganges–Brahmaputra–Meghna floodplain alone, approximately 2.1 million ha of wetlands has been lost to flood control and a million ha of wetlands has been lost to drainage and irrigation development (Khan et al., 1994). Therefore, wetlands are facing serious challenges from environmental changes and anthropogenic impacts (Sarkar, 1993; Nair, 2004; Ahmed and Falk, 2008).

There are four categories of wetland values that have been recognized in the society in Bangladesh. However, they are categorized as environmental values, economic values (direct, indirect, option, existence, and bequest values), and social and cultural values. Based on this category, wetland ecosystem offers different types of services for the communities. The wetland ecosystem types in Bangladesh have been categorized in Table 29.2. Wetlands in three (Bera, Santhia, and Sujanagar) upazilas in the Pabna district and the Sundarbans coastal mangrove wetlands contain very rich components of biodiversity of local, national, and regional significance (Khan et al., 1994; Rahman, 1995).

The wetland environment unites the inhabitants into a society, which has a definite shape, culture, and livelihood pattern. Due to the availability of a wide variety of harvestable products, the people in and around the wetlands are traditionally self-reliant and have subsistence-oriented economy and livelihoods. Despite all these support to human livelihoods, many parts of the country have experienced loss or degradation of wetlands on a huge scale because of agricultural use, urbanization, and excessive exploitation by the local population. The wetlands in Bera, Santhia, and Sujanagar are also facing similar kinds of threats mainly due to overexploitation, which ultimately deteriorates the wetland ecology. Owing to continuous degradation of the wetland ecology, the respondents of the wetlands give emphasis about the importance of conserving the wetlands in the case areas in Pabna district. So there is a clear interlinkage between the economic and ecological functions of wetlands in Pabna district. The development of integrated management plan with a provision of the participation of local stakeholders may become the possible way of conserving this wetland ecology. Figure 29.5 illustrates the freshwater wetland distribution in the Bera, Santhia, and Sujanagar upazilas of Pabna district in Bangladesh. The freshwater wetlands and community people in Pabna district are under threats due to anthropogenic influences, natural calamities such as devastating floods, riverbank erosion, and climate change impacts. Figure 29.5 shows the critical geographical location and the characteristics of the freshwater wetlands in Bera, Santhia, and Sujanagar upazilas of Pabna district in Bangladesh.

The Bera, Santhia, and Sujanagar upazilas together cover an area of 914.16 km^2. The total area is a rich and unique wetland biodiversity in the northwestern part of Bangladesh. There are 118,232 households that cover the total population of 706,492 (2007) in these three upazilas (Table 29.1). The groundwater of the Bera upazila is gradually polluted by arsenic (As) contamination. The groundwater of this upazila was focused to contain arsenic above 10 μg/L, which is a threat and risk for human health. A long duration of drinking of Arsenic contaminated ground water as a result of accumulation in the human body. The riverbank erosion of the braided Jamuna River in Bera upazila is one of the major natural disasters and a severe problem in Bangladesh. Thousands of hectares of floodplain land are eroded each year, leaving many people homeless with damaging or destroying infrastructures and wetlands by the mighty Jamuna River (Mansour et al., 2011).

Table 29.1 shows the dynamic poor wetland communities in Bera, Santhia, and Sujanagar upazilas, which are more vulnerable to climate change, because they tend to be located in geographically vulnerable areas, such as flood-prone, riverbank erosion-prone, and tornado-prone areas of Bangladesh. Most of the community people are dependent on wetland natural resources for their livelihoods (Reid et al., 2010). Climate change refers to short-, medium-, and long-term changes in weather patterns and temperature that are predicted to happen or already happening as a result of anthropogenic emissions of greenhouse gases such as carbon dioxide. These changes include extreme weather, temperature increase, floods, and long duration of monsoonal rainfalls, and dryness character in the case area is recognized as a new threat for wetland ecosystem (IPCC, 2007; Reid et al., 2010). Vulnerability to floods and riverbank erosion and climate change is not just a function of geography or dependence on natural resources; it also has social, economic, and political dimensions that influence how climate change affects different groups (Action Aid, 2005). Based on the climatic and geographical condition, the wetlands in Bangladesh have categories in different groups (Table 29.2).

According to Sarkar and Nishat (1993), Bangladesh is more than 68,722 km^2 of closed freshwater areas and 6,100 km^2 of brackish water and mangrove swamp areas that are recognized as wetlands in Bangladesh. Inlands including haors, baors, and beels are vanishing or ecologically degraded due to various natural and anthropogenic causes, such as population growth, demand for wetland products in markets, unplanned floodplain conservation, and maintaining their ecological and hydrologic

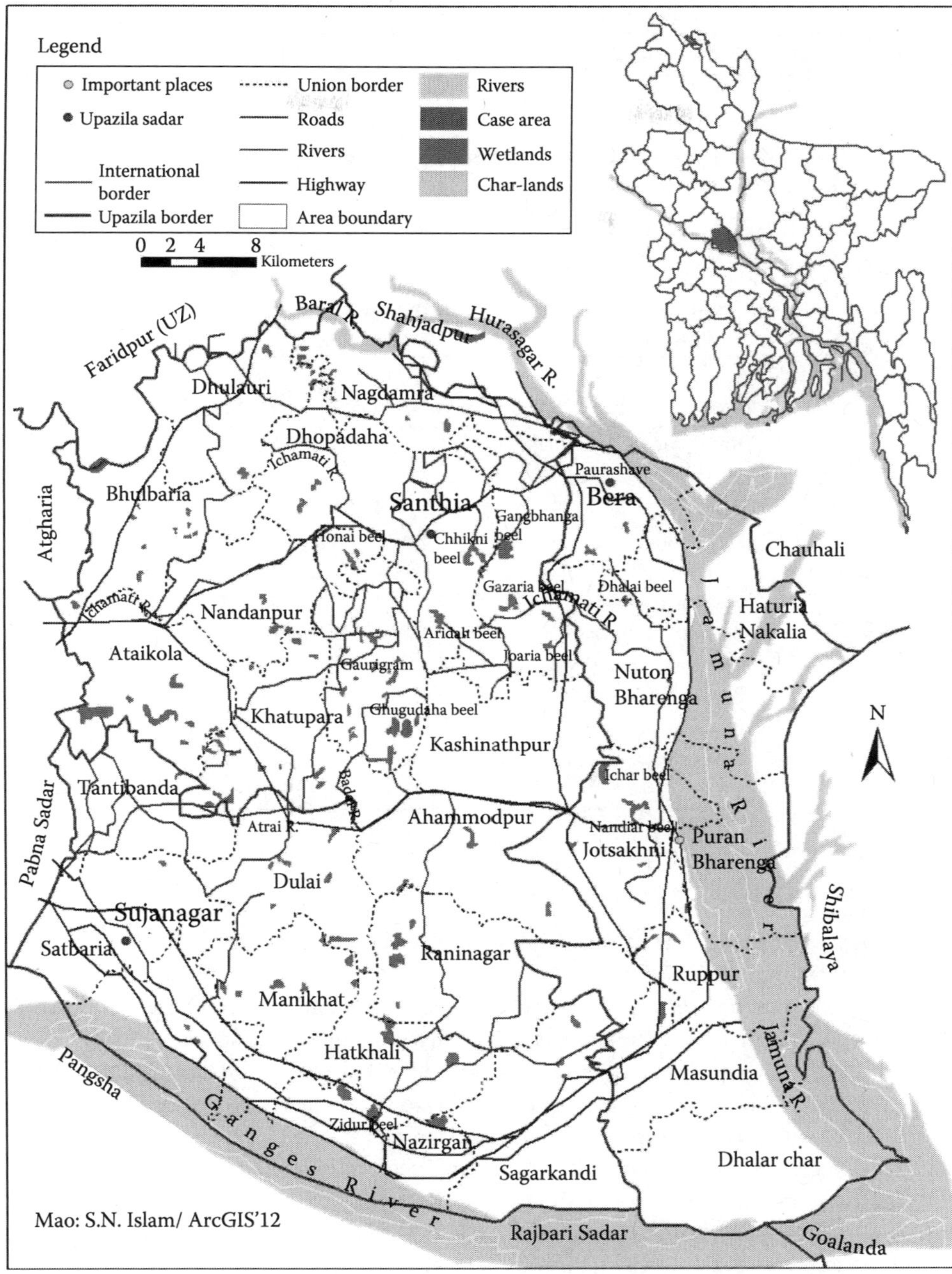

FIGURE 29.5 The freshwater case area (Bera, Santhia, and Sujanagar upazilas) in Bangladesh.

functions (Thompson and Sultana, 1996; Sultana and Thompson, 2008). Presently, there are 43 sites that have been identified in Bangladesh as wetlands and protected areas, and most of them are ecologically sensitive sites (Figure 29.6) (Rahman and Ahmed, 2002; Talukdar et al., 2008). The major wetlands and ecologically sensitive sites in Bangladesh are shown in Figure 29.6, where 15 sites are declared as protected area by the Ministry of Environment and Forest of Bangladesh. These sites are not well protected, and at present there is no management and conservation policy. As a whole, the management strategies for the natural constructed wetlands in Bangladesh are not well structured.

TABLE 29.1 Scenarios of Bera, Santhia, and Sujanagar Upazilas of Pabna District

Upazilas	Area km^2	Population	Male (%)	Female (%)	Density	Household	Village	Literacy (%)
Bera	248.6	208,897	52.11	47.89	840	35351	157	24
Santhia	331.56	283463	52.2	47.8	855	46745	253	23
Sujanagar	334.4	214132	51.86	48.14	640	36136	195	23
Total	914.16	706492	52.05 (Average)	47.95 (Average)	778 (Average)	118232	605	23.66 (Average)

TABLE 29.2 Different Wetland Types and Areas in Bangladesh

Wetland Types	Area Coverage/km^2	Percentage of Area (%)
Rivers, canals, and estuaries	10,300.00	13.77
Natural depressions	1,141.69	1.52
Ponds	1,619.43	2.16
Oxbow lakes	544.88	0.73
Reservoirs	688.00	0.92
Seasonal flood lands	28,000.00	37.42
Brackish water farms	873.00	1.17
Mangrove wetlands	6,100.00	8.15
Beel and haor	5,000.00	6.68
Peatland	155.00	0.21
Swamp forest	20,400.00	27.27
Total	74,822.00	100.00

29.3.1 Geography and Geological History

Bangladesh is located in South Asia between 20′ 34′ to 26′ 38′ N latitude and 88′ 1′ to 92′ 42′ E longitude, with a national territory of 147,570 km^2. Located in the southeastern part of the Indian subcontinent, it is bordered by India on the west, north, and east, except for a small portion in the east by Myanmar (Burma), sharing a 288 km border. The south is a highly irregular deltaic coastline of about 710 km long, which is dissected by many rivers and streams flowing into the Bay of Bengal. It has a population of about 150 million until 2009, with a very low per capita gross national product (GNP) of US $ 690. Bangladesh consists mainly of floodplains, except terraces in the Madhupur, Barind tract, and hills in Sylhet and Chittagong hill tracts. The major types of landscapes in Bangladesh are floodplains (80% of total land), terraces (8%), and hills (12%) (Hassan and Mulamootti, 1994). Except the eastern hilly region, the country mostly lies in the delta of the three active rivers, that is, the Ganges, Brahmaputra, and Meghna.

The alluvial sediments of the three mighty rivers formed the deltaic floodplain and the wetlands in Bera, Santhia, and Sujanagar upazilas of Pabna district where the Brahmaputra (Jamuna) and the Ganges Rivers have joint. On the other hand, the south coastal region developed accordingly through alluvial sedimentation. The Sundarbans mangrove wetland region enjoys humid tropical monsoon climate with proximity to sea as an added advantage. The wetland area has an average elevation of 0.9 to 2.1 above mean sea level. Considering the geological statements, the mangrove wetlands in Bangladesh originated at least 53,000–7,000 years ago, which is a relatively short age compacted with the global age of mangrove (Islam and Gnauck, 2008; Islam, 2010). Wetlands in Bangladesh are also cherished for their ecotourism and cultural importance to the nation. In terms of aesthetic values, wetlands of Bangladesh have many potential utilities (Khan et al., 1994; Nishat, 2003). They contain very rich components of biodiversity of local, national, and regional significance (Khan, 1993; Bennett et al., 1995; Talukdar et al., 2008). In the

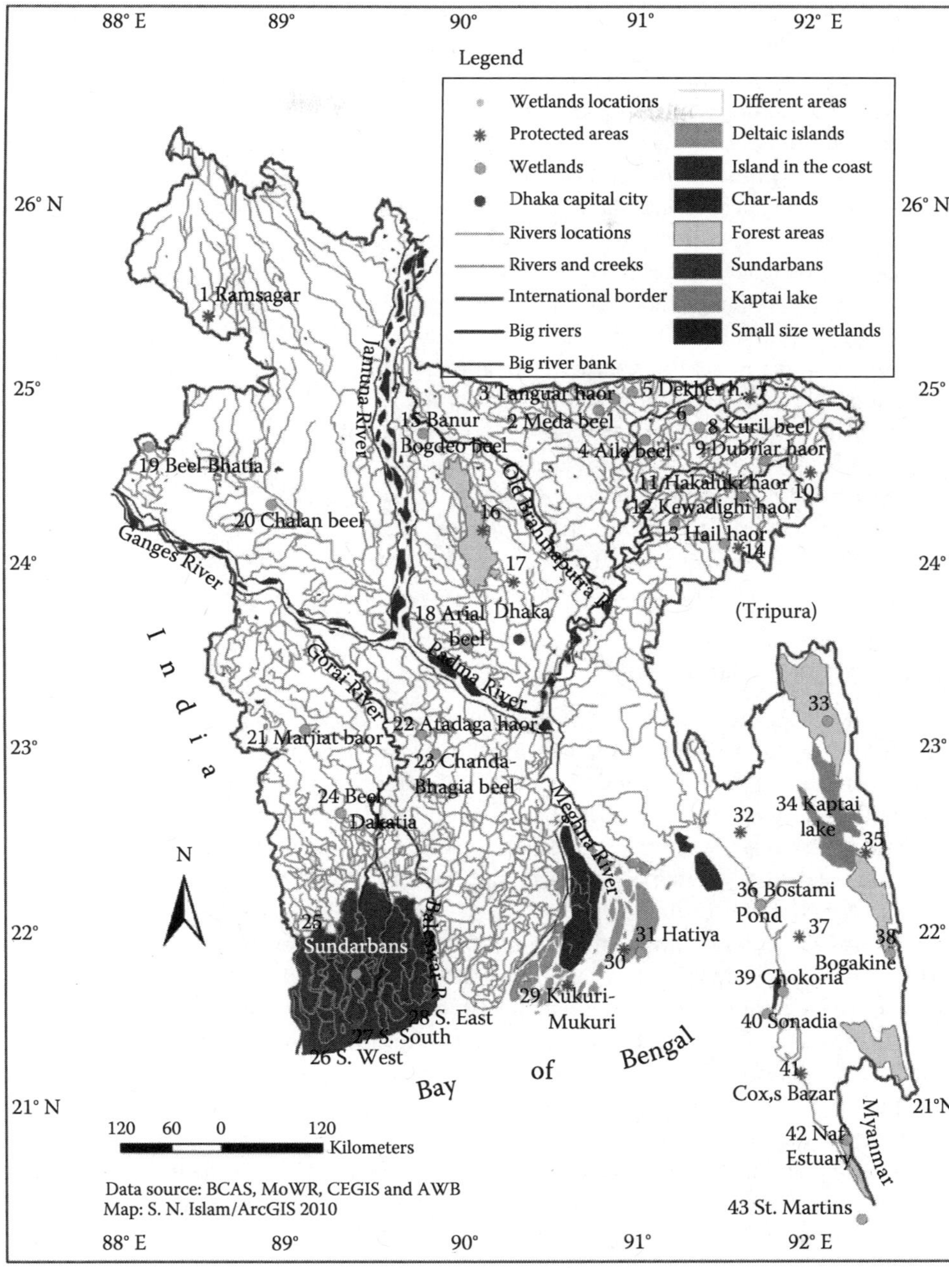

FIGURE 29.6 The coastal and the Sundarbans mangrove wetlands in Bangladesh.

early nineteenth century, Chalan beel used to cover an area of about 1085 km^2, but it was reduced to 368 km^2 in 1909, of which only 85 km^2 remains underwater throughout the year. It has since shrunk to only 26 km^2. This is the very common scenario in Bangladesh. The wetlands in Bera, Santhia, and Sujanagar upazilas have been reduced at the similar rate as Chalan beel reduced.

Global climate variability and increased population pressure in Bangladesh have resulted in large-scale demand of cultivable land for extra food production (Ali, 1988; Rasid, 1993; Zaman, 1993; Bennett et al., 1995). Conversions into other uses have resulted in net loss of wetlands in Bangladesh (Khan, 1993;

Hughes et al., 1994). Despite the threats to wetland sites and species in Bangladesh, very few studies of the functions, uses, values, and causes of degradation and of wetlands in Bangladesh are yet to be undertaken (Nokashima and Khan, 1993; Chowdhury, 1988). In order to sustain wetland resources, there is the need to develop appropriate strategies for extensive study involving both physical and environmental conditions deemed necessary for the optimization of resource utilization of wetlands in Bangladesh (Ali, 1990; Khan et al., 1994). However, many stakeholders are involved in wetland activities, and it is not easy to solve the challenges without proper policy model. It is therefore important to build up the awareness of people who are directly related to the wetlands of Bangladesh (Islam and Kitazawa, 2012).

The wetlands of Bangladesh have distinct characteristics, and that is permanent wetland to seasonal wetland and seasonal wetland to flooded wetland or agricultural land. The transformation of wetlands is accelerated by new human settlements, roads, and highways. Physical wetland changes fragment and isolate fauna and flora, eventually leading to endangerment or extinction. Because of push and pull factors, rural people move from rural to urban areas (Rana et al., 2009; Islam and Kitazawa, 2012). These poorly planned structures create waterlogs and had serious impact on wetlands. Although the Government of Bangladesh (GOB) takes some initiatives such as signing international environmental treaties, the situation of wetland management has not changed yet. Nevertheless, in Bangladesh, it is difficult to declare wetlands as a protected area because many poor people are directly or indirectly dependent on wetlands. Without proper alternatives for the wetland-dependent poor people, it is not possible to declare critical wetland as protected area (Islam and Kitazawa, 2012).

29.3.2 Data and Methodology

The study has been conducted based on primary and secondary data sources. The primary data on freshwater and coastal mangrove wetland and livelihood information were collected from the local people using nonstructural and structural and semistructural questionnaire around the Sundarbans and in Bera, Santhia, and Sujanagar upazilas in Pabna district in Bangladesh. The participatory rural appraisal (PRA) practices were arranged in the Sundarbans region in greater Khulna region, Gangbhanga beel (Santhia), Ghugudaha beel, (Santhia), Hatkhali and Raninagar beel area (Sujanagar), Char beel, and Nadiar beel in Bera upazila of Pabna district.

This study was also conducted with the author's field visit and personal observation for the past 10 years in the study area. Geomorphologic anthropological village-settled study and questionnaire survey for benchmark data collection were conducted with local community and stakeholders. The households covering the hand-holding pattern, tenancy, agricultural cropping systems and crop marketing, occupation, demographic character, literacy, etc., were collected in 2003, 2008, and 2011. A number of approaches for information and data collection and analysis were used including literature review, which helps obtain the basic understanding of wetlands, degradation, and natural livelihoods.

The secondary data inputs were collected from different publications of the government agencies, NGO reports, and research organization in Bangladesh and other countries. Collected data and information were analyzed and visualized using Microsoft Excel, Origin 8, and ArcGIS 10.

29.3.3 Surface Water Supply

The Bengal delta is the world's largest delta comprising 100,000 km^2 of riverine floodplain and deltaic plain (Goodbred et al., 2003; Hori and Saito, 2007). The high-tide energy results in tide-dominated deltas like the Meghna estuary, where distributary channels with linear river mouth bars are present (Hori and Saito, 2007). Sediment deposition occurs only by river flushing in the river-dominated delta, while in tide-dominated deltas, sediments are reworked and redeposited (Hori and Saito, 2007). Every year, about 1×10^{12} m^3/s of freshwater is brought into the Meghna estuary by the three major rivers, the Ganges, the Jamuna, and the upper Meghna. These rivers are major sources of water supply to the wetlands and balance the wetland ecosystems. The largest rivers in Asia are created from the Himalayas and

Tibetan plateau. There are 32 large rivers that are located around the Himalayas range and run through the long way and join with the sea. The water source of these large rivers is supplying surface freshwater and protecting the Asian wetlands. Accordingly in Bangladesh, the major rivers are supplying freshwater and balance the wetlands in Bangladesh. As a whole, the Brahmaputra (Jamuna) River carries water flow 60,000–100,000 m^3/s and about 600 million tons of sediment particles; most part of this particle is sand and clay in every year (FAP 24, 1993) [19]. The Ganges flow was 37,000 m^3/s in 1962, whereas it has reduced to 364 m^3/s in 2006. These two river flows are supplying sufficient freshwater and protecting the freshwater wetlands in Bera, Santhia, and Sujanagar upazilas in Pabna district. The reduction of the Ganges freshwater in the upstream area is the root cause of salinity intrusion in the southwestern region, which is affecting the Sundarbans coastal mangrove saline wetlands in Bangladesh.

29.3.4 Indigenous Management and Conservation of Wetlands

The public and private institutes tried to coordinate with the Ramsar regime on wetlands (1971–1998) and the Conservation of Biodiversity Development (CBD) regime (1992–1998). Besides, there are initiatives that have been taken by the different government periods. The legislation on wetland management has been considered by the government, and some initiatives have been taken as well as the Haor Development Board Ordinance, 1997; the Bangladesh Environmental Conservation Act, 1995; the East Bengal Protection of Conservation of Fish Act, 1950; the East Bengal State Acquisition and Tenancy Act, 1950; the Land Reform Board Act, 1989; the Bangladesh Wildlife (Preservation) (Amendment) Act, 1974; the Environment Policy 1992; and the Bangladesh Environmental Conservation Act, 1995, which give due importance to wetlands, and related issues are as follows: rivers, canal, roads, lakes, haor, beels, baors, and all other water bodies should be kept free from pollution. Besides the national laws, Bangladesh is signatory of Ramsar convention, CITES–(Convention on International Trade in Endangered Species of Wild Fauna and Flora), CBD–Convention on Biological Diversity), UNESCO–(United Nations Educational Scientific and Cultural Organization), IUCN–(International Union for Conservation of Nature), etc., but unfortunately, the scenarios of wetland degradation in Bangladesh did not change yet due to unplanned policy implementation; lack of knowledge, skills, and capacity; and less research activities.

Local regimes have three classes of parties as country government, civil society, and the secretariats. Many challenges and uncertainty build up a regime for better wetland management in Bangladesh. These include (1) collective action and free rider problems, (2) poverty and education break, (3) lack of good governance, (4) policy and legislation gap, and (5) natural challenge and uncertainty. Before independence, which was in 1971, these regions had two wetland regimes. These include British colonial regime and postindependent Bangladesh regime. Unfortunately, due to lack of political stability of the country, policymakers are not able to build an international standard wetland management policy. Finally, political instability plays a role in any kind of management including wetland management in the Pabna district as well as in the whole Bangladesh (Islam and Kitazawa, 2012).

The National Water Policy (NWPo) of Bangladesh (1999) also gives due importance on research and development of knowledge and capacity building for sustainable management. In Article 3, the objectives of NWPo are to develop a state of knowledge and capacity that will enable the country to design water resource management plans by itself with economic efficiency, gender equity, social justice, and environmental awareness. In Article 4.15, the following specific objectives are not ensued such as appropriate technologies, development and promotion of water management techniques, and production of skilled professionals for water resource management. The NWPo of Bangladesh is a guideline framework for the nation but unfortunately not any strong guideline direction for wetland management over there (Akter et al., 2010). A study of the Economic and Social Council for Asia and the Pacific (ESCAP) in 1988 and GOB has referred five sets of constraints to the development of a strategy for coastal wetland resource management in Bangladesh such as policymaking, planning for coastal resources, integrated resource management, coastal wetland and marine resource sustainability, local environmental ecological perspective, and lack in knowledge of coastal environment and understanding (Islam and Gnauck, 2009, 2009a).

The National Wetland Policy has been drafted by the Ministry of Environment and Forestry, but it still remains in the preliminary stage and not yet implemented. The main features of this policy include the following: maintenance of biodiversity and landscape protection, maintenance of ecosystem functions and ensuring socioeconomic benefits, and promotion of economic development and establishment of principles for sustainable resource utilization. Considering the present hostile situation of wetland biodiversity, a long-term wetland management plan for the sustainable use of wetland natural resources is urgently needed. The current ongoing development project like Wetland Biodiversity Conservation issues in Bera, Santhia, and Sujanagar upazilas in Pabna district and coastal wetland region in Bangladesh could be a new hope for the new strategies for the future development and sustainability of wetland biodiversity conservation in Bangladesh.

29.4 Results and Discussions

Wetland is the most important issue in nature and natural resource effectiveness and management. Most of the wetlands are very complex regarding their hydrologic, biological, and geomorphological functioning and dynamics. These wetlands have been utilized by farmers, fisherman, and other stakeholders in terms of drainage and annual banning for stock farming for hundreds or even thousands of years. Wetlands are used as the driving force of ecosystem services, biodiversity conservation, and community livelihood sustainability. In general, millions of people are directly dependent on wetland resources and its services and goods. Unfortunately, a million ha of wetlands is declined due to natural and anthropogenic impacts. In the case of Bangladesh, there are five types of wetlands that are used substantially, and the area is gradually reducing. The major freshwater wetlands and saline water wetlands are polluted due to climate change impacts and salinity intrusion in the coastal wetlands and arsenic contamination in groundwater in the freshwater wetland region in Bangladesh. The present management of wetlands in Bangladesh is not a sophisticated management system; the GOB has developed a draft for wetland management policy and conservation.

29.4.1 Water Quality in the Wetland Regions

Freshwater quality in the wetland areas has been degraded owing to devastating floods and lowland inundation. Floods and sedimentation, industrial waste dumping, chemical fertilizer used in crop production, riverbank erosion, and accretion causes are responsible for water quality degradation. On the other hand, eugenic and biogenic causes are also potential issues for water quality damage. However, the net cropped area in the wetland region of the coastal zone in Bangladesh has been decreasing over the year due to various purposes, and the most common one is the land inundation and salinity intrusion by tidal water. Freshwater reduction along with intrusion of saline water is perhaps the most devastating consequence of climate change in the coast of Bangladesh. The freshwater quality of the wetlands in the Bera, Santhia, and Sujanagar upazilas is gradually degrading because of unplanned and substantial use of water resources; besides, annual flood inundation is the main reason of quality degradation.

The dominant wetland and land use in the coastal region of Bangladesh is about 144,085 and 83,416 ha, respectively (Islam and Gnauck, 2009). It has been estimated that 830,000 million ha of coastal wetlands has been identified, which are affected by soil salinity at different degrees. It is estimated that a net reduction of 0.5 million metric tons of rice production will take place due to a 0.3 m sea level rise in coastal areas of Bangladesh (IPCC, 2007). The quality of water in the coastal wetlands and Sundarbans region is also degraded especially in the dry season when high-salinity intrusion occurs. The CEGIS study result shows that water quality has degraded in the coastal rivers where 60% water is of poor quality, considering the parameter of EC dS/m is over 4.0; that means 40% are of good quality, considering the EC dS/m rate is less than 2.0 (EGIS, 2000). Where water quality is damaging at that case, the other elements of ecosystem like soil, wildlife, and vegetation are also damaging due to high salinity, siltation, and shortage of freshwater.

29.4.2 Water and Soil Salinity in the Wetlands

The wetland areas in the coastal region of Bangladesh are affected due to the high rate of water and soil salinity intrusion. In general, the annual pattern of salinity changes in the Sundarbans region is also related with the changes of freshwater flow from upstream rivers. The adverse effects of salinity intrusion on ecosystem of the Sundarbans mangrove wetlands are manifested in the drying of tops of Sundari (Heritiera fomes) trees retrogression of forest types, slow forest growth, and reduced productivity of forest sites (MPO, 1986). The peak salinity was found to be about 56,186 dS/m in 2001 and 2002, and minimum salinity during postmonsoon was found to be about 10,805 dS/m (IWM, 2003). This salinity rate has crossed the salinity threshold value for the mangrove wetland ecosystems in the coastal region. Some mangrove species have been dried and displaced due to high-salinity penetration and intrusion in the coastal area in Bangladesh.

Salinity in the southern part of the bay remains less than 10,805 dS/m during monsoon and starts to increase at a steady rate up to about 32,415 dS/m during the dry season (IWM, 2003). Salinity in the western part is not reduced to a low-salinity range even during monsoon period; salinity increases at a steady rate during the dry periods. Almost 265 km^2 *H. fomes* type of forest moderately and 210 km^2 areas are severely affected, which is one of the main threats for a sustainable mangrove forest management and its ecosystems.

The highest soil salinity levels measured were ECs 41.2 dS/m at Nilkamal and ECs 40 dS/m at Mirganj, and third highest rate of soil salinity is ECs 24 dS/m at Munchiganj point in the northwestern Sundarbans mangrove coastal wetland region. The increasing salinity levels are major threats for both biotic and abiotic factors of mangrove wetland ecosystems in the region (Islam and Gnauck, 2009). The coastal region covers an area of about 29,000 km^2 or about 20% of the country. The coastal areas of Bangladesh cover more than 30% of the cultivable lands of the country. About 53% of the coastal areas are affected by salinity. Salinity causes unfavorable environment and hydrologic situations that restrict the normal crop production throughout the year. The factors that contribute significantly to the development of saline soil are tidal flooding during wet season, direct inundation by saline water, and upward or lateral movement of saline groundwater during dry season. The severity of salinity problem in the coastal wetlands increases with the desiccation of the soil.

29.4.3 Degraded Wetland Ecosystem Services

The freshwater wetlands in Bera, Santhia, and Sujanagar upazila wetland sites in Pabna district and the coastal mangrove wetlands in the southwestern region in Bangladesh are playing a potential role in socioeconomic improvement in the local community. It is estimated that over 6 million people are benefiting either directly or indirectly on the Sundarbans mangrove wetland resources in the coastal region (Anon, 1995). An area of 121,000 ha of the mangrove wetlands has experienced catastrophic losses over the past 100 years. Since the diversion of upstream freshwater, salinity has penetrated up to 173 km north from the coast. The change of salinity affected mangrove wetland plant species and animals in a variety of ways (Islam, 2010). Almost 45% mangrove wetlands are destroyed due to shrimp cultivation, construction, and other developmental activities in Bangladesh.

Considering the water quality in the Sundarbans coastal wetland region, the average values of DO concentration ranges (2002) were found to be high at the eastern rivers while low in the western part where DO ranges from 4.90 to 6.90 mg/L with an average value of 5.99 mg/L, and BOD is found to be 8.03 mg/L in the coast and 12.98 mg/L in the west site. The values of COD were found to be 26.28 mg/L in the east and 215.29 mg/L in the west, and the values of NH_3-N were found to be 5.61 mg/L in the west during the dry season (March–June). All these values of water quality parameters including NaCl, DO, BOD, COD, and NH_3-N have crossed the considerable limits of water quality in the lower tidally active delta in the coastal region. The ecosystem process and functions are the biological, chemical, and physical interactions between ecosystem components (IWM, 2003; Islam and Gnauck, 2007, 2008,

2009). The four types of coastal mangrove wetland services are functioning in the lower tidally active deltaic region. Therefore, the water quality maintenance is necessary in the coastal saline water wetland ecosystem in the Bengal coast in Bangladesh.

29.4.4 Threats for Wetland Biodiversity

A social–ecological system consists of a biogeophysical unit and its associated social actors and institutions. The management of agro-wetland biodiversity can be observed in the case of wetlands, medicinal plants, multifunctional farms, and organic farms. The biodiversity policies in Bangladesh and the "protection of flora and fauna varieties and community rights frame the legal situation and the governance structure for many issues relevant for the management of wetland biodiversity."

The baseline biodiversity survey report in 2011 conducted by the Flood Hazard Research Centre, Bangladesh (fhrc), stated that there were 105 river bird species that were available in three upazilas in Pabna district. There are 94 bird species that were available in 2009–2010, whereas only 82 birds were available in 2010–2011 wet seasons. The report also stated that 63 fish species are available in Bera, Santhia, and Sujanagar upazilas in Pabna district. That means the report finding shows that the species are gradually reducing due to time and seasons. The fishing yields have declined sharply and the number of fish species has been greatly reduced. About 20 species are regarded as officially endangered species. Within the project area, there are 19 national birds that are endangered bird species in the world context. In the Sundarbans mangrove wetland case, a number of species like the Javan rhinoceros (*Rhinoceros sondaicus*), water buffalo (*Bubalus bubalis*), swamp deer (*Cervus duvauceli*), gaur (*Bos gaurus*), hog deer (*Axis porcinus*), and marsh crocodile (*Crocodylus palustris*) became extinct in the Sundarbans mangrove wetland region. In the coastal wetlands in Bangladesh, there are 15 plant species that are listed in the International Union for Conservation of Nature (IUCN) red book list. Most of these species cannot tolerate the high water and soil salinity in the coastal region. The effect of climate change is a threat to biodiversity and natural resources of the increase of local population. The main reasons are a lack of cross-sectoral cooperation in the management of wetland and floodplain disciplines involved with several ministries and weak decentralized specialized authorities. The negative effects are the substantial use of natural resources and degradation of wetland ecosystem, which is the reason for loss of biodiversity and a growing decline in fish stocks. Therefore, it can be stated that there are some negative influences that are disturbing the species settlement in the case areas. Although the baseline report did not work on flora and other fauna, the medicinal plant species are very potential issues for the sustainable natural wetland resource management and conservation point of view. The report of Flood Hazard Research Centre (2011) also mentioned about the primary initiatives of the GOB.

29.4.5 Approach for Wetland Sustainability and Natural Resource Management

The study findings illustrated minor initiatives undertaken by the Ministry of Environment and Forest and the Ministry of Water Resources of the Bangladesh Government. In general, the freshwater and saline water wetlands are controlled by the local elite groups. Actions that aimed to retain water resources have been undertaken to a lesser extent. The raising groundwater level in wetland areas may increase ET, and as a result, it can reduce runoff in the summer period. The freshwater wetland sites in Pabna district are losing their hydrologic cycle, and the scarcity of freshwater is creating ecological damage in the Bera, Santhia, and Sujanagar upazila wetland sites. On the other hand, the mangrove wetland sites of the southern coastal region of Bangladesh are affected by saline water intrusion and water quality degradation. The anthropogenic influences on wetland resources in 15 sites have already been marked with a red list. The study result also reveals that the authority does not follow any policy guideline or appropriate management approaches for the wetland conservation in Bangladesh (Islam, 2010).

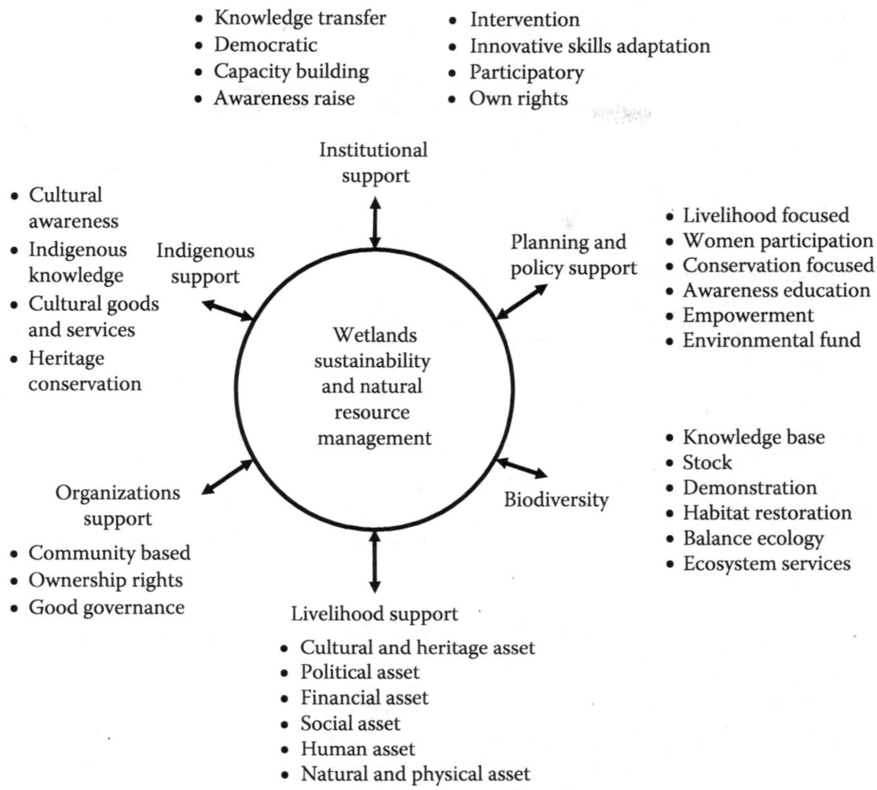

FIGURE 29.7 The wetland sustainability and natural resource management.

Figure 29.7 illustrates the conceptual model of wetland and natural resource management that could ensure sustainability.

The fundamental elements of natural resource management and sustainability such as institutional support, planning and policy support, biodiversity support, livelihood support, organizational support, and indigenous support have made the policy framework of wetland sustainability and natural resource management for freshwater and saline water wetland sites. The natural resource management situation and its ecological, socioeconomic, cultural, and environmental impacts are tragic in all freshwater and saline water wetland sites in Bangladesh (Islam and Ganuck, 2007; Islam et al., 2012). The National Wetland Policy has been drafted by the Ministry of Environment and Forestry, but it is not in implementation stage. The main features of this policy include the following issues:

- Maintenance of biodiversity and landscape protection
- Maintenance of ecosystem functions and ensuring socioeconomic benefits
- Promotion of economical development and establishment of principles for sustainable resource utilization

For better planning and sustainable management of wetland resources, the fundamental model structure and policy framework and its prime factors and subfactors should follow properly. The conceptual model (Figure 29.7) should be implemented in freshwater and saline water wetland sites in Bangladesh that could achieve a good result for better management and that could ensure livelihood sustainability (Cambers, 1992; DFID, 1999). The finding of this study could become an important guideline for planning and sustainable management of wetland and site conservation in Bangladesh as well as similar categories of wetland sites in other parts of the world.

29.5 Summary and Conclusions

Wetlands can be a driving force for community social and economic development. They have the ability to focus tremendous energy and to generate significant creative and economic betterment. In general, the country's wetland natural resources as well as the ecosystems are degrading due to anthropogenic influences and natural calamities. The freshwater and saline water wetlands in Bangladesh are in alarming situation. The hydrologic cycle of the wetland areas is losing the balance. The country needs an adequate interdisciplinary policy and strategies and political wills to implement it for sustainable management and protection of wetlands and ecologically sensitive wetland ecosystems in Bangladesh. The methods for the generation of transformation knowledge to achieve and maintain natural resources, agro-farming system development, fishing development, and, in general, wise use of wetland natural resources could protect the biodiversity in the freshwater wetlands and coastal saline mangrove wetland areas. The following recommendations could be followed in the implementation and management stages:

- To increase people's awareness, a yearly local cultural festival on international wetland day in different parts of the world as well as in the case areas in Bangladesh. The local culture and Heritage, wetland ecosystem services and goods could be demonstrated and marketed to enforce the local capacity of income generation could be arranged.
- Capacity building training for the stakeholders and resource users could be arranged through this local-based wetland project implementation for the local government authorities for skills development and good governance.
- The requirement to ensure the provision of adequate institutional capacity for policy development, delivery, and monitoring; the importance of considering wetlands as water management infrastructure rather than nature reserves; and the need to consider the wise use of wetlands both within and beyond urban boundaries and understand the interconnectivity of watershed-scale issues.
- Traditionally, frameworks and spatial plans are area based and structured on administrative boundaries. This approach fails to address or in many cases even recognize that these boundaries are not functional or commensurate with the environment or boundaries required for the protection of wetland or ecosystem services.
- There is a clear need to develop wetland policies and plans around the protection of ecosystem services; however, there is limited awareness of this concept within both national governments and local administrations.
- In the developing world, considerable issues related to unplanned or informal settlement usefully fall outside of the normal planning framework; however, they are often hot spots of environmental degradation.

For time series analysis, the satellite and remote sensing imageries could be used to compare the historical changes of agricultural cropping systems, land-use pattern, shrinking trends of wetlands, and present scenario measures. Special data analysis result shows services of wetland ecosystems in three upazilas of Pabna district and the southwestern coastal mangrove wetlands in Bangladesh. As a whole, the developed wetland biodiversity conservation planning and policy framework could be the fundamental output of applied research that could be implemented in the protected and sensitive wetlands in Bangladesh as well as other parts of the world.

References

Action Aid International. 2005. *Participatory Vulnerability Analysis: A Step-by Guide for Field Staff.* AAI, London, U.K.

Ahmed, R. and Falk, G.C. 2008. Bangladesh: Environment under pressure. *Geographische Rundschau International Edition*, 4(1): 13–19.

Akonda, A.W. 1989. Bangladesh. In: Scott, D.A. (ed.), *A Directory of Asian Wetlands*. IUCN, Gland, Switzerland, pp. 541–581.

Akter, J., Islam, S.N., and Gnauck, A. 2010. Water resources management in the coastal region of Bangladesh. In: *Modelling and Simulation of Ecosystems, Workshop Kölpinsee 2009*, Shaker Verlag, Aachen, Germany, pp. 167–185.

Ali, M.Y. 1988. Open water fisheries and environmental changes. In: Rahman, A.A., Huq, S., and Conway, G.R. (eds.), *Environmental Aspects of Surface Water Systems of Bangladesh*. University Press Ltd, Dhaka, Bangladesh.

Ali, S.I. 1990. Haor basin ecosystem. In: Rahman, A.A., Huq, S., and Gony, G.R. (eds.), *Environmental Aspect of Surface Water System of Bangladesh*. University Press Ltd, Dhaka, Bangladesh, pp. 38–72.

Anon, 1995. Integrated resource management plan of the Sundarbans reserved forest, Vol. 1, Draft Final Report of FAO/UNEP Project BGD/84/056-Khulna, Bangladesh.

Bennett, R.L., Steinhaus, K.A., Uhrich, S.B., O'Sullivan, C., Resta, R.G., and Doyle, D.L. 1995. Recommendations for standardized pedigree nomenclature. *American Journal of Human Genetics*, 56: 745–752.

BGS and DPHE (British Geological Survey and Dept. of Public Health Engineering), 2001. Arsenic contamination of groundwater in Bangladesh. BGS Technical Report, WC/00/19.Vol.2 Final Report, Dhaka.

Cambers, R. 1992. *Rural Appraisal; Rapid Relaxed and Participatory*. IDS Discussion Paper 311. International Development Studies, Brighton, U.K.

Cannon, T. 2002. Gender and climate hazards in Bangladesh. *Gender and Development*, 10(2): 45–50.

Carter, V. (2010), Technical aspects of wetlands—Wetland hydrology, water quality, and associated functions. *US Geological Survey* Water Supply Paper 2425, Washington DC.

Chowdhury, A.M.R. 1988. The 1987 flood in Bangladesh: An estimate of damage in twelve villages. *Disasters*, 12(4): 294–300.

Chowdhury, N.T. 2009. Water management in Bangladesh: An analytical review. *Journal of Water Policy*, IWA Publishing, 12(1): 32–51.

Davies, J. and Claridge, G. 1993. *Wetland Benefits the Potential for Wetlands to Support and Maintain Development*. Asian Ecology Bureau, Kuala Lumpur, Malaysia, pp. 46–52.

DFID—Department of Foreign Aid and International Development, 1999. Sustainable livelihoods guideline sheets (Framework) British Government, London, U.K., pp. 1–26.

EGIS—(Environmental Geographical Information Services), 2000. *Environmental Baseline of Gorai River Restoration Project*. Environment and GIS support project for water sector planning (EGIS-II), Ministry of Water Resources, Government of Bangladesh and Government of the Netherlands, Dhaka, the Netherlands, p. 150.

Enarson, E. and Morrow, B.H. 1998. *The Gendered Terrain of Disaster: Through Women's Eyes*. Praeger, Westport, CT.

FAP (Flood Action Plan) 24, 1993. River survey project- morphology of Gorai off-take. Special Report No.10, WARPO, Dhaka, Bangladesh.

Goodbred, S.L., Kuehl, S.A., Stecler, M.S., and Sarker, M.H. 2003. Controls on facies distribution and stratigraphic presentation in the Ganges-Brahmaputra delta sequence. *Sedimentary Geology*, 155: 301–316.

Gopal, B. 1999. Natural and constructed wetlands for wastewater treatment: potentials and problems. *Water Science and Technology*, 40(3): 27–35.

Gopal, B. and Wetzel, R.G. 1995. Limnology in developing countries, Indian Association for Limnology, Vol. 1:1-230.

Hassan, S. and Mulamoottil, G. 1994. Environmental problem in Dhaka city: A study of mismanagement. *Cities*, 11(3): 195–200.

Hori, K. and Saito, Y. 2007. Classification, architecture, and evolution of large-river deltas. In: Gupta, A. (ed.), *Large Rivers: Geomorphology and Management*. John Wiley & Sons Ltd, London, U.K., pp. 75–96.

Hughes, R., Adnan, S., and Clayton, B.D. 1994. *Floodplains or Flood Plans? A Review of Approaches to Water Management in Bangladesh.* IIED, London, U.K., pp. 1–12.

IPCC—(Intergovernmental Panel on Climate Change), 2007. Summary for policymakers. In: Parry, M.L. et aL (eds.,) *Climate Change 2007: Impact Adaptation and Vulnerability. Contribution of Working Group II to the Fourth Assessment Report of the IPCC.* Cambridge University Press, Cambridge, U.K., p. 1000.

Islam, M., Saha, N., and Rahman, M. 2012. *Impact of Livelihood Dependence on Ecological Functions of Wetlands—An Investigation on Hakaluki haor: The Largest Freshwater Ecosystem in Bangladesh.* LAP Lambert Academic Publishing, Saarbrucken, Germany.

Islam, M.N. and Kitazawa, D. 2012. Modelling of freshwater wetland management strategies for building the public awareness at local level in Bangladesh. *Mitigation and Adaptation Strategies for Global Change,* 18: 869–888.

Islam, S.N. 2010. Threatened wetlands and ecologically sensitive ecosystems management in Bangladesh. *Frontiers of Earth Science in China,* 4(4): 438–448.

Islam, S.N. and Gnauck, A. 2007. Effects of salinity intrusion in mangrove wetlands ecosystems in the Sundarbans: An alternative approach for sustainable management. In: Okruszko, T., Jerecka, M., and Kosinski, K. (eds.), *Wetlands: Monitoring Modelling and Management.* Taylor & Francis Group, London, U.K., pp. 315–322.

Islam, S.N. and Gnauck, A. 2008. Mangrove wetland ecosystems in Ganges-Brahmaputra delta in Bangladesh. *Frontiers of Earth Science in China,* 2(4): 439–448.

Islam, S.N., and Gnauck, A. 2009. Threats to the Sundarbans mangrove wetland ecosystems from transboundary water allocation in the Ganges basin: A preliminary problem analysis. In: S. Eslamian (ed.), *International Journal of Ecological Economics and Statistics (IJEES),* 13(9): 64–78.

Islam, S.N. and Gnauck, A. 2009a. The coastal mangrove wetland ecosystems in the Ganges Delta: A case study on the Sundarbans in Bangladesh. In: *Proceeding of American Association of Petroleum Geologist—AAPG Hedberg Conference on Variations in Fluvial-Deltaic and Coastal Reservoirs Deposited in Tropical Environments, from 29th April to 2nd May, 2009,* Jakarta, Indonesia, pp. 26–29.

IWM (Institute of Water Modelling), 2003. Sundarbans biodiversity conservations project-surface water modeling TA No. 3158 BAN, Final Report, Vol. 1, IWM, Dhaka, Bangladesh.

Khan, A.A. 1993. Freshwater wetlands in Bangladesh: Opportunities and options. In: *Freshwater Wetlands in Bangladesh-Issues and Approaches for Management.* IUCN, Dhaka, Bangladesh, pp. 1–7.

Khan, S.M., Haq, E., Huq, S., Rahman, A.A., Rashid, S.M.A., and Ahmed, H. 1994. *Wetlands of Bangladesh.* Holyday Printers, Dhaka, Bangladesh, pp. 1–88.

Kim, G.K., Park, Y.M., and Choi, S.M. 2006. Developing a wetland type classification system in the Republic of Korea. *Landscape and Ecological Engineering,* 2: 93–110.

Kundzewicz, Z.W. 2003. *Ecohydrology for Sustainable Wetlands Under Global Change-Data, Models, Management Techniques and Data Assessment in Wetlands Hydrology.* Warsaw Agricultural University Press, Warsaw, Poland, pp. 25—35.

Mansour, M.R., Akrami, A., Ghobadi, H.S., Denji, A.K., and Ezatrahimi, A.G. 2011. Effect of dietary mannan oligosaccharide (MOS) on growth performance, survival, body composition, and some hematological parameters in giant sturgeon juvenile (Huso huso linnaeus, 1754). *Fish Physiology and Biochemistry,* Springer Science and Business Media 1–7.

Mather, R.G., Montgomery, W.I., and Portig, A.A. 1993. Exploitation of intertidal Zostera species by Brent geese (Branta bernicla hrota) Why dig for your dinner? *Biology and Environment: Proceedings of the Royal Irish Academy,* 98B(3): 147–152.

Mitsch, W.I. and Gosselink, I.G. 1986. *Wetlands.* Van Nostand Reinhold, New York, pp. 1–220.

Moore, M.T., Rodgers, V.H., Cooper, C.M., and Smith, Jr., S. 2000. Constructed wetlands for mitigation of atrazine associated agricultural runoff. *Environmental Pollution,* 110(3): 393–399.

MPO, 1986. Groundwater resources of Bangladesh. Technical Report No. 5. Master Plan Organization, Hazra, Dhaka, Bangladesh.

Nair, K.S. 2004. Wetlands management to meet the food and water crisis- example from a tropical region. In: *Abstracts Book of 7th INTECOL International Conference on Wetlands,* July 25–30, 2004, Utrecht, the Netherlands, p. 217.

Nishat, A. 1993. Freshwater wetlands in Bangladesh: Status and issues. In: Nishat, A., Hussain, Z., Ray, M.K., and Karim, A. (eds.), *Freshwater Wetlands in Bangladesh—Issues and Approaches for Management.* IUCN, Dhaka, Bangladesh, pp. 9–22.

Nishat, A. 2003. Management of freshwater wetlands in Bangladesh: Issues and strategy. http://www.doe-bd.org/cwbmp/inception_report/on 03.03. 2012.

Nokashima, S. and Khan, M.H. 1993. A basic guide book to understanding the environmental impacts of rural roads on the wetlands. CARE International in Bangladesh, Dhaka.

Rahman, M.M. 1995. Wetlands and biodiversity: A case study of common property resources in Bangladesh. *Conference Paper at the 5th Annual Common Property,* Conference held from May 24–28, 1995, Bodo, Norway, pp. 1–10.

Rahman, S. and Ahmed, K.K. 2002. Performance of fisheries sector planning in Bangladesh. *Outlook on Agriculture,* 31(4): 243–251.

Rana, M.P., Chowdhury, M.S.H., Sohel, M.S.I., Akhter, S., and Koike, M. 2009. Status and socio-economic significance of wetland in the tropics: A study from Bangladesh. *IForest,* 2: 172–177.

Rasid, H. 1993. Preventing flooding or regulating flood level? Case studies on perception of flood alleviation in Bangladesh. *Natural Hazards,* 8(1): 39–57.

Rashid, H.E. 1991. *Geography of Bangladesh.* University Press Limited, Dhaka, Bangladesh.

Reid, H., Alam, M., Berger, R., Cannon, T., Huq, S., and Milligan, A. 2010. *Community-Based Adaptation to Climate Change: An Overview.* IIED, Russell Press, London, U.K., pp. 11–30.

Sarkar, S.U. 1993. Faunal diversity and their conservation in freshwater wetlands. In: Nishat, A., Hussain, Z., Roy, M.K., and Wold Conservation Union (eds.), *Freshwater Wetlands in Bangladesh: Issues and Approaches for Management.* IUCN, Gland, Switzerland, pp. 1–364.

Sarker, M.H. 2008. Morphological response of the Brahmaputra-Padma-lower Meghna River system to the Assam Earthquake of 1950. Unpublished PhD Thesis, School of Geography, University of Nottingham, Nottingham, U.K., pp. 1–296.

Smedley, P.L. and Kinniburgh, D.G. 2002. A review of the source, behavior and distribution of arsenic in natural waters. *Applied Geochemistry,* 17: 517–568.

Sultana, F. 2010. Living in hazardous waterscapes: Gendered vulnerabilities and experiences of floods and disasters. *Environmental Hazards-* Human and Policy Dimensions, New York, pp 43–53.

Sultana, P. and Thompson, P. 2008. Gender and local floodplain management institutions: A case study from Bangladesh. *Journal of International Development,* 20: 53–68.

Talukdar, B., Nakagoshi, N., and Rashid, M.S. 2008. State and management of wetlands in Bangladesh. *Landscape Ecology Engineering,* 5(1): 81–90.

Thompson, P. and Sultana, P. 1996. Distributional and social impacts of flood control in Bangladesh. *The Geographical Journal,* 162(1): 1–13.

U.S. Department of Agriculture—(USDA). 2003. Natural Resources Conservation Service (NRCS)-Global distribution of wetlands map, USDA-NRCS, Soil Science division, World Soil Resources, Washington, DC.

Zaman, S.M.H. 1993. Agricultural development and sustainability of wetlands. In: Nishat, A. et al. (eds.), *Freshwater Wetlands in Bangladesh: Issues and Approaches for Management.* IUCN, Dhaka, Bangladesh, pp. 283.

Index

A

B

C

E

H

N

O

P

Q

R

T

U

V

W

Z